UNITED STATES DEPARTMENT OF THE INT

BUREAU OF RECLAMATION

DESIGN OF GRAVITY DAMS

DESIGN MANUAL FOR CONCRETE GRAVITY DAMS

A Water Resources Technical Publication

Denver, Colorado
1976

As the Nation's principal conservation agency, the Department of the Interior has responsibility for most of our nationally owned public lands and natural resources.

This includes fostering the wisest use of our land and water resources, protecting our fish and wildlife, preserving the environmental and cultural values of our national parks and historical places, and providing for the enjoyment of life through outdoor recreation.

The Department assesses our energy and mineral resources and works to assure that their development is in the best interests of all our people.

The Department also has a major responsibility for American Indian reservation communities and for people who live in Island Territories under U.S. administration.

UNITED STATES GOVERNMENT PRINTING OFFICE

DENVER : 1976

Frontispiece.—Grand Coulee Dam and Powerplants.—CN 222-117-14091—July 22, 1975

Preface

This manual presents instructions, examples, procedures, and standards for use in the design of concrete gravity dams. It serves as a guide to sound engineering practices in the design of concrete gravity dams and provides the technically trained, qualified design engineer with specialized and technical information that can be readily used in the design of such a dam.

The manual came into being because of the numerous requests made to the Bureau for its latest concepts on the design of concrete dams. A companion Bureau manual "Design of Arch Dams" is also being prepared and will soon be published.

"Design of Gravity Dams" was prepared to cover all heights of concrete gravity dams except small dams under 50 feet which are covered in "Design of Small Dams." Foundations for the design of dams discussed in this book are assumed to be rock.

The material used in this book from "Design of Small Dams" has been revised to make it applicable to larger concrete gravity dams. Although most of this text is related exclusively to the design of dams and appurtenant structures, it is important that the designer be familiar with *the purpose of the project of which the dam is a part, the considerations influencing its justification, and the manner of arriving at the size and type of structure to be built.* Factors which affect the selection of the type of dam and its location are discussed in chapter II, "Design Considerations." Chapter XV discusses the ecological and environmental considerations required in constructing a dam. The integrity of the structural design requires strict adherence to specifications for the concrete and to the practice of good workmanship in concrete production. Therefore, a summary of Bureau of Reclamation concrete construction practices or methods is included in chapter XIV, "Concrete Construction."

The manual should be of service to all concerned with the planning and designing of water storage projects, but it cannot relieve the agency or person using it of the responsibility for a safe and adequate design. The limitations stated in the design procedures should be heeded.

This book was prepared by engineers of the Bureau of Reclamation, U.S. Department of the Interior, at the Engineering and Research Center, Denver, Colorado, under the direction of H. G. Arthur, Director of Design and Construction, and Dr. J. W. Hilf,* Chief, Division of Design. The text was written by members of the Concrete Dams Section, Hydraulic Structures Branch, Division of Design, except for Appendix G "Inflow Design Flood Studies," which was written by D. L. Miller,* of the Flood and Sedimenation Section, Water and Management Planning Branch, Division of Planning Coordination. Members of the Concrete Dams Section who made substantial contributions to the text include: M. D. Copen,* J. Legas, E. A. Lindholm, G. S. Tarbox, F. D. Reed,* C. L.

---

*Retired

Townsend,* J. S. Conrad,* R. O. Atkinson, R. R. Jones, M. A. Kramer, C. W. Jones,* J. L. Von Thun, G. F. Bowles, and J. T. Richardson.* The major editing and coordinating of the test was done by E. H. Larson,* and the final preparation of the text for printing was done by R. E. Haefele and J. M. Tilsley, all of the Publications Section, Technical Services and Publications Branch, Division of Engineering Support. The authors and editors wish to express their appreciation to the personnel in the General Services Branch for their contributions and to the technicians of Concrete Dams Section and Drafting Branch who prepared charts, tables, and drawings for use in the text.

The methods of design and analysis were developed through the efforts of dedicated Bureau engineers during the many years the Bureau of Reclamation has been designing and constructing concrete gravity dams. Their efforts are gratefully acknowledged.

There are occasional references to proprietary materials or products in this publication. These must not be construed in any way as an endorsement of the Bureau of Reclamation since such endorsement cannot be made for proprietary products or processes of manufacturers or the services of commercial firms for advertising, publicity, sales, or other purposes.

Contents

Chapter II—Design Considerations—Continued

J. CONSTRUCTION ASPECTS

K. MISCELLANEOUS CONSIDERATIONS

L. BIBLIOGRAPHY

Chapter III—Design Data and Criteria

A. INTRODUCTION

B. CONCRETE

C. FOUNDATION

D. LOADS

Chapter III—Design Data and Criteria—Continued

D. LOADS–Continued

E. LOADING COMBINATIONS

F. FACTORS OF SAFETY

G. BIBLIOGRAPHY

Chapter IV—Layout and Analysis

A. LAYOUT

B. THE GRAVITY METHOD OF STRESS AND STABILITY ANALYSIS

Chapter IV—Layout and Analysis—Continued

B. THE GRAVITY METHOD OF STRESS AND STABILITY ANALYSIS—Continued

C. TRIAL-LOAD METHODS OF ANALYSIS

1. Trial-Load Twist Method of Analysis, Joints Ungrouted

Chapter IV—Layout and Analysis—Continued

C. TRIAL-LOAD METHODS OF ANALYSIS—Continued

2. Trial-Load Twist Method of Analysis, Joints Grouted

3. Analysis of Curved Gravity Dams

D. DYNAMIC ANALYSIS

E. THE FINITE ELEMENT METHOD

1. Two-Dimensional Finite Element Program

Chapter IV—Layout and Analysis—Continued

E. THE FINITE ELEMENT METHOD—Continued

F. FOUNDATION ANALYSIS

Chapter IV—Layout and Analysis—Continued

F. FOUNDATION ANALYSIS—Continued

G. BIBLIOGRAPHY

Chapter V—River Diversion

A. DIVERSION REQUIREMENTS

B. METHODS OF DIVERSION

C. SPECIFICATIONS REQUIREMENTS

Chapter VI—Foundation Treatment

A. EXCAVATION

B. GROUTING

C. DRAINAGE

D. BIBLIOGRAPHY

Chapter VII—Temperature Control of Concrete

A. INTRODUCTION

B. METHODS OF TEMPERATURE CONTROL

Chapter VII—Temperature Control of Concrete— Continued

B. METHODS OF TEMPERATURE CONTROL—Continued

C. TEMPERATURE STUDIES

D. DESIGN CONSIDERATIONS

E. CONSTRUCTION OPERATIONS

Chapter VII—Temperature Control of Concrete—Continued

E. CONSTRUCTION OPERATIONS—Continued

F. BIBLIOGRAPHY

Chapter VIII—Joints in Structures

Chapter IX—Spillways

A. GENERAL DESIGN CONSIDERATIONS

Chapter IX—Spillways—Continued

A. GENERAL DESIGN CONSIDERATIONS—Continued

B. DESCRIPTION OF SPILLWAYS

C. CONTROL STRUCTURES

Chapter IX—Spillways— Continued

C. CONTROL STRUCTURES—Continued

D. HYDRAULICS OF DISCHARGE CHANNELS

E. HYDRAULICS OF TERMINAL STRUCTURES

Chapter IX—Spillways— Continued

E. HYDRAULICS OF TERMINAL STRUCTURES—Continued

F. HYDRAULICS OF MORNING GLORY (DROP INLET) SPILLWAYS

G. STRUCTURAL DESIGN

H. BIBLIOGRAPHY

Chapter X—Outlet Works and Power Outlets

A. INTRODUCTION

B. OUTLET WORKS OTHER THAN POWER OUTLETS

Chapter X—Outlet Works and Power Outlets— Continued

Chapter X—Outlet Works and Power Outlets— Continued

Chapter XI—Galleries and Adits

Chapter XI—Galleries and Adits—Continued

Chapter XII—Miscellaneous Appurtenances

Chapter XIII—Structural Behavior Measurements

Chapter XIII—Structural Behavior Measurements—Continued

Chapter XIV—Concrete Construction

Chapter XV—Ecological and Environmental Considerations

A. INTRODUCTION

B. FISH AND WILDLIFE CONSIDERATIONS

Chapter XV—Ecological and Environmental Considerations—Continued

C. RECREATIONAL CONSIDERATIONS

D. DESIGN CONSIDERATIONS

E. BIBLIOGRAPHY

Appendix A—The Gravity Method of Stress and Stability Analysis

Appendix B—Trial-Load Twist Analysis—Joints Grouted

Appendix B—Trial-Load Twist Analysis—Joints Grouted—Continued

Appendix C—Finite Element Method of Analysis

A. TWO-DIMENSIONAL FINITE ELEMENT ANALYSIS

B. THREE-DIMENSIONAL FINITE ELEMENT ANALYSIS

Appendix D—Special Methods of Nonlinear Stress Analysis

Appendix D—Special Methods of Nonlinear Stress Analysis—Continued

Appendix E—Comparison of Results by Gravity and Trial-Load Methods

Appendix F—Hydraulic Data and Tables

Appendix F—Hydraulic Data and Tables—Continued

Appendix G—Inflow Design Flood Studies

A. COLLECTION OF HYDROLOGIC DATA FOR USE IN ESTIMATING FLOODFLOWS

B. ANALYSES OF BASIC HYDROLOGIC DATA

Appendix G—Inflow Design Flood Studies—Continued

B. ANALYSES OF BASIC HYDROLOGIC DATA—Continued

C. SYNTHETIC UNIT HYDROGRAPH

Appendix G—Inflow Design Flood Studies—Continued

C. SYNTHETIC UNIT HYDROGRAPH—Continued

D. STREAMFLOW ROUTING

E. DESIGN STORM STUDIES

Appendix G—Inflow Design Flood Studies—Continued

Appendix G—Inflow Design Flood Studies—Continued

Appendix G—Inflow Design Flood Studies—Continued

Appendix H—Sample Specifications for Concrete

Appendix H—Sample Specifications for Concrete—Continued

Appendix H—Sample Specifications for Concrete—Continued

Appendix H—Sample Specifications for Concrete—Continued

Appendix I—Sample Specifications for Controlling Water and Air Pollution

TABLES

TABLES IN APPENDICES

FIGURES

FIGURES IN APPENDICES

Chapter I

Introduction

1-1. *Scope.*—A concrete gravity dam, as discussed in this manual, is a solid concrete structure so designed and shaped that its weight is sufficient to ensure stability against the effects of all imposed forces. Other types of dams exist which also maintain their stability through the principle of gravity, such as buttress and hollow gravity dams, but these are outside the scope of this book. Further, discussions in this manual are limited to dams on rock foundations and do not include smaller dams generally less than 50 feet high which are discussed in the Bureau of Reclamation publication "Design of Small Dams"[1][1].

The complete design of a concrete gravity dam includes not only the determination of the most efficient and economical proportions for the water impounding structure, but also the determination of the most suitable appurtenant structures for the control and release of the impounded water consistent with the purpose or function of the project. This manual presents the basic assumptions, design considerations, methods of analysis, and procedures used by designers within the Engineering and Research Center, Bureau of Reclamation, for the design of a gravity dam and its appurtenances.

1-2. *Classifications.*—Gravity dams may be classified by plan as straight gravity dams and curved gravity dams, depending upon the axis alinement. The principal difference in these two classes is in the method of analysis. Whereas a straight gravity dam would be analyzed by one of the gravity methods discussed in this manual (ch. IV), a curved gravity dam would be analyzed as an arch dam structure, as discussed in the Bureau's manual "Design of Arch Dams"[2]. For statistical purposes, gravity dams are classified with reference to their structural height. Dams up to 100 feet high are generally considered as low dams, dams from 100 to 300 feet high as medium-height dams, and dams over 300 feet high as high dams.

1-3. *General Dimensions.*—For uniformity within the Bureau of Reclamation, certain general dimensions have been established and are defined as follows:

The *structural height* of a concrete gravity dam is defined as the difference in elevation between the top of the dam and the lowest point in the excavated foundation area, exclusive of such features as narrow fault zones. The top of the dam is the crown of the roadway if a roadway crosses the dam, or the level of the walkway if there is no roadway. Although curb and sidewalk may extend higher than the roadway, the level of the crown of the roadway is considered to be the top of the dam.

The *hydraulic height*, or height to which the water rises behind the structure, is the difference in elevation between the lowest point of the original streambed at the axis of the dam and the maximum controllable water surface.

The *length* of the dam is defined as the distance measured along the axis of the dam at the level of the top of the main body of the dam or of the roadway surface on the crest, from abutment contact to abutment contact, exclusive of abutment spillway; provided that, if the spillway lies wholly within the dam and

---

[1]Numbers in brackets refer to items in the bibliography, sec. 1-5.

not in any area especially excavated for the spillway, the length is measured along the axis extended through the spillway to the abutment contacts.

The *volume* of a concrete dam should include the main body of the dam and all mass concrete appurtenances not separated from the dam by construction or contraction joints. Where a powerplant is constructed on the downstream toe of the dam, the limit of concrete in the dam should be taken as the downstream face projected to the general excavated foundation surface.

1-4. Gravity Dam Definitions. —Terminology relating to the design and analysis of gravity dams and definitions of the parts of gravity dams as used in this manual are as follows:

A *plan* is an orthographic projection on a horizontal plane, showing the main features of a dam and its appurtenant works with respect to the topography and available geological data. A plan should be oriented so that the direction of streamflow is toward the top or toward the right of the drawing.

A *profile* is a developed elevation of the intersection of a dam with the original ground surface, rock surface, or excavation surface along the axis of the dam, the upstream face, the downstream face, or other designated location.

The *axis* of the dam is a vertical reference plane usually defined by the upstream edge of the top of the dam.

A *section* is a representation of a dam as it would appear if cut by a plane. A *beam section* is taken horizontally through the dam. A *cantilever section* is a vertical section taken normal to the axis and usually oriented with the reservoir to the left.

A *beam element*, or *beam*, is a portion of a gravity dam bounded by two horizontal planes 1 foot apart. For purposes of analysis the edges of the elements are assumed to be vertical.

A *cantilever element,* or *cantilever,* is a portion of a gravity dam bounded by two vertical planes normal to the axis and 1 foot apart.

A *twisted structure* consists of vertical elements with the same structural properties as the cantilevers, and of horizontal elements with the same properties as the beams. The twisted structure resists torsion in both the vertical and horizontal planes.

The *height* of a *cantilever* is the vertical distance between the base elevation of the cantilever section and the top of the dam.

The *thickness* of a dam at any point is the distance between upstream and downstream faces along a line normal to the axis through the point.

The *abutment* of a *beam element* is the surface, at either end of the beam, which contacts the rock of the canyon wall.

The *crest* of a dam is the top of the dam.

1-5. Bibliography.

[1] "Design of Small Dams," second edition, Bureau of Reclamation, 1973.
[2] "Design of Arch Dams," first edition, Bureau of Reclamation, 1976.

Chapter II

Design Considerations

A. LOCAL CONDITIONS

2-1. General.—Although not of immediate concern to the designer of a dam and its appurtenances, the early collection of data on local conditions which will eventually relate to the design, specifications, and construction stages is advisable. Local conditions are not only needed to estimate construction costs, but may be of benefit when considering alternative designs and methods of construction. Some of these local conditions will also be used to determine the extent of the project designs, including such items as access roads, bridges, and construction camps.

2-2. Data to be Submitted.—Local conditions should be described and submitted as part of the design data as follows:

(1) The approximate distance from the nearest railroad shipping terminal to the structure site; load restrictions and physical inadequacies of existing roads and structures and an estimate of improvements to accommodate construction hauling; an estimate of length and major structures for access roads; and possible alternative means for delivering construction materials and equipment to the site.

(2) Local freight or trucking facilities and rates.

(3) Availability of housing and other facilities in the nearest towns; requirements for a construction camp; and need for permanent buildings for operating personnel.

(4) Availability or accessibility of public facilities or utilities such as water supply, sewage disposal, electric power for construction purposes, and telephone service.

(5) Local labor pool and general occupational fields existing in the area.

B. MAPS AND PHOTOGRAPHS

2-3. General.—Maps and photographs are of prime importance in the planning and design of a concrete dam and its appurtenant works. From these data an evaluation of alternative layouts can be made preparatory to determining the final location of the dam, the type and location of its appurtenant works, and the need for restoration and/or development of the area.

2-4. Survey Control.—Permanent horizontal and vertical survey control should be established at the earliest possible time. A grid coordinate system for horizontal control should be established with the origin located so that all of the features (including borrow areas) at a major structure will be in one quadrant. The coordinate system should be related to a State or National coordinate system, if practicable. All previous survey work, including topography and location and ground surface elevation of subsurface exploration holes, should be corrected to agree with the permanent control system; and all subsequent survey work, including location and ground

3

surface elevations, should be based on the permanent control.

2-5. Data to be Submitted.—A general area map should be obtained locating the general area within the State, together with county and township lines. This location map should show existing towns, highways, roads, railroads, and shipping points. A vicinity map should also be obtained using such a scale as to show details on the following:

(1) The structure site and alternative sites.

(2) Public utilities.

(3) Stream gaging stations.

(4) Existing manmade works affected by the proposed development.

(5) Locations of potential construction access roads, sites for a Government camp and permanent housing area, and sites for the contractor's camp and construction facilities.

(6) Sources of natural construction materials.

(7) Existing or potential areas or features having a bearing on the design, construction, operation, or management of project features such as recreational areas, fish and wildlife areas, building areas, and areas of ecological interest.

The topography of the areas where the dam and any of its appurtenant works are to be located is of prime concern to the designer.

Topography should be submitted covering an area sufficient to accommodate all possible arrangements of dam, spillway, outlet works, diversion works, construction access, and other facilities; and should be based on the permanently established horizontal and vertical survey control. A scale of 1 inch equals 50 feet and a contour interval of 5 feet will normally be adequate. The topography should extend a minimum of 500 feet upstream and downstream from the estimated positions of the heel and toe of the dam and a sufficient distance beyond each end of the dam crest to include road approaches. The topography should also cover the areas for approach and exit channels for the spillway. The topography should extend to an elevation sufficiently high to permit layouts of access roads, spillway structures, and visitor facilities.

Ground and aerial photographs are beneficial and can be used in a number of ways. Their principal value is to present the latest data relating to the site in such detail as to show conditions affecting the designs. Close-up ground photographs, for example, will often give an excellent presentation of local geology to supplement that obtained from a topographic map. Where modifications are to be made to a partially completed structure, such photographs will show as-constructed details which may not show on any drawings.

C. HYDROLOGIC DATA

2-6. Data to be Submitted.—In order to determine the potential of a site for storing water, generating power, or other beneficial use, a thorough study of hydrologic conditions must be made. Necessary hydrologic data will include the following:

(1) Streamflow records, including daily discharges, monthly volumes, and momentary peaks.

(2) Streamflow and reservoir yield.

(3) Project water requirements, including allowances for irrigation and power, conveyance losses, reuse of return

flows, and stream releases for fish; and dead storage requirements for power, recreation, fish and wildlife, etc.

(4) Flood studies, including inflow design floods and floods to be expected during periods of construction.

(5) Sedimentation and water quality studies, including sediment measurements, analysis of dissolved solids, etc.

(6) Data on ground-water tables in the vicinity of the reservoir and damsite.

(7) Water rights, including interstate compacts and international treaty effects,

and contractual agreements with local districts, power companies, and individuals for subordination of rights, etc.

Past records should be used as a basis for predicting conditions which will develop in the future. Data relating to streamflow may be obtained from the following sources:

(1) Water supply papers—U.S. Department of the Interior, Geological Survey, Water Resources Division.

(2) Reports of state engineers.

(3) Annual reports—International Boundary and Water Commission, United States and Mexico.

(4) Annual reports—various interstate compact commissions.

(5) Water right filings, permits—state engineers, county recorders.

(6) Water right decrees—district courts.

Data on sedimentation may be obtained from:

(1) Water supply papers—U.S. Department of the Interior, Geological Survey, Quality of Water Branch.

(2) Reports—U.S. Department of the Interior, Bureau of Reclamation; and U.S. Department of Agriculture, Soil Conservation Service.

Data for determining the quality of the water may be obtained from:

(1) Water supply papers—U.S. Department of the Interior, Geological Survey, Quality of Water Branch.

(2) Reports—U.S. Department of Health, Education, and Welfare, Public Health Service, and Environmental Protection Agency, Federal Water Control Administration.

(3) Reports—state public health departments.

2-7. *Hydrologic Investigations.*—Hydrologic investigations which may be required for project studies include the determination of the following: yield of streamflow, reservoir yield, water requirements for project purposes, sediment which will be deposited in the reservoir, floodflows, and ground-water conditions.

The most accurate estimate possible must be prepared of the portion of the streamflow yield that is surplus to senior water rights, as the basis of the justifiable storage. Reservoir storage will supplement natural yield of streamflow during low-water periods. Safe reservoir yield will be the quantity of water which can be delivered on a firm basis through a critical low-water period with a given reservoir capacity. Reservoir capacities and safe reservoir yields may be prepared from mass curves of natural streamflow yield as related to fixed water demands or from detailed reservoir operation studies, depending upon the study detail which is justified. Reservoir evaporation and other incidental losses should be accounted for before computation of net reservoir yields.

The critical low-water period may be one drought year or a series of dry years during the period of recorded water history. Water shortages should not be contemplated when considering municipal and industrial water use. For other uses, such as irrigation, it is usually permissible to assume tolerable water shortages during infrequent drought periods and thereby increase water use during normal periods with consequent greater project development. What would constitute a tolerable irrigation water shortage will depend upon local conditions and the crops to be irrigated. If the problem is complex, the consulting advice of an experienced hydrologist should be secured.

The annual rate at which sediment will be deposited in the reservoir should be ascertained to ensure that sufficient sediment storage is provided in the reservoir so that the useful functions of the reservoir will not be impaired by sediment deposition within the useful life of the project or the period of economic analysis, say 50 to 100 years. The expected elevation of the sediment deposition may also influence the design of the outlet works, necessitating a type of design which will permit raising the intake of the outlet works as the sediment is deposited.

Water requirements should be determined for all purposes contemplated in the project. For irrigation, consideration should be given to climatic conditions, soil types, type of crops, crop distribution, irrigation efficiency and conveyance losses, and reuse of return flows.

For municipal and industrial water supplies, the anticipated growth of demand over the life of the project must be considered. For power generation, the factors to be considered are load requirements and anticipated load growth.

Knowledge of consumptive uses is important in the design and operation of a large irrigation project, and especially for river systems as a whole. However, of equal and perhaps more importance to an individual farm or project is the efficiency with which the water is conveyed, distributed, and applied. The losses incidental to application on the farm and the conveyance system losses and operational waste may, in many instances, exceed the water required by the growing crops. In actual operation, the amount of loss is largely a matter of economics. In areas where water is not plentiful and high-value crops are grown, the use of pipe or lined conveyance systems and costly land preparation or sprinkler systems can be afforded to reduce losses to a minimum. A part of the lost water may be consumed nonbeneficially by nonproductive areas adjacent to the irrigated land or in drainage channels. Usually most of this water eventually returns to a surface stream or drain and is referred to as return flow.

In planning irrigation projects, two consumptive use values are developed. One, composed of monthly or seasonal values, is used with an adjustment for effective precipitation and anticipated losses mentioned above to determine the total water requirement for appraising the adequacy of the total water supply and determining reservoir storage requirements. The other, a peak use rate, is used for sizing the canal and lateral system.

Evapotranspiration, commonly called consumptive use, is defined as the sum of evaporation from plant and soil surfaces and transpiration from plants and is usually expressed in terms of depth (volume per unit area). Crop consumptive use is equal to evapotranspiration plus water required for plant tissue, but the two are usually considered the same. Predictions or estimates of evapotranspiration are basic parameters for the engineer or agronomist involved in planning and developing water resources. Estimates of evapotranspiration are also used in assessing the disposition of water in an irrigation project, evaluating the irrigation water-management efficiency, and projecting drainage requirements.

Reliable rational equations are available for estimating evapotranspiration when basic meteorological parameters such as net radiation, vapor pressure and temperature gradients, wind speed at a prescribed elevation above the crops or over a standard surface, and soil heat flux are available. When information on these parameters is not available, which is the usual case, recourse is made to empirical methods. Numerous equations, both empirical and partially based on theory, have been developed for estimating potential evapotranspiration. Estimates from these methods are generally accepted as being of suitable accuracy for planning and developing water resources. Probably the methods most widely used at this time are the Blaney-Criddle method shown in reference [1][1] and the Soil Conservation Service adaptation of the Blaney-Criddle method, shown in reference [2].

A more recent method, nearly developed sufficiently for general usage, is the Jensen-Haise solar radiation method shown in reference [3]. In general terms, these methods utilize climatic data to estimate a climatic index. Then coefficients, reflecting the stage of growth of individual crops and their actual water requirement in relationship to the climatic index, are used to estimate the consumptive use requirements for selected crops.

Project studies must include estimates of floodflows, as these are essential to the determination of the spillway capacity. Consideration should also be given to annual minimum and mean discharges and to the magnitudes of relatively common floods having 20-, 10-, and 4-percent chances of occurrence, as this knowledge is essential for construction purposes such as diverting the stream, providing cofferdam protection, and scheduling

---

[1]Numbers in brackets refer to items in the bibliography, sec. 2-31.

operations. Methods of arriving at estimates of floodflows are discussed in appendix G. If the feasibility studies are relatively complete, the flood determination may be sufficient for design purposes. If, however, floodflows have been computed for purposes of the feasibility study without making full use of all available data, these studies should be carefully reviewed and extended in detail before the actual design of the structure is undertaken. Frequently, new data on storms, floods, and droughts become available between the time the feasibility studies are made and construction starts. Where such changes are significant, the flood studies should be revised and brought up to date.

Project studies should also include a ground-water study, which may be limited largely to determining the effect of ground water on construction methods. However, some ground-water situations may have an important bearing on the choice of the type of dam to be constructed and on the estimates of the cost of foundations. Important ground-water information sometimes can be obtained in connection with subsurface investigations of foundation conditions.

As soon as a project appears to be feasible, steps should be taken in accordance with State water laws to initiate a project water right.

D. RESERVOIR CAPACITY AND OPERATION

2-8. *General.*—Dam designs and reservoir operating criteria are related to the reservoir capacity and anticipated reservoir operations. The loads and loading combinations to be applied to the dam are derived from the several standard reservoir water surface elevations. Reservoir operations are an important consideration in the safety of the structure and should not be overlooked in the design. Similarly, the reservoir capacity and reservoir operations are used to properly size the spillway and outlet works. The reservoir capacity is a major factor in flood routings and may determine the size and crest elevation of the spillway. The reservoir operation and reservoir capacity allocations will determine the location and size of outlet works for the controlled release of water for downstream requirements and flood control.

Reservoir area-capacity tables should be prepared before the final designs and specifications are completed. These area-capacity tables should be based upon the best available topographic data and should be the official document for final design and administrative purposes until superseded by a reservoir resurvey. Electronic computer programs are an aid in preparation of reservoir area and capacity data. These computers enable the designer to quickly have the best results

obtainable from the original field data.

2-9. *Reservoir Allocation Definitions.*—To ensure uniform reporting of data for design and construction, the following standard designations of water surface elevations and reservoir capacity allocations are used by the Bureau of Reclamation:

(a) *General.* Dam design and reservoir operation utilize reservoir capacity and water surface elevation data. To ensure uniformity in the establishment, use, and publication of these data, the following standard definitions of water surface elevations and reservoir capacities shall be used. Reservoir capacity as used here is exclusive of bank storage capacity.

(b) *Water Surface Elevation Definitions.* (Refer to fig. 2-1.)

(1) *Maximum Water Surface* is the highest 'acceptable water surface elevation with all factors affecting the safety of the structure considered. Normally, it is the highest water surface elevation resulting from a computed routing of the inflow design flood through the reservoir on the basis of established operating criteria. It is the top of surcharge capacity.

(2) *Top of Exclusive Flood Control Capacity* is the reservoir water surface elevation at the top of the reservoir capacity allocated to exclusive use for regulation of

RESERVOIR CAPACITY ALLOCATIONS

TYPE OF DAM			REGION	STATE	
OPERATED BY					RESERVOIR
CREST LENGTH FT; CREST WIDTH FT					DAM
VOLUME OF DAM CU YD					PROJECT
CONSTRUCTION PERIOD					DIVISION
STREAM					UNIT
RES AREA ACRES AT EL					STATUS OF DAM
ORIGINATED BY:			APPROVED BY:		
(Initials) (Code) (Date)			(Initials) (Code) (Date)		

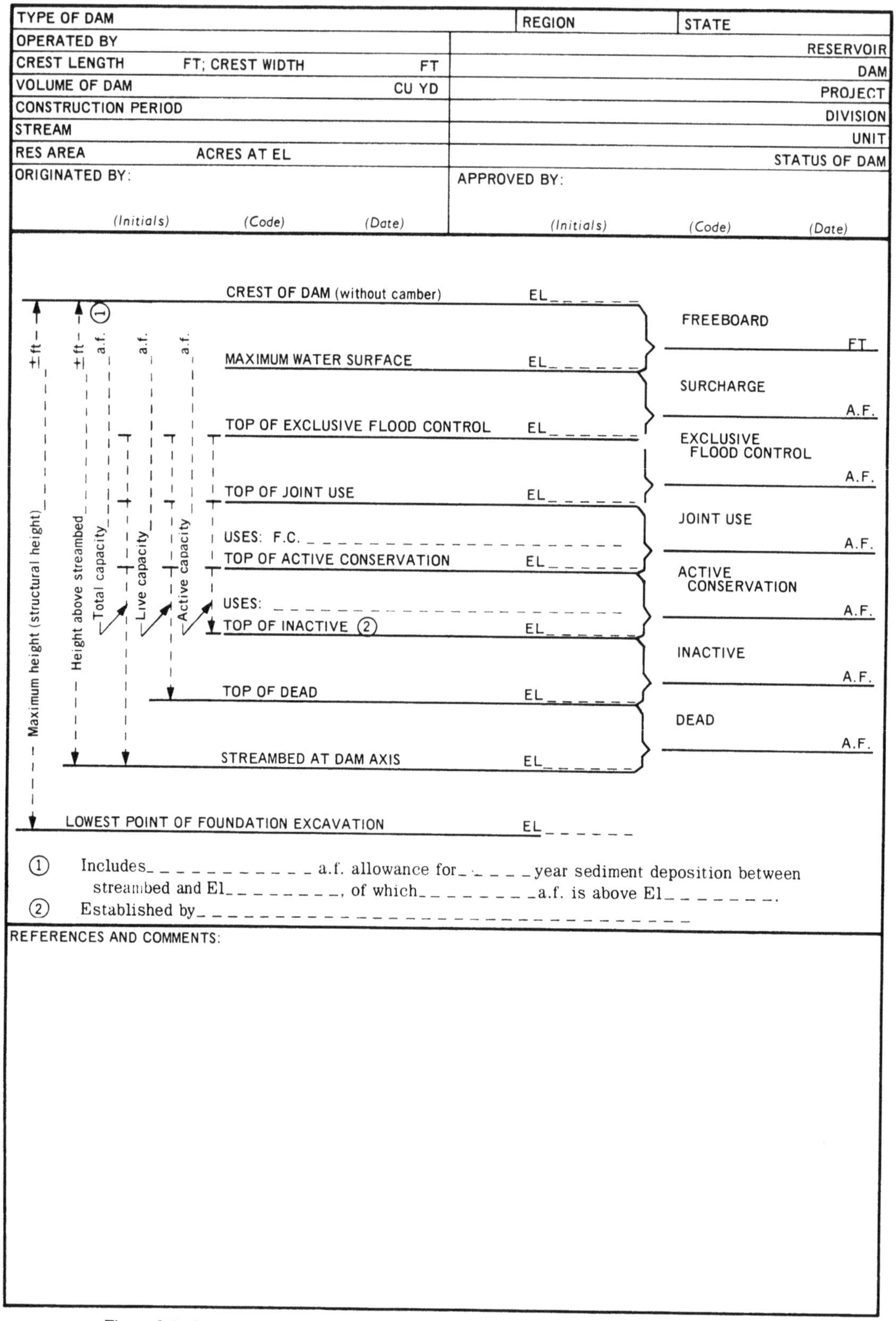

Figure 2-1. Reservoir capacity allocation sheet used by Bureau of Reclamation.

flood inflows to reduce damage downstream.

(3) *Maximum Controllable Water Surface Elevation* is the highest reservoir water surface elevation at which gravity flows from the reservoir can be completely shut off.

(4) *Top of Joint Use Capacity* is the reservoir water surface elevation at the top of the reservoir capacity allocated to joint use, i.e., flood control and conservation purposes.

(5) *Top of Active Conservation Capacity* is the reservoir water surface elevation at the top of the capacity allocated to the storage of water for conservation purposes only.

(6) *Top of Inactive Capacity* is the reservoir water surface elevation below which the reservoir will not be evacuated under normal conditions.

(7) *Top of Dead Capacity* is the lowest elevation in the reservoir from which water can be drawn by gravity.

(8) *Streambed at the Dam Axis* is the elevation of the lowest point in the streambed at the axis of the dam prior to construction. This elevation normally defines the zero for the area-capacity tables.

(c) *Capacity Definitions.*

(1) *Surcharge Capacity* is reservoir capacity provided for use in passing the inflow design flood through the reservoir. It is the reservoir capacity between the maximum water surface elevation and the highest of the following elevations:

    a. Top of exclusive flood control capacity.

    b. Top of joint use capacity.

    c. Top of active conservation capacity.

(2) *Total Capacity* is the reservoir capacity below the highest of the elevations representing the top of exclusive flood control capacity, the top of joint use capacity, or the top of active conservation capacity. In the case of a natural lake which has been enlarged, the total capacity includes the dead capacity of the lake. If the dead capacity of the natural lake has not been measured, specific mention of this fact should be made. Total capacity is used to express the total quantity of water which can be impounded and is exclusive of surcharge capacity.

(3) *Live Capacity* is that part of the total capacity from which water can be withdrawn by gravity. It is equal to the total capacity less the dead capacity.

(4) *Active Capacity* is the reservoir capacity normally usable for storage and regulation of reservoir inflows to meet established reservoir operating requirements. Active capacity extends from the highest of the top of exclusive flood control capacity, the top of joint use capacity, or the top of active conservation capacity, to the top of inactive capacity. It is the total capacity less the sum of the inactive and dead capacities.

(5) *Exclusive Flood Control Capacity* is the reservoir capacity assigned to the sole purpose of regulating flood inflows to reduce flood damage downstream. In some instances the top of exclusive flood control capacity is above the maximum controllable water surface elevation.

(6) *Joint Use Capacity* is the reservoir capacity assigned to flood control purposes during certain periods of the year and to conservation purposes during other periods of the year.

(7) *Active Conservation Capacity* is the reservoir capacity assigned to regulate reservoir inflow for irrigation, power, municipal and industrial use, fish and wildlife, navigation, recreation, water quality, and other purposes. It does not include exclusive flood control or joint use capacity. The active conservation capacity extends from the top of the active conservation capacity to the top of the inactive capacity.

(8) *Inactive Capacity* is the reservoir capacity exclusive of and above the dead capacity from which the stored water is normally not available because of operating agreements or physical restrictions. Under abnormal conditions, such as a shortage of water or a requirement for structural repairs, water may be evacuated from this space

after obtaining proper authorization. The highest applicable water surface elevation described below usually determines the top of inactive capacity.

a. The lowest water surface elevation at which the planned minimum rate of release for water supply purposes can be made to canals, conduits, the river, or other downstream conveyance. This elevation is normally established during the planning and design phases and is the elevation at the end of extreme drawdown periods.

b. The established minimum water surface elevation for fish and wildlife purposes.

c. The established minimum water surface elevation for recreation purposes.

d. The minimum water surface elevation as set forth in compacts and/or agreements with political subdivisions.

e. The minimum water surface elevation at which the powerplant is designed to operate.

f. The minimum water surface elevation to which the reservoir can be drawn using established operating procedures without endangering the dam, appurtenant structures, or reservoir shoreline.

g. The minimum water surface elevation or the top of inactive capacity established by legislative action.

(9) *Dead Capacity* is the reservoir capacity from which stored water cannot be evacuated by gravity.

2-10. *Data to be Submitted.*—To complete the designs of the dam and its appurtenant works, the following reservoir design data should be submitted:

(1) Area-capacity curves and/or tables computed to an elevation high enough to allow for storage of the spillway design flood.

(2) A topographic map of the reservoir site prepared to an appropriate scale.

(3) Geological information pertinent to reservoir tightness, locations of mines or mining claims, locations of oil and natural gas wells.

(4) Completed reservoir storage allocations and corresponding elevations.

(5) Required outlet capacities for respective reservoir water surfaces and any required sill elevations. Give type and purpose of reservoir releases and the time of year these must be made. Include minimum releases required.

(6) Annual periodic fluctuations of reservoir levels shown by tables or charts summarizing reservoir operation studies.

(7) Method of reservoir operation for flood control and maximum permissible releases consistent with safe channel capacity.

(8) Physical, economic, or legal limitations to maximum reservoir water surface.

(9) Anticipated occurrence and amounts of ice (thickness) and floating debris, and possible effect on reservoir outlets, spillway, and other appurtenances.

(10) Extent of anticipated wave action, including a discussion of wind fetch.

(11) Where maintenance of flow into existing canals is required, determine maximum and probable carrying capacity of such canal, and time of year when canals are used.

E. CLIMATIC EFFECTS

2-11. *General.*—The climatic conditions which are to be encountered at the site affect the design and construction of the dam. Measures which should be employed during the construction period to prevent cracking of concrete must be related to the ambient temperatures encountered at the site. Construction methods and procedures may also be dependent upon the weather conditions, since weather affects the rate of construction and the overall construction schedule. Accessibility of the site during periods of inclement weather affects the construction schedule and should be investigated.

2-12. *Data to be Submitted.*—The following data on climatic conditions should be submitted as part of the design data:

(1) Weather Service records of mean

monthly maximum, mean monthly minimum, and mean monthly air temperatures for the nearest station to the site. Data on river water temperatures at various times of the year should also be obtained.

(2) Daily readings of maximum and minimum air temperatures should be submitted as soon as a station can be established at the site.

(3) Daily readings of maximum and minimum river water temperatures should be submitted as soon as a station can be established at the site.

(4) Amount and annual variance in rainfall and snowfall.

(5) Wind velocities and prevailing direction.

F. CONSTRUCTION MATERIALS

2-13. *Concrete Aggregates*.—The construction of a concrete dam requires the availability of suitable aggregates in sufficient quantity to construct the dam and its appurtenant structures. Aggregates are usually processed from natural deposits of sand, gravel, and cobbles. However, if it is more practical, they may be crushed from suitable rock. For small dams, the aggregates may be obtained from existing commercial sources. If the aggregates are obtained from borrow pits or rock quarries, provisions should be made to landscape and otherwise restore the areas to minimize adverse environmental effects. If aggregates are available from the reservoir area, particularly below minimum water surface, their adverse effects would be minimized. However, any early storage in the reservoir, prior to completion of the dam, may rule out the use of aggregate sources in the reservoir.

2-14. *Water for Construction Purposes*.—For large rivers, this item is relatively unimportant except for quality of the water. For small streams and offstream reservoirs, water for construction purposes may be difficult to obtain. An adequate supply of water for construction purposes such as washing aggregates and cooling and batching concrete should be assured to the contractor, and the water rights should be obtained for him. If necessary to use ground water, information on probable sources and yields should be obtained. Information on locations and yields of existing wells in the vicinity, restrictions if any on use of ground water, and necessary permits should also be obtained.

2-15. *Data to be Submitted*.—In addition to the data on concrete aggregates and water for construction purposes, the following data on construction materials should be obtained:

(1) An earth materials report containing information on those potential sources of soils, sand, and gravel which could be used for backfill and bedding materials.

(2) Information on riprap for protection of slopes.

(3) Information on sources and character of acceptable road surfacing materials, if required.

(4) References to results of sampling, testing, and analysis of construction materials.

(5) Photographs of sources of construction materials.

(6) Statement of availability of lumber for structural work.

G. SITE SELECTION

2-16. *General*.—A water resources development project is designed to perform a certain function and to serve a particular area. Once the purpose and the service area are defined, a preliminary site selection can be made.

Following the determination of the adequacy of the water supply as discussed in subchapter C, the two most important considerations in selecting a damsite are: (1) the site must be adequate to support the dam and the appurtenant structures, and (2) the area upstream from the site must be suitable for a reservoir. There are often several suitable sites along a river where the dam can be located.

The site finally selected should be that where the dam and reservoir can be most economically constructed with a minimum of interference with local conditions and still serve their intended purpose. An experienced engineer can usually eliminate some of the sites from further consideration. Cost estimates may be required to determine which of the remaining sites will provide the most economical structure.

2-17. *Factors in Site Selection.*—In selecting a damsite the following should be considered:

Topography — A narrow site will minimize the amount of material in the dam thus reducing its cost, but such a site may be adaptable to an arch dam and this possibility should be investigated.

Geology — The foundation of the dam should be relatively free of major faults and shears. If these are present, they may require expensive foundation treatment to assure an adequate foundation.

Appurtenant Structures — While the cost of these structures is usually less than the cost of the dam, economy in design may be obtained by considering their effect at the time of site selection. For example, if a river has a large flow, a large spillway and diversion works will be required. Selecting a site which will better accommodate these appurtenances will reduce the overall cost.

Local Conditions — Some sites may have roads, railroads, powerlines, canals, etc., which have to be relocated, thus increasing the overall costs.

Access — Accessibility of the site has a very definite effect on the total cost. Difficult access may require the construction of expensive roads. An area suitable for the contractor's plant and equipment near the site will reduce the contractor's construction costs.

H. CONFIGURATION OF DAM

2-18. *Nonoverflow Section.*—A gravity dam is a concrete structure designed so that its weight and thickness insure stability against all the imposed forces. The downstream face will usually be a uniform slope which, if extended, would intersect the vertical upstream face at or near the maximum reservoir water level. The upper portion of the dam must be thick enough to resist the shock of floating objects and to provide space for a roadway or other required access. The upstream face will normally be vertical. This concentrates most of the concrete weight near the upstream face where it will be most effective in overcoming tensile stresses due to the reservoir water loading. The thickness is also an important factor in resistance to sliding and may dictate the slope of the downstream face. Thickness may also be increased in the lower part of the dam by an upstream batter.

2-19. *Overflow Section.*—The spillway may be located either in the abutment or on the dam. If it is located on a portion of the dam, the section should be similar to the abutment section but modified at the top to accommodate the crest and at the toe to accommodate the energy dissipator. The elevation of the crest and its shape will be determined by hydraulic requirements, and the shaping at the toe by the energy dissipator. Stability requirements for the overflow section may involve some changes from the theoretical hydraulic shapes. Hydraulic design of the overflow section is discussed fully in chapter IX. For structural design of the dam see chapters III and IV.

I. FOUNDATION INVESTIGATIONS

2-20. *Purpose.*—The purpose of a foundation investigation is to provide the data necessary to properly evaluate a foundation. A properly sequenced and organized foundation investigation will provide the data necessary to evaluate and analyze the foundation at any stage of investigation.

2-21. *Field Investigations.*—The collection, study, and evaluation of foundation data is a continuing program from the time of the appraisal investigation to the completion of construction. The data collection begins with an appraisal and continues on a more detailed basis through the design phase. Data are also collected continuously during construction to correlate with previously obtained information and to evaluate the need for possible design changes.

(a) *Appraisal Investigation.*—The appraisal investigation includes a preliminary selection of the site and type of dam. All available geologic and topographic maps, photographs of the site area, and data from field examinations of natural outcrops, road cuts, and other surface conditions should be utilized in the selection of the site and preliminary evaluation of the foundation.

The amount of investigation necessary for appraisal will vary with the anticipated difficulty of the foundation. In general, the investigation should be sufficient to define the major geologic conditions with emphasis on those which will affect design. A typical geologic map and profile are shown on figures 2-2 and 2-3.

The geologic history of a site should be thoroughly studied, particularly where the geology is complex. Study of the history may assist in recognizing and adequately investigating hidden but potentially dangerous foundation conditions.

Diamond core drilling during appraisal investigations may be necessary in more complex foundations and for the foundations for larger dams. The number of drill holes required will depend upon the areal extent and complexity of the foundation. Some foundations may require as few as three or four drill holes to define an uncertain feature. Others may require substantially more drilling to determine foundation treatment for a potentially dangerous foundation condition.

Basic data that should be obtained during the appraisal investigation, with refinement continuing until the construction is complete, are:

(1) Dip, strike, thickness, composition, and extent of faults and shears.

(2) Depth of overburden.

(3) Depth of weathering.

(4) Joint orientation and continuity.

(5) Lithology throughout the foundation.

(6) Physical properties tests of the foundation rock. Tests performed on similar foundation materials may be used for estimating the properties in the appraisal phase.

(b) *Feasibility Investigation.*—During the feasibility phase, the location of the dam is usually finalized and the basic design data are firmed up. The geologic mapping and sections are reviewed and supplemented by additional data such as new surveys and additional drill holes. The best possible topography should be used. In most cases, the topography is easily obtained by aerial photogrammetry to almost any scale desired.

The drilling program is generally the means

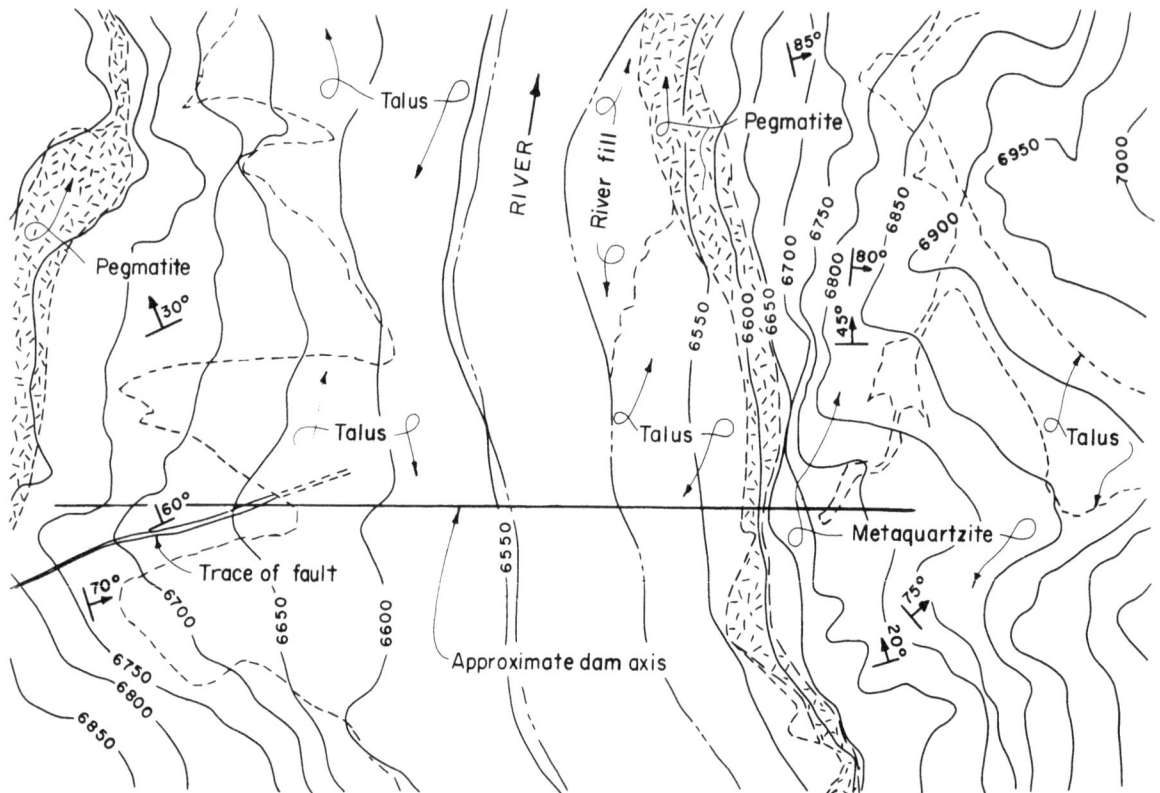

$\top$ 60° Indicates strike and dip of fault $\downarrow$ 10°   Indicates strike and dip of joints

Figure 2-2. A typical geologic map of a gravity damsite.—288-D-2952

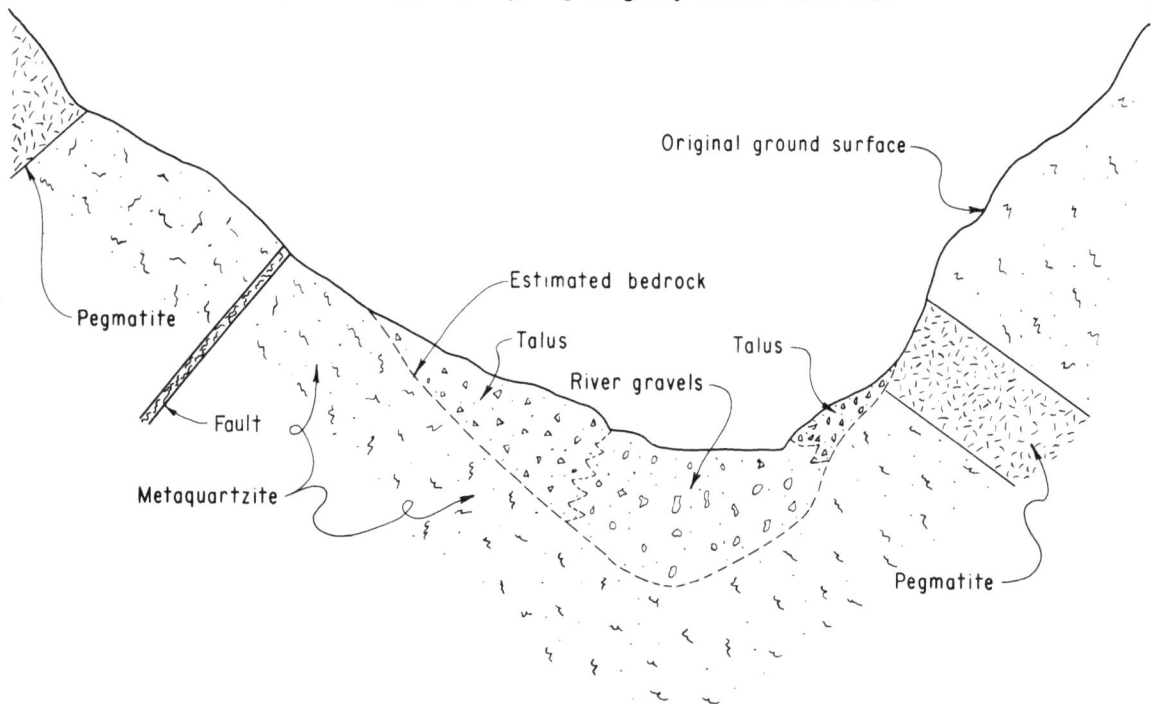

Figure 2-3. A typical geologic profile of a damsite.—288-D-2954

of obtaining the additional data required for the feasibility stage. The program takes advantage of any knowledge of special conditions revealed during the appraisal investigation. The drill holes become more specifically oriented and increased in number to better define the foundation conditions and determine the amount of foundation treatment required.

The rock specimens for laboratory testing during the feasibility investigations are usually nominal, as the actual decision for construction of the dam has not yet been made. Test specimens should be obtained to determine more accurately physical properties of the foundation rock and for petrographic examination. Physical properties of joint or fault samples may be estimated by using conservative values from past testing of similar materials. The similarity of materials can be judged from the cores retrieved from the drilling.

(c) *Final Design Data.*—Final design data are required prior to the preparation of the specifications. A detailed foundation investigation is conducted to obtain the final design data. This investigation involves as many drill holes as are necessary to accurately define the following items:

(1) Strike, dip, thickness, continuity, and composition of all faults and shears in the foundation.

(2) Depth of overburden.

(3) Depth of weathering throughout the foundation.

(4) Joint orientation and continuity.

(5) Lithologic variability.

(6) Physical properties of the foundation rock, including material in the faults and shears.

The foundation investigation may involve, besides diamond core drilling, detailed mapping of surface geology and exploration of dozer trenches and exploratory openings such as tunnels, drifts, and shafts. The exploratory openings can be excavated by contract prior to issuing final specifications. These openings provide the best possible means of examining the foundation. In addition, they provide

excellent in situ testing locations and areas for test specimen collection.

In addition to test specimens for determining the physical properties, specimens may be required for final design for use in determining the shear strength of the rock types, healed joints, and open joints. This information may be necessary to determine the stability of the foundation and is discussed as the shear-friction factor in subchapter F of chapter III.

Permeability tests should be performed as a routine matter during the drilling program. The information obtained can be utilized in establishing flow nets which will aid in studying uplift conditions and establishing drainage systems. The permeability testing methods presently used by the Bureau of Reclamation are described in designation E-18 of the Earth Manual [4] and the report entitled "Drill Hole Water Tests—Technical Instructions," published by the Bureau of Reclamation in July 1972.

2-22. *Construction Geology.*—The geology as encountered in the excavation should be defined and compared with the preexcavation geology. Geologists and engineers should consider carefully any geologic change and check its relationship to the design of the structure.

As-built geology drawings should be developed even though revisions in design may not be required by changed geologic conditions, since operation and maintenance problems may develop requiring detailed foundation information.

2-23. *Foundation Analysis Methods.*—In most instances, a gravity dam is keyed into the foundation so that the foundation will normally be adequate if it has enough bearing capacity to resist the loads from the dam. However, a foundation may have faults, shears, seams, joints, or zones of inferior rock that could develop unstable rock masses when acted on by the loads of the dam and reservoir. The safety of the dam against sliding along a joint, fault, or seam in the foundation can be determined by computing the shear-friction factor of safety. This method of analysis is explained in subchapter F of chapter III. If there are several joints, faults, or seams along which failure can occur, the potentially unstable rock mass can be analyzed by a method called rigid block analysis. This method is explained in detail in subchapter F of chapter IV. These methods of analysis may also be applied to slope stability problems.

The data required for these two methods of analysis are:

(1) Physical properties.

(2) Shearing and sliding strengths of the discontinuities and the rock.

(3) Dip and strike of the faults, shears, seams, and joints.

(4) Limits of the potentially unstable rock mass.

(5) Uplift pressures on the failure surfaces.

(6) Loads to be applied to the rock mass.

When a foundation is interspersed by many faults, shears, joints, seams, and zones of inferior rock, the finite element method of analysis can be used to determine the bearing capacity and the amount of foundation treatment required to reduce or eliminate areas of tension in the foundation. The description of this method can be found in subchapter E of chapter IV. In addition to the data required for the rigid block analysis, the finite element analysis requires the deformation moduli of the various parts of the foundation.

2-24. *In Situ Testing*.—In situ shear tests [5] are more expensive than similar laboratory tests; consequently, comparatively few can be run. The advantage of a larger test surface may require that a few in situ tests be supplemented by a greater number of laboratory tests. The shearing strength relative to both horizontal and vertical movement should be obtained by either one or a combination of both methods.

Foundation permeability tests may be run in conjunction with the drilling program or as a special program. The tests should be performed according to designation E-18 of the Earth

Manual [4] and the report entitled "Drill Hole Water Tests—Technical Instructions," published by the Bureau of Reclamation in July 1972.

2-25. *Laboratory Testing*.—The following laboratory tests are standard and the methods and test interpretations should not vary substantially from one laboratory to another. A major problem involved with laboratory testing is obtaining representative samples. Sample size is often dictated by the laboratory equipment and is a primary consideration. Following is a list of laboratory tests:

Physical Properties Tests

(1) Compressive strength
(2) Elastic modulus
(3) Poisson's ratio
(4) Bulk specific gravity
(5) Porosity
(6) Absorption

Shear Tests

(1) Direct shear ⎫ Perform on intact
                 ⎬ specimens and
(2) Triaxial shear ⎭ those with healed
                     joints
(3) Sliding friction   Perform on open
                       joints

Other Tests

(1) Solubility
(2) Petrographic analysis

2-26. *Consistency of Presentation of Data*.—It is important that the design engineers, laboratory personnel, and geologists be able to draw the same conclusions from the information presented in the investigations. The standardization of the geologic information and laboratory test results is therefore essential and is becoming increasingly so with the newer methods of analysis.

J. CONSTRUCTION ASPECTS

2-27. *General.*—The construction problems that may be encountered by the contractor in constructing the dam and related features should be considered early in the design stage. One of the major problems, particularly in narrow canyons, is adequate area for the contractor's construction plant and equipment and for storage of materials in the proximity of the dam. Locating the concrete plant to minimize handling of the concrete and the aggregates and cement can materially reduce the cost of the concrete.

Permanent access roads should be located to facilitate the contractor's activities as much as practicable. This could minimize or eliminate unsightly abandoned construction roads. Structures should be planned to accommodate an orderly progression of the work. The length of the construction season should be considered. In colder climates and at higher elevations it may be advantageous to suspend all or part of the work during the winter months. Adequate time should be allowed for construction so that additional costs for expedited work are not encountered.

2-28. *Construction Schedule.*—The contractor's possible methods and timing of construction should be considered at all times during the design of the dam and its appurtenant structures. Consideration of the problems which may be encountered by the contractor can result in significant savings in the cost of construction. By developing an anticipated construction schedule, potential problems in the timing of construction of the various parts can be identified. If practicable, revisions in the design can be made to eliminate or minimize the effect of the potential problems. The schedule can be used to program supply contracts and other construction contracts on related features of the project. It is also useful as a management tool to the designer in planning his work so that specifications and construction drawings can be provided when needed.

The construction schedule can be made by several methods such as Critical Path Method (CPM), Program Evaluation and Review Technique (PERT), and Bar Diagram. Figure 2-4 shows a network for a portion of a hypothetical project for a CPM schedule. Data concerning the time required for various parts of the work and the interdependencies of parts of the work can be programmed into a computer which will calculate the critical path. It will also show slack time or areas which are not critical. In this example, there are two paths of activities. The path which is critical is the preparation of specifications, awarding of contract, and the construction of "A," "B," and "D". The second path through construction of "C" and "E" is not critical. As the work progresses, the current data on the status of all the phases of work completed and in progress can be fed back into the computer. The computer will then recompute the critical path, thus establishing a new path if another phase of the work has become critical, and will point out any portion of the work that is falling behind the required schedule.

Figure 2-5 shows the construction schedule for the hypothetical project on a bar diagram. This diagram is made by plotting bars to the length of time required for each portion of the work and fitting them into a time schedule, checking visually to make sure interrelated activities are properly sequenced.

K. MISCELLANEOUS CONSIDERATIONS

2-29. *Data to be Submitted.*—Many items not covered above affect the design and construction of a dam. Some of these are noted below. In securing and preparing design data, the adequacy and accuracy of the data should contemplate their possible subsequent utility for expansion into specifications design data.

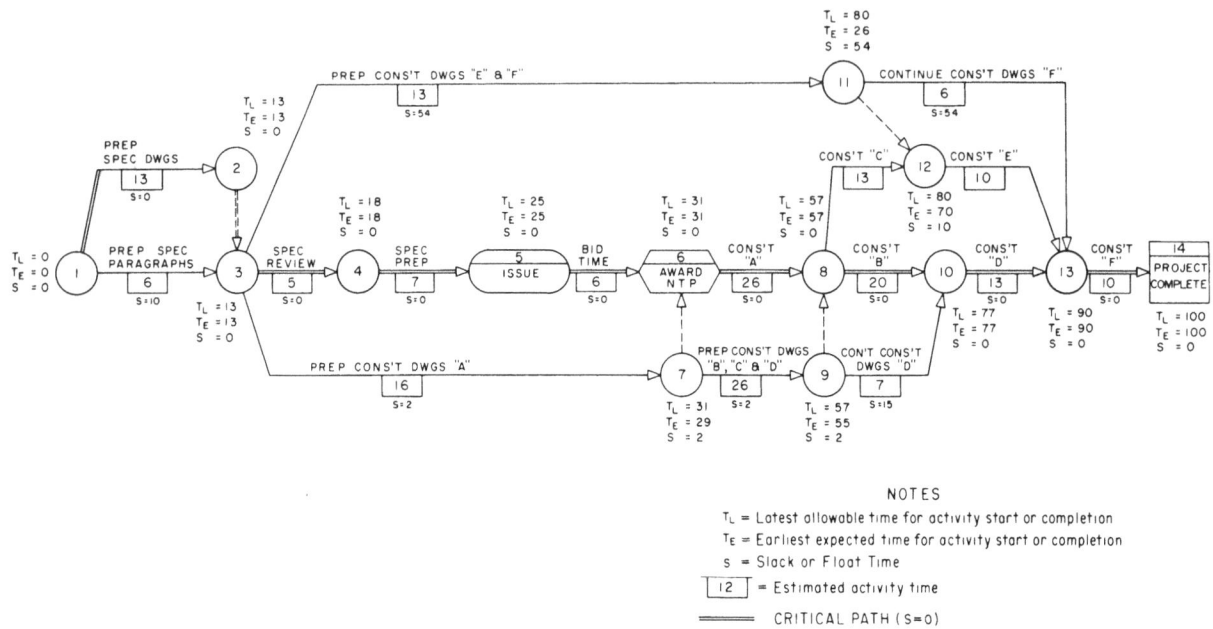

Figure 2-4. **Typical construction schedule using Critical Path Method (CPM).—288-D-2955**

(1) Details of roadway on crest of dam (and approaches) if required.

(2) Present or future requirement for highway crossing on dam.

(3) Details on fishways and screens, with recommendations of appropriate fish authorities.

(4) Existing works to be replaced by incorporation into dam.

(5) Future powerplant or power development.

(6) Navigation facilities.

(7) Possibility of raising crest of dam in future.

(8) Anticipated future river channel improvement or other construction which might change downstream river regimen.

(9) Recreational facilities anticipated to be authorized, and required provisions for public safety.

(10) Recommended period of construction.

(11) Commitments for delivery of water or power.

(12) Designation of areas within right-of-way boundaries for disposal of waste materials.

2-30. *Other Considerations*.—Design consideration must take into account construction procedures and costs. An early evaluation and understanding of these is necessary if a rapid and economical construction of the dam is to be attained.

Designs for mass concrete structures and their appurtenances should be such that sophisticated and special construction equipment will not be required. Thin, curved walls with close spacing of reinforcement may be desirable for several reasons, and may represent the minimum cost for materials such as cement, flyash, admixtures, aggregates, and reinforcing steel. However, the cost of forming and labor for construction of this type and the decreased rate of concrete placement may result in a much higher total cost than would result from a simpler structure of greater dimensions.

Design and construction requirements should permit and encourage the utilization of machine power in place of manpower wherever practicable. Any reduction in the requirement for high-cost labor will result in a significant cost savings in the completed structure. Work

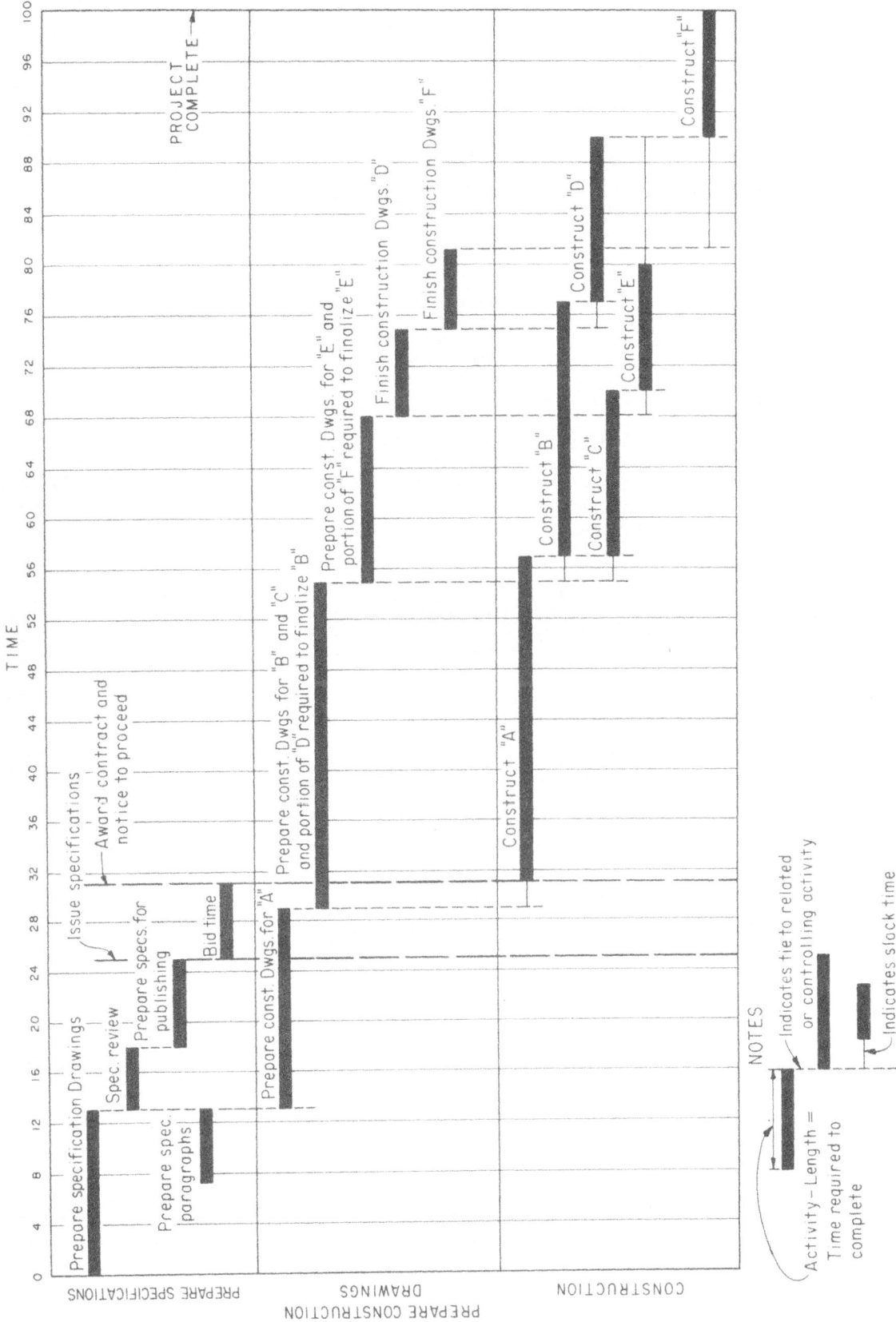

Figure 2-5. Typical construction schedule using a bar diagram.—28S-D-2956

areas involved in a high labor use include the placing, compaction, and curing of the concrete, the treatment and cleanup of construction joints, and the repair and finishing of the concrete surfaces.

Forming is a significant cost in concrete structures. Designs should permit the simpler forms to be used, thus facilitating fabrication,

installation, and removal of the forms. Repetitive use of forms will materially reduce forming costs. Although wooden forms are lower in initial cost, they can only be used a limited number of times before they warp and fail to perform satisfactorily. The reuse of steel forms is limited only by the designs and the demands of the construction schedule.

L. BIBLIOGRAPHY

2-31. *Bibliography*

[1] U.S. Department of Agriculture, Agricultural Research Service, "Determining Consumptive Use and Irrigation Water Requirements," Technical Bulletin No. 1275, December 1962.

[2] U.S. Department of Agriculture, Soil Conservation Service, "Irrigation Water Requirements," Technical Release No. 21, April 1967.

[3] Jensen, M. E., "Water Consumption by Agricultural Plants," Water Deficits and Plant Growth, vol. II, Academic Press, New York, N.Y., 1968, pp. 1-22.

[4] "Field Permeability Tests in Boreholes," Earth Manual, Designation E-18, Bureau of Reclamation, 1974.

[5] "Morrow Point Dam Shear and Sliding Friction Tests," Concrete Laboratory Report No. C-1161, Bureau of Reclamation, 1965.

[6] Wallace, G. B., Slebir, E. J., and Anderson, F. A., "Radial Jacking Test for Arch Dams," Tenth Rock Mechanics Symposium, University of Texas, Austin, Tex., 1968.

[7] Wallace, G. B., Slebir, E. J., and Anderson, F. A., "In Situ Methods for Determining Deformation Modulus Used by the Bureau of Reclamation," Winter Meeting, American Society for Testing and Materials, Denver, Colo., 1969.

[8] Wallace, G. B., Slebir, E. J., and Anderson, F. A., "Foundation Testing for Auburn Dam," Eleventh Symposium on Rock Mechanics, University of California, Berkeley, Calif., 1969.

Chapter III

Design Data and Criteria

A. INTRODUCTION

3-1. *Basic Assumptions.*—Computational methods require some basic assumptions for the analysis of a gravity dam. The assumptions which cover the continuity of the dam and its foundation, competency of the concrete in the dam, adequacy of the foundation, and variation of stresses across the sections of the dam are as follows:

(1) Rock formations at the damsite are, or will be after treatment, capable of carrying the loads transmitted by the dam with acceptable stresses.

(2) The dam is thoroughly bonded to the foundation rock throughout its contact with the canyon.

(3) The concrete in the dam is homogeneous, uniformly elastic in all directions, and strong enough to carry the applied loads with stresses below the elastic limit.

(4) Contraction joints that are keyed and grouted may be considered to create a monolithic structure, and loads may be transferred horizontally to adjacent blocks by both bending and shear. If the joints are keyed but not grouted, loads may be transferred horizontally to adjacent blocks by shear across the keys. Where joints are neither keyed nor grouted, the entire load on the dam will be transferred vertically to the foundation. If joints are grouted, they will be grouted before the reservoir loads are applied so that the structure acts monolithically.

(5) Horizontal and vertical stresses vary linearly from the upstream face to the downstream face.

(6) Horizontal shear stresses have a parabolic variation from the upstream face to the downstream face.

B. CONCRETE

3-2. *Concrete Properties.*—A gravity dam must be constructed of concrete which will meet the design criteria for strength, durability, permeability, and other properties. Although mix proportions are usually controlled by strength and/or durability requirements, the cement content should be held to an acceptable minimum in order to minimize the heat of hydration. Properties of concrete vary with age and with proportions and types of ingredients.

Tests must be made on specimens using the full mass mix and the specimens must be of sufficient age to adequately evaluate the strength and elastic properties which will exist for the concrete in the dam [1][1].

(a) *Strength.*—The strength of concrete should satisfy early load and construction requirements, and at some specific age should

---

[1]Numbers in brackets refer to items in the bibliography, sec. 3-23.

have the specified compressive strength as determined by the designer. This specific age is often 365 days but may vary from one structure to another.

Tensile strength of the concrete mix should be determined as a companion test series using the direct tensile test method.

Shear strength is a combination of internal friction, which varies with the normal compressive stress, and cohesive strength. Companion series of shear strength tests should be conducted at several different normal stress values covering the range of normal stresses to be expected in the dam. These values should be used to obtain a curve of shear strength versus normal stress.

(b) *Elastic Properties.*—Concrete is not a truly elastic material. When concrete is subjected to a sustained load such as may be expected in a dam, the deformation produced by that load may be divided into two parts—the elastic deformation, which occurs immediately due to the instantaneous modulus of elasticity; and the inelastic deformation, or creep, which develops gradually and continues for an indefinite time. To account for the effects of creep, the sustained modulus of elasticity is used in the design and analysis of a concrete dam.

The stress-strain curve is, for all practical purposes, a straight line within the range of usual working stresses. Although the modulus of elasticity is not directly proportional to the strength, the high strength concretes usually have higher moduli. The usual range of the instantaneous modulus of elasticity for concrete at 28-day age is between 2.0×10^6 and 6.0×10^6 pounds per square inch.

(c) *Thermal Properties.*—The effects of temperature change on a gravity dam are dependent on the thermal properties of the concrete. Thermal properties necessary for the evaluation of temperature effects are the coefficient of thermal expansion, thermal conductivity, and specific heat [7]. The coefficient of thermal expansion is the length change per unit length per degree temperature change. Thermal conductivity is the rate of heat conduction through a unit thickness over a unit area of the material subjected to a unit

temperature difference between faces. The specific heat is defined as the amount of heat required to raise the temperature of a unit mass of the material 1 degree. Diffusivity of concrete is an index of the facility with which concrete will undergo temperature change. Diffusivity is a function of the values of specific heat, thermal conductivity, and density.

(d) *Dynamic Properties.*—Concrete, when subjected to dynamic loadings, may exhibit characteristics unlike those occurring during static loadings. Testing is presently underway in the Bureau's laboratory to determine the properties of concrete when subjected to dynamic loading. Until sufficient test data are available, static strengths and the instantaneous modulus of elasticity should be used.

(e) *Other Properties.*—In addition to the strength, elastic modulus, and thermal properties, several other properties of concrete should be evaluated during the laboratory testing program. These properties, which must be determined for computations of deformations and stresses in the concrete structures, are Poisson's ratio, unit weight, and any autogenous growth or drying shrinkage.

(f) *Average Concrete Properties.*—For preliminary studies until laboratory test data are available, the necessary values may be estimated from published data [2] for similar tests. Until long-term load tests are made to determine the effects of creep, the sustained modulus of elasticity should be taken as 60 to 70 percent of the laboratory value of the instantaneous modulus of elasticity.

If no tests or published data are available, the following may be assumed for preliminary studies:

Specified compressive strength = 3,000 to 5,000 p.s.i.

Tensile strength = 4 to 6 percent of the compressive strength

Shear strength:

Cohesion = 10 percent of the compressive strength

Coefficient of internal friction = 1.0

Sustained modulus of elasticity = 3.0×10^6 p.s.i. (static load including effects of creep)

Instantaneous modulus of elasticity = 5.0 x 10^6 p.s.i. (dynamic or short time load)

Coefficient of thermal expansion = 5.0 x 10^{-6} per degree F.

Poisson's ratio = 0.20

Unit weight of concrete = 150 pounds per cubic foot.

C. FOUNDATION

3-3. *Introduction.*—Certain information concerning the foundation is required for design of the gravity dam section. The design of the dam and any treatment to the foundation (see sec. 6-3) to improve its properties are considered separate problems. If treatments are applied to the foundation, the data used for the design of the dam should be based on the properties of the foundation after treatment. A geologic investigation is required to determine the general suitability of the site and to identify the types and structures of the materials to be encountered. After these identifications have been made the following three parameters should be determined:

(1) For each material the shear strengths of intact portions, the sliding friction strengths of discontinuities, and the shear strength at each interface with a different material (including the strength at the interface of concrete and the material exposed on the completed excavated surface).

(2) The permeability of each material.

(3) The deformation modulus of the foundation.

The discussion of foundation investigation in chapter II (secs. 2-20 through 2-26) lists the physical properties normally required and the samples desired for various foundation materials.

3-4. *Foundation Deformation.*—Accurate knowledge of the modulus of deformation of the foundation of a gravity dam is required to:

(1) Determine the extent of relative deformation between locations where physical properties vary along the foundation in the vertical or horizontal directions.

(2) Determine the stress concentrations in the dam or foundation due to local low modulus regions adjacent to or below the dam.

(3) Determine the stress distribution to be used in detailed stability studies.

The foundation investigation should provide information related to or giving deformation moduli and elastic moduli. (Deformation modulus is the ratio of stress to elastic plus inelastic strain. Elastic modulus is the ratio of stress to elastic strain.) The information includes elastic modulus of drill core specimens, elastic modulus and deformation modulus from in situ jacking tests, deformation modulus of fault or shear zone material, and logs of the jointing occurring in recovered drill cores. Knowledge of the variation in materials and their relative prevalence at various locations along the foundation is provided by the logs of drill holes and by any tunnels in the foundation.

When the composition of the foundation is nearly uniform over the extent of the dam contact, has a regular jointing pattern, and is free of low modulus seams, the three conditions listed above do not exist and thus an accurate deformation modulus is not required. An estimate based on reduction of the elastic modulus of drill core specimens will suffice. However, when a variation of materials, an irregular jointing pattern, and fault and shear zones exist, the deformation moduli of each type of material in the foundation will be required for design. The analysis of the interaction of the dam and foundation may be accomplished by using finite element analysis.

The moduli values are determined by laboratory or in situ testing, and if necessary are modified to account for factors not included in the initial testing. The modification to the modulus value of rock may be determined according to the rock quality index [3] or joint shear index [4]. Modification of the moduli values for shear or fault zone material may be required if the geometry of the zone is quite variable. An example of such a modification is given in reference [4].

3-5. Foundation Strength. —Compressive strength of the foundation rock can be an important factor in determining thickness requirements for a dam at its contact with the foundation. Where the foundation rock is nonhomogeneous, a sufficient number of tests, as determined by the designer, should be made to obtain compressive strength values for each type of rock in the loaded part of the foundation.

A determination of tensile strength of the rock is seldom required because discontinuities such as unhealed joints and shear seams cannot transmit tensile stress within the foundation.

Resistance to shear within the foundation and between the dam and its foundation results from the cohesion and internal friction inherent in the materials and at the concrete-rock contact. These properties are found from laboratory and in situ testing as discussed in sections 2-24 and 2-25. However, when test data are not available, values of the properties may be estimated (subject to the limitations discussed below) from published data [2, 5, 6] and from tests on similar materials.

The results of laboratory triaxial and direct shear tests, as well as in situ shear tests, will typically be reported in the form of the Coulomb equation,

$$R = CA + N \tan \phi \qquad (1)$$

where:

R = shear resistance,
C = unit cohesion,
A = area of section,
N = effective normal force, and
$\tan \phi$ = tangent of angle of friction,

which defines a linear relationship between shear resistance and normal load. Experience has shown that such a representation of shear resistance is usually realistic for most intact rock. For other materials, the relationship may not be linear and a curve of shear strength versus normal load should be used as discussed

later for the condition of an existing joint. Also, it may be very difficult to differentiate between cohesive and friction resistance for materials other than intact rock.

In the case of an existing joint in rock, the shear strength is derived basically from sliding friction and usually does not vary linearly with the normal load. Therefore, the shear resistance should be represented by a curve of shear resistance versus normal load, as shown by the curve OA in figure 3-1. If a straight line, BC, had been used, it would have given values of shear resistance too high where it is above the curve OA, and values too low where it is below. A linear variation may be used to represent a portion of the curve. Thus, the line DE can be used to determine the shear resistance for actual normal loads between N_1 and N_2 without significant error. However, for normal loads below N_1 or above N_2, its use would give a shear resistance which is too high and the design would therefore be unsafe.

Other potential sliding planes, such as shear zones and faults, should be checked to determine if the shear resistance should be linear or curvilinear. As with the jointed rock, a linear variation can be assumed for a limited range of normal loads if tests on specimens verify this type of variation for that range of normal loads.

The specimens tested in the laboratory or in situ are usually small with respect to the planes analyzed in design. Therefore, the scale effect should be carefully considered in determining the shear resistance to be used in design.

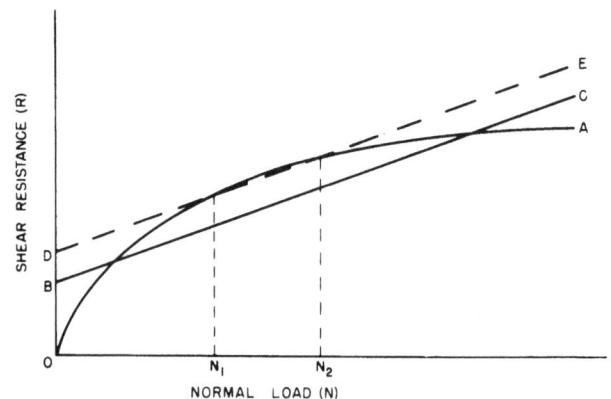

Figure 3-1. Shear resistance on an existing joint in rock.—288-D-2957

Among the factors to be considered in determining the scale effect at each site are the following:

(1) Comparisons of tests of various sizes.

(2) Geological variations along the potential sliding planes.

(3) Current research on scale effect.

When a foundation is nonhomogeneous, the potential sliding surface may be made up of different materials. The total resistance can be determined by adding the shear resistances offered by the various materials, as shown in the following equation:

$$R_t = R_1 + R_2 + R_3 + \cdots \cdots R_n \qquad (2)$$

where:

$$R_t = \text{total resistance, and}$$
$$R_1, R_2, R_3, \text{etc.} = \text{resistance offered by the various materials.}$$

When determining the shear resistance offered by the various materials, the effect of deformation should be considered. The shear resistance given by the Coulomb equation or the curves of shear resistance versus normal load are usually the maximum for the test specimen without regard to deformation. Some materials obtain their maximum resistance with less deformation than others. For example, intact rock will not deform as much as a joint in rock or a sheared zone when maximum shear resistance of the material is reached.

The following example illustrates the importance of including the effect of deflection in determining the resistance offered by each material in nonhomogeneous foundations. This example has only 5 percent intact rock to emphasize that a small quantity of high-strength intact rock can make a significant contribution to the total resistance. Such a situation is not normally encountered but can and has occurred.

Example: Determine the shear resistance on a potential sliding plane which is 1,000 square feet in area for the following conditions:

(1) Normal load, $N = 10,000$ kips.

(2) The plane is 5 percent intact rock ($A_r = 50$ square feet), 20 percent sheared material

($A_s = 200$ square feet), and 75 percent joint ($A_j = 750$ square feet).

(3) The values of cohesion and tan ϕ for each material are as follows:

Material	Cohesion (p.s.f.)	Tan ϕ
Intact rock	200,000	1.80
Sheared material	3,000	0.30
Joint	0	0.75

(4) The normal load on each material is:

Intact rock	$N_r = 2,000$ kips
Sheared material	$N_s = 1,000$ kips
Joint	$N_j = 7,000$ kips

The shear resistance is determined as follows:

$$R_r = \frac{200,000(50)}{1,000} + 1.8(2,000)$$

$$= 13,600 \text{ kips}$$

$$R_s = \frac{3,000(200)}{1,000} + 0.3(1,000)$$

$$= 900 \text{ kips}$$

$$R_j = \frac{0(750)}{1,000} + 0.75(7,000)$$

$$= 5,250 \text{ kips}$$

$$R_t = 13,600 + 900 + 5,250$$

$$= 19,750 \text{ kips}$$

For this example, an analysis of the shear strength versus deflection shows that the movement of the intact rock at failure is 0.02 inch. At this deflection the sheared material will have developed only 50 percent of its strength and the joint only 5 percent. Therefore, the actual developed strength at the time the rock would fail is:

$$13,600 + 900 \times 0.50 + 5,250 \times 0.05$$
$$= 14,312 \text{ kips.}$$

This is about 70 percent of the maximum shear strength computed above without considering deformation.

In some situations, the potential sliding surface comprised of several different materials may exhibit greater total shear resistance after any intact materials are sheared. For example, if the cohesive strength of intact rock is low but the normal load acting on the total surface is large, the sliding friction strength of the combined materials can exceed the shear resistance determined before the rock is sheared. For this reason, a second analysis should be performed which considers only the sliding friction strength of the surfaces.

3-6. *Foundation Permeability*.—The design of a gravity dam and its foundation requires a knowledge of the hydrostatic pressure distribution throughout the foundation. The exit gradient for shear zone materials that surface near the downstream toe of the dam should also be determined to check against the possibility of piping (see sec. 6-4).

The laboratory values for permeability of sample specimens are applicable only to that portion or portions of the foundation which they represent. The permeability is controlled by a network of geological features such as joints, faults, and shear zones. The permeability of the geologic features can be determined best by in situ testing. Pressure distributions for design should include the appropriate influences of the permeability and extent of all the foundation materials and geologic features. Such a determination may be made by several methods including two- and three-dimensional physical models, two- and three-dimensional finite element models, and electric analogs.

D. LOADS

3-7. *Reservoir and Tailwater*.—Reservoir and tailwater loads to be applied to the structure are obtained from reservoir operation studies and tailwater curves. These studies are based on operating and hydrologic data such as reservoir capacity, storage allocations, streamflow records, flood hydrographs, and reservoir releases for all purposes. A design reservoir can be derived from these operation studies which will reflect a normal high water surface.

The hydrostatic pressure at any point on the dam is equal to the hydraulic head at that point times the unit weight of water (62.4 lb. per cu. ft.).

The normal design reservoir elevation is the highest elevation that water is normally stored. It is the *Top of Joint Use Capacity,* if joint use capacity is included. If not, it is the *Top of Active Conservation Capacity.* For definitions of reservoir capacities, see section 2-9.

Maximum design reservoir elevation is the highest anticipated water surface elevation and usually occurs in conjunction with the routing of the inflow design flood through the reservoir.

The tailwater elevation used with a particular reservoir elevation should be the minimum that can be expected to occur with that reservoir elevation.

3-8. *Temperature.*—Volumetric changes due to temperature change [7] will transfer load across transverse contraction joints if the joints are grouted. These horizontal thrusts will then result in twist effects and in additional loading of the abutments. These effects may or may not be beneficial from a stress and stability standpoint and should be investigated using the "Trial-Load Twist Method of Analysis" discussed in chapter IV (secs. 4-25 through 4-29).

When making studies to determine concrete temperature loads, varying weather conditions can be applied. Similarly, a widely fluctuating reservoir water surface will affect the concrete temperatures. In determining temperature loads, the following conditions and temperatures are used:

(1) *Usual weather conditions.*—The combination of daily air temperatures, a 1-week cycle representative of the cold

(hot) periods associated with barometric pressure changes, and the mean monthly air temperatures. This condition will account for temperatures which are halfway between the mean monthly air temperatures and the minimum (maximum) recorded air temperatures at the site.

(2) *Usual concrete temperatures.*—The average concrete temperatures between the upstream and downstream faces which will result from usual air temperatures, reservoir water temperatures associated with the design reservoir operation, and solar radiation.

Secondary stresses can occur around openings and at the faces of the dam due to temperature differentials. These temperature differentials are caused by differences in the temperature of the concrete surfaces due to ambient air and water temperature variations, solar radiation, temperature of air or water in openings, and temperature of the concrete mass. These secondary stresses are usually localized near the faces of the dam and may produce cracks which give an unsightly appearance. If stress concentrations occur around openings, cracking could lead to progressive deterioration. Openings filled with water, such as outlets, are of particular concern since cracks, once formed, would fill with water which could increase the uplift or pore pressure within the dam.

3-9. *Internal Hydrostatic Pressures.*—Hydrostatic pressures from reservoir water and tailwater act on the dam and occur within the dam and foundation as internal pressures in the pores, cracks, joints, and seams. The distribution of pressure through a horizontal section of the dam is assumed to vary linearly from full hydrostatic head at the upstream face to zero or tailwater pressure at the downstream face, provided the dam has no drains or unlined water passages. When formed drains are constructed, the internal pressure should be modified in accordance with the size, location, and spacing of the drains. Large unlined penstock transitions or other large openings in dams will require special modification of internal pressure patterns. Pressure distribution

in the foundation may be modified by the ground water in the general area.

The internal pressure distribution through the foundation is dependent on drain size, depth, location, and spacing; on rock porosity, jointing, faulting; and to some extent on the grout curtain. Determination of such pressure distribution can be made from flow nets computed by several methods including two- and three-dimensional physical models, two- and three-dimensional finite element models, and electric analogs. Such a flow net, modified by effects of drainage and grouting curtains, should be used to determine internal pressure distribution. However, the jointing, faulting, variable permeability, and other geologic features which may further modify the flow net should be given full consideration.

The component of internal hydrostatic pressure acting to reduce the vertical compressive stresses in the concrete on a horizontal section through the dam or at its base is referred to as uplift or pore pressure. Records are kept of the pore pressure measurements in most Bureau of Reclamation dams. Figure 3-2 illustrates actual measured uplift pressures at the concrete-rock contact as compared with design assumptions for Shasta Dam.

Laboratory tests indicate that for practical purposes pore pressures act over 100 percent of the area of any section through the concrete. Because of possible penetration of water along construction joints, cracks, and the foundation contact, internal pressures should be considered to act throughout the dam. It is assumed that the pressures are not affected by earthquake acceleration because of the transitory nature of such accelerations.

Internal hydrostatic pressures should be used for analyses of the foundation, the dam, and overall stability of the dam at its contact with the foundation.

For preliminary design purposes, uplift pressure distribution in a gravity dam is assumed to have an intensity at the line of drains that exceeds the tailwater pressure by one-third the differential between headwater and tailwater levels. The pressure gradient is then extended to headwater and tailwater

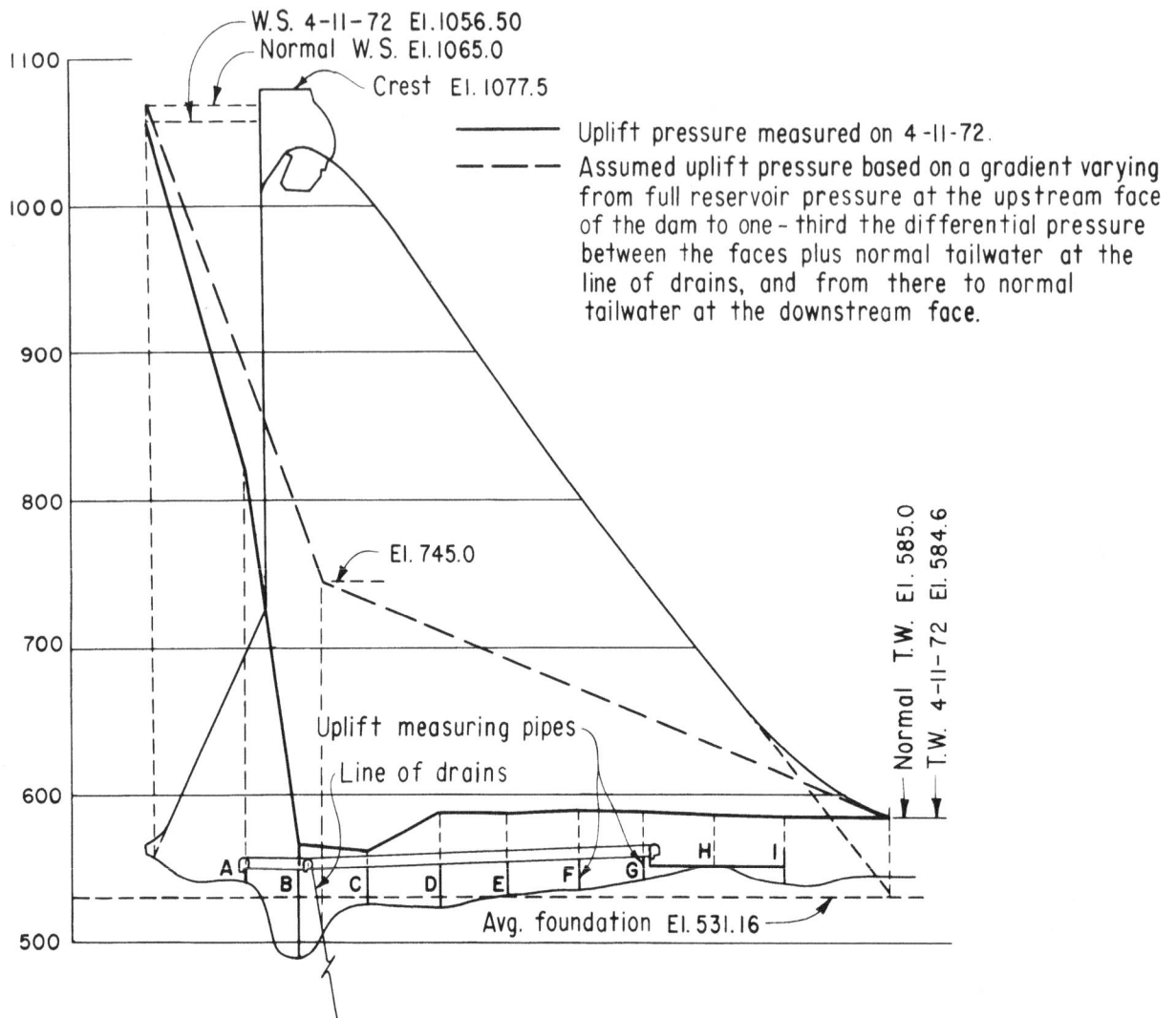

Figure 3-2. Comparison of assumed and uplift pressures on a gravity dam (Shasta Dam in California).—288-D-2959

levels, respectively, in straight lines. If there is no tailwater, the downstream end of a similar pressure diagram is zero at the downstream face. The pressure is assumed to act over 100 percent of the area.

In the final design for a dam and its foundation, the internal pressures within the foundation rock and at the contact with the dam will depend on the location, depth, and spacing of drains as well as on the joints, shears, and other geologic structures in the rock. Internal pressures within the dam depend on the location and spacing of the drains. These internal hydrostatic pressures should be

determined from flow nets computed by electric analogy analysis, three-dimensional finite element analysis, or other comparable means.

3-10. Dead Load.—The magnitude of dead load is considered equal to the weight of concrete plus appurtenances such as gates and bridges. For preliminary design the unit weight of concrete is assumed to be 150 pounds per cubic foot. For final design the unit weight of concrete should be determined by laboratory tests.

3-11. Ice.—Existing design information on ice pressure is inadequate and somewhat approximate. Good analytical procedures exist for computing ice pressures, but the accuracy of results is dependent upon certain physical data which must come from field and laboratory tests [8].

Ice pressure is created by thermal expansion of the ice and by wind drag. Pressures caused by thermal expansion are dependent on the temperature rise of the ice, the thickness of the ice sheet, the coefficient of expansion, the elastic modulus, and the strength of the ice. Wind drag is dependent on the size and shape of the exposed area, the roughness of the surface, and the direction and velocity of the wind. Ice loads are usually transitory. Not all dams will be subjected to ice pressure, and the designer should decide after consideration of the above factors whether an allowance for ice pressure is appropriate. The method of Monfore and Taylor [9] may be used to determine the anticipated ice pressure. An acceptable estimate of ice load to be expected on the face of a structure may be taken as 10,000 pounds per linear foot of contact between the ice and the dam, for an assumed ice depth of 2 feet or more when basic data are not available to compute pressures.

3-12.   *Silt*.—Not all dams will be subjected to silt pressure, and the designer should consider all available hydrologic data before deciding whether an allowance for silt pressure is necessary. Horizontal silt pressure is assumed to be equivalent to that of a fluid weighing 85 pounds per cubic foot. Vertical silt pressure is determined as if silt were a soil having a wet density of 120 pounds per cubic foot, the magnitude of pressure varying directly with depth. These values include the effects of water within the silt.

3-13.   *Earthquake*.—Concrete dams are elastic structures which may be excited to resonance when subjected to seismic disturbances. Two steps are necessary to obtain loading on a concrete dam due to such a disturbance. First, an estimate of magnitude and location must be made of the earthquake to which the dam will be subjected and the resulting rock motions at the site determined. The second step is the analysis of the response of the dam to the earthquake by either the response spectrum or time-history method.

Most earthquakes are caused by crustal movements of the earth along faults. Geologic examinations of the area should be made to locate any faults, determine how recently they have been active, and estimate the probable length of fault. Seismological records should also be studied to determine the magnitude and location of any earthquakes recorded in the area. Based on these geological and historical data, hypothetical earthquakes usually of magnitudes greater than the historical events are estimated for any active faults in the area. These earthquakes are considered to be the most severe earthquakes associated with the faults and are assumed to occur at the point on the fault closest to the site. This defines the *Maximum Credible Earthquake* and its location in terms of Richter Magnitude M and distance d to the causative fault.

Methods of determining a design earthquake that represents an operating-basis event are under development. These methods should consider historical records to obtain frequency of occurrence versus magnitude, useful life of the structure, and a statistical approach to determine probable occurrence of various magnitude earthquakes during the life of the structure. When future developments produce such methods, suitable safety factors will be included in the criteria.

The necessary parameters to be determined at the site using attenuation methods [10] are acceleration, predominant period, duration of shaking, and frequency content.

Attenuation from the fault to the site is generally included directly in the formulas used to compute the basic data for response spectra. A response spectrum graphically represents the maximum response of a structure with one degree of freedom having a specific damping and subjected to a particular excitation. A response spectrum should be determined for each magnitude-distance relationship by each of three methods as described in appendix D of reference [10]. The design response spectrum of a structure at a site is the composite of the above spectra.

Time-history analyses of a dam are sometimes desirable. The required accelerograms may be produced by appropriate adjustment of existing or artificially generated accelerograms. The previously mentioned parameters are necessary considerations in the development of synthetic accelerograms or in the adjustment of actual recorded accelerograms.

The analytical methods used to compute material frequencies, mode shapes, and structural response are discussed in chapter IV.

E. LOADING COMBINATIONS

3-14. *General.*—Designs should be based on the most adverse combination of probable load conditions, but should include only those loads having reasonable probability of simultaneous occurrence. Combinations of transitory loads, each of which has only remote probability of occurrence at any given time, have negligible probability of simultaneous occurrence and should not be considered as a reasonable basis for design. Temperature loadings should be included when applicable (see sec. 3-8).

Gravity dams should be designed for the appropriate loading combinations which follow, using the safety factors prescribed in sections 3-19 through 3-22.

3-15. *Usual Loading Combination.*—

(1) Normal design reservoir elevation, with appropriate dead loads, uplift, silt, ice, and tailwater. If temperature loads are applicable, use minimum usual temperatures.

3-16. *Unusual and Extreme Loading Combinations.*—

(1) *Unusual Loading Combination.*—Maximum design reservoir elevation, with appropriate dead loads, uplift, silt, minimum temperatures occurring at that time if applicable, and tailwater.

(2) *Extreme Loading Combination.*—Normal design reservoir elevation, with appropriate dead loads, uplift, silt, ice, usual minimum temperatures if applicable, and tailwater, plus the effects of the *Maximum Credible Earthquake.*

3-17. *Other Studies and Investigations.*—

(1) Maximum design reservoir elevation, with appropriate dead loads, silt, minimum temperature occurring at that time if applicable, and tailwater, plus uplift with drains inoperative.

(2) Dead load.

(3) Any of the above loading combinations for foundation stability.

(4) Any other loading combination which, in the designer's opinion, should be analyzed for a particular dam.

F. FACTORS OF SAFETY

3-18. *General.*—All design loads should be chosen to represent as nearly as can be determined the actual loads which will act on the structure during operation. Methods of determining load-resisting capacity of the dam should be the most accurate available. All uncertainties regarding loads or load-carrying capacity must be resolved as far as practicable by field or laboratory tests, thorough exploration and inspection of the foundation, good concrete control, and good construction practices. On this basis, the factor of safety will be as accurate an evaluation as possible of the capacity of the structure to resist applied loads. All safety factors listed are minimum values.

Dams, like other important structures, should be frequently inspected. In particular, where uncertainties exist regarding such factors as loads, resisting capacity, or characteristics of the foundation, it is expected that adequate observations and measurements will be made of the structural behavior of the dam and its foundation to assure that the structure is at all times behaving as designed.

The factors of safety for the dam are based on the "Gravity Method of Stress and Stability Analysis" (secs. 4-5 through 4-10). Although lower safety factors may be permitted for limited local areas within the foundation, overall safety factors for the dam and its foundation (after beneficiation) should meet the requirements for the loading combination being analyzed. Somewhat higher safety factors should be used for foundation studies because of the greater amount of uncertainty involved in assessing foundation load resisting capacity. For other loading combinations where safety factors are not specified, the designer is responsible for selection of safety factors consistent with those for loading combination categories discussed in sections 3-14 through 3-17.

3-19. Allowable Stresses.—The maximum allowable compressive stress in the concrete for the Usual Loading Combinations should be not greater than the specified compressive strength divided by a safety factor of 3.0. Under no circumstances should the allowable compressive stress exceed 1,500 pounds per square inch for Usual Loading Combinations. In the case of Unusual Loading Combinations, the maximum allowable compressive stress should be determined by dividing the specified compressive strength by a safety factor of 2.0. The maximum allowable compressive stress for the Unusual Loading Combinations should, in no case, exceed 2,250 pounds per square inch. The allowable compressive stress for the Extreme Loading Combination shall be determined in the same way using a factor of safety greater than 1.0.

In order not to exceed the allowable tensile stress, the minimum allowable compressive stress computed without internal hydrostatic pressure should be determined from the following expression, which takes into account the tensile strength of the concrete at lift surfaces:

$$\sigma_{z_u} = pwh - \frac{f_t}{s} \qquad (3)$$

where:

σ_{z_u} = minimum allowable compressive stress at the upstream face,

p = a reduction factor to account for drains,

w = unit weight of water,

h = depth below reservoir surface,

f_t = tensile strength of concrete at lift surfaces, and

s = safety factor.

All parameters must be specified using consistent units.

The value of p should be 1.0 if drains are not present and 0.4 if drains are used. The value of s should be 3.0 for Usual and 2.0 for Unusual Loading Combinations. The allowable value of σ_{z_u} for the usual loading combination should never be less than zero. Cracking should be assumed to occur if the stress at the upstream face is less than σ_{z_u} computed from the above equation with a value for s of 1.0 for the Extreme Loading Combination. The structure may be deemed safe for this loading if, after cracking has been included, stresses in the structure do not exceed specified strengths and sliding stability is maintained.

The maximum allowable compressive stress in the foundation shall be less than the compressive strength of the foundation material divided by safety factors of 4.0, 2.7, and 1.3 for the Usual, Unusual, and Extreme Loading Combinations, respectively.

3-20. Sliding Stability.—The shear-friction factor of safety, Q, as computed using equation (4), is a measure of the safety against sliding or shearing on any section. It applies to any section of the structure or its contact with the foundation. For gravity dams the shear-friction factor of safety should be greater than 3.0 for Usual Loading Combinations, 2.0 for Unusual Loading Combinations, and 1.0 for the Extreme Loading Combination.

The shear-friction factor of safety, Q, is the ratio of resisting to driving forces as computed by the expression:

$$Q = \frac{CA + (\Sigma N + \Sigma U)\tan\phi}{\Sigma V} \qquad (4)$$

where:

C = unit cohesion,
A = area of the section considered,
ΣN = summation of normal forces,
ΣU = summation of uplift forces,
$\tan\phi$ = coefficient of internal friction,
     and
ΣV = summation of shear forces.

All parameters must be specified using consistent units and with proper signs according to the convention shown in figure 4-1.

Values of cohesion and internal friction should be determined by actual tests of the foundation materials and the concrete proposed for use in the dam.

3-21. Cracking.—Cracking is assumed to occur in a gravity dam if the vertical normal stress (computed without uplift) at the upstream face is less than the minimum required stress as computed by equation (3). Such cracking is not permitted in new designs except for the Extreme Loading Combination. However, for existing dams, cracking may be permitted for the condition of maximum water surface with drains inoperative in addition to the Extreme Loading Combination.

When checking the stability of an existing dam for the loading condition of maximum water surface with drains inoperative, the uplift (or internal hydrostatic) pressure is assumed to vary linearly from full reservoir level at the upstream face to tailwater level at the downstream face. If cracking occurs the foundation pressure diagram is assumed to be as shown in figure 3-3(D). The foundation pressure diagram is determined by the following procedure:

(1) A horizontal crack is assumed to extend from the upstream face to a point where the vertical stress is equal to the uplift pressure at the upstream face, point 4 on figure 3-3(D).

(2) From figures 3-3(A) and (D), taking moments about the center of gravity of the base, the following equations are obtained:

$$e' = \frac{\Sigma M}{\Sigma W - \overline{A3} \cdot T} \qquad (5)$$

and

$$T_1 = 3\left(\frac{T}{2} - e'\right) \qquad (6)$$

where:

e' = eccentricity of the stress diagram after cracking,
ΣM = summation of moments of all forces,
ΣW = summation of vertical forces,
$\overline{A3}$ = internal hydrostatic pressure at the upstream face,
T = thickness of section, and
T_1 = remaining uncracked portion of the thickness.

Therefore the stress at the downstream face, $\overline{B5}$, is:

$$\overline{B5} = \frac{2(\Sigma W - \overline{A3} \cdot T)}{T_1} + \overline{A3} \qquad (7)$$

Because of the rapidly cycling changes in stress during earthquakes, it should be assumed that the internal hydrostatic pressures are zero in the cracks caused by the extreme loading. Equations (5) and (7) should be revised to account for the zero internal hydrostatic pressure in the crack. For these computations, T should be taken as the thickness of the uncracked portion shown as T_1 in equations (6) and (7). The value of $\overline{A'4}$ should be the uplift pressure at the end of the crack in the uncracked portion and ΣM should include the moment of the altered uplift pressures taken about the center of gravity of the original section. The T in equation (6) is the full thickness of the original section. The value of T_1 to be used as T in equation (5) must be estimated for the first computation. Thereafter, for succeeding computations of e', the value obtained for T_1 in equation (6) should be used. Several cycles of computation

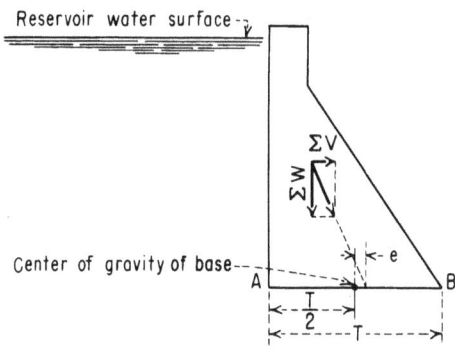

(A) VERTICAL CROSS-SECTION

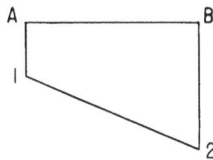

(B) PRESSURE DIAGRAM WITHOUT UPLIFT

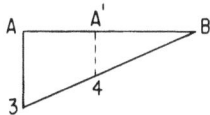

(C) UPLIFT PRESSURE DIAGRAM

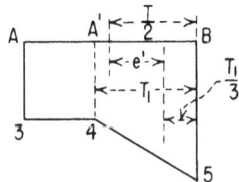

(D) COMBINED PRESSURE DIAGRAM

DIAGRAMS OF BASE PRESSURES
ACTING ON A GRAVITY DAM

Figure 3-3. Foundation base pressures for a gravity
dam.—288-D-2510

using equations (5) and (6) may be required to obtain adequate agreement between the value used for T in equation (5) and the value

computed for T_1 in equation (6).

The uncracked area of the base is substituted for A in equation (4). The section is considered satisfactory for any of these loading conditions if the stress at the downstream face, from equation (7), does not exceed the allowable stress, and the shear-friction factor of safety is sufficient to ensure stability. A shear-friction factor of safety greater than 2.0 would be considered satisfactory for the Unusual Loading Combination and greater than 1.0 for the Extreme Loading Combination.

A gravity dam should be considered safe against overturning if $\overline{B5}$, the ordinate in figure 3-3(D), is less than the allowable stresses in the concrete and the foundation rock for the appropriate loading combinations.

3-22. *Foundation Stability*.—Joints, shears, and faults which form identifiable blocks of rock are often present in the foundation. Effects of such planes of weakness on the stability of the foundation should be carefully evaluated. Methods of analysis for foundation stability under these circumstances are discussed in section 4-50. The determination of effective shear resistance for such foundation conditions is given in detail in section 3-5.

The factor of safety against sliding failure of these foundation blocks, as determined by the shear-friction factor, Q, using equation (4), should be greater than 4.0 for the Usual Loading Combination, 2.7 for the Unusual Loading Combination, and 1.3 for the Extreme Loading Combination. If the computed safety factor is less than required, foundation treatment can be included to increase the safety factor to the required value.

Treatment to accomplish specific stability objectives such as prevention of differential displacements (see sec. 4-51) or stress concentrations due to bridging (see sec. 4-52) should be designed to produce the safety factor required for the loading combination being analyzed.

G. BIBLIOGRAPHY

3-23. *Bibliography*.—

[1] "Concrete Manual," Bureau of Reclamation, eighth edition, 1975.

[2] "Properties of Mass Concrete in Bureau of Reclamation Dams," Concrete Laboratory Report No. C-1009, Bureau of Reclamation, 1961.

[3] Stagg, K. G., and Zienkiewicz, O. C., "Rock Mechanics in Engineering Practice," John Wiley & Sons, London, England, 1968.

[4] Von Thun, J. L., and Tarbox, G. S., "Deformation Moduli Determined by Joint Shear Index and Shear Catalog," Proceedings, International Symposium on Rock Mechanics, Nancy, France, 1971.

[5] "Physical Properties of Some Typical Foundation Rocks," Concrete Laboratory Report No. SP-39, Bureau of Reclamation, 1953.

[6] Link, Harald, "The Sliding Stability of Dams," Water Power–Part I, March 1969; Part II, April 1969; Part III, May 1969, London, England.

[7] Townsend, C. L., "Control of Cracking in Mass Concrete Structures," Engineering Monograph No. 34, Bureau of Reclamation, 1965.

[8] Monfore, G. E., "Experimental Investigations by the Bureau of Reclamation," Trans. ASCE, vol. 119, 1954, p. 26.

[9] Monfore, G. E., and Taylor, F. W., "The Problem of an Expanding Ice Sheet," Bureau of Reclamation Memorandum, March 18, 1948.

[10] Boggs, H. L., Campbell, R. B., Klein, I. E., Kramer, R. W., McCafferty, R. M., and Roehm, L. H., "Methods for Estimating Design Earthquake Rock Motions," Bureau of Reclamation, April 1972.

[11] "Design of Small Dams," Bureau of Reclamation, second edition, 1973.

Chapter IV

Layout and Analysis

4-1. Introduction. —A brief discussion of guidelines for making a gravity dam layout is given in sections 4-2 through 4-4. The layout represents the initial step in the design procedure for a new structure. After a layout is completed, a stress and stability analysis of the structure must be made to determine the stress distributions and magnitudes and the stability factor. If the analytical results do not fall within the established allowable limits or the stress distributions are not satisfactory because of stress concentrations, modifications to improve the design must be made by reshaping the structure. The design of a gravity dam is accomplished by making successive layouts, each one being progressively improved based on the results of a stress analysis. It is difficult to discuss layouts without discussing analysis and vice versa, because each operation is essential to the other.

Stress analyses of gravity dams fall into two classifications—those analyses based on gravity action and those based on the trial-load method. (See also sec. 4-30.) The "Gravity Method of Analysis," which is discussed in considerable detail in sections 4-5 through 4-10, provides a two-dimensional solution for straight gravity dams. The method is based on the assumptions that a straight gravity dam is comprised of a number of vertical elements, each of which carries its load to the foundation without any transfer of the load from or to adjacent vertical elements and that vertical stresses vary linearly. It is usually sufficient to compute stresses and stability factors at the base elevation and selected elevations above the base for both a maximum overflow section and a maximum nonoverflow section. This method

of analysis is used for designing straight gravity concrete dams in which the transverse contraction joints are neither keyed nor grouted.

The stress analysis of a straight gravity dam in which the transverse contraction joints are keyed, whether grouted or not, is a three-dimensional problem. One method used by the Bureau is the "Trial-Load Method of Analysis" in which it is assumed that the dam is comprised of three systems of elements each occupying the entire volume of the structure and independent of the others. These systems are the vertical cantilevers, the horizontal beams, and the twisted elements. The loads on the dam are divided between these systems in such a manner as to produce equal deflections and rotations at conjugate points.

The more recently developed "Finite Element Method", which can be used for two-dimensional studies to determine the stress distributions and for the three-dimensional studies for grouted joints, is discussed in sections 4-36 through 4-48. An example of its use is presented in appendix C.

Analytical methods of determining the response of gravity dams to earthquake ground accelerations are presented in sections 4-31 through 4-35. The response of a structure is defined as its behavior as a result of an earthquake disturbance. The response is usually represented as a measure of the structure's displacement acceleration or velocity. Either the time variation of a particular response or its maximum value during the disturbance may be of interest. The determination of natural frequencies and mode shapes is a fundamental part of dynamic analysis. Dynamic analyses are used in stress analysis methods to determine loadings for computing stresses due to earthquake.

A discussion of foundation analyses is given in sections 4-49 through 4-52. A dam is no better than its foundation, and therefore an evaluation of the foundation behavior is necessary to ensure a competent load-bearing system consisting of the dam and the foundation. Analytical methods are presented to evaluate foundation stability and local overstressing due to foundation deficiencies.

Certain special, rigorous methods of analysis, such as the "Slab Analogy Method" [1][1] and "Lattice Analogy Method" [2], which may be used for the determination of nonlinear stress distributions are included in lesser detail in appendix D along with photoelastic model studies.

(a) *Level of Design.*—The level of design for a gravity dam, whether appraisal, feasibility, or final, differs only by the level of investigation used to determine design data. Details of these levels of investigation in the field are discussed in section 2-21. The levels of investigations in the laboratory are usually dependent on the levels of field investigations.

A. LAYOUT

4-2. *Nonoverflow Section.*—The shape of the maximum nonoverflow section is determined by the prescribed loading conditions, the shear resistance of the rock, and the height of the maximum section. The upstream face of a gravity dam is usually made vertical to concentrate the concrete weight at the upstream face where it acts to overcome the effects of the reservoir waterload. Except where additional thickness is required at the crest, as discussed below, the downstream face will usually have a uniform slope which is determined by both stress and stability requirements at the base. This slope will be adequate to meet the stress and stability requirements at the higher elevations unless a large opening is included in the dam. The crest thickness may be dictated by roadway or other access requirements, but in any case it should be adequate to withstand possible ice pressures and the impact of floating objects. When additional crest thickness is used, the downstream face should be vertical from the downstream edge of the crest to an intersection with the sloping downstream face.

A batter may be used on the lower part of the upstream face to increase the thickness at the base to improve the sliding safety of the base. However, unacceptable stresses may develop at the heel of the dam because of the change in moment arm for the concrete weight about the center of gravity of the base. If a batter is used, stresses and stability should be checked where the batter intersects the vertical upstream face. The dam should be analyzed at any other changes in slope on either face.

4-3. *Spillway Section.*—The overflow or spillway section should be designed in a similar manner to the nonoverflow section. The curves describing the spillway crest and the junction of the slope with the energy dissipator are designed to meet hydraulic requirements discussed in chapter IX. The slope joining these curves should be tangent to each curve and, if practicable, parallel to the downstream slope on the nonoverflow section. The spillway section should be checked for compliance with stress and stability requirements. An upstream batter may be used on the spillway section under the same conditions as for the nonoverflow section. Figure B-1 in appendix B is a typical layout drawing of a gravity dam showing a nonoverflow section, a typical spillway section, a plan, and a profile.

4-4. *Freeboard.*—Current Bureau practice is to allow the maximum water surface elevation to be coincident with the top of the nonoverflow section of the dam, and to consider that the standard 3.5-foot-high solid parapet acts as a freeboard. Exceptional cases may point to a need for more freeboard, depending on the anticipated wave height.

---

[1]Numbers in brackets refer to items in the bibliography, sec. 4-55.

B. THE GRAVITY METHOD OF STRESS AND STABILITY ANALYSIS

4-5. *Description and Use.*—The "Gravity Method of Stress and Stability Analysis" is used a great deal for preliminary studies of gravity dams, depending on the phase of design and the information required. The gravity method is also used for final designs of straight gravity dams in which the transverse contraction joints are neither keyed nor grouted. For dams in which the transverse joints are keyed and grouted, the "Trial-Load Twist Analsysis" including the beam structure should be used. If the joints are keyed but left ungrouted, the "Trial-Load Twist Analysis" should omit the beam structure.

The gravity method provides an approximate means for determination of stresses in a cross section of a gravity dam. It is applicable to the general case of a gravity section with a vertical upstream face and with a constant downstream slope and to situations where there is a variable slope on either or both faces. Equations are given with standard forms and illustrations showing calculation of normal and shear stresses on horizontal planes, normal and shear stresses on vertical planes, and principal stresses, for both empty-reservoir and full-reservoir conditions, including the effects of tailwater and earthquake shock. Uplift pressures on a horizontal section are usually not included with the contact pressures in the computation of stresses, and are considered separately in the computation of stability factors.

The formulas shown for calculating stresses are based on the assumption of a trapezoidal distribution of vertical stress and a parabolic distribution of horizontal shear stress on horizontal planes. These formulas provide a direct method of calculating stresses at any point within the boundaries of a transverse section of a gravity dam. The assumptions are substantially correct, except for horizontal planes near the base of the dam where the effects of foundation yielding are reflected in the stress distributions. At these locations the stress changes which occur due to foundation yielding are usually small in dams of low or medium height but they may be important in high dams. Stresses near the base of a high masonry dam should therefore be checked by the "Finite Element Method" or other comparable methods of analysis.

The analysis of overflow sections presents no added difficulties. Usually, the dynamic effect of overflowing water is negligible and any additional head above the top of the section can be included as an additional vertical load on the dam. In some cases some increase in horizontal load may be justified for impact. An example of the gravity method of analysis is given in appendix A.

4-6. *Assumptions.*—Design criteria are given in chapter III. However, those assumptions peculiar to the gravity analysis are listed below:

(1) The concrete in the dam is a homogeneous, isotropic, and uniformly elastic material.

(2) There are no differential movements which occur at the damsite due to waterloads on the reservoir walls and floors.

(3) All loads are carried by the gravity action of vertical, parallel-side cantilevers which receive no support from the adjacent elements on either side.

(4) Unit vertical pressures, or normal stresses on horizontal planes, vary uniformly as a straight line from the upstream face to the downstream face.

(5) Horizontal shear stresses have a parabolic variation across horizontal planes from the upstream face to the downstream face of the dam.

4-7. *Notations for Normal Reservoir Loading.*—Symbols and definitions for normal reservoir loading are given below. This loading includes full-reservoir load and usual tailwater loads on the dam, as shown on figure 4-1. The "section" referred to is one formed by a horizontal plane through the cantilever element, except when otherwise specified.

Properties and Dimensions.

O = origin of coordinates, at downstream edge of section.

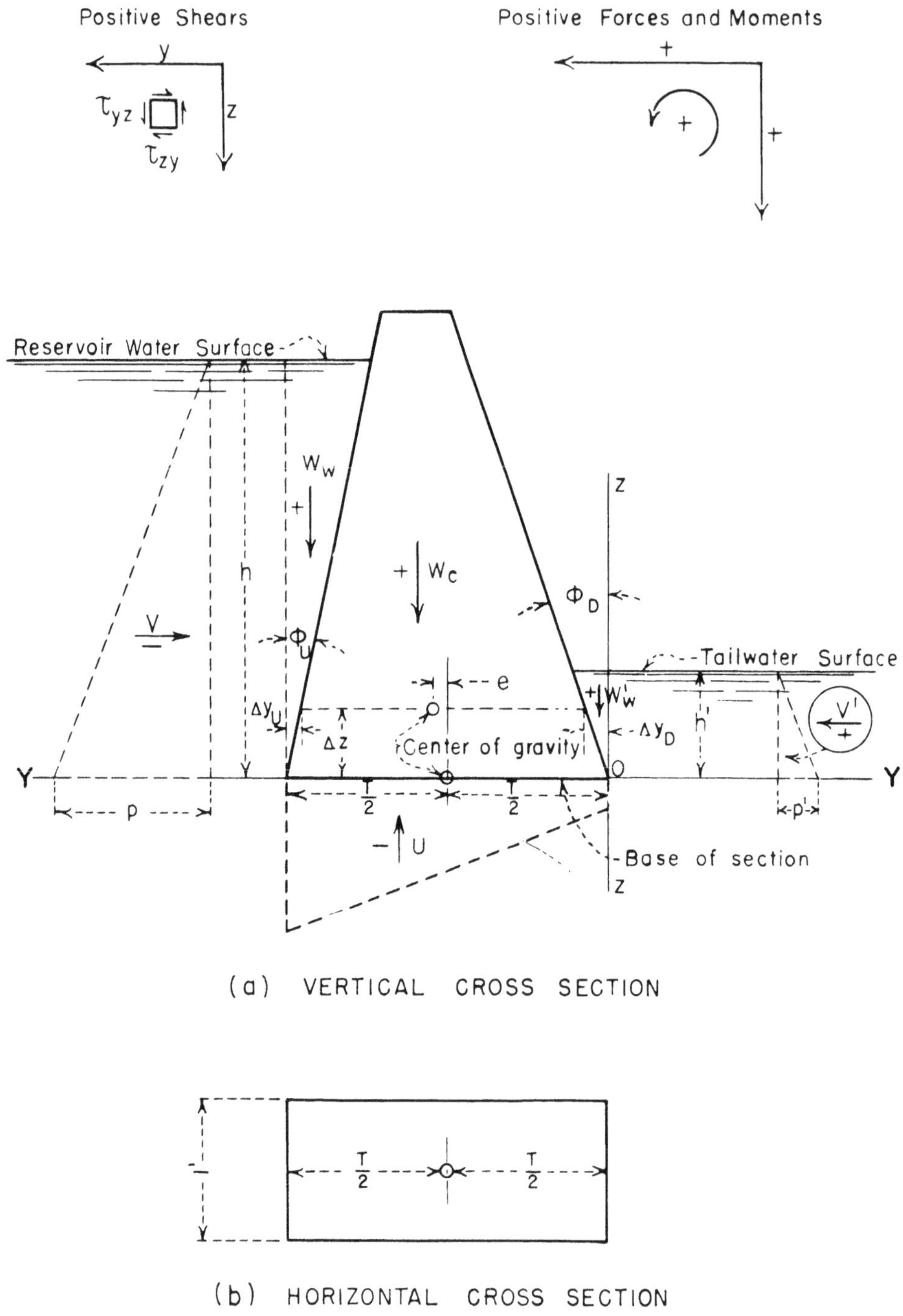

Positive Shears

τ_{yz} τ_{zy}

Positive Forces and Moments

$+$

$+$

Reservoir Water Surface

W_w

$+$

h

V

ϕ_U

W_c

$+$

ϕ_D

e

Tailwater Surface

Δy_U

$+$ W_w'

V'

Δz

Center of gravity

Δy_D

h'

Y

$\frac{T}{2}$

$\frac{T}{2}$

O

Y

p

$-p'$

$-U$

Base of section

(a) VERTICAL CROSS SECTION

$\frac{T}{2}$

$\frac{T}{2}$

(b) HORIZONTAL CROSS SECTION

Figure 4-1. Cross section of a parallel-side cantilever showing usual loading
combination.—DS2-2(1)

ϕ = angle between face of element and the vertical.

T = horizontal distance from upstream edge to downstream edge of section.

c = horizontal distance from center of gravity of section to either upstream or downstream edge, equal to $T/2$.

A = area of section, equal to T.

I = moment of inertia of section about center of gravity, equal to $T^3/12$.

ω_c = unit weight of concrete or masonry.

ω = unit weight of water.

h or h' = vertical distance from reservoir or tailwater surface, respectively, to section.

p or p' = reservoir water or tailwater pressure, respectively, at section. It is equal to ωh or $\omega h'$.

Forces and Moments.

W_c = dead-load weight above base of section under consideration.

M_c = moment of W_c about center of gravity of section.

W_w or W_w' = vertical component of reservoir or tailwater load, respectively, on face above section.

M_w or M_w' = moment of W_w or W_w' about center of gravity of section.

V or V' = horizontal component of reservoir or tailwater load, respectively, on face above section. This is equal to $\dfrac{\omega h^2}{2}$ or $\dfrac{\omega(h')^2}{2}$.

M_p or M_p' = moment of V or V' about center of gravity of section, equal to $\dfrac{\omega h^3}{6}$ or $\dfrac{\omega(h')^3}{6}$.

ΣW = resultant vertical force above section, equal to $W_c + W_w + W_w'$.

ΣV = resultant horizontal force above section, equal to $V + V'$.

ΣM = resultant moment of forces above section about center of gravity. It is equal to $M_c + M_w + M_w' + M_p + M_p'$. A positive moment produces compression on the section at the upstream face. All positive normal stresses are compressive.

U = total uplift force on horizontal section.

Positive horizontal forces act in the upstream direction.

Stresses.

σ_z = normal stress on horizontal plane.

σ_y = normal stress on vertical plane.

$\tau_{zy} = \tau_{yz}$ = shear stress on vertical or horizontal plane. A positive shear stress is shown by the sketch in the upper left of figure 4-1.

$a, a_1, a_2, b, b_1, b_2, c_1, c_2, d_2$ = constants.

σ_{p1} = first principal stress.

σ_{p2} = second principal stress.

ϕ_{p1} = angle between σ_{p1} and the vertical. It is positive in a clockwise direction.

Subscripts.

U = upstream face.

D = downstream face.

w = vertical water component.

p = horizontal water component.

4-8.   Notations for Horizontal Earthquake.—The hydrodynamic pressures due to horizontal rock motions from earthquakes are computed using the method introduced in section 4-31. The notation is as follows:

p_E = pressure normal to face.

$\alpha = \dfrac{\text{horizontal earthquake acceleration}}{\text{acceleration of gravity}}$, varies by elevation as computed in section 4-33.

z = depth of reservoir at section being studied.

h = vertical distance from the reservoir surface to the elevation in question.

C_m = a dimensionless pressure coefficient obtained from figure 4-18.

W_{WE} or W'_{WE} = change in vertical component of reservoir water load or tailwater load on face above section due to horizontal earthquake loads.

M_{WE} or M'_{WE} = moment of W_{WE} or W'_{WE} about center of gravity of section.

V_E = horizontal inertia force of concrete weight above section.

M_E = moment of V_E about center of gravity of section.

V_{pE} or V'_{pE} = change in horizontal component of reservoir or tailwater load on face above section due to horizontal earthquake loads.

M_{pE} or M'_{pE} = moment of V_{pE} or V'_{pE} about center of gravity of section.

ΣW = resultant vertical force above section, equal to $W_c + W_w + W'_w$ $\pm W_{WE} \pm W'_{WE}$.

ΣV = resultant horizontal force above section, equal to $V + V' \pm V_E$ $\pm V_{pE} \pm V'_{pE}$.

ΣM = resultant moment of forces above horizontal section about center of gravity. It is equal to $M_c + M_w$ $+ M'_w + M_p + M'_p \pm M_E$ $\pm M_{WE} \pm M'_{WE} \pm M_{pE}$ $\pm M'_{pE}$.

The algebraic signs of the terms with subscript E in the earthquake equations for ΣW, ΣV, and ΣM depend upon the direction assumed for the horizontal earthquake acceleration of the foundation.

No notation is given for vertical earthquake shock. A vertical earthquake shock is assumed to produce an acceleration of the mass of the dam and water in direct proportion to the value of α. This is equivalent to increasing or decreasing the density of concrete and water, depending on direction of shock.

4-9. *Forces and Moments Acting on Cantilever Element.*—Forces acting on the cantilever element, including uplift, are shown for normal loading conditions on figure 4-1. Reservoir and tailwater pressure diagrams are shown for the portion of the element above the horizontal cross section $O\ Y$. Positive forces, moments, and shears are indicated by the directional arrows.

Hydrodynamic and concrete inertia forces acting on the cantilever element for a horizontal earthquake shock are in addition to those forces shown on figure 4-1. The forces are negative for a foundation acceleration acting in an upstream direction and positive for a foundation acceleration acting in a downstream direction.

Forces and moments for static loads are easily computed for each section by determining areas and moment arms of the triangular pressure diagrams and the area and eccentricity of vertical sections. However, to evaluate the quantities V_{pE}, V'_{pE}, W_{WE}, and W'_{WE} for hydrodynamic effects of earthquake shock, it is necessary to use the procedure outlined in sections 4-31 through 4-35.

4-10. *Stress and Stability Equations.*—A summary of the equations for stresses computed by the gravity method is given on figures 4-2 and 4-3. These equations include those for normal stresses on horizontal planes, shear stresses on horizontal and vertical planes, normal stresses on vertical planes, and direction and magnitude of principal stresses, for any point within the boundaries of the cantilever element. In the equations of figure 4-3, terms that are due to earthquake are preceded by the algebraic signs, plus or minus (±). The correct sign to use is indicated by accompanying notes.

Effects of earthquake shocks on stresses may be excluded merely by omitting the terms preceded by the plus or minus (±) signs from the equations, which results in the correct equations for computing stresses for normal loading conditions.

DERIVATION OF STRESS FORMULAE
(HORIZONTAL EARTHQUAKE UPSTREAM)

From Figure a:

To maintain **rotational** equilibrium about A,

$$(\tau_{zyu}\,dy)\frac{dz}{2} = (\tau_{yzu}\,dz)\frac{dy}{2}$$

$$\tau_{zyu} = \tau_{yzu}$$

To maintain vertical equilibrium,

$$(p + p_E)\,dy - \sigma_{zu}\,dy - \tau_{yzu}\,dz = 0$$

$$\tau_{yzu}\,dz = (p + p_E)\,dy - \sigma_{zu}\,dy$$

$$\tau_{yzu} = \frac{(p + p_E)\,dy - \sigma_{zu}\,dy}{dz}$$

$$\frac{dy}{dz} = \tan\phi_U$$

$$\tau_{yzu} = \tau_{zyu} = -\left[\sigma_{zu} - p - p_E\right]\tan\phi_U$$

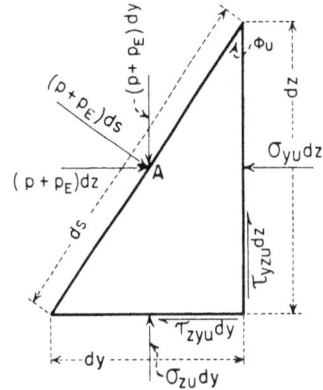

(a)- FORCES ACTING ON
DIFFERENTIAL ELEMENT AT
UPSTREAM FACE

From Figure b:

To maintain rotational equilibrium about B,

$$(-\tau_{zyD}\,dy)\frac{dz}{2} = (-\tau_{yzD}\,dz)\frac{dy}{2}$$

$$\tau_{zyD} = \tau_{yzD}$$

To maintain vertical equilibrium,

$$(p' - p'_E)\,dy - \sigma_{zD}\,dy - (-\tau_{yzD})\,dz = 0$$

$$\tau_{yzD}\,dz = \sigma_{zD}\,dy - (p' - p'_E)\,dy$$

$$\tau_{yzD} = \frac{\left[\sigma_{zD} - (p' - p'_E)\right]dy}{dz}$$

$$\frac{dy}{dz} = \tan\phi_D$$

$$\tau_{yzD} = \tau_{zyD} = \left[\sigma_{zD} - p' + p'_E\right]\tan\phi_D$$

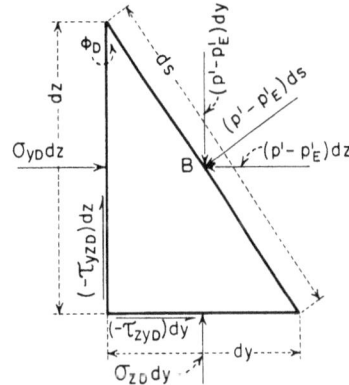

(b)- FORCES ACTING ON
DIFFERENTIAL ELEMENT AT
DOWNSTREAM FACE

From Figure c

To maintain rotational equilibrium about C,

$$(\tau_{IU}\,dr)\frac{ds}{2} = 0$$

$$\tau_{IU} = 0$$

To maintain vertical equilibrium,

$$\sigma_{zu}\,dy - \left[(p + p_E)\,ds\right]\sin\phi_U - (\sigma_{IU}\,dr)\cos\phi_U + (\tau_{IU}\,dr)\sin\phi_U = 0$$

$$\sigma_{IU} = \frac{\sigma_{zu}\,dy - \left[(p + p_E)\,ds\right]\sin\phi_U}{dr\,\cos\phi_U}$$

$$ds = dy\,\sin\phi_U$$
$$dr = dy\,\cos\phi_U$$

$$\sigma_{IU} = \frac{\sigma_{zu}\,dy - (p + p_E)\,dy\,\sin^2\phi_U}{dy\,\cos^2\phi_U}$$

Then:

$$\sigma_{IU} = \sigma_{zu}\,\sec^2\phi_U - (p + p_E)\tan^2\phi_U$$

In a similar manner:

$$\sigma_{ID} = \sigma_{zD}\,\sec^2\phi_D - (p' - p'_E)\tan^2\phi_D$$

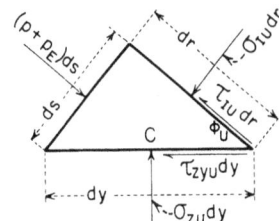

(c)- DIFFERENTIAL ELEMENT
AT UPSTREAM FACE

Figure 4-2. Derivation of stress formulae for a concrete gravity dam.—DS2-2(4)

STRESSES IN STRAIGHT GRAVITY DAMS, INCLUDING EFFECTS OF TAILWATER AND HORIZONTAL EARTHQUAKE

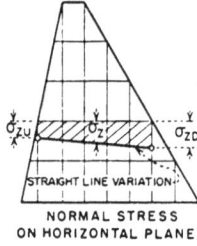

NORMAL STRESS ON HORIZONTAL PLANE

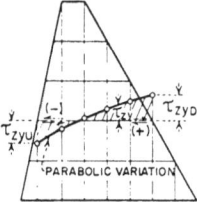

SHEAR STRESS ON HORIZONTAL PLANE

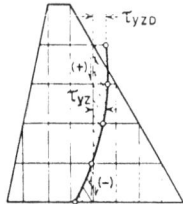

SHEAR STRESS ON VERTICAL PLANE

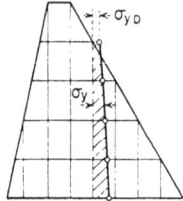

NORMAL STRESS ON VERTICAL PLANE

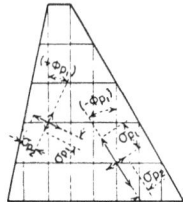

PRINCIPAL STRESS

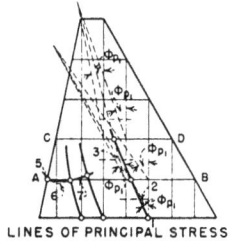

LINES OF PRINCIPAL STRESS

NORMAL STRESS ON HORIZONTAL PLANE, σ_z

I. $\sigma_z = a + by$

$$a = \sigma_{ZD} = \frac{\Sigma W}{T} - \frac{6\Sigma M}{T^2} \qquad b = \frac{12\Sigma M}{T^3}$$

SHEAR STRESS ON HORIZONTAL OR VERTICAL PLANE, $\tau_{yz} = \tau_{zy}$

II. $\tau_{zy} = \tau_{yz} = a_1 + b_1 y + c_1 y^2$

$$a_1 = \tau_{zyD} = (\sigma_{ZD} - p' \pm^{\ddagger} p'_E) \tan \phi_D$$

$$b_1 = -\frac{T}{T}\left(\frac{6\Sigma V}{T} + 2\tau_{zyU} + 4\tau_{zyD}\right)$$

$$c_1 = \frac{1}{T^2}\left(\frac{6\Sigma V}{T} + 3\tau_{zyU} + 3\tau_{zyD}\right)$$

Check at face for $y = T$, $\tau_{zyU} = -(\sigma_{ZU} - p \pm^{\ddagger} p_E) \tan \phi_U$

Note: ($\ddagger$ Use (+) sign if horizontal acceleration of foundation is upstream.)

     ($*$ Use (−) sign if horizontal acceleration of foundation is upstream.)

NORMAL STRESS ON VERTICAL PLANE, σ_y

III. $\sigma_y = a_2 + b_2 y + c_2 y^2 + d_2 y^3$

$$a_2 = \sigma_{YD} = a_1 \tan \phi_D + p' \pm^{\ddagger} p'_E; \qquad b_2 = b_1 \tan \phi_D + \frac{\partial a_1}{\partial z} \pm^{\ddagger} \lambda \omega_c$$

$$\frac{\partial a_1}{\partial z} = \tan \phi_D\left(\frac{\partial \sigma_{ZD}}{\partial z} - \omega^{\circledast} \pm^{\ddagger} \frac{\partial p'_E}{\partial z}\right) + \frac{\partial \tan \phi_D}{\partial z}\left(\sigma_{ZD} - p' \pm^{\ddagger} p'_E\right)$$

   ($\circledast$ ω to be omitted if tailwater is absent)

$$\frac{\partial \sigma_{ZD}}{\partial z} = \omega_c + \tan \phi_U\left(\frac{12\Sigma M}{T^3} + \frac{2\Sigma W}{T^2} - \frac{2p}{T} \pm^{\ddagger} \frac{2p_E}{T}\right) + \tan \phi_D\left(\frac{12\Sigma M}{T^3} - \frac{4\Sigma W}{T^2} + \frac{4p'}{T} + {}^{\ddagger}\frac{4p'_E}{T}\right) - \frac{6\Sigma V}{T^2}$$

$$\frac{\partial p'_E}{\partial z} = \frac{(p'_E - p'^{*}_E)}{\Delta z} \qquad \frac{\partial \tan \phi_D}{\partial z} = \frac{\tan \phi_D - \tan \phi^{*}_D}{\Delta z}$$

$$c_2 = c_1 \tan \phi_D + \frac{1}{2}\frac{\partial b_1}{\partial z}$$

$$\frac{\partial b_1}{\partial z} = -\frac{1}{T^2}\left[6\left(\frac{\partial \Sigma V}{\partial z}\right) - \frac{\partial T}{\partial z}\left(\frac{12\Sigma V}{T} + 2\tau_{zyU} + 4\tau_{zyD}\right)\right] - \frac{1}{T}\left[2\left(\frac{\partial \tau_{zyU}}{\partial z}\right) + 4\left(\frac{\partial \tau_{zyD}}{\partial z}\right)\right]$$

$$\frac{\partial \Sigma V}{\partial z} = -(p - p' \pm^{\ddagger} \lambda \omega_c T \pm^{\ddagger} p_E \pm^{\ddagger} p'_E) \qquad \frac{\partial T}{\partial z} = \tan \phi_U + \tan \phi_D$$

$$\frac{\partial \tau_{zyU}}{\partial z} = \tan \phi_U\left(\omega^{\circledast} - \frac{\partial \sigma_{ZU}}{\partial z} \pm^{\ddagger} \frac{\partial p_E}{\partial z}\right) + \frac{\partial \tan \phi_U}{\partial z}\left(p \pm^{\ddagger} p_E - \sigma_{ZU}\right)$$

   ($\circledast$ ω to be omitted if reservoir water is absent)

$$\frac{\partial \sigma_{ZU}}{\partial z} = \omega_c + \tan \phi_U\left(\frac{4p}{T} \pm^{\ddagger}\frac{4p_E}{T} - \frac{4\Sigma W}{T^2} - \frac{12\Sigma M}{T^3}\right) + \tan \phi_D\left(\frac{2\Sigma W}{T} \pm^{\ddagger}\frac{2p'_E}{T} - \frac{2p'}{T} - \frac{12\Sigma M}{T^3}\right) + \frac{6\Sigma V}{T^2}$$

$$\frac{\partial \tau_{zyD}}{\partial z} = \frac{\partial a_1}{\partial z} \qquad \frac{\partial p_E}{\partial z} = \frac{(p_E - p^{*}_E)}{\Delta z} \qquad \frac{\partial \tan \phi_U}{\partial z} = \frac{\tan \phi_U - \tan \phi^{*}_U}{\Delta z}$$

$$d_2 = \frac{1}{3}\frac{\partial c_1}{\partial z}$$

$$\frac{\partial c_1}{\partial z} = \frac{1}{T^3}\left[6\left(\frac{\partial \Sigma V}{\partial z}\right) - \frac{\partial T}{\partial z}\left(\frac{18\Sigma V}{T} + 6\tau_{zyU} + 6\tau_{zyD}\right)\right] + \frac{1}{T^2}\left[3\left(\frac{\partial \tau_{zyU}}{\partial z}\right) + 3\left(\frac{\partial \tau_{zyD}}{\partial z}\right)\right]$$

Check at face for $y = T$, $\sigma_{yU} = (p \pm^{\ddagger} p_E - \tau_{zyU} \tan \phi_U)$

MAGNITUDE OF PRINCIPAL STRESSES, $\sigma_{p_1}, \sigma_{p_2}$

IV. $\sigma_{p_1} = \frac{\sigma_z + \sigma_y}{2} \pm \sqrt{\left(\frac{\sigma_z - \sigma_y}{2}\right)^2 + (\tau_{zy})^2}$ $\begin{cases} \text{If } (\sigma_z - \sigma_y) > 0, \text{ use}(+) \\ \text{If } (\sigma_z - \sigma_y) < 0, \text{ use}(-) \end{cases} \begin{cases} \text{Alternate sign gives } \sigma_{p_2} \text{ which} \\ \text{is perpendicular to } \sigma_{p_1}. \end{cases}$

Check at upstream face either σ_{p_1} or $\sigma_{p_2} = \sigma_{zU} \sec^2 \phi_U - (p \pm^{\ddagger} p_E) \tan^2 \phi_U$

Check at downstream face either σ_{p_1} or $\sigma_{p_2} = \sigma_{ZD} \sec^2 \phi_D - (p' \pm^{\ddagger} p'_E) \tan^2 \phi_D$

DIRECTION OF FIRST PRINCIPAL STRESS, ϕ_{p_1}

V. $\phi_{p_1} = \frac{1}{2} \arctan\left(-\frac{\tau_{zy}}{\frac{\sigma_z - \sigma_y}{2}}\right)$ $\begin{cases} \text{If } \tan 2\phi_{p_1} = (+), \ 0 < \phi_{p_1} < (+45°) \\ \text{If } \tan 2\phi_{p_1} = (-), \ (-45°) < \phi_{p_1} < 0 \end{cases} \begin{array}{l}\text{Measured from Vertical.} \\ \text{Clockwise positive.}\end{array}$

Check at upstream and downstream face, $\phi_{p_1} = \phi$, or $(90° - \phi)$

METHOD OF CONSTRUCTING LINES OF PRINCIPAL STRESS

1. From the intersection of a chosen vertical plane and the base of the cantilever cross-section, measure ϕ_{p_1} at that point and draw tangent 1-2 half way to the horizontal section A-B next above.
2. At the two points on section A-B between which the line must pass, lay off the angles which give the direction of the principal stress being considered and prolong the lines to their point of intersection.
3. Between this point of intersection and point 2 of the tangent 1-2, draw tangent 2-3 half way to the horizontal section C-D next above A-B.
4. Continue the interpolation until the succession of tangents has reached either of the faces or the top of the cantilever cross-section. (It may be necessary to interpolate between angles at the intersections of successive horizontal sections and a vertical plane.)
5. Through the points of intersection of these tangents and the horizontal sections (or vertical planes), draw the curves or lines of principal stress. Draw sufficient lines to include the whole section.
6. From the intersection of a horizontal section and the upstream face (or from the intersection of a vertical plane and the base), measure the angle of the stress complementary to the one just drawn and draw tangent 5-6 from the upstream face half way to the first line of principal stress already drawn.
7. From 6, draw tangent 6-7 perpendicular to the first line of principal stress and extending half way to the next line of principal stress.
8. Continue the construction of the tangents perpendicular to the lines of principal stress already drawn until the top, base, or downstream face of the cantilever cross-section is reached.
9. Connect the points of intersection of these tangents and the lines of principal stress first drawn with curves or lines of complementary principal stress. Draw sufficient curves to include the whole section.

NOTE

The figures on this sheet are for illustrative purposes only. They do not represent results for any specific condition of loading.

$*$ The value of the quantity is to be determined at a horizontal plane Δz distance above the horizontal section under consideration

Figure 4-3. Stresses in straight gravity dams.—288-D-3152

The shear-friction factor, Q, for horizontal planes, is the ratio of resisting to driving forces as computed by the expression:

$$Q = \frac{CA + (\Sigma N + \Sigma U)\tan\phi}{\Sigma V} \qquad (4)$$

where:

C = unit cohesion,
A = area of horizontal section considered,
ΣN = summation of normal forces,
ΣU = summation of uplift forces,
tan ϕ = coefficient of internal friction, and
ΣV = summation of shear forces.

All parameters must be specified using consistent units and with proper signs according to the convention shown in figure 4-1.

Shear-friction factors are computed for each respective elevation for which stresses are calculated in the cantilever element for the same condition of loading. All possible conditions of loading should be investigated. It should be noted that high stability is indicated by high shear-friction factors. The allowable minimum value for this factor for use in design is given in section 3-20.

The factor of safety for overturning is not usually tabulated with other stability factors for Bureau dams, but may be calculated if desired by dividing the total resisting moments by the total moments tending to cause overturning about the downstream toe. Thus:

$$\left[\begin{array}{c}\text{Overturning safety}\\\text{factor}\end{array}\right] = \frac{\text{moments resisting}}{\text{moments overturning}}$$

Before bodily overturning of a gravity dam can take place, other failures may occur such as crushing of the toe material, and cracking of the upstream material with accompanying increases in uplift pressure and reduction of the shear resistance. However, it is desirable to provide an adequate factor of safety against the overturning tendency. This may be accomplished by specifying the maximum allowable stress at the downstream face of the dam. Because of their oscillatory nature, earthquake forces are not considered as contributing to the overturning tendency.

C. TRIAL-LOAD METHODS OF ANALYSIS

1. Trial-Load Twist Method of

Analysis, Joints Ungrouted

4-11. Introduction.—A gravity dam may be considered to be made up of a series of vertical cantilever elements from abutment to abutment. If the cross-canyon profile is narrow with steep sloping walls, each cantilever from the center of the dam towards the abutments will be shorter than the preceding one. Consequently, each cantilever will be deflected less by the waterload than the preceding one and more than the succeeding one. If the transverse contraction joints in the dam are keyed, and regardless of whether grouted or ungrouted, the movements of each cantilever will be restrained by the adjacent ones. The longer cantilever will tend to pull the adjacent shorter cantilever forward and the shorter cantilever will tend to hold it back. This interaction between adjacent cantilever elements causes torsional moments, or twists, which materially affect the manner in which the waterload is distributed between the cantilever elements in the dam. This changes the stress distribution from that found by the ordinary gravity analysis in which the effects of twist, as well as deformation of the foundation rock, are neglected. All straight gravity dams having keyed transverse contraction joints should therefore be treated as three-dimensional structures and designed on that basis.

If the canyon is wide and flat, the cantilevers in the central portion of the dam are of about the same length and the effects of twist are usually negligible. However, twist effects may be important in the abutment regions where the length of the cantilevers changes rapidly.

This twist action tends to twist the cantilevers from their seats on the sloping canyon walls, thus tending to develop cracks in the dam in these regions.

A sharp break in the cross-canyon profile will result in an abrupt change in the length of cantilevers in that region. This has the effect of introducing an irregular wedge of rock in the dam and causes a very marked change in stresses and stability factors. Such conditions should be eliminated, if possible, by providing additional rock excavation so that a smooth profile is obtained.

4-12. *Theory*.—The use of a twisted structure for analyzing dams was suggested in a review of the trial-load method of analysis made for the Bureau of Reclamation in 1930. It was proposed that instead of replacing the arch dam by two systems of structural elements, one of arches and the other of cantilevers, one could replace it by three systems each occupying the whole volume of the dam. These three systems would consist of vertical cantilevers, horizontal arches, and a twisted structure. The new system, designated the twisted structure, would have the primary purpose of resisting the twisting moments in horizontal and vertical sections. Thus the vertical cantilevers would be considered as having no torsional resistance, since the twisting moments would be assigned to the twisted structure. Subsequently, the conception of the twisted structure was applied to the analysis of a gravity dam.

A gravity dam is constructed as a series of vertical blocks fixed at the base, as shown on figure 4-4. These blocks may be assumed to be capable of resisting torsions and shears in horizontal planes and bending in vertical planes for the condition of joints ungrouted. This assumption is based upon the theory that the contraction joints, through the action of the keyways, can transfer load horizontally to the abutments by means of shear in the horizontal elements and torsion in the vertical element, but cannot transfer twisting moment horizontally to the abutment. For joints ungrouted, it is obvious that the contraction joints cannot be assumed to provide resistance to bending in either vertical or horizontal planes.

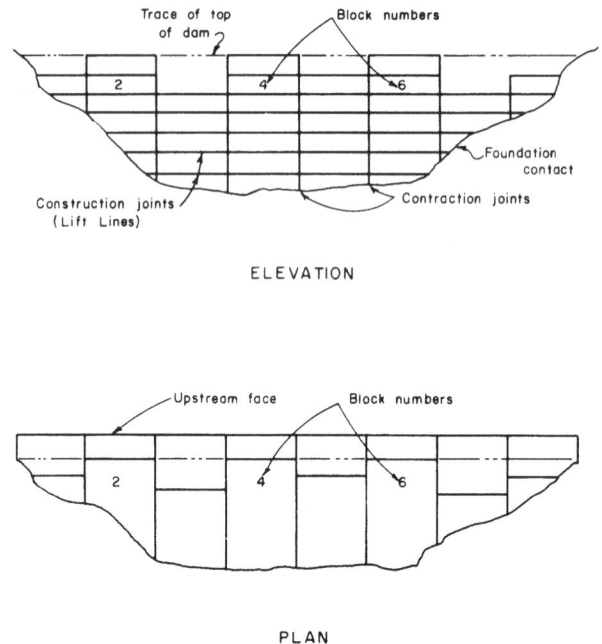

Figure 4-4. Schematic view simulating partial construction of a gravity dam in vertical blocks.—288-D-3113

In the "Trial-Load Twist Method of Analysis," the dam is assumed to be divided into a number of vertical and horizontal elements, the vertical elements of which are usually considered to be 1 foot in width and the horizontal elements 1 foot in height. Further, in the manner described below, these elements are assumed to make up two structural systems, called the cantilever structure and the twisted structure, each of which occupies the entire volume of the dam.

The cantilever structure consists of a series of vertical cantilever elements as described above, which abut on the foundation and transfer thereto the dead load and a portion of the total external loads by gravity action only. These cantilevers carry external loads downward to the foundation by bending and shear along horizontal planes without being restrained by the twisted-structure elements. The cantilever is subjected to two types of deformation: bending caused by flexure, and detrusion caused by shear in each cantilever element.

The twisted structure consists of vertical twisted elements with the same structural properties as the cantilevers in the cantilever structure; and of horizontal elements which are subjected to shear only. The vertical twisted elements resist no bending and shear, but resist only twisting moments produced by the shear due to loads on the horizontal elements of the twisted structure. A major part of the deflection of the horizontal twisted-structure elements is caused by the angular rotations of the cantilevers. Also included in the total twisted-structure deflection are the movement of the abutment due to forces brought down by the cantilever which joins the foundation at a common point with the horizontal element, and the shear deflection in the horizontal element due to loads on the twisted structure. The horizontal elements are segmental and are incapable of resisting bending moments if the joints are not grouted. The cantilever structure and twisted structure are illustrated on figure 4-5.

In order to make a twist analysis it is necessary to load the cantilever structure and twist structure by trial. For convenience, only a limited number of selected elements are analyzed which will provide satisfactory results for representative points in the dam. After selection of the elements, the total waterload on the dam is divided between the two structures by trial. The deflections of the cantilevers and twisted structure are then determined at conjugate points. For the first trial there will be little agreement in deflection at these points, but the process is repeated until the continuity of the structures is restored. Stresses may then be computed from known forces and moments and are assumed to represent the true stresses within the dam.

The following sections show the equations and procedure for analyzing a dam for joints ungrouted. Detailed computations are not given for the complete procedure, since another analysis, which is given later for joints grouted, shows the calculations which demonstrate the principles involved in the present analysis.

4-13. *Notations.—*

x, y, z = coordinates along X axis, Y axis, and Z axis, respectively.

M_x = bending moment in plane parallel to YZ plane, foot-pounds.

M_z = bending moment in plane parallel to XY plane, foot-pounds.

M_{xy} = twisting moment in horizontal XY plane, foot-pounds.

M_{zy} = twisting moment in vertical ZY plane, foot-pounds.

V = horizontal thrust of waterload, pounds.

V_c = shear in cantilever due to horizontal component of waterload carried by cantilever, pounds.

V_T = shear in horizontal element of twisted structure due to horizontal component of waterload carried by the twisted structure, pounds.

Subscript $_A$ such as in $_A M_{xy}$ and $_A V_C$ indicates abutment value for twisting moment in XY plane and shear at base of cantilever, respectively.

T = thickness of dam at a given elevation, feet.

E_c = modulus of elasticity of concrete in tension or compression, pounds per square foot.

E_r = modulus of elasticity of abutment material in tension or compression, pounds per square foot.

G = modulus of elasticity of concrete in shear, pounds per square foot.

μ = Poisson's ratio.

θ_z = angular rotation in horizontal plane, radians.

Δy = deflection normal to axis of dam, feet.

Z = vertical distance from base of cantilever, feet.

I = moment of inertia for a vertical cantilever of unit width or a horizontal beam of unit height of cross-section, feet[4].

J = a factor used in computing angular rotations of cantilevers due to torsions—joints ungrouted.

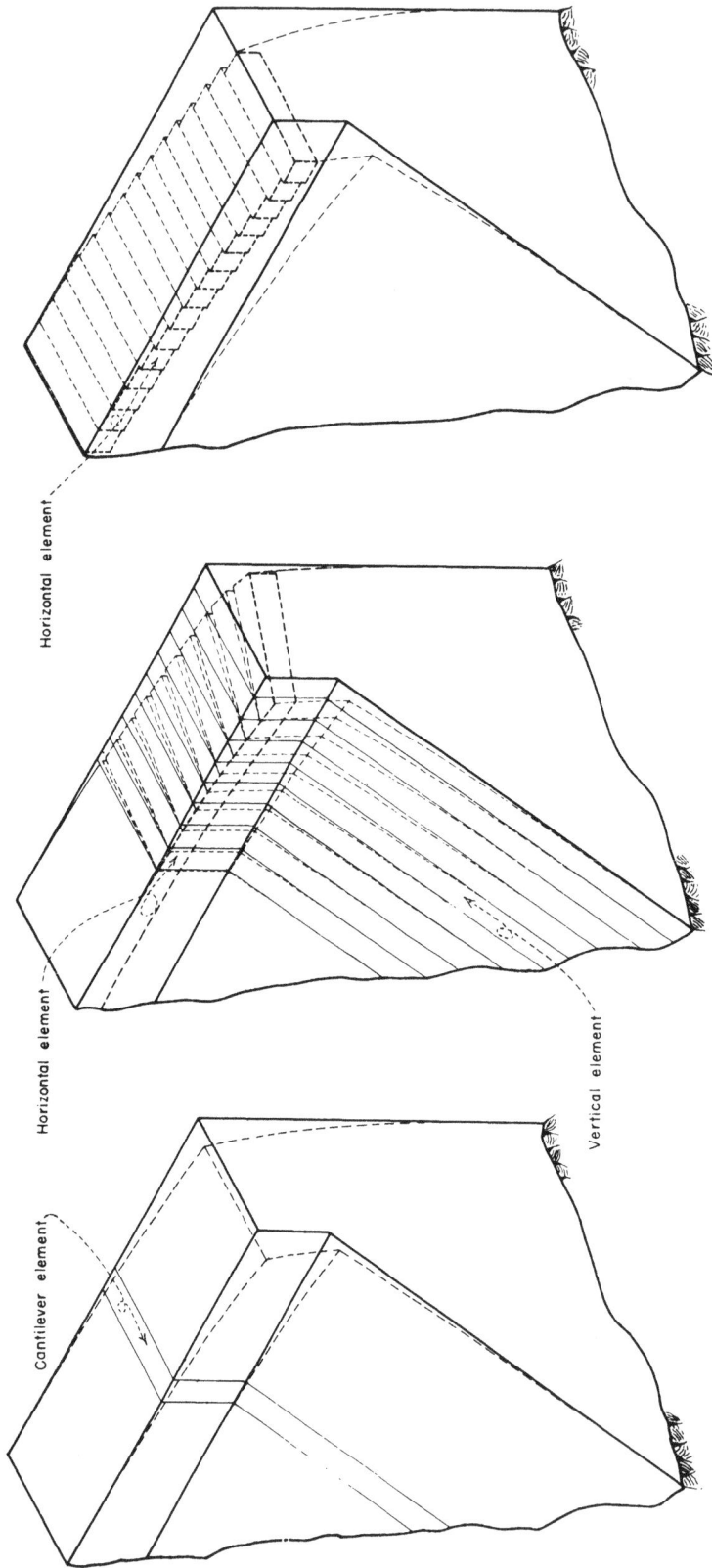

DEFLECTION OF HORIZONTAL ELEMENT DUE
TO SHEAR DETRUSION

THE TWISTED-STRUCTURE SYSTEM

Note: Horizontal element incapable of resisting
bending moment or torsional shear.

Horizontal element

DEFLECTION OF HORIZONTAL ELEMENT DUE
TO TWIST OF VERTICAL ELEMENTS

Horizontal element

Vertical element

THE CANTILEVER STRUCTURE SYSTEM

Cantilever element

Figure 4-5. Cantilever and twisted-structure systems—joints ungrouted.—DS2-2(23)

A = area of cross-section of a cantilever or beam, square feet. For a unit width or height, $A = T$.

L = length of a horizontal beam, feet.

p = external pressure at depth h, pounds per square foot.

P = unit load ordinate, pounds per square foot.

K = a constant which depends upon the ratio of the actual shear distribution to a shear distributed uniformly. In these analyses $K/G = 3/E$ and $K = 1.25$.

ψ = angle between canyon wall at cantilever base and the vertical, degrees.

α = angular movement of abutment in vertical plane due to unit bending moment $_A M_x$, radians.

γ = abutment movement normal to axis of dam due to unit shear force at the abutment, feet.

a_2 $\begin{cases} \text{= angular movement of abutment in vertical plane due to unit horizontal shear force at the abutment, radians.} \\ \text{= abutment movement normal to axis of dam due to unit bending moment } _A M_x, \text{ feet.} \end{cases}$

δ = angular movement of abutment in horizontal plane due to unit twisting moment $_A M_{xy}$, radians.

The convention of signs to be used is shown on figure 4-6.

4-14. *Foundation Constants*.—Rotation and deformations of the foundation surface for moments and forces of unity, per unit length, are given by the following formulas, in which k is a function of μ and b/a, and T is equal to a' (see figs. 4-7 to 4-10).

$$\alpha' = \frac{k_1}{E_r T^2} \tag{1}$$

$$\alpha'' = \frac{k_5}{E_r T} \tag{2}$$

$$\gamma' = \frac{k_3}{E_r} \tag{3}$$

$$\delta' = \frac{k_4}{E_r T^2} \tag{4}$$

The above equations contain elastic constants, E_r and μ, which are usually determined by direct experimental methods. The curves shown on figures 4-7 to 4-10 provide an easy means for determining values of k_1 to k_5, inclusive, after the ratio b/a has been determined by means described below. It is impossible to obtain a definite value of b/a for an irregular foundation surface. An approximation of some kind is necessary, and at present the following method is used. The surface of contact between the dam and foundation is developed and plotted as shown on figure 4-11. This surface is replaced by a rectangle of the same area and approximately the same proportions, called the equivalent developed area. The ratio of length to width of the rectangle is taken as the ratio b/a for the foundation in question. The value of b/a is therefore a constant for a particular dam. In computing deformations for a particular element, the width a' is made equal to T, the thickness of the dam at the element considered, making $T/b' = a/b$, or $b' = (b/a) T$.

The final equations for the foundation movements of a unit horizontal element at either abutment of the dam are shown below. The algebraic signs are as used for the left abutment, and the asterisk (*) indicates that signs are to be reversed for movements at the right abutment.

$$*\theta_z = M_z \, \alpha + V \, \alpha_2 \tag{5}$$

$$\Delta y = V \gamma + M_z \, \alpha_2 \tag{6}$$

for which:

$$\alpha = \alpha' \cos^3 \psi + \delta' \sin^2 \psi \cos \psi \tag{7}$$

NOTES:

Normal loads are applied to the faces.

Twist loads are applied at the cantilever ₵.

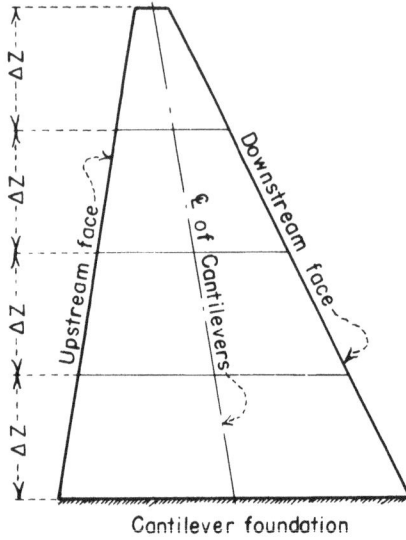

(a) VERTICAL CROSS-SECTION

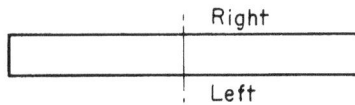

(b) HORIZONTAL CROSS-SECTION

MAXIMUM CANTILEVER AND CANTILEVER TO LEFT OF MAXIMUM VERTICAL CROSS-SECTION (L) (LOOKING UPSTREAM)						CANTILEVER TO RIGHT OF MAXIMUM VERTICAL CROSS-SECTION (R) (LOOKING UPSTREAM)				
DIRECTION OF POSITIVE MOVEMENTS	DIRECTION OF POSITIVE FORCES AND MOMENTS	DIRECTION OF POSITIVE LOADS	DIRECTION OF FORCES AND MOMENTS DUE TO POSITIVE LOADS	DIRECTION OF MOVEMENTS DUE TO POSITIVE LOADS		DIRECTION OF POSITIVE MOVEMENTS	DIRECTION OF POSITIVE FORCES AND MOMENTS	DIRECTION OF POSITIVE LOADS	DIRECTION OF FORCES AND MOMENTS DUE TO POSITIVE LOADS	DIRECTION OF MOVEMENTS DUE TO POSITIVE LOADS
ALL DIRECTIONS REFER TO FIGURES						ALL DIRECTIONS REFER TO FIGURES				
FIGURE (a)						FIGURE (a)				
Δy ← +	V ← + ⟋ M + ↺	NORMAL ± →	{ V → M − ↻	Δy →		Δy ← +	V ← + ⟋ M + ↺	NORMAL ± →	{ V → M − ↻	Δy →
FIGURE (b)						FIGURE (b)				
θ + ↺	₩ + ↺	TWIST + ↻	₩ − ↻	θ − ↻		θ + ↺	₩ + ↻	TWIST + ↻	₩ − ↻	θ + ↺

Figure 4-6. Direction of positive movements, forces, moments, and loads; and direction of forces, moments, and movements due to positive loads.—DS2-2(24)

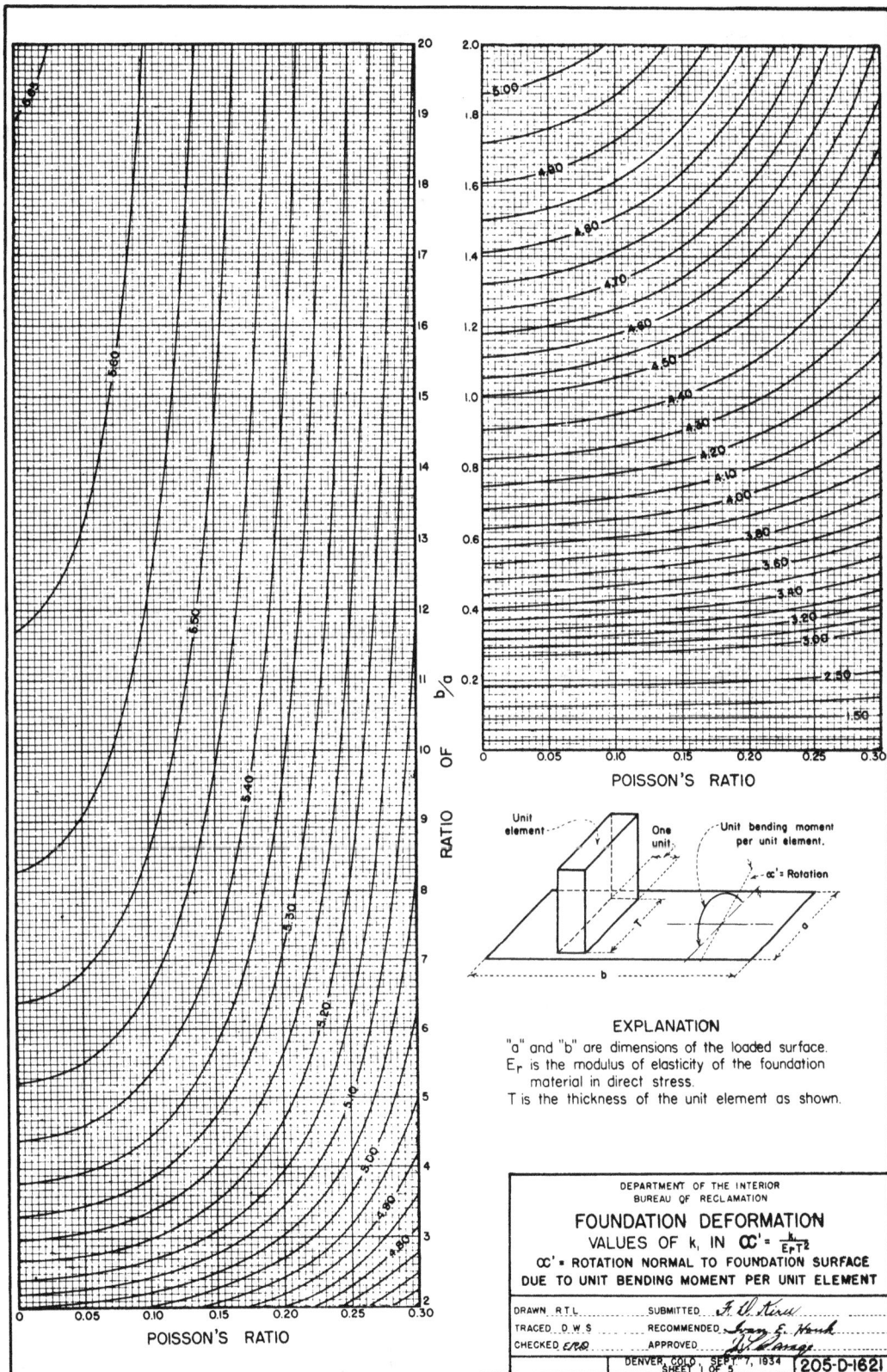

Figure 4-7. Foundation deformation—values of k_1 in equation (1).

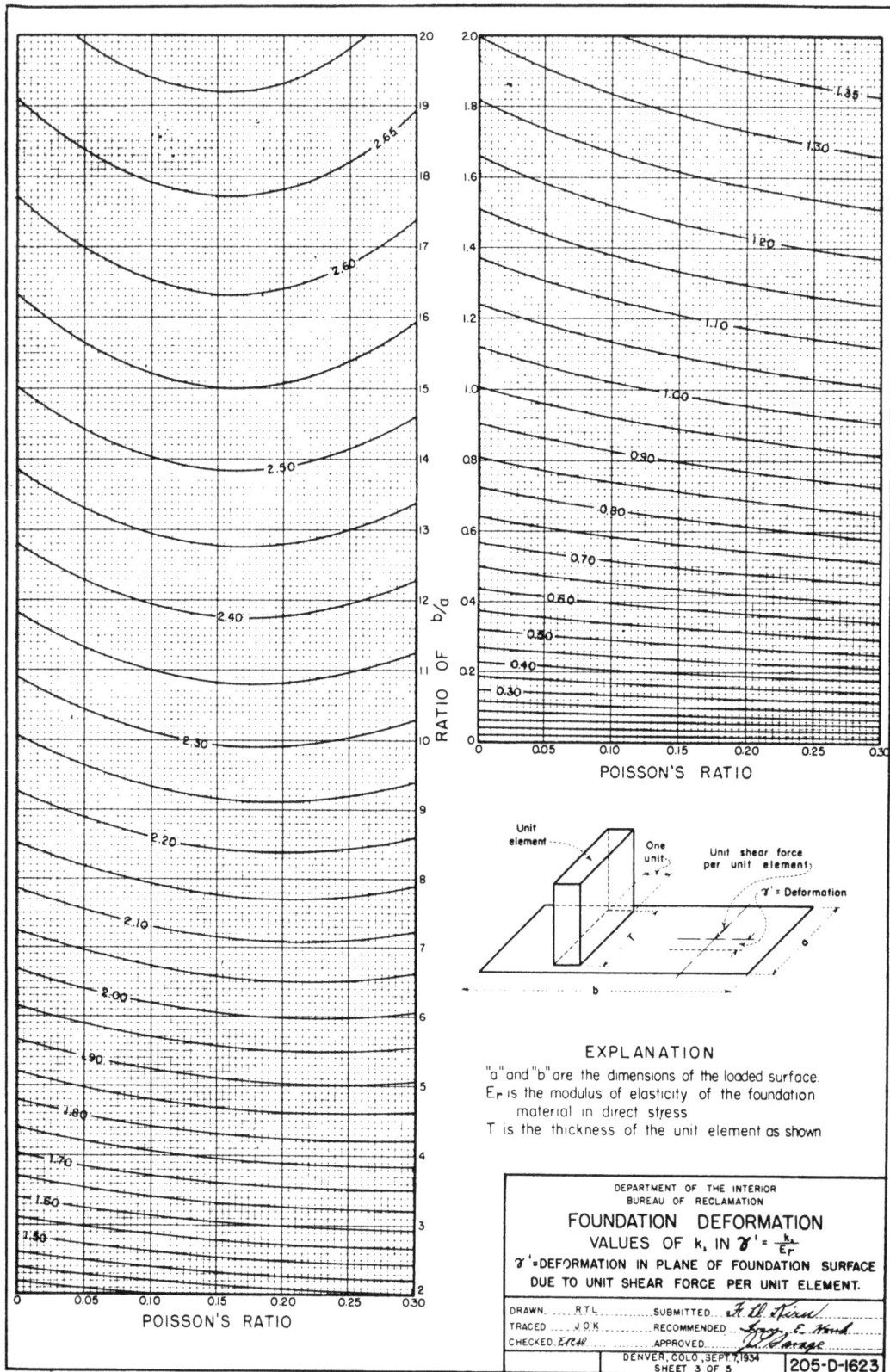

Figure 4-8. Foundation deformation–values of k_3 in equation (3).

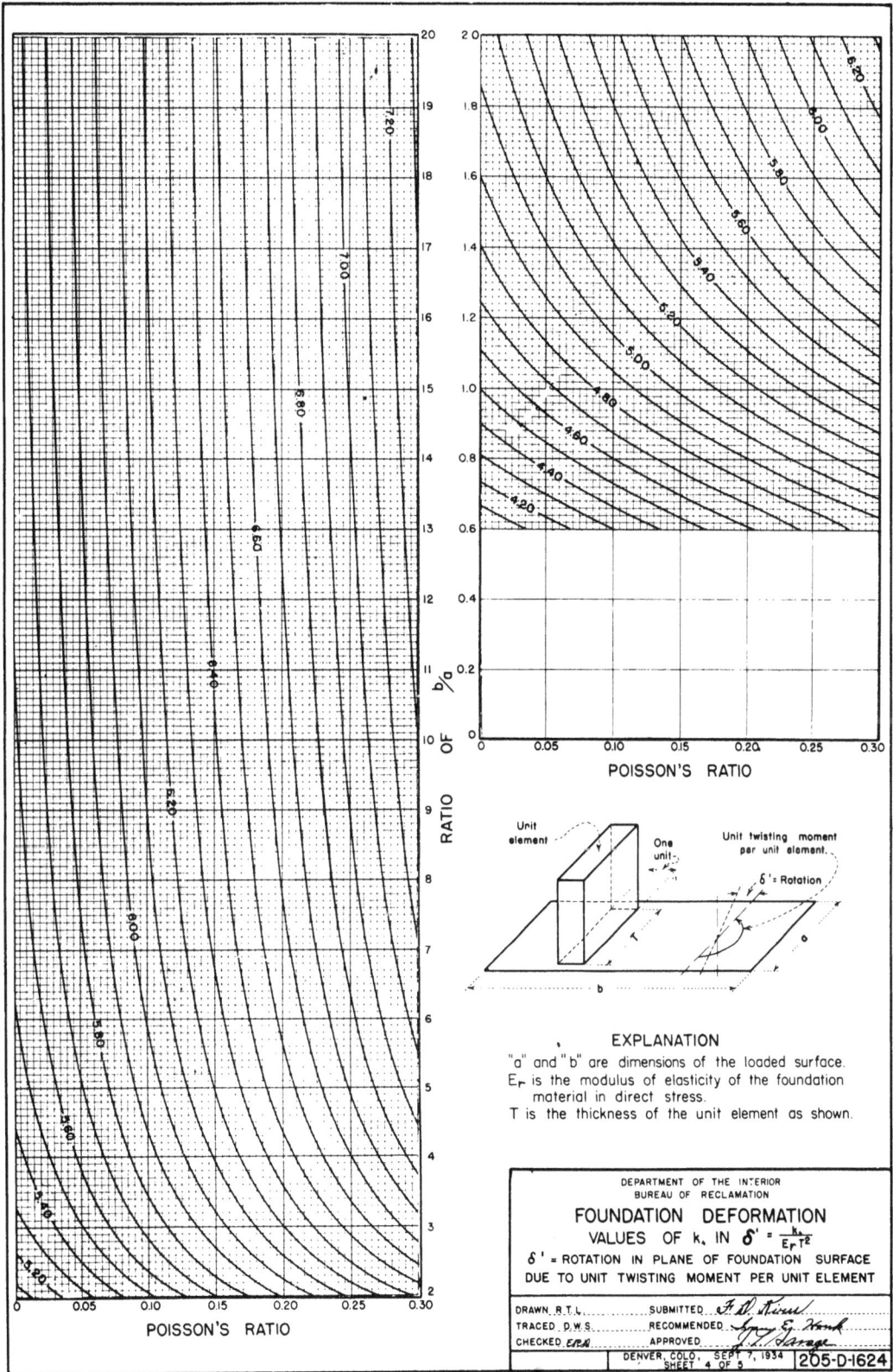

Figure 4-9. Foundation deformation—values of k_4 in equation (4).

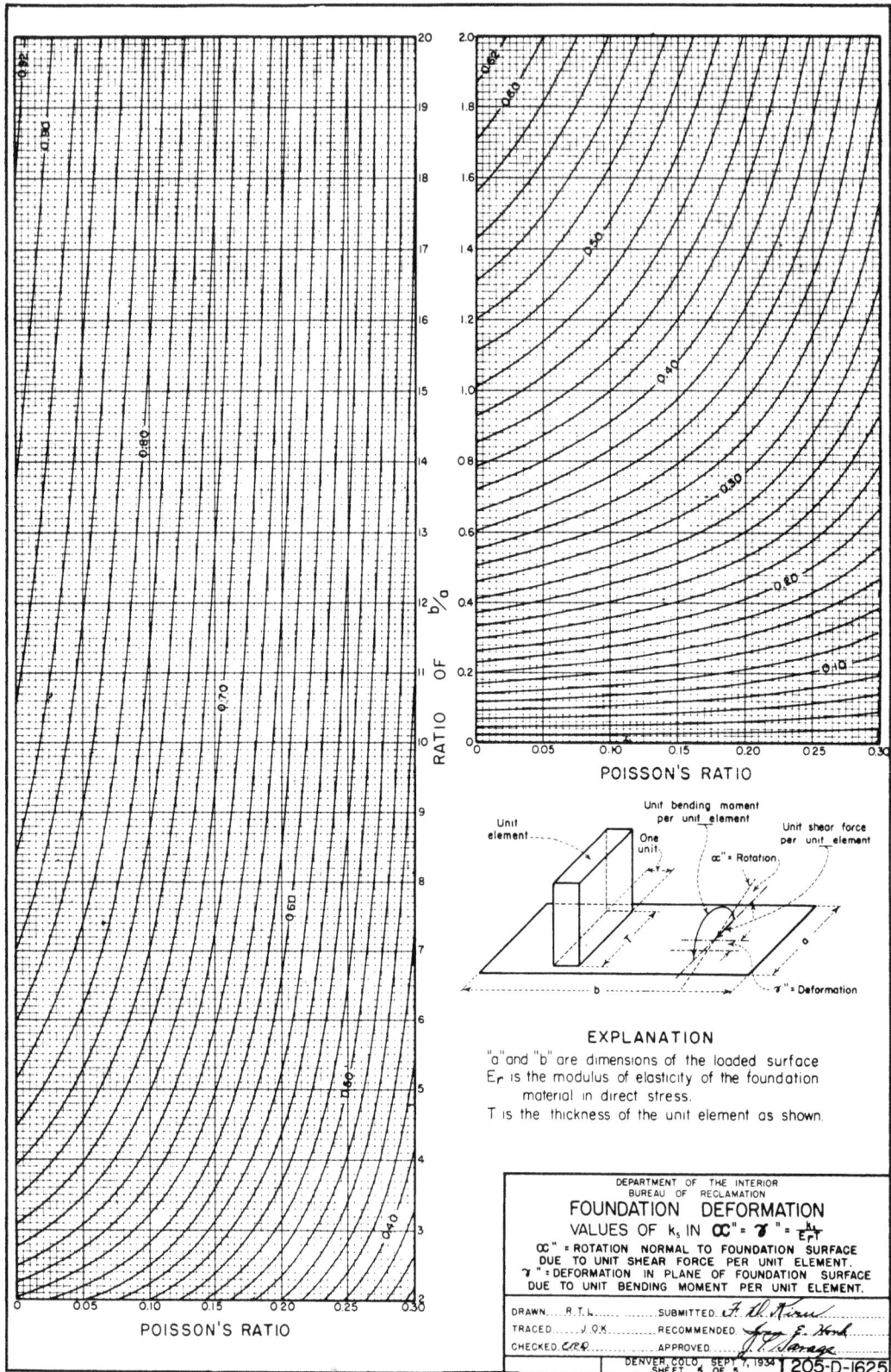

Figure 4-10. Foundation deformation—values of k_5 in equation (2).

(a) PLAN

(b) DEVELOPMENT

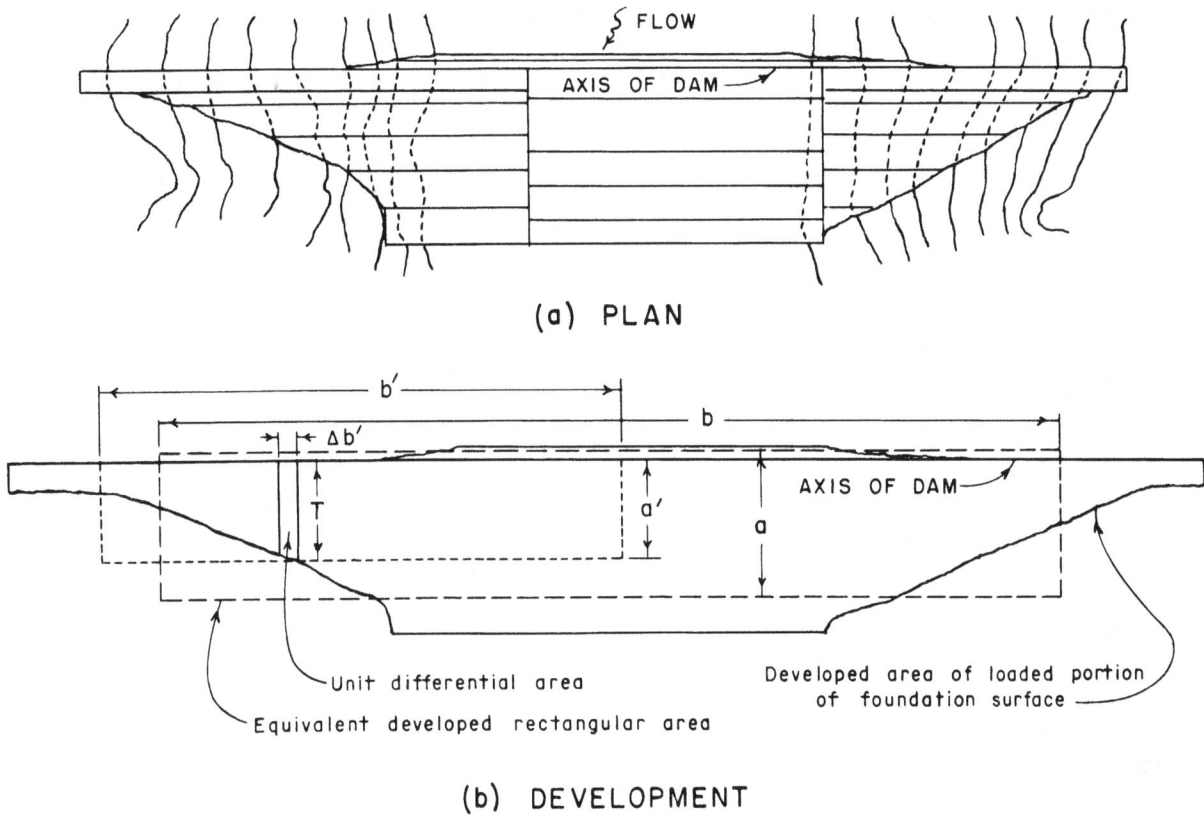

Figure 4-11. **Loaded area of a foundation surface.—288-D-3153**

$$\alpha_2 = \alpha'' \cos^2 \psi \qquad (8)$$

$$\gamma = \gamma' \cos \psi \qquad (9)$$

It is customary to require that δ' for a unit differential area on one side of the dam be an average value for the equivalent developed area of that side of the dam. If the damsite is approximately symmetrical about the maximum section, dimensions of the equivalent developed area for either or both sides of the dam are a and $b/2$. For this reason, ratios $\dfrac{b/2}{a}$ and $\dfrac{a}{b/2}$ are substituted for the ratio b/a in some cases in obtaining values from the curves on figures 4-7 through 4-10. These substitutions are indicated below:

For α', α'', and γ' use ratio $\dfrac{b}{a}$.

For δ' use ratio $\dfrac{b/2}{a}$.

The final equations for movements of a unit vertical element at either abutment of the dam are shown below. As before, the algebraic signs are as used for the left abutment and the asterisk (*) indicates that signs are to be reversed for movements at the right abutment.

$$\theta_x = M_x \, \alpha + V \, \alpha_2 \qquad (10)$$

$${}^*\theta_z = M_{xy} \, \delta \qquad (11)$$

$$\Delta y = V \, \gamma + M_x \, \alpha_2 \qquad (12)$$

for which:

$$\alpha = \alpha' \sin^3 \psi + \delta' \sin \psi \cos^2 \psi \qquad (13)$$

$$\alpha_2 = \alpha'' \sin^2 \psi \qquad (14)$$

$$\delta = \delta' \sin^3 \psi + \alpha' \sin \psi \cos^2 \psi \qquad (15)$$

$$\gamma = \gamma' \sin \psi \qquad (16)$$

4-15. *Selection of Elements.* —If the dam is symmetrical, only half of the dam need be analyzed. If it is not symmetrical, the dam is divided at a convenient plane near the center where the canyon floor is relatively flat, or where it is expected that little twist action is likely to exist. Each part of a nonsymmetrical dam is analyzed separately and continuity is established by bringing deflections into agreement at the dividing plane. For analyzing one-half of a symmetrical dam, usually five to seven horizontal elements and from four to seven vertical elements are selected to represent the structure. For a nonsymmetrical dam, usually nine to eleven vertical elements are required. Along a steeply sloping abutment and at points of irregularity, additional vertical elements may be required. Horizontal and vertical elements should be selected so that they have common abutments and foundations. Occasionally, this may not be possible for one or two vertical elements which must be placed at critical locations. The closest spacing of elements should be in the region of greatest twist.

4-16. *Loads, Forces, and Moments.* —Forces and moments due to dead load, waterload, earthquake shock, and other loads, and notations are determined as indicated in sections 4-7, 4-8, and 4-9. The concrete weight is assigned to the cantilevers entirely, since it is assumed that deflections due to weight take place gradually during construction of the dam prior to grouting of contraction joints. The position assumed by the cantilevers due to concrete weight is the zero position from which subsequent movements of the structure are measured. Stresses due to concrete weight are added to those determined from the trial-load adjustment. The calculation of deflections due to concrete weight is not required in the analysis.

It is convenient to assign certain loads initially to the cantilevers. These include horizontal and vertical earthquake concrete-inertia loads, vertical silt load, superstructure load, horizontal ice load, and static and hydrodynamic vertical waterloads. The deflections caused by these initial loads must be considered subsequently along with trial-load deflections, as explained later.

4-17. *Initial and Unit Deflections of Cantilevers.* —Prior to starting an adjustment, it is necessary to determine the properties of the cantilevers and calculate initial and unit deflections of cantilevers due to initial loads and unit normal loads, respectively.

The calculation of unit forces and moments due to unit normal loads is illustrated on figure 4-12. These loads are a system of triangular loads having a unit value of P, generally 1,000 pounds per square foot, at each respective elevation and decreasing uniformly to zero at elevations above and below the point of application.

Normal deflections of each cantilever due to initial loads and those due to unit normal loads are calculated by the equation,

$$\Delta y = \Sigma \left[\Sigma \frac{M_x}{EI} \Delta Z + {}_A M_x \, \alpha + {}_A V_c \, \alpha_2 \right] \Delta Z$$

$$+ {}_A M_x \, \alpha_2 + (\Sigma \frac{K V_c}{G A} \Delta Z + {}_A V_c \, \gamma) \quad (17)$$

in which the symbols have the meanings given in section 4-13. Unit abutment movements α, α_2, and γ, for use in the above equation, are determined for each cantilever by means of equations and curves given in section 4-14. The *underlined* portion of equation (17) represents deflection due to shear, while the remainder represents deflection due to bending. No special attention need be given the underlined portion of the equation when used for the analysis with joints ungrouted, but it will be referred to in the explanation for the analysis with joints grouted.

4-18. *Unit Rotations of Vertical Elements of Twisted Structure Due to Unit Twisting Couple.* —Unit rotations of vertical twisted elements due to unit triangular twisting-couple loads are illustrated on figure 4-13. As previously stated, the vertical elements have the same structural characteristics as the cantilevers.

As shown on the figure, each twisting-couple load has a unit value of P, usually 1,000 foot-pounds per square foot, at the elevation at

UNIT CANTILEVER ELEMENT — PARALLEL SIDES

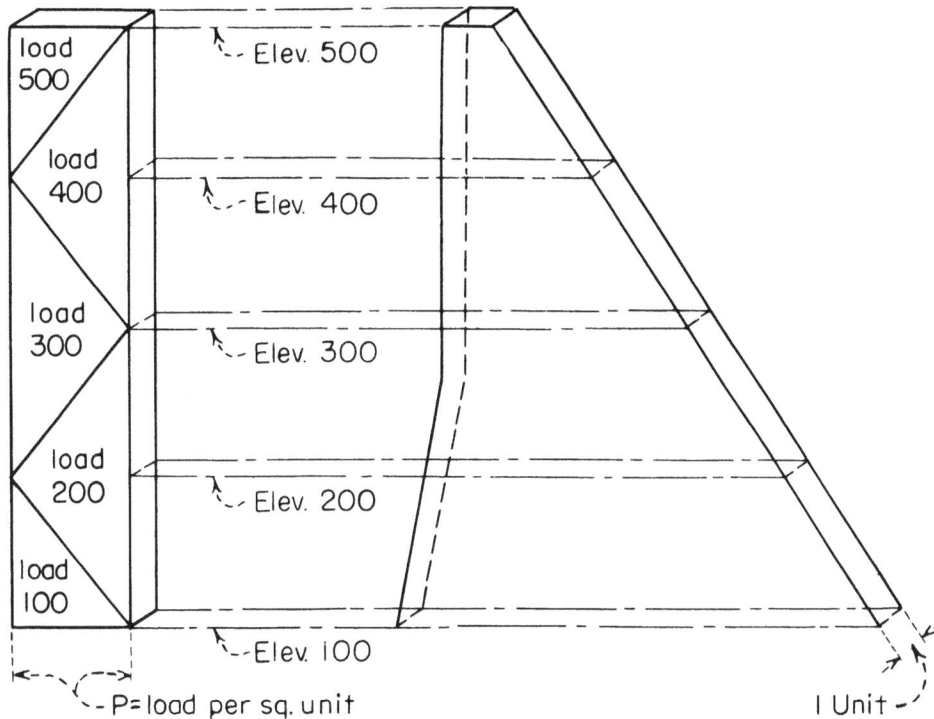

SHEARS AND MOMENTS DUE TO UNIT NORMAL LOADS

At Elevation 400

Load 500 (No.1)

$$\triangle V = \frac{P}{2} \cdot 100 = 50 \, P$$

$$\triangle M_x = \frac{P}{2} \cdot 100 \cdot \frac{2}{3} \cdot 100 = 3333 \, P$$

Load 400 (No.2)

$$\triangle V = \frac{P}{2} \cdot 100 = 50 \, P$$

$$\triangle M_x = \frac{P}{2} \cdot 100 \cdot \frac{1}{3} \cdot 100 = 1667 P$$

At Elevation 300

Load 500 (No.1)

$$\triangle V = 50 \, P$$

$$\triangle M_x = \frac{P}{2} \cdot 100 \left(\frac{2}{3} 100 + 100 \right) = 8333 P$$

Load 400 (No.2)

$$\triangle V = \frac{P}{2} \cdot 200 = 100 \, P$$

$$\triangle M_x = \frac{P}{2} 200 \cdot 100 = 10{,}000 \, P$$

Load 300 (No.3)

$$\triangle V = \frac{P}{2} \cdot 100 = 50 \cdot P$$

$$\triangle M_x = \frac{P}{2} \cdot 100 \, \frac{1}{3} \cdot 100 = 1667 \, P$$

Other Elevations Similar

$\triangle V =$ Horizontal force of portion of load above an elevation.

$\triangle M =$ Moment of $\triangle V$ about an elevation.

Note: In accordance with sign convention, ΔV and ΔM are negative.

Figure 4-12. **Unit normal loads on a cantilever.**—DS2-2(25)

UNIT CANTILEVER ELEMENT – PARALLEL SIDES

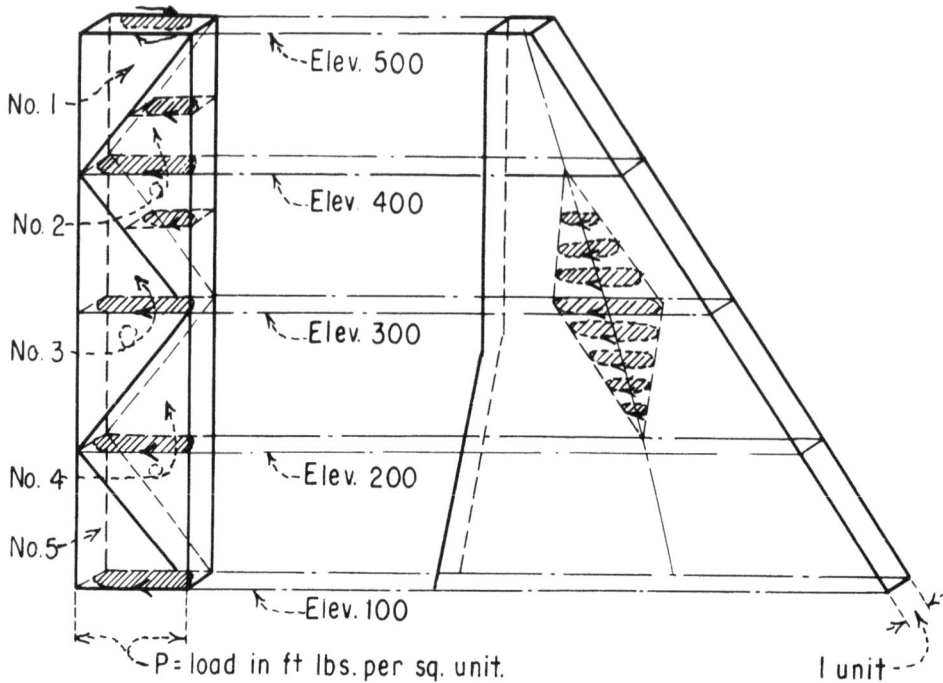

TWISTING MOMENTS DUE TO UNIT TWIST LOADS

At Elevation 400

Load 500 (No.1)

$\triangle M_{xy_1} = \frac{P}{2} 100 = 50P$

Load 400 (No.2)

$\triangle M_{xy_2} = \frac{P}{2} 100 = 50P$

At Elevation 300

Load 500 (No.1)

$\triangle M_{xy_1} = \frac{P}{2} 100 = 50P$

Load 400 (No.2)

$\triangle M_{xy_2} = \frac{P}{2} 200 = 100P$

Load 300 (No.3)

$\triangle M_{xy_3} = \frac{P}{2} 100 = 50P$

At Elevation 200

$\triangle M_{xy_1} = 50P$ $\qquad$ $\triangle M_{xy_3} = 100P$

$\triangle M_{xy_2} = 100P$ $\qquad$ $\triangle M_{xy_4} = 50P$

Other Elevations Similar

Note: In accordance with sign convention, M_{xy} is negative.

Figure 4-13. **Unit twist loads on a cantilever.**–DS2-2(26)

which the load is applied, which decreases uniformly to zero at elevations of horizontal elements above and below the point of application. The value of the twisting moment at any elevation due to a given unit load is equal to the volume of the portion of the wedge representing that load, above the given elevation, as may be seen by the calculations given on figure 4-13. Unit rotations of a vertical element on the left side of the dam are calculated by substitution of the above twisting moments in the equation which follows, where the symbols have the meanings given in section 4-13.

$$\theta_Z = \Sigma \frac{M_{xy}}{GJ} \Delta Z + {}_A M_{xy} \, \delta \qquad (18)$$

In the above, J is a factor for determining twist in a shaft of uniform cross section. The values of J are computed [3] from the equation,

$$J = \beta \, b \, c^3 \qquad (19)$$

where:

 b = longer side of horizontal cross section of element. (In this case the element is the block between two ungrouted contraction joints.)
 c = shorter side of horizontal cross section of element.

The following tabulation gives values of β for various ratios of b/c:

b/c	1.00	1.50	1.75	2.00	2.50	3.00
β	0.141	0.196	0.214	0.229	0.249	0.263

b/c	4.00	6.00	8.00	10.0	∞	
β	0.281	0.299	0.307	0.313	0.333	

To facilitate determining the proper values of J for the different elevations of each vertical element, the data in the above tabulation have been plotted and a curve drawn as shown on figure 4-14. The ordinates of the curve are the values of β and the abscissas are the corresponding ratios of b/c from the above table. Using the computed ratio of b/c for the

elevation at which J is to be computed, the value of β is determined from the curve. This value of β, together with the values of b and c, is then substituted in equation (19) and J computed for that elevation. This procedure is repeated for each elevation analyzed. The values of J thus computed are for a block having a width equal to the distance between the ungrouted contraction joints. To determine J for an element 1 foot wide, the computed values are divided by the distance between the contraction joints.

The values of J determined by the above method will hold for any unit element within this block or similar blocks having the same distance between the contraction joints and the same thickness at each elevation. However, if there are other blocks in the dam that have different distances between the contraction joints, thus changing the ratio of b/c at different beam elevations, values of J for these blocks must also be determined. Care should be taken to assign the proper values to b and c in each computation, b being the longer side of the rectangular cross section of the block at the elevation under consideration and c the shorter side.

Equation (19) was developed for shafts or beams of uniform cross section, and the values of J computed from this equation for the vertical elements in the dam are therefore only approximately correct since the cross sections are not uniform.

4-19. *Unit Deflections of Horizontal Elements of Twisted Structure.*—Unit deflections of horizontal elements due to shear are used in calculating deflections for each adjustment. Unit loads are applied to the horizontal elements by means of triangular loads which have a value of P pounds per square foot at the abutment and vary as a straight line to zero at the intersections of each respective vertical element with the horizontal element. The shear deflections due to a uniform load and those due to a unit concentrated load at the vertical dividing plane are also computed (see fig. 4-15). The concentrated load is used to provide deflection agreement of the two portions of the dam. The general equation used to compute the above deflection is:

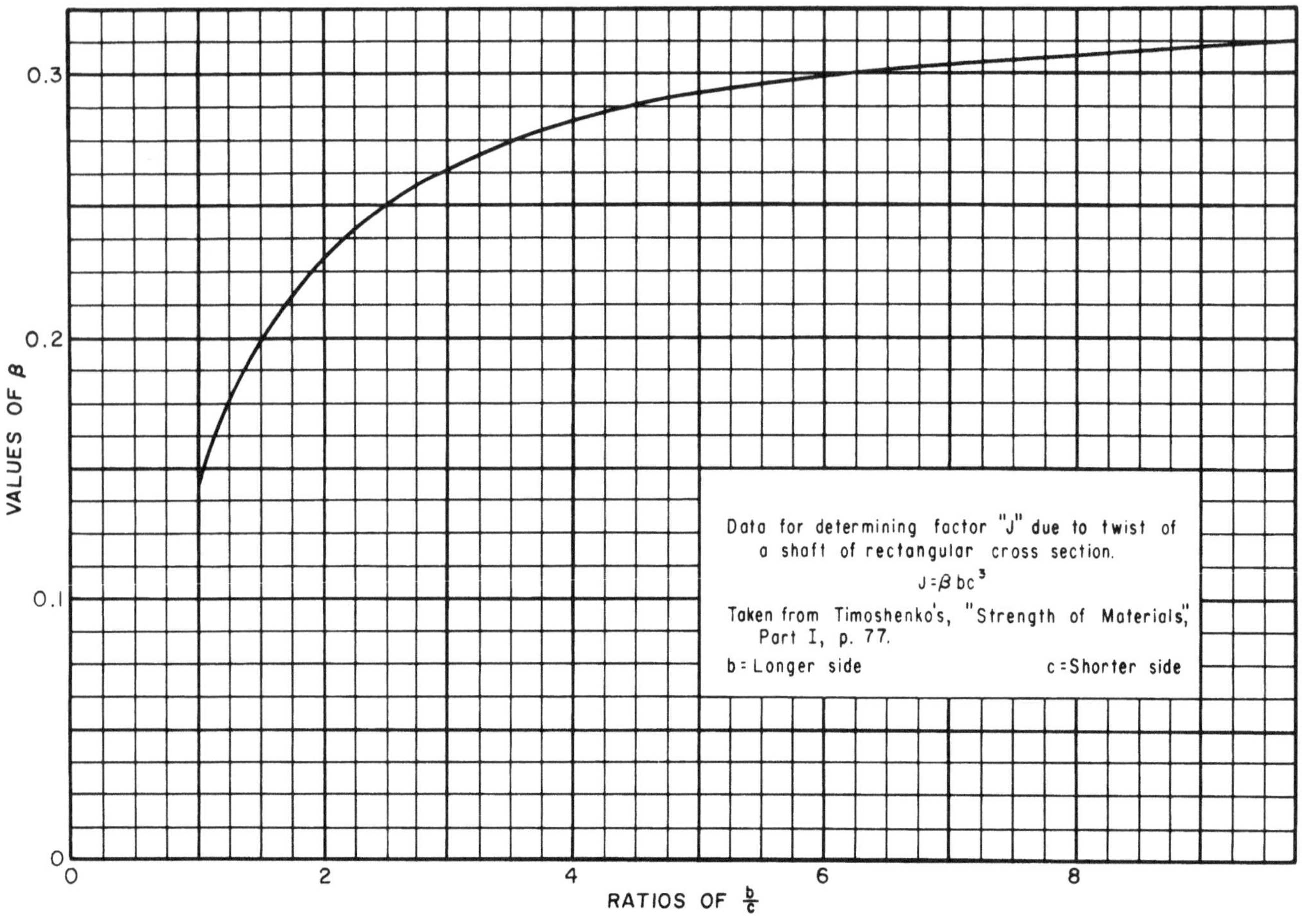

Figure 4-14. Graph for determining *J* factor due to twist of a shaft of rectangular cross section.—288-D-3154

$$\Delta y = \int \frac{K\,V_T}{G\,A}\,dx + {}_A V_T\,\gamma \qquad (20)$$

in which the symbols have the meanings given in section 4-13.

Using $K = 1.25$ and $G = E/2(1 + \mu)$, the general equation reduces to the following, where L is the length of the half-element, L' is the length of the loaded portion, measured from the abutment, and x is the distance from the abutment to the point where deflection is desired. For a unit triangular load,

$$\Delta y = -\frac{P}{2\,E\,A\,L'}\left[3\,(L')^2\,x - 3\,L'\,x^2 + x^3\right]$$

$$+ V_A\,\gamma \qquad (21)$$

For a uniform load,

$$\Delta y = -\frac{3\,P}{2\,E\,A}\left[2\,L\,x - x^2\right] + V_A\,\gamma \quad (22)$$

For a unit concentrated load,

$$\Delta y = -\frac{3\,P\,x}{E\,A} + V_A\,\gamma \qquad (23)$$

Shear forces are equal to the area under the unit-load diagram from the dividing plane of the dam to the cantilever points under consideration, and are negative in sign.

Values for unit deflections due to shear in horizontal elements are tabulated for convenient use in the adjustments.

4-20. Trial Loads.—Following the computation of unit-load deflections and

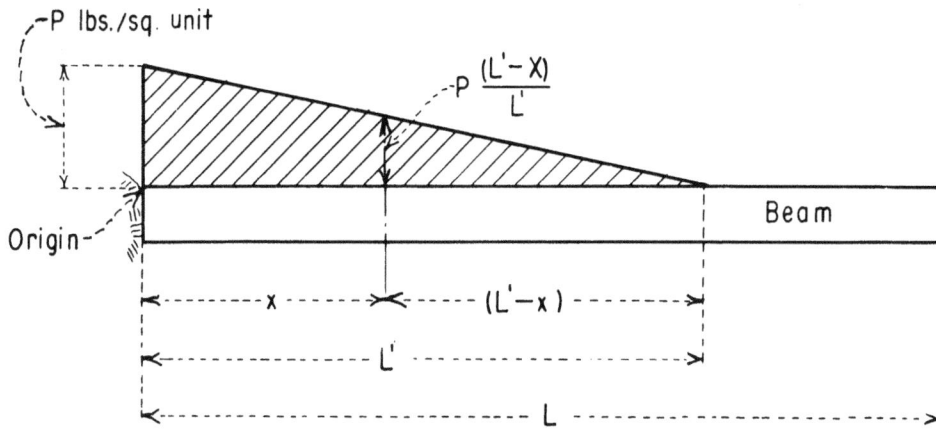

(a) TRIANGULAR LOAD ON HORIZONTAL ELEMENT

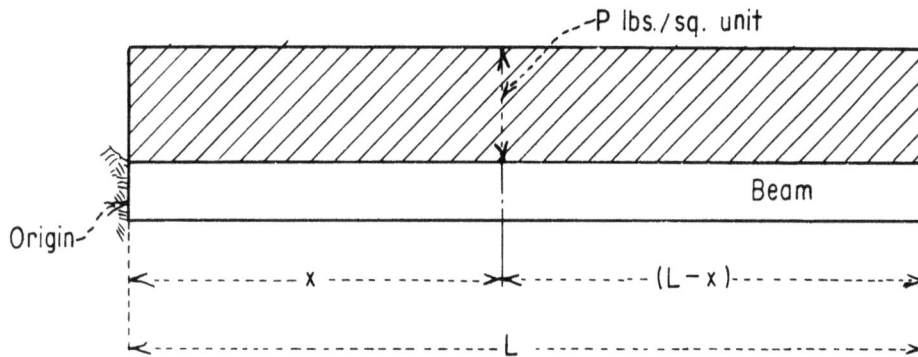

(b) UNIFORM LOAD ON HORIZONTAL ELEMENT

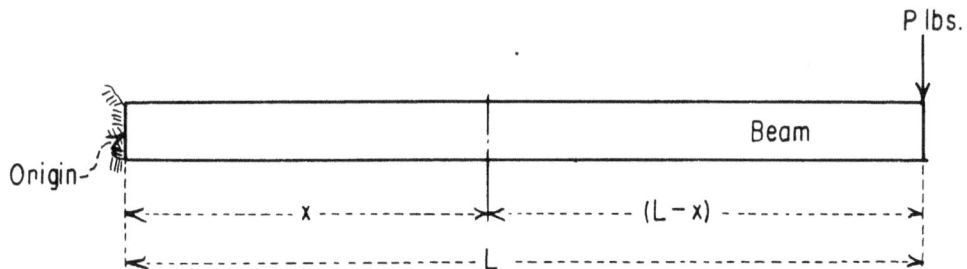

(c) CONCENTRATED LOAD ON HORIZONTAL ELEMENT

Figure 4-15. Loads on a horizontal element.—DS2-2(28)

rotations of horizontal and vertical elements, the next step in the analysis is the trial division of horizontal waterload between the cantilever structure and the twisted structure. This depends a great deal on experience and judgment. An examination of unit deflections is frequently of help in revealing the relative elasticity of the elements. Generally the cantilevers carry the greater proportion of load, especially in the middle portion of the dam. Near the abutments, however, the twisted structure usually carries the greater proportion of load, depending on the size of dam, shape of canyon, and elastic properties of the concrete and rock. One part of the total waterload is placed on the cantilever structure and the other part on the horizontal elements of the twisted structure. No external waterload is placed directly on the vertical twisted elements.

4-21. Angular Rotation of Vertical Twisted Elements due to Trial Loads on Horizontal Elements.—The shear force due to trial normal loads on a horizontal element, at the location of any vertical element, is equal to the area under the load diagram from the dividing plane to the vertical element under consideration. A twisting couple is produced in the vertical element as a result of the shear force in the horizontal element at that point. Since the width of the vertical element is unity, the twisting couple in the cantilever is numerically equal to the shear in the horizontal element. Since couple loads are in thousands of foot-pounds per square foot, shear forces must be expressed in thousands of pounds units. The general sign convention must be followed. By summating the respective products of these couple loads and the rotations of vertical twisted elements due to unit twist loads, the angular rotations in the vertical elements due to shear forces on the horizontal elements are obtained.

4-22. Deflections of Twisted Structure.—The deflections of the twisted structure due to angular rotations of vertical elements are obtained by integrating the angular rotations along the horizontal elements from the abutment to the dividing plane of the dam. This, however, gives only one component

of the twisted-structure deflection. The other part is shear detrusion due to trial loads on the horizontal elements, which is computed by summating the respective products of the loads and the unit deflections of horizontal elements due to shear (see sec. 4-19). The sum of these two parts is the total twisted-structure deflection due to trial loads. However, for comparison with cantilever deflections as to agreement at conjugate points, there must be added to the twisted-structure deflections the abutment movement of the particular vertical element of the cantilever structure (herein termed the conjugate vertical element) which has an abutment common with the horizontal element of the twisted structure.

4-23. Deflections of Cantilever Structure.—Cantilever deflections due to trial loads are calculated by summating the respective products of these loads and the cantilever deflections due to unit normal loads (see sec. 4-17). To these deflections are added algebraically the deflections due to initial loads, and, for comparison with the total twisted-structure deflections, are also added the movement at the base of the cantilever due to shear at the abutment of the conjugate horizontal element of the twisted structure. For cantilevers in a relatively flat canyon bottom where the foundations do not coincide with the ends of horizontal elements, this latter abutment movement is usually assumed equal to zero.

4-24. Stresses and Stability Factors.—After satisfactory continuity of the structure, or agreement of deflections at conjugate points, has been obtained by trial, the total shears and moments at various points in the cantilevers may be computed from the established trial loads. This is done by summating the products of the trial loads and the unit moments and shears, respectively, due to 1,000-pound unit loads. Stresses are then calculated for trial loads and added algebraically to stresses calculated for concrete weight at the faces of the dam. Stress equations are given in a later section. Stability factors are calculated in the usual manner.

2. Trial-Load Twist Method of Analysis, Joints Grouted

4-25. *Description of Method.*—The grouting of contraction joints welds the vertical blocks of the dam into a monolithic structure. In this case the dam has a different action under load than when joints are ungrouted. Grouting reduces the deflections of the structure for a given loading, since both horizontal and vertical elements of the dam are subject to bending and twisting in both horizontal and vertical planes. For very small dams, say under 50 feet in height, the effect of bending in the horizontal elements is very small and may be neglected; for higher dams, however, it is usually included. The general procedure is similar to that used for joints ungrouted, but an additional structure, designated the beam structure, is introduced for resistance to bending in horizontal elements. For this analysis, then, we have the cantilever structure, the twisted structure, and the beam structure, or three structures instead of two, the twisted structure being composed of both vertical and horizontal elements as previously described.

If the dam acts as a monolith, as assumed, the deflections of the cantilevers, horizontal beams, and twisted structure—due to trial divisions of waterload between the three systems—must be brought into agreement in all parts of the dam. Furthermore, for complete continuity, the longitudinal slopes of the cantilever must equal the transverse slopes of horizontal elements, and the longitudinal slopes of the horizontal elements must equal the transverse slopes of the cantilevers. A general slope adjustment is not necessary, however, since the adjustment of deflections in the horizontal elements of the twisted structure produces agreement of the longitudinal slopes of the horizontal elements with the transverse slopes of the cantilevers.

The adjustment is more complicated than for joints ungrouted, since three structures are used instead of two. As an initial step in the analysis, the waterload is divided by trial between the three structures, as illustrated diagrammatically for a cantilever element on figure 4-16(b). It should be noted that one-half of the twisted-structure load is carried to the foundation by the vertical twisted elements and one-half to the abutment by the horizontal twisted elements. That such a distribution may be assumed for the twisted-structure load was shown by H. M. Westergaard in 1930 in his review of the trial-load analysis of arch dams. The principle will be explained by illustration. Figure 4-17(a) shows a triangular dam, 5 units high and 10 units along the crest from the dividing plane of symmetry to the left abutment. The vertical cross sections in planes normal to the plane of the paper are assumed to be of unit uniform thickness from the top to the base. Rigid foundations are assumed, hence abutment and foundation rotations are omitted from the analysis. A twisted-structure load applied at a point x_5, z_{10} produces angular rotations in the beams and cantilevers of the twisted structure. The calculations on figure 4-17 show that if these rotations are integrated along their respective planes from beam abutment and cantilever foundation to the point of application of the load, the resulting deflections at the latter point are equal [4], from which it can be concluded that equal amounts of load are transferred vertically and horizontally by the twisted structure. It should be noted that, while the assumption of an equal load distribution is correct for a uniform-thickness section, it is only approximately true for a variable-thickness section.

By hypothesis, the beam and cantilever structures can resist only bending and shear, while the twisted structure can resist only twist and shear. Figure 4-17(b) illustrates an element of the beam structure subjected to load. Any portion of this element, *ABCD*, which may also be considered as part of a cantilever element, is seen to be in equilibrium due to moments and shears set up by load *P*. The total clockwise moment acting on the element is *P* multiplied by the arm of 10 feet, plus a couple consisting of the shear *P* multiplied by an arm of 1 foot, which is balanced by a counterclockwise resisting moment of *P* multiplied by 11 feet. Therefore, the load on the beam does not require a resisting twist in the cantilever

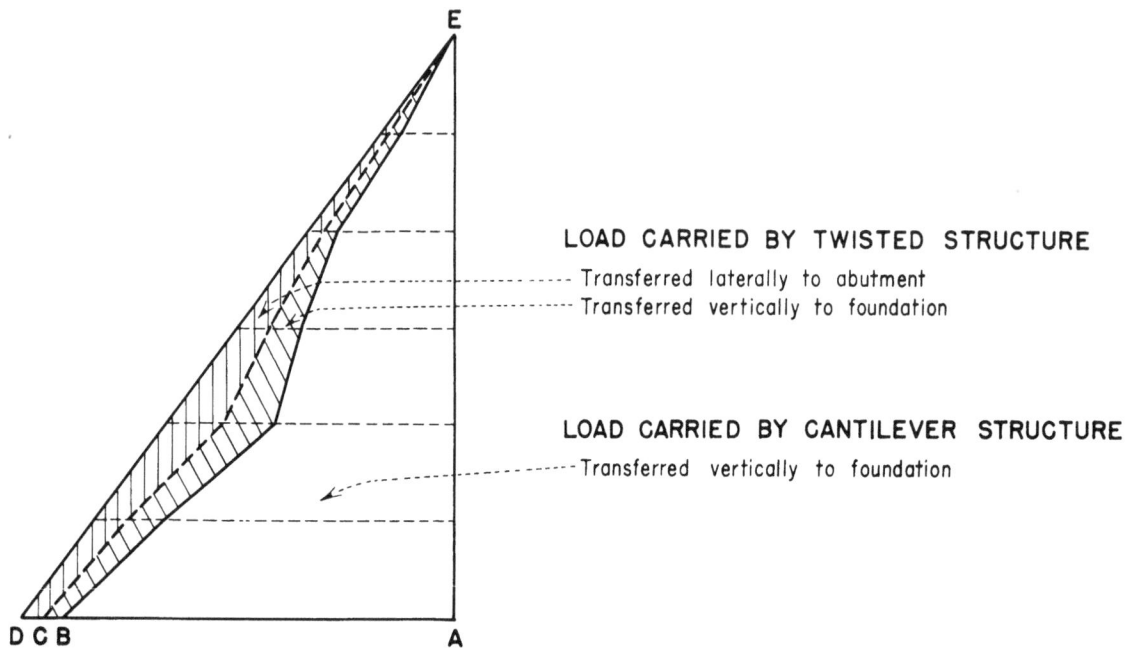

(a) NEGLECTING EFFECTS OF HORIZONTAL
BEAM ACTION IN BENDING

LOAD CARRIED BY TWISTED STRUCTURE
---- Transferred laterally to abutment
---- Transferred vertically to foundation

LOAD CARRIED BY CANTILEVER STRUCTURE
---- Transferred vertically to foundation

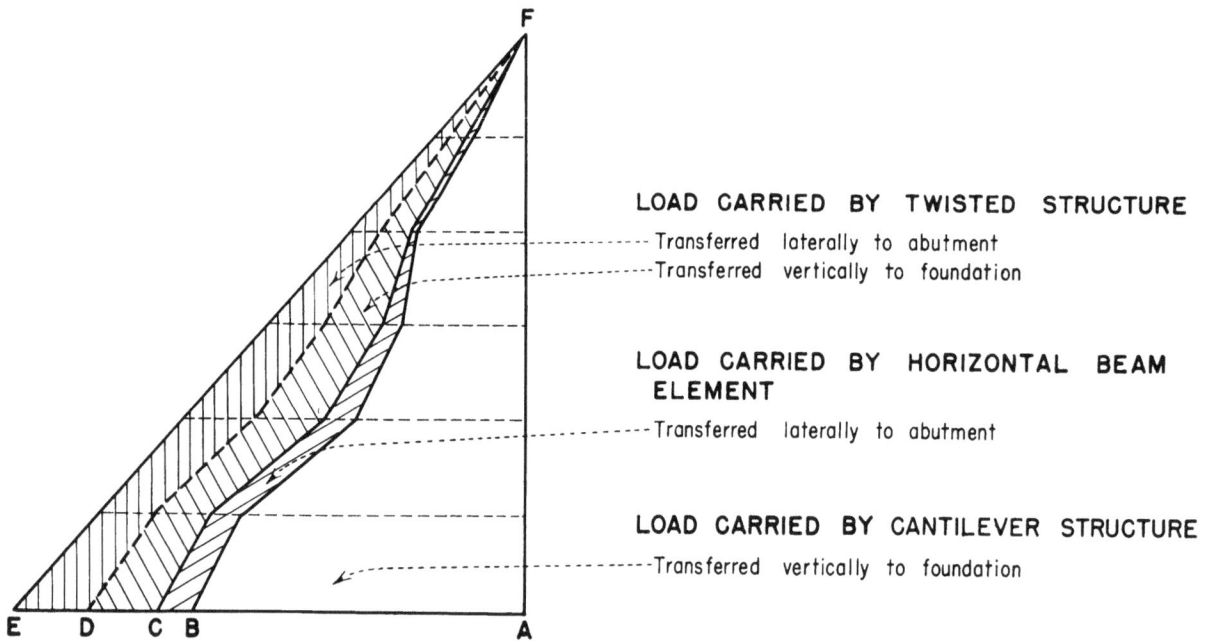

(b) INCLUDING EFFECTS OF HORIZONTAL
BEAM ACTION IN BENDING

LOAD CARRIED BY TWISTED STRUCTURE
---- Transferred laterally to abutment
---- Transferred vertically to foundation

LOAD CARRIED BY HORIZONTAL BEAM
ELEMENT
---- Transferred laterally to abutment

LOAD CARRIED BY CANTILEVER STRUCTURE
---- Transferred vertically to foundation

Figure 4-16. Trial-load twist analysis for a straight gravity dam—joints grouted. Division of external horizontal loads is shown.—103-D-275

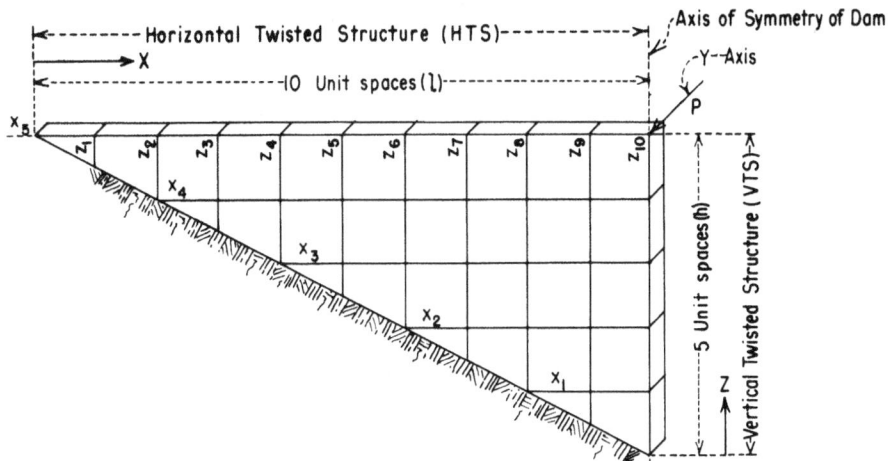

(a) DISTRIBUTION OF TWISTED-STRUCTURE LOAD

BASIC ASSUMPTIONS

Dam is symmetrical in triangular site about axis z_{10}. Vertical cross-sections in planes normal to plane of paper are of unit uniform thickness from top to base of dam. $2GI$ = unity. Half length of dam, l, = twice height, h. Abutment and foundation deformations not included, shear detrusions omitted.

Let a unit twisted-structure load of P intensity be applied at a point x_5, z_{10} in the dam. Let the angular rotations in both the xy and yz planes due to unit twisting moments = 1 per unit length. Assume one-half the twisted-structure load is carried horizontally to abutments and one-half carried vertically to foundation, by twist action.

In HTS $\quad M_{yz} = \dfrac{P}{2}, \quad ----- \theta_{yz} = \dfrac{P}{2} \cdot x, \quad ----- \Delta Y = \int \theta_{yz}\, dz$

In VTS $\quad M_{xy} = \dfrac{P}{2}, \quad ----- \theta_{xy} = \dfrac{P}{2} \cdot z, \quad ----- \Delta Y = \int \theta_{xy}\, dx$

For HTS $\quad \Delta Y$ at point $x_5 z_{10}$, $Z = \dfrac{X}{2}$

$$\therefore \Delta Y = \int_{X=0}^{X=10} \frac{P}{2} \cdot \frac{X}{2}\ dx = \left[\frac{PX^2}{8}\right]_0^{10} = 12.5P$$

For VTS $\quad \Delta Y$ at point $x_5 z_{10}$, $X = 2Z$

$$\therefore \Delta Y = \int_{Z=0}^{Z=5} \frac{P}{2} \cdot 2z\ dz = \left[\frac{PZ^2}{2}\right]_0^5 = 12.5P$$

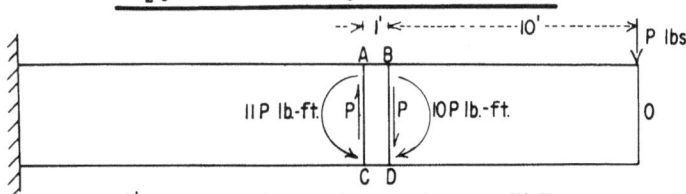

(b) BEAM OR CANTILEVER ELEMENT

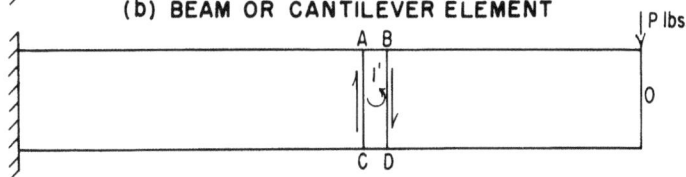

(c) TWISTED-STRUCTURE ELEMENT

Figure 4-17. Twisted-structure loads.—DS2-2(30)

element for equilibrium. Similarly, the load carried by a cantilever requires no resisting twist in the beam element.

Figure 4-17(c) illustrates an element of the twisted structure subjected to load. Any portion of this element, *ABCD*, which may also be considered as part of a vertical twisted-structure element, is subjected to a resultant shear couple of amount *P* with an arm of 1 foot. Therefore, for equilibrium, a load on the horizontal element of the twisted structure requires a resisting twist in the vertical element, and similarly a load on the vertical element requires a resisting twist in the horizontal element.

4-26. *Assumptions.*—From the foregoing considerations, the structural action of the elements of the dam may be assumed as follows:

(1) The cantilever elements resist shears in horizontal planes and bending in vertical planes.

(2) The horizontal beam elements resist shears in vertical planes and bending in horizontal planes.

(3) The twisted structure resists twisting moments and shears in horizontal and vertical planes.

4-27. *Horizontal Beam Elements.*—The horizontal beam elements are assumed to be 1 foot in vertical thickness with horizontal top and bottom faces and vertical upstream and downstream faces. Calculations of deflections are made by the ordinary theory of flexure for beams, with contributions from abutment yielding included. The same types of unit loads as described in section 4-19 for horizontal elements of the twisted structure are used for calculating unit deflections of each horizontal

beam. In addition, however, there is included a concentrated moment load at the "free end" or "crown" of the beam, that is, at the dividing plane, and also a concentrated normal load at the same free end. Unit slopes are calculated for the crown and the abutment. Unit slopes of the beams at the abutment are used to obtain the effect of beam abutment forces on rotation of the base of the conjugate vertical twisted element, and unit slopes at the crown are used to establish slope agreement at the dividing plane between beams in the left half and beams in the right half of the dam.

4-28. *Notations.*—In addition to nomenclature given in section 4-13, the following terms are given for equations used for movement in horizontal beams. Figure 4-15, used for illustrating horizontal elements of the twist structure, is also illustrative of horizontal beams.

x = distance from abutment to any point under load.

x_p = distance from abutment to any point on beam where deflection is desired.

L' = distance from abutment to end of load.

L = length of beam from abutment to crown.

M_L, V_L = moment and shear, respectively, at any point due to external load to right of point, for left half of beam. For right half of beam use subscript $_R$.

Subscript:

$_T$ = twisted structure.

4-29. *Equations.*—Equation (17) is used for computing cantilever deflections due to initial loads and due to unit normal loads. The underlined portions of equation (17) and subsequent equations give cantilever deflections due to unit shear loads. The deflection of the cantilever due to the portion of load carried by vertical elements of the twisted structure may be determined by use of these unit-shear deflection equations. The following equation is used in place of equation (18) for calculating the angular rotations of cantilevers in horizontal planes:

$$\theta_z = \Sigma \frac{M_{xy}}{2\,G\,I} \Delta z + {}_A M_{xy}\, \delta \tag{24}$$

The general equations for rotation and deflection at any point in a horizontal beam element, including effects of bending, shear, and abutment movement, are as follows:

$$\theta = \int \frac{M \, dx}{E \, I} + M_A \, \alpha + V_A \, \alpha_2 \tag{25}$$

$$\Delta y = \int \frac{M \, x \, dx}{E \, I} + 3 \int \frac{V \, dx}{E \, A} + M_A \, \alpha \, x + M_A \, \alpha_2 + V_A \, \gamma + V_A \, \alpha_2 \, x \tag{26}$$

(a) *Triangular Load.*–Slopes and deflections due to a triangular normal load may be calculated at any point along the centerline of a beam in the left half of the dam by means of the equations given in this subsection. Equations for the right half of the dam are the same except for a reversal in the sign of slopes. Equations for moment and shear are:

$$M = - \frac{P \, (L' - x)^3}{6 \, L'} = - \frac{P}{6 \, L'} \left[(L')^3 - 3 \, (L')^2 \, x + 3 \, L' \, x^2 - x^3 \right] \tag{27}$$

$$V = - \frac{P \, (L' - x)^2}{2 \, L'} = - \frac{P}{2 \, L'} \left[(L')^2 - 2 \, L' \, x + x^2 \right] \tag{28}$$

The equation for slope at any point is:

$$\theta = \int_0^{x_p} \frac{M \, dx}{E \, I} = - \frac{P}{6 \, E \, I \, L'} \left[(L')^3 \int_0^{x_p} dx - 3 \, (L')^2 \int_0^{x_p} x \, dx \right.$$

$$\left. + 3 \, L' \int_0^{x_p} x^2 \, dx - \int_0^{x_p} x^3 \, dx \right]$$

$$= - \frac{P}{24 \, E \, I \, L'} \left[4 \, (L')^3 \, x_p - 6 \, (L')^2 \, x_p^2 + 4 \, L' x_p^3 - x_p^4 \right]$$

$$+ M_A \, \alpha + V_A \, \alpha_2 \tag{29}$$

For

$$x_p = L' \text{ to } x_p = L,$$

$$\theta = - \frac{P \, (L')^3}{24 \, E \, I} + M_A \, \alpha + V_A \, \alpha_2 \tag{30}$$

The equation for deflection at any point is:

$$\Delta y = \int_0^{x_p} \frac{M(x_p - x)}{EI} \, dx + 3 \int_0^{x_p} \frac{V \, dx}{EA}$$

$$= -\frac{P}{120 \, EIL'} \left[10 \, (L')^3 \, x_p^2 - 10 \, (L')^2 \, x_p^3 + 5 \, L' x_p^4 - x_p^5 \right]$$

$$- \frac{P}{2 \, EAL'} \left[3 \, (L')^2 \, x_p - 3 \, L' x_p^2 + x_p^3 \right] + V_A \, \gamma + M_A \, \alpha_2$$

$$+ \left[M_A \, \alpha + V_A \, \alpha_2 \right] \, x_p \tag{31}$$

When

$$x_p = L',$$

$$\Delta y = -\frac{P \, (L')^4}{30 \, EI} - \frac{P \, (L')^2}{2 \, EA} + V_A \, \gamma + M_A \, \alpha_2 + \left[M_A \, \alpha + V_A \, \alpha_2 \right] \, x_p \tag{32}$$

For

$$x_p > L' \text{ to } x_p = L,$$

$$\Delta y = -\frac{P \, (L')^4}{30 \, EI} - \frac{P \, (L')^2}{2 \, EA} - \frac{P \, (L')^3}{24 \, EI} (x_p - L') + V_A \, \gamma + M_A \, \alpha_2 + \left[M_A \, \alpha + V_A \, \alpha_2 \right] \, x_p \tag{33}$$

(b) *Uniform Load.*—The equations for moment and shear due to a uniform normal load on a horizontal beam are:

$$M = -\frac{P \, (L - x)^2}{2} = -\frac{P}{2} \left[L^2 - 2 \, L \, x + x^2 \right] \tag{34}$$

$$V = -P \, (L - x) \tag{35}$$

The equation for slope at any point is:

$$\theta = \int_0^{x_p} M \, dx \tag{36}$$

$$\theta = -\frac{P}{6 \, EI} \left[3 \, L^2 \, x_p - 3 \, L \, x_p^2 + x_p^3 \right] + M_A \, \alpha + V_A \, \alpha_2 \tag{37}$$

When

$$x_p = L,$$

$$\theta = -\frac{P L^3}{6 E I} + M_A \, \alpha + V_A \, \alpha_2 \tag{38}$$

The equation for deflection at any point is:

$$\Delta y = \int_0^{x_p} \frac{M (x_p - x) \, dx}{E I} + 3 \int_0^{x_p} \frac{V \, dx}{E A}$$

$$= -\frac{P}{24 E I} \left[6 L^2 x_p^2 - 4 L x_p^3 + x_p^4 \right] - \frac{3 P}{2 E A} \left[2 L x_p - x_p^2 \right]$$

$$+ V_A \, \gamma + M_A \, \alpha_2 + \left[M_A \, \alpha + V_A \, \alpha_2 \right] \, x_p \tag{39}$$

When

$$x_p = L,$$

$$\Delta y = -\frac{P L^4}{8 E I} - \frac{3 P L^2}{2 E A} + V_A \, \gamma + M_A \, \alpha_2 + \left[M_A \, \alpha_2 + V_A \, \alpha_2 \right] \, L \tag{40}$$

(c) *Concentrated Moment at Free End of Beam.*—The equations for moment, shear, slope, and deflection for this condition are:

$$M = - P \qquad\qquad V = 0$$

$$\theta = -\frac{P x_p}{E I} + M_A \, \alpha \tag{41}$$

$$\Delta y = -\frac{P x_p^2}{2 E I} + M_A \, \alpha_2 + M_A \, \alpha \, x_p \tag{42}$$

For $x_p = L$, the latter value is substituted in the above equations.

(d) *Concentrated Normal Load at Free End of Beam.*—The equations for moment, shear, and slope for this condition are:

$$M = - P (L - x) \qquad\qquad V = - P$$

$$\theta = -\frac{P}{2 E I} (2 L x_p - x_p^2) + M_A \, \alpha + V_A \, \alpha_2 \tag{43}$$

When $x_p = L$,

$$\theta = -\frac{PL^2}{2EI} + M_A \, \alpha + V_A \, \alpha_2 \qquad (44)$$

The equation for deflection is:

$$\Delta y = -\frac{P}{6EI}(3Lx_p{}^2 - x_p{}^3) - \frac{3Px_p}{EA} + \left[M_A \, \alpha + V_A \, \alpha_2\right] \, x_p + M_A \, \alpha_2 + V_A \, \gamma \qquad (45)$$

When $x_p = L$,

$$\Delta y = -\frac{PL^3}{3EI} - \frac{3PL}{EA} + \left[M_A \, \alpha + V_A \, \alpha_2\right] \, L + M_A \, \alpha_2 + V_A \, \gamma \qquad (46)$$

The *underlined portions* of the preceding equations are equivalent to expressions for unit cantilever deflections obtained by equations (21) to (23), inclusive. Therefore, by keeping separate the underlined portions of equations (33), (40), and (46), shear deflections due to unit shear loads on horizontal elements are obtained at the same time as beam deflections due to unit normal loads. An example of a twist analysis of a gravity dam with joints grouted is shown in appendix B.

3. Analysis of Curved Gravity Dams

4-30. *Method of Analysis.*—If a gravity dam is curved in plan only for convenience in locating the structure on the existing topography and contraction joints are not grouted, the analyses should be made as described for straight gravity dams. However, if the joints are grouted and the dam is curved, arch action is an important factor in the reliability of the structure. Under these circumstances, it is desirable to analyze such a structure by an arch dam analysis method rather than by the gravity method described earlier. The arch dam analysis, including computerized application, is described in the Bureau of Reclamation publication "Design of Arch Dams" [17].

D. DYNAMIC ANALYSIS

4-31. *Introduction.*—The following method for dynamic analysis of concrete gravity dams can be described as a lumped mass, generalized coordinate method using the principle of mode superposition [5]. Application of the method is done by computer, and matrix methods of structural analysis are used. The method is similar to that proposed by Chopra [6].

4-32. *Natural Frequencies and Mode Shapes.*—The section analyzed is a two-dimensional cross section of the dam. The section is represented by finite elements [7] with the concrete mass lumped at the nodal points. The natural frequencies f_1, f_2, f_3, etc., and the corresponding mode shapes $(\phi_i)_1$, $(\phi_i)_2$, $(\phi_i)_3$ (where i indicates the assigned number of the mass point) are found by the simultaneous solution of equations of dynamic equilibrium for free vibration. There is one equation for each lumped mass. This problem is known as an eigenvalue problem. There are standard computer solutions available for the eigenvalue problem.

The input that will be required for the solution of the eigenvalue problem will be the stiffness matrix $[K]$ and the mass matrix $[M]$. A typical element in the stiffness matrix, K_{ij}, represents the force at i due to a unit deflection of j with all other points remaining fixed. The mass matrix is a diagonal matrix of the lumped masses. Each lumped mass includes the mass of the concrete associated with that point. To represent the effect of the water against the dam on the frequencies and mode shapes, a mass of water is divided appropriately between the mass points. The volume of water assumed to be vibrating with the dam is given by an equation developed by Westergaard [8].

The equation is:

$$b = 7/8 \sqrt{h\,z} \tag{47}$$

where:

b = the dimension of the water measured horizontally from the upstream face,
z = the depth of water at the section being studied, and
h = the distance from the water surface to the point in question.

4-33. *Response to an Earthquake.*—Given the natural frequencies, mode shapes, and an acceleration record of an earthquake, the following equation expresses the acceleration of point i in mode n, $\ddot{x}_{in}$, as a function of time:

$$\ddot{x}_{in} = \phi_{in} \frac{\Sigma_i M_i \phi_{in}}{\Sigma_i M_i \phi_{in}^2} \frac{2\pi}{T_n} \int_0^t \ddot{x}_g(\tau) e^{\left[\frac{\lambda 2\pi(t-\tau)}{T_n}\right]} \sin\frac{2\pi}{T_n}(t-\tau)\,d\tau \tag{48}$$

where:

$$T_n = \frac{1}{f_n},$$

λ = viscous damping factor,
$\ddot{x}_g(\tau)$ = the acceleration of the ground as a function of time, digitized for the numerical evaluation of the integral,
ϕ = nodal displacement,
M = mass,
τ = time, and
t = a particular time, $\tau = t$.

Little data are available on the damping in concrete gravity dams, expressed as λ in equation (48). Chopra [9] indicates that a reasonable assumption for λ in a concrete gravity structure is 0.05.

Equation (48) is evaluated at chosen increments of time. An increment of 0.01 second has been used. At the end of each of these time increments, the accelerations for all node points in all the modes being considered are summed as in the following equation:

$$\ddot{x}_{i_{TOTAL}} = \underset{n}{\Sigma}\ \ddot{x}_{in} \tag{49}$$

The response history is scanned for the time of the maximum value of acceleration at the crest. The $\ddot{x}_{i_{TOTAL}}$ at this time is the acceleration for the dam. These values can be divided by the acceleration of gravity to give α_i.

After the acceleration ratios, α_i, are determined for the necessary elevations in the dam, the resulting loads on the structure should be calculated as described in the following paragraphs.

4-34. Loads Due to Horizontal Earthquake Acceleration.—For dams with vertical or sloping upstream faces, the variation of hydrodynamic earthquake pressure with depth is given by the equations below [10]:

$$p_E = C \, \alpha \, wz \tag{50}$$

$$C = \frac{C_m}{2} \left[\frac{h}{z}(2 - \frac{h}{z}) + \sqrt{\frac{h}{z}(2 - \frac{h}{z})} \right] \tag{51}$$

where:

$\quad p_E$ = pressure normal to the face,
$\quad\ C$ = a dimensionless pressure coefficient,
$\quad\ \alpha = \dfrac{\text{horizontal earthquake acceleration}}{\text{acceleration of gravity}}$,
$\quad\ w$ = unit weight of water,
$\quad\ z$ = depth of reservoir at section being studied,
$\quad\ h$ = vertical distance from the reservoir surface to the elevation in question, and
$\quad C_m$ = the maximum value of C for a given slope, as obtained from figure 4-18.

For dams with combination vertical and sloping face, if the height of the vertical portion of the upstream face of the dam is equal to or greater than one-half the total height of dam, analyze as if vertical throughout.

If the height of the vertical portion of the upstream face of the dam is less than one-half the total height of the dam, use the pressure which would occur assuming that the upstream face has a constant slope from the water surface elevation to the heel of the dam.

Values of V_{pE} or V_{pE}' and M_{pE} or M_{pE}' should be computed for each increment of elevation selected for the study and the totals obtained by summation because of the nonlinear response. The inertia forces for concrete in the dam should be computed for each increment of height, using the average acceleration factor for that increment. The inertia forces to be used in considering an elevation in the dam are the summation of all the incremental forces above that elevation and the total of their moments about the center of gravity at the elevation being considered.

The horizontal concrete inertia force (V_E) and its moment (M_E) can be calculated using Simpson's rule.

4-35. Effects of Vertical Earthquake Accelerations.—The effects of vertical accelerations may be determined using the appropriate forces, moments, and the vertical acceleration factor. The forces and moments due to water pressure normal to the faces of the dam and those due to the dead loads should be multiplied by the appropriate acceleration factors to determine the increase (or decrease) caused by the vertical accelerations.

E. THE FINITE ELEMENT METHOD

4-36. Introduction.—The finite element method utilizes the idea that a continuous body may be considered an assemblage of distinct elements connected at their corners.

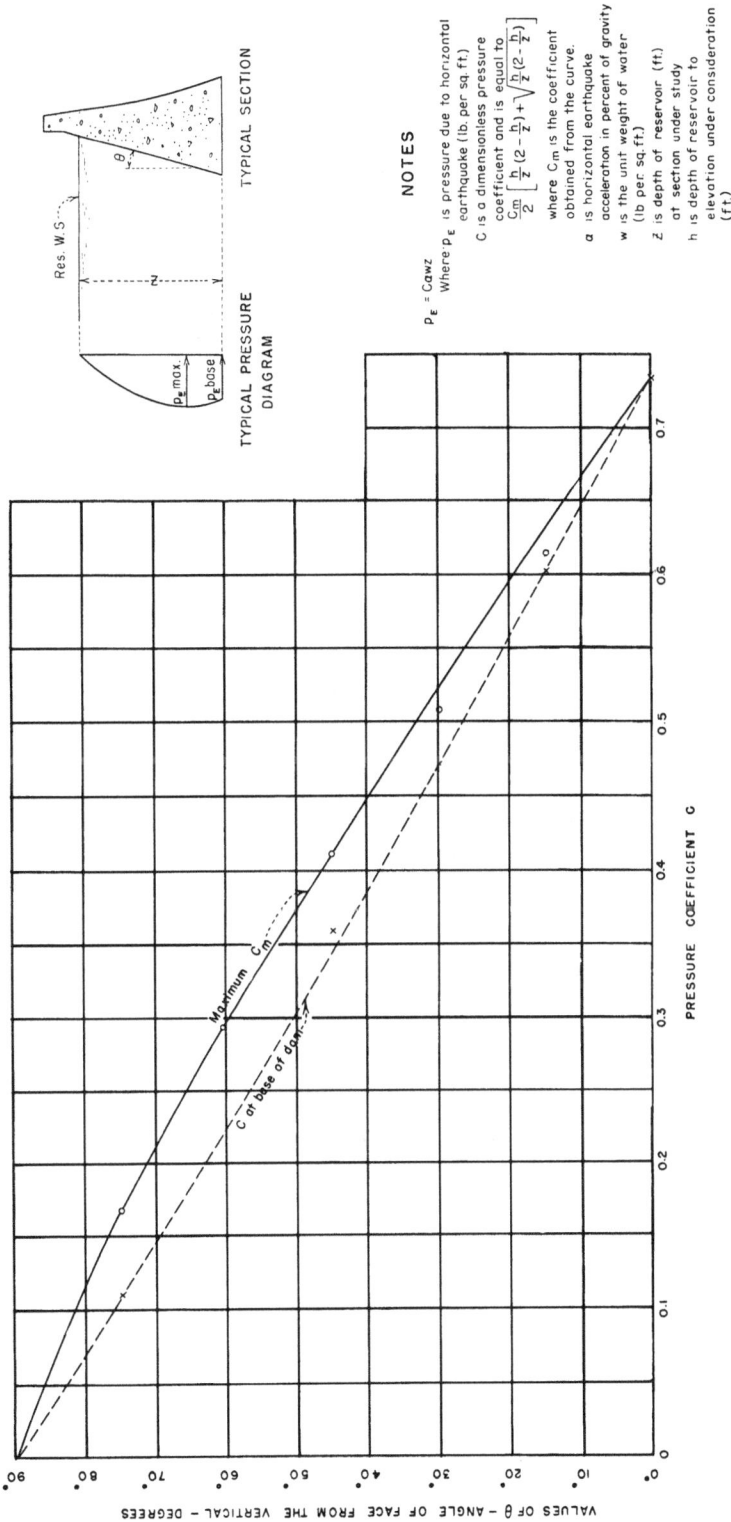

Figure 4-18. Hydrodynamic pressures upon the sloping face of a dam due to horizontal earthquake effect.—288-D-3155

This computerized method has become a widely used and accepted means of stress analysis in the last decade. The literature of the past few years contains numerous examples of specialized uses of the finite element method. The reason for the ready acceptance and tremendous amount of use of this method is that it made possible the approximate solution of many problems which engineers had been neglecting, overdesigning, or grossly approximating. The inclusion of complex geometrical and physical property variations prior to adaption of the finite element method and the modern high-speed digital computer was simply beyond the realm of reality. The finite element method permits a very close approximation of the actual geometry and extensive variations of material properties simply and inexpensively. The formulation and theory of the finite element method are given in several publications including those by Clough [11] and Zienkiewicz [12].

Because of the ability of the method to analyze special situations, this is the area in which the most application has been made. The two-dimensional finite element method is capable of analyzing the majority of problems associated with variations in the geometry of sections of the dam. Three-dimensional effects can be approximated by making a two-dimensional analysis in more than one plane. The two-dimensional finite element method is capable of solving for stresses economically even when great detail is necessary to attain sufficient accuracy.

When the structure or loading is such that plane stress or strain conditions may not be assumed, the three-dimensional finite element method may be used. The applicability of this method to problems with extensive detail is limited by computer storage capacity and economics. However, the method is often used for problems with near uniform cross section or where only the general state of stress is desired. Additionally, the three-dimensional method finds application when the effect of an eccentric load or member is to be found.

Many two- and three-dimensional finite element programs with varying accuracy and capability have been written. The programs used by the Bureau for analyses connected with gravity dams are discussed below.

I. Two-Dimensional Finite Element Program

4-37. *Purpose.*—The purpose of this computer program is to determine deformations and stresses within two-dimensional plane stress structures of arbitrary shape. The structure may be loaded by concentrated forces, gravity, and temperature, or by given displacements. Materials whose properties vary in compression and tension may be included by successive approximations.

4-38. *Method.*—The structure is divided into elements of arbitrary quadrilateral or triangular shape. The verticies of these shapes form nodal points. The deflections at the nodal points due to various stresses applied to each element are a function of the element geometry and material properties. The coefficient matrix relating this deflection of the element to the load applied is the individual element stiffness matrix. These stiffnesses are combined with the stiffnesses of all the other elements to form a global stiffness matrix. The loads existing at each node are determined. The deflections of each node in two directions are unknown. The same number of equations relating stiffness coefficients times unknown deflections to existing loads (right hand members) have been generated. The very large coefficient matrix is banded and symmetric. Advantage of this fact is taken into account in the storage of this matrix. The equations are solved by Gauss elimination.

In this method each unknown is progressively solved for in terms of the other unknowns existing in the equation. This value is then substituted into the next equation. The last equation then is expressible in only one unknown. The value of this unknown is determined and used in the solution of the previous equation which has only two unknowns. This process of back substitution continues until all unknowns are evaluated. The known deflections, the stiffness of the individual elements, and the equations relating strain and stress for the element are then used to calculate the stress condition for the element.

4-39. Input.—The problem is defined by a card input that describes the geometry and boundary conditions of the structure, the material properties, the loads, the control information for plotting, and the use of options in the program. Mesh generation, load generation, and material property generation are incorporated in the system.

4-40. Output.—The output of this program consists primarily of a print of the input data and the output of displacements at each node and stresses within each element. In addition, a microfilm display of the mesh and of portions of the mesh with stresses plotted on the display is available. Some punched card output is also available for special purposes of input preparation or output analysis.

4-41. Capabilities.—

(1) *Loading.*—External forces, temperature, and known displacements are shown, and accelerations given as a percentage of the acceleration due to gravity in the X and Y directions.

(2) *Physical property variations.*—The program allows reading-in changes in modulus, density, reference temperature, and accelerations after each analysis. Stresses and displacements may then be computed with the new properties and loading without redefinition of the structure.

(3) *Plotting.*—A microfilm plot of the entire grid or details of it may be obtained. The detailed plot may be blank or can be given with principal, horizontal, vertical, and shear stresses. Either plot may also be obtained with the material number identification given within each element.

(4) *Bilinear material properties.*—The program allows for input of a modulus in compression and in tension. The tension modulus is included in successive approximations after the determination of tension in an element has been made.

(5) *Openings.*—An opening may be simulated in the structure by assigning a material number of zero to any element or by actually defining the structure with the opening not included in the definition. The former method allows for optimum use of mesh generation and allows for considerably more flexibility.

(6) *Checking and deck preparation.*—Several options exist that allow for checking and facilitating input preparation.

(7) *Shear stiffness.*—The effect of shear stiffness in the third dimension may be included.

(8) *Units.*—The program output units match the input units. In general, these units are not shown on the output. The option exists, however, that allows units to be given on the output in feet and pounds per square inch provided that the input was in feet and kips.

(9) *Normal stress and shear stress on a plane.*—The normal stress and the shear stress on any given plane can be computed. In addition, given the angle of internal friction and the cohesion for the plane, the factor of safety against sliding can be computed.

(10) *Reference temperature.*—Temperature loads are applied with respect to a given reference temperature for the entire problem. If certain portions of the problem have different reference temperatures, these may be input on the material properties card and would override the overall reference temperature for that material only.

(11) *External forces* may be applied using boundary pressures. The program calculates concentrated loads at the nodes based on these pressures.

(12) *The input coordinates* may be prepared by digitizing a scale drawing of the problem. The actual scale can be adjusted for within the program by inputting a scale factor on the control card. The coordinates used by the program are the input coordinates times the scale factor. If no scale factor is involved, the coordinates are used as they are given.

4-42. Limitations.—

(1) *Nodes,* 999; *elements,* 949; *materials,* 100.

(2) *Bandwidth* (maximum difference between nodes of any element) = 42.

(3) *Maximum number of rows* in a detailed plot section = 25.

4-43. Approximations.—

(1) *Linear deflection distribution* between nodes.

(2) *Curved surface* has to be approximated

by a series of straight lines.

(3) *Points of fixity* must be established on the boundaries.

(4) *Two-dimensional* plane stress.

4-44. *Application to Gravity Dams.*—Two-dimensional finite element analysis is adaptable to gravity dam analysis when the assumption of planarity is used. The stress results for loading of typical transverse sections (perpendicular to the axis) are directly applicable. Sections including auxiliary works can be analyzed to determine their stress distribution. Both transverse and longitudinal sections should be prepared and analyzed for local areas with extensive openings. The results of the stress distributions are combined to approximate the three-dimensional state of stress.

The two-dimensional finite element analysis allows the foundation with its possible wide variation in material properties to be included with the dam in the analysis. Zones of tension cracks and weak seams of material can be included in the foundation. The internal hydrostatic pressure can be included as loads on the section to be analyzed.

Foundation treatment requirements for achieving suitable stresses and deformations can be determined with acceptable accuracy using this two-dimensional finite element program.

An example is given in appendix C which illustrates the application of the two-dimensional finite element method to analysis of a gravity dam and its foundation.

2. Three-Dimensional Finite Element Program

4-45. *Application.*—This computer program, which was developed by the University of California at Berkeley, uses the Zienkiewicz-Irons isoparametric eight-nodal-point (hexahedron) element to analyze three-dimensional elastic solids [12]. The elements use the local or natural coordinate system which is related to the X-Y-Z system by a set of linear interpolation functions. These local coordinates greatly

simplify the stiffness formulation for the element. The displacements are also assumed to vary linearly between the nodes. Thus the same interpolation functions can be used for displacements. This common relationship of geometry and displacement is the reason for the name isoparametric element.

Once the displacement functions have been established, the element strains can be formulated. The nodal point displacements are related to the element strains in the strain-displacement relations. The element stress is related to strain using the stress-strain relations for an elastic solid. Energy considerations (either minimum potential energy or virtual work) are used to establish the relationship between nodal point displacements and nodal point forces. The relationship is a function of the stress-strain and the strain-displacement characteristics. This function, by definition, is the element stiffness.

The element stiffness is the key feature in the finite element solution. Each element stiffness is combined into a global stiffness matrix. In this matrix the stiffness at each node is obtained by summing the contribution from each element which contains that node. A set of equations for the entire system is obtained by equating the products of the unknown displacements times the stiffnesses to the known forces at each nodal point.

Nodal displacements are determined by solving this set of equations. Stresses are computed at the nodes of each element, using the same strain-displacement and stress-strain relations used in the formulation of the element stiffness. The stresses at a node are taken as the average of the contributions from all the elements meeting at that node.

4-46. *Capabilities and Limitations.*—The program is able to analyze any three-dimensional elastic structure. The linear displacement assumption, however, limits the efficient use of the program to problems where bending is not the primary method of load resistance. Accurate modeling of bending requires the use of several elements (three have been shown to work fairly well) across the bending section. When acceptance of load is by shear and/or normal displacement along with bending, the element is capable of modeling the displacement efficiently. A comparison of the accuracy of elements by Clough [13] demonstrates this point with several sample problems.

The program capacity for a 65,000-word-storage computer is 900 elements, 2,000 nodal points, and a maximum bandwidth of 264. The bandwidth is defined as three times the maximum difference between any two node numbers on an element plus 3. On a CDC 6400 electronic computer the time for analysis in seconds is approximately:

$$0.024 \begin{bmatrix} \text{number of} \\ \text{nodal points} \end{bmatrix} + 0.45 \begin{bmatrix} \text{number of} \\ \text{elements} \end{bmatrix}$$

$$+ 100 \begin{bmatrix} \dfrac{\text{number of}}{\text{nodal points}} \\ \dfrac{}{775} \end{bmatrix} \times 3 \times \begin{bmatrix} \dfrac{\text{bandwidth}}{96} \end{bmatrix}^2 \quad (52)$$

For a problem which uses the full program capacity, this is equal to:

$$0.024(2,000) + 0.45(900)$$

$$+ \frac{100(2,000)}{775} \times 3 \times \left(\frac{264}{96}\right)^2 = 6,308 \text{ seconds,}$$

or about 105 minutes.

The cost of operating increases approximately as the square of the bandwidth. This economic consideration often restricts the user to a relatively coarse mesh.

Capability for use of mesh generation, concentrated loads, automatic uniform or hydrostatic load application, and varying material properties exists in the program.

The elements (see fig. 4-19) are arbitrary six-faced solids formed by connecting the appropriate nodal points by straight lines. Nonrectangular solid elements, however, require additional time for stiffness formulation because of the necessity of increased numerical integration.

4-47. Input.—The structure to be analyzed is approximated by an assemblage of elements. The finest mesh (smallest sized elements) are located in the region of greatest stress change

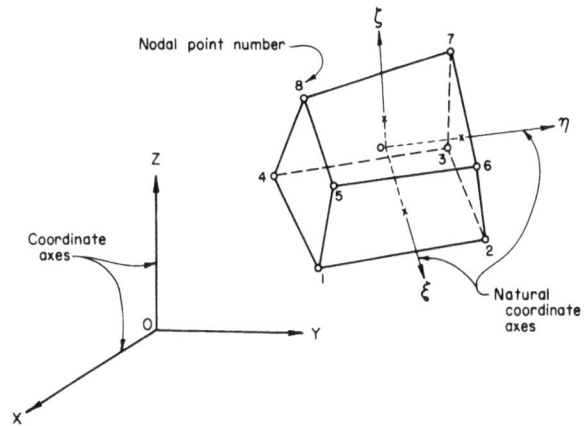

Figure 4-19. **A finite element with nodal point numbers and coordinate axes.**—288-D-2994

to allow for accurate modeling of deformations. The division is also made such that the minimum bandwidth is possible, and the nodes and elements are numbered with this consideration in mind. The program requires the following basic information:

(1) Operational data such as title, number of jobs, number of elements, maximum bandwidth, number of materials, etc.

(2) The conditions of restraint on the boundary.

(3) The material description of the elements.

(4) The accuracy of integration required for each element.

(5) The X-Y-Z coordinates of each nodal point and the eight nodal point numbers forming each element (mesh generation can be used to accomplish these functions).

(6) Applied forces (it is possible to use automatic load generation).

4-48. Output.—The program output consists of:

(1) A reprint of all input information including the information automatically generated.

(2) The displacements in the X, Y, and Z directions for each of the nodal points.

(3) The normal stress in the X, Y, and Z directions and the shear stress in the

XY, *YZ*, and *XZ* planes at each nodal point.

A sample problem showing mesh

formulation, input data, and output is given in appendix C.

F. FOUNDATION ANALYSIS

4-49. *Purpose.*—The foundation or portions of it must be analyzed for stability whenever the rock against which the dam thrusts has a configuration such that direct shear failure is possible or whenever sliding failure is possible along faults, shears, and joints. Associated with stability are problems of local overstressing in the dam due to foundation deficiencies. The presence of such weak zones can cause problems under either of two conditions: (1) when differential displacement of rock blocks occurs on either side of weak zones, and (2) when the width of a weak zone represents an excessive span for the dam to bridge over. To prevent local overstressing, the zones of weakness in the foundation must be strengthened so that the applied forces can be distributed without causing excessive differential displacements, and so that the dam is not overstressed due to bridging over the zone. Analyses can be performed to determine the geometric boundaries and extent of the necessary replacement concrete to be placed in weak zones to limit overstressing in the dam.

1. Stability Analyses

4-50. *Methods Available.*—Methods available for stability analysis are:
  (a) Two-Dimensional Methods.
    (1) Rigid section method.
    (2) Finite element method.
  (b) Three-Dimensional Methods.
    (1) Rigid block method.
    (2) Partition method.
    (3) Finite element method.
Each of these analyses produces a shearing force and a normal force. The normal force can be used to determine the shearing resistance as described in section 3-5. The factor of safety against sliding is then computed by dividing the shear resistance by the shearing force.

4-51. *Two-Dimensional Methods.*—A problem may be considered two dimensional if the geological features creating the questionable stability do not vary in cross section over a considerable length so that the end boundaries have a negligible contribution to the total resistance, or when the end boundaries are free faces offering no resistance. The representation of such a problem is shown on figure 4-20.

(a) *Rigid Section Method.*—The rigid section method offers a simple method of analysis. The assumption of no deformation of the section allows a solution according to statics and makes the method comparable to the three-dimensional rigid block method. As shown on figure 4-20, the resultant of all loads on the section of mass under investigation are resolved into a shearing force, *V*, parallel to the potential sliding plane and a normal force, *N*. The normal force is used in determining the amount of resistance as discussed in section 3-5. The factor of safety or shear friction factor is determined by dividing the resisting force by the sliding force.

This method may also be used when two or more features combine to form the potential sliding surface. For this case each feature can be assumed to form a section. Load which cannot be carried by one section is then transferred to the adjacent one as an external load. This procedure is similar to the method of slices in soil mechanics, except that the surface may have abrupt changes in direction.

(b) *Finite Element Method.*—The finite element method, discussed in sections 4-36 through 4-48, allows deformations to occur and permits more accurate placement of loads. The analysis gives the resulting stress distribution in the section. This distribution allows the variation in normal load to be considered in determination of the resisting force and shearing force along the potential

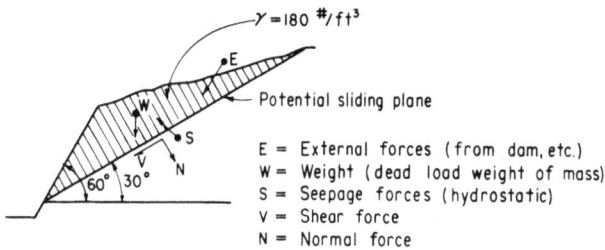

Figure 4-20. Sketch illustrating the two-dimensional stability problem.—288-D-2996

sliding plane. The shear friction factor can then be computed along the plane to determine the stability. When the stress distribution along the plane is known, a check can be made to determine if stress concentrations may cause failure of the material in localized areas.

It should be noted that this distribution can be approximated without using the finite element method if the potential sliding mass and underlying rock are homogeneous. The finite element method is very useful if there are materials with significantly different properties in the section.

4-52. *Three-Dimensional Methods.*—A typical three-dimensional stability problem is a four-sided wedge with two faces exposed and the other two faces offering resistance to sliding. The wedge shown on figure 4-21 is used in the discussion to illustrate the various methods.

(a) *Rigid Block Method* [14].—The following assumptions are made for this method:

(1) All forces may be combined into one resultant force.

(2) No deformation within the block mass can take place.

(3) Sliding on a single plane can occur only if the shear force on the plane is directed toward an exposed (open or free) face.

(4) Sliding on two planes can occur only in the direction of the intersection of the two planes and toward an exposed face.

(5) No transverse shear forces are developed (that is, there is no shear on the planes normal to the sliding direction).

The rigid block analysis proceeds in the following manner:

(1) The planes forming the block are defined.

(2) The intersections of the planes form the edges of the block.

(3) The areas of the faces of the block and the volume of the block are computed.

(4) The hydrostatic forces, if applicable, are computed normal to the faces.

(5) The resultant of all forces is computed.

(6) The possibility for sliding on one or two planes is checked.

(7) The factor of safety against sliding is computed for all cases where sliding is possible.

To determine whether the rock mass could slide on one or two planes, a test is applied to each possible resisting plane. If the resultant vector of all forces associated with the rock mass has a component normal to and directed into a plane, it will offer resistance to sliding. If only one plane satisfies the criterion, the potential sliding surface will be one plane; and if two planes satisfy the criterion the potential sliding surface will be the two planes.

Sliding on three planes is impossible according to the assumptions of rigid block. If an analysis of a block with many resistant faces is desired according to rigid block procedure, several blocks will need to be analyzed with any excess shear load from each block applied to the next.

The resultant force for the case of a single sliding plane is resolved into one normal and one shear force. For the case of sliding on two planes, the resultant is divided into a shear force along the intersection line and a resultant force normal to the intersection line. Forces normal to the two planes are then computed such that they are in equilibrium with the resultant normal force. As a result of assumption (5) at the beginning of this subsection, these normal loads are the maximum that can occur and the resulting shear resistance developed is a maximum. Figures 4-22(a) and 4-22(b) show a section

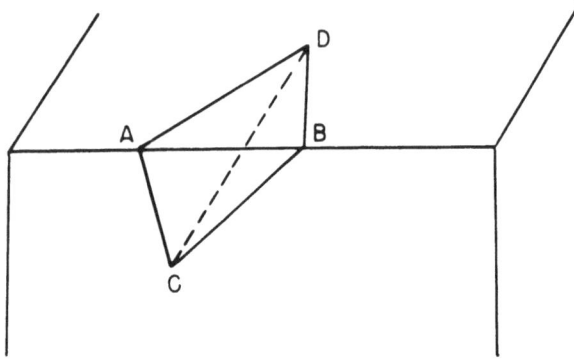

Figure 4-21. Four-sided failure wedge for three-dimensional stability analysis.—288-D-2997

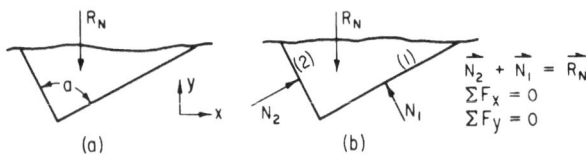

R_N = The portion of the resultant normal to the direction of potential movement.

N = The normal load on the face indicated by the subscript.

Figure 4-22. Section through a sliding mass normal to the intersection line of two planes.—288-D-2998

through the potential sliding mass normal to the intersection line of the two planes with the resultant normal load balanced by normals to the two potential sliding planes.

The shearing resistance developed for either a single plane or two planes is computed using the normal forces acting on the planes and the methods discussed in section 3-5.

(b) *Partition Method.*—The rigid block method permits no deformation of the mass of the block. Because of this restriction no shear load is developed in the potential sliding planes transverse to the direction of sliding. The development of shear in the transverse direction decreases the normal load and consequently the developable shear resistance [15]. An approximation to the minimum developable shear resistance is made by the partition method. In this method the planes (normal to the sliding direction) are parted according to the dead load associated with each plane as shown on figures 4-23(a) and 4-23(b). The component of the external load

perpendicular to the sliding direction is then proportionately assigned to each plane according to the ratio of projected areas of the planes with respect to the direction of loading as shown on figures 4-23(c) and 4-23(d). (Note: If the external load is parallel to one of the planes, the load assignment may have to be assumed differently depending on the point of load application.) All the forces on each plane are then combined to form a resultant on that plane (fig. 4-23(e)). This resultant is assumed to be balanced by a normal force and a shear force on that plane (fig. 4-23(f)). The normal force is then used in determining the resistance of the block to sliding.

Although it is recognized that the developable shear force is probably less than that required to balance the resultant, the assumption that this strength is developed allows computation of the minimum developable strength. The shear resistance developed by using N_1 and N_2 (fig. 4-23(f)) is considered the minimum possible.

The shearing force tending to drive the block in the direction of sliding is determined as described for the rigid block method. The computation of the resistance according to the partition method utilizes the information obtained for the rigid block analysis, and therefore requires very little additional computation. The shear resistance determined by the rigid block method is an upper bound and that determined by the partition method is considered a lower bound. As the angle between the planes (see fig. 4-22(a)) increases, the results obtained from the two methods converge. The correct shear resistance lies between the upper and lower bounds and is a function of the deformation properties of the sliding mass and host mass of rock, and even more importantly of the sliding and deformation characteristics of the joint or shear material forming the surface. The effect of these properties on the resistance developed can be approximated by using a three-dimensional finite element program with planar sliding zone elements. This method is discussed in the next subsection.

The partition method can be extended to multifaced blocks very readily. Just as the section normal to the direction of sliding is

$$\overrightarrow{W_1'} + \overrightarrow{W_2'} = \overrightarrow{W'}$$

$$E_1' = \frac{aE'}{a+b}$$

$$E_2' = \frac{bE'}{a+b}$$

$$\overrightarrow{E_1'} + \overrightarrow{E_2'} = \overrightarrow{E'}$$

(a)   (b)   (c)   External Load=E'   (d)

$$\overrightarrow{R_1'} = \overrightarrow{W_1'} + \overrightarrow{E_1'}$$

$$\overrightarrow{R_2'} = \overrightarrow{W_2'} + \overrightarrow{E_2'}$$

$$\overrightarrow{R_1'} + \overrightarrow{R_2'} = \overrightarrow{R'}$$

$$\overrightarrow{R'} = \overrightarrow{E'} + \overrightarrow{W'}$$

(e)   (f)

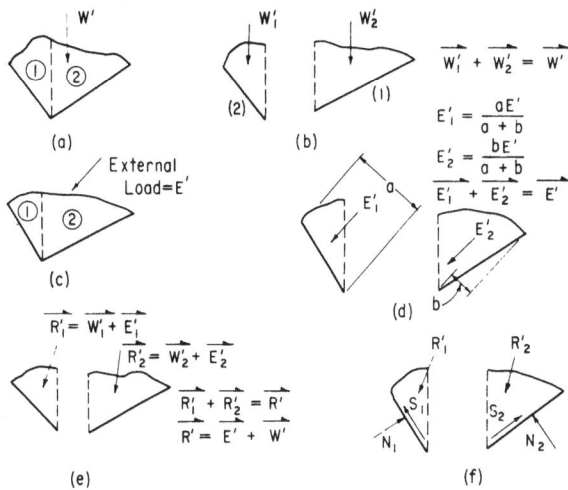

Where: W = DEAD LOAD
        E = EXTERNAL LOADS
        R = RESULTANT LOAD ON MASS
Subscripts refer to the appropriate portions of
the mass. No subscript implies that the entire mass
is being considered.
Planes are normal to the direction of potential sliding
Loads resolved into the plane normal to direction of potential
sliding are indicated with a prime.

Figure 4-23. **Partition method of determining shear
resistance of a block.—288-D-2999**

partitioned, so can a section along the direction of sliding be divided as shown on figure 4-24. Excess shear load from one partition, *A*, must be applied to the adjacent one, *B*, as an external loading as shown on the figure.

(c) *Finite Element Method.*—A program developed by Mahtab [16] allows representation of the rock masses by three-dimensional solid elements and representation of the potential sliding surface by two-dimensional planar elements. The planar elements are given properties of deformation in compression (normal stiffness) and in shear (shear stiffness) in two directions.

The ratio of the normal stiffness to the shear stiffness influences greatly the amount of load which will be taken in the normal direction and in the transverse shear direction. If the normal stiffness is much greater than the shear stiffness, as is the case for a joint with a slick coating, the solution approaches that given by the rigid block method. However, as the shear stiffness increases with respect to the normal stiffness, more load is taken by transverse shear

and the solution given by the partition method is approached.

The three-dimensional finite element method allows another important refinement in the solution of stability problems. Since deformations are allowed, the stress state on all planes of a multifaced block can be computed rather than approximated and stress concentrations located.

The refinements available in the analysis by the three-dimensional finite element method should be used when the upper and lower bounds determined by the other methods are significantly different. The method should also be used if there is considerable variation in material properties either in the potential sliding planes or in the rock masses.

A more detailed discussion of the finite element method is given in sections 4-36 through 4-48.

2. Other Analyses

4-53. *Differential Displacement Analysis.*— The problem of relative deflection or differential displacement of masses or blocks within the foundation arises due to variations in the foundation material. Methods that approximate or compute the displacement of masses or zones within the foundation are required to analyze problems of this nature. Typical problems that may occur are as follows:

(1) Displacement of a mass whose stability depends on sliding friction.

(2) Displacement of a mass sliding into a low modulus zone.

(3) Displacement of a mass with partial intact rock continuity.

(4) Displacement of zones with variable loading taken by competent rock in two directions but cut off from adjacent rock by weak material incapable of transmitting shear load.

The displacements may be approximated by: (1) extension of shear-displacement data obtained from specimen testing in situ or in the laboratory; (2) model testing; (3) development of an analytical model which can be solved

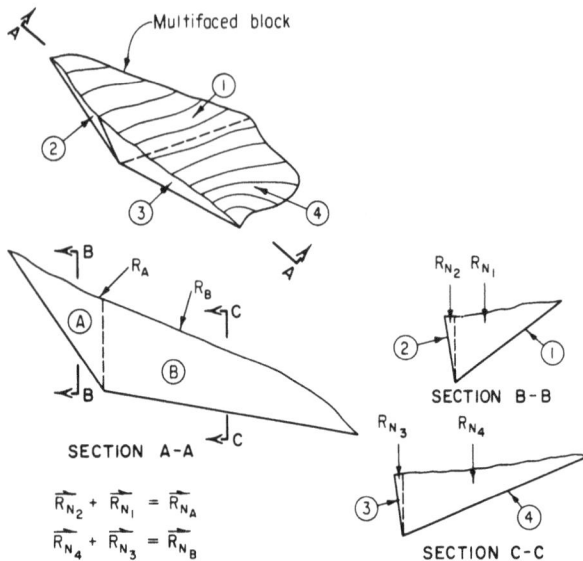

NOTE: Circled numbers refer to faces.
Circled letters refer to blocks.
R_{N_i} = The portion of the resultant assigned to a face.
The sub-subscript indicates the face number.
R_{N_A} = The portion of the resultant normal to the
direction of potential movement of a block.
The sub-subscript refers to the block.
R_A = Resultant external load acting on block A.

Figure 4-24. **Partition method extended to multifaced blocks.—288-D-3000**

manually; or (4) two- or three-dimensional finite element methods.

Although the method used depends on the particular problem, it should be noted that the finite element method offers considerable advantage over the other procedures. The finite element method allows accurate material property representation, gives stress distribution, and permits representation of treatment necessary to obtain acceptable displacements.

4-54. *Analysis of Stress Concentrations Due to Bridging.*—A stress concentration may occur in the dam due to the presence of a low-modulus zone within the foundation as shown on figure 4-25. To minimize the buildup of stress in the dam, a portion of the weak material in the low-modulus zone may be replaced with concrete. The depth of replacement required is determined as the depth when stresses in the dam and foundation are within allowable limits. The two-dimensional finite element method, discussed in sections 4-37 through 4-44, is an excellent method for solving this problem.

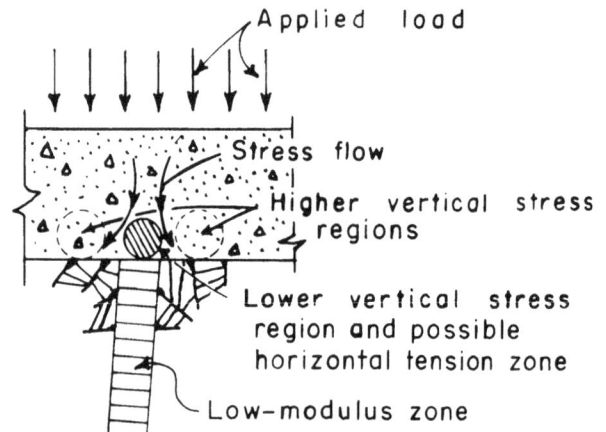

Figure 4-25. **Stress distribution near a low-modulus zone.—288-D-3001**

G. BIBLIOGRAPHY

4-55. *Bibliography*

[1] Westergaard, H. M., "Computations of Stresses in Bridge Slabs Due to Wheel Loads," Public Roads, vol. II, March 1930, pp. 1-23.
[2] McHenry, Douglas, "A Lattice Analogy for the Solution of Stress Problems," Institution of Civil Engineers, Paper 5350, vol. 21, December 1943, pp. 59-82.
[3] Timoshenko, S., "Strength of Materials," Part I, p. 270, 1956.
[4] Timoshenko, S., "Theory of Elastic Stability," Chapter 6, 1961.
[5] Clough, R. W., "Earthquake Response of Structures," Chapter 12 of Earthquake Engineering (R. L. Wiegel, coordinating editor), Prentice-Hall, Englewood Cliffs, N.J., 1970.
[6] Chopra, A. K., and Chakrabarti, P., "A Computer Solution for Earthquake Analysis of Dams," Report No. EERC70-5, Earthquake Engineering Research Center, University of California, Berkeley, Calif., 1970.
[7] Morgan, E. D., and Anderson, H. W., "Stress Analysis Using Finite Elements," Report No. SA-1, Bureau of Reclamation, 1969.
[8] Westergaard, H. M., "Water Pressures on Dams During Earthquakes," Transactions, American Society of Civil Engineers, vol. 98, 1933.

[9] Chopra, A. K., and Chakrabarti, P., "The Koyna Earthquake of December 11, 1967, and the Performance of Koyna Dam," Report No. EERC-71-1, Earthquake Engineering Research Center, University of California, Berkeley, Calif., p. 28, 1971.

[10] Zanger, C. N., "Hydrodynamic Pressures on Dams Due to Horizontal Earthquake Effects," Bureau of Reclamation, Special Assignments Section Report No. 21, October 18, 1950.

[11] Clough, Ray W., "The Finite Element Method in Plane Stress Analysis," ASCE Conference Papers (Second Conference on Electronic Computation, September 1960).

[12] Zienkiewicz, O. C., "The Finite Element in Structural and Continuum Mechanics," McGraw-Hill, London, 1967.

[13] Clough, R. W., "Comparison of Three-Dimensional Finite Elements," Proceedings of the Symposium on the Application of Finite Element Methods in Civil Engineering, Vanderbilt University, Nashville, Tenn., November 13-14, 1969.

[14] Londe, P., (1965), Une Methode d'Analyze o'trois dimensions de la stabilite d'une rive rocheme, Annls Ponts Chaus. No. 1 37-60.

[15] Guzina, Bosko, and Tucovic, Ignjat, "Determining the Minimum Three-Dimensional Stability of a Rock Wedge," Water Power, London, October 1969.

[16] Mahtab, M. A., and Goodman, R. E., "Three-Dimensional Finite Element Analysis of Jointed Rock Slopes," Final Report to Bureau of Reclamation, contract No. 14-06-D-6639, December 31, 1969.

[17] "Design of Arch Dams," Bureau of Reclamation, 1976.

Chapter V

River Diversion

A. DIVERSION REQUIREMENTS

5-1. *General.*—The design for a dam which is to be constructed across a stream channel must consider diversion of the streamflow around or through the damsite during the construction period. The extent of the diversion problem will vary with the size and flood potential of the stream; at some damsites diversion may be costly and time-consuming and may affect the scheduling of construction activities, while at other sites it may not offer any great difficulties. However, a diversion problem will exist to some extent at all sites except those located offstream, and the selection of the most appropriate scheme for handling the flow of the stream during construction is important to obtain economy in the cost of the dam. The scheme selected ordinarily will represent a compromise between the cost of the diversion facilities and the amount of risk involved. The proper diversion plan will minimize serious potential flood damage to the work in progress at a minimum of expense. The following factors should be considered in a study to determine the best diversion scheme:

(1) Characteristics of streamflow.
(2) Size and frequency of diversion flood.
(3) Regulation by existing upstream dam.
(4) Methods of diversion.
(5) Specifications requirements.
(6) Turbidity and water pollution control.

5-2. *Characteristics of Streamflow.*— Streamflow records provide the most reliable information regarding stream characteristics, and should be consulted whenever available.

Depending upon the geographical location of the drainage area, floods on a stream may be the result of snowmelt, rain on snow, seasonal rains, or cloudbursts. Because these types of runoff have their peak flows and their periods of low flow at different times of the year, the nature of runoff will influence the selection of the diversion scheme. A site subject mainly to snowmelt or rain on snow floods will not have to be provided with elaborate measures for use later in the construction season. A site where seasonal rains may occur will require only the minimum of diversion provisions for the rest of the year. A stream subject to cloudbursts which may occur at any time is the most unpredictable and probably will require the most elaborate diversion scheme, since the contractor must be prepared to handle both the low flows and floodflows at all times during the construction period.

5-3. *Selection of Diversion Flood.*—It is not economically feasible to plan on diverting the largest flood that has ever occurred or may be expected to occur at the site, and consequently some lesser requirement must be decided upon. This, therefore, brings up the question as to how much risk to the partially completed work is involved in the diversion scheme under consideration. Each site is different and the extent of damage done by flooding is dependent upon the area of foundation and structure excavation that would be involved, and the time and cost of cleanup and reconstruction that would be required.

In selecting the flood to be used in the diversion designs, consideration should be given to the following:

(1) How long the work will be under

construction, to determine the number of flood seasons which will be encountered.

(2) The cost of possible damage to work completed or still under construction if it is flooded.

(3) The cost of delay to completion of the work, including the cost of forcing the contractor's equipment to remain idle while the flood damage is being repaired.

(4) The safety of workmen and possibly the safety of downstream inhabitants in case the failure of diversion works results in unnatural flooding.

After an analysis of these factors is made, the cost of increasing the protective works to handle progressively larger floods can be compared to the cost of damages resulting if such floods occurred without the increased protective work. Judgment can then be used in determining the amount of risk that is warranted. Figure 5-1 shows a view from the right abutment of Monticello Dam with a major flood flowing over the low blocks and flooding the construction site. This flood did not damage the dam and caused only nominal damage to the contractor's plant.

The design diversion flood for each dam is dependent upon so many factors that rules cannot be established to cover every situation. Generally, however, for small dams which will be constructed in a single season, only the floods which may occur for that season need be considered. For most small dams, involving at the most two construction seasons, it should be sufficiently conservative to provide for a flood with a probability of occurrence of 20 percent. For larger dams involving more than a 2-year construction season, a design diversion flood with a probability of occurrence of anywhere between 20 and 4 percent may be established depending on the loss risk and the completion time for the individual dam.

Floods may be recurrent; therefore, if the diversion scheme involves temporary storage of cloudburst-type runoff, facilities must be provided to evacuate such storage within a reasonable period of time, usually a few days.

5-4. *Regulation by an Existing Upstream Dam.*—If the dam is to be built on a stream below an existing dam or other control structure, it is sometimes possible to modify the characteristics of the streamflow by planned operation of the existing structure. During the construction period, a modified program of operation of the existing structure may be used to reduce the peak of the flood outflow hydrograph and reduce the diversion requirements at the construction site. Upstream control can also be utilized to reduce flow during the construction of cofferdams, plugging of diversion systems and the removal of cofferdams.

5-5. *Turbidity and Water Pollution Control.*—One of the more important factors to be considered in determining the diversion scheme is how the required construction work affects the turbidity and pollution of the stream. A scheme that limits the turbidity, present in all diversion operations, to the shortest practicable period and creates less total effect on the stream should be given

Figure 5-1. View from right abutment of partially completed Monticello Dam in California, showing water flowing over low blocks.—SO-1446-R2

much consideration. Factors which contribute to turbidity in the stream during diversion are the construction and removal of cofferdams, required earthwork in or adjacent to the stream, pile driving, and the dumping of waste material. Therefore, all diversion schemes should be reviewed for the effect of pollution and turbidity on the stream during construction and removal of the diversion works, as well as the effect on the stream during the time construction is carried on between the cofferdams. Sample specifications for the control of turbidity and pollution are shown in appendix I.

B. METHODS OF DIVERSION

5-6. *General.*—The method or scheme of diverting floods during construction depends on the magnitude of the flood to be diverted; the physical characteristics of the site; the size and shape of dam to be constructed; the nature of the appurtenant works, such as the spillway, penstocks, and outlet works; and the probable sequence of construction operations. The objective is to select the optimum scheme considering practicability, cost, turbidity and pollution control, and the risks involved. The diversion works should be such that they may be incorporated into the overall construction program with a minimum of loss, damage, or delay.

Diverting streams during construction utilizes one or a combination of the following provisions: tunnels driven through the abutments, flumes or conduits through the dam area, or multiple-stage diversion over the tops of alternate construction blocks of the dam. On a small stream the flow may be bypassed around the site by the installation of a temporary wood or metal flume or pipeline, or the flow may be impounded behind the dam during its construction, pumps being used if necessary to control the water surface. In any case, barriers are constructed across or along the stream channel in order that the site, or portions thereof, may be unwatered and construction can proceed without interruption.

A common problem is the meeting of downstream requirements when the entire flow of the stream is stopped following closure of the diversion works. Downstream requirements may demand that a small flow be maintained at all times. In this case the contractor must provide the required flow by pumping or by other means (bypasses or siphons) until water is stored in the reservoir to a sufficient elevation so that releases may be made through the outlet works.

Figure 5-2 shows how diversion of the river was accomplished during the construction of Folsom Dam and Powerplant on the American River in California. This photograph is included because it illustrates many of the diversion principles discussed in this chapter. The river, flowing from top to bottom in the picture, is being diverted through a tunnel; "a" and "b" mark the inlet and outlet portals, respectively. Construction is proceeding in the original river channel between earthfill cofferdams "c" and "d." Discharge from pipe "e" at the lower left in the photograph is from unwatering of the foundation. Since it was impracticable to provide sufficient diversion tunnel capacity to handle the large anticipated spring floods, the contractor made provisions to minimize damage that would result from overtopping of the cofferdam. These provisions included the following:

(1) Placing concrete in alternate low blocks in the dam "f" to permit overflowing with a minimum of damage.

(2) Construction of an auxiliary rockfill and cellular steel sheet-piling cofferdam "g" to protect the powerplant excavation "h" from being flooded by overtopping of the cofferdam.

(3) Early construction of the permanent training wall "i" to take advantage of the protection it affords.

5-7. *Tunnels.*—It is usually not feasible to do a significant amount of foundation work in a narrow canyon until the stream is diverted. If the lack of space or a planned powerplant or other feature eliminates diversion through the construction area by flume or conduit, a tunnel

Figure 5-2. Diversion of the river during construction of Folsom Dam and Powerplant in California.–AR-1627-CV.

may prove the most feasible means of diversion. The streamflow may be bypassed around the construction area through tunnels in one or both abutments. A diversion tunnel should be of a length that it bypasses the construction area. Where suitable area required by the contractor for shops, storage, fabrication, etc., is not readily available, it may be advantageous to lengthen the tunnel to provide additional work area in the streambed. However, the tunnel should be kept as short as practicable for economic and hydraulic reasons. Figure 5-3 shows such a tunnel which was constructed at Flaming Gorge Dam site, a relatively narrow canyon, to permit diversion through the abutment.

The diversion system must be designed to bypass, possibly also contain part of, the design diversion flood. The size of the diversion tunnel will thus be dependent on the magnitude of the diversion flood, the height of the upstream cofferdam (the higher the head, the smaller the tunnel needs to be for a given discharge), and the size of the reservoir formed by the cofferdam if this is appreciable. An economic study of cofferdam height versus tunnel size may be involved to establish the most economical relationship.

The advisability of lining the diversion tunnel will be influenced by the cost of a lined tunnel compared with that of a larger unlined tunnel of equal carrying capacity; the nature of the rock in the tunnel, as to whether it can stand unsupported and unprotected during the passage of the diversion flows; and the permeability of the material through which the tunnel is carried, as it will affect the amount of leakage through or around the abutment.

If tunnel spillways are provided in the design, it usually proves economical to utilize

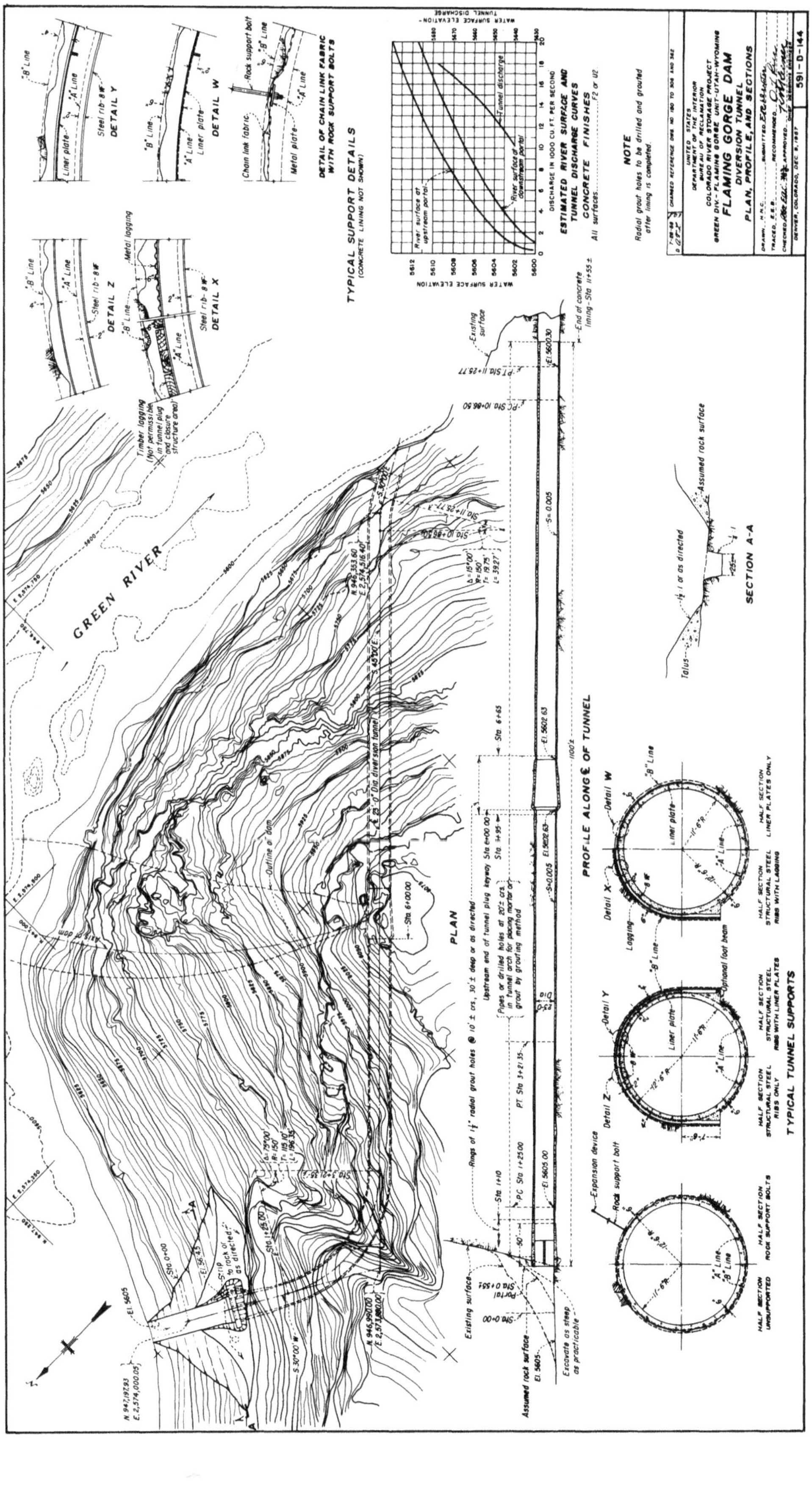

Figure 5-3. Diversion tunnel for Flaming Gorge Dam, a large concrete dam in Utah—plan, profile, and sections.

them in the diversion plan. When the proposed spillway tunnel consists of a high intake and a sloping tunnel down to a near horizontal portion of tunnel close to streambed elevation, a diversion tunnel can be constructed between the near horizontal portion of tunnel and the channel elevation upstream to effect a streambed bypass. Figure 5-4 shows such a typical diversion tunnel which will permit diversion through the lower, nearly horizontal portion of the spillway tunnel. Provisions for the final plugging, such as excavation of keyways, grouting, etc., should be incorporated into the initial construction phase of the diversion tunnel.

Some means of shutting off diversion flows must be provided; in addition, some means of regulating the flow through the diversion tunnel may be necessary. Closure devices may consist of a timber, concrete, or steel bulkhead gate; a slide gate; or stoplogs. Regulation of flow to satisfy downstream needs after storage of water in the reservoir has started can be accomplished by the use of a slide gate on a temporary bypass until the water surface in the reservoir reaches the level of the outlet works intake. Figure 5-5 shows the closure structure constructed at Flaming Gorge Dam, which was incorporated in the upstream 50-foot length of the diversion tunnel. A high-pressure slide gate on a small conduit was provided in the left side of the closure structure to bypass required flows while filling the reservoir to the elevation of the river outlet.

Permanent closure of the diversion tunnel is made by placing a concrete plug in the tunnel. If the tunnel passes close to and under the dam, the plug should be located near the upstream face in line with the grout curtain cutoff or it may extend entirely under the dam, depending on the stresses from the dam and the condition of the foundation. If the diversion tunnel joins a permanent tunnel, the plug is usually located immediately upstream from the intersection as indicated in figure 5-4. Keyways may be excavated into rock or formed into the lining to insure adequate shear resistance between the plug and the rock or lining. After the plug has been placed and the concrete cooled, grout is forced through previously installed grout connections into the contact between the plug and the surrounding rock or concrete lining to insure a watertight joint.

5-8. Conduits Through Dam.—Diversion conduits at stream level are sometimes provided through a dam. These conduits may be constructed solely for the purpose of diversion or they may be conduits which later will form part of the outlet works or power penstock systems. As with tunnels, some means of shutting off the flow at the end of the diversion period and a method of passing downstream water requirements during the filling of the reservoir must be incorporated into the design of the conduit. The most common procedure for closing the diversion conduit before the placement of the permanent plug is by lowering bulkheads down the upstream face of the dam which will seal against the upstream face. Figure 5-6 shows typical details of a conduit through a dam.

After serving their purpose, all diversion conduits must be filled with concrete for their entire length. This is accomplished with the bulkheads in place. The conduit should be provided with keyways, metal seals, and grouting systems within the initial construction to assure a satisfactory permanent seal. The shrinkage and temperature of the plug concrete should be controlled by the installation of a cooling system.

5-9. Flumes.—In a wide canyon, an economical method of diversion may be the use of a flume to carry the streamflow around the construction area. A flume may also be used to carry the streamflow over a low block and through the construction area. The flume should be designed to accommodate the design diversion flood, or a portion thereof if the flume is used in conjunction with another method of diversion. The most economical scheme can be found by comparing costs of various cofferdam heights versus the corresponding flume capacity. Large flumes may be of steel or timber frame with a timber lining, and smaller flumes may be of timber or metal construction, pipe, etc.

The flume is usually constructed around one side or the other of the damsite or over a low

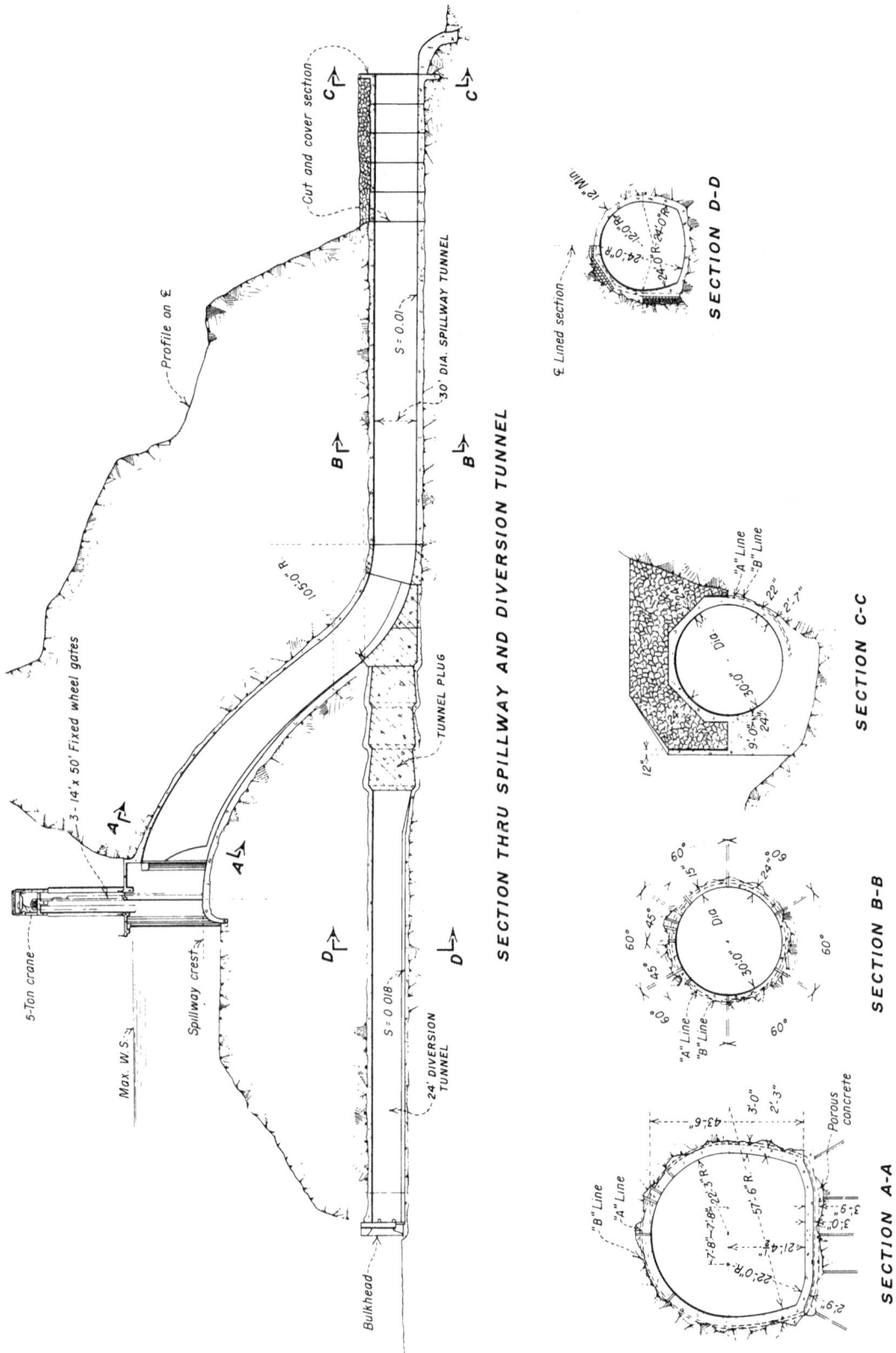

Figure 5-4. Typical arrangement of diversion tunnel with spillway tunnel.—288-D-3002

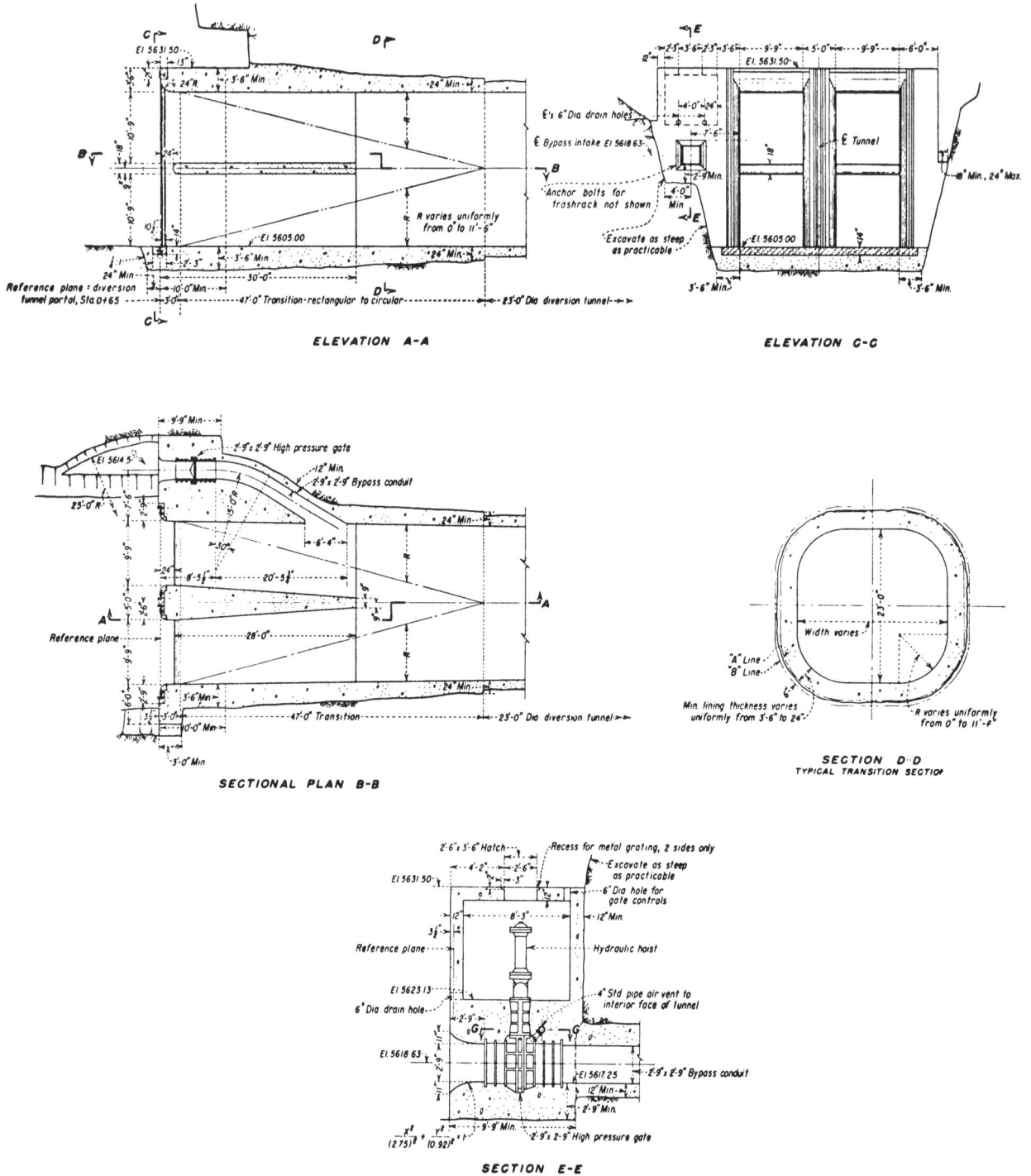

Figure 5-5. Diversion tunnel closure structure for a large concrete dam (Flaming Gorge Dam in Utah).—288-D-3003

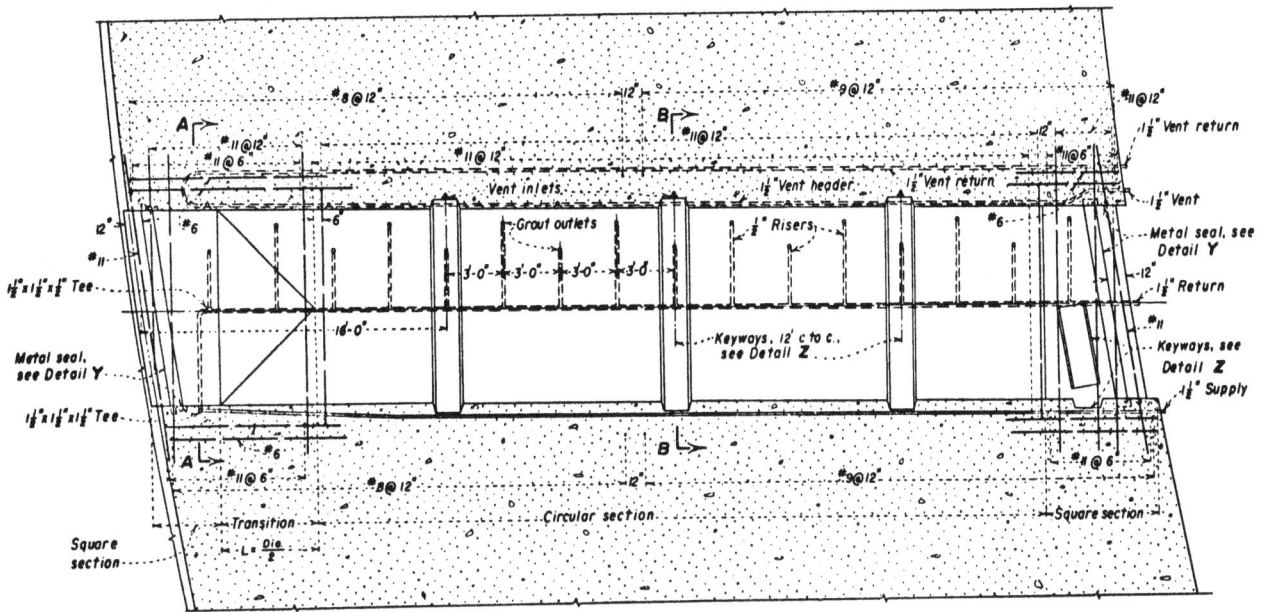

SECTION ALONG ℄ OF CONDUIT

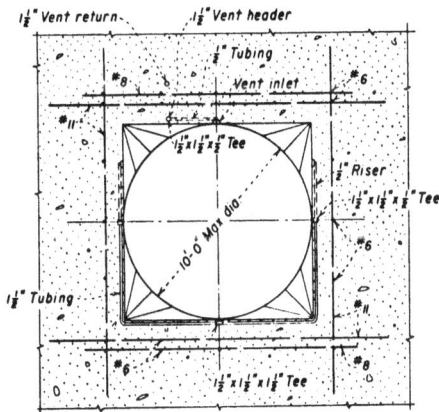

SECTION A-A

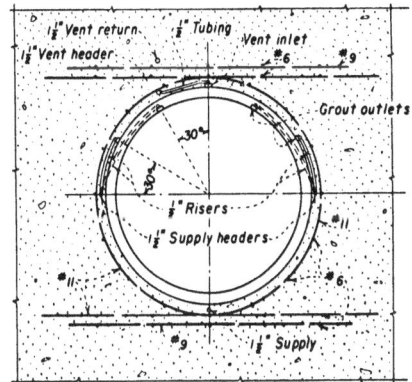

SECTION B-B

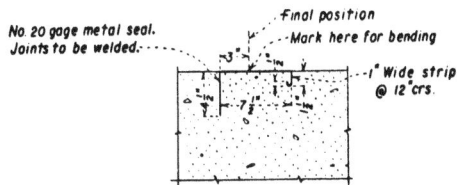

DETAIL Y

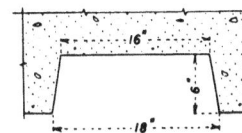

DETAIL Z

Figure 5-6. Diversion conduit through Morrow Point Dam, a thin arch structure in Colorado—plan and sections.—288-D-3004

block. The flume can then be moved to other areas as the work progresses and stage construction can be utilized. During the construction of Canyon Ferry Dam, a steel-framed, timber-lined flume was constructed along the right bank of the river to be used as the first stage of diversion. The flume was designed for a capacity of approximately 23,000 cubic feet per second. The completed flume can be seen in figure 5-7 and a view of the flume in use can be seen in figure 5-8.

5-10. Multiple-Stage Diversion.—The multiple-stage method of diversion over the tops of alternate low construction blocks or through diversion conduits in a concrete dam requires shifting of the cofferdam from one side of the river to the other during construction. During the first stage, the flow is restricted to one portion of the stream channel while the dam is constructed to a safe elevation in the remainder of the channel. In the second stage, the cofferdam is shifted and the stream is carried over low blocks or through diversion conduits in the constructed section of the dam while work proceeds on the unconstructed portion of the dam. The dam is then carried to its final height, with diversion ultimately being made through the spillway, penstock, or permanent outlets. Figure 5-9 shows diversion through a conduit in a concrete dam, with excess flow over the low blocks.

5-11. Cofferdams.—A cofferdam is a temporary dam or barrier used to divert the stream or to enclose an area during construction. The design of an adequate

Figure 5-7. Completed diversion flume at Canyon Ferry damsite in Montana. Note the large size of flume required to pass the design flow, amounting to 23,000 cubic feet per second.–P-584–MRBP

Figure 5-8. Completed diversion flume at Canyon Ferry damsite in use for first-stage diversion.—P-591—MRBP

cofferdam involves the problem of construction economics. Where the construction is timed so that the foundation work can be executed during the low water season, use of cofferdams can be held to a minimum. Where the streamflow characteristics are such that this is not practicable, the cofferdam must be so designed that it is not only safe, but also of the optimum height. The height to which a cofferdam should be constructed involves an economic study of cofferdam height versus diversion works capacity, including routing studies of the diversion design flood. This is particularly true when the outlet works requirements are small. It should be remembered that floodwater accumulated behind the cofferdam must be evacuated in time to accommodate a recurrent storm. The maximum height to which it is feasible to construct the cofferdam without encroaching upon the area to be occupied by the dam must also be considered. Furthermore, the design of the cofferdam must take into consideration the effect that excavation and unwatering of the foundation of the dam will have on the cofferdam stability, and must anticipate removal, salvage, and other factors.

When determining the type and location of the cofferdams, the effects on the stream as related to water pollution and turbidity should be examined for each scheme under consideration. Unwatering work for structural

Figure 5-9. Flows passing through diversion opening and over low blocks of a concrete and earth dam (Olympus Dam in Colorado).—EPA-PS-330-CBT

foundations, constructing and removing cofferdams, and earthwork operations adjacent to or encroaching on streams or watercourses should be conducted in such a manner as to prevent muddy water and eroded materials from entering the channel. Therefore, the cofferdams should be placed in such a location that earthwork near the stream will be kept to a minimum, by containing as much of the excavation and work area within the confines of the cofferdams as practicable. During the construction and removal of the cofferdams, mechanized equipment should not be operated in flowing water except where necessary to perform the required work, and this should be restricted as much as possible.

Generally, cofferdams are constructed of materials available at the site. The two types normally used in the construction of dams are earthfill cofferdams and rockfill cofferdams,

the design considerations of which closely follow those for permanent small dams of the same type. Other types, although not as common, include timber or concrete cribs filled with earth or rock, and cellular steel cofferdams filled with pervious material. Cribs and cellular steel cofferdams can be used when space for a cofferdam is limited or material is scarce. Cellular cofferdams are especially adaptable to confined areas where currents are swift and normal cofferdam construction would be difficult.

In many situations, a combination of several types of cofferdams may be used to develop the diversion scheme in the most economical and practical manner. The type of cofferdam would be determined for each location depending upon such factors as the materials available, required height, available space, swiftness of water, and ease of removal.

C. SPECIFICATIONS REQUIREMENTS

5-12. *Contractor's Responsibilities.*—It is general practice to require the contractor to assume responsibility for the diversion of the stream during construction of the dam and appurtenant structures. The requirement should be defined by appropriate paragraphs in the specifications which describe the contractor's responsibilities and inform him as to what provisions, if any, have been incorporated in the design to facilitate construction. Usually the specifications should not prescribe the capacity of the diversion works, nor the details of the diversion method to be used; but hydrographs prepared from streamflow records, if available, should be included. Also, the specifications usually require that the contractor's diversion plan be subject to the owner's approval.

In some cases the entire diversion scheme might be left in the contractor's hands, with the expectation that the flexibility afforded to the contractor's operations by allowing him to choose the scheme of diversion will be reflected in low bids. Since various contractors will usually present different schemes, the schedule of bids in such instances should require diversion of the river to be included as a lump-sum bid. Sometimes a few pertinent paragraphs are appropriate in the specifications giving stipulations which affect the contractor's construction procedures. For example, restriction from certain diversion schemes may be specified because of safety requirements, geology, ecology, or time and space limitations. The contractor may also be required to have the dam constructed to a certain elevation or have the channel or other downstream construction completed before closure of the diversion works is permitted.

These or similar restrictions tend to guide the contractor toward a safe diversion plan. However, to further define the contractor's responsibility, other statements should be made to the effect that the contractor shall be responsible for and shall repair at his expense any damage to the foundation, structures, or any other part of the work caused by flood, water, or failure of any part of the diversion or protective works. The contractor should also be cautioned concerning the use of the hydrographs, by a statement to the effect that the contracting authority does not guarantee the reliability or accuracy of any of the hydrographs and assumes no responsibility for any deductions, conclusions, or interpretations that may be made from them.

5-13. *Designer's Responsibilities.*—For difficult and/or hazardous diversion situations, it may prove economical for the owner to assume the responsibility for the diversion plan. One reason for this is that contractors tend to increase bid prices for diversion of the stream if the specifications contain many restrictions and there is a large amount of risk involved. A definite scheme of cofferdams and tunnels might be specified where the loss of life and property damage might be heavy if a cofferdam built at the contractor's risk were to fail.

Another consideration is that many times the orderly sequence of constructing various stages of the entire project depends on a particular diversion scheme being used. If the responsibility for diversion rests on the contractor, he may pursue a different diversion scheme, with possible delay to completion of the entire project. This could result in a delay in delivery of irrigation water or in generation of power, or both, with a subsequent loss in revenue.

If the owner assumes responsibility for the diversion scheme, it is important that the diversion scheme be realistic in all respects, and compatible with the probable ability and capacity of the contractor's construction plant.

Chapter VI

Foundation Treatment

A. EXCAVATION

6-1. General.—The entire area to be occupied by the base of the concrete dam should be excavated to firm material capable of withstanding the loads imposed by the dam, reservoir, and appurtenant structures. Considerable attention must be given to blasting operations to assure that excessive blasting does not shatter, loosen, or otherwise adversely affect the suitability of the foundation rock. All excavations should conform to the lines and dimensions shown on the construction drawings where practicable; however, it may be necessary or even desirable to vary dimensions or excavation slopes due to local conditions.

Foundations such as shales, chalks, mudstones, and siltstones may require protection against air and water slaking, or in some environments, against freezing. Such excavations can be protected by leaving a temporary cover of several feet of unexcavated material, by immediately applying a minimum of 12 inches of pneumatically applied mortar to the exposed surfaces, or by any other method that will prevent damage to the foundation.

6-2. Shaping.—If the canyon profile for a damsite is relatively narrow with steep sloping walls, each vertical section of the dam from the center towards the abutments is shorter in height than the preceding one. Consequently sections closer to the abutments will be deflected less by the reservoir load and sections closer toward the center of the canyon will be deflected more. Since most gravity dams are keyed at the contraction joints, the result is a torsional effect in the dam that is transmitted to the foundation rock.

A sharp break in the excavated profile of the canyon will result in an abrupt change in the height of the dam. The effect of the irregularity of the foundation rock causes a marked change in stresses in both the dam and foundation, and in stability factors. For this reason, the foundation should be shaped so that a uniformly varying profile is obtained free of sharp offsets or breaks.

Generally, a foundation surface will appear as horizontal in the transverse (upstream-downstream) direction. However, where an increased resistance to sliding is desired, particularly for structures founded on sedimentary rock foundations, the surface can be sloped upward from heel to toe of the dam. The foundation excavation for Pueblo Dam (fig. 6-1), a massive head buttress-type gravity dam, is an example of an excavation sloped in the transverse direction. Figure 6-1 also represents a special type of situation wherein the foundation excavation is shaped to the configuration of the massive head buttress.

6-3. Dental Treatment.—Very often the exploratory drilling or final excavation uncovers faults, seams, or shattered or inferior rock extending to such depths that it is impracticable to attempt to clear such areas out entirely. These conditions require special treatment in the form of removing the weak material and backfilling the resulting excavations with concrete. This procedure of reinforcing and stabilizing such weak zones is frequently called "dental treatment."

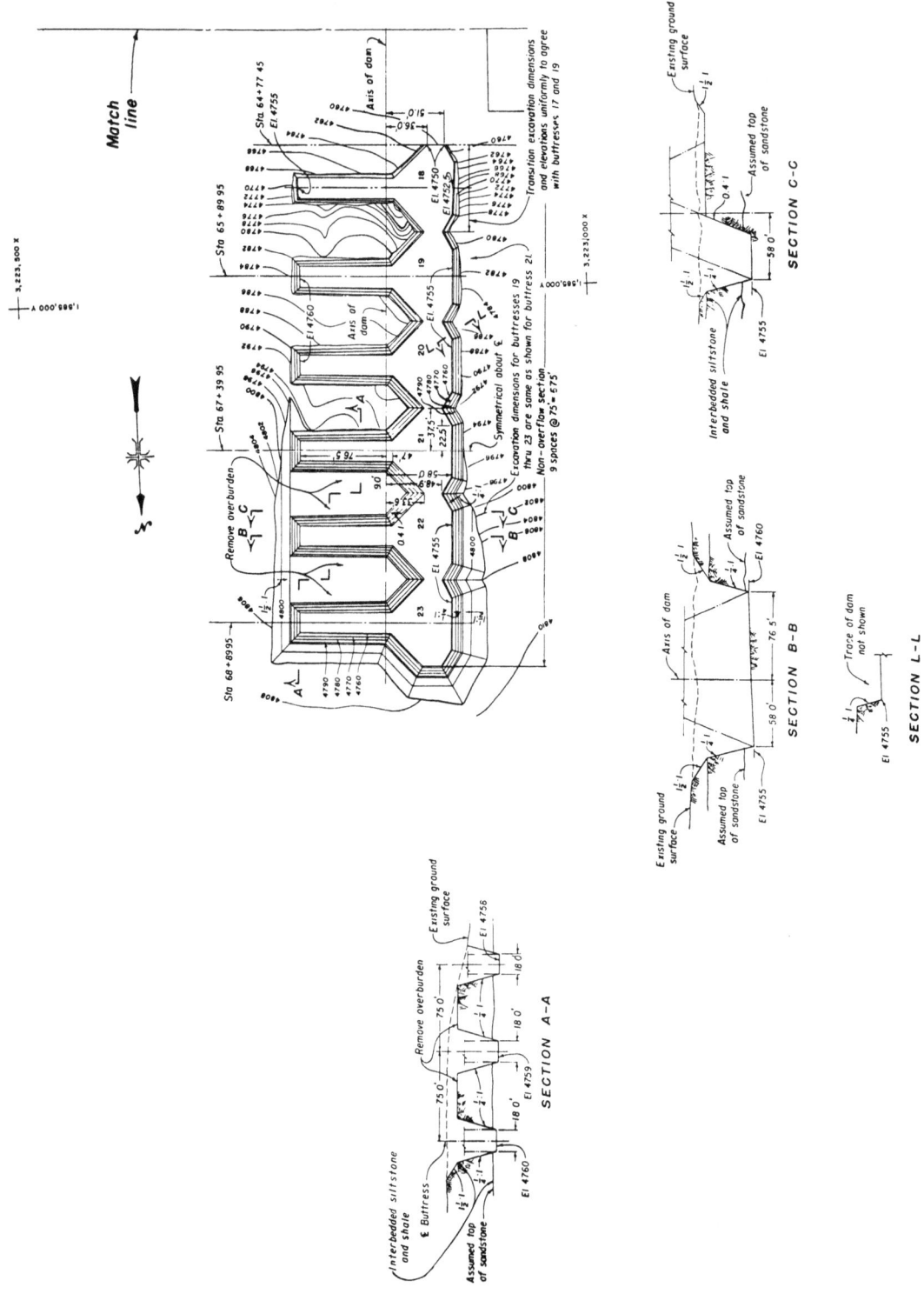

Figure 6-1. Excavation layout for Pueblo Dam and spillway in Colorado—a concrete buttress-type structure (sheet 1 of 2).—288-D-3006(1/2)

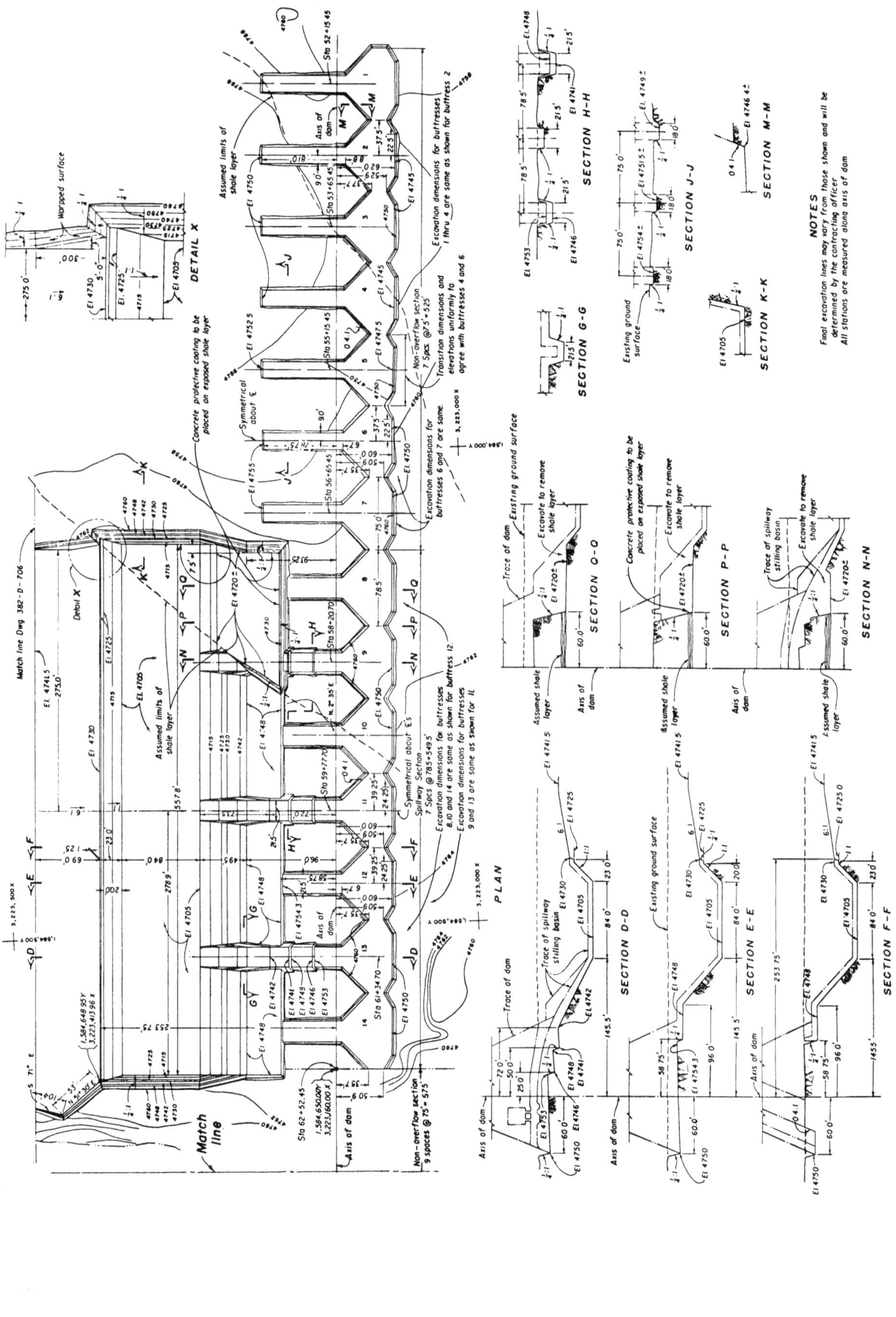

Figure 6-1. Excavation layout for Pueblo Dam and spillway in Colorado—a concrete buttress-type structure (sheet 2 of 2).–288-D-3006(2/2)

Theoretical studies have been made to develop general rules for guidance as to how deep transverse seams should be excavated. These studies, based upon foundation conditions and stresses at Shasta and Friant Dams, have resulted in the development of the following approximate formulas for determining the depth of dental treatment:

$$d = 0.002\, bH + 5 \text{ for } H \geqq 150 \text{ feet}$$

$$d = 0.3\, b + 5 \qquad \text{for } H < 150 \text{ feet}$$

where:

H = height of dam above general foundation level in feet,
b = width of weak zone in feet, and
d = depth of excavation of weak zone below surface of adjoining sound rock in feet.
(In clay gouge seams, d should not be less than $0.1\,H$.)

These rules provide a means of approach to the question of how much should be excavated, but final judgment must be exercised in the field during actual excavation operations.

Although the preceding rules are suitable for application to foundations with a relatively homogeneous rock foundation with nominal faulting, some damsites may have several distinct rock types interspersed with numerous faults and shears. The effect of rock-type anomalies complicated by large zones of faulting on the overall strength and stability of the foundation requires a definitive analysis. Such a study was performed for Coulee Forebay Dam wherein the finite element method of analysis was used in evaluating the foundation. (See subchapter E of chapter IV and also appendix C.) This method provides a way to combine the physical properties of various rock types, and geologic discontinuities such as faults, shears, and joint sets into a value representative of the stress and deformation in a given segment of the foundation. The method also permits substitution of backfill concrete in

faults, shears, and zones of weak rock, and thus evaluates the degree of beneficiation contributed by the "dental concrete."

Data required for the finite element method of analysis are: dimensions and composition of the lithologic bodies and geologic discontinuities, deformation moduli for each of the elements incorporated into the study, and the loading pattern imposed on the foundation by the dam and reservoir. Methods for obtaining data related to the rock and discontinuities are discussed in the sections on foundation investigations in chapter II.

"Dental treatment" may also be required to improve the stability of rock masses. By inputting data related to the shearing strength of faults, shear, joints, intact rock, pore water pressures induced by the reservoir and/or ground water, the weight of the rock mass, and the driving forces induced by the dam and reservoir, a safety factor for a particular rock mass can be calculated.

Methods of rock stability analysis are discussed in chapter IV in the sections on finite element method and foundation analysis.

6-4. *Protection Against Piping.*—The approximate and analytical methods discussed above will satisfy the stress, deformation, and stability requirements for a foundation, but they may not provide suitable protection against piping. Faults and seams may contain material conducive to piping and its accompanying dangers, so to mitigate this condition upstream and downstream cutoff shafts should be excavated in each fault or seam and backfilled with concrete. The dimension of the shaft perpendicular to the seam should be equal to the width of the weak zone plus a minimum of 1 foot on each end to key the concrete backfill into sound rock. The shaft dimension parallel with the seam should be at least one-half of the other dimension. In any instance a minimum shaft dimension of 5 feet each way should be used to provide working space.

The depth of cutoff shafts may be computed by constructing flow nets and computing the cutoff depths required to eliminate piping effects, or by the methods outlined by Khosla in reference [1].[1] These two methods are particularly applicable for medium to high dams. For low head dams, the weighted creep method for determining cutoff depths as shown in chapter VIII of "Design of Small Dams" [2] may be used.

Other adverse foundation conditions may be due to horizontally bedded clay and shale seams, caverns, or springs. Procedures for treating these conditions will vary and will depend upon field studies of the characteristics of the particular condition to be remedied.

B. GROUTING

6-5. *General*.—The principal objectives of grouting in a rock foundation are to establish an effective barrier to seepage under the dam and to consolidate the foundation. Spacing, length, and orientation of grout holes and the procedure to be followed in grouting a foundation are dependent on the height of the structure and the geologic characteristics of the foundation. Since the characteristics of a foundation will vary for each site, the grouting plan must be adapted to suit field conditions.

Grouting operations may be performed from the surface of the excavated foundation, from the upstream fillet of the dam, from the top of concrete placements for the dam, from galleries within the dam, and from tunnels driven into the abutments, or any combination of these locations.

The general plan for grouting the foundation rock of a dam provides for preliminary low-pressure, shallow consolidation grouting to be followed by high-pressure, deep curtain grouting. As used here, "high pressure" and "low pressure" are relative terms. The actual pressures used are usually the maximum that will result in filling the cracks and voids as completely as practicable without causing any uplift or lateral displacement of foundation rock.

6-6. *Consolidation Grouting*.—Low-pressure grouting to fill voids, fracture zones, and cracks at and below the surface of the excavated foundation is accomplished by drilling and grouting relatively shallow holes, called "B" holes. The extent of the area grouted and the depth of the holes will depend on local conditions.

Usually for structures 100 feet and more in height, a preliminary program will call for lines of holes parallel to the axis of the dam extending from the heel to the toe of the dam and spaced approximately 10 to 20 feet apart. Holes are staggered on alternate lines to provide better coverage of the area. The depths of the holes vary from 20 to 50 feet depending on local conditions and to some extent on the height of the structure. For structures less than 100 feet in height and depending on local conditions, "B" hole grouting has been applied only in the area of the heel of the dam. In this case the upstream line of holes should lie at or near the heel of the dam to furnish a cutoff for leakage of grout from the high-pressure holes drilled later in the same general location. "B" holes are drilled normal to the excavated surface unless it is desired to intersect known faults, shears, fractures, joints, and cracks. Drilling is usually accomplished from the excavated surface, although in some cases drilling and grouting to consolidate steep abutments has been accomplished from the tops of concrete placements in the dam to prevent "slabbing" of the rock. In rarer cases, consolidation grouting has been performed from foundation galleries within the dam after the concrete placement has reached a certain elevation. This method of consolidation grouting requires careful control of grouting pressures and close inspection of the foundation to assure that the structure is not being disbonded from the foundation. Figure 6-2 illustrates a typical spacing and length pattern for "B" hole grouting.

In the execution of the consolidation grouting program, holes with a minimum diameter of 1½ inches are drilled and grouted 40 to 80 feet apart before split-spaced

---

[1]Numbers in brackets refer to items in the bibliography, sec. 6-9.

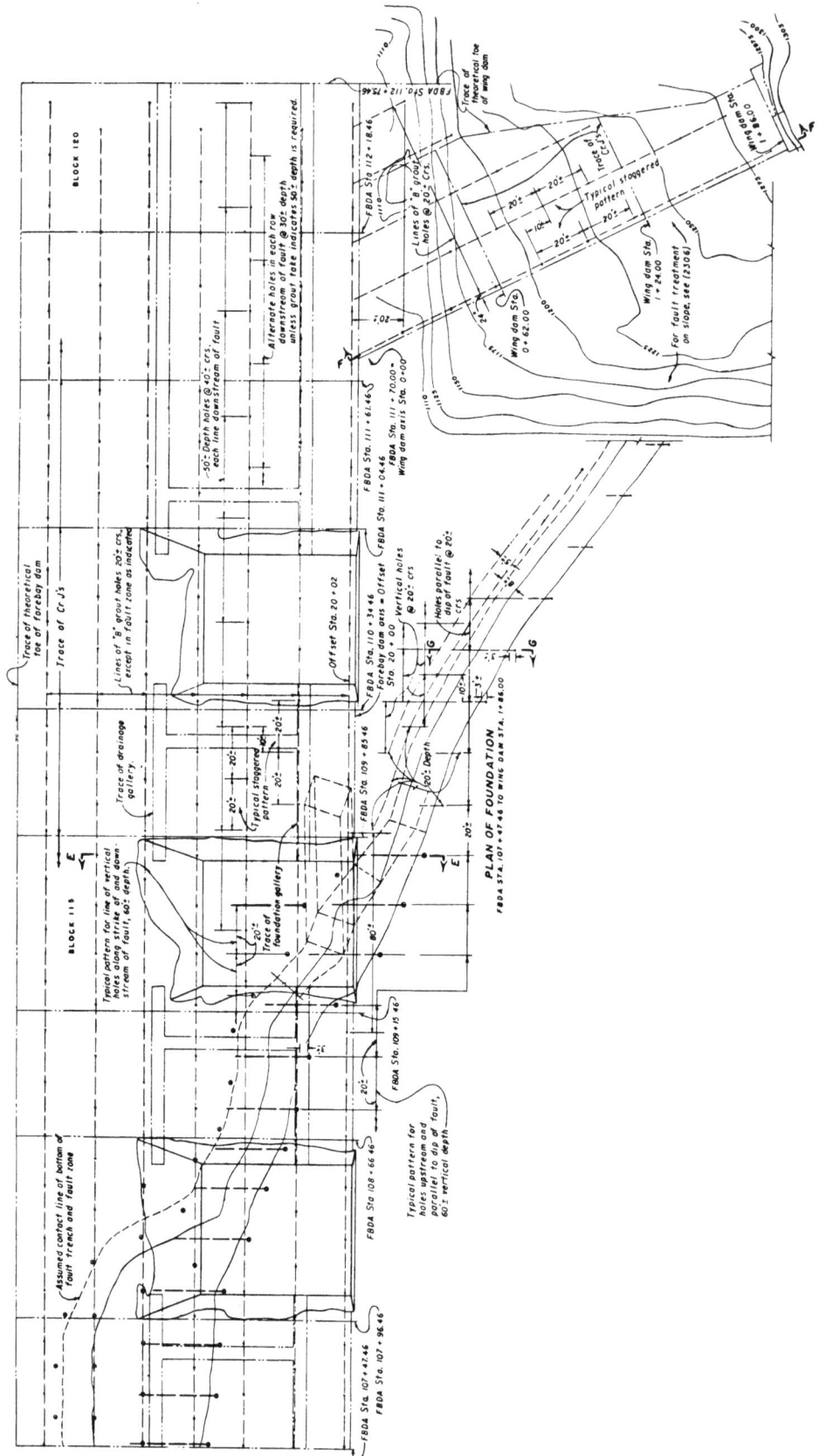

Figure 6-2. Foundation treatment for Grand Coulee Forebay Dam in Washington (sheet 1 of 2).–288-D-3005(1/2)

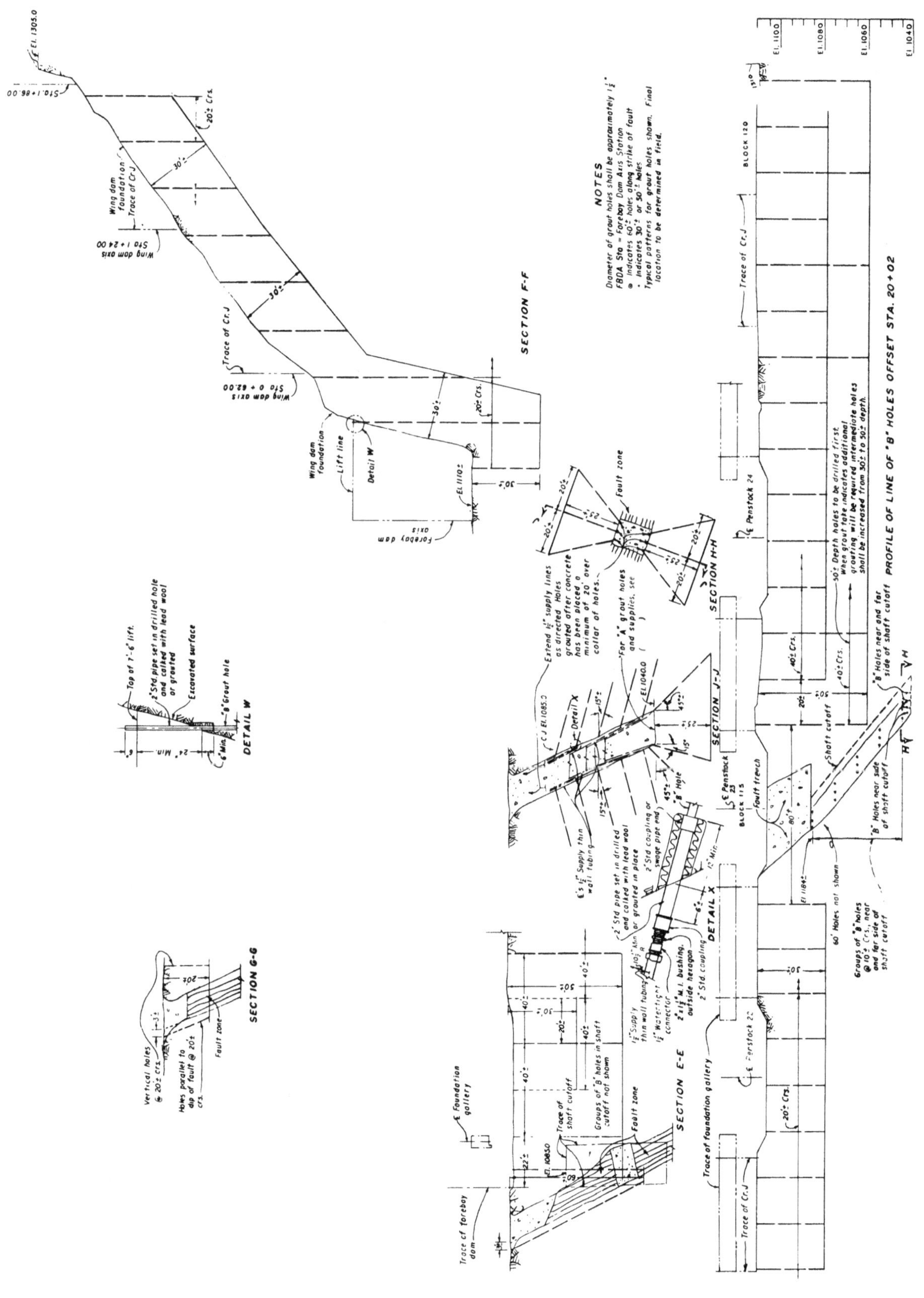

Figure 6-2. Foundation treatment for Grand Coulee Forebay Dam in Washington (sheet 2 of 2).—288-D-3005(2/2)

intermediate holes are drilled. The amount of grout which the intermediate holes accept determines whether additional intermediate holes should be drilled. This split-spacing process is continued until grout "take" for the final closure holes is negligible and it is reasonably assured that all groutable seams, fractures, cracks, and voids have been filled.

Water-cement ratios for grout mixes may vary widely depending on the permeability of the foundation rock. Starting water-cement ratios usually range from 8:1 to 5:1 by volume. Most foundations have an optimum mix that can be injected which should be determined by trial in the field by gradually thickening the starting mix. An admixture such as sand or clay may be added if large voids are encountered.

Consolidation grouting pressures vary widely and are dependent in part on the characteristics of the rock, i.e., its strength, tightness, joint continuity, stratification, etc.; and on the depth of rock above the stage being grouted. In general, grout pressures as high as practicable but which, as determined by trial, are safe against rock displacement, are used in grouting. These pressures may vary from a low of 10 pounds per square inch to a high range of 80 to 100 pounds per square inch. A common rule of thumb is to increase the above minimum collar pressure by 1 pound per square inch per foot of depth of hole above the packer, as a trial. If the take is small the pressure may be increased.

6-7. *Curtain Grouting.*—Construction of a deep grout curtain near the heel of the dam to control seepage is accomplished by drilling deep holes and grouting them using higher pressure. These holes are identified as "A" holes when drilled from a gallery. Tentative designs will usually specify a single line of holes drilled on 10-foot centers, although wider or closer spacing may be required depending on the rock condition. To permit application of high pressures without causing displacement in the rock or loss of grout through surface cracks, this grouting procedure is carried out subsequent to consolidation grouting and after some of the concrete has been placed. Usually, grouting will be accomplished from galleries within the dam and from tunnels driven into the abutments especially for this purpose. However, when no galleries are provided, as is the case for most low gravity dams, high-pressure grouting is done from curtain holes located in the upstream fillet of the dam before reservoir storage is started. Such grouting holes are identified as "C" holes.

The alinement of holes should be such that the base of the grout curtain will be located on the vertical projection of the heel of the dam. If drilled from a gallery that is some distance from the upstream face, the holes may be inclined as much as 15° upstream from the plane of the axis. If the gallery is near the upstream face, the holes will be nearly vertical. Holes drilled from foundation tunnels may be inclined upstream or they may be vertical depending on the orientation of the tunnel with the axis of the dam. When the holes are drilled from the upstream fillet, they are usually inclined downstream. Characteristics of the foundation seams may also influence the amount of inclination.

To facilitate drilling, pipes of 2-foot minimum length are embedded in the floor of the gallery or foundation tunnel, or in the upstream fillet. When the structure has reached an elevation that is sufficient to prevent movement of concrete, the grout holes are drilled through these pipes and into the foundation. Although the tentative grouting plan may indicate holes to be drilled on 10-foot centers, the usual procedure will be first to drill and grout holes approximately 40 feet apart, or as far apart as necessary to prevent grout from one hole leaking into another drilled but ungrouted hole. Also, leakage into adjacent contraction joints must be prevented by prior grouting of the joints. Intermediate holes, located midway between the first holes, will then be drilled and grouted. Drilling and grouting of additional intermediate holes, splitting the spaces between completed holes, will continue until the desired spacing is reached or until the amount of grout accepted by the last group of intermediate holes indicates no further grouting is necessary.

The depth to which the holes are drilled will vary greatly with the characteristics of the foundation and the hydrostatic head. In a hard, dense foundation, the depth may vary from 30 to 40 percent of the head. In a poor foundation the holes will be deeper and may reach as deep as 70 percent of the head. During the progress of the grouting, local conditions may determine the actual or final depth of grouting. Supplementary grouting may also be required after the waterload has come on the dam and observations have been made of the rate of seepage and the accompanying uplift.

For high dams where foundation galleries are located at a relatively long distance from the upstream face, as at Grand Coulee Dam, "A" hole grouting may be augmented by a line of "C" holes, drilled from the upstream face of the dam and inclined downstream in order to supplement the main grout curtain. The depth of these holes is usually about 75 feet and their spacing is usually the same as for the "A" holes. The supplementary grout curtain formed by grouting this line of holes serves as an upstream barrier for subsequent "A" hole grouting, permitting higher "A" hole grout pressures with less chance of excessive upstream grout travel.

Usually the foundation will increase in density and tightness of seams as greater depths are reached, and the pressure necessary to force grout into the tight joints of the deep planes may be sufficient to cause displacements of the upper zones. Two general methods of grouting are used, each permitting the use of higher pressures in the lower zones.

(1) *Descending stage grouting* consists of drilling a hole to a limited depth or to its intersection with an open seam, grouting to that depth, cleaning out the hole after the grout has taken its initial set, and then drilling and grouting the next stage. To prevent backflow of grout during this latter operation, a packer is seated at the bottom of the previously grouted stage. This process is repeated, using higher pressures for each succeeding stage until the final depth is reached.

(2) *Ascending stage grouting* consists of drilling a hole to its final depth and grouting the deepest high-pressure stage first by use of a packer which is seated at the top of this stage. The packer limits grout injection to the desired stage and prevents the grout from rising into the hole above the packer. After grouting this stage, the grout pipe is raised so that the packer is at the top of the next stage which is subsequently grouted using somewhat lower pressure. This stage process is repeated, working upward until the hole is completely grouted. Ascending stage grouting is becoming more generally used, as it reduces the chances for displacement of the foundation rock, gives better control as to the zones of injection, and expedites the drilling.

The discussion in section 6-6 concerning grout pressures applies in general to curtain grouting. An exception is that higher initial collar pressures are permitted, depending on the height of concrete above the hole.

C. DRAINAGE

6-8. *Foundation   Drainage.*—Although   a well-executed grouting program may materially reduce the amount of seepage, some means must be provided to intercept the water which will percolate through and around the grout curtain, and, if not removed, may build prohibitive hydrostatic pressures on the base of the structure. Drainage is usually accomplished by drilling one or more lines of holes downstream from the high-pressure grout curtain. The size, spacing, and depth of these holes are assumed on the basis of judgment of the physical characteristics of the rock. Holes are usually 3 inches in diameter (NX size). Spacing, depth, and orientation are all influenced by the foundation conditions. Usually the holes are spaced on 10-foot centers with depths dependent on the grout curtain and reservoir depths. As a general rule, hole depths vary from 20 to 40 percent of the reservoir depth and 35 to 75 percent of the deep curtain grouting depth.

Drain holes should be drilled after all foundation grouting has been completed in the area. They can be drilled from foundation and drainage galleries within the dam, or from the downstream face of the dam if no gallery is provided. Frequently drainage holes are drilled from foundation grouting and drainage tunnels excavated into the abutments.

In some instances where the stability of a rock foundation may be beneficiated by reducing the hydrostatic pressure along planes of potentially unstable rock masses, drainage holes have been introduced to alleviate this condition. A collection system for such drainages should be designed so that flows can be gathered and removed from the area.

D. BIBLIOGRAPHY

6-9. *Bibliography*.

[1] Khosla, A. N., "Design of Dams on Permeable Foundations," Central Board of Irrigation, India, September 1936.

[2] "Design of Small Dams," Bureau of Reclamation, second edition, 1973.

Chapter VII

Temperature Control of Concrete

A. INTRODUCTION

7-1. *Purposes.*—Temperature control measures are employed in mass concrete dams to (1) facilitate construction of the structure, (2) minimize and/or control the size and spacing of cracks in the concrete, and (3) permit completion of the structure during the construction period. The measures and degree of temperature control to be employed are determined by studies of the structure, its method of construction, and its temperature environment.

Cracking in mass concrete structures is undesirable because it affects the watertightness, internal stresses, durability, and appearance of the structures. Cracking will occur when tensile stresses are developed which exceed the tensile strength of the concrete. These tensile stresses may occur because of imposed loads on the structure, but more often occur because of restraint against volumetric change. The largest volumetric change in mass concrete results from change in temperature. The cracking tendencies which occur as a result of temperature changes and temperature differentials can be reduced to acceptable levels, in most instances, by the use of appropriate design and construction procedures.

Temperature control measures which minimize volumetric changes make possible the use of larger construction blocks, thereby resulting in a more rapid and economical construction. One of these measures, post cooling, is also necessary if contraction joint grouting is to be accomplished. A gravity dam with longitudinal joints must have a monolithic section in an upstream-downstream direction. Therefore, provision for the construction of gravity dams with longitudinal contraction joints must include measures by which the concrete is cooled and contraction joints are closed by grouting before the reservoir loads are applied.

Complete temperature treatments, over and above the use of precooling measures and embedded pipe cooling systems, have been used in some structures. In these instances, reductions were made in the amount of cement used, low-heat cements were specified, and effective use was made of pozzolan to replace a part of the cement. Glen Canyon Dam, because of the size of the construction blocks and the relatively low grouting temperature, was constructed with a 50° F. maximum placing temperature, embedded cooling coils, a type II cement, and a mix containing 2 sacks of cement and 1 sack of pozzolan per cubic yard of concrete.

7-2. *Volumetric Changes.*—Mass concrete structures undergo volumetric changes which, because of the dimensions involved, are of concern to the designer. The changes in volume due to early-age temperature changes can be controlled within reasonable limits and incorporated into the design of the structure. The final state of temperature equilibrium depends upon site conditions, and little if any degree of control over the subsequent periodic volumetric changes can be effected.

The ideal condition would be simply to eliminate any temperature drop. This could be achieved by placing concrete at such a low temperature that the temperature rise due to hydration of the cement would be just sufficient to bring the concrete temperature up to its final stable state. Most measures for the prevention of temperature cracking, however, can only approach this ideal condition. The degree of success is related to site conditions, economics, and the stresses in the structure.

The volumetric changes of concern are those caused by the temperature drop from the peak temperature, occurring shortly after placement, to the final stable temperature of the structure. A degree of control over the peak temperature can be attained by limiting the placing temperature of the fresh concrete and by minimizing the temperature rise after placement. The placing temperature can be varied, within limits, by precooling measures which lower the temperatures of one or more of the ingredients of the mix before batching. The temperature rise in newly placed concrete can be restrained by use of embedded pipe cooling systems, placement in shallow lifts with delays between lifts, and the use of a concrete mix designed to limit the heat of hydration. These measures will reduce the peak temperature which otherwise would have been attained. Proportionately, this reduction in peak temperature will reduce the subsequent volumetric change and the accompanying crack-producing tendencies.

7-3. *Factors to be Considered.*—The methods and degree of temperature control should be related to the site conditions and the structure itself. Such factors as exposure conditions during and after construction, final stable temperature of the concrete mass, seasonal temperature variations, the size and type of structure, composition of the concrete, construction methods, and rate of construction should be studied and evaluated in order to select effective, yet economical, temperature control measures. The construction schedule and design requirements must also be studied to determine those procedures necessary to produce favorable temperature conditions during construction. Such factors as thickness of lifts, time interval between lifts, height differentials between blocks, and seasonal limitations on placing of concrete should be

evaluated. Study of the effect of these variables will permit the determination of the most favorable construction schedules consistent with the prevention of cracking from temperature stresses.

Some structures favor the use of a particular method of temperature control. Since open longitudinal contraction joints would prevent a block from carrying its load as a monolith, gravity dams with longitudinal joints must provide for contraction joint grouting of the longitudinal joints. This normally requires cooling by means of an embedded pipe cooling system and grouting of the joints before the reservoir load is applied. The gravity-type dam with no longitudinal contraction joints requires only that degree of temperature control necessary to prevent structural cracking circumferentially across the block as the block cools and approaches its final stable temperature. Precooling of aggregates and the use of low-heat cements, reduced cement content, and pozzolans are normally adopted as temperature control measures for gravity dams containing no longitudinal joints.

While longitudinal contraction joints must always be grouted, any decision to grout transverse contraction joints in straight gravity dams depends upon the magnitude of load transfer across the joint. Since this load transfer depends largely upon the height and axis profile shape of the dam, no specific criteria can be made for all straight gravity dams. If these transverse joints are to be grouted, an embedded pipe cooling system will normally be required.

7-4. *Design Data.*—The collection of design data should start at the inception of the project and should be continued through the construction period. Data primarily associated with the determination of temperature control measures include the ambient air temperatures at the site, river water temperatures, anticipated reservoir and tailwater temperatures, and the diffusivity of the concrete in the dam.

The estimate of air temperatures which will occur in the future at a given site is based on air temperatures which have occurred in the past, either at that location or one in the near vicinity. The U.S. Weather Service has collected climatological data at a great number of locations, and long-time records from one or more of these nearby locations may be selected and adjusted to the site. For this adjustment, an increase of 250 feet in elevation can be assumed to decrease the air temperature 1° F. Similarly, an increase of 1.4° in latitude can be assumed to decrease the temperature 1° F. River water temperatures and streamflow data can be obtained from various hydrometeorological and water supply reports and papers. A program for obtaining actual maximum and minimum daily air and river water temperatures at the site should be instituted as soon as possible to verify or adjust the data assumed for early studies. Representative wet- and dry-bulb temperatures should also be obtained throughout the year.

The best estimate of the future reservoir water temperatures would be one based on water temperatures recorded at nearby reservoirs of similar depth and with similar inflow and outflow conditions. The Bureau of Reclamation has obtained reservoir water temperatures over a period of several years in a number of reservoirs. From these data, maximum ranges of temperature for the operating conditions encountered were determined. When no data are available on nearby reservoirs, the next best estimate of the reservoir temperatures can be obtained by the principle of heat continuity. This method takes into consideration the quantity and temperature of the water entering and leaving the reservoir, and the heat transfer across the reservoir surface. These heat budget computations, though accurate in themselves, are based on estimates of evaporation, conduction, absorption and reflection of solar radiation, and reradiation—which in turn are related to cloud cover, air temperatures, wind velocities, and relative humidity. Because of these variables, any forecast of temperature conditions in a reservoir based on the principle of heat continuity can only be considered as an estimate.

The diffusivity of concrete, h^2, is an index of the facility with which concrete will undergo temperature change. Although desirable from the heat standpoint, it is not practicable to select aggregate, sand, and cement for a concrete on the basis of heat characteristics. The thermal properties of the concrete must therefore be accepted for what they are. The value of the diffusivity of concrete is expressed in square feet per hour, and can be determined from the relationship,

$$h^2 = \frac{K}{C\rho}$$

where:

K = conductivity in B.t.u. per foot per hour per ° F.,
C = specific heat in B.t.u. per pound per ° F., and
ρ = density in pounds per cubic foot.

Values of the diffusivity for a given concrete are determined from laboratory tests, although they must normally be estimated for early studies. As the thermal characteristics of the coarse aggregate largely govern the thermal characteristics of the concrete, the earliest of these estimates can be based upon the probable type of coarse aggregate to be used in the concrete. Table 7-1 gives the thermal properties of concretes in Bureau of Reclamation dams and representative values for several rock types.

7-5. *Cracking.*—Temperature cracking in mass concrete occurs as tensile stresses are developed when a temperature drop takes place in the concrete and some degree of restraint exists against this volumetric change. The stresses developed are related to the amount and rate of the temperature drop, the age of the concrete when the temperature drop takes place, and the elastic and inelastic properties of the particular concrete. The restraint may be external, such as the restraint exerted by the foundation of a structure; or it may be internal, such as the restraint exerted by a mass upon its surface. Tensile stresses also occur when a nonlinear temperature variation occurs across a section of the structure. Because of the inelastic properties of concrete, the stresses developed are related to the temperature history of the structure.

Table 7-1.–*Thermal properties of concrete for various dams.*

Dam	Density (saturated) lb./cu. ft.	Conductivity K B.t.u./ft.-hr.-°F.			Specific heat C B.t.u./lb.-°F.			Diffusivity h² ft.²/hr.		
		50°	70°	90°	50°	70°	90°	50°	70°	90°
East Canyon . . . . (predominately quartz and quartzite)	152.9	2.56	2.53	2.50	0.208	0.213	0.217	0.081	0.078	0.075
Glen Canyon	148.4	2.02	2.01	2.01	.211	.216	.222	.065	.063	.061
Seminoe	155.3	1.994	1.972	1.951	.204	.213	.222	.063	.060	.057
Norris	160.6	2.120	2.105	2.087	.234	.239	.247	.056	.055	.053
Wheeler	145.5	1.815	1.800	1.785	.223	.229	.236	.056	.054	.052
Flaming Gorge . . . . (limestone and sandstone)	150.4	1.78	1.77	1.76	.221	.226	.232	.054	.052	.050
Kortes mixes: 1 bbl. cement/cu. yd. and 0.0-percent air	157.6	1.736	1.724	1.711	.210	.215	.221	.052	.051	.049
0.85 bbl. cement/cu. yd. and 0.0-percent air	158.1	1.715	1.710	1.705	.209	.215	.220	.052	.050	.049
Hungry Horse	150.1	1.72	1.72	1.71	.217	.223	.229	.053	.051	.050
Hoover	156.0	1.699	1.688	1.677	.212	.216	.221	.051	.050	.049
Gibson	155.2	1.676	1.667	1.657	.218	.222	.229	.050	.048	.047
Canyon Ferry	151.3	1.63	1.62	1.61	.214	.218	.222	.050	.049	.048
Swift . . . . (limestone)	158.2	1.82	1.79	1.76	.237	.242	.246	.049	.047	.041
Altus	149.7	1.578	1.579	1.580	.225	.229	.234	.047	.046	.045
Monticello	153.1	1.57	1.56	1.55	.225	.230	.235	.046	.044	.043
Yellowtail	152.8	1.57	1.56	1.55	.219	.223	.227	.047	.046	.045
Angostura mixes: 0.9 bbl. cement/cu. yd. and 3.0-percent air	151.2	1.491	1.484	1.478	.221	.228	.234	.045	.043	.042
1.04 bbl. cement/cu. yd. and 0.0-percent air	152.6	1.571	1.554	1.537	.227	.234	.240	.045	.044	.042
Hiwassee	155.7	1.505	1.491	1.478	.218	.225	.233	.044	.042	.041
Parker	155.1	1.409	1.402	1.395	.213	.216	.221	.043	.042	.041
Owyhee	152.1	1.376	1.373	1.369	.208	.214	.222	.044	.042	.041
O'Shaughnessy	152.8	1.316	1.338	1.354	.217	.218	.223	.040	.040	.040
Friant mixes: Portland cement	153.6	1.312	1.312	1.312	.214	.214	.217	.040	.040	.039
20-percent pumicite	153.8	1.229	1.232	1.234	.216	.221	.227	.037	.036	.035
Shasta	157.0	1.299	1.309	1.319	.222	.229	.235	.037	.037	.036
Bartlett	156.3	1.293	1.291	1.289	.216	.222	.230	.038	.037	.036
Morris	156.9	1.290	1.291	1.293	.214	.216	.222	.039	.038	.037
Chickamauga	156.5	1.287	1.277	1.266	.225	.229	.233	.037	.036	.035
Morrow Point . . . . (andesite-basalt)	145.5	0.99	0.97	0.94	.212	.217	.222	.032	.031	.029
Grand Coulee	158.1	1.075	1.077	1.079	.219	.222	.227	.031	.031	.030
Ariel	146.2	0.842	0.884	0.915	.228	.235	.244	.025	.026	.026
Bull Run	159.1	0.835	0.847	0.860	.215	.225	.234	.024	.024	.023

Thermal Properties of Coarse Aggregate

	Density (saturated) lb./cu. ft.	Conductivity K B.t.u./ft.-hr.-°F.			Specific heat C B.t.u./lb.-°F.			Diffusivity h² ft.²/hr.		
Quartzite	151.7	2.052	2.040	2.028	.209	.217	.226	.065	.062	.059
Dolomite	156.2	1.948	1.925	1.903	.225	.231	.238	.055	.053	.051
Limestone	152.8	1.871	1.842	1.815	.221	.224	.230	.055	.054	.052
Granite	150.9	1.515	1.511	1.588	.220	.220	.224	.046	.045	.045
Basalt	157.5	1.213	1.212	1.211	.226	.226	.230	.034	.034	.033
Rhyolite	146.3	1.197	1.203	1.207	.220	.226	.232	.037	.036	.036

The most common cracking in mass concrete occurs when large blocks of concrete are placed on the foundation in the fall of the year, after which concreting is stopped for the winter. Under these conditions, foundation restraint is high, large drops in temperature are possible because concrete placing temperatures and peak temperatures are relatively high, and concrete temperatures will be dropping quite rapidly due to exposure conditions. For blocks not larger than 50 by 50 feet, cracking under these conditions has no particular pattern. In larger blocks, and where the length-to-width ratio is over 2, cracking under the above conditions often occurs at or near the third points of the longer side. Generally, if the blocks are not placed more than 10 or 15 feet off the foundation, cracking will start at the exposed top edge of the block and progress into the block and down the side to within a few feet of the foundation. Such cracks vary from extremely small or hairline surface cracks which penetrate only a few inches into the mass, to irregular structural cracks of varying width which completely cross the construction block. The maximum crack width is at the top edge and normally will be from 1/32 to 1/64 inch in width.

Similar cracking across the full width of a block can occur during the colder months of the year in a high block which has been constructed well off the foundation and which is 25 to 50 feet higher than the adjacent blocks. In this instance, the upper part of the block will cool at a relatively fast rate while that part of the block below the elevation of the adjacent blocks may remain at the same temperature or may possibly rise in temperature depending upon its age.

Surface cracking which occurs because of internal restraint seldom follows any particular pattern. The most general cracking is along the horizontal construction joints where the tensile strength is low. Such cracking normally occurs when wood or insulated steel forms are used and then removed when exposure temperatures are low. Upon removal of the forms, the surface is subjected to a thermal drop which sets up a severe temperature gradient between the surface and the interior. Practically all of these cracks are from hairline width to 1/64 inch in thickness. Aside from the horizontal construction joints, most other surface cracking is evidenced by vertical or near-vertical cracks associated with surface irregularities such as openings, reentrant corners, or construction discontinuities which occurred during placement. Most of these cracks do not progress beyond the one placement lift, but those that do often are the beginning of the cracks described above.

B. METHODS OF TEMPERATURE CONTROL

7-6. *Precooling.*—One of the most effective and positive temperature control measures is that which reduces the placing temperature of the concrete. Methods of reducing the placing temperature which would otherwise be obtained at a site can be varied from restricting concrete placement during the hotter part of the day or the hotter months of the year, to a full treatment of refrigerating the various parts of the concrete mix to obtain a predetermined, maximum concrete placing temperature.

The method or combination of methods used to reduce concrete placing temperatures will vary with the degree of cooling required and the contractor's equipment and previous experience. For some structures, sprinkling and shading of the coarse aggregate piles may be the only precooling measures required. The benefits of sprinkling depend largely on the temperature of the applied water and on the contractor's operations at the stockpile. A secondary benefit, evaporative cooling, can also be obtained but is restricted to areas with a low relative humidity. Insulating and/or painting the surfaces of the batching plant, water lines, etc., with reflective paint can also be beneficial.

Mixing water can be cooled to varying degrees, the more common temperatures being from 32° to 40° F. Adding slush or crushed ice to the mix is an effective method of cooling because it takes advantage of the latent heat of fusion of ice. The addition of large amounts of ice, however, may not be very practical in some instances. For example, if the coarse aggregate and sand both contain appreciable amounts of free water, the amount of water to be added to the mix may be so small that replacement of part of the added water with ice would not be appreciable.

Cooling of the coarse aggregates to about 35° F. can be accomplished in several ways. One method is to chill the aggregate in large tanks of refrigerated water for a given period of time. Relatively effective cooling of coarse aggregate can also be attained by forcing refrigerated air through the aggregate while the aggregate is draining in stockpiles, while it is on a conveyor belt and while it is passing through the bins of the batching plant. Spraying with cold water will also cool the aggregate. Sand may be cooled by passing it through vertical tubular heat exchangers. Cold air jets directed on the sand as it is transported on conveyor belts can also be used. Immersion of sand in cold water is not practical because of the difficulty in removing the free water from the sand after cooling.

Cooling of the cement is seldom practicable. Bulk cement in the quantities used for dams is almost always obtained at relatively high temperatures, generally from 140° to 180° F. Seldom will it cool naturally and lose a sizable portion of the excess heat before it is used.

Use of the above treatments has resulted in concrete placing temperatures of 50° F. in a number of instances. Concrete placing temperatures as low as 45° F. have been attained, but these can usually be achieved only at a considerable increase in cost. The temperature of the concrete at the mixing plant should be 3° to 4° F. lower than the desired placing temperature. This will compensate for the heat developed and absorbed by the concrete during mixing and transporting.

7-7. *Postcooling*.—Postcooling of mass concrete in gravity dams is used primarily to prevent cracking during construction. It is also required where longitudinal contraction joints are used and where grouting of transverse contraction joints is required, in order to reduce the temperature of the concrete to the desired value prior to grouting. The layout of embedded cooling systems used in postcooling mass concrete is described in section 7-20.

Postcooling is an effective means of crack control. Artificially cooling mass concrete by circulating cold water through embedded cooling coils on the top of each construction lift will materially reduce the peak temperature of the concrete below that which would otherwise be attained. However, these embedded coils will not actually prevent a temperature rise in the concrete, because of the high rate of heat development during the first few days after placement and the relatively low conductivity of the concrete. The use of an embedded pipe system affords flexibility in cooling through operation of the system. Any desired degree of cooling may be accomplished at any place at any time. This can minimize the formation of large temperature gradients from the warm interior to the colder exterior. The formation of such gradients in the fall and winter is particularly conducive to cracking.

7-8. *Amount and Type of Cement*.—Mass concrete structures require lesser amounts of cement than the ordinary size concrete structures because of a lower strength requirement. Because of their dimensions, however, less heat is lost to the surfaces and a greater maximum temperature is attained. Since the heat generated within the concrete is directly proportional to the amount of cement used per cubic yard, the mix selected should be that one which will provide the required strength and durability with the lowest cement content. The cement content in mass concrete structures has varied in the past from 4 to 6 sacks of cement per cubic yard, but present-day structures contain as low as 2 sacks of cement plus other cementing materials.

The heat-producing characteristics of cement play an important role in the amount of temperature rise. Although cements are classified by type as type I, type II, etc., the heat generation within each type may vary widely because of the chemical compounds in the cement. Types II and IV were developed for use in mass concrete construction. Type II cement is commonly referred to as modified cement, and is used where a relatively low heat generation is desirable. Type IV cement is a low-heat cement characterized by its low rate of heat generation during early age.

Specifications for portland cement generally

do not state within what limits the heat of hydration shall be for each type of cement. They do, however, place maximum percentages on certain chemical compounds in the cement. They further permit the purchaser to specifically request maximum heat of hydration requirements of 70 or 80 calories per gram at ages 7 and 28 days, respectively, for type II cement; and 60 or 70 calories per gram at ages 7 and 28 days, respectively, for type IV cement.

In most instances, type II cement will produce concrete temperatures which are acceptable. In the smaller structures, type I cement will often be entirely satisfactory. Other factors being equal, type II cement should be selected because of its better resistance to sulfate attack, better workability, and lower permeability. Type IV cement is now used only where an extreme degree of temperature control is required. For example, it would be beneficial near the base of long blocks where a high degree of restraint exists. Concrete made with type IV cement requires more curing than concrete made with other types of cement, and extra care is required at early ages to prevent damage to the concrete from freezing during cold weather. Often, the run-of-the-mill cement from a plant will meet the requirements of a type II cement, and the benefits of using this type of cement can be obtained at little or no extra cost. Type IV cement, because of its special composition, is obtained at premium prices.

7-9. *Use of Pozzolans.*—Pozzolans are used in concrete for several reasons, one of which is to reduce the peak temperature due to heat of hydration from the cementing materials in the mix. This is possible because pozzolans develop heat of hydration at a much lower rate than do portland cements. Pozzolans can also be used as a replacement for part of the portland cement to improve workability, effect economy, and obtain a better quality concrete. The more common pozzolans used in mass concrete include calcined clays, diatomaceous earth, volcanic tuffs and pumicites, and fly ash. The actual type of pozzolan to be used is normally determined by cost and availability.

7-10. *Miscellaneous Measures.*—(a) *Shallow*

Construction Lifts.—Shallow construction or placement lifts can result in a greater percentage of the total heat generated in the lift being lost to the surface. Such a temperature benefit exists only during periods of time when the exposure temperatures are lower than the concrete temperature as described in section 7-22. Unless the site conditions are such that a sizable benefit can be obtained, shallow placement lifts are generally limited to placements over construction joints which have experienced prolonged exposure periods, or over foundation irregularities where they are helpful in the prevention of settlement cracks.

(b) *Water Curing.*—Water curing on the top and sides of each construction lift will reduce the temperature rise in concrete near the surfaces as described in section 7-29. Proper application of water to the surfaces will cause the surface temperature to approximate the curing water temperature instead of the prevailing air temperatures. In areas of low humidity, the effect of evaporative cooling may result in a slightly lower surface temperature than the temperature of the curing water.

(c) *Retarding Agents.*—Retarding agents added to the concrete mix will provide a temperature benefit when used in conjunction with pipe cooling. The retarding agents reduce the early rate of heat generation of the cement, so that the total temperature rise during the first 2 or 3 days will be 2^0 or perhaps 3^0 F. lower than for a similar mix without retarder. The actual benefit varies with the type and amount of retarder used. The percentage of retarder by weight of cement is generally about one-fourth to one-third of 1 percent. Percentages higher than this may give added temperature benefit but can create construction problems such as delay in form removal, increased embedment of form ties required, etc.

(d) *Surface Treatments.*—If the near-surface concrete of a mass concrete structure can be made to set at a relatively low temperature and can be maintained at this temperature during the early age of the concrete, say, for the first 2 weeks, cracking at the surface can be minimized. Under this condition, tensions at the surface are reduced or the surface may even be put into compression when the interior mass of the concrete subsequently drops in temperature. Such surface cooling can be accomplished by circulating water in closely spaced embedded cooling-pipe coils placed adjacent to and parallel with the exposed surfaces, by use of cold water sprays on noninsulated steel forms and on the exposed concrete surfaces, or by use of special refrigerated forms.

(e) *Rate of Temperature Drop.*— Temperature stresses and the resultant tendency to crack in mass concrete can be minimized by controlling the rate of temperature drop and the time when this drop occurs. In thick sections with no artificial cooling, the temperature drop will normally be slow enough as to present no problem. In thin sections with artificial cooling, however, the temperature can drop quite rapidly and the drop may have to be controlled. This can be accomplished by reducing the amount of cooling water circulated through the coils or by raising the cooling water temperature. The operation of the cooling systems, and the layout of the header systems to supply cooling water to the individual cooling coils, should be such that each coil can be operated independently. No-cooling periods should also be utilized where necessary. In thin sections where no artificial cooling is employed, the temperature drop during periods of cold weather can be controlled by the use of insulated forms and insulation placed on exposed surfaces. Such measures not only reduce the rate of change, but also reduce the temperature gradients near the surface resulting in a definite reduction in cracking.

C. TEMPERATURE STUDIES

7-11. *General Scope of Studies.*—The measures required to obtain a monolithic structure and the measures necessary to reduce cracking tendencies to a minimum are determined by temperature control studies. In addition to the climatic conditions at the site, the design requirements of the structure and the probable construction procedures and schedules require study to determine the methods and degree of temperature control for the structure.

Early design studies and specification requirements are based on existing data and on a possible construction schedule. The ambient temperatures and probable concrete temperatures are then related to the dimensions of the structure, the conditions arising during construction, and the desired design stresses. As a result of these studies, a maximum concrete placing temperature may be determined, measures taken to limit the initial temperature rise within the concrete, and protective measures planned to alleviate cracking conditions arising during the construction period. Actual exposure conditions, water temperatures, and construction progress may vary widely from the assumed conditions, and adjustments should be made during the construction period to obtain the best structure possible consistent with economy and good construction practices.

The following discussions cover the more common temperature investigations and studies. In all of these studies, certain conditions must be assumed. Since any heat flow computation is dependent on the validity of the assumed exposure conditions and concrete properties, experience and good judgment are essential.

7-12. *Range of Concrete Temperatures.*— The ranges or amplitudes of the mean concrete temperature at various elevations of a gravity dam are used in several studies of stresses within the dam. This range of mean concrete temperature is determined from the air and water temperatures at the site, as modified by the effects of solar radiation. For preliminary studies, the range of mean concrete temperature can be obtained in a short computation by applying the air and water exposure temperatures as sinusoidal waves with applicable periods of 1 day, 1 week or 2 weeks depending upon the severity of the weather to be used for the design, and 1 year. Solar radiation is then added to obtain the final range of mean concrete temperature.

For average (mean) weather conditions, the ambient air temperatures are obtained from a plotting of the mean monthly air temperatures on a year scale. For usual and extreme weather conditions, the above ambient air temperatures are adjusted for a 7-day period and a 14-day period, respectively, at the high and low points of the annual curve. The amount of the adjustment for these weather conditions is described in subsection (a) below.

The thickness of section for these studies is measured along lines normal to the exposed surfaces, the intersection of the normals being equidistant from the two faces.

(a) *Ambient Air Temperatures.*—When computing the range of mean concrete temperature, mean daily, mean monthly, and mean annual air temperatures are used. The theory applies the daily and annual air temperatures as sinusoidal variations of temperature, even though the cycles are not true sine waves. The annual and daily amplitudes are assumed to be the same for all weather conditions.

To account for the maximum and minimum recorded air temperatures, a third and somewhat arbitrary temperature cycle is assumed. This temperature variation is associated with the movements of barometric pressures and storms across the country. Plots throughout the western part of the United States show from one to two cycles per month. Arbitrarily, this third temperature variation is assumed as a sine wave with either a 7-day or 14-day period for usual weather conditions and extreme weather conditions, respectively. For extreme weather conditions, the amplitudes of the arbitrary cycle are assigned numerical values which, when added to the amplitudes of the daily and annual cycles, will account for the actual maximum and minimum recorded air temperatures at the site. For usual weather conditions, these amplitudes are assigned values which account for temperatures halfway

between the mean monthly maximum (minimum) and the maximum (minimum) recorded. When computing the mean concrete temperature condition, no third cycle is used.

(b) *Reservoir Water Temperatures.*—The reservoir water temperatures used in determining the range of mean concrete temperature for a proposed dam are those temperatures which will occur after the reservoir is in operation. These reservoir water temperatures vary with depth, and for all practical purposes can be considered to have only an annual cycle. For preliminary studies, the range of mean concrete temperature with full reservoir is the normal condition. For final designs, stage construction should be taken into consideration and the design reservoir operation used. When the reservoir is to be filled or partially filled before concrete temperatures have reached their final stage of temperature equilibrium, further studies are needed for the particular condition.

(c) *Solar Radiation Effect.*—The downstream face of a dam, and the upstream face when not covered by water, receives an appreciable amount of radiant heat from the sun. This has the effect of warming the concrete surface above the surrounding air temperature. The amount of this temperature rise above the air temperature was recorded on the faces of several dams in the western portion of the United States. These data were then correlated with theoretical studies which took into consideration varying slopes, orientation of the exposed faces, and latitudes. The results of these studies are presented in reference [1].[1] These theoretical temperature rises due to solar radiation should be corrected by a terrain factor obtained from an east-west profile of the site terrain. This is required because the theoretical computations assumed a horizontal plane at the base of the structure, and the effect of the surrounding terrain is to block out some hours of sunshine. This terrain factor will vary with elevation and from abutment to abutment.

---

[1]Numbers in brackets refer to items in bibliography, sec. 7-31.

(d) *Amplitudes of Concrete Temperatures.*—
The range or amplitude of concrete
temperatures is determined by applying the
above-described external sinusoidal air and
water temperatures to the edges of a
theoretical flat slab, the width of the slab being
equal to the thickness of the dam at the
elevation under consideration. The problem is
idealized by assuming that no heat flows in a
direction normal to the slab. The law of
superposition is used in that the final
amplitude in the concrete slab is the sum of the
amplitudes obtained from the different
sinusoidal variations.

To apply the theoretical heat flow in a
practical manner, unit values are assumed for
the several variables and a curve is drawn to
show the ratio of the variation of the mean
temperature of the slab to the variation of the
external temperature. Figure 7-1 shows the
relationship thus derived for temperature
variations in flat slabs exposed to sinusoidal
variations for $h^2 = 1.00$ square foot per day, a
period of 1 day, and a thickness of slab of l_1. A
correlation equation is given to take into
account the actual thickness of dam, diffusivity
constant, and period of time. The
computations are shown in figures 7-2 and
7-3.[2] For the actual thickness of dam, l_2, a
value of l_1 is obtained from the correlation
equation for each of the air temperature cycles.
For each value of l_1, a ratio of the variation of
mean concrete temperature to the variation of
external temperature is obtained. The sums of
the products of these ratios and their respective
amplitudes are algebraically added to and
subtracted from the mean annual air
temperature to obtain mean concrete
temperatures for the condition of air on both
faces. Mean concrete temperatures are then
obtained in the same manner for a fictitious
condition of water on both faces, and the two
conditions are simply averaged together to
obtain the condition of air on one face and
water on the other. Solar radiation values are
then added to obtain the final range of mean
concrete temperatures.

7-13. *Temperature Gradients.*—Temperature
distributions in a mass where boundary
conditions vary with time are easily determined
by the Schmidt method. (See references [1],
[2], [3], [4].) This method is generally used
for temperature studies of mass concrete
structures when the temperature gradient or
distribution across the section is desired. The
depth of freezing, and temperature distribution
after placement are typical of the solutions
which can be obtained by this step-by-step
method. Different exposure temperatures on
the two faces of the theoretical slab and the
autogenous heat of hydration are easily taken
into consideration.

An early objection to the Schmidt method
of temperature computation was the time
required to complete the step-by-step
computation. This has been overcome by the
use of electronic data processing machines
which save many man-hours of work. Programs
have been developed which will take into
consideration any thickness of section, varying
exposures on the two faces of the slab, variable
initial temperatures, a varying heat of
hydration with respect to time, and increasing
the thickness of slab at regular intervals as
would occur when lifts of concrete are placed
on previously placed lifts.

A second method of temperature
computation in mass concrete which is
particularly adaptable to thick walls and
placement lifts near the rock foundation was
devised by R. W. Carlson. This method is
described in reference [5]. It, like the Schmidt
method, is essentially a step-by-step integration
which can be simplified by selection of certain
variables. Conditions such as initial
temperature distributions, diffusivity, and
adiabatic temperature rise must be known or
assumed. Carlson's method can also be
modified to take into account the flow of heat
between different materials. This would be the
case where insulated or partially insulated
forms are used, or where concrete lifts are
placed on rock foundations.

The variation in temperature in a
semi-infinite solid at any particular point can
also be estimated from figure 7-4. This
illustration gives the ratio of the temperature

---

[2]These and several other figures and tables in this chapter
were reprinted from Bureau of Reclamation Engineering
Monograph No. 34, listed as reference [1] in the bibliography,
sec. 7-31.

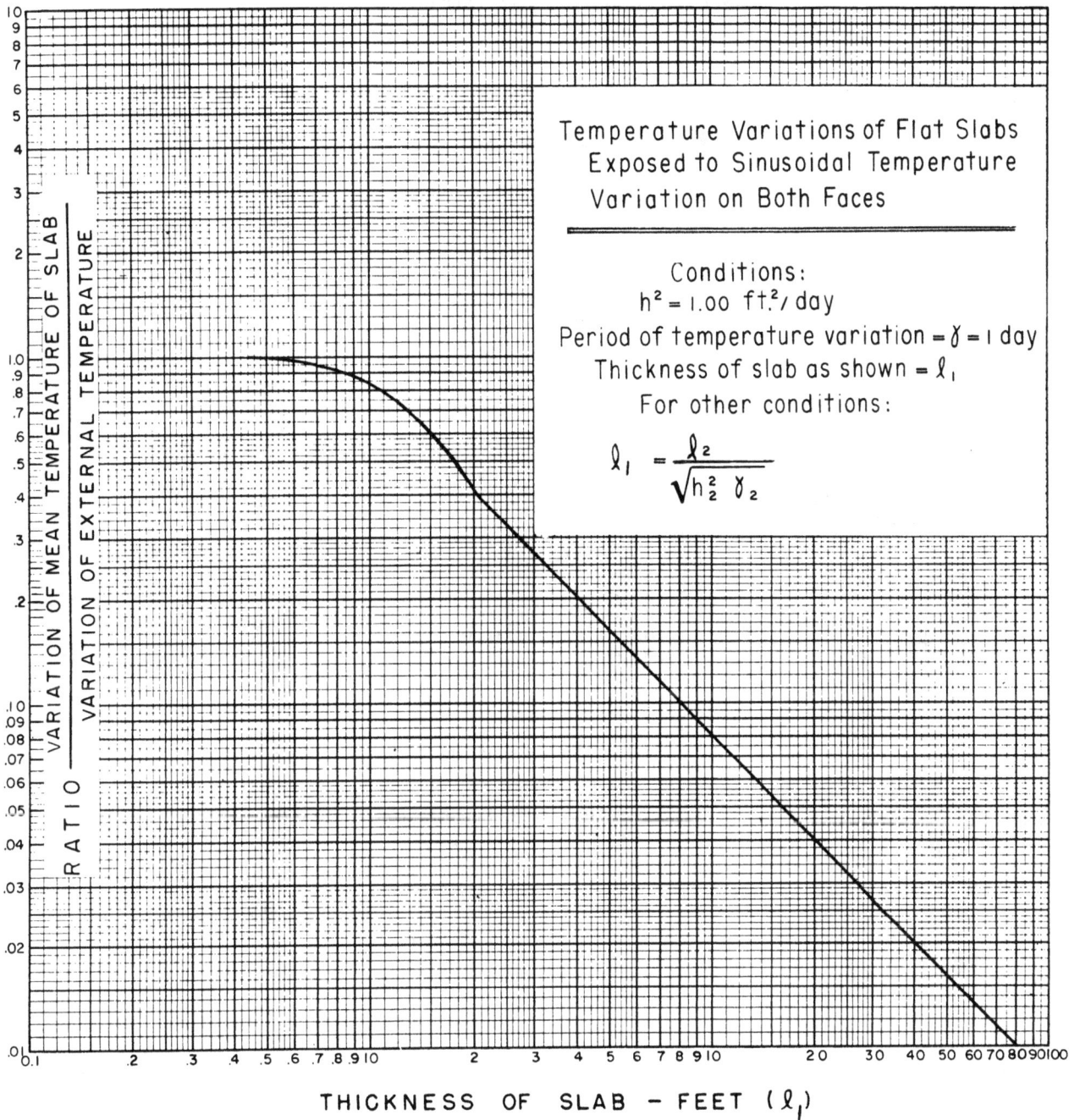

Figure 7-1. Temperature variations of flat slabs exposed to sinusoidal temperature variations on both faces.—288-D-3008

range in the concrete at the particular point, to the temperature range at the surface for daily, 15-day, and annual cycles of temperature.

Stresses due to temperature gradients may

be of concern not only during the construction period but during the life of the structure. Stresses across a section due to temperature gradients can be obtained from the expression

RANGE OF MEAN CONCRETE TEMPERATURES

Hungry Horse DAM

(Effect of solar radiation not included)

For yearly change $\quad \ell_1 = \dfrac{\ell_2}{\sqrt{h_1^2\, r_2}} = \dfrac{\ell_2}{\sqrt{0.053 \times 8760}} = 0.0464\,\ell_2$

For 365-hr. change $\quad \ell_1 = \dfrac{\ell_2}{\sqrt{0.053 \times 365}} = 0.227\,\ell_2$

For daily change $\quad \ell_1 = \dfrac{\ell_2}{\sqrt{0.053 \times 24}} = 0.887\,\ell_2$

Remarks: Air temperatures taken from 34-year record at Columbia Falls, Montana. $h^2 = 0.053$ from laboratory data

Avg. Annual Air Temperature 43.2°F.	Extreme Weather Conditions		Usual Weather Conditions	
	Above Mean	Below Mean	Above Mean	Below Mean
Yearly	22.0	21.4	22.0	21.4
365-hr.	30.6	49.6	20.4	25.2
Daily	7.2	7.2	7.2	7.2

Elevation	Thickness of Dam ℓ_2	Due to Yearly Range ℓ_1	Ratio from Curve	Due to 365-hr. Range ℓ_1	Ratio from Curve	Due to Daily Range ℓ_1	Ratio from Curve	Ext. Wea. Cond. Amp. Above Mean	Ext. Wea. Cond. Amp. Below Mean	Ext. Temp. Max.	Ext. Temp. Min.	Usual Wea. Cond. Amp. Above Mean	Usual Wea. Cond. Amp. Below Mean	Usual Temp. Max.	Usual Temp. Min.	Water Max.	Water Min.	Water Avg.	Water Amp.	Mean Conc. Amp.	Mean Conc. Max.	Mean Conc. Min.	Ext. Wea. Max.	Ext. Wea. Min.	Usual Wea. Max.	Usual Wea. Min.
3565	40	1.86	0.460	9.08	0.087	35.48	0.022	12.9	14.3	56.1	28.9	12.1	12.2	55.3	31.0	69.0	32.0	50.5	18.5	8.5	59.0	42.0	57.5	35.5	57.2	36.5
3550	39	1.81	.480	8.85	.089	34.59	.023	13.4	14.9	56.6	28.3	12.5	12.7	55.7	30.5	69.0	33.0	51.0	18.0	8.6	59.6	42.4	58.1	35.4	57.6	36.4
3500	56	2.60	.310	12.71	.063	49.67	.016	8.9	9.9	52.1	33.3	8.2	8.3	51.4	34.9	54.5	36.8	45.6	8.8	2.7	48.3	42.9	50.2	38.1	49.9	38.9
3450	81	3.76	.215	18.39	.043	71.85	.011	6.1	6.8	49.3	36.4	5.7	5.8	48.9	37.4	47.5	38.5	43.0	4.5	1.2	44.2	41.8	46.7	39.1	46.6	39.6
3400	111	5.15	.156	25.20	.032	—	—	4.4	4.9	47.6	38.3	4.1	4.1	47.3	39.1	45.0	39.0	42.0	3.0	0.5	42.5	41.5	45.0	39.7	44.9	40.3
3350	141	6.54	.122	32.01	.025			3.4	3.9	46.6	39.3	3.2	3.2	46.4	40.0	43.0	39.0	41.0	2.0	0.2	41.2	40.8	43.9	40.0	43.8	40.4
3300	171	7.93	.100	38.82	.021			2.8	3.2	46.0	40.0	2.6	2.7	45.8	40.5	43.0	39.0	41.0	2.0	0.2	41.2	40.8	43.6	40.4	43.5	40.6
3250	201	9.33	.085	45.63	.018			2.4	2.7	45.6	40.5	2.2	2.3	45.4	40.9	43.0	39.0	41.0	2.0	0.2	41.2	40.8	43.4	40.6	43.3	40.8
3200	231	10.72	.075	52.44	.015			2.1	2.3	45.3	40.9	2.0	2.0	45.2	41.2	43.0	39.0	41.0	2.0	0.2	41.2	40.8	43.2	40.8	43.2	41.0
3150	274	12.71	.063	62.20	.013			1.8	2.0	45.0	41.2	1.7	1.7	44.9	41.5	43.0	39.0	41.0	2.0	0.1	41.1	40.9	43.0	41.0	43.0	41.2
3100	292	13.55	.060	66.28	.012			1.7	1.9	44.9	41.3	1.6	1.6	44.8	41.6	43.0	39.0	41.0	2.0	0.1	41.1	40.9	43.0	41.1	42.9	41.3
3067	366	16.98	.047	83.08	—			1.0	1.0	44.2	42.2	1.0	1.0	44.2	42.2	43.0	39.0	41.0	2.0	0.1	41.1	40.9	42.7	41.5	42.7	41.5

NOTES: ℓ_2 = Thickness of dam, ft. Curve referred to is "Temperature Variations of Flat Slabs Exposed to Sinusoidal Temperature Variations on Both Faces."

Figure 7-2. Computation form, sheet 1 of 2—range of mean concrete temperatures.—288-D-3009

$$\sigma_y = \frac{eE}{b^3(1-\mu)}\left[\, b^2 \int_0^b T(x)\,dx + 3(2x-b)\int_0^b (2x-b)T(x)\,dx - b^3\,T(x)\,\right]$$

where:

e = thermal coefficient of expansion,

E = modulus of elasticity,

μ = Poisson's ratio, and

b = thickness of section with a temperature distribution $T(x)$.

Where the temperature variation, $T(x)$, cannot be expressed analytically, the indicated integrations can be performed numerically by the use of Simpson's rule. For example, using b = 30 feet, $e = 6.0 \times 10^{-6}$, $E = 2,500,000$ pounds per square inch, $\mu = 0.20$, and an assumed $T(x)$, the stresses would be computed as shown in table 7-2.

The above expression for stress is not valid in all essentials for those temperature gradients which occur during the first few days after placement, because the extreme creep characteristics of the concrete during this age result in a highly indeterminate condition of stress. The expression is also not valid where external restraints occur such as near the foundation of a block or structure.

7-14. *Temperature Rise.*—Newly placed concrete undergoes a rise in temperature due to the exothermic reaction of the cementing materials in the concrete. Early temperature rise studies may be based on past experience records with the type of cement to be used.

RANGE OF MEAN CONCRETE TEMPERATURES
Hungry Horse DAM

(Effect of Solar Radiation included)

Latitude 48°N

Remarks: _____
_____
_____
_____
_____
_____
_____

Elevation	Thickness of Dam	Effects of Solar Radiation				MEAN CONCRETE TEMPERATURES							
						Exposed to air on both faces				Air on D.S. Face Water on U.S. Face			
						Ext. Wea. Conditions		Usual Wea. Conditions		Ext. Wea. Conditions		Usual Wea. Conditions	
		U.S.	D.S.	Avg.	½DS	Max.	Min.	Max.	Min.	Max	Min	Max	Min.
3565	40	6.1	4.3	5.2	2.2	61.3	34.1	60.5	36.2	59.7	37.6	59.4	38.7
3550	39	6.0	5.0	5.5	2.5	62.1	33.8	61.2	36.0	60.6	37.9	60.1	38.9
3500	56	5.8	5.6	5.7	2.8	57.8	39.0	57.1	40.6	53.0	40.9	52.7	41.7
3450	81	5.6	5.6	5.6	2.8	54.9	42.0	54.5	43.0	49.5	41.9	49.4	42.4
3400	111	5.4	5.4	5.4	2.7	53.0	43.7	52.7	44.5	47.7	42.4	47.6	43.0
3350	141	5.2	5.5	5.3	2.7	51.9	44.6	51.7	45.3	46.6	42.7	46.5	43.1
3300	171	5.0	5.4	5.2	2.7	51.2	45.2	51.0	45.7	46.3	43.1	46.1	43.3
3250	201	5.0	6.0	5.5	3.0	51.1	46.0	50.9	46.4	46.4	43.6	46.3	43.8
3200	231	4.8	4.7	4.8	2.3	50.1	45.7	50.0	46.0	45.5	43.1	45.5	43.3
3150	274	4.6	4.7	4.6	2.3	49.6	45.8	49.5	46.1	45.3	43.3	45.3	43.5
3100	292	4.4	4.7	4.5	2.3	49.4	45.8	49.3	46.1	45.3	43.4	45.2	43.6

Elevation	Terrain Factor %	SOLAR RADIATION VALUES												Average Temp. Rise	
		131° Point 1 49°				106° Point 2 74°				81° Point 3 99°					
		Upstream		Downstream		Upstream		Downstream		Upstream		Downstream			
		Normal angle		Normal angle		Normal angle		Normal angle		Normal angle		Normal angle			
		Slope	Temp. Rise 100% / Actual	Slope	Temp. Rise 100% / Actual	Slope	Temp. Rise 100% / Actual	Slope	Temp. Rise 100% / Actual	Slope	Temp. Rise 100% / Actual	Slope	Temp. Rise 100% / Actual	U.S.	D.S.
3565	100	0	7.3 / 7.3	0	2.7 / 2.7	0	6.3 / 6.3	0	4.3 / 4.3	0	4.8 / 4.8	0	6.0 / 6.0	6.1	4.3
3550	97	0	7.3 / 7.1	0.25	3.4 / 3.3	0	6.3 / 6.1	0.25	5.1 / 4.9	0	4.8 / 4.7	0.25	7.1 / 6.9	6.0	5.0
3500	94	0	7.3 / 6.9	.5	4.0 / 3.8	0	6.3 / 5.9	.5	6.0 / 5.6	0	4.8 / 4.5	.5	7.9 / 7.4	5.8	5.6
3450	91	0	7.3 / 6.6	.6	4.2 / 3.8	0	6.3 / 5.7	.6	6.2 / 5.6	0	4.8 / 4.4	.6	8.1 / 7.4	5.6	5.6
3400	88	0	7.3 / 6.4	.6	4.2 / 3.7	0	6.3 / 5.6	.6	6.2 / 5.5	0	4.8 / 4.2	.6	8.1 / 7.1	5.4	5.4
3350	85	0	7.3 / 6.2	.87	4.9 / 4.2	0	6.3 / 5.4	.6	6.2 / 5.3	0	4.8 / 4.1	.6	8.1 / 6.9	5.2	5.5
3300	82	0	7.3 / 6.0	.87	4.9 / 4.0	0	6.3 / 5.2	.6	6.2 / 5.1	0	4.8 / 3.9	.82	8.6 / 7.1	5.0	5.4
3250	79					0	6.3 / 5.0	.6	6.2 / 4.9	0	4.8 / 3.8	.97	8.8 / 7.0	5.0	6.0
3200	76					0	6.3 / 4.8	.6	6.2 / 4.7					4.8	4.7
3150	73					0	6.3 / 4.6	.7	6.5 / 4.7					4.6	4.7
3100	70					0	6.3 / 4.4	.8	6.7 / 4.7					4.4	4.7

Figure 7-3. Computation form, sheet 2 of 2—range of mean concrete temperatures.—288-D-3010

Figure 7-5 shows typical temperature rise curves for the various types of cement. The temperature rise curves are based on 1 barrel (4 sacks) of cement per cubic yard of concrete, a diffusivity of 0.050 square foot per hour, and no embedded pipe cooling. These curves should be used only for preliminary studies because there are wide variations of heat generation within each type of cement and of diffusivity in concrete. (See reference [6].) Where less than 4 sacks of cement per cubic yard is to be used, the temperature rise can be estimated by direct proportion since the heat generation is directly proportional to the amount of cement.

As with cements, the heat-development characteristics of pozzolans vary widely. When a pozzolan is to be used to replace a part of the cement, the heat of hydration of the pozzolan, for early studies, can be assumed to be about 50 percent of that developed by an equal amount of cement. For final temperature control studies, the heat generation for a particular concrete mix should be obtained by laboratory tests using the actual cement, pozzolan, concrete mix proportions, and mass-cure temperature cycle for the concrete to be placed in the structure.

The above heat of hydration relates to the adiabatic temperature rise in the concrete. Because the surfaces of a structure are exposed

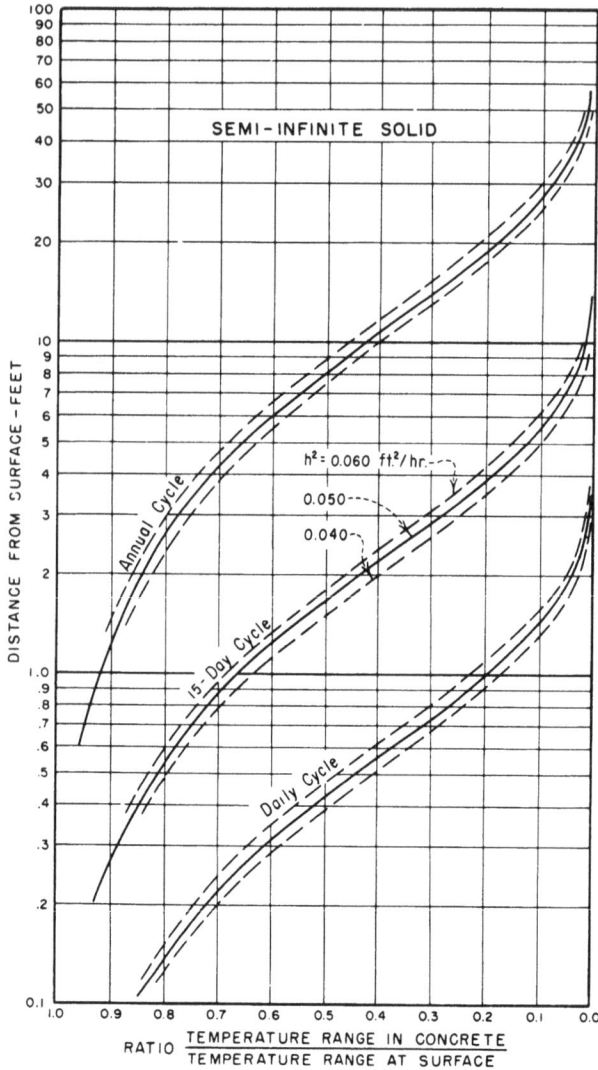

Figure 7-4. Temperature variations with depth in semi-infinite solid.—288-D-3011

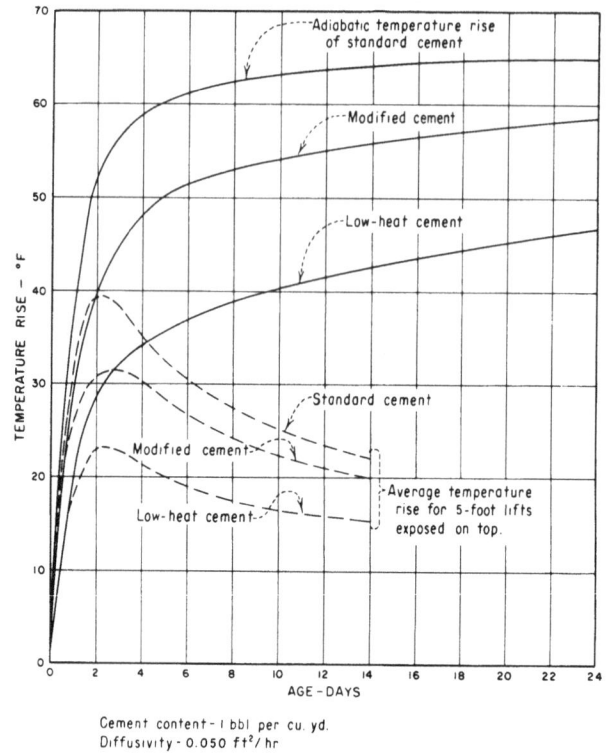

Cement content - 1 bbl per cu. yd.
Diffusivity - 0.050 ft²/ hr

Figure 7-5. Temperature rise in mass concrete for various types of cement.—288-D-3013

or in contact with inert or near-inert bodies, a flow of heat will take place through these surfaces and the actual temperature rise in the concrete will be affected accordingly. The loss or gain of heat to the surface due to exposure conditions, and the loss or gain of heat from an underlying lift or to the foundation are illustrated in reference [7]. Schmidt's method or Carlson's method can also be used to determine the actual temperature rise.

Several difficulties are encountered in the conditions given in reference [7]. For example, the theoretical equation for the adiabatic temperature rise is given as $T = T_o (1 - e^{-mt})$,

and T_o and m are selected to make the theoretical curve fit the laboratory data. Any variance between the theoretical and actual curves will result in some error in the theoretical heat loss in the heat-generating lift. The loss from the inert lift does not take into consideration a varying surface temperature, which also introduces an error. A third error may be introduced when a new lift is placed on an older lift which is still generating heat. Depending upon the age of the older lift, the heat generated may still be enough to be considered.

7-15. *Artificial Cooling.*—The design of an artificial cooling system requires a study of each structure, its environment, and the maximum temperatures which are acceptable from the standpoint of crack control. The temperature effects of various heights of placement lifts and such layout variables as size, spacing, and length of embedded coils should be investigated. Variables associated with the operation of the cooling systems, such as rate of water circulation and the

Table 7-2.—*Computation of temperature stress.*

x ft.	$T(x)$ °F.	$(2x-30)\,T(x)$ ft.–°F.		σ_y lb./ft.²	σ_y lb./in.²
0	0.0	0	For the given conditions:	10,584	74
3	8.3	−199		4,125	29
6	15.8	−284	$\dfrac{eE}{b^3(1-\mu)}=0.1$	−174	−1
9	22.7	−272		−2,853	−20
12	29.1	−175	$\sigma_y=0.1[(900)(1003.8)$	−4,182	−29
15	35.1	0	$+3(2x-30)(8862)$	−4,431	−31
18	40.7	244	$-(30)^3\,T(x)]$	−3,600	−25
21	46.0	552		−1,959	−14
24	50.9	916		762	5
27	55.6	1334	Simplifying:	4,023	28
30	60.0	1800	$\sigma_y=5317x-2700\,T(x)$	8,094	56
$\displaystyle\int_0^b$	1003.8	8862	$+10{,}584$		

temperature differential between the cooling water and the concrete being cooled, are studied concurrently. All of these factors should be considered in arriving at an economical cooling system which can achieve the desired temperature control.

The theory for the removal of heat from concrete by embedded cooling pipes was first developed for use in Hoover Dam. (See reference [7].) From these studies, a number of curves and nomographs were prepared for a vertical spacing (height of placement lift) of 5 feet. The concrete properties and a single rate of flow of water were also used as constants. Subsequent to the earlier studies, the theory was developed using dimensionless parameters. Nomographs were then prepared on the basis of a ratio of b/a of 100, where b is the radius of the cooled cylinder and a is the radius of the cooling pipe. Actual cooling pipe spacings are nominal spacings and will seldom result in a b/a ratio of 100. In order to take the actual horizontal and vertical spacings into consideration, a fictitious diffusivity constant can be used which is based on tests of concrete made with similar aggregates. Table 7-3 gives the values of D, D^2, and h^2_f for various spacings of cooling pipe. The b/a ratios of the spacings shown vary from about 34 to 135. Within these limits, the values of h^2_f may be used with sufficient accuracy.

Figures 7-6 and 7-7 are used for pipe cooling computations. In these illustrations,

$$X = \frac{\left[\begin{array}{c}\text{Difference between mean temper-}\\\text{ature of the concrete and temper-}\\\text{ature of the cooling water}\end{array}\right]}{\left[\begin{array}{c}\text{Initial temperature difference}\\\text{between the concrete and the}\\\text{cooling water}\end{array}\right]}$$

$$Y = \frac{\left[\begin{array}{c}\text{Temperature rise of the}\\\text{cooling water}\end{array}\right]}{\left[\begin{array}{c}\text{Initial temperature difference}\\\text{between the concrete and the}\\\text{cooling water}\end{array}\right]}$$

K = conductivity of the concrete,
L = length of cooling coil,
c_w = specific heat of water,
ρ_w = density of water,
q_w = volume of water flowing through the coil,
t = time from start of cooling,
D = diameter of the cooling cylinder, and
h^2 = diffusivity of the concrete.

Consistent units of time and distance must be used throughout.

The curves in figures 7-6 and 7-7 are used in a straight-forward manner as long as no appreciable heat of hydration is occurring in the concrete during the period of time under consideration. When the effect of artificial cooling is desired during the early age of the

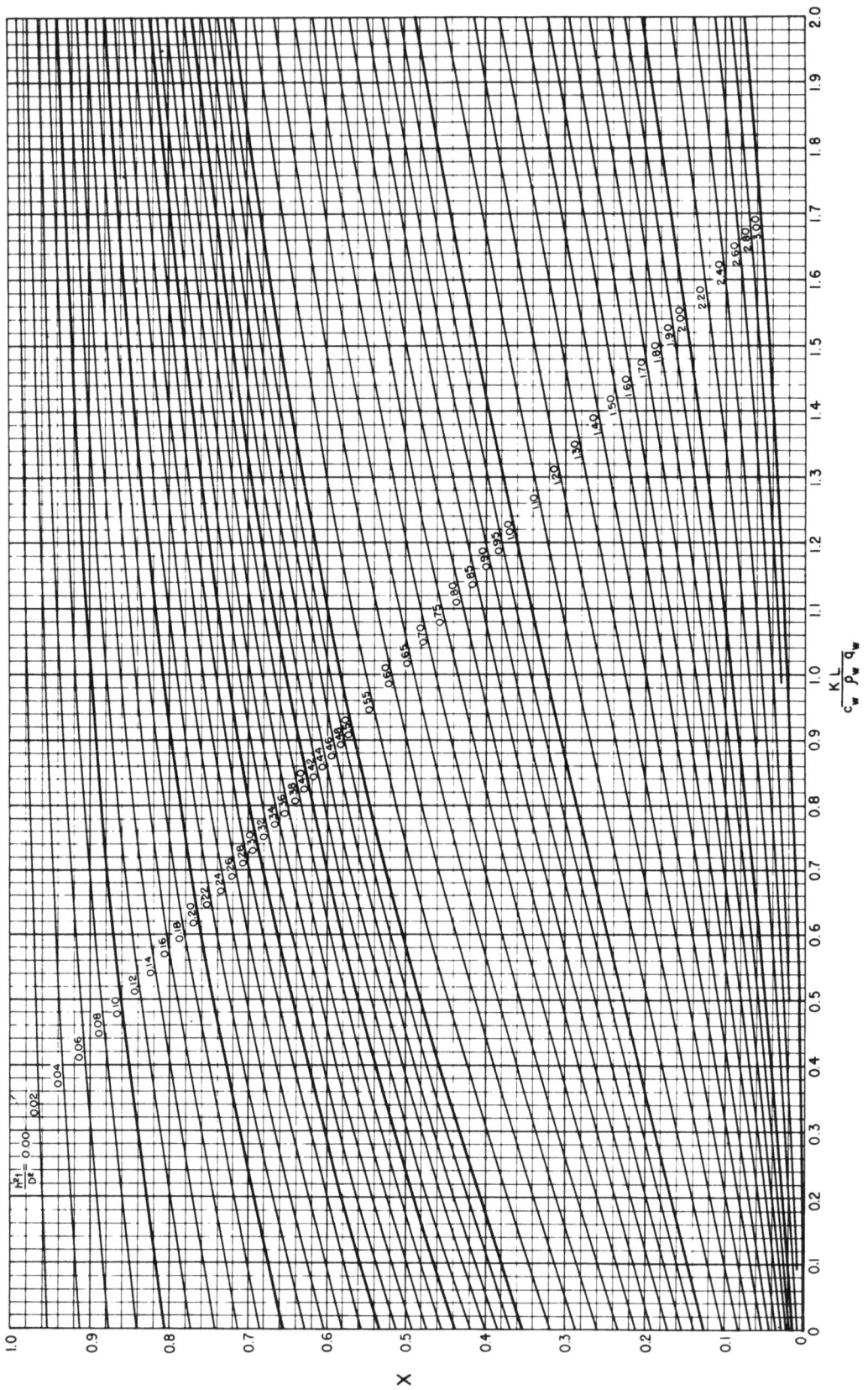

Figure 7-6. Pipe cooling of concrete—values of X.—288-D-3015

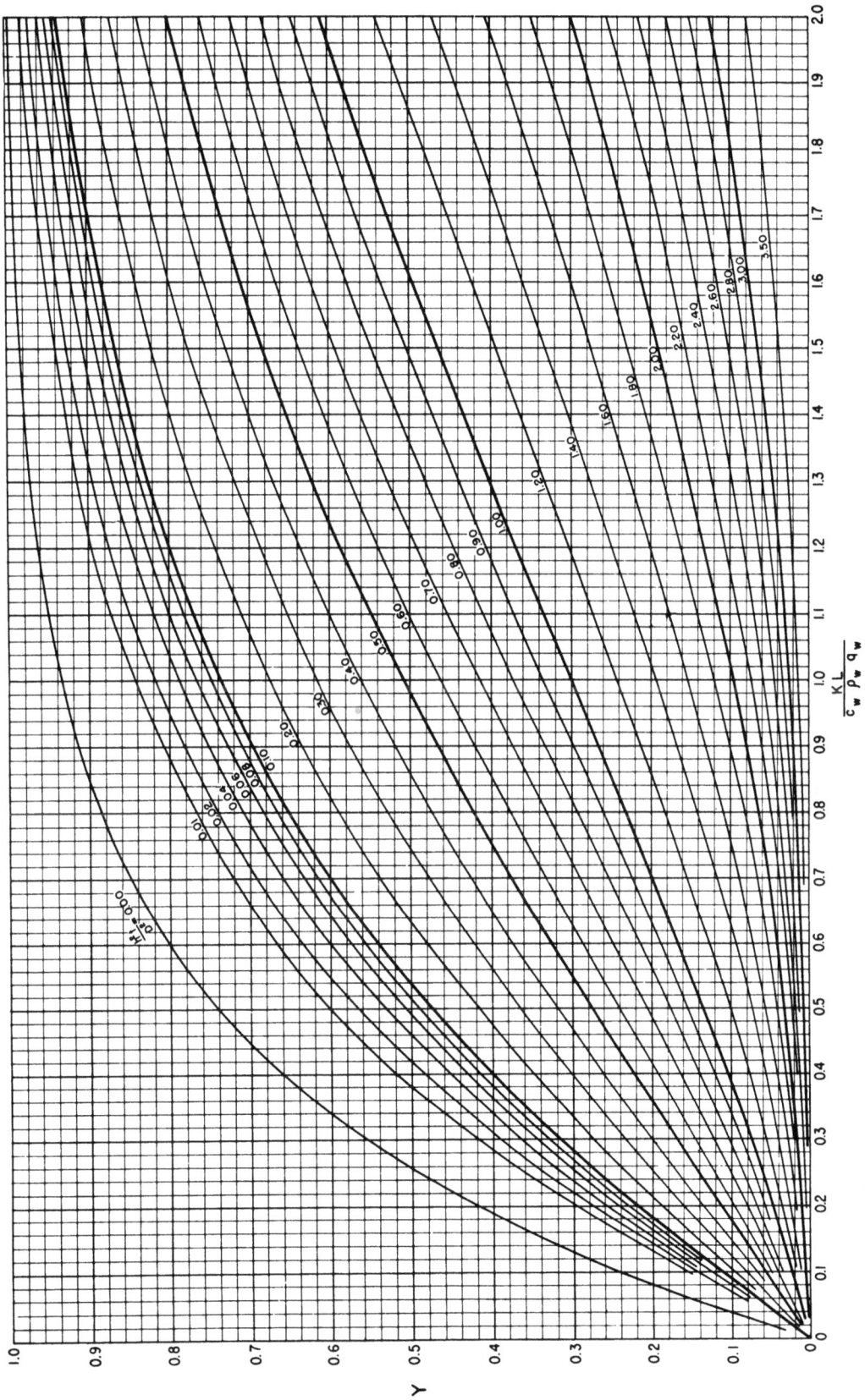

Figure 7-7. Pipe cooling of concrete–values of Y.–288-D-3016

Table 7-3.—*Values of D, D^2, and h^{2}f for pipe cooling.*

Spacing		D	D²	h²f
Vertical (feet)	Horizontal (feet)			
2½	2½	2. 82	7. 95	1. 31h²
5	2½	3. 99	15. 92	1. 19h²
5	3	4. 35	18. 92	1. 16
5	4	5. 02	25. 20	1. 12
5	5	5. 64	31. 81	1. 09
5	6	6. 18	38. 19	1. 07
7½	2½	4. 88	23. 81	1. 13h²
7½	4	6. 15	37. 82	1. 07
7½	5	6. 86	47. 06	1. 04
7½	6	7. 54	56. 85	1. 02
7½	7½	8. 46	71. 57	1. 00
7½	9	9. 26	85. 75	0. 98
10	10	11. 284	127. 33	0. 94h²

concrete, a step-by-step computation is required which takes into consideration heat increments added at uniform time intervals during the period.

Varying the temperature of the water circulated through the coil, the length of the embedded coil, and the horizontal spacing of the pipe are effective means of varying the cooling operation to obtain the desired results. Figures 7-8, 7-9, and 7-10[3] show how these variables affect the concrete temperatures. These studies were made using 4 sacks of type II cement per cubic yard, a diffusivity of 0.050 square foot per hour, a flow of 4 gallons per minute through 1-inch outside-diameter pipe, 5-foot placement lifts, and a 3-day exposure of each lift. Figures 7-9 and 7-10 were derived using the adiabatic temperature rise shown in figure 7-8. In general, cooling coil lengths of 800 to 1,200 feet are satisfactory. Spacings varying from 2½ feet on the rock foundation to 6 feet on tops of 7½-foot lifts have been used. The temperature of the cooling water has varied from a refrigerated brine at about 30° F. to river water with temperatures as high as 75° F.

[3] These three illustrations are reprinted from an article "Control of Temperature Cracking in Mass Concrete," by C. L. Townsend, published in ACI Publication SP-20, "Causes, Mechanism, and Control of Cracking in Concrete," 1968.

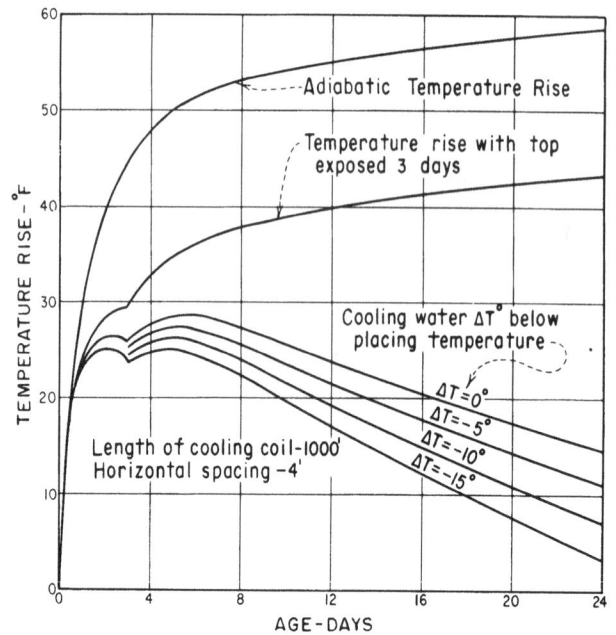

Figure 7-8. Artificial cooling of concrete—effect of cooling water temperature. (From ACI Publication SP-20.)—288-D-3017

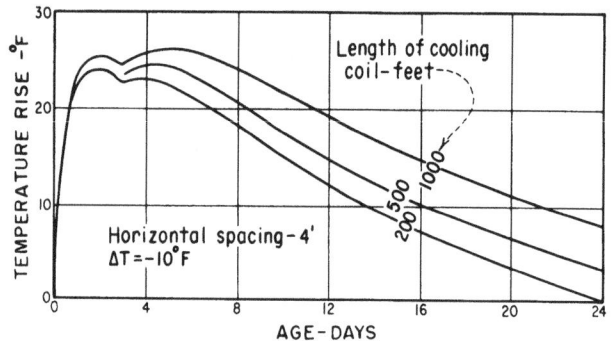

Figure 7-9. Artificial cooling of concrete—effect of coil length. (From ACI Publication SP-20.)—288-D-3018

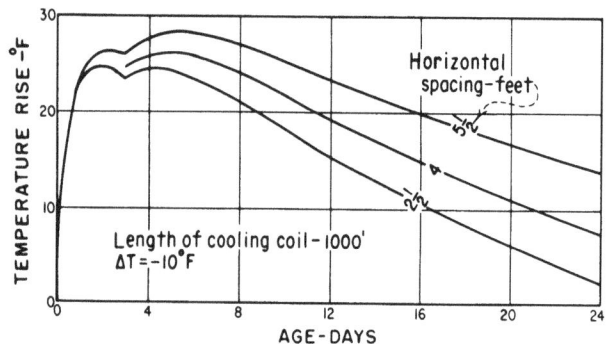

Figure 7-10. Artificial cooling of concrete—effect of horizontal spacing of pipe. (From ACI Publication SP-20.)—288-D-3019

Varying the size of the embedded pipe will affect the cooling results but is uneconomical as compared to the other methods of varying the cooling. The use of 1-inch outside-diameter metal pipe or tubing is common practice. Although black steel pipe is cheaper in material cost, aluminum tubing has been used in many instances because it can be furnished in coils and will result in a lower installation cost. Increasing the rate of flow through 1-inch pipe will give a marked improvement of performance up to a rate of 4 gallons per minute. However, doubling the flow to 8 gallons per minute decreases the time required for cooling by only 20 to 25 percent for average conditions, whereas it doubles the capacity requirements, increases the friction losses, and more than doubles the power costs.

7-16. Miscellaneous Studies.—Solutions for idealized heat flow problems associated with the design and construction of mass concrete dams are given in reference [7]. Illustrative examples are given which demonstrate the use of the theory in practical applications. Temperature distributions and gradients in semi-infinite solids are given for both constant exposure and variable exposure temperature conditions. Natural cooling of slabs, cylinders, and spheres is discussed using initial uniform temperature distributions, uniformly varying initial temperatures, constant exposure temperatures, and variable exposure temperatures.

Studies for the insulation requirements on concrete structures as a protection against freezing and to minimize the formation of extreme temperature gradients are discussed in reference [8].

Although specific methods of cooling are normally left to the contractor, the requirements for cooling the various parts of a concrete mix to obtain a predetermined placing temperature can require a detailed study. The various considerations for such an operation are discussed in an article by F. B. Kinley in reference [9].

D. DESIGN CONSIDERATIONS

7-17. Placing Temperatures.—The maximum temperature attained in mass concrete is determined to a large extent by the temperature of the concrete as it is placed in the structure. This makes the placing temperature of the concrete of concern because (1) lower concrete temperatures will minimize temperature differentials near the surface, and (2) a measure of control over the subsequent temperature drop from the maximum concrete temperature to the grouting or final stable state temperature can be achieved.

When no special provisions are employed, concrete placing temperatures will approximate the mean monthly air temperature, ranging from $4°$ to $6°$ F. higher than the mean air temperature in the wintertime and this same amount lower than the mean air temperature in the summertime. The actual temperature of the concrete mix depends upon the temperatures, batch weights, and specific heats of the separate materials going into the concrete mix.

The placing temperature of the concrete may be lowered by reducing the temperatures of one or more of the separate materials. The computation for determining the temperature of a mix, both with and without precooling measures, is illustrated in references [1] and [9].

Minimal tensile stresses at the base of a placement lift will be developed if the placing temperature of the concrete is at or slightly below the temperature of the foundation and if the temperature rise is minimized. These tensile stresses resulting after placement will be lower if successive lift placements in a block are made at regular, periodic intervals with the shortest practicable time between lifts. Form removal and lifting of forms, installation of required metalwork, and construction joint cleanup will normally require a minimum of almost 3 days between lifts.

7-18. Closure Temperature.—One design consideration related to temperature control is the grouting or closure temperature of the contraction joints in the dam. Normally, the closure temperature in a gravity dam is the minimum mean concrete temperature during operation, but the actual temperature may be influenced by practical or economic considerations. The designer often has to make a design decision whether to use only the river water available to cool the concrete, thereby losing the benefit of 2° to 5° F. additional cooling which could be obtained by artificial methods, or to obtain the desired temperature reduction by requiring mechanically refrigerated water to perform the cooling.

From the practical standpoint, it is possible to cool the concrete by means of an embedded pipe cooling system to within 4° or 5° F. of the mean temperature of the cooling water. Concrete temperatures as low as 35° F. have been obtained with a refrigerating plant using brine as the coolant. Where cooling is accomplished with river water, concrete temperatures attainable depend on the mean river water temperature. At Hungry Horse Dam, river water at 32° to 34° F. was available during the colder months of the year, and final cooling was accomplished to 38° F. with this river water. Where river water is limited in quantity and is relatively warm, refrigeration of the cooling water will be required.

7-19. Size of Construction Block.— Temperature cracking in mass concrete structures is related to the dimensions and shape of the construction blocks in the structure and to the climatic conditions occurring during the construction period. Generally, a block with a length of 50 feet or less can be placed with only a minimum of control. Likewise, blocks up to 200 feet long can be placed with normal temperature control measures and have no more than nominal cracking. The location of appurtenances generally controls the spacing between transverse contraction joints, but this spacing should be guided to some extent by the shape of the block as it progresses from the foundation to the top of the dam.

(a) *Length of Construction Block.*—For a given site and given loading conditions, the thickness of a dam is determined by gravity analyses. Where this thickness is large, the section can be broken into two or more construction blocks separated by longitudinal joints, or it can be constructed as a single block by applying rigid temperature control measures. Normally, a 25° to 30° F. temperature drop can be permitted in blocks of the size commonly used before tensile stresses are developed which will be great enough to cause cracking across the block. In low temperature climates, special precautions are needed to avoid high differential temperatures caused by sudden temperature drops.

The length of a construction block is not governed by the capacity of the concrete mixing plant, since each block is first constructed to its full width and height at the downstream end of the block and then progressively placed to the upstream face. More generally, the length of block is related to the tensile stresses which tend to develop within the block between the time the block is placed and the time it reaches its final temperature. The stresses are subject to some degree of control by operations affecting the overall temperature drop from the maximum temperature to the final or closure temperature, the rate of temperature drop, the thermal coefficient of expansion, and the age of the concrete when it is subjected to the temperature change. Factors in addition to temperature which affect the stresses in the block are the effective modulus of elasticity between the block and its foundation, the elastic and inelastic properties of the concrete, and the degree of external restraint.

The actual stresses will further vary between rather wide limits because of conditions occurring during the construction period which introduce localized stress conditions. Tensile stresses and resulting cracks may occur because the larger blocks, by reason of their greater area, will have a greater number of stress concentrations arising from the physical irregularities and variable composition of the foundation. Cracks may also occur because of delays in the construction schedule and construction operations. Longer blocks are more likely to have cold joints created during placement of the concrete, and these cold joints are definite planes of weakness. A special problem exists with respect to the longer blocks at the base of the dam. These will normally be exposed for longer periods of time because concrete placement is always slow at the start of a job. Under this condition, extreme temperature gradients may form near the surfaces. The stresses caused by these steep temperature gradients may then cause cracks to form along any planes of weakness which exist as a result of construction operations.

Unlike ordinary structural members undergoing temperature change, the stresses induced in mass concrete structures by temperature changes are not capable of being defined with any high degree of accuracy. The indeterminate degree of restraint and the varying elastic and inelastic properties of the concrete, particularly during the early age of the concrete, make such an evaluation an estimate at best. Field experiences on other jobs should guide the designer to a great extent. Such experiences are reflected in table 7-4 which can be used as a guide during the early stages of design.

(b) *Width of Construction Block.*— Contraction joints are normally spaced about 50 feet apart, but may be controlled in some parts of the dam by the spacing and location of penstocks and river outlets, or by definite breaks and irregularities of the foundation. Although a uniform spacing of joints is not necessary, it is desirable so that the contraction joint openings will be essentially uniform at the time of contraction joint grouting. Spacings have varied from 30 to 80 feet as measured along the axis of the dam. When the blocks are 30 feet or less in width, a larger temperature drop than would otherwise be necessary may be required to obtain a groutable opening of the contraction joint. This temperature drop should be compatible with the permissible drop for the long dimension of the block.

A further consideration is the maximum length-to-width ratio of the blocks which will exist as construction of a block progresses from its foundation to the top of the dam. If the ratio of the longer dimension to the shorter dimension is much over 2½, cracking at

Table 7-4.—*Temperature treatment versus block length.*

Block length	Treatment		
Over 200 feet	Use longitudinal joint. Stagger longitudinal joints in adjoining blocks by minimum of 30 feet		
	Temperature drop from maximum concrete temperature to grouting temperature—°F.		
	Foundation to $H=0.2L$ [1]	$H=0.2L$ to $0.5L$ [1]	Over $H=0.5L$ [1]
150 to 200 feet__	25	35	40
120 to 150 feet__	30	40	45
90 to 120 feet___	35	45	No restriction
60 to 90 feet____	40	No restriction	No restriction
Up to 60 feet___	45	No restriction	No restriction

[1] $H=$ height above foundation; $L=$ block length.

approximate third points of the block can be expected. Ratios of 2 to 1 or less are desirable, if practicable.

7-20. *Concrete Cooling Systems.*—The layout of the concrete cooling systems consists of pipe or tubing placed in grid-like coils over the top surface of each lift of concrete after the concrete has hardened. Coils are formed by joining together lengths of thin-wall metal pipe or tubing. The number of coils in a block depends upon the size of the block and the horizontal spacing. Supply and return headers, with manifolds to permit individual connections to each coil, are normally placed on the downstream face of the dam. In some instances, cooling shafts, galleries, and embedded header systems can be used to advantage. Figures 7-11 and 7-12 show cooling details for Glen Canyon Dam.

The velocity of flow of the cooling water through the embedded coils is normally required to be not less than 2 feet per second, or about 4 gallons per minute for the commonly used 1-inch pipe or tubing. Cooling water is usually pumped through the coils, although a gravity system has at times been used. When river water is used, the warmed water is usually wasted after passing through the coils. River water having a high percentage of solids should be avoided as it can clog the cooling systems. When refrigerated water is used, the warmed water is returned to the

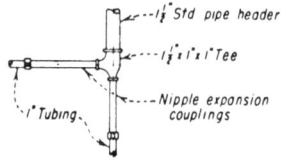

DETAIL Z

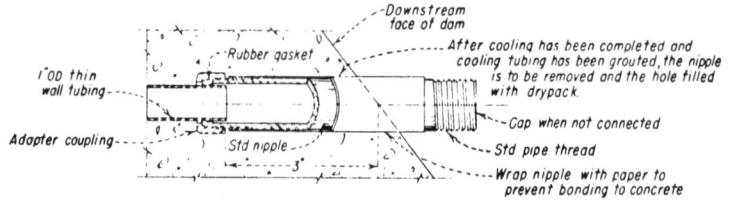

SURFACE CONNECTION DETAIL
I"O.D. COOLING AND THERMOMETER TUBING AT FACE OF DAM

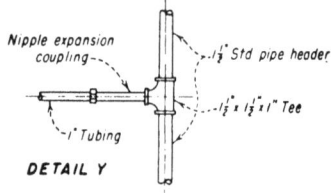

DETAIL Y

EXPLANATION

Thermocouple wire
1½" Std pipe header
I" O.D thin wall tubing

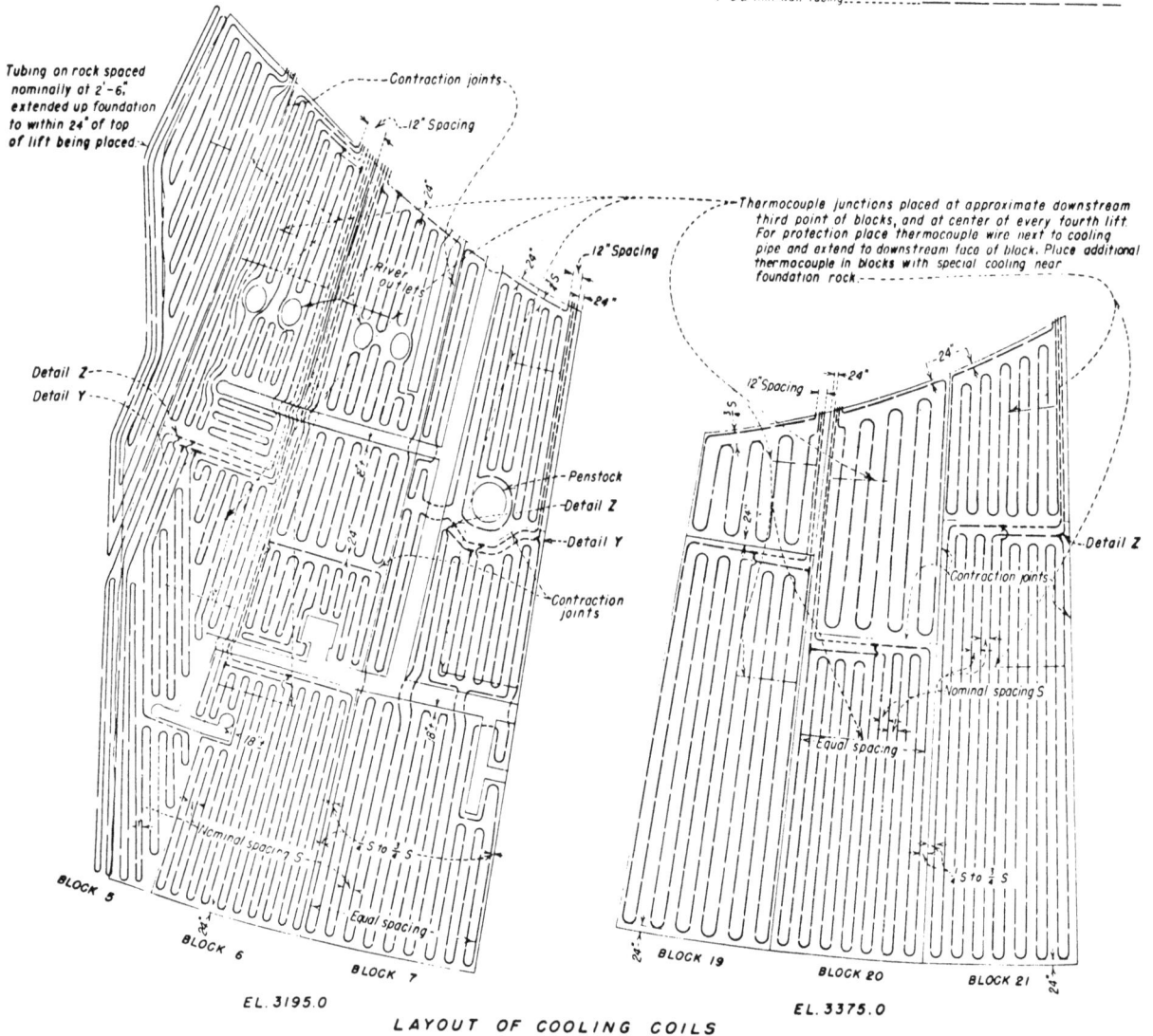

LAYOUT OF COOLING COILS

Figure 7-11. Glen Canyon Dam—cooling pipe layout.—288-D-3021

Extend header or tubing 6" min thru high block Thermocouple wire to be temporarily coiled and suspended on header or tubing for protection until embedment

Steel sheathing

High

1½" Std pipe header, or 1" OD tubing

Top of 7'-6" lift

Block

INITIAL INSTALLATION

Wrap exposed header or tubing with paper to prevent bonding to concrete.

Place loop in thermocouple wire under wrapping paper

Expansion coupling

High block

1½" Std pipe header, or 1" OD tubing

Low block

Extend thermocouple wire to downstream face

Contraction joint

FINAL INSTALLATION

TYPICAL EXPANSION COUPLING
SHOWING CONTRACTION JOINT CROSSING

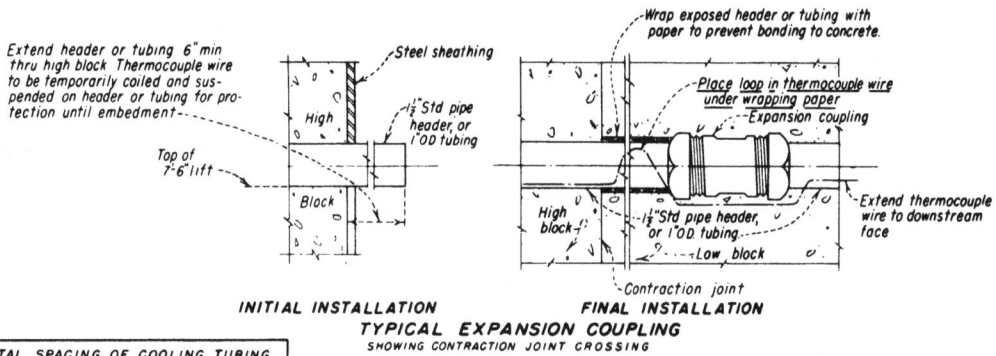

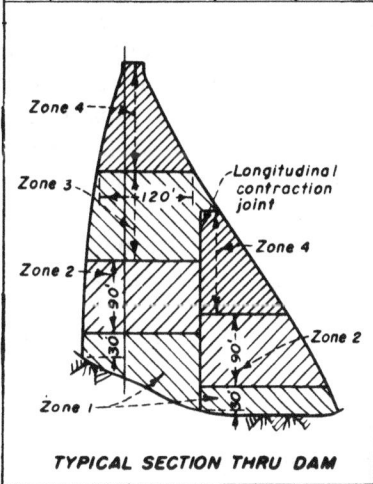

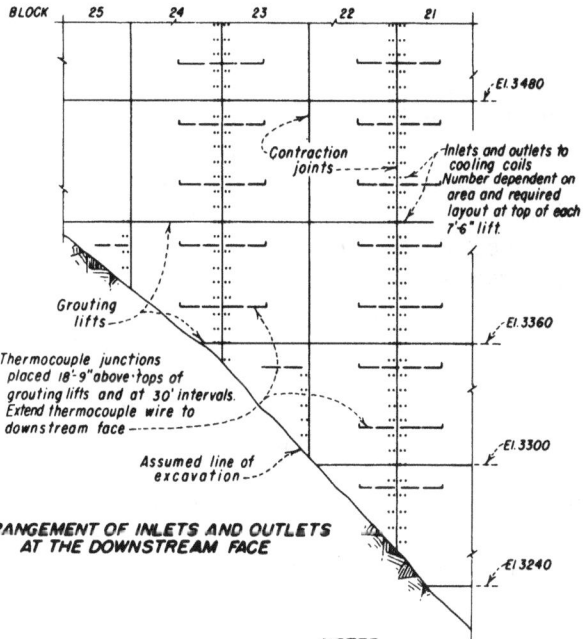

HORIZONTAL SPACING OF COOLING TUBING
(CRITERIA TO BE APPLIED TO EACH BLOCK)

ZONE	DESCRIPTION	MAXIMUM NOMINAL SPACING S	
		CONCRETE PLACED	
		NOV DEC JAN FEB	ALL OTHER MONTHS
1	Between the foundation and 30' above the highest point of the foundation	3'-0" on all concrete, 2'-6" along foundation rock	3'-0" on all concrete, 2'-6" along foundation rock
2	Between 30' and 120' above the highest point of the foundation	5'-0"	4'-0"
3	Greater than 120' above the highest point of the foundation and where the 7'-6" lift is longer than 120'	6'-0"	5'-0"
4	Greater than 120' above the highest point of the foundation and where the 7'-6" lift is 120' in length or less	6'-0"	6'-0"

Zone 4

Zone 3

Longitudinal contraction joint

120'

Zone 2

Zone 4

90'

Zone 2

30'

90'

Zone 1

TYPICAL SECTION THRU DAM

BLOCK 25 24 23 22 21

El. 3480

Contraction joints

Inlets and outlets to cooling coils Number dependent on area and required layout at top of each 7'-6" lift.

Grouting lifts

El. 3360

Thermocouple junctions placed 18'-9" above tops of grouting lifts and at 30' intervals. Extend thermocouple wire to downstream face

Assumed line of excavation

El. 3300

El. 3240

ARRANGEMENT OF INLETS AND OUTLETS
AT THE DOWNSTREAM FACE

NOTES

Actual required foundations may differ widely from assumed excavation lines shown

Cooling coils shall be placed on top of each 7'-6" concrete lift.

Place tubing on all rock surfaces to within 24" of the top of the lift being placed.

Cooling tubing to be placed to clear openings in dam a min of 18" or as directed.

Expansion couplings shall be used at contraction joint crossings.

Where tubing is installed for thermometer wells, the embedded end of the tubing is to be flattened and crimped to seal against grout leakage.

Arrangement of tubing may vary from that shown. The actual arrangement of the tubing in the structure shall be as directed.

Where a block is bounded by the downstream face and requires two or more coils, the contractor may elect to terminate all coils at downstream face in lieu of using 1½ headers. Upstream blocks requiring two or more coils will require headers.

Tubing placed on rock to be spaced at 2'-6"; tubing at top of each 7'-6" lift to be spaced according to zones as shown in table.

Coils placed in Zone 1 shall be approximately 800' in length with no coil longer than 900' in length; all other coils shall be approximately 1200' in length with no coil longer than 1300' in length. Adjacent coils served by the same header shall be as nearly the same length as possible.

Thermometer wells will be used to determine concrete temperatures at locations directed to supplement or replace thermocouple wire.

Cooling tubing placed within 25 to 30 feet of the foundation rock shall, where practicable, be placed as separate coils or with separate headers to facilitate special cooling in this area.

Each block shall have an independent cooling system at each concrete lift.

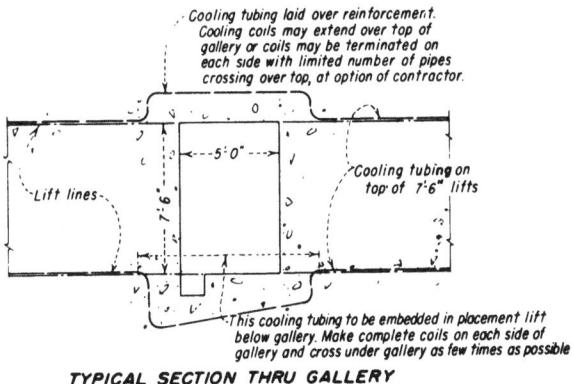

Cooling tubing laid over reinforcement. Cooling coils may extend over top of gallery or coils may be terminated on each side with limited number of pipes crossing over top, at option of contractor.

5'-0"

7'-6"

Lift lines

Cooling tubing on top of 7'-6" lifts

This cooling tubing to be embedded in placement lift below gallery. Make complete coils on each side of gallery and cross under gallery as few times as possible

TYPICAL SECTION THRU GALLERY

Figure 7-12. Glen Canyon Dam—concrete cooling details.—288-D-3022

water coolers in the refrigerating plant, recooled, and recirculated.

For control of· the cooling operations, electrical resistance-type thermometers can be embedded at midlift and the electrical cable extended to a terminal board where readings can be taken whenever desired. Thermometer tubes can also be embedded in the concrete. Insert-type thermometers are inserted into these tubes when readings are desired. In many installations thermocouples have been used and are not as costly as the thermometer installations. The thermocouples are placed in the fresh concrete at midlift and at least 10 feet from an exposed face, with the lead wires from the thermocouples carried to readily accessible points on the downstream face.

Varying the length of the embedded coil, the horizontal spacing of the pipe, and the temperature of the water circulated through the coil can be done during the construction period to meet changed conditions. The effect of these variables is given in section 7-15.

Specification requirements for the installation and operation of the cooling systems should provide for the cooling systems to be water tested prior to embedment to assure the operation of each individual coil. The arrangement of the pipe headers and connections to the individual cooling coils should be such as to insure dependable and continuous operation. Provisions should be made in the pumping or header systems for reversing the flow of water in the individual coils once each day. This is necessary to obtain a uniform cooling across the block. Because of varying construction schedules and progress and varying climatic conditions, the specifications should also provide that the times when cooling is to be performed in the individual cooling coils be as directed by the contracting officer. This will permit the operation of the cooling systems to be such as to minimize adverse conditions of temperature drops and temperature gradients which could lead to undesirable cracking.

7-21. *Height Differentials.*—A maximum height differential between adjacent blocks is normally specified in construction specifications for concrete dams. From a temperature standpoint, an even temperature distribution throughout the structure will be obtained when all blocks in the dam are placed in a uniform and continuous manner. This even temperature distribution is desirable because of the subsequent uniform pattern of contraction joint openings. Extreme temperature gradients on the exposed sides of blocks will also be lessened when each lift is exposed for a minimum length of time.

Minimizing the overall height differential between the highest and lowest blocks in the dam will cause construction of the dam to progress uniformly up from the bottom of the canyon. Contraction joints can then be grouted in advance of a rising reservoir, thus permitting storage at earlier times than would be possible if construction progress were concentrated in selected sections of the dam.

The height differential specified is a compromise between the uniform temperature conditions and construction progress desired, and the contractor's placement program. In practice, the maximum height differential between adjacent blocks is usually 25 feet when 5-foot lifts are used or 30 feet when 7½-foot lifts are used. The maximum differential between the highest block in the dam and the lowest block is usually limited to 40 feet when 5-foot lifts are used and 52.5 feet when 7½-foot lifts are used.

If cold weather is to be expected during any part of the construction period, height differentials between adjacent blocks should be limited to those needed for construction. If concrete placement is to be discontinued during winter months, the height differentials should be reduced to practical minimums before the shutdown period.

7-22. *Lift Thickness.*—Economy of construction should be considered in determining the heights of placement lifts in mass concrete. Shallow lifts not only slow up construction but result in increased construction joints which have to be cleaned and prepared for the next placement lift. Secondarily, the thickness of lift should be considered and related to the temperature control measures proposed for the structure.

When no precooling measures are used, the

placing temperature of the concrete will approximate the ambient temperature at the site. With this condition, a considerable portion of the total heat of hydration in a placement lift can be lost through the top exposed surface before the next lift is placed. Shallow lifts and longer delays between placement lifts will result in the minimum temperature rise in the concrete under these conditions. The opposite condition may occur, and should be studied, when precooling measures are used. During the summer months, the ambient temperatures will normally be higher than the concrete temperatures for the first few days after placement and a heat gain will result. Under these conditions, higher placement lifts and minimum periods of time between placements would be beneficial.

7-23. Delays Between Placements.— Construction of mass concrete blocks by placement lifts incurs periodic time delays between lifts. Depending upon ambient temperatures, these delays can be beneficial or harmful. The minimum elapsed time between placing of successive lifts in any one block is usually restricted to 72 hours, but temperature studies should be made to relate heat loss or heat gain to the placement lifts. These studies should take into account the anticipated temperature control measures and the seasonal effects to be met during the construction period. Delays between placements, and lift thicknesses should be studied simultaneously to take these variables into consideration as discussed in section 7-22.

The size and number of construction blocks in the dam will influence the time between placement lifts. Normal construction operations will require a minimum of 2 or 3 days between lifts. On the larger dams, however, an average placement time of about 6 or 7 days between successive lifts in a block will elapse because of the large number of construction blocks and the concrete yardage involved.

7-24. Closure Slots.— Closure slots are 2- to 4-foot-wide openings left in the dam between adjacent blocks during construction. Closure is made by filling the slot with concrete at a time when temperature conditions are favorable, usually during the late winter months of the construction period when the adjacent blocks are at minimum temperature. The use of closure slots will often expedite construction and will result in economy of labor and materials. Adverse stress conditions resulting from an unusual valley profile or undesirable temperature effects may be noted during the design or construction phases of a dam, which can often be overcome or reduced to safe proportions by the use of open joints or slots during the construction period.

E. CONSTRUCTION OPERATIONS

7-25. Temperature Control Operations.— The typical temperature history of artificially cooled concrete is shown on figure 7-13. Owing to hydration of the cement, a temperature rise will take place in the concrete after placement. After the peak temperature is reached, the temperature will decline depending upon the thickness of section, the exposure conditions, the rate and amount of continued heat of hydration, and whether or not artificial cooling is continued. The peak temperature is generally reached between ages 7 and 20 days in massive concrete sections where no artificial cooling is employed. These sections may maintain this maximum temperature for several weeks, after which the temperature will drop slowly over a period of several years. In thin structures or when artificial cooling is employed, the peak temperature is generally reached at about age 2½ to 6 days, after which the temperature can drop at a fairly rapid rate. With artificial cooling, the rate of temperature drop is usually limited to ½° to 1° F. per day, exposure conditions permitting. In thin structures exposed to very low air temperatures, the exposure conditions alone may cause temperatures to decline as much as 3° to 4° F. per day.

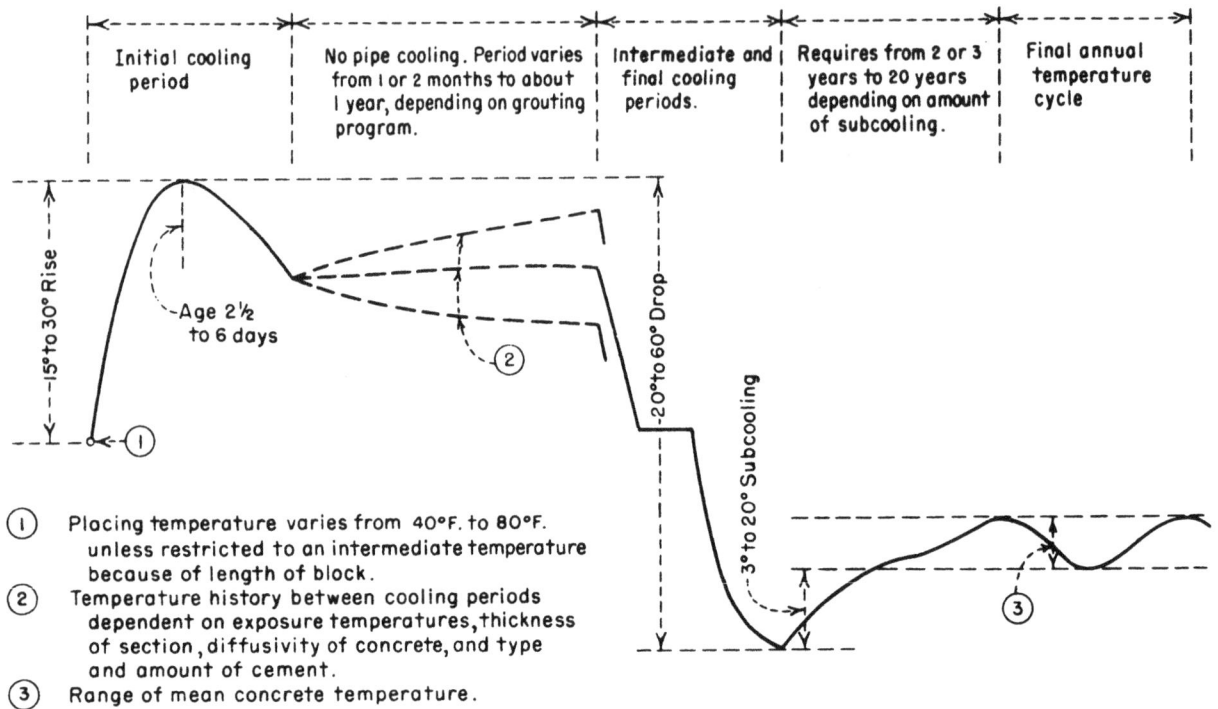

Initial cooling period

No pipe cooling. Period varies from I or 2 months to about I year, depending on grouting program.

Intermediate and final cooling periods.

Requires from 2 or 3 years to 20 years depending on amount of subcooling.

Final annual temperature cycle

① Placing temperature varies from 40°F. to 80°F. unless restricted to an intermediate temperature because of length of block.

② Temperature history between cooling periods dependent on exposure temperatures, thickness of section, diffusivity of concrete, and type and amount of cement.

③ Range of mean concrete temperature.

Figure 7-13. Temperature history of artificially cooled concrete.—288-D-3024

Initial cooling is normally accomplished with water not warmer than that obtainable from the river. Intermediate and final cooling may be accomplished with either river water or refrigerated water, depending upon the temperatures involved. River water will usually be sufficient if its temperature is 4° to 5° F. below the grouting temperature and if such a temperature persists for a minimum of about 2 months. The main objection to refrigerated water is its high cost. Advantages, however, include its availability at any time of the year and the wide range of temperatures possible.

Timely operation of the embedded cooling system will reduce the tendency of the concrete to crack during the construction period. The effects of unanticipated changes such as a change in the type or amount of cement used or the curing method employed, exposure temperatures varying from those assumed, or any other factor which influences concrete temperatures are normally taken into account by varying the period of flow and the temperature and rate of flow of the cooling water. Intermittent cooling periods can be used

to lower interior temperatures prior to exposure of the concrete to cold weather. During cold weather placement, the normal period of initial cooling may be shortened considerably to prevent forcing too rapid a drop in temperature. Depending upon the dimensions of the structure and the exposures expected, insulating the exposed surfaces while artificially cooling the interior may be necessary to control temperature cracking. This is especially true for areas near corners of the construction blocks where temperatures can drop very rapidly.

(a) *Initial Cooling.*—Artificial cooling is employed for a limited period of time initially. Upon completion of this initial cooling period, temperatures within the concrete may continue to drop but at a slower rate, they may hold steady at about the same temperature, or they may start rising again. This part of the temperature history is primarily dependent upon the thickness of section and the exposure conditions existing at the time. Continued heat of hydration at this age may also affect the concrete but would be of lesser importance.

The normal initial cooling period is from 10 to 16 days. During this initial cooling period, the concrete temperatures are reduced from the maximum concrete temperature to such a value that, upon stoppage of the flow of water through the cooling system, the continued heat of hydration of the cement will not result in temperatures higher than the maximum previously obtained. The rate of cooling is controlled so that the tensions in the concrete caused by the drop in temperature will not exceed the tensile strength of the concrete for that age of concrete.

In the early spring and late fall months when exposure temperatures may be low, the length of the initial cooling period and the rate of temperature drop can be critical in thin concrete sections. In these sections, pipe cooling, combined with the low exposure temperatures, can cause the concrete temperature to drop too fast. During these seasons, artificial cooling should be stopped shortly after the peak temperature is reached and the concrete then allowed to cool in a natural manner. In structures with thicker sections, the exposure temperatures have less effect on the immediate temperature drop, and the initial cooling period can be continued with the primary purpose of controlling the differential temperature between the exposed faces and the interior.

(b) *Intermediate and Final Cooling.*— Subsequent to the initial cooling period, intermediate and final cooling periods are employed to obtain desired temperature distributions or desired temperatures prior to contraction joint grouting. Final cooling for contraction joint grouting is normally accomplished just prior to grouting the contraction joints, the program of cooling being dictated by construction progress, method of cooling, season of the year, and any reservoir filling criteria.

As indicated in figure 7-13, cooling prior to grouting the contraction joints is normally started after the concrete has attained an age of 2 months to 1 year. Cooling is normally performed by grout lifts. In the smaller construction blocks, final cooling may be accomplished in a single, continuous cooling period. In the larger blocks, however, the final cooling should be performed in two steps to reduce the vertical temperature gradient between grout lifts. The first of these steps is commonly referred to as the intermediate cooling period and the second step as the final cooling period.

In practice, the intermediate cooling period for a grout lift lowers the temperature of the concrete in that lift to approximately halfway between the temperature existing at the start of the cooling period and the desired final temperature. Each grout lift, in succession, undergoes this intermediate cooling period before the final cooling of the next lower grout lift is undertaken.

Depending upon the temperature drop and final temperature to be obtained, the season of the year when this cooling is accomplished, and the temperature of the cooling water, the intermediate and final cooling periods will require a total of from 30 to 60 days. The rate of temperature drop should be held to not more than 1° F. per day, and a rate of $\frac{1}{2}^\circ$ to $\frac{3}{4}^\circ$ F. per day is preferable.

It is theoretically possible to compute the required temperature drop to obtain a desired joint opening. The theoretical joint opening does not occur, however, because some compression is built up in the block as the temperature increases during the first few days after placement. A temperature drop of 4° to 8° F. from the maximum temperature, depending on the creep properties of the concrete, may be required to relieve this compression before any contraction joint opening will occur. Measured joint openings in Hungry Horse Dam averaged 75 percent of the theoretical. Other experiences with arch dams having block widths of approximately 50 feet have indicated that a minimum temperature drop of 25° F. from the maximum temperature to the grouting temperature is desirable, and will result in groutable contraction joint openings of 0.06 to 0.10 inch. For the wider blocks with 70 feet or more between contraction joints, a temperature drop of 20° F. will usually be sufficient.

(c) *Warming Operations.*—Prolonged

exposure of horizontal construction joints will often result in poor bond of the construction lifts. Horizontal leafing cracks may occur between the older and newer concretes, extending from the face of the structure into the interior. Cracks of this type quite often lead to freezing and thawing deterioration of the concrete. Preventive steps should be directed toward obtaining a better than average bond between the old concrete and the new concrete. This includes minimizing the temperature differential between the old and the new concrete. Several shallow placement lifts placed over the cold construction joint may be sufficient. For lifts exposed over a winter season, treatment may include warming the top 10 to 15 feet of the old concrete to the placing temperature of the new concrete. This will reduce the temperature gradient which will occur. The warming operation can be performed by circulating warm water through the embedded cooling coils. Warming operations should immediately precede the placement of the new concrete. If exposure temperatures are extremely low at the time placement is to be resumed, insulation should be placed over the tops of the lifts during the warming operations.

7-26. *Foundation Irregularities.*—Although the designs assume relatively uniform foundation and abutment excavations, the final excavation may vary widely from that assumed. Faults or crush zones are often uncovered during excavation, and the excavation of the unsound rock leaves depressions or holes which must be filled with concrete. Unless this backfill concrete has undergone most of its volumetric shrinkage at the time overlying concrete is placed, cracks can occur in the overlying concrete near the boundaries of the backfill concrete as loss of support occurs due to continuing shrinkage of the backfill concrete. Where the area of such dental work is extensive, the backfill concrete should be placed and cooled before additional concrete is placed over the area.

Similar conditions exist where the foundation has abrupt changes in slope. At the break of slope, cracks often occur because of the differential movement which takes place between concrete held in place by rock, and concrete held in place by previously placed concrete which has not undergone its full volumetric shrinkage. A forced cooling of the concrete adjacent to and below the break in slope, and a delay in placement of concrete over the break in slope, can be employed to minimize cracking at these locations. If economical, the elimination of these points of high stress concentration is worthwhile. Such cracks in lifts near the abutments very often develop leakage and lead to spalling and deterioration of the concrete.

7-27. *Openings in Dam.*—Because openings concentrate stresses at their corners, all possible means should be used to minimize stresses at the surfaces of such openings. Proper curing methods should be used at all times. The entrances to such openings should be bulkheaded and kept closed, with self-closing doors where traffic demands, to prevent the circulation of air currents through the openings. Such air currents not only tend to dry out the surfaces but can cause the formation of extreme temperature gradients during periods of cold weather.

7-28. *Forms and Form Removal.*—The time of removal of forms from mass concrete structures is important in reducing the tendency to crack at the surface. This is especially true when wooden forms or insulated steel forms are used. If exposure temperatures are low and if the forms are left in place for several days, the temperature of the concrete adjacent to the form will be relatively high when the forms are stripped, and the concrete will be subjected to a thermal shock which may cause cracking. From the temperature standpoint, these forms should either be removed as early as practicable or should remain in place until the temperature of the mass has stabilized. In the latter case, a uniform temperature gradient will be established between the interior mass and the surface of the concrete, and removal of the forms, except in adverse exposure conditions, will have no harmful results.

When the ordinary noninsulated steel form is used, the time of form removal may or may not be important. The use of steel forms which are kept cool by continuous water sprays will tend to cause the near-surface concrete to set at a lower temperature than the interior of the mass. Form removal can then be accomplished with no detrimental effects. If, however, water sprays are not used to modify the temperature of the steel forms, the early-age temperature variation of the face concrete may be even greater than the daily cycle of air temperature because of absorbed heat from solar radiation and reradiation.

7-29. Curing.—Drying shrinkage can cause, as a skin effect, hairline cracks on the surface of a mass concrete structure. The primary objection to these random hairline cracks of limited depth is that they are usually the beginning of further and more extensive cracking and spalling under adverse exposure conditions. Following the removal of forms, proper curing is important if drying shrinkage and resulting surface cracking are to be avoided. Curing compounds which prevent the loss of moisture to the air are effective in this respect, but lack the cooling benefit which can be obtained by water curing. In effect, water curing obtains a surface exposure condition more beneficial than the fluctuating daily air temperature. With water curing, the daily exposure cycle is dampened because the daily variation of the water temperature is less than that of the air temperature.

A benefit also occurs from the evaporative cooling effect of the water on the surface. The evaporative cooling effect is maximized by intermittent sprays which maintain the surface of the concrete in a wet to damp condition with some free water always available.

In general, water curing should be used instead of membrane curing on mass concrete structures. Where appearance is of prime importance, other methods of curing may be considered because water curing will often result in stains on the faces. Water curing during periods of cold weather also can be a safety problem because of icing hazards.

7-30. Insulation.—During the fall of the year when placing temperatures are still relatively high, and during periods of cold weather, the temperature of the surface concrete tends to drop rapidly to the exposure temperature. This may occur while the interior concrete is still rising in temperature. Such conditions will cause high tensile stresses to form at the surface. Surface treatments previously described can reduce these temperature gradients, particularly when used in conjunction with artificial cooling, but the use of insulation will give greater protection. Such insulation may be obtained by measures varying from simply leaving wooden or insulated forms in place, to the use of commercial-type insulation applied to the forms or to the surfaces of the exposed concrete. Tops of blocks can be protected with sand or sawdust when an extended exposure period is anticipated.

Unless required immediately after placement to prevent surface freezing, the insulation should be placed after the maximum temperature is reached in the lift. This permits loss of heat to the surface and will cause the near-surface concrete to set at a relatively low temperature. Normally, during periods of cold weather, the insulation is removed at such time as required for placement of the next lift. Otherwise, it may be removed when the cold weather has abated or when interior temperatures have been reduced substantially below the peak temperatures.

Whatever the type of insulation, measures should be taken to exclude as much moisture from the insulation as practicable. The insulation should also be as airtight as possible. For a short period of exposure, small space heaters may be used, either by themselves or in conjunction with work enclosures. Care should be taken when using space heaters in enclosed areas to avoid drying out the concrete surfaces.

F. BIBLIOGRAPHY

7-31. Bibliography.

[1] "Control of Cracking in Mass Concrete Structures," Engineering Monograph No. 34, Water Resources Technical Publication, Bureau of Reclamation, 1965.

[2] Schack, Alfred, "Industrial Heat Transfer," John Wiley & Sons, New York, N.Y., 1933.

[3] Jakob, Max, "Heat Transfer," vol. I, pp. 373-375, John Wiley & Sons, New York, N.Y., 1949.

[4] Grinter, L. E., "Numerical Methods of Analysis in Engineering," p. 86, Macmillan Co., New York, N.Y., 1949.

[5] "A Simple Method for the Computation of Temperatures in Concrete Structures," ACI Proceedings, vol. 34 (November-December 1937 ACI Journal).

[6] "Thermal Properties of Concrete," Part VII, Bulletin No. 1, Boulder Canyon Project Final Reports, Bureau of Reclamation, 1940.

[7] "Cooling of Concrete Dams," Part VII, Bulletin No. 3, Boulder Canyon Project Final Reports, Bureau of Reclamation, 1949.

[8] "Insulation Facilitates Winter Concreting," Engineering Monograph No. 22, Bureau of Reclamation, 1955.

[9] Kinley, F. B., "Refrigeration for Cooling Concrete Mix," Air Conditioning, Heating and Ventilating, March 1955.

Chapter VIII

Joints in Structures

8-1. *Purpose.*—Cracking in concrete dams is undesirable because cracking in random locations can destroy the monolithic nature of the structure, thereby impairing its serviceability and leading to an early deterioration of the concrete. Joints placed in mass concrete dams are essentially designed cracks, located where they can be controlled and treated to minimize any undesirable effects. The three principal types of joints used in concrete dams are contraction, expansion, and construction joints.

Contraction and expansion joints are provided in concrete structures to accommodate volumetric changes which occur in the structure after placement. Contraction joints are provided in a structure to prevent the formation of tensile cracks as the structure undergoes a volumetric shrinkage due to a temperature drop. Expansion joints are provided in a unit-structure to allow for the expansion (a volumetric increase due to temperature rise) of the unit in such a manner as not to change the stresses in, or the position of, an adjacent unit or structure. Construction joints are placed in concrete structures to facilitate construction, to reduce initial shrinkage stresses, to permit installation of embedded metalwork, or to allow for the subsequent placing of other concrete, including backfill and second-stage.

8-2. *Contraction Joints.*—In order to control the formation of cracks in mass concrete dams, current practice is to construct the dam in blocks separated by transverse contraction joints. These contraction joints are vertical and normally extend from the foundation to the top of the dam. Transverse joints are normal to the axis of the dam and are continuous from the upstream face to the downstream face.

Depending upon the size of the structure, it may also be necessary to provide longitudinal contraction joints in the blocks formed by the transverse contraction joints. If longitudinal contraction joints are provided, construction of the dam will consist of placing a series of adjoining columnar blocks, each block free to undergo its own volume change without restraint from the adjoining blocks. The longitudinal contraction joints are also vertical and parallel to the axis of the dam. The joints are staggered a minimum of 25 feet at the transverse joints. Generally, both transverse and longitudinal joints pass completely through the structure. As the longitudinal joint nears the sloping downstream face, and in the upstream sections of dams with sloping upstream faces, either the direction of the joint is changed from the vertical to effect a perpendicular intersection with the face, with an offset of 3 to 5 feet, or the joint is terminated at the top of a lift when it is within 15 to 20 feet of the face. In the latter case, strict temperature control measures will be required to prevent cracking of the concrete directly above the termination of the joint.

Typical transverse contraction joints can be seen on figures 8-1 and 8-2, and a typical longitudinal contraction joint can be seen on figure 8-3.

Contraction joints should be constructed so that no bond exists between the concrete blocks separated by the joint. Reinforcement should not extend across a contraction joint. The intersection of the joints with the faces of the dam should be chamfered to give a desirable appearance and to minimize spalling. In order to standardize block identification on all future dams, a criterion has recently been established which calls for the designation of blocks in the longitudinal direction by number, starting with block 1 on the right abutment (looking downstream). The blocks in each transverse row are to be designated by letter starting with the upstream block as the "A" block.

8-3. *Expansion Joints.*—Expansion joints are provided in concrete structures primarily to accommodate volumetric change due to temperature rise. In addition, these joints frequently are installed to prevent transferal of stress from one structure to another. Notable examples are: (1) powerplants constructed adjacent to the toe of a dam, wherein the powerplant and the mass of the dam are separated by a vertical expansion joint; and (2) outlet conduits encased in concrete and extending downstream from the dam, in which case an expansion joint is constructed near the toe of the dam separating the encasement concrete from the dam.

Like contraction joints, previously discussed, expansion joints are constructed so that no bond exists between the adjacent concrete structures. A corkboard, mastic, sponge rubber, or other compressible-type filler usually separates the joint surfaces to prevent stress or load transferal. The thickness of the compressible material will depend on the magnitude of the anticipated deformation induced by the load.

8-4. *Construction Joints.*—A construction joint in concrete is defined as the surface of previously placed concrete upon or against which new concrete is to be placed and to which the new concrete is to adhere when the previously placed concrete has attained its initial set and hardened to such an extent that the new concrete cannot be incorporated integrally with the earlier placed concrete by vibration. Although most construction joints are planned and made a part of the design of the structure, some construction joints are expedients used by a contractor to facilitate construction. Construction joints may also be required because of inadvertent delays in concrete placing operations. Treatment and preparation of construction joints are discussed in chapter XIV.

8-5. *Spacing of Joints.*—The location and spacing of transverse contraction joints should be governed by the physical features of the damsite, details of the structures associated with the dam, results of temperature studies, placement methods, and the probable concrete mixing plant capacity.

Foundation defects and major irregularities in the rock are conducive to cracking and this can sometimes be prevented by judicious location of the joints. Although cracks may develop normal to the canyon wall, it is not practicable to form inclined joints. Consideration should be given to the canyon profile in spacing the joints so that the tendency for such cracks to develop is kept to a minimum.

Outlets, penstocks, spillway gates, or bridge piers may affect the location of joints and consequently influence their spacing. Consideration of other factors, however, may lead to a possible relocation of these appurtenances to provide a spacing of joints which is more satisfactory to the dam as a whole. Probably the most important of these considerations is the permissible spacing of the joints determined from the results of concrete temperature control studies. If the joints are too far apart, excessive shrinkage stresses will produce cracks in the blocks. On the other hand, if the joints are too close together, shrinkage may be so slight that the joints will not open enough to permit effective grouting. Data on spacing of joints as related to the degree of temperature control are discussed in chapter VII.

Contraction joints should be spaced close enough so that, with the probable placement methods, plant capacity, and the type of concrete being used, batches of concrete placed in a lift can always be covered while the concrete is still plastic. For average conditions, a spacing of 50 feet has proved to be satisfactory. In dams where pozzolan and retarders are used, spacings up to 80 feet have been acceptable. An effort should be made to keep the spacing uniform throughout the dam.

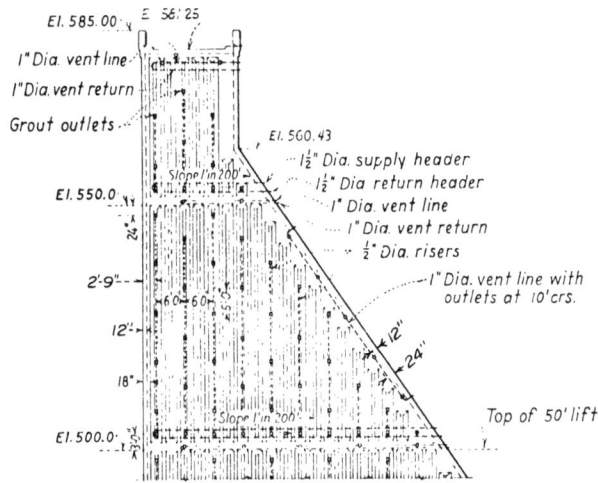

TYPICAL ELEVATION OF CONTRACTION JOINT

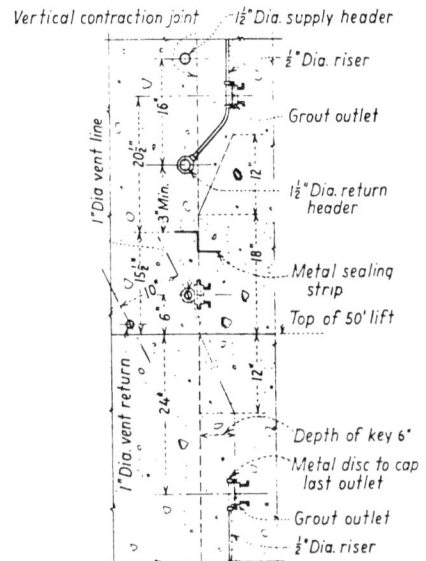

SECTION B-B

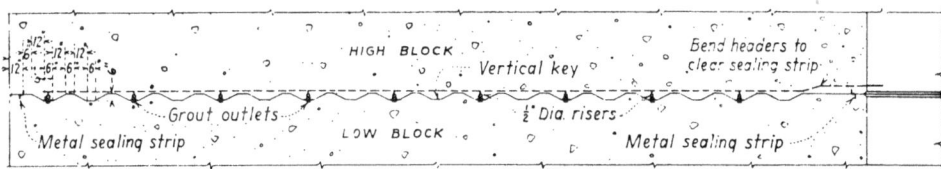

SECTION A-A

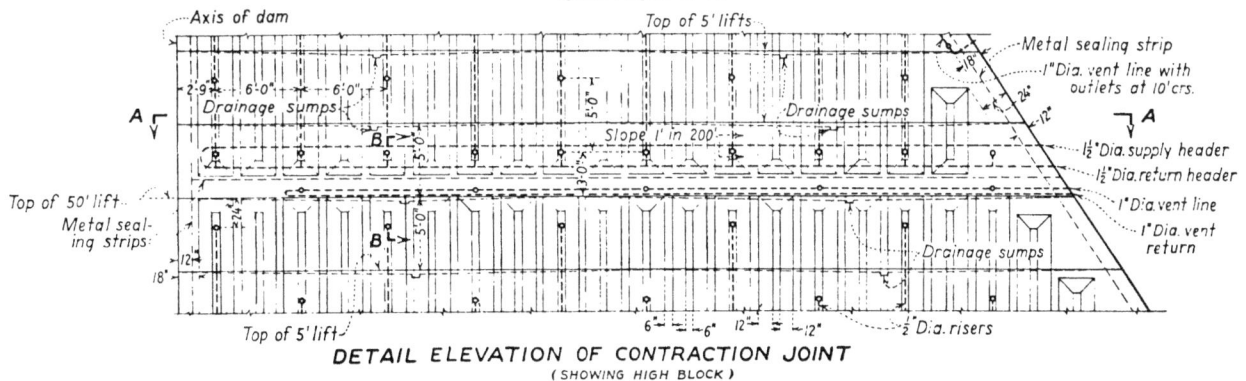

DETAIL ELEVATION OF CONTRACTION JOINT
(SHOWING HIGH BLOCK)

Figure 8-1. Typical keyed transverse contraction joint for a concrete gravity dam (Friant Dam in California).—288-D-3030

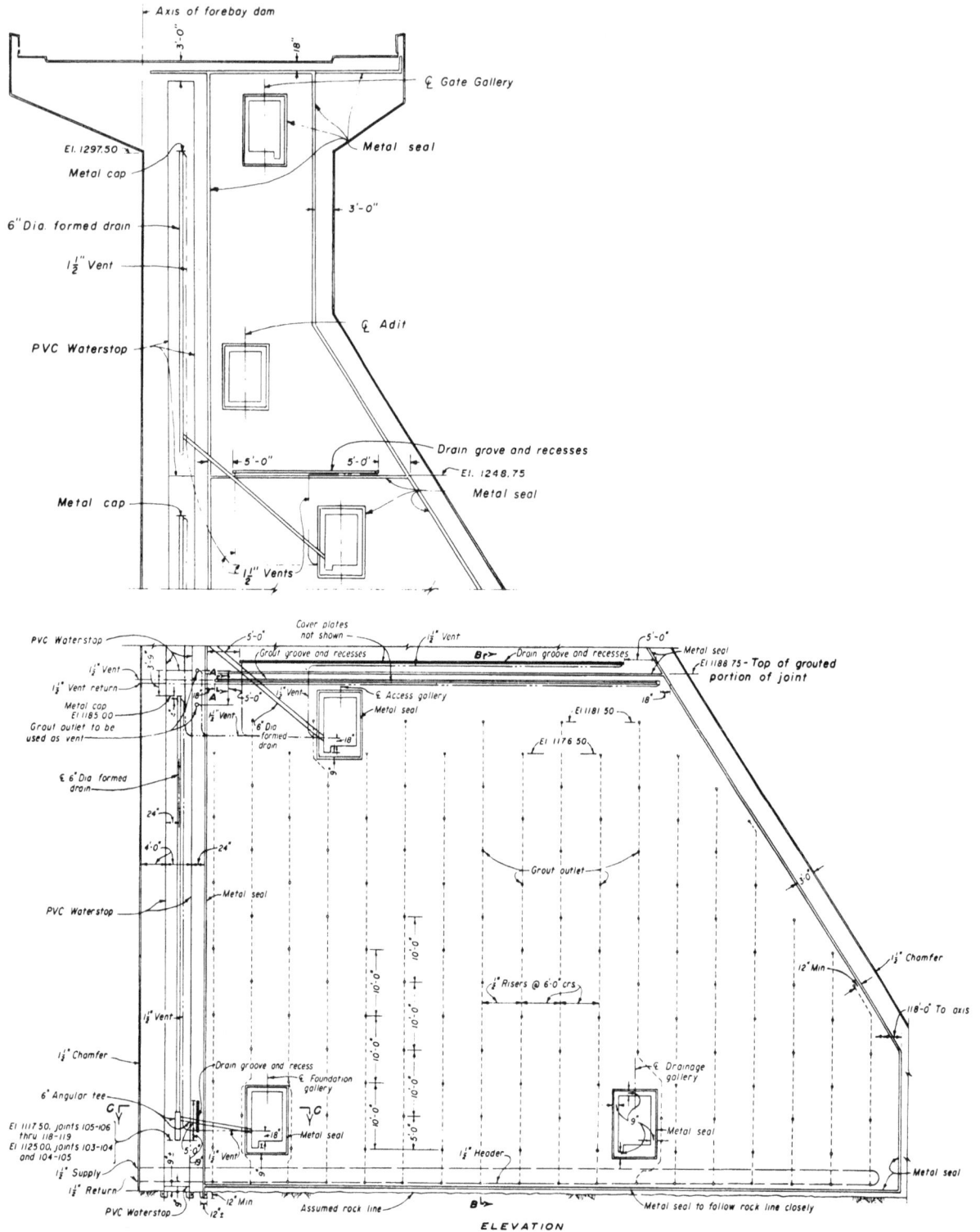

Figure 8-2. Typical unkeyed transverse contraction joint (Grand Coulee Forebay Dam in Washington). (sheet 1 of 2).–288-D-3032(1/2)

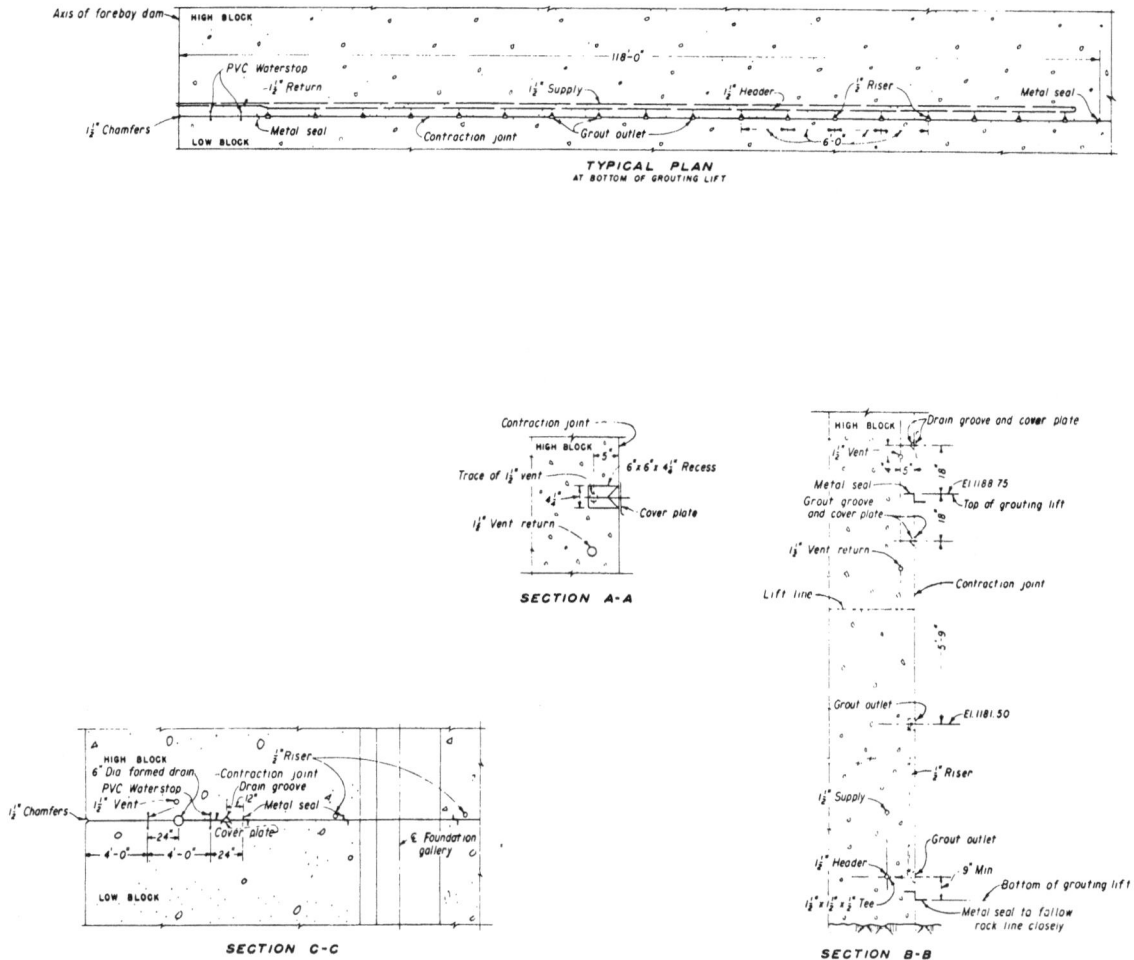

Figure 8-2. Typical unkeyed transverse contraction joint (Grand Coulee Forebay Dam in Washington). (sheet 2 of 2).—288-D-3032(2/2)

The practice of spacing longitudinal joints follows, in general, that for the transverse joints, except that the lengths of the blocks are not limited by plant capacity. Depending on the degree to which artificial temperature control is exercised, spacings of 50 to 200 feet may be employed.

8-6. Keys.—Vertical keys in transverse joints are used primarily to provide increased shearing resistance between blocks; thus, when the joints and keys are grouted, a monolithic structure is created which has greater rigidity and stability because of the transfer of load from one block to another through the keys. A secondary benefit of the use of keys is that they minimize water leakage through the joints. The keys increase the percolation distance through joints and, by forming a series of constrictions, are beneficial in hastening the sealing of the joints with mineral deposits.

Keys are not always needed in the transverse contraction joints of concrete gravity dams. Because the requirement for keys adds to form and labor costs, the need for keys and the benefits which would be attained from their use should be investigated and determined for each dam. Keys may be used to transfer horizontal loads to the abutments, thereby obtaining a thinner dam than would otherwise be possible. Foundation irregularities may be such that a bridging action over certain portions of the foundation would be desirable. Keys can be used to lock together adjacent blocks to help accomplish this bridging action.

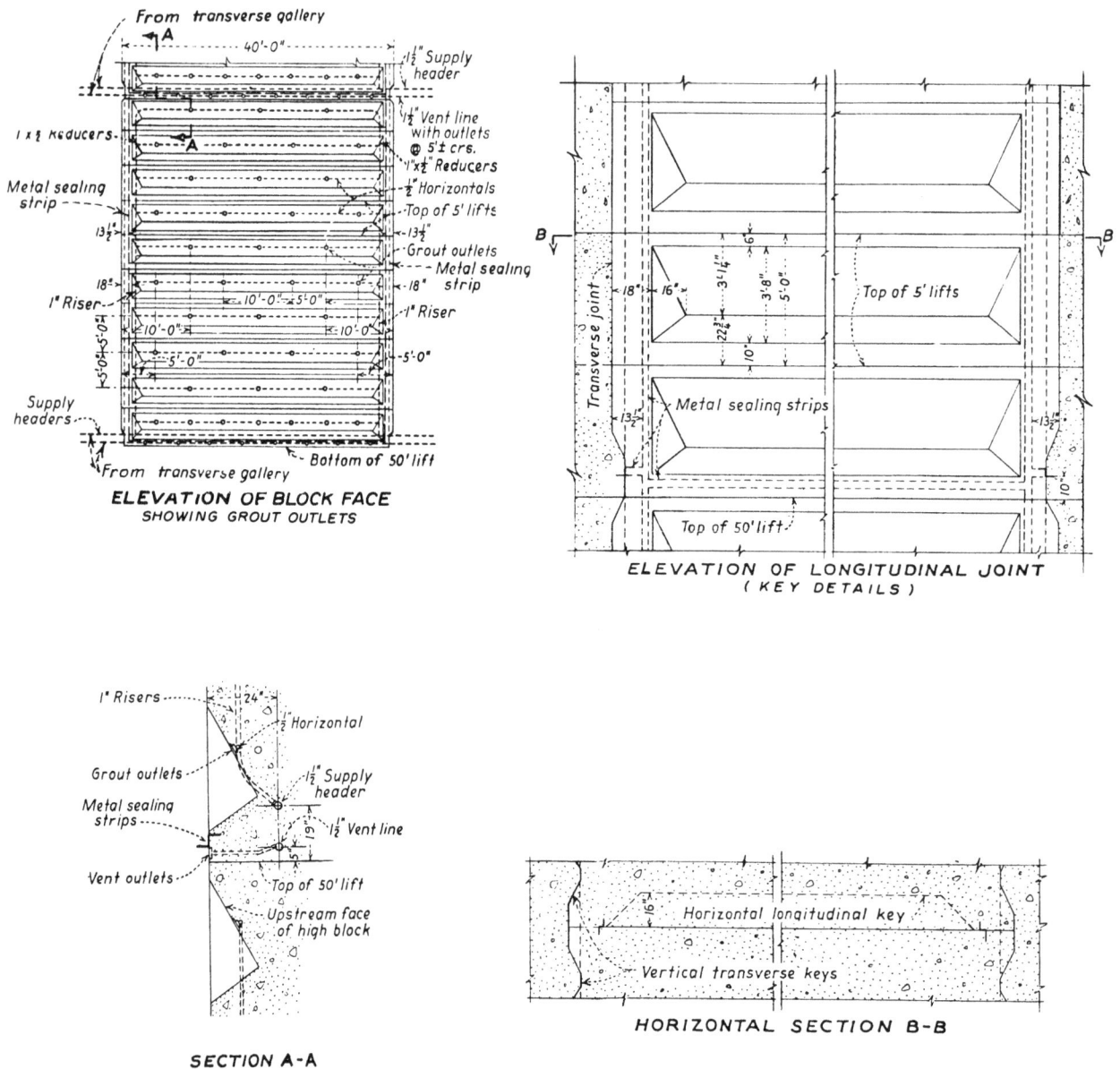

ELEVATION OF BLOCK FACE
SHOWING GROUT OUTLETS

ELEVATION OF LONGITUDINAL JOINT
(KEY DETAILS)

SECTION A-A

HORIZONTAL SECTION B-B

Figure 8-3. Typical longitudinal contraction joint for a concrete gravity dam (Grand Coulee Dam in Washington).—288-D-3034

In blocks where large openings are provided for penstocks, gate chambers, or other large features, keys can be used to improve the stability of the block.

The transverse joint key developed by the Bureau has been standardized. The standard key offers minimum obstruction to the flow of grout, provides a good theoretical shear value, eliminates sharp corners which commonly crack upon removal of forms, improves the

reentrant angles conducive to crack development associated with volume changes, and is well adapted to the construction of forms. Figure 8-1 shows the shape and dimensions of the standard key on the face of a typical transverse contraction joint.

Shear keys are important accessories in longitudinal contraction joints and are provided to maintain stability of the dam by increasing the resistance to vertical shear. The key faces are inclined to make them conform approximately with the lines of principal stress for full waterload. Inasmuch as the direction of principal stresses varies from the upstream face to the downstream face of the dam and from the foundation to the crest, an unlimited number of key shapes with resulting high forming costs would be required if close conformity were considered necessary. In order to simplify keyway forms, a single key shape, determined largely by the general direction of the lines of principal stress in the lower, downstream portion of the dam where the vertical shear is at a maximum, has been adopted for standard use. Details of the shape and dimensions of longitudinal keys used on Grand Coulee Dam are shown on figure 8-3. These keys are proportioned to accommodate the 5-foot concrete placement lifts used on that dam.

8-7. Seals.—The opening of transverse contraction joints between construction blocks provides passages through the dam which, unless sealed, would permit the leakage of water from the reservoir to the downstream face. To prevent this leakage, seals are installed in the joints adjacent to the upstream face. Seals are also required on both transverse and longitudinal joints during grouting operations to confine the fluid grout in the joint. Figure 8-4 illustrates typical seals used in contraction joints.

For seals to be effective in the contraction joints of concrete dams, installation is of greater importance than shape or material. Good workmanship in making connections, adequate protection to keep them from becoming torn prior to embedment, and careful placement and consolidation of the concrete around the seals are of primary importance.

(a) *Metal Seals.*—The most common type of seal used in concrete dams has been a metal seal embedded in the concrete across the joint. Metal seals are similar in design whether used as water or grout seals. Bureau practice has standardized two shapes—the Z-type and the M-type. The Z-type seal is of simpler design, is easily installed and spliced, but will accommodate only small lateral movements.

Such a seal is well adapted to joints which are to be grouted, since grouting tends to consolidate the two blocks and restrict any movement. The M-type seal is more difficult to splice, but its shape accommodates greater movement of the joint. This shape is well adapted for use as a water seal in ungrouted joints. Figure 8-4 shows the general dimensions and connections for the Z- and M-type seals.

Metal seals are made from a 12- or 15-inch strip of corrosion-resistant metal, usually copper or stainless steel. No. 20 gage United States Standard (0.0375-inch thick) stainless steel has proved satisfactory. The stainless steel is more rigid and will stay in position during embedment better than the more ductile copper. It is harder to weld, however, and is generally higher in initial cost. Copper strip can be furnished in rolls and will minimize the number of connections which have to be made.

(b) *Polyvinyl Chloride Seals.*—Recent advancements in the specifications for and manufacture of materials have resulted in the acceptance of polyvinyl chloride (PVC) as a suitable material for joint seals. This material can be manufactured in a number of shapes and sizes. The 12-inch seal having a ½-inch thickness, serrations, and a center bulb is acceptable for high dams. The 9-inch similar seal is satisfactory for low dams.

(c) *Other Seals.*—Rubber seals have been used in special joints in concrete sections of dams and appurtenant works where it is desired to provide for greater movement at the joint than can be accommodated by metal seals. Rubber seals have been used successfully in contraction joints between piers and the cantilevers of drum gate crests, to permit unrestrained deflection of the cantilevers and prevent leakage from the reservoir into the drum gate chamber. They can also be used in expansion and contraction joints of thin cantilever walls in stilling basins to prevent objectionable leakage caused by unequal deflection and settlement of the walls. A similar use would be in ungrouted contraction joints of low diversion dams to prevent excessive leakage caused by differential settlement.

Asphalt seals have not proved satisfactory

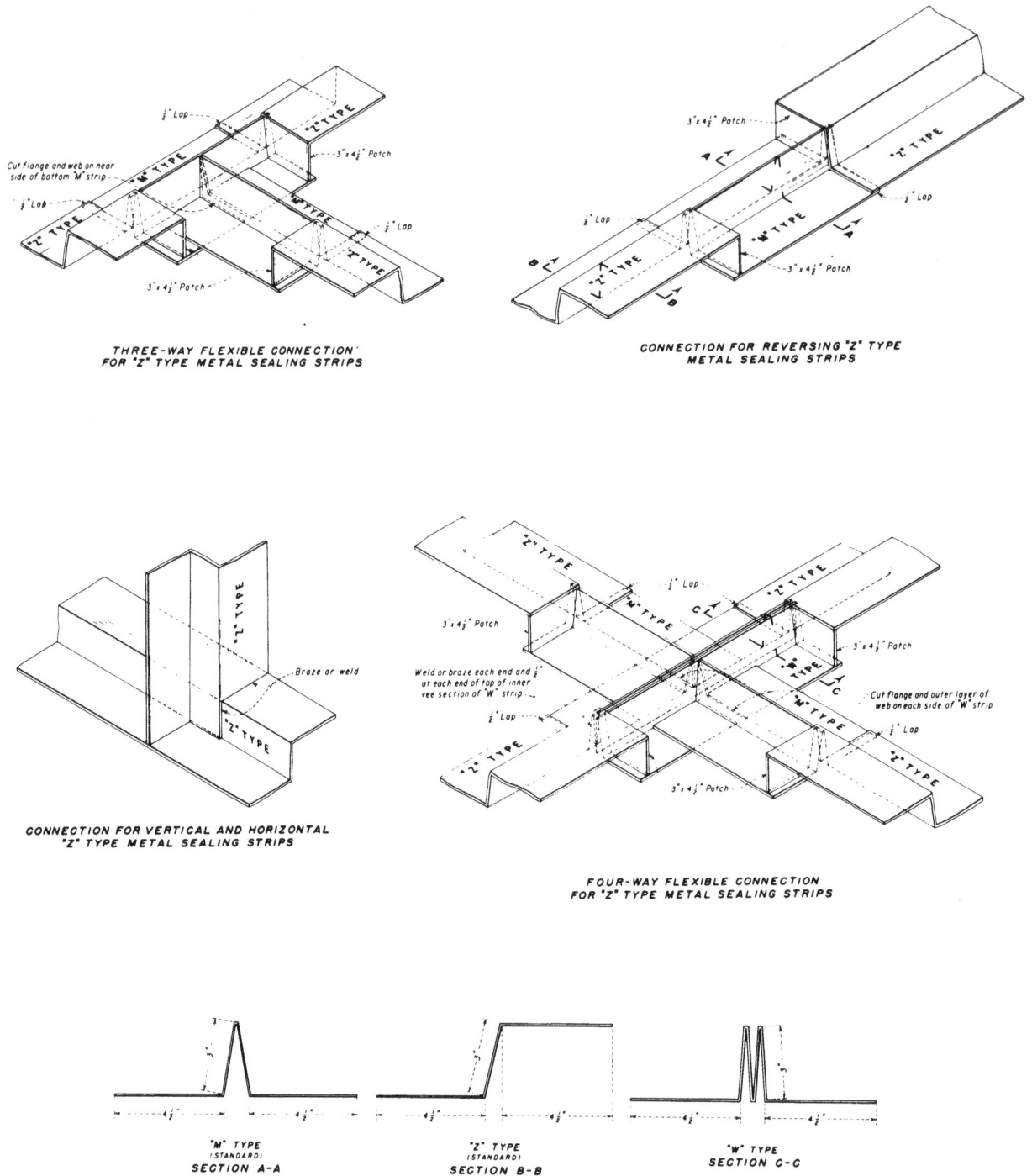

Figure 8-4. Metal seals and connections at contraction joints.—288-D-3200

for sealing contraction joints in concrete dams, and they are no longer used.

8-8. *Joint Drains*.—Drainage of contraction joints is desirable to prevent development of excessive pressure in the joints during the construction period and seepage of reservoir water through the joints during operation. Where contraction joints are to be ungrouted, 5- or 6-inch-diameter formed joint drains are constructed on the joints. These joint drains discharge the seepage water into the gallery drainage system. Where joint grouting systems are installed, the joints can be drained effectively during the construction period by utilizing the piping for the grouting system. Effective grouting when the joints are opened their widest will normally obviate any further need for drainage of the joint. Since provision for open joint drains makes effective grouting difficult, joint drains are usually omitted on Bureau dams where contraction joint grouting is to be performed.

8-9. *Grouting Systems*.—The purpose of contraction joint grouting is to bind the blocks together so that the structure will act as a monolithic mass. In some cases, the stability of the dam does not require the entire mass to act as a monolith and the transverse contraction joints need not be grouted. Longitudinal contraction joints must be grouted so that blocks in a transverse row act monolithically. Also, grouting of transverse construction joints may be required only in the lower portion of the joint as shown on figure 8-2.

In order to make the individual blocks act as a monolith, a grout mixture of portland cement and water is forced into each joint under pressure. Upon setting, the mixture will form a cement mortar which fills the joint. The means of introducing grout into the joint is through an embedded pipe system. Typical pipe systems are shown on figures 8-1, 8-2, and 8-3.

In order to insure complete grouting of a contraction joint before the grout begins to set, and to prevent excessive pressure on the seals, the joint is normally grouted in lifts 50 to 60 feet in height, although heights to about 75 feet have been used. Such a grouting lift in a transverse joint consists of an area bounded on the sides by seals adjacent to the upstream and downstream faces of the dam, and on the top and bottom by seals normally 50 to 60 feet apart. Since the longitudinal joints are staggered, the grouting area of a longitudinal joint is bounded by vertical seals placed close to the adjacent transverse joints and horizontal seals placed at 50- to 60-foot intervals in elevation. Each area of a transverse or longitudinal joint is sealed off from adjacent areas and has its own piping system independent of all other systems.

The layout of a piping system for transverse joints is illustrated on figures 8-1 and 8-2. A horizontal 1½-inch-diameter looped supply-header-return is embedded in the concrete adjacent to the lower boundary of the lift. One-half-inch-diameter embedded vertical risers take off from the header at approximately 6-foot intervals and terminate near the top of the lift or near the downstream face of the dam. Grout outlets are connected to the risers at 10-foot staggered intervals to give better coverage of the joint. The looped supply-header-return permits the delivery of grout to the various ½-inch riser pipes from either or both ends of the header, as may be desired, and provides reasonable assurance that grout will be admitted to all parts of the joint area. The top of each grout lift is vented to permit the escape of air, water, and thin grout which rises in the joint as grouting proceeds. A triangular grout groove can be formed in the face of the high block and covered with a metal plate which serves as a form for the concrete when the adjacent low block is placed. Vent pipes are connected to each end of the groove, thereby providing venting in either direction which will allow venting to continue if an obstruction is formed at any one point in the system. In some cases, a row of vent outlets may be used in lieu of a grout groove as shown on figure 8-1.

The piping arrangement for longitudinal joints is illustrated on figure 8-3. A horizontal 1½-inch-diameter looped supply-header-return line from either the downstream face or the gallery system is embedded in the concrete adjacent to the lower boundary of the lift. The 1½-inch supply line conveys the grout to the piping at each longitudinal joint. At each side of the grouting lift, a 1-inch-diameter riser takes off from the header and extends nearly to the top of the lift. The return line aids in the release of entrapped air and water in the system, and may be used for grouting the joint in the event the supply line becomes plugged. One-half-inch-diameter horizontal distribution pipes are connected between these risers spaced at 5 feet or 7 feet 6 inches, conforming to the height of the placement lifts. Grout outlets are attached to the horizontal distribution pipes at approximately staggered 10-foot intervals. As in the case of transverse joints, grout grooves or vent outlets are provided at the top of each lift and are connected to 1½-inch-diameter vent pipes which lead to the downstream face or to a gallery.

The location of the inlets and outlets of the supply-header-return and vents varies with conditions. Normally, these piping systems terminate at the downstream face of the dam. Under some conditions, these systems can be arranged to terminate in galleries. In order that the exposed ends of these systems will not be exposed after grouting operations have been completed, the pipes are terminated with a protruding pipe nipple which is wrapped with paper to prevent bonding to the concrete. This nipple is removed when no longer needed and the holes thus formed are dry-packed with mortar.

Typical grout outlets are shown on figure 8-5. The metal fitting alternative consists of two conduit boxes connected to the riser by a standard pipe tee. The blockout alternative is a blockout with a galvanized sheet steel cover. The riser goes through the blockout and a 2-inch section of the pipe is cut out. In erection, the box or blockout is placed in the high or first placed block and secured to the form. After the concrete has hardened and the forms have been removed, the cover box or sheet steel cover is placed in position and firmly held in place by wire or nails. A metal strap fastened to the cover serves as an anchor to fasten the cover to the second or low block so that the cover moves with it. When the two blocks contract upon cooling, the covers and the box or blockout are pulled apart and an

opening equal to the joint opening is provided for grout injection.

The grout grooves, formed in contraction joints and used for venting air, water, and thin grout, are covered with metal cover plates which act as forms when the concrete is placed in the low block. Details of the installation of the metal cover plates and the grout grooves are shown on figure 8-5. Before the cover plates are placed, the grooves are cleaned thoroughly of all concrete, dirt, and other foreign substances. At the upper edges of the cover plates, the joint between the cover plate and the concrete is covered with dry cement mortar or with asphalt emulsion to prevent mortar from the concrete from plugging the groove.

8-10. *Grouting Operations.*—Before any lift of a joint is grouted, the lift is washed thoroughly with air and water under pressure, the header and vent systems are tested to determine that they are unobstructed, and the joint is allowed to remain filled with water for a period of 24 hours. Immediately prior to being grouted, the water is drained from the joint lifts to be grouted. During the grouting operations, the lifts in two or more ungrouted adjacent joints at the same level are filled with water to the level of the top of the lift being grouted. As the grouting of the lift of the joint nears completion, the grouting lift of the joint immediately above the lift being grouted is filled with water. Immediately after a grouting operation is completed, the water is drained from the joints in the lift above, but the water is not drained from the adjacent ungrouted joint lifts at the same level until 6 hours after completion of the grouting operation.

The material used in grouting contraction joints is a mixture of cement and water, the consistency of which varies from thin to thick as the operation proceeds. Usually, a 2 to 1 mixture by volume of water and cement is used at the start of the grouting operation to assure grout travel and the filling of small cracks. As the grouting proceeds the mixture is thickened to a 1 to 1 water-cement ratio to fill the grout system and joint. If the joint is wide and accepts grout readily, grout of 0.7 to 0.8 water-cement ratio by volume may be used to

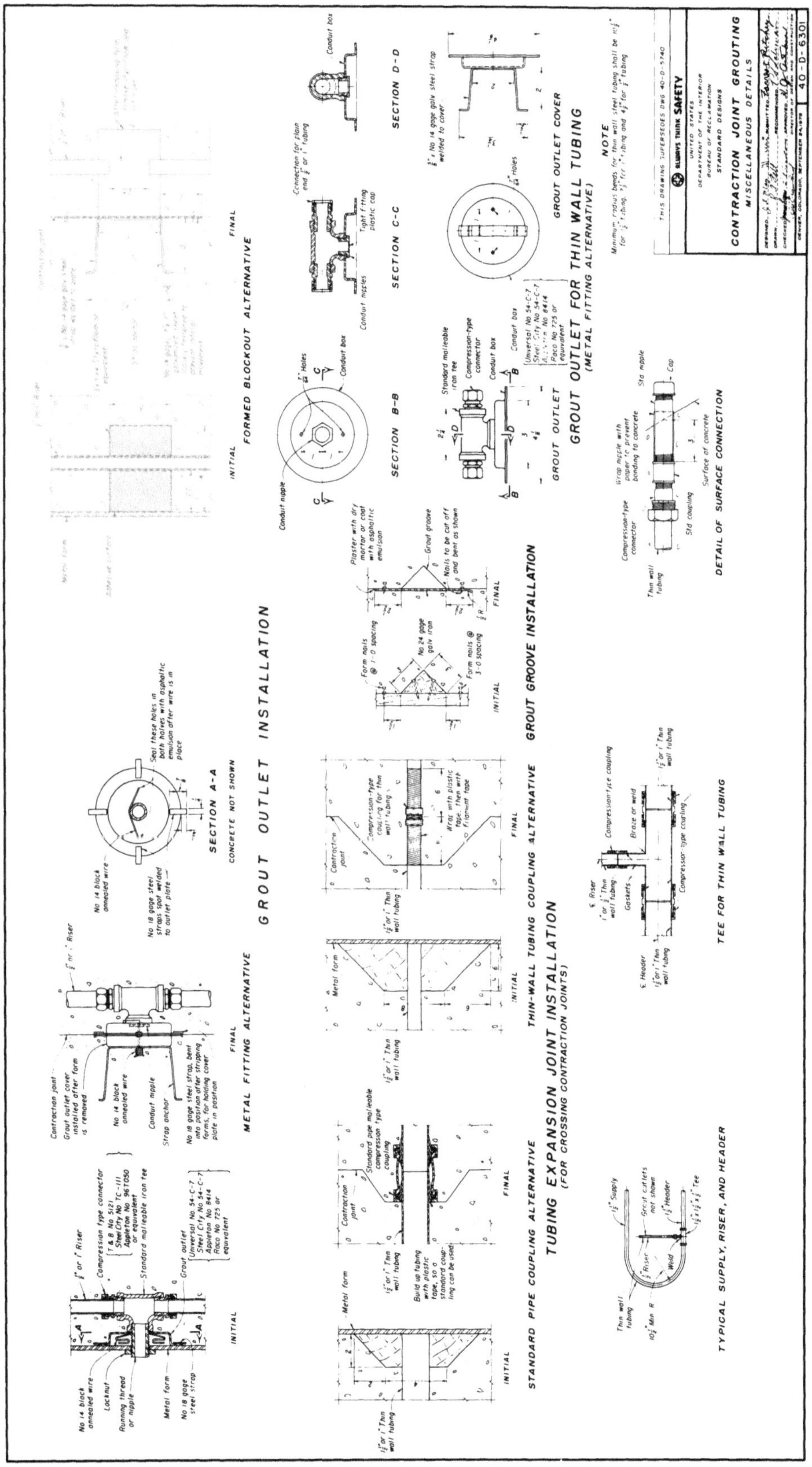

Figure 8-5. Grouting system details.

finish the operation. Normally, the supply line from the grout pump is connected to the supply so that grout first enters the joint through outlets in the most remote riser pipe, thereby setting up conditions most favorable for the expulsion of air, water, and diluted grout as the grouting operations proceed. If the grout introduced in the normal way makes a ready appearance at the return, the indications are that the header system is unobstructed and the return header can be capped. Grout from the header is forced up the risers and into the joint through the grout outlets, while air and water is forced up to the vent groove above.

Grouting of contraction joints in a dam is normally done in groups and in separate successive lifts, beginning at the foundation and finishing at the top of the dam. The grout is applied in rotation from joint to joint by batches in such quantities and with such time delays as necessary to allow the grout to settle in the joint. Each joint is filled at approximately the same rate. The grouting of each joint lift is completed before the grout

takes its set in the grouting system, but the lift is not grouted so rapidly that the grout will not settle in the joint. In no case is the time consumed in filling any lift of a joint less than 2 hours.

When thick grout flows from the vent outlets, injection is stopped for awhile to allow the grout to settle. After several repetitions of a showing of thick grout, the valves on the outlets are closed. The pressure on the supply line is then increased to the allowable limit for the particular joint to force grout into all small openings of the joint and to force the excess water into the pores of the concrete, leaving a grout film of lower water-cement ratio and higher density in the joint. The limiting pressure, usually from 30 to 50 pounds per square inch as measured at the vent, must be low enough to avoid deflecting the block excessively or causing opening of the grouted portion of the joint below. This maximum pressure is maintained until no more grout can be forced into the joint, and the system is then sealed off.

Chapter IX

Spillways

A. GENERAL DESIGN CONSIDERATIONS

9-1. *Function*.—Spillways are provided at storage and detention dams to release surplus or floodwater which cannot be contained in the allotted storage space, and at diversion dams to bypass flows exceeding those which are turned into the diversion system. Ordinarily, the excess is drawn from the top of the pool created by the dam and released through a spillway back to the river or to some natural drainage channel. Figure 9-1 shows the spillway at Grand Coulee Dam in operation.

The importance of a safe spillway cannot be overemphasized; many failures of dams have been caused by improperly designed spillways or by spillways of insufficient capacity. However, concrete dams usually will be able to withstand moderate overtopping. Generally, the increase in cost of a larger spillway is not directly proportional to increase in capacity. Very often the cost of a spillway of ample capacity will be only moderately higher than that of one which is obviously too small.

In addition to providing sufficient capacity, the spillway must be hydraulically and structurally adequate and must be located so that spillway discharges will not erode or undermine the downstream toe or abutments of the dam. The spillway's flow surfaces must be erosion resistant to withstand the high scouring velocities created by the drop from the reservoir surface to tailwater, and usually some device will be required for dissipation of energy at the bottom of the drop.

The frequency of spillway use will be determined by the runoff characteristics of the drainage area and by the nature of the development. Ordinary riverflows are usually stored in the reservoir, used for power generation, diverted through headworks, or released through outlets, and the spillway is not required to function. Spillway flows will result during floods or periods of sustained high runoff when the capacities of other facilities are exceeded. Where large reservoir storage is provided, or where large outlet or diversion capacity is available, the spillway will be utilized infrequently. Where storage space is limited and outlet releases or diversions are relatively small compared to normal riverflows, the spillway will be used frequently.

9-2. *Selection of Inflow Design Flood*.—(a) *General Considerations*.—When floods occur in an unobstructed stream channel, it is considered a natural event for which no individual or group assumes responsibility. However, when obstructions are placed across the channel, it becomes the responsibility of the sponsors either to make certain that hazards to downstream interests are not appreciably increased or to obligate themselves for damages resulting from operation or failure of such structures. Also, the loss of the facility and the loss of project revenue occasioned by a failure should be considered.

If danger to the structures alone were involved, the sponsors of many projects would prefer to rely on the improbability of an extreme flood occurrence rather than to incur the expense necessary to assure complete safety. However, when the risks involve downstream interests, including widespread damage and loss of life, a conservative attitude

Figure 9-1. Drumgate-controlled ogee-type overflow spillway in operation at Grand Coulee Dam in Washington. Note Third Powerplant construction in left background.—P1222-142-13418

is required in the development of the inflow design flood. Consideration of potential damage should not be confined to conditions existing at the time of construction. Probable future development in the downstream flood plain, encroachment by farms and resorts, construction of roads and bridges, etc., should be evaluated in estimating damages and hazards to human life that would result from failure of a dam.

Dams impounding large reservoirs and built on principal rivers with high runoff potential unquestionably can be considered to be in the high-hazard category. For such developments, conservative design criteria are selected on the basis that failure cannot be tolerated because

of the possible loss of life and because of the potential damages which could approach disaster proportions. However, dams built on isolated streams in rural areas where failure would neither jeopardize human life nor create damages beyond the sponsor's financial capabilities can be considered to be in a low-hazard category. For such developments design criteria may be established on a much less conservative basis. There are numerous instances, however, where failure of dams of low heights and small storage capacities have resulted in loss of life and heavy property damage. Most dams will require a reasonable conservatism in design, primarily because of the criterion that a dam failure must not present a serious hazard to human life.

(b) *Inflow Design Flood Hydrograph.*—Concrete dams are usually built on rivers from major drainage systems and impound large reservoirs. Because of the magnitude of the damage which would result from a failure of the dam, the probable maximum flood is used as the inflow design flood. The hydrograph for this flood is based on the hydrometeorological approach, which requires estimates of storm potential and the amount and distribution of runoff. The derivation of the probable maximum flood is discussed in appendix G.

The probable maximum flood is based on a rational consideration of the chances of simultaneous occurrence of the maximum of the several elements or conditions which contribute to the flood. Such a flood is the largest that reasonably can be expected and is ordinarily accepted as the inflow design flood for dams where failure of the structure would increase the danger to human life. The inflow design flood is determined by evaluating the hydrographs of the following situations to ascertain the most critical flood:

(1) A probable maximum rainstorm in conjunction with a severe, but not uncommon, antecedent condition.

(2) A probable maximum rainstorm in conjunction with a major snowmelt flood somewhat smaller than the probable maximum.

(3) A probable maximum snowmelt flood in conjunction with a major rainstorm less severe than the probable maximum for that season.

9-3. *Relation of Surcharge Storage to Spillway Capacity.* —The inflow design flood is normally represented in the form of a hydrograph, which charts the rate of flow in relation to time. A typical hydrograph representing a storm runoff is illustrated in figure 9-2, curve A. The flow into a reservoir at any time and the momentary peak can be read from this curve. The area under the curve is the volume of the inflow, since it represents the product of rate of flow and time.

Where no surcharge storage is allowed in the reservoir, the spillway capacity must be sufficiently large to pass the peak of the flood. The peak rate of inflow is then of primary interest and the total volume in the flood is of lesser importance. However, where a relatively large storage capacity above normal reservoir level can be made available economically by constructing a higher dam, a portion of the flood volume can be retained temporarily in reservoir surcharge space and the spillway capacity can be reduced considerably.

In many projects involving reservoirs, economic considerations will necessitate a design utilizing surcharge. The most economical combination of surcharge storage and spillway capacity requires flood routing studies and economic studies of the costs of spillway-dam combinations, subsequently described.

9-4. *Flood Routing.* —The storage accumulated in a reservoir depends on the difference between the rates of inflow and outflow. For an interval of time Δt, this relationship can be expressed by the equation:

$$\Delta S = Q_i \Delta t - Q_o \Delta t \qquad (1)$$

where:

ΔS = storage accumulated during Δt,
Q_i = average rate of inflow during Δt, and
Q_o = average rate of outflow during Δt.

Referring to figure 9-2, the rate of inflow at any time, t, is shown by the inflow design flood hydrograph; the rate of outflow may be obtained from the curve of spillway discharge versus reservoir water surface elevation; and storage is shown by the curve of reservoir capacity versus reservoir water surface elevation.

The quantity of water a spillway can discharge depends on the size and type of spillway. For a simple overflow crest the flow will vary with the head on the crest, and the surcharge will increase with an increase in spillway discharge. For a gated spillway, however, outflow can be varied with respect to reservoir head by operation of the gates. For example, one assumption for an operation of a gate-controlled spillway might be that the gates will be regulated so that inflow and outflow are

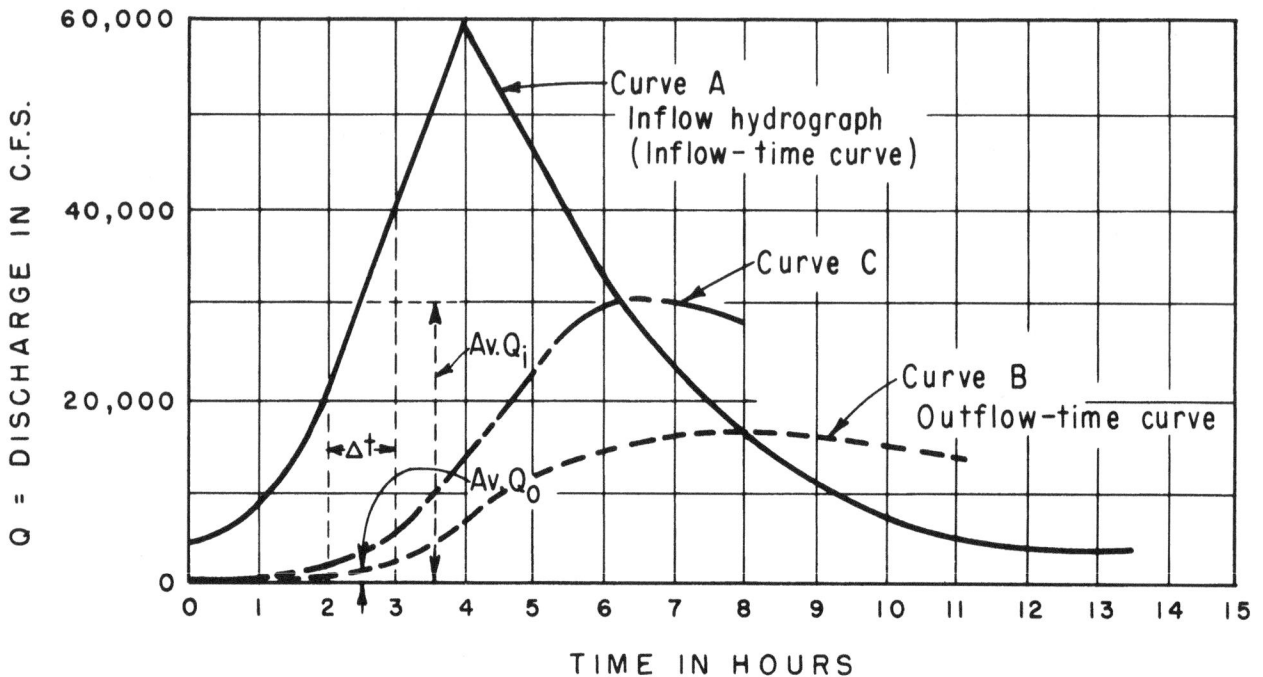

Figure 9-2. Typical inflow and outflow hydrographs.—288-D-3035

equal until the gates are wide open; or an assumption can be made to open the gates at a slower rate so that surcharge storage will accumulate before the gates are wide open.

Outflows need not necessarily be limited to discharges through the spillway but might be supplemented by other releases such as through river outlets, irrigation outlets, and powerplant turbines. In all such cases the size, type, and method of operation of the spillway and other releases with reference to the storage and/or to the inflow must be predetermined in order to establish an outflow-elevation relationship.

If simple equations could be established for the inflow design flood hydrograph curve, the outflow (as may be modified by operational procedures), and the reservoir capacity curve, a solution of flood routing could be made by mathematical integration. However, simple equations usually cannot be written for these variables, and such a solution is not practical. Many techniques of flood routing have been devised, each with its advantages and disadvantages. These techniques vary from a strictly arithmetical method to an entirely graphical solution.

Electronic computers are being used to make flood routing computations. The computer programs were developed using an iteration technique. For simplicity, an arithmetical trial and error tabular method is illustrated in this manual. Data required for the routing, which is the same regardless of the method used, are as follows:

(1) Inflow hydrograph, figure 9-2.

(2) Reservoir capacity, figure 9-3.

(3) Outflow, figure 9-4. (Spillway discharge only was assumed in this illustration.)

The flood routing computations are shown in table 9-1. The procedure for making the computations is as follows:

(1) Select a time interval, Δt, column (2).

(2) Obtain column (3) from the inflow hydrograph, figure 9-2.

(3) Column (4) represents average inflow for Δt in c.f.s. (cubic feet per second).

(4) Obtain column (5) by converting column (4) values of c.f.s. for Δt to acre-feet (1 c.f.s. for 12 hours = 1 acre-foot).

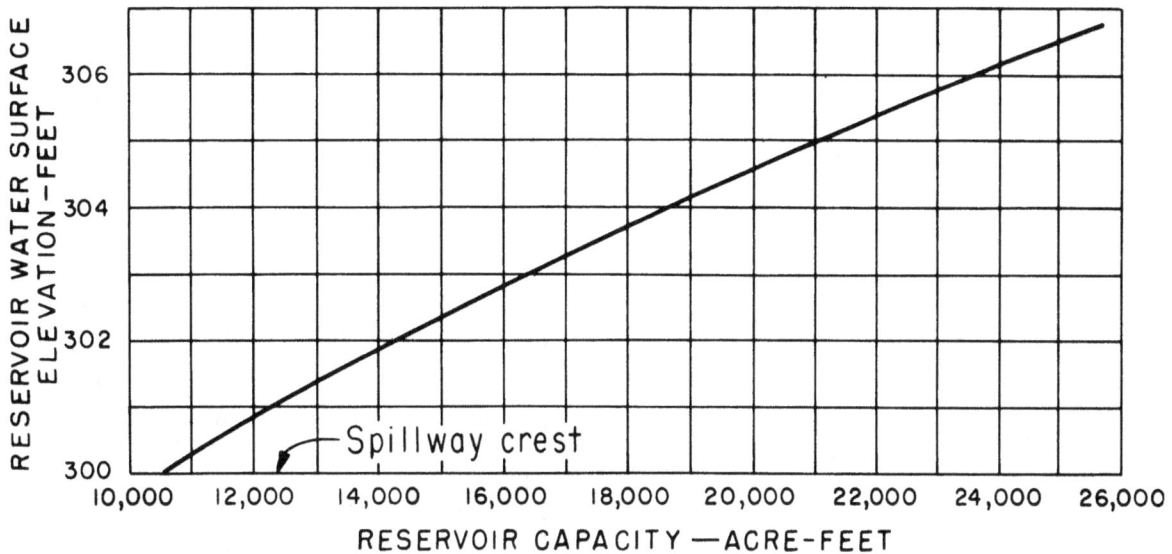

Figure 9-3. Typical reservoir capacity curve.—288-D-3036

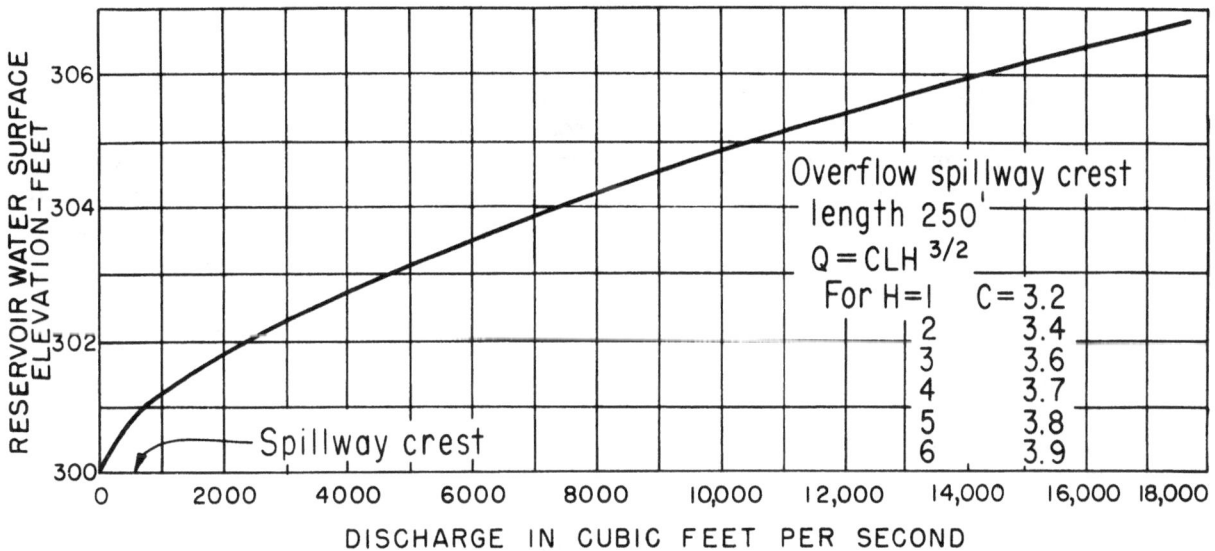

Figure 9-4. Typical spillway discharge curve.—288-D-3037

(5) Assume trial reservoir water surface in column (6), determine the corresponding rate of outflow from figure 9-4, and record in column (7).

(6) Average the rate of outflow determined in step (5) and the rate of outflow for the reservoir water surface which existed at the beginning of the period and enter in column (8).

(7) Obtain column (9) by converting column (8) values of c.f.s. for Δt to acre-feet, similar to step (4).

(8) Column (10) = column (5) minus column (9).

(9) The initial value in column (11) represents the reservoir storage at the beginning of the inflow design flood. Determine subsequent values by adding ΔS values from column (10) to the previous column (11) value.

Table 9-1.—*Flood routing computations.*

(1) Time t, hours	(2) Δt, hours	(3) Inflow at time, t, c.f.s.	(4) Average rate of inflow Q_i for Δt, c.f.s.	(5) Inflow, acre-feet	(6) Trial reservoir storage elevation at time t,	(7) Outflow at time t, c.f.s.	(8) Average rate of outflow Q_o for Δt, c.f.s.	(9) Outflow, acre-feet	(10) Incremental storage Δs, acre-feet	(11) Total storage, acre-feet	(12) Reservoir elevation, end of Δt, feet	(13) Remarks
0		4,000				0				10,500	300.0	
	1		6,000	500	~~300.2~~	~~50~~	~~25~~	~~2~~	~~498~~	~~10,998~~	~~300.3~~	~~HIGH~~
1		8,000			300.3	100	50	4	496	10,996	300.3	OK
	1		14,000	1,167	~~300.8~~	~~540~~	~~320~~	~~27~~	~~1,140~~	~~12,136~~	~~301.0~~	~~HIGH~~
2		20,000			301.0	800	450	38	1,129	12,125	301.0	OK
	1		30,000	2,500	~~302.3~~	~~3,000~~	~~1,900~~	~~158~~	~~2,342~~	~~14,467~~	~~302.1~~	~~LOW~~
3		40,000			302.1	2,600	1,700	142	2,358	14,483	302.1	OK
	1		50,000	4,167	~~303.9~~	~~7,100~~	~~4,850~~	~~404~~	~~3,763~~	~~18,246~~	~~303.8~~	~~LOW~~
4		60,000			303.8	6,900	4,750	396	3,771	18,254	303.8	OK
	1		53,500	4,458	~~305.0~~	~~10,600~~	~~8,750~~	~~729~~	~~3,729~~	~~21,983~~	~~305.3~~	~~HIGH~~
5		47,000			305.3	11,600	9,250	771	3,687	21,941	305.3	OK
	1		40,000	3,333	~~306.3~~	~~15,300~~	~~13,450~~	~~1121~~	~~2,212~~	~~24,153~~	~~306.2~~	~~LOW~~
6		33,000			306.2	15,100	13,350	1113	2,220	24,161	306.2	OK
	1		28,500	2,375	306.6	16,800	15,950	1329	1,046	25,207	306.6	OK
7		24,000										
	1		20,000	1,667	306.7	17,200	17,000	1417	250	25,457	306.7	OK
8		16,000										
	1		13,500	1,125	306.6	16,800	17,000	1417	-292	25,165	306.6	OK
9		11,000										
	2		8,000	1,333	~~306.0~~	~~14,300~~	~~15,550~~	~~2592~~	~~-1259~~	~~23,906~~	~~306.1~~	~~HIGH~~
11		5,000			306.1	14,700	15,750	2,625	-1292	23,873	306.1	OK

(10) Determine reservoir elevation in column (12) corresponding to storage in column (11) from figure 9-3.

(11) Compare reservoir elevation in column (12) with trial reservoir elevation in column (6). If they do not agree within 0.1 foot, make a second trial elevation and repeat procedure until agreement is reached.

The outflow-time curve resulting from the flood routing shown in table 9-1 has been plotted as curve B on figure 9-2. As the area under the inflow hydrograph (curve A) indicates the volume of inflow, so will the area under the outflow hydrograph (curve B) indicate the volume of outflow. It follows then that the volume indicated by the area between the two curves will be the surcharge storage. The surcharge storage computed in table 9-1 can, therefore, be checked by comparing it with the measured area on the graph.

A rough approximation of the relationship of spillway size to surcharge volume can be obtained without making an actual flood routing, by arbitrarily assuming an approximate outflow-time curve and then measuring the area between it and the inflow hydrograph. For example, if the surcharge volume for the problem shown on figure 9-2 is sought where a 30,000-c.f.s. spillway would be provided, an assumed outflow curve represented by curve C can be drawn and the area between this curve and curve A can be planimetered. Curve C will reach its apex of 30,000 c.f.s. where it crosses curve A. The volume represented by the area between the two curves will indicate the approximate surcharge volume necessary for this capacity spillway.

9-5. *Selection of Spillway Size and Type.*—(a) *General Considerations.*—In determining the best combination of storage and spillway capacity to accommodate the selected inflow design flood, all pertinent factors of hydrology, hydraulics, geology, topography, design requirements, cost, and benefits should be considered. These considerations involve such factors as (1) the characteristics of the flood hydrograph; (2) the damages which would result if such a flood occurred without the dam; (3) the damages which would result if such a flood occurred with the dam in place; (4) the damages which would occur if the dam or spillway should fail; (5) effects of various dam and spillway combinations on the probable increase or decrease of damages above or below the dam (as indicated by reservoir backwater curves and tailwater curves); (6) relative costs of increasing the capacity of spillways; and (7) use of combined outlet facilities to serve more than one function, such as control of releases and control or passage of floods. Other outlets, such as river outlets, irrigation outlets, and powerplant turbines, should be considered in passing part of the inflow design flood when such facilities are expected to be available in time of flood.

The outflow characteristics of a spillway depend on the particular device selected to control the discharge. These control facilities may take the form of an overflow crest or orifice. Such devices can be unregulated or they can be equipped with gates or valves to regulate the outflow.

After the overflow characteristics have been selected, the maximum spillway discharge and the maximum reservoir water level can be determined by flood routing. Other components of the spillway can then be proportioned to conform to the required capacity and to the specific site conditions, and a complete layout of the spillway can be established. Cost estimates of the spillway and dam can then be made. Estimates of various combinations of spillway capacity and dam height for an assumed spillway type, and of alternative types of spillways, will provide a basis for selection of the economical spillway type and the optimum relation of spillway capacity to height of dam. Figures 9-5 and 9-6 illustrate the results of such a study. The relationships of spillway capacities to maximum reservoir water surfaces obtained from the flood routings is shown on figure 9-5 for two spillways. Figure 9-6 illustrates the comparative costs for different combinations of spillway and dam, and indicates a combination which results in the least total cost.

To make such a study as illustrated requires many flood routings, spillway layouts, and spillway and dam estimates. Even then, the study is not necessarily complete since many other spillway arrangements could be considered. A comprehensive study to

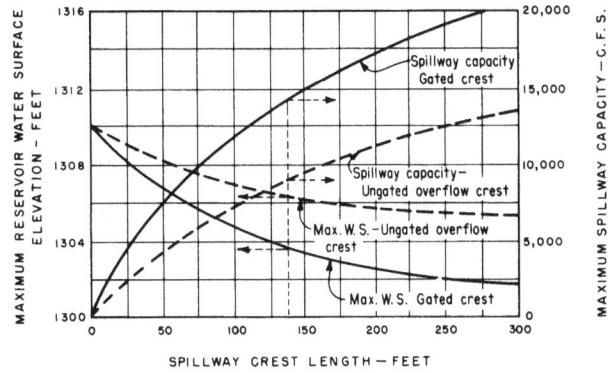

Figure 9-5. Spillway capacity—surcharge relationship.—288-D-3039

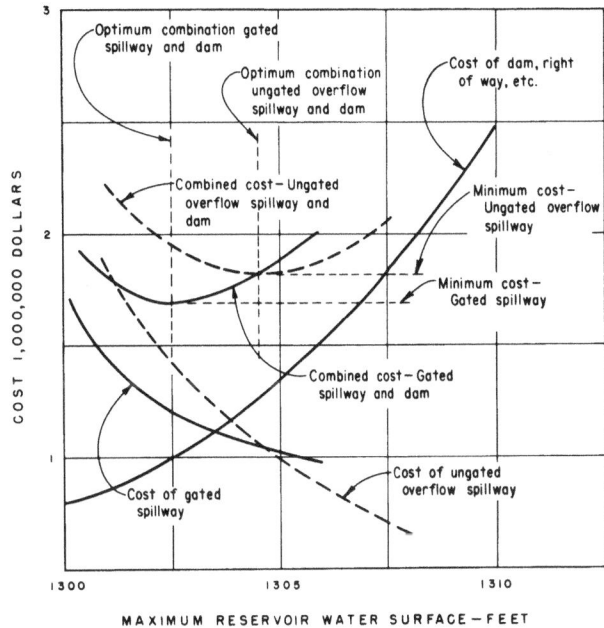

Figure 9-6. Comparative cost of spillway-dam combinations.—288-D-3040

determine alternative optimum combinations and minimum costs may not be warranted for the design of some dams. Judgment on the part of the designer would be required to select for study only the combinations which show definite advantages, either in cost or adaptability. For example, although a gated spillway might be slightly cheaper overall than an ungated spillway, it may be desirable to adopt the latter because of its less complicated construction, its automatic and trouble-free operation, its ability to function without an attendant, and its less costly maintenance.

(b) *Combined Service and Auxiliary Spillways.*—Where site conditions are favorable, the possibility of gaining overall economy by utilizing an auxiliary spillway in conjunction with a smaller service-type structure should be considered. In such cases the service spillway should be designed to pass floods likely to occur frequently and the auxiliary spillway control set to operate only after such small floods are exceeded. In certain instances the outlet works may be made large enough to serve also as a service spillway. Conditions favorable for the adoption of an auxiliary spillway are the existence of a saddle or depression along the rim of the reservoir which leads into a natural waterway, or a gently sloping abutment where an excavated channel can be carried sufficiently beyond the dam to avoid the possibility of damage to the dam or other structures.

Because of the infrequency of use, it is not necessary to design the entire auxiliary spillway for the same degree of safety as required for other structures; however, at least the control portion must be designed to forestall failure, since its breaching would release large flows from the reservoir. For example, concrete lining may be omitted from an auxiliary spillway channel excavated in rock which is not easily eroded. Where the channel is excavated through less competent material, it might be lined but terminated above the river channel with a cantilevered lip rather than extending to a stilling basin at river level. The design of auxiliary spillways is often based on the premise that some damage to portions of the structure from passage of infrequent flows is permissible. Minor damage by scour to an unlined channel, by erosion and undermining at the downstream end of the channel, and by creation of an erosion pool downstream from the spillway might be tolerated.

An auxiliary spillway can be designed with a fixed crest control, or it can be stoplogged or gated to increase the capacity without additional surcharge head.

B. DESCRIPTION OF SPILLWAYS

9-6. *Selection of Spillway Layout*.—The design of a spillway, including all of its components, can be prepared by properly considering the various factors influencing the spillway size and type, and correlating alternatively selected components. Many combinations of components can be used in forming a complete spillway layout. After the hydraulic size and outflow characteristics of a spillway are determined by routing of the design flood, the general dimensions of the control can be selected. Then, a specific spillway layout can be developed by considering the topography and foundation conditions, and by fitting the control structure and the various components to the prevailing conditions.

Site conditions greatly influence the selection of location, type, and components of a spillway. Factors that must be considered in the selection are the possibility of incorporating the spillway into the dam, the steepness of the terrain that would be traversed by a chute-type spillway, the amount of excavation required and the difficulty of its disposal, the chances of scour of the flow surfaces and the need for lining the spillway channel, the permeability and bearing capacity of the foundation, the stability of the excavated slopes, and the possible use of a tunnel-type spillway.

The adoption of a particular size or arrangement for one of the spillway components may influence the selection of other components. For example, a wide control structure with the crest placed normal to the centerline of the spillway would require a long converging transition to join it to a narrow discharge channel or to a tunnel; a better alternative might be the selection of a narrower gated control structure or a side channel control arrangement. Similarly, a wide stilling basin may not be feasible for use with a cut-and-cover conduit or tunnel, because of the long, diverging transition needed.

A spillway may be an integral part of a dam such as an overflow section of a concrete dam, or it may be a separate structure. In some instances, it may be integrated into the river diversion plan for economy. Thus, the location, type, and size of other appurtenances are factors which may influence the selection of a spillway location or its arrangement. The final plan will be governed by overall economy, hydraulic sufficiency, and structural adequacy.

The components of a spillway and common types of spillways are described and discussed herein. Hydraulic design criteria and procedures are discussed in sections 9-10 through 9-29.

9-7. Spillway Components.—(a) *Control Structure.*—A major component of a spillway is the control device, since it regulates and controls the outflows from the reservoir. This control limits or prevents outflows below fixed reservoir levels, and it also regulates releases when the reservoir rises above these levels. The control structure is usually located at the upstream end of the spillway and consists of some form of overflow crest or orifice. Sometimes the configuration of the spillway downstream from the control structure is such that with higher discharges the structure no longer controls the flow. For example, with the morning glory spillway, shown on figure 9-44, the tunnel rather than the crest or orifice usually controls the flow at higher discharges (see sec. 9-25).

Control structures may take various forms in both positioning and shape. In plan, overflow crests can be straight, curved, semicircular, U-shaped, or circular. Figure 9-7 shows the circular crest for the morning glory spillway at Hungry Horse Dam. Orifice controls can be placed in a horizontal, inclined, or vertical position. The orifice can be circular, square, rectangular, triangular, or varied in shape.

An overflow can be sharp crested, ogee shaped, broad crested, or of varied cross section. Orifices can be sharp edged, round edged, or bellmouth shaped, and can be placed so as to discharge with a fully contracted jet or with a suppressed jet. They may discharge freely or discharge partly or fully submerged.

Figure 9-7. Circular crest for morning glory spillway at Hungry Horse Dam in Montana.—P447-105-5587

(b) *Discharge Channel.*—Flow released through the control structure usually is conveyed to the streambed below the dam in a discharge channel or waterway. Exceptions are where the discharge falls free from an arch dam crest or where the flow is released directly along the abutment hillside to cascade down the abutment face. The conveyance structure may be the downstream face of a concrete dam, an open channel excavated along the ground surface on one abutment, or a tunnel excavated through an abutment. The profile may be variably flat or steep; the cross section may be variably rectangular, trapezoidal, circular, or of other shape; and the discharge channel may be wide or narrow, long or short.

Discharge channel dimensions are governed primarily by hydraulic requirements, but the selection of profile, cross-sectional shape, width, length, etc., is influenced by the geologic and topographic characteristics of the site. Open channels excavated in the abutment usually follow the ground surface profile; steep canyon walls usually make a tunnel desirable. In plan, open channels may be straight or curved, with sides parallel, convergent, divergent, or a combination of these. Discharge channels must be cut through or lined with material which is resistant to the scouring action of the high velocities, and which is structurally adequate to withstand the forces from backfill, uplift, waterloads, etc.

(c) *Terminal Structure.*—When water flows in a spillway from reservoir pool level to downstream river level, the static head is converted to kinetic energy. This energy manifests itself in the form of high velocities which if impeded result in large pressures. Means of returning the flow to the river without serious scour or erosion of the toe of the dam or damage to adjacent structures must usually be provided.

In some cases the discharge may be delivered at high velocities directly to the stream where the energy is absorbed along the streambed by impact, turbulence, and friction. Such an arrangement is satisfactory where erosion-resistant bedrock exists at shallow depths in the channel and along the abutments or where the spillway outlet is sufficiently removed from the dam or other appurtenances to avoid damage by scour, undermining, or abutment sloughing. The discharge channel may be terminated well above the streambed level or it may be continued to or below streambed.

Upturned deflectors, cantilevered extensions, or flip buckets can be provided to project the jet some distance downstream from the end of the structure. Often, erosion of the streambed in the area of impact of the jet can be minimized by fanning the jet into a thin sheet by the use of a flaring deflector.

Where severe scour at the point of jet impingement is anticipated, a plunge pool can be excavated in the river channel and the sides and bottom lined with riprap or concrete. It may be expedient to perform a minimum of excavation and to permit the flow to erode a natural pool; protective riprapping or concrete lining may be later provided to halt the scour if necessary. In such arrangements an adequate cutoff or other protection must be provided at the end of the spillway structure to prevent it from being undermined.

Where serious erosion to the streambed is to be avoided, the high energy of the flow must be dissipated before the discharge is returned to the stream channel. This can be accomplished by the use of an energy dissipating device, such as a hydraulic jump basin, a roller bucket, an apron, a basin incorporating impact baffles and walls, or some similar energy absorber or dissipator.

(d) *Entrance and Outlet Channels.*— Entrance channels serve to draw water from the reservoir and convey it to the control structure. Where a spillway draws water immediately from the reservoir and delivers it directly back into the river, as in the case with an overflow spillway over a concrete dam, entrance and outlet channels are not required. However, in the case of spillways placed through abutments or through saddles or ridges, channels leading to the spillway control and away from the spillway terminal structure may be required.

Entrance velocities should be limited and channel curvatures and transitions should be made gradual, in order to minimize head loss through the channel (which has the effect of reducing the spillway discharge) and to obtain uniformity of flow over the spillway crest. Effects of an uneven distribution of flow in the entrance channel might persist through the spillway structure to the extent that undesirable erosion could result in the downstream river channel. Nonuniformity of head on the crest may also result in a reduction in the discharge.

The approach velocity and depth below crest level each have important influences on the discharge over an overflow crest. As discussed in section 9-11(b), a greater approach depth with the accompanying reduction in approach velocity will result in a larger discharge coefficient. Thus, for a given head over the crest, a deeper approach will permit a shorter crest length for a given discharge. Within the limits required to secure satisfactory flow conditions and nonscouring velocities, the determination of the relationship of entrance channel depth to channel width is a matter of economics. When the spillway entrance channel is excavated in material that will be eroded by the approach velocity, a zone of riprap is often provided immediately upstream from the inlet lining to prevent scour of the channel floor and side slope adjacent to the spillway concrete.

Outlet channels convey the spillway flow from the terminal structure to the river channel below the dam. An outlet channel should be excavated to an adequate size to pass the anticipated flow without forming a control which will affect the tailwater stage in the stilling device.

The outlet channel dimensions and its need for protection by lining or riprap will depend on the nature of the material through which the channel is excavated and its susceptibility to scouring. Although stilling devices are provided, it may be impossible to reduce resultant velocities below the natural velocity in the original stream; and some scouring of the riverbed, therefore, may not be avoidable. Further, under natural conditions the beds of many streams are scoured during the rising stage of a flood and filled during the falling stage by deposition of material carried by the flow. After creation of a reservoir the spillway will normally discharge clear water and the material scoured by the high velocities will not be replaced by deposition. Consequently, there will be a gradual retrogression of the downstream riverbed, which will lower the tailwater stage-discharge relationship. Conversely, scouring where only a pilot channel is provided may build up bars and islands downstream, thereby effecting an aggradation of the downstream river channel which will raise the tailwater elevation with respect to discharges. The dimensions and erosion-protective measures at the outlet channel may be influenced by these considerations.

9-8. *Spillway Types.*—Spillways are ordinarily classified according to their most distinguishing feature, either as it pertains to the control, to the discharge channel, or to some other component. Spillways often are referred to as controlled or uncontrolled, depending on whether they are gated or ungated. Common types are the free fall, ogee (overflow), side channel, chute or open channel, tunnel, and morning glory spillways.

(a) *Free Fall Spillways.*—A free fall spillway is one in which the flow drops freely, usually into the streambed. Flows may be free discharging, as with a sharp-crested weir or orifice control, or they may be supported part way down the face of the dam and then trajected away from the dam by a flip bucket.

Where no artificial protection is provided at the base, scour will occur in some streambeds and will form a deep plunge pool. The volume and depth of the hole are related to the range of discharges, the height of the drop, and the depth of tailwater. The erosion-resistant properties of the streambed material including bedrock have little influence on the size of the hole, the only effect being the time necessary to scour the hole to its full depth. Probable depths of scour are discussed in section 9-24. Where erosion cannot be tolerated, a plunge pool can be created by constructing an auxiliary dam downstream from the main structure, or by excavating a basin which is then provided with a concrete apron or bucket.

If tailwater depths are sufficient, a hydraulic jump will form when a free fall jet falls upon a flat apron. It has been demonstrated that the momentum equation for the hydraulic jump may be applied to the flow conditions at the base of the fall to determine the elements of the jump.

(b) *Ogee (Overflow) Spillways.*—The ogee spillway has a control weir which is ogee- or S-shaped in profile. The upper curve of the ogee ordinarily is made to conform closely to the profile of the lower nappe of a ventilated sheet of water falling from a sharp-crested weir. Flow over the crest is made to adhere to the face of the profile by preventing access of air to the underside of the sheet. For discharges at designed head, the flow glides over the crest with minimum interference from the boundary surface and attains near-maximum discharge efficiency. The profile below or downstream of the upper curve of the ogee is continued tangent along a slope to support the flowing sheet on the face of the weir. A reverse curve at the bottom of the slope turns the flow onto the apron of a stilling basin, into a flip bucket, or into the spillway discharge channel. Figure 9-1 shows this type of spillway in operation at Grand Coulee Dam.

The upper curve at the crest may be made either broader or sharper than the nappe profile. A broader shape will support the sheet and positive hydrostatic pressure will occur along the contact surface. The supported sheet thus creates a backwater effect and reduces the efficiency of discharge. For a sharper shape, the sheet tends to pull away from the crest and to produce subatmospheric pressure along the contact surface. This negative pressure effect increases the effective head, and thereby increases the discharge.

An ogee crest and apron may comprise an entire spillway, such as the overflow portion of a concrete gravity dam, or the ogee crest may be only the control structure for some other type of spillway. Because of its high discharge efficiency, the nappe-shaped profile is used for most spillway control crests.

(c) *Side Channel Spillways.*—The side channel spillway is one in which the control weir is placed along the side of and approximately parallel to the upper portion of the spillway discharge channel. Flow over the crest falls into a narrow trough behind the weir, turns an approximate right angle, and then continues into the main discharge channel. The side channel design is concerned only with the hydraulic action in the upstream reach of the discharge channel and is more or less independent of the details selected for the other spillway components. Flows from the side channel can be directed into an open discharge channel or into a closed conduit or inclined tunnel. Flow into the side channel might enter on only one side of the trough in the case of a steep hillside location, or on both sides and over the end of the trough if it is located on a knoll or gently sloping abutment. Figure 9-8 shows the Arizona spillway at Hoover Dam which consists of a side channel discharging into a large tunnel.

Discharge characteristics of a side channel spillway are similar to those of an ordinary overflow and are dependent on the selected profile of the weir crest. However, for maximum discharges the side channel flow may

Figure 9-8. Drumgate-controlled side channel spillway in operation at Hoover Dam on the Colorado River.—BC P5492

differ from that of the overflow spillway in that the flow in the trough may be restricted and may partly submerge the flow over the crest. In this case the flow characteristics will be controlled by the channel downstream from the trough.

Although the side channel is not hydraulically efficient nor inexpensive, it has advantages which make it adaptable to certain spillway layouts. Where a long overflow crest is desired in order to limit the surcharge head and the abutments are steep and precipitous, or where the control must be connected to a narrow discharge channel or tunnel, the side channel is often the best choice.

(d) *Chute Spillways.*—A spillway whose discharge is conveyed from the reservoir to the downstream river level through an open channel, placed either along a dam abutment or through a saddle, is called a chute or open channel spillway. These designations can apply regardless of the control device used to regulate the flow. Thus, a spillway having a chute-type discharge channel, though controlled by an overflow crest, a gated orifice, a side channel crest, or some other control device, might still be called a chute spillway. However, the name is most often applied when the spillway control is placed normal or nearly normal to the axis of an open channel, and where the streamlines of flow both above and below the control crest follow in the direction of the axis.

Chute spillways ordinarily consist of an entrance channel, a control structure, a discharge channel, a terminal structure, and an outlet channel. The simplest form of chute spillway has a straight centerline and is of uniform width. Often, either the axis of the entrance channel or that of the discharge channel must be curved to fit the alinement of the chute to the topography. In such cases, the curvature is confined to the entrance channel if possible, because of the low approach velocities. Where the discharge channel must be curved, its floor is sometimes superelevated to guide the high-velocity flow around the bend, thus avoiding a piling up of flow toward the outside of the chute.

Chute spillway profiles are usually influenced by the site topography and by subsurface foundation conditions. The control structure is generally placed in line with or upstream from the dam. Usually the upper portion of the discharge channel is carried at minimum grade until it "daylights" along the downstream hillside to minimize excavation. The steep portion of the discharge channel then follows the slope of the abutment.

Flows upstream from the crest are generally at subcritical velocity, with critical velocity occurring when the water passes over the control. Flows in the chute are ordinarily maintained at supercritical stage, either at constant or accelerating rates, until the terminal structure is reached. For good hydraulic performance, abrupt vertical changes or sharp convex or concave vertical curves in the chute profile should be avoided. Similarly, the convergence or divergence in plan should be gradual in order to avoid cross waves, "ride-up" on the walls, excessive turbulence, or

uneven distribution of flow at the terminal structure.

Figure 9-9 shows the chute-type structure at Elephant Butte Dam in New Mexico.

(e) *Tunnel Spillways.*—Where a tunnel is used to convey the discharge around a dam, the spillway is called a tunnel spillway. The spillway tunnel usually has a vertical or inclined shaft, a large-radius elbow, and a horizontal tunnel at the downstream end. Most forms of control structures, including overflow crests, vertical or inclined orifice entrances, and side channel crests can be used with tunnel spillways.

With the exception of morning glory spillways, discussed later, tunnel spillways are designed to flow partly full throughout their length. To guarantee free flow in the tunnel, the ratio of the flow area to the total tunnel area is often limited to about 75 percent. Air vents may be provided at critical points along the tunnel to insure an adequate air supply which will avoid unsteady flow through the spillway.

Tunnel spillways may present advantages for damsites in narrow canyons with steep abutments or at sites where there is danger to open channels from snow or rock slides.

(f) *Morning Glory Spillways.*—A morning glory spillway (sometimes called a drop inlet spillway) is one in which the water enters over a horizontally positioned lip, which is circular in plan, drops through a vertical or sloping shaft, and then flows to the downstream river channel through a horizontal or near horizontal tunnel. The structure may be considered as being made up of three elements; namely, an overflow control weir, an orifice control section, and a closed discharge channel.

Discharge characteristics of the morning glory spillway usually vary with the range of head. The control will shift according to the relative discharge capacities of the weir, the orifice, and the tunnel. For example, as the head increases, the control will shift from weir flow over the crest to orifice flow in the throat and then to full tunnel flow in the downstream portion of the spillway. Full tunnel flow design for spillways, except those with extremely low drops, is not recommended, as discussed in

Figure 9-9. Chute type spillway (left) at Elephant Butte Dam in New Mexico.—P24-500-1250

section 9-29.

A morning glory spillway can be used advantageously at damsites in narrow canyons where the abutments rise steeply or where a diversion tunnel is available for use as the downstream leg. Another advantage of this type of spillway is that near maximum capacity is attained at relatively low heads; this characteristic makes the spillway ideal for use where the maximum spillway outflow is to be limited. This characteristic also may be considered disadvantageous, in that there is little increase in capacity beyond the designed heads, should a flood larger than the selected inflow design flood occur. This would not be a disadvantage if this type of spillway were used as a service spillway in conjunction with an auxiliary spillway.

9-9. *Controls for Crests.*—The simplest form of control for a spillway is the free or uncontrolled overflow crest which

automatically releases water whenever the reservoir water surface rises above crest level. The advantages of the uncontrolled crest are the elimination of the need for constant attendance and regulation of the control device by an operator, and the freedom from maintenance and repairs of the device.

A regulating device or movable crest must be employed if a sufficiently long uncontrolled crest or a large enough surcharge head cannot be obtained for the required spillway capacity. Such control devices will also be required if the spillway is to release storages below the normal reservoir water surface. The type and size of the selected control device may be influenced by such conditions as discharge characteristics of a particular device, climate, frequency and nature of floods, winter storage requirements, flood control storage and outflow provisions, the need for handling ice and debris, and special operating requirements. Whether an operator will be in attendance during periods of flood, and the availability of electricity, operating mechanisms, operating bridges, etc., are other factors which will influence the type of control device employed.

Many types of crest control have been devised. The type selected for a specific installation should be based on a consideration of the factors noted above as well as economy, adaptability, reliability, and efficiency. In the classification of movable crests are such devices as flashboards and stoplogs. Regulating devices include vertical and inclined rectangular lift gates, radial gates, drum gates, and ring gates. These may be controlled manually or automatically. Automatic gates may be either mechanical or hydraulic in operation. The gates are often raised automatically to follow a rising water surface, then lowered if necessary to provide sufficient spillway capacity for larger floods.

(a) *Flashboards and Stoplogs.*—Flashboards and stoplogs provide a means of raising the reservoir storage level above a fixed spillway crest level, when the spillway is not needed for releasing floods. Flashboards usually consist of individual boards or panels supported by vertical pins or stanchions anchored to the crest; stoplogs are boards or panels spanning horizontally between grooves recessed into supporting piers. In order to provide adequate spillway capacity, the flashboards or stoplogs must be removed before the floods occur, or they must be designed or arranged so that they can be removed while being overtopped.

Various arrangements of flashboards have been devised. Some must be placed and removed manually, some are designed to fail after being overtopped, and others are arranged to drop out of position either automatically or by being manually triggered after the reservoir exceeds a certain stage. Flashboards provide a simple economical type of movable crest device, and they have the advantage that an unobstructed crest is provided when the flashboards and their supports are removed. They have numerous disadvantages, however, which greatly limit their adaptability. Among these disadvantages are the following: (1) They present a hazard if not removed in time to pass

floods, especially where the reservoir area is small and the stream is subject to flash floods; (2) they require the attendance of an operator or crew to remove them, unless they are designed to fail automatically; (3) if they are designed to fail when the water reaches certain stages their operation is uncertain, and when they fail they release sudden and undesirably large outflows; (4) ordinarily they cannot be restored to position while flow is passing over the crest; and (5) if the spillway functions frequently the repeated replacement of flashboards may be costly.

Stoplogs are individual beams or girders set one upon the other to form a bulkhead supported in grooves at each end of the span. The spacing of the supporting piers will depend on the material from which the stoplogs are constructed, the head of water acting against the stoplogs, and the handling facilities provided for installing and removing them. Stoplogs which are removed one by one as the need for increased discharge occurs are the simplest form of a crest gate.

Stoplogs may be an economical substitute for more elaborate gates where relatively close spacing of piers is not objectionable and where removal is required only infrequently. Stoplogs which must be removed or installed in flowing water may require such elaborate hoisting mechanisms that this type of installation may prove to be as costly as gates. A stoplogged spillway requires the attendance of an operating crew for removing and installing the stoplogs. Further, the arrangement may present a hazard to the safety of the dam if the reservoir is small and the stream is subject to flash floods, since the stoplogs must be removed in time to pass the flood.

(b) *Rectangular Lift Gates.*—Rectangular lift gates span horizontally between guide grooves in supporting piers. Although these gates may be made of wood or concrete, they are often made of metal (cast iron or steel). The support guides may be placed either vertically or inclined slightly downstream. The gates are raised or lowered by an overhead hoist. Water is released by undershot orifice flow for all gate openings.

For sliding gates the vertical side members of

the gate frame bear directly on the guide members; sealing is effected by the contact pressure. The size of this type of installation is limited by the relatively large hoisting capacity required to operate the gate because of the sliding friction that must be overcome.

Where larger gates are needed, wheels can be mounted along each side of the rectangular lift gates to carry the load to a vertical track on the downstream side of the pier groove. The use of wheels greatly reduces the amount of friction and thereby permits the use of a smaller hoist.

(c) *Radial Gates.*—Radial gates are usually constructed of steel. They consist of a cylindrical segment which is attached to supporting bearings by radial arms. The face segment is made concentric to the supporting pins so that the entire thrust of the waterload passes through the pins; thus, only a small moment need be overcome in raising and lowering the gate. Hoisting loads then consist of the weight of the gate, the friction between the side seals and the piers, and the frictional resistance at the pins. The gate is often counterweighted to partially counterbalance the effect of its weight, which further reduces the required capacity of the hoist.

The small hoisting effort needed to operate radial gates makes hand operation practical on small installations which otherwise might require power. The small hoisting forces involved also make the radial gate more

adaptable to operation by relatively simple automatic control apparatus. Where a number of gates are used on a spillway, they might be arranged to open automatically at successively increasing reservoir levels, or only one or two might be equipped with automatic controls, while the remaining gates would be operated by hand or power hoists.

(d) *Drum Gates.*—Drum gates are constructed of steel plate and, since they are hollow, are buoyant. Each gate is triangular in section and is hinged to the upstream lip of a hydraulic chamber in the weir structure, in which the gate floats. Water introduced into or drawn from the hydraulic chamber causes the gate to swing upwards or downwards. Controls governing the flow of water into and out of the hydraulic chamber are located in the piers adjacent to the chambers. Figure 9-8 shows the drum gates on the Arizona spillway at Hoover Dam, which are automatic in operation.

(e) *Ring Gates.*—A ring gate consists of a full-circle hollow steel ring with streamlined top surface which blends with the surface of a morning glory inlet structure. The bottom portion of the ring is contained within a circular hydraulic chamber. Water admitted to or drawn from the hydraulic chamber causes the ring to move up or down in the vertical direction. Figure 9-7 shows the morning glory spillway for Hungry Horse Dam with the ring gate in the closed position.

C. CONTROL STRUCTURES

9-10. Shape for Uncontrolled Ogee Crest.—Crest shapes which approximate the profile of the under nappe of a jet flowing over a sharp-crested weir provide the ideal form for obtaining optimum discharges. The shape of such a profile depends upon the head, the inclination of the upstream face of the overflow section, and the height of the overflow section above the floor of the entrance channel (which influences the velocity of approach to the crest).

A simple scheme suitable for most dams with a vertical upstream face is to shape the

upstream surface (in section) to an arc of a circle and the downstream surface to a parabola. The necessary information for defining the shape is shown on figure 9-10. This method will define a crest which approximates the more refined shape discussed below. It is suitable for preliminary estimates and for final designs when a refined shape is not required.

Crest shapes have been studied extensively in the Bureau of Reclamation hydraulic laboratories, and data from which profiles for overflow crests can be obtained have been

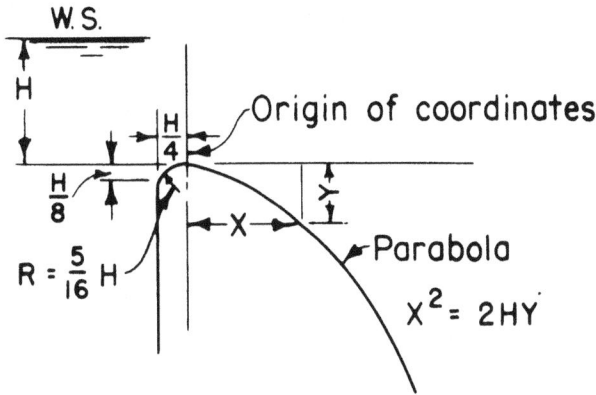

Figure 9-10. **A simple ogee crest shape with
a vertical upstream face.—288-D-3041**

published [1].[1] For most conditions the data
can be summarized according to the form
shown on figure 9-11(A), where the profile is
defined as it relates to axes at the apex of the
crest. That portion upstream from the origin is
defined as either a single curve and a tangent or
as a compound circular curve. The portion
downstream is defined by the equation:

$$\frac{y}{H_o} = -K\left(\frac{x}{H_o}\right)^n \qquad (2)$$

in which K and n are constants whose values
depend on the upstream inclination and on the
velocity of approach. Figure 9-11 gives values
of these constants for different conditions.

The approximate profile shape for a crest
with a vertical upstream face and negligible
velocity of approach is shown on figure 9-12.
The profile is constructed in the form of a
compound circular curve with radii expressed
in terms of the design head, H_o. This definition
is simpler than that shown on figure 9-11, since
it avoids the need for solving an exponential
equation; further, it is presented in a form
easily used by a layman for constructing forms
or templates. For ordinary conditions of design
of spillways where the approach height, P (fig.
9-11(A)), is equal to or greater than one-half
the maximum head on the crest, this profile is
sufficiently accurate to avoid seriously reduced

---

[1]Numbers in brackets refer to items in the bibliography,
sec. 9-31.

crest pressures and does not materially alter the
hydraulic efficiency of the crest. When the
approach height is less than one-half the
maximum head on the crest, the profile should
be determined from figure 9-11.

In some cases, it is necessary to use a crest
shape other than that indicated by the above
design. Information from model studies
performed on many spillways has been
accumulated and a compilation of the
coefficient data has been made. This
information is shown in Engineering
Monograph No. 9 [2]. In this monograph, the
crests are plotted in a dimensionless form with
the design head, H_o, equal to 1. By plotting
other crests to the same scale, comparisons
with model-tested crest shapes can be made.

**9-11. *Discharge Over an Uncontrolled
Overflow Ogee Crest.*—**The discharge over an
ogee crest is given by the formula:

$$Q = CLH_e^{3/2} \qquad (3)$$

where:

Q = discharge,
C = a variable coefficient of
        discharge,
L = effective length of crest, and
H_e = total head on the crest, including
        velocity of approach head, h_a.

The total head on the crest, H_e, does not
include allowances for approach channel
friction losses or other losses due to curvature
of the upstream channel, entrance loss into the
inlet section, and inlet or transition losses.
Where the design of the approach channel
results in appreciable losses, they must be
added to H_e to determine reservoir elevations
corresponding to the discharges given by the
above equation.

(a) *Coefficient of Discharge.*—The discharge
coefficient, C, is influenced by a number of
factors, such as (1) the depth of approach, (2)
relation of the actual crest shape to the ideal
nappe shape, (3) upstream face slope, (4)
downstream apron interference, and (5)
downstream submergence. The effect of these
various factors is discussed in subsections (b)

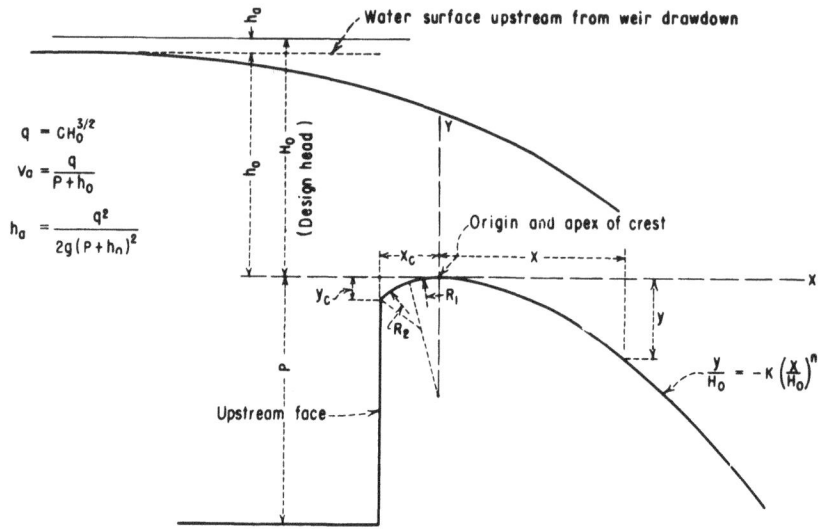

$$q = CH_0^{3/2}$$

$$v_a = \frac{q}{P + h_0}$$

$$h_a = \frac{q^2}{2g(P + h_0)^2}$$

$$\frac{y}{H_0} = -K\left(\frac{x}{H_0}\right)^n$$

(A) ELEMENTS OF NAPPE-SHAPED CREST PROFILES

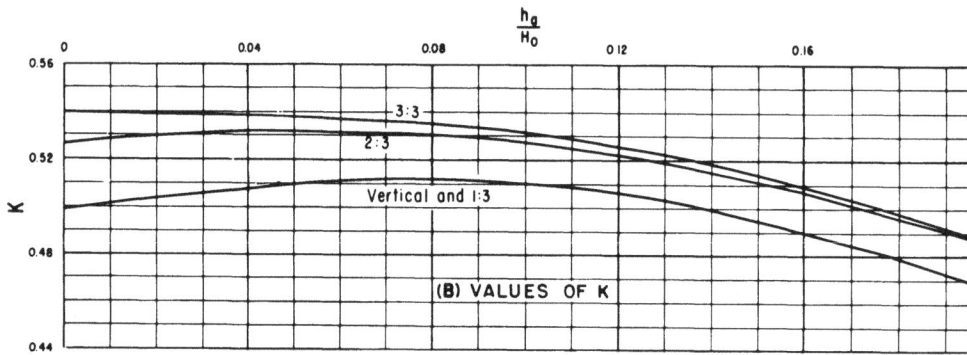

(B) VALUES OF K

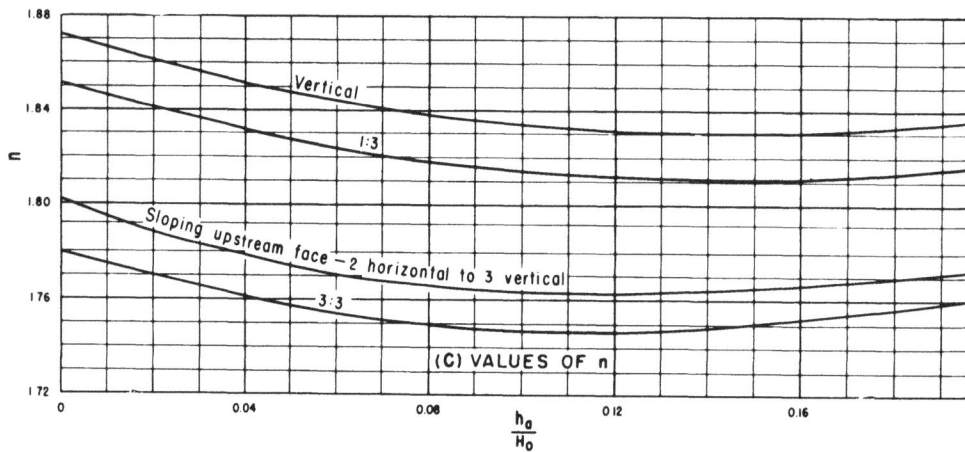

(C) VALUES OF n

Figure 9-11. Factors for definition of nappe-shaped crest profiles (sheet 1 of 2).—288-D-2406

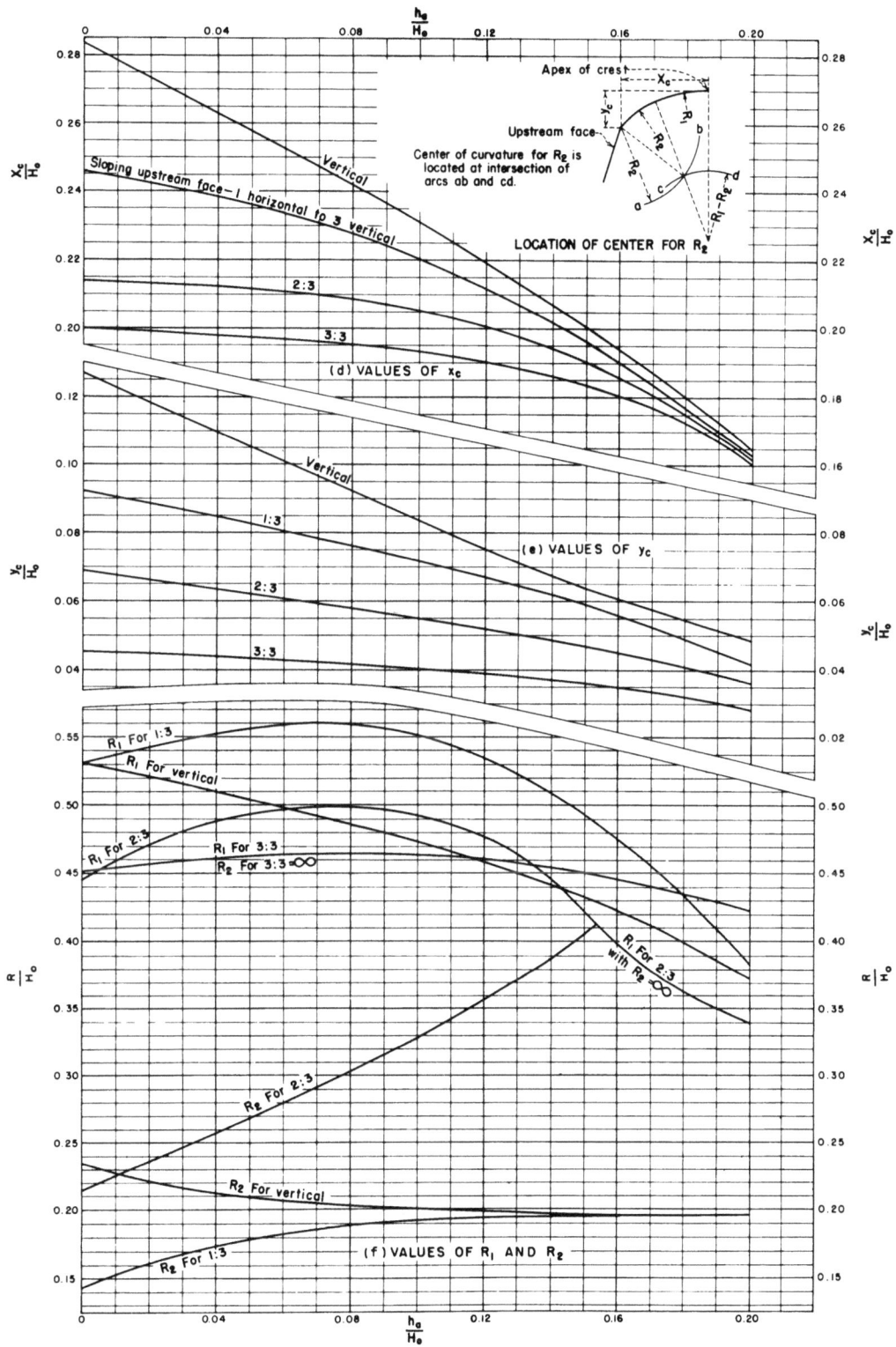

Figure 9-11. Factors for definition of nappe-shaped crest profiles (sheet 2 of 2).—288-D-2407

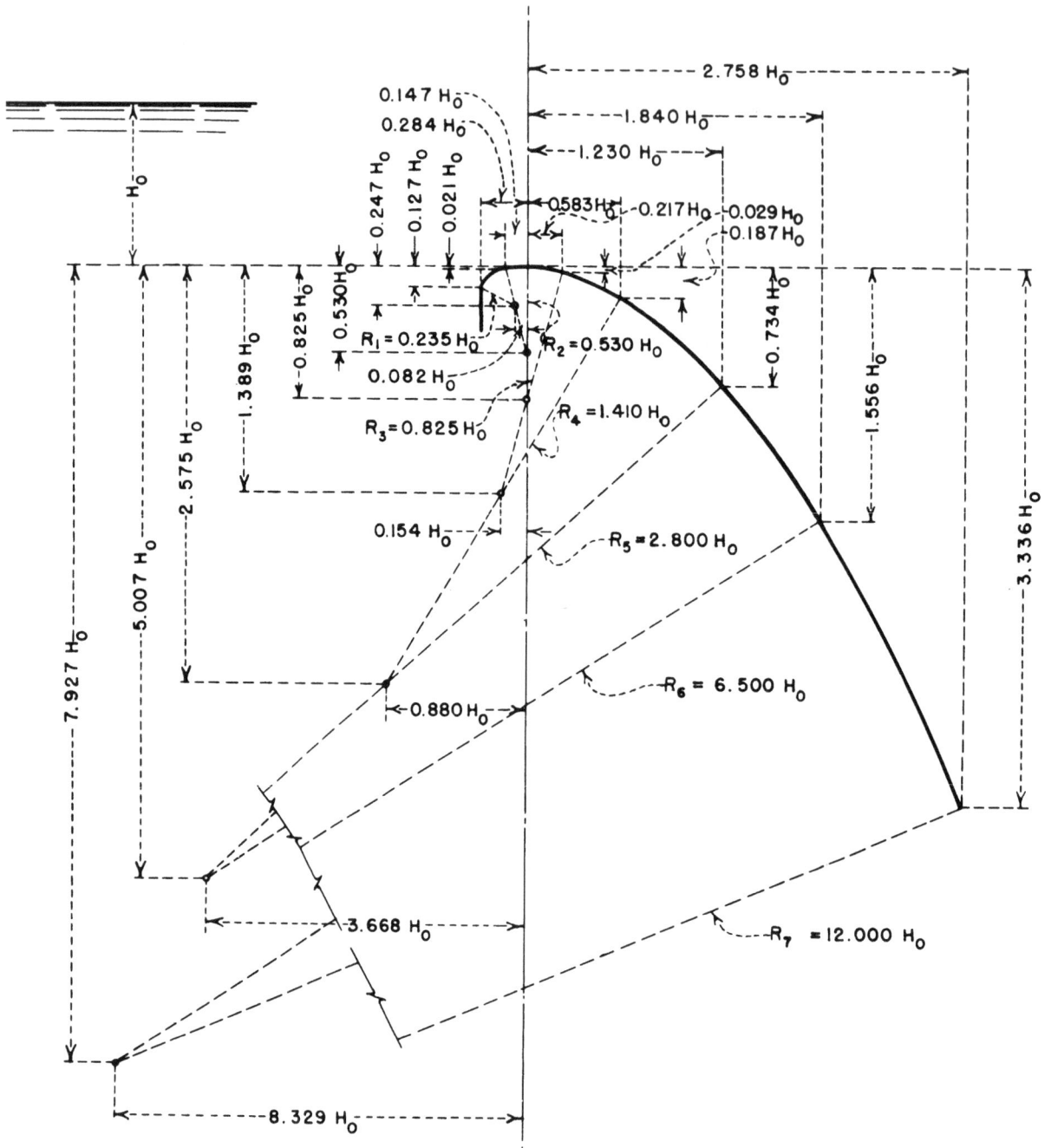

Figure 9-12. Ogee crest shape defined by compound curves.—288-D-2408

through (d). The effect of the discharge coefficient for heads other than the design head is discussed in subsection (e). The discharge coefficient for various crest profiles can be determined from Engineering Monograph No. 9 [2] or it may be approximated by finding the design shape it most nearly matches.

(b) *Effect of Depth of Approach.*—For a high sharp-crested weir placed in a channel, the velocity of approach is small and the under side of the nappe flowing over the weir attains maximum vertical contraction. As the approach depth is decreased, the velocity of approach increases and the vertical contraction diminishes. When the weir height becomes zero, the contraction is entirely suppressed and the overflow weir becomes in effect a channel or a broad-crested weir, for which the theoretical coefficient of discharge is 3.087. If the sharp-crested weir coefficients are related to the head measured from the point of maximum contraction instead of to the head above the sharp crest, coefficients applicable to ogee crests shaped to profiles of under nappes for various approach velocities can be established. The relationship of the ogee crest coefficient, C_o, to various values of $\dfrac{P}{H_o}$ is shown on figure 9-13. These coefficients are valid only when the ogee is formed to the ideal nappe shape, that is when $\dfrac{H_e}{H_o} = 1$.

(c) *Effect of Upstream Face Slope.*—For small ratios of the approach depth to head on the crest, sloping the upstream face of the overflow results in an increase in the coefficient of discharge. For large ratios the effect is a decrease of the coefficient. Within the range considered in this text, the coefficient of discharge is reduced for large ratios of $\dfrac{P}{H_o}$ only for relatively flat upstream slopes. Figure 9-14 shows the ratio of the coefficient for an overflow ogee crest with a sloping face to the coefficient for a crest with a vertical upstream face as obtained from figure 9-13, as related to values of $\dfrac{P}{H_o}$.

(d) *Effect of Downstream Apron Interference and Downstream Submergence.*—When the water level below an overflow weir is high enough to affect the discharge, the weir is said to be submerged. The vertical distance from the crest of the overflow weir to the downstream apron and the depth of flow in the downstream channel, as it relates to the head pool level, are factors which alter the coefficient of discharge.

Five distinct characteristic flows can occur below an overflow crest, depending on the relative positions of the apron and the downstream water surface: (1) Flow will continue at supercritical stage; (2) a partial or incomplete hydraulic jump will occur immediately downstream from the crest; (3) a true hydraulic jump will occur; (4) a drowned jump will occur in which the high-velocity jet will follow the face of the overflow and then continue in an erratic and fluctuating path for a considerable distance under and through the slower water; and (5) no jump will occur—the jet will break away from the face of the overflow and ride along the surface for a short distance and then erratically intermingle with the slow-moving water underneath. Figure 9-15 shows the relationship of the floor positions and downstream submergences which produce these distinctive flows.

Where the downstream flow is at supercritical stage or where the hydraulic jump occurs, the decrease in the coefficient of discharge is due principally to the back-pressure effect of the downstream apron and is independent of any submergence effect due to tailwater. Figure 9-16 shows the effect of downstream apron conditions on the coefficient of discharge. It will be noted that this curve plots the same data represented by the vertical dashed lines on figure 9-15 in a slightly different form. As the downstream apron level nears the crest of the overflow $\left(\dfrac{h_d + d}{H_e} \text{ approaches } 1.0 \right)$, the coefficient of discharge is about 77 percent of that for unretarded flow. On the basis of a coefficient of 3.98 for unretarded flow over a high weir, the coefficient when the weir is submerged will be about 3.08, which is virtually' the coefficient for a broad-crested weir.

From figure 9-16 it can be seen that when the $\dfrac{h_d + d}{H_e}$ values exceed about 1.7, the downstream floor position has little effect on the coefficient, but there is a decrease in the coefficient caused by tailwater submergence. Figure 9-17 shows the ratio of the coefficient of discharge where affected by tailwater conditions, to the coefficient for free flow conditions. This curve plots the data

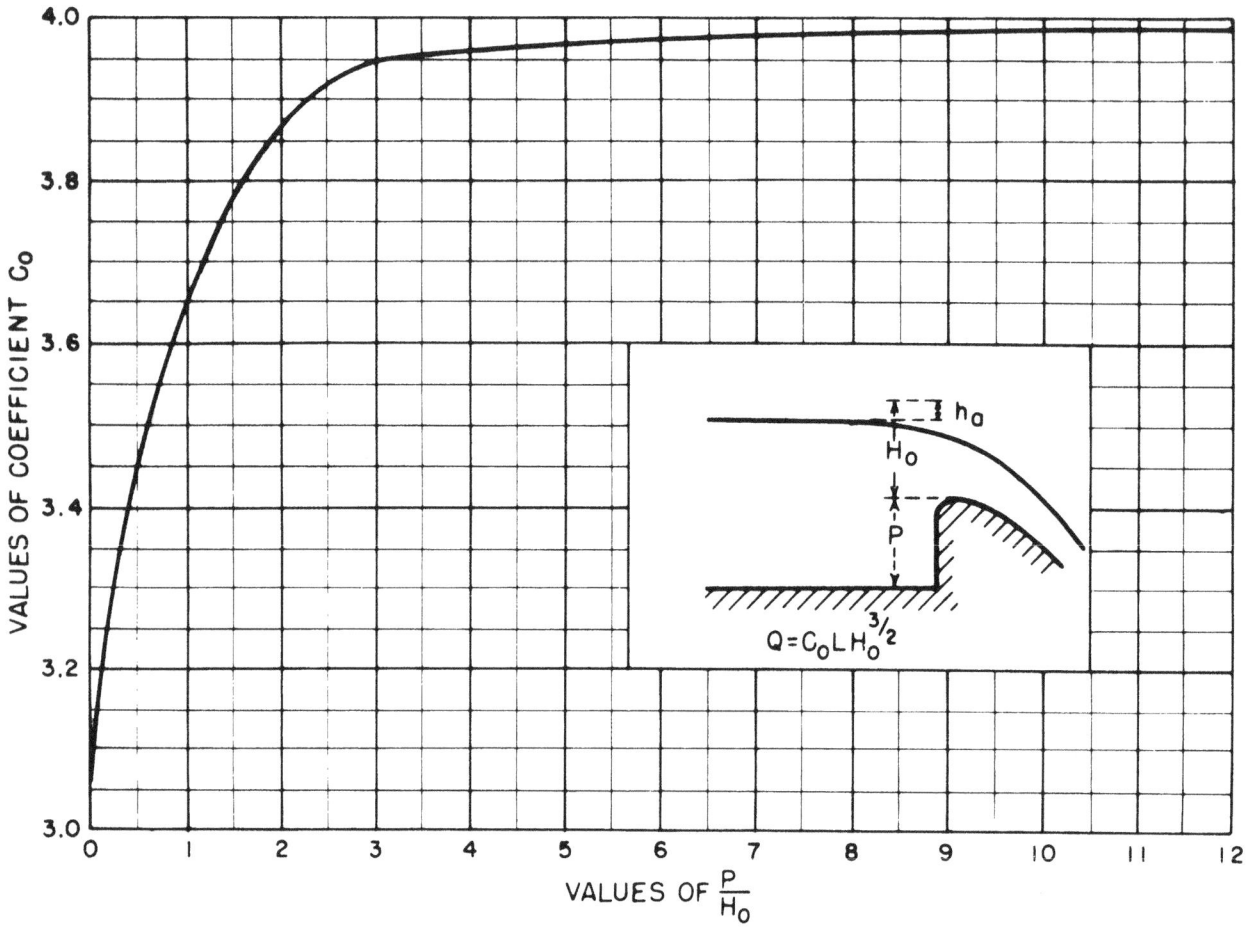

Figure 9-13. Coefficient of discharge for ogee-shaped crest with vertical upstream face.—288-D-3042

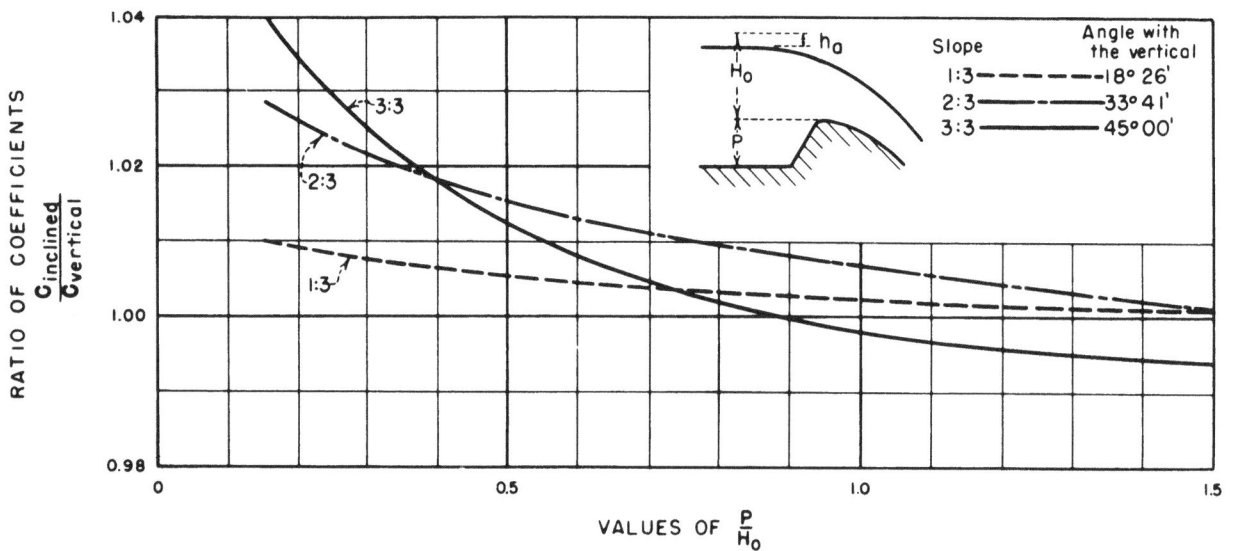

Figure 9-14. Coefficient of discharge for ogee-shaped crest with sloping upstream face.—288-D-2411

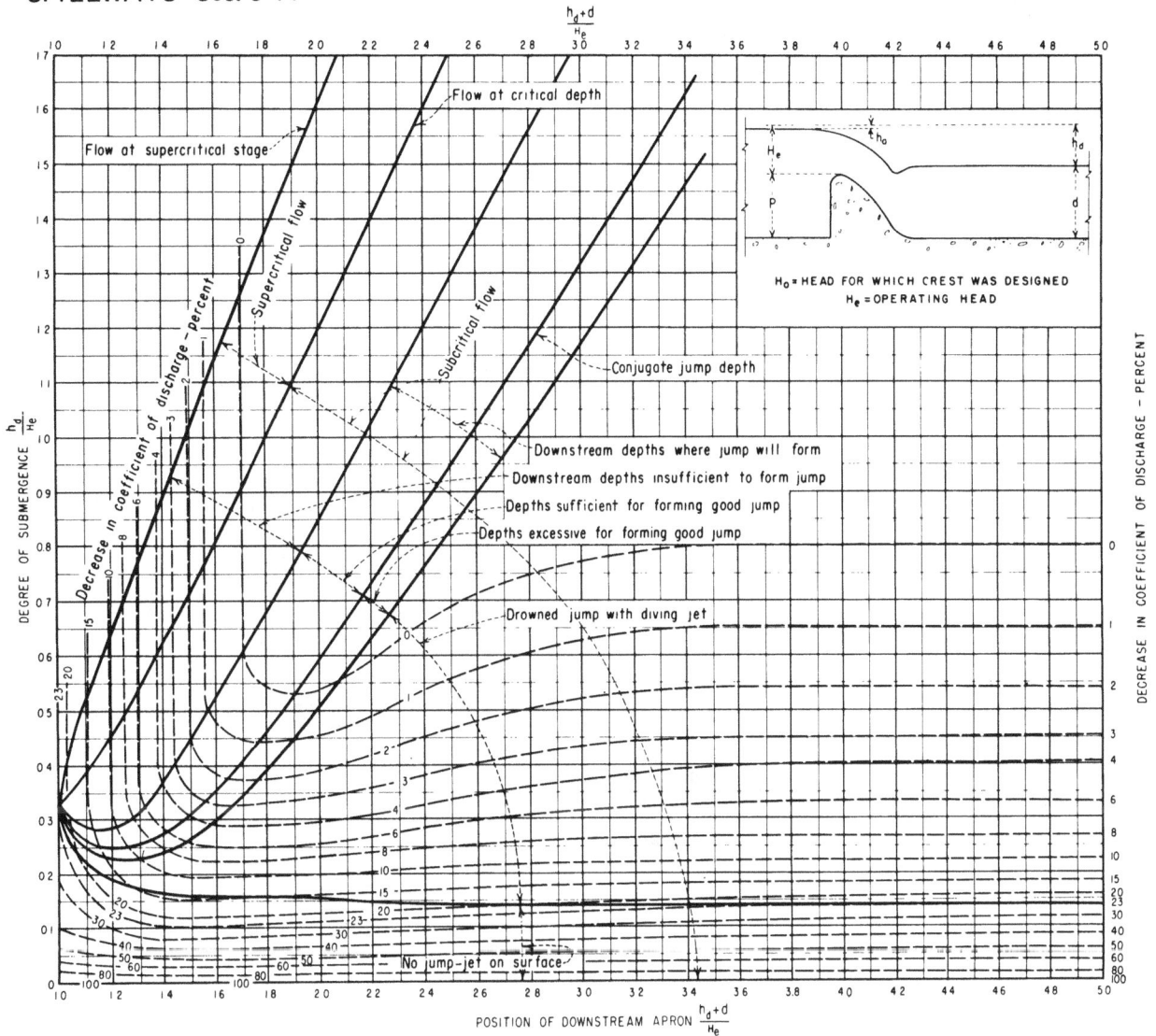

Figure 9-15. Effects of downstream influences on flow over weir crests.—288-D-2412

represented by the horizontal dashed lines on figure 9-15 in a slightly different form. Where the dashed lines on figure 9-15 are curved, the decrease in the coefficient is the result of a combination of tailwater effects and downstream apron position.

(e) *Effect of Heads Differing from Design Head.*—When the crest has been shaped for a head larger or smaller than the one under consideration, the coefficient of discharge, C, will differ from that shown on figure 9-13. A widened shape will result in positive pressures along the crest contact surface, thereby reducing the discharge; with a narrower crest

shape negative pressures along the contact surface will occur, resulting in an increased discharge. Figure 9-18 shows the variation of the coefficient as related to values of $\dfrac{H_e}{H_o}$, where H_e is the actual head being considered. The adjusted coefficient can be used for preparing a discharge-head relationship.

(f) *Pier and Abutment Effects.*—Where crest piers and abutments are shaped to cause side contractions of the overflow, the effective length, L, will be less than the net length of the crest. The effect of the end contractions may

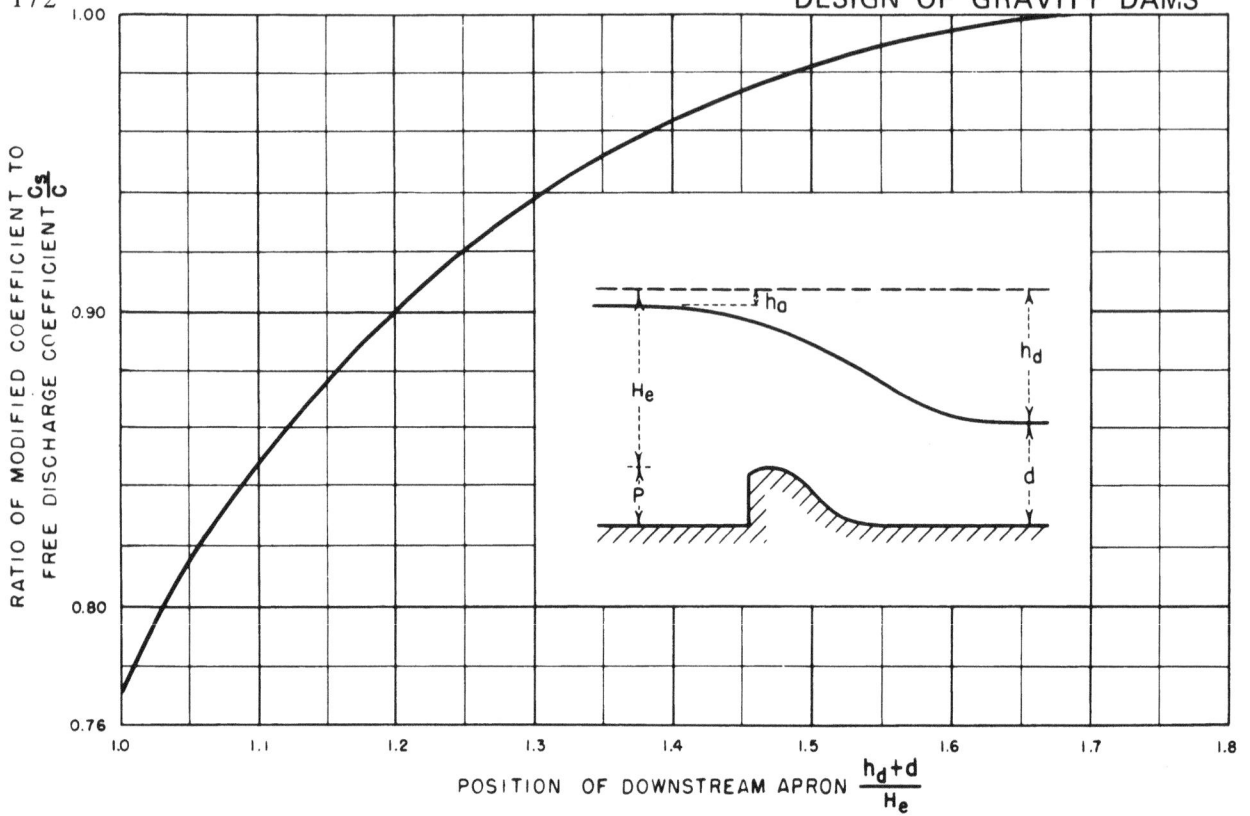

Figure 9-16. Ratio of discharge coefficients due to apron effect.—288-D-2413

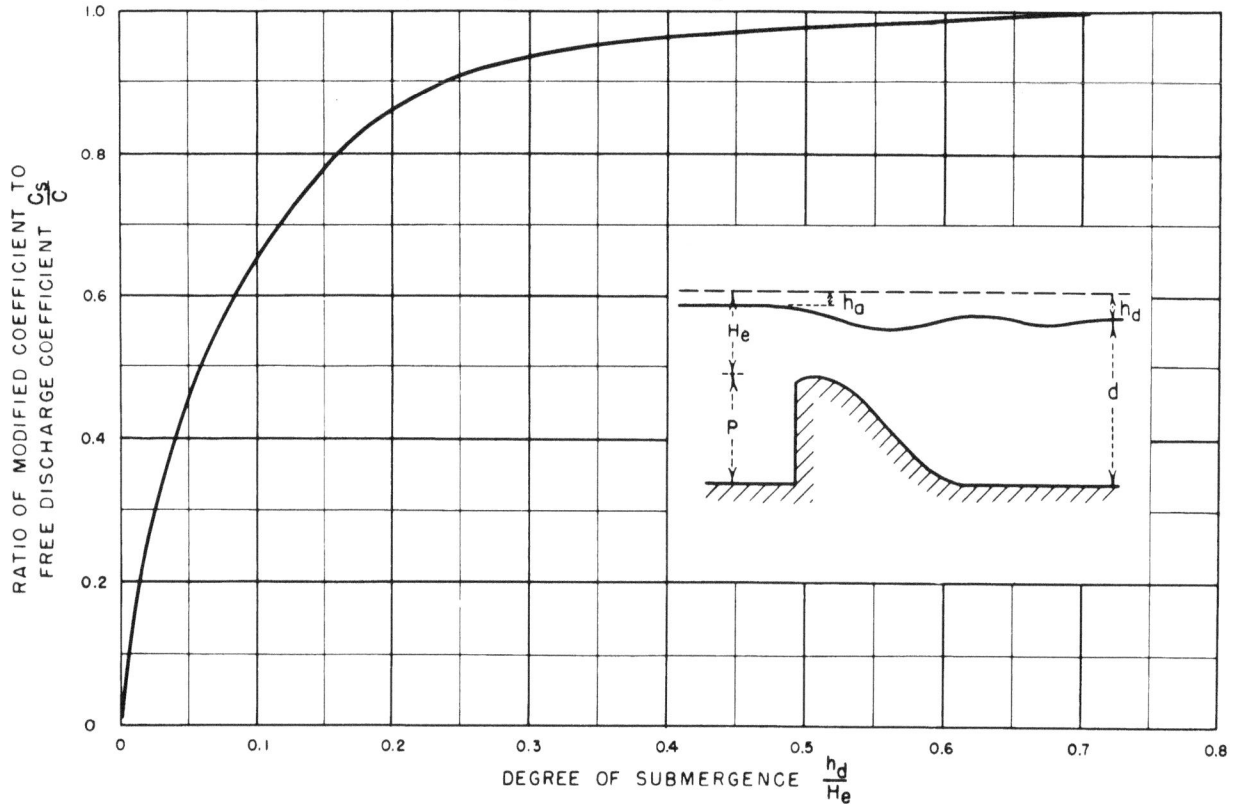

Figure 9-17. Ratio of discharge coefficients due to tailwater effect.—288-D-2414

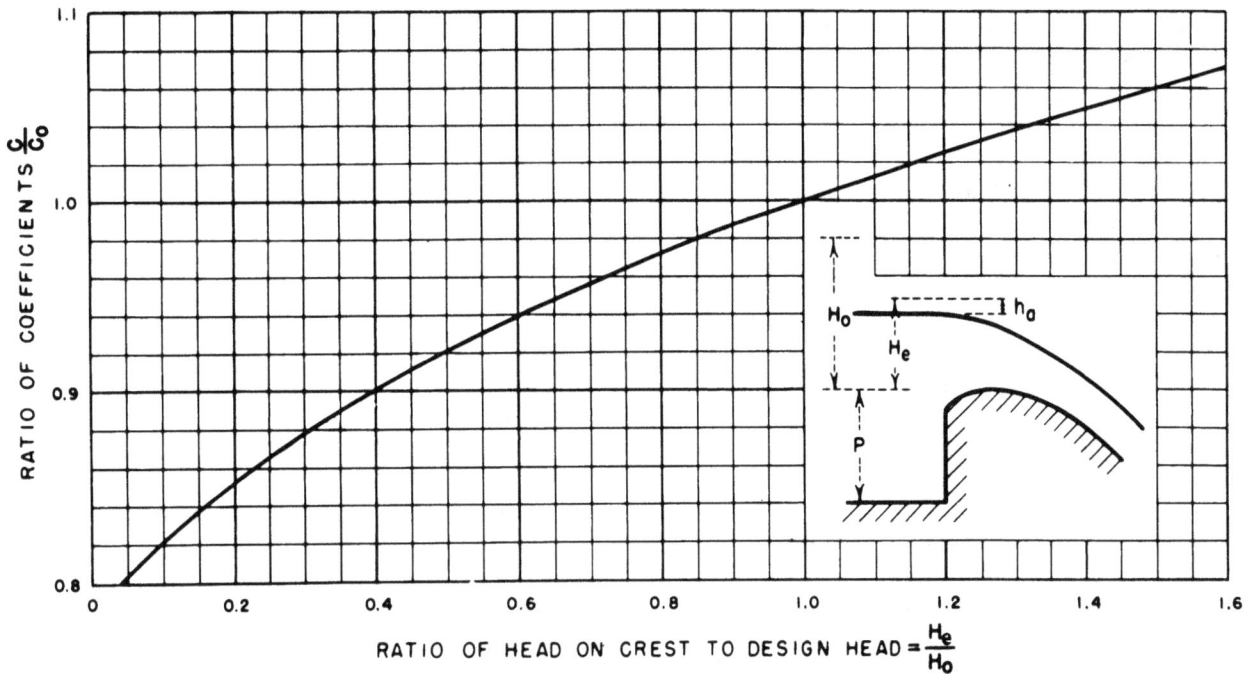

Figure 9-18. Coefficient of discharge for other than the design head.—288-D-2410

be taken into account by reducing the net crest length as follows:

$$L = L' - 2 \left(N K_p + K_a \right) H_e \qquad (4)$$

where:

    L = effective length of crest,
    L' = net length of crest,
    N = number of piers,
    K_p = pier contraction coefficient,
    K_a = abutment contraction coefficient, and
    H_e = total head on crest.

The pier contraction coefficient, K_p, is affected by the shape and location of the pier nose, the thickness of the pier, the head in relation to the design head, and the approach velocity. For conditions of design head, H_o, average pier contraction coefficients may be assumed as follows:

	K_p
For square-nosed piers with corners rounded on a radius equal to about 0.1 of the pier thickness	0.02
For round-nosed piers	0.01
For pointed-nose piers	0

The abutment contraction coefficient is affected by the shape of the abutment, the angle between the upstream approach wall and the axis of flow, the head in relation to the design head, and the approach velocity. For conditions of design head, H_o, average coefficients may be assumed as follows:

	K_a
For square abutments with headwall at 90° to direction of flow	0.20
For rounded abutments with headwall at 90° to direction of flow, when $0.5H_o \geqq r \geqq 0.15H_o$	0.10
For rounded abutments where $r > 0.5H_o$ and headwall is placed not more than 45° to direction of flow	0

where r = radius of abutment rounding.

9-12. *Uncontrolled Ogee Crests Designed for Less Than Maximum Head.* —Economy in the design of an ogee crest may sometimes be effected by using a design head less than the maximum expected for determining the ogee profile. Use of a smaller head for design results in increased discharges for the full range of heads. The increase in capacity makes it possible to achieve economy by reducing either the crest length or the maximum surcharge head.

Tests have shown that the subatmospheric pressures on a nappe-shaped crest do not exceed about one-half the design head when the design head is not less than about 75 percent of the maximum head. As long as these subatmospheric pressures do not approach pressures which might induce cavitation, they can be tolerated. Care must be taken, however, in forming the surface of the crest where these negative pressures will occur, since unevenness caused by abrupt offsets, depressions, or projections will amplify the negative pressures to a magnitude where cavitation conditions can develop.

The negative pressure on the crest may be resolved into a system of forces acting both upward and downstream. These forces should be considered in analyzing the structural stability of the crest structure.

An approximate force diagram of the subatmospheric pressures when the design head used to determine the crest shape is 75 percent of the maximum head, is shown on figure 9-19. These data are based on average results of tests made on ideal shaped weirs with negligible velocities of approach. Pressures for intermediate head ratios can be assumed to vary linearly, considering that no subatmospheric pressure prevails when $\frac{H_o}{H_e}$ is equal to 1.

9-13. *Gate-Controlled Ogee Crests.* — Releases for partial gate openings for gated crests will occur as orifice flow. With full head on the gate and with the gate opened a small amount, a free discharging trajectory will follow the path of a jet issuing from an orifice. For a vertical orifice the path of the jet can be expressed by the parabolic equation:

$$-y = \frac{x^2}{4H} \qquad (5)$$

where H is the head on the center of the opening. For an orifice inclined an angle of θ from the vertical, the equation will be:

$$-y = x \tan \theta + \frac{x^2}{4H \cos^2 \theta} \qquad (6)$$

If subatmospheric pressures are to be avoided along the crest contact, the shape of the ogee downstream from the gate sill must conform to the trajectory profile.

Gates operated with small openings under high heads produce negative pressures along the crest in the region immediately below the gate if the ogee profile drops below the trajectory profile. Tests have shown that subatmospheric pressures would be equal to about one-tenth of the head when the gate is operated at small opening and the ogee is shaped to the nappe profile as defined by equation (2) for maximum head H_o. The force diagram for this condition is shown on figure 9-20.

The adoption of a trajectory profile rather than a nappe profile downstream from the gate sill will result in a wider ogee, and reduced discharge efficiency for full gate opening. Where the discharge efficiency is unimportant and where a wider ogee shape is needed for structural stability, the trajectory profile may be adopted to avoid subatmospheric pressure zones along the crest. Where the ogee is shaped to the ideal nappe profile for maximum head, the subatmospheric pressure area can be minimized by placing the gate sill downstream from the crest of the ogee. This will provide an orifice which is inclined downstream for small gate openings, and thus will result in a steeper trajectory more nearly conforming to the nappe-shaped profile.

9-14. *Discharge Over Gate-Controlled Ogee Crests.* —The discharge for a gated ogee crest at partial gate openings will be similar to flow through an orifice and may be computed by the equation:

$$Q = \frac{2}{3}\sqrt{2g} \, CL \left(H_1^{3/2} - H_2^{3/2} \right) \qquad (7)$$

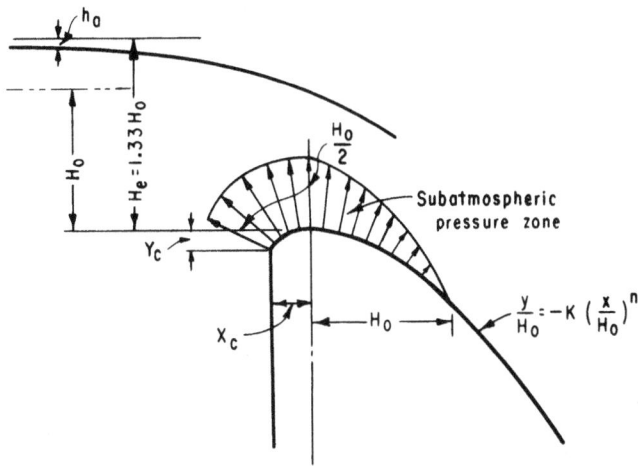

Figure 9-19. Subatmospheric crest pressures for a 0.75 ratio of H_o to H_e.--288-D-3043

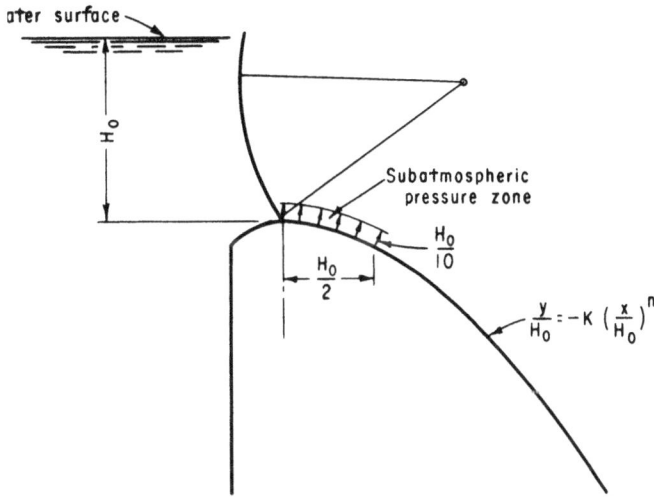

Figure 9-20. Subatmospheric crest pressures for undershot gate flow.--288-D-3044

where H_1 and H_2 are the total heads (including the velocity head of approach) to the bottom and top of the orifice, respectively. The coefficient, C, will differ with different gate and crest arrangements; it is influenced by the approach and downstream conditions as they affect the jet contractions. Thus, the top contraction for a vertical leaf gate will differ from that for a curved, inclined radial gate; the upstream floor profile will affect the bottom contraction of the issuing jet; and the downstream profile will affect the back pressure and consequently the effective head.

Figure 9-21 shows coefficients of discharge for orifice flows for various ratios of gate opening to total head. The curve represents averages determined for the various approach and downstream conditions described and is sufficiently reliable for determining discharges for most spillway structures. The curve is for a gate at the apex of the ogee crest, and so long as the bottom of the gate when closed is less than 0.03 H_o vertically from the apex the coefficient should not change significantly.

9-15. *Orifice Control Structures.*--Orifice control structures are often incorporated into a concrete gravity dam, one or more orifices being formed through the dam. If the invert of the orifice is below normal water surface the orifice must be gated. If the invert is at or above normal water surface the orifice may be either gated or ungated. Figure 9-22 shows typical orifice control structures.

(a) *Shape.*--The entrance to the orifice must be streamlined to eliminate negative pressures. Portions of ellipses are used to streamline the entrances. The major axis of the ellipse is equal to the height or width of the orifice H or W in figure 9-22, and the minor axis is one-third of this amount. Orifices may be horizontal or they may be inclined downward to change the location of the impingement area in the case of a free fall spillway, or to provide improved alinement into a discharge channel. If inclined downward, the bottom of the orifice should be shaped similar to an ogee crest to eliminate negative pressures. The top should be made parallel to or slightly converging with the bottom.

(b) *Hydraulics*--The discharge characteristics of an orifice flowing partially full with the upper nappe not in contact with the orifice are similar to those of an ogee crest, and the discharge can be computed by use of equation (3). The discharge coefficient, C, can be determined as described in section 9-10 for an overflow crest. Where practicable, a model study should be made to confirm the value of the coefficient. An orifice flowing full will function similar to a river outlet, and the discharge can be determined using the same procedures as for river outlets discussed in chapter X.

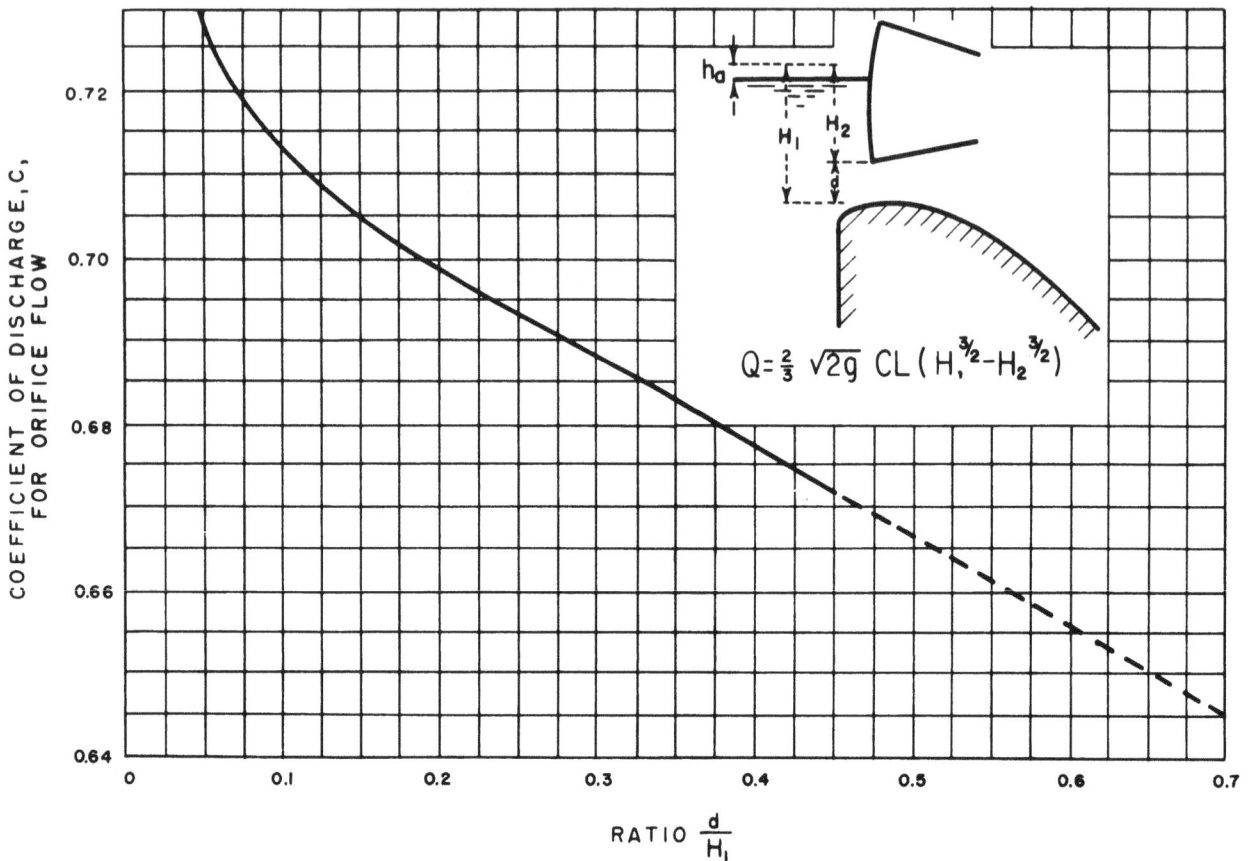

Figure 9-21. Coefficient of discharge for flow under a gate (orifice flow).—288-D-3045

9-16. *Side Channel Control Structures.* —The side channel control structure consists of an ogee crest to control releases from the reservoir, and a channel immediately downstream of and parallel to the crest to carry the water to the discharge channel.

(a) *Layout.*—The ogee crest is designed by the methods in section 9-10 if the crest is uncontrolled or section 9-13 if it is controlled.

The cross-sectional shape of the side channel trough will be influenced by the overflow crest on the one side and by the bank conditions on the opposite side. Because of turbulences and vibrations inherent in side channel flow, a side channel design is ordinarily not considered except where a competent foundation such as rock exists. The channel sides will, therefore, usually be a concrete lining placed on a slope and anchored directly to the rock. A trapezoidal cross section is the one most often employed for the side channel trough. The

width of such a channel in relation to the depth should be considered. If the width to depth ratio is large, the depth of flow in the channel will be shallow, similar to that depicted by the cross section *abfg* on figure 9-23. It is evident that for this condition a poor diffusion of the incoming flow with the channel flow will result. A cross section with a minimum width-depth ratio will provide the best hydraulic performance, indicating that a cross section approaching that depicted as *adj* on the figure would be the ideal choice both from the standpoint of hydraulics and economy. Minimum bottom widths are required, however, to avoid construction difficulties due to confined working space. Furthermore, the stability of the structure and the hillside which might be jeopardized by an extremely deep cut in the abutment must also be considered. Therefore, a minimum bottom width must be selected which is commensurate

TYPICAL HORIZONTAL ORIFICE   SEC. A-A

TYPICAL INCLINED ORIFICE

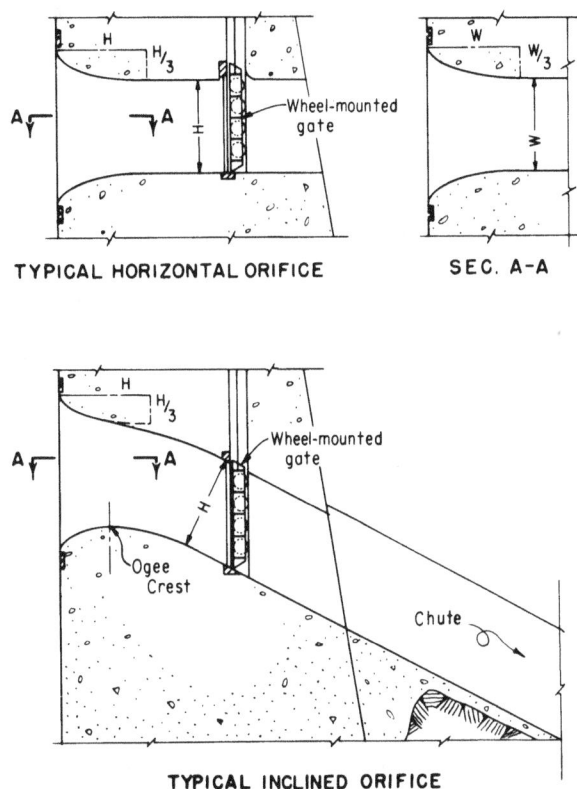

Figure 9-22. Typical orifice control
structures.—288-D-3046

with both the practical and structural aspects of the problem.

The slope of the channel profile is arbitrary; however, a relatively flat slope will provide greater depths and slower velocities and consequently will ensure better intermingling of flows at the upstream end of the channel and avoid the possibility of accelerating or supercritical flows occurring in the channel for smaller discharges.

A control section is usually constructed downstream from the side channel trough. It is achieved by constricting the channel sides or elevating the channel bottom to produce a point of critical flow. Flows upstream from the control will be at the subcritical stage and will provide a maximum of depth in the side channel trough. The side channel bottom and control dimensions are then selected so that flow in the trough immediately downstream from the crest will be at the greatest depth possible without excessively submerging the

flow over the crest. Flow in the discharge channel downstream from the control will be the same as that in an ordinary channel or chute type spillway. If a control section is not provided, the depth of water and its velocity in the side channel will depend upon either the slope of the side channel trough floor or the backwater effect of the discharge channel.

Figure 9-24(A) illustrates the effect of a control section and the slope of the side channel trough floor on the water surface profile. When the bottom of the side channel trough is selected so that its depth below the hydraulic gradient is greater than the minimum specific energy depth, flow will be either at the subcritical or supercritical stage, depending on the relation of the bottom profile to critical slope or on the influences of a downstream control section. If the slope of the bottom is greater than critical and a control section is not established below the side channel trough, supercritical flow will prevail throughout the length of the channel. For this stage, velocities will be high and water depths will be shallow, resulting in a relatively high fall from the reservoir water level to the water surface in the trough. This flow condition is illustrated by profile B′ on figure 9-24(A). Conversely, if a control section is established downstream from the side channel trough to increase the upstream depths, the channel can be made to flow at the subcritical stage. Velocities at this stage will be less than critical and the greater depths will result in a smaller drop from the reservoir water surface to the side channel water surface profile. The condition of flow for subcritical depths is illustrated on figure 9-24(A) by water surface profile A′.

The effect of the fall distance from the reservoir to the channel water surface for each type of flow is depicted on figure 9-24(B). It can be seen that for the subcritical stage, the incoming flow will not develop high transverse velocities because of the low drop before it meets the channel flow, thus effecting a good diffusion with the water bulk in the trough. Since both the incoming velocities and the channel velocities will be relatively slow, a fairly complete intermingling of the flows will take place, thereby producing a comparatively

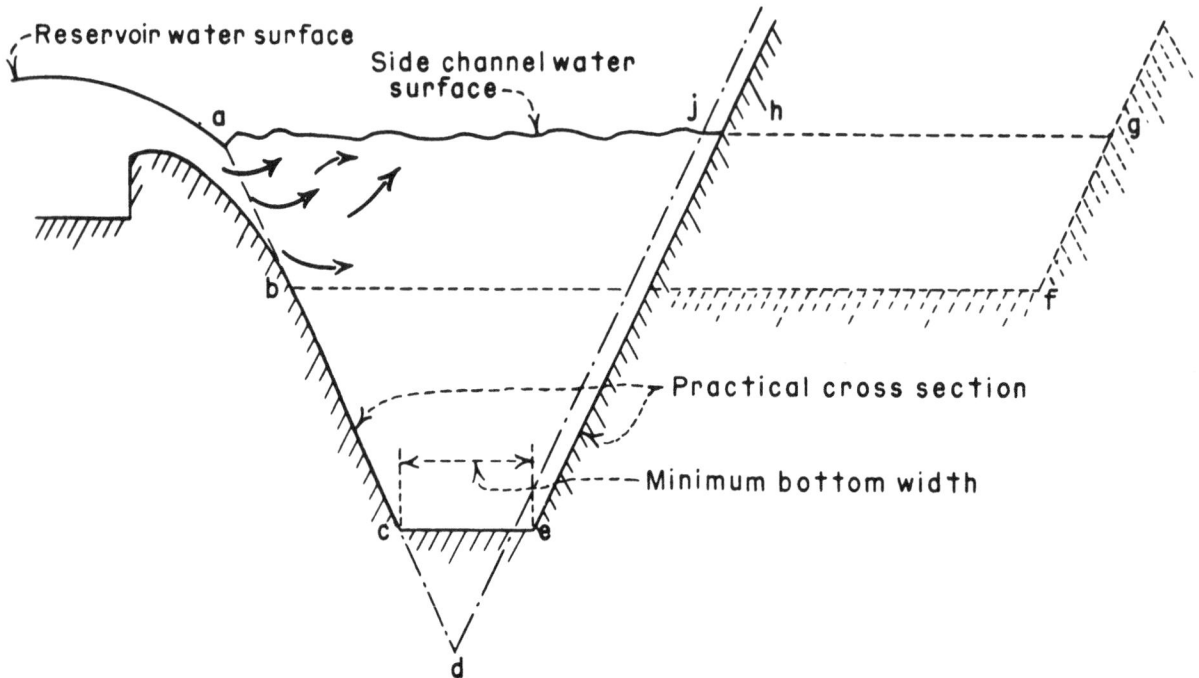

Figure 9-23. Comparison of side channel cross sections.—288-D-2419

smooth flow in the side channel. Where the channel flow is at the supercritical stage, the channel velocities will be high, and the intermixing of the high-energy transverse flow with the channel stream will be rough and turbulent. The transverse flows will tend to sweep the channel flow to the far side of the channel, producing violent wave action with attendant vibrations. It is thus evident that flows should be maintained at subcritical stage for good hydraulic performance. This can be achieved by establishing a control section downstream from the side channel trough.

Variations in the design can be made by assuming different bottom widths, different channel slopes, and varying control sections. A proper and economical design can usually be achieved after comparing several alternatives.

(b) *Hydraulics.*—The theory of flow in a side channel [3] is based principally on the law of conservation of linear momentum, assuming that the only forces producing motion in the channel result from the fall in the water surface in the direction of the axis. This premise assumes that the entire energy of the flow over the crest is dissipated through its intermingling with the channel flow and is therefore of no assistance in moving the water along the channel. Axial velocity is produced only after the incoming water particles join the channel stream.

For any short reach of channel Δx, the change in water surface, Δy, can be determined by either of the following equations:

$$\Delta y = \frac{Q_1 (v_1 + v_2)}{g (Q_1 + Q_2)} \left[(v_2 - v_1) + \frac{v_2 (Q_2 - Q_1)}{Q_1} \right] \quad (8)$$

$$\Delta y = \frac{Q_2 (v_1 + v_2)}{g (Q_1 + Q_2)} \left[(v_2 - v_1) + \frac{v_1 (Q_2 - Q_1)}{Q_2} \right] \quad (9)$$

where Q_1 and v_1 are values at the beginning of the reach and Q_2 and v_2 are the values at the end of the reach. The derivation of these formulas can be found in reference [3].

By use of equation (8) or (9), the water surface profile can be determined for any particular side channel by assuming successive short reaches of channel once a starting point is found. The solution of equation (8) or (9) is obtained by a trial-and-error procedure. For a reach of length Δx in a specific location, Q_1 and Q_2 will be known.

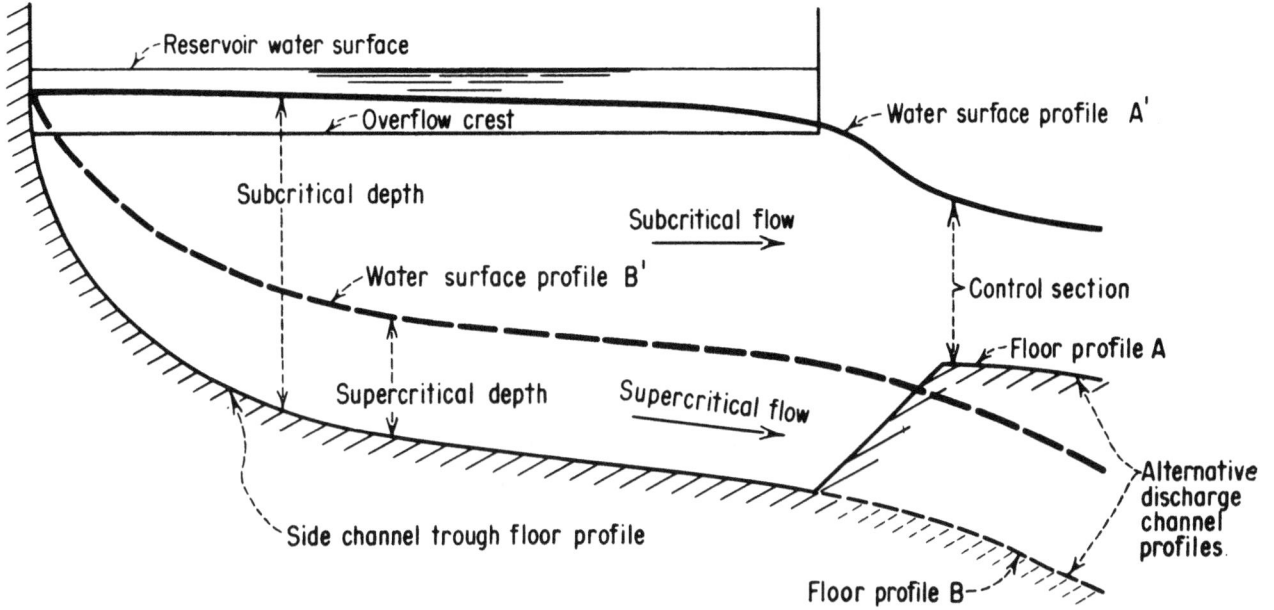

(A) SIDE CHANNEL PROFILE

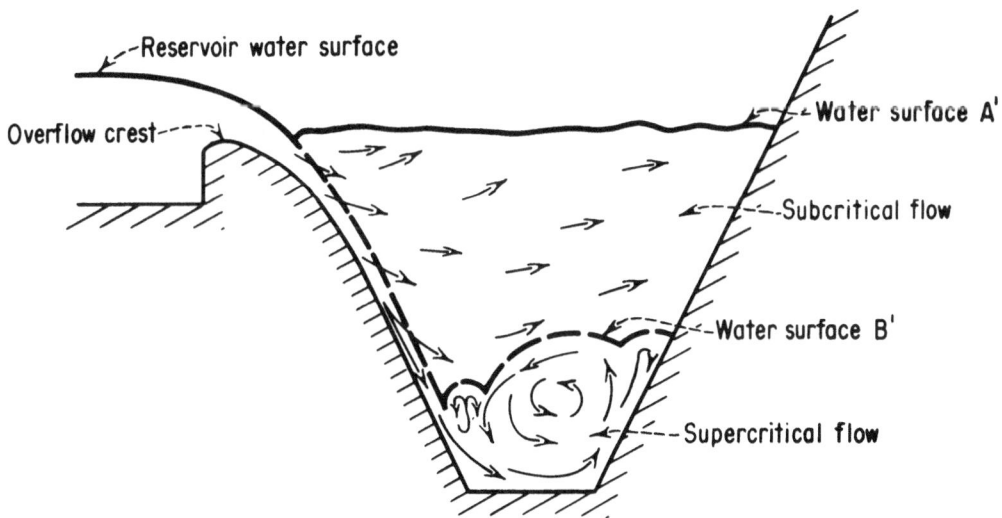

(B) SIDE CHANNEL CROSS SECTION

Figure 9-24. Side channel flow characteristics.—288-D-2418

As in other water surface profile determinations, the depth of flow and the hydraulic characteristics of the flow will be affected by backwater influences from some control point, or by critical conditions along the reach of the channel under consideration. A control section is usually constructed at the downstream end of the side channel. After determining the depth of water at the control section, the water surface at the downstream end of the side channel can be determined by routing the water between the two points. With the depth of water at the downstream point known, depths for successive short reaches can be computed as previously described. It is assumed that a maximum of two-thirds submergence of the crest can be tolerated without affecting the water surface profile.

D. HYDRAULICS OF DISCHARGE CHANNELS

9-17. *General.*—Discharge generally passes through the critical stage in the spillway control structure and enters the discharge channel as supercritical or shooting flow. To avoid a hydraulic jump below the control, the flow must remain at the supercritical stage throughout the length of the channel. The flow in the channel may be uniform or it may be accelerated or decelerated, depending on the slopes and dimensions of the channel and on the total drop. Where it is desired to minimize the grade to reduce excavation at the upstream end of a channel, the flow might be uniform or decelerating, followed by accelerating flow in the steep drop leading to the downstream river level. Flow at any point along the channel will depend upon the specific energy, $(d + h_v)$, available at that point. This energy will equal the total drop from the reservoir water level to the floor of the channel at the point under consideration, less the head losses accumulated to that point. The velocities and depths of flow along the channel can be fixed by selecting the grade and the cross-sectional dimensions of the channel.

The velocities and depths of free surface flow in a channel, whether an open channel or a tunnel, conform to the principle of the conservation of energy as expressed by the Bernoulli's theorem, which states: "The absolute energy of flow at any cross section is equal to the absolute energy at a downstream section plus intervening losses of energy." As applied to figure 9-25 this relationship can be expressed as follows:

$$\Delta Z + d_1 + h_{v_1} = d_2 + h_{v_2} + \Delta h_L \qquad (10)$$

When the channel grades are not too steep, for practical purposes the normal depth d_n can be considered equal to the vertical depth d, and ΔL can be considered to be the horizontal distance. The term Δh_L includes all losses which occur in the reach of channel, such as friction, turbulence, impact, and transition losses. Since in most channels changes are made gradually, ordinarily all losses except those due to friction can be neglected. The friction loss can then be expressed as:

$$\Delta h_L = s \cdot \Delta L \qquad (11)$$

where s is the average friction slope expressed by either the Chezy or the Manning formula. For the reach ΔL, the head loss can be expressed as $\Delta h_L = \left(\dfrac{s_1 + s_2}{2} \right) \Delta L$. From the Manning formula, as given in section F-2(c) of appendix F,

$$s = \left(\frac{vn}{1.486 r^{2/3}} \right)^2$$

The coefficient of roughness, n, will depend on the nature of the channel surface. For conservative design the frictional loss should be maximized when evaluating depths of flow and minimized when evaluating the energy content of the flow. For a concrete-lined channel, a conservative value of n, varying from 0.014 for a channel with good alinement and a smooth finish to 0.018 for a channel with poor

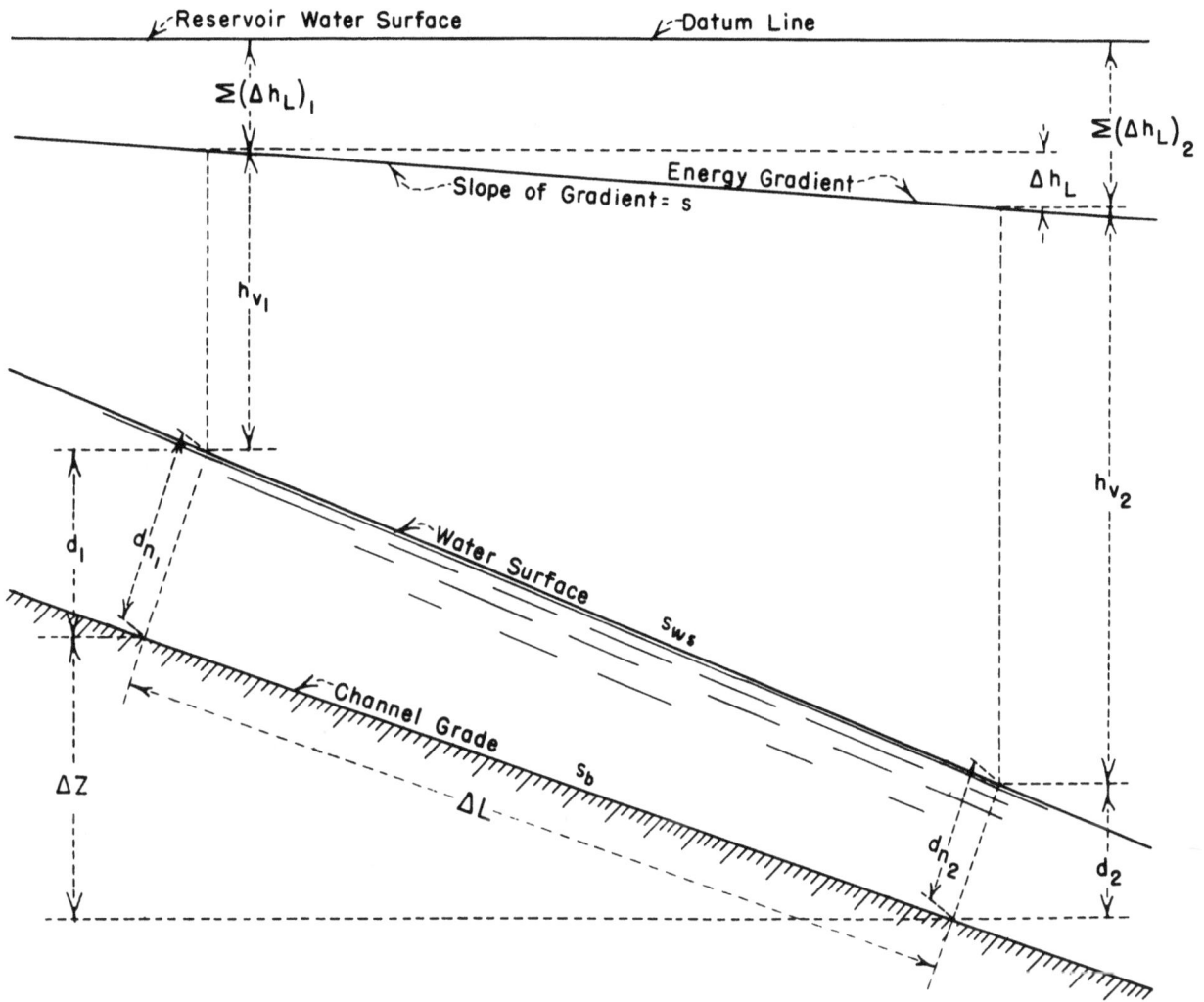

Figure 9-25. Sketch illustrating flow in open channels.—288-D-2421

alinement and some unevenness in the finish, should be used in estimating the depth of flow. For determining specific energies of flow needed for designing the dissipating device, a value of n of about 0.008 should be assumed.

Where only rough approximations of depths and velocities of flow in a discharge channel are desired, the total head loss $\Sigma(\Delta h_L)$ to any point along the channel might be expressed in terms of the velocity head. Thus, at any section the relationship can be stated: Reservoir water surface elevation minus floor grade elevation = $d + h_v + Kh_v$. For preliminary spillway layouts, K can be assumed as approximately 0.2 for determining depths of flow and 0.1 or less for evaluating the energy of flow. Rough

approximations of losses can also be obtained from figure 9-26. The assumptions used in determining the losses in figure 9-26 are discussed in section F-2(f) of appendix F.

9-18. *Open Channels.*—(a) *Profile*—The profile of an open channel is usually selected to conform to topographic and geologic site conditions. It is generally defined as straight reaches joined by vertical curves. Sharp convex and concave vertical curves should be avoided to prevent unsatisfactory flows in the channel. Convex curves should be flat enough to maintain positive pressures and thus avoid the tendency for separation of the flow from the floor. Concave curves should have a sufficiently long radius of curvature to minimize the

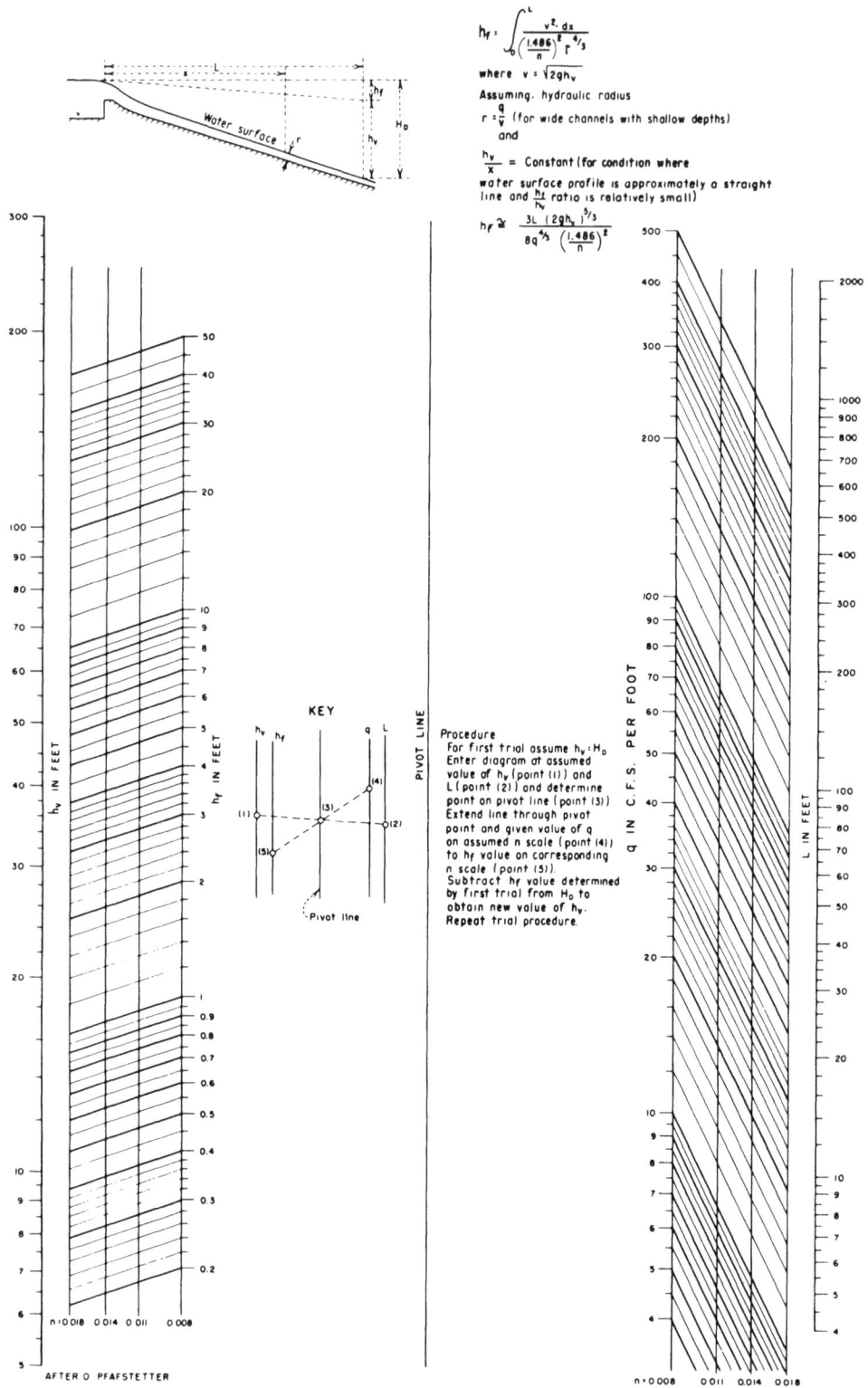

Figure 9-26. Approximate losses in chutes for various values of water surface drop and channel length.—288-D-3047

dynamic forces on the floor brought about by the centrifugal force which results from a change in the direction of flow.

To avoid the tendency for the water to spring away from the floor and thereby reduce the surface contact pressure, the floor shape for convex curvature should be made substantially flatter than the trajectory of a free-discharging jet issuing under a head equal to the specific energy of flow as it enters the curve. The curvature should approximate a shape defined by the equation:

$$-y = x \tan \theta + \frac{x^2}{K[4(d + h_v) \cos^2 \theta]} \qquad (12)$$

where θ is the slope angle of the floor upstream from the curve. Except for the factor K, the equation is that of a free-discharging trajectory issuing from an inclined orifice. To assure positive pressure along the entire contact surface of the curve, K should be equal to or greater than 1.5.

For the concave curvature, the pressure exerted upon the floor surface by the centrifugal force of the flow will vary directly with the energy of the flow and inversely with the radius of curvature. An approximate relationship of these criteria can be expressed in the equations:

$$R = \frac{2qv}{p} \text{ or } R = \frac{2dv^2}{p} \qquad (13)$$

where:

    R = the minimum radius of curvature
        measured in feet,
    q = the discharge in c.f.s. per foot of
        width,
    v = the velocity in feet per second,
    d = the depth of flow in feet, and
    p = the normal dynamic pressure exerted
        on the floor, in pounds per square
        foot.

An assumed value of $p = 100$ will normally produce an acceptable radius; however, a minimum radius of $10d$ is usually used. For the reverse curve at the lower end of the ogee crest,

radii of not less than $5d$ have been found acceptable.

(b) *Convergence and Divergence.*—The best hydraulic performance in a discharge channel is obtained when the confining sidewalls are parallel and the distribution of flow across the channel is maintained uniform. However, economy may dictate a channel section narrower or wider than either the crest or the terminal structure, thus requiring converging or diverging transitions to fit the various components together. Sidewall convergence must be made gradual to avoid cross waves, "ride ups" on the walls, and uneven distribution of flow across the channel. Similarly, the rate of divergence of the sidewalls must be limited or else the flow will not spread to occupy the entire width of the channel uniformly, which may result in undesirable flow conditions at the terminal structure.

The inertial and gravitational forces of streamlined kinetic flow in a channel can be expressed by the Froude number parameter, $\frac{v}{\sqrt{gd}}$. Variations from streamlined flow due to outside interferences which cause an expansion or a contraction of the flow also can be related to this parameter. Experiments have shown that an angular variation of the flow boundaries not exceeding that produced by the equation,

$$\tan \alpha = \frac{1}{3F} \qquad (14)$$

will provide an acceptable transition for either a contracting or an expanding channel. In this equation, $F = \frac{v}{\sqrt{gd}}$ and α is the angular variation of the sidewall with respect to the channel centerline; v and d are the averages of the velocities and depths at the beginning and at the end of the transition. Figure 9-27 is a nomograph from which the tangent of the flare angle or the flare angle in degrees may be obtained for known values of depth and velocity of flow.

(c) *Channel Freeboard.*—In addition to using a conservative value for n in determining

DEPTH OF FLOW IN FEET

TANGENT OF FLARE ANGLE

FLARE ANGLE IN DEGREES

VELOCITY IN FEET PER SECOND

FLOW

$$\text{TAN FLARE} = \frac{1}{3F}$$

$$F = \frac{V}{\sqrt{gd}}$$

After C. Freeman

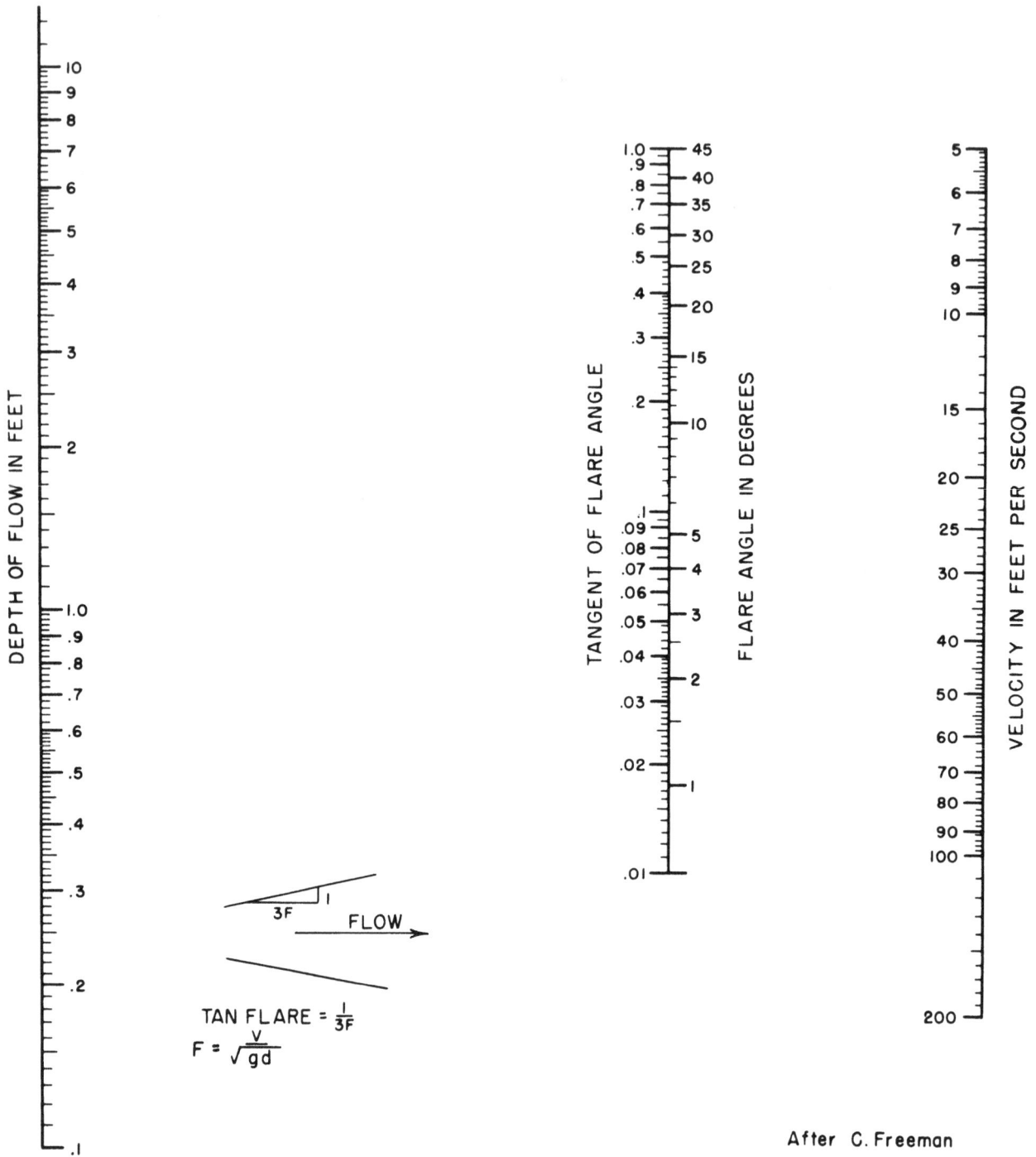

Figure 9-27. Flare angle for divergent or convergent channels.–288-D-2422

the depth of water, a freeboard of 3 to 6 feet is usually provided to allow for air bulking, wave action, etc. When the channel is constructed on the downstream face of the dam and some overtopping of the wall will not cause damage,

a minimal freeboard can be permitted. Where damage can occur, such as when the channel is located on an abutment, the higher freeboard is needed for safety. Engineering judgment should be used in setting the height of freeboard by comparing the cost of additional wall height against the possible damage due to overtopping of the channel walls. Wherever practicable, a hydraulic model should be used in determining the wall height.

In some cases, a minimum wall height of about 10 feet is used since there is very little increase in cost of a 10-foot wall over a lower wall. Also, the fill behind the wall provides a berm for catching material sloughing off the excavation slope, thus preventing it from getting into the channel.

9-19. Tunnel Channels.—(a) *Profile.*—Figure 9-28 shows a typical tunnel spillway channel. The profile at the upper end is curved to coincide with the profile of the control structure. The inclined portion is usually sloped at 55° from the horizontal. Steeper slopes increase the total length of the tunnel. On flatter slopes the blasted rock tends to stay on the slope during excavation rather than falling to the bottom where it can be easily removed from the tunnel.

The radius of the elbow at the invert may be determined by using equation (13); however, a radius of about 10 tunnel diameters is usually satisfactory. From the elbow, the tunnel is usually excavated on a slight downslope to the downstream portal.

(b) *Tunnel Cross Section.*—In the transition, the cross section changes from that required at the control structure to that required for the tunnel downstream from the elbow. This transition may be accomplished in one or more stages and is usually completed upstream of the elbow. Because a circular shape better resists the external loadings, it is usually desirable to attain a circular shape as soon as practicable.

The transition should be designed so that a uniform flow pattern is maintained and no negative pressures are developed which could lead to cavitation damage. No criteria have been established for determining the shape of the transition. Preliminary layouts are made using experience gained from previous tunnels. The layout should be checked using equation (10) so that no portion of the transition will flow more than 75 percent full (in area). This will allow for air bulking of the water and avoid complete filling of the tunnel. If the

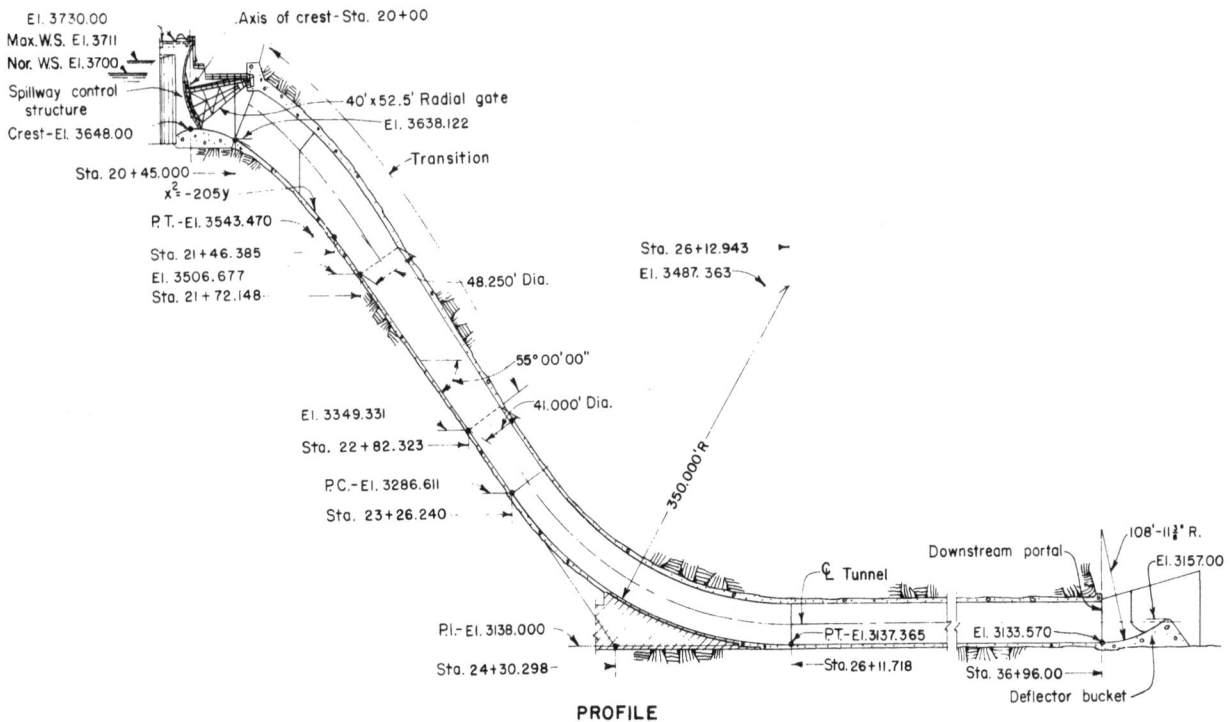

Figure 9-28. Profile of typical tunnel spillway channel.—288-D-3048

tunnel were to flow full, the control could move from the control structure and cause surging in the tunnel.

Downstream of the elbow, generally the slope of the energy gradient (equivalent to the friction slope, s) is greater than the slope of the tunnel invert (see fig. 9-25). This condition causes the velocity of the water to decrease and the depth of the water to increase. Usually it is not economically feasible to change the tunnel size downstream of the elbow, and therefore the conditions at the downstream end of the tunnel determine the size of the tunnel. This portion of the tunnel is frequently used as a part of the diversion scheme. If diversion flows are large, it may be economical to make the tunnel larger than required for the spillway flows. Because proper function of the spillway is essential, consideration should be given in these instances to checking of the final layout in a hydraulic model.

9-20. *Cavitation Erosion of Concrete Surfaces.* —Concrete surfaces adjacent to high-velocity flow must be protected from cavitation erosion. Cavitation will occur when, due to some irregularities in the geometry of the flow surface, the pressure in the flowing water is reduced to the vapor pressure, about 0.363 pound per square inch absolute at 70° F. As the vapor cavities move with the flowing water into a region of higher pressure, the cavities collapse causing instantaneous positive water pressures of many thousands of pounds per square inch. These extremely high localized pressures will cause damage to any flow surface adjacent to the collapsing cavities (reference [4]).

Protection against cavitation damage may include (1) use of surface finishes and alinements devoid of irregularities which might produce cavitation, (2) use of construction materials which are resistant to cavitation damage, or (3) admission of air into the flowing water to cushion the damaging high pressures of collapsing cavities. (See reference [5].)

E. HYDRAULICS OF TERMINAL STRUCTURES

9-21. *Hydraulic Jump Stilling Basins.* — Where the energy of flow in a spillway must be dissipated before the discharge is returned to the downstream river channel, the hydraulic jump basin is an effective device for reducing the exit velocity to a tranquil state. Figure 9-29 shows a hydraulic-jump stilling basin in operation at Canyon Ferry Dam in Montana.

The jump which will occur in a stilling basin has distinctive characteristics and assumes a definite form, depending on the energy of flow which must be dissipated in relation to the depth of the flow. Comprehensive tests have been performed by the Bureau of Reclamation [6] in connection with the hydraulic jump. The jump form and the flow characteristics can be related to the Froude number parameter, $\dfrac{v}{\sqrt{gd}}$. In this context v and d are the velocity and depth, respectively, before the hydraulic jump occurs, and g is the acceleration due to gravity. Forms of the hydraulic jump phenomena for various ranges of the Froude number are illustrated on figure 9-30. The depth d_2, shown on the figure, is the downstream conjugate depth, or the minimum tailwater depth required for the formation of a hydraulic jump. The actual tailwater depth may be somewhat greater than this, as discussed in subsection (d).

When the Froude number of the incoming flow is equal to 1.0, the flow is at critical depth and a hydraulic jump cannot form. For Froude numbers from 1.0 up to about 1.7, the incoming flow is only slightly below critical depth, and the change from this low stage to the high stage flow is gradual and manifests itself only by a slightly ruffled water surface. As the Froude number approaches 1.7, a series of small rollers begin to develop on the surface, which become more intense with increasingly higher values of the number. Other than the surface roller phenomena, relatively smooth flows prevail throughout the Froude number

Figure 9-29. Overflow gate-controlled spillway on Canyon Ferry Dam in Montana.—P296-600-883

range up to about 2.5. Stilling action for the range of Froude numbers from 1.7 to 2.5 is designated as form A on figure 9-30.

For Froude numbers between 2.5 and 4.5 an oscillating form of jump occurs, the entering jet intermittently flowing near the bottom and then along the surface of the downstream channel. This oscillating flow causes objectionable surface waves which carry considerably beyond the end of the basin. The action represented through this range of flows is designated as form B on figure 9-30.

For the range of Froude numbers for the incoming flow between 4.5 and 9, a stable and well-balanced jump occurs. Turbulence is confined to the main body of the jump, and the water surface downstream is comparatively smooth. As the Froude number increases above

9, the turbulence within the jump and the surface roller becomes increasingly active, resulting in a rough water surface with strong surface waves downstream from the jump. Stilling action for the range of Froude numbers between 4.5 and 9 is designated as form C on figure 9-30 and that above 9 is designated as form D.

Figure 9-31 plots relationships of conjugate depths and velocities for the hydraulic jump in a rectangular channel or basin. Also indicated on the figure are the ranges for the various forms of hydraulic jump described above.

(a) *Hydraulic Design of Stilling Basins.*—Stilling basins are designed to provide suitable stilling action for the various forms of hydraulic jump previously discussed. Type I basin, shown on figure 9-32, is a rectangular

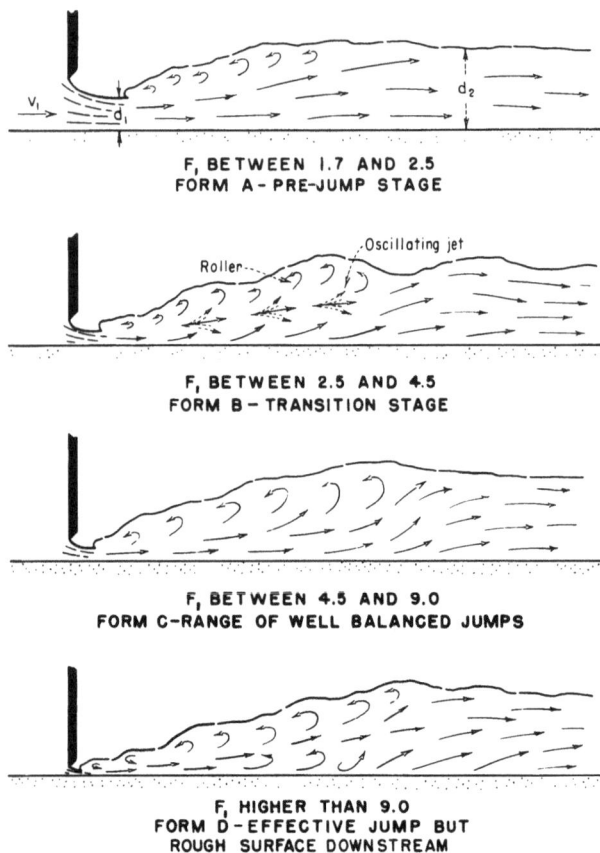

Figure 9-30. Characteristic forms of hydraulic jump related to the Froude number.—288-D-2423

channel without any accessories such as baffles or sills and is designed to confine the entire length of the hydraulic jump. Seldom are stilling basins of this type designed since it is possible to reduce the length and consequently the cost of the basin by the installation of baffles and sills, as discussed later for types II, III, and IV basins. The type of basin best suited for a particular situation will depend upon the Froude number.

(1) *Basins for Froude numbers less than 1.7.*—For a Froude number of 1.7 the conjugate depth d_2 is about twice the incoming depth, or about 40 percent greater than the critical depth. The exit velocity v_2 is about one-half the incoming velocity, or 30 percent less than the critical velocity. No special stilling basin is needed to still flows where the incoming flow Froude factor is less than 1.7, except that the channel lengths beyond the point where the depth starts to change should be not less than about $4d_2$. No baffles or other dissipating devices are needed.

(2) *Basins for Froude numbers between 1.7 and 2.5.*—Flow phenomena for basins where the incoming flow factors are in the Froude number range between 1.7 and 2.5 will be in the form designated as the prejump stage, as illustrated on figure 9-30. Since such flows are not attended by active turbulence, baffles or sills are not required. The basin should be a type I basin as shown on figure 9-32 and it should be sufficiently long and deep to contain the flow prism while it is undergoing retardation. Depths and lengths shown on figure 9-32 will provide acceptable basins.

(3) *Basins for Froude numbers between 2.5 and 4.5.*—Jump phenomena where the incoming flow factors are in the Froude number range between 2.5 and 4.5 are designated as transition flow stage, since a true hydraulic jump does not fully develop. Stilling basins to accommodate these flows are the least effective in providing satisfactory dissipation, since the attendant wave action ordinarily cannot be controlled by the usual basin devices. Waves generated by the flow phenomena will persist beyond the end of the basin and must often be dampened by means of wave suppressors.

Where a stilling device must be provided to dissipate flows for this range of Froude number, the basin shown on figure 9-33 which is designated as type IV basin, has proved to be relatively effective for dissipating the bulk of the energy of flow. However, the wave action propagated by the oscillating flow cannot be entirely dampened. Auxiliary wave dampeners or wave suppressors must sometimes be employed to provide smooth surface flow downstream. Because of the tendency of the jump to sweep out and as an aid in suppressing wave action, the water depths in the basin should be about 10 percent greater than the computed conjugate depth.

Often the need for utilizing the type IV basin in design can be avoided by selecting stilling basin dimensions which will provide flow conditions which fall outside the range of transition flow. For example, with an 800-c.f.s.

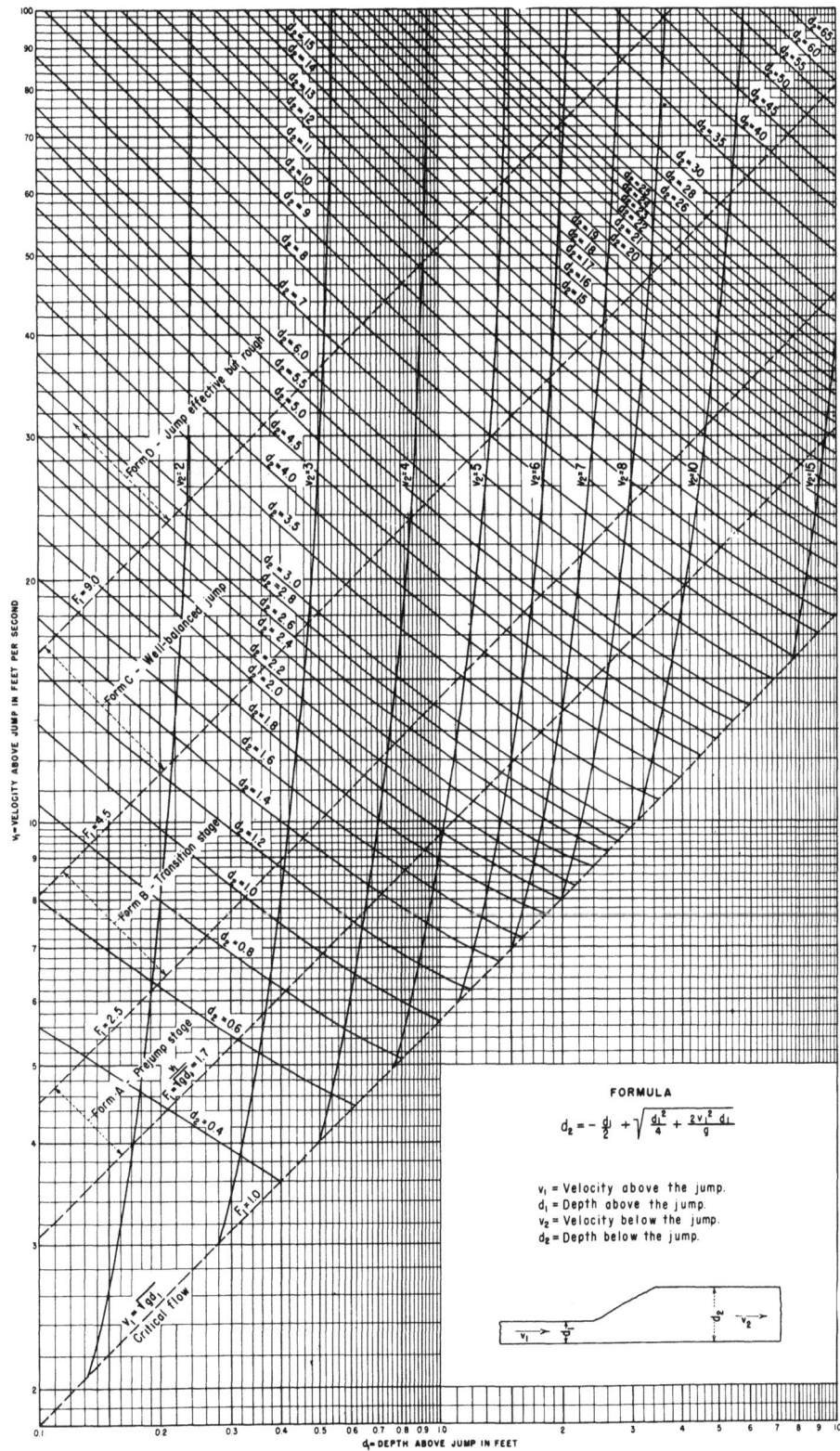

Figure 9-31. Relations between variables in hydraulic jumps for rectangular channels.—288-D-2424

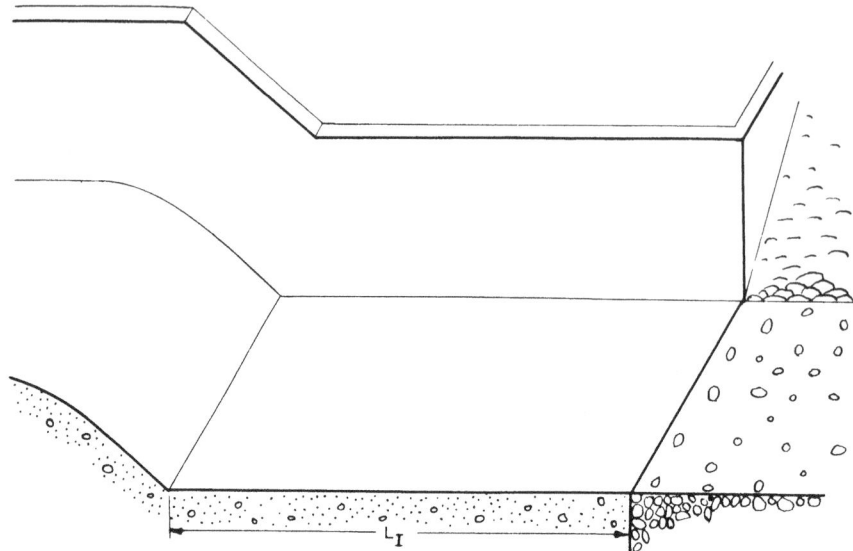

(A) TYPE I BASIN DIMENSIONS

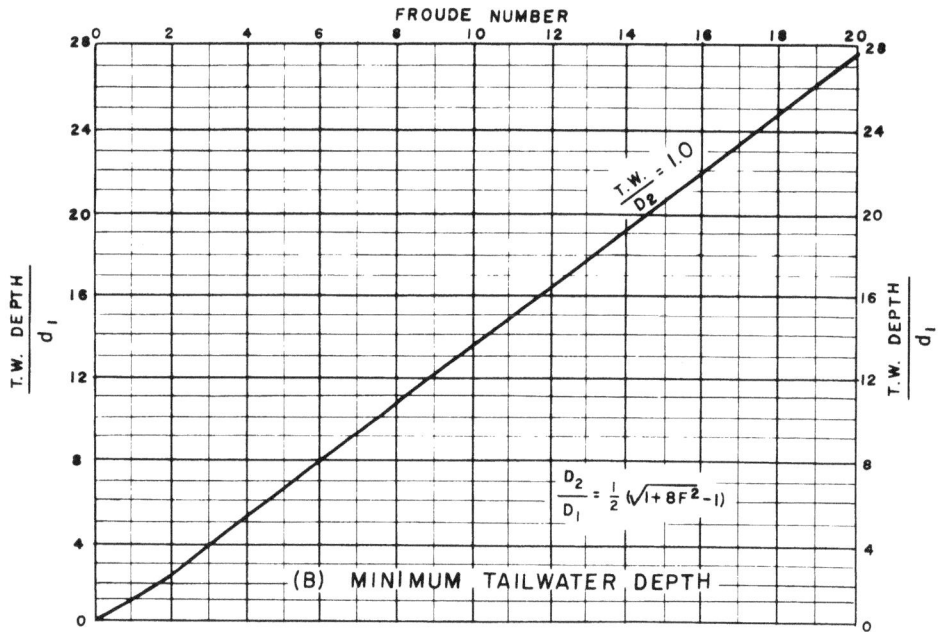

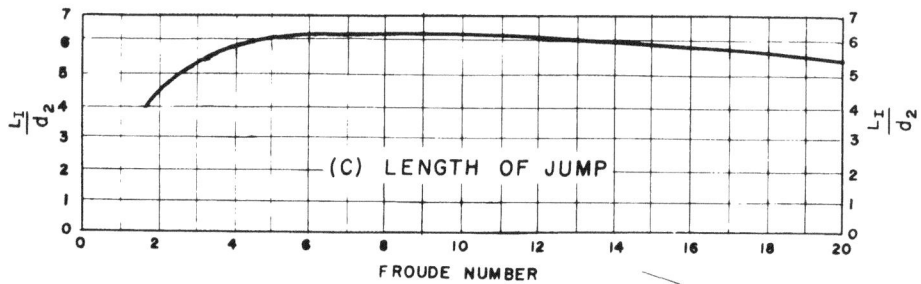

$$\frac{T.W.}{D_2} = 1.0$$

$$\frac{D_2}{D_1} = \frac{1}{2} \left(\sqrt{1+8F^2} - 1 \right)$$

(B) MINIMUM TAILWATER DEPTH

(C) LENGTH OF JUMP

Figure 9-32. Type I stilling basin characteristics.—288-D-3049

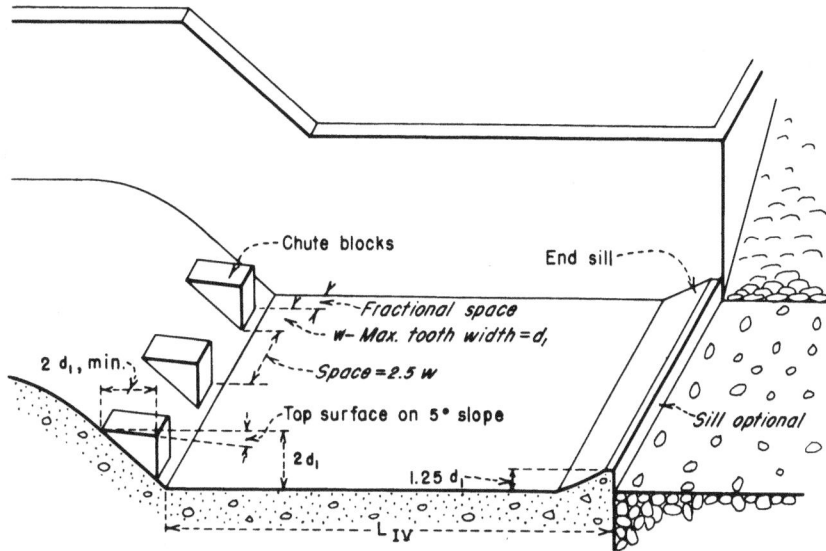

(A) TYPE IV BASIN DIMENSIONS

(B) MINIMUM TAILWATER DEPTHS

(C) LENGTH OF JUMP

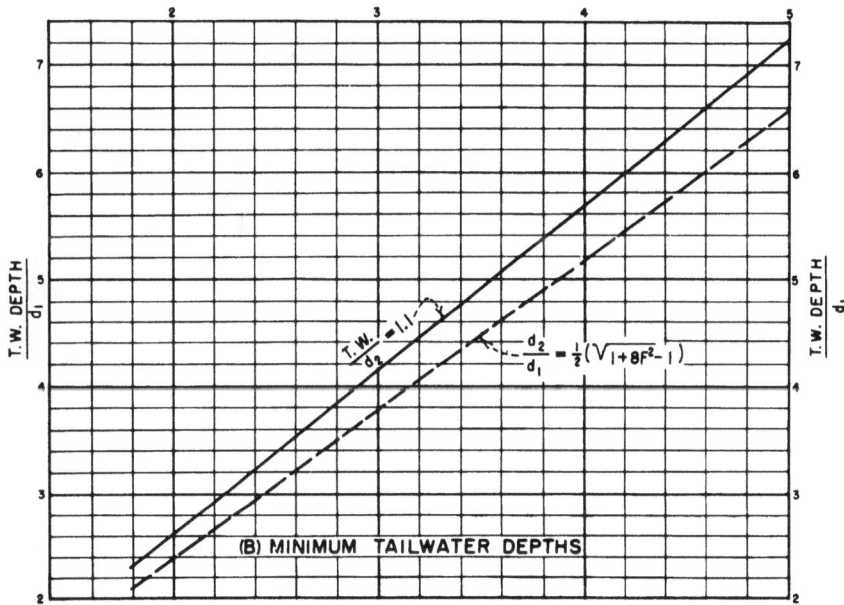

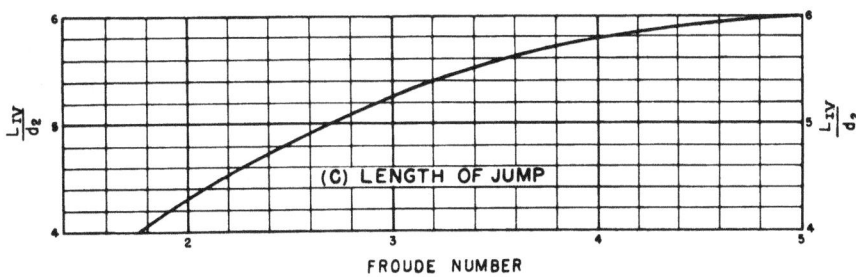

Figure 9-33. Stilling basin characteristics for Froude numbers between 2.5 and 4.5.—288-D-3050

capacity spillway where the specific energy at the upstream end of the basin is about 15 feet and the velocity into the basin is about 30 feet per second, the Froude number will be 3.2 for a basin width of 10 feet. The Froude number can be raised to 4.6 by widening the basin to 20 feet. The selection of basin width then becomes a matter of economics as well as hydraulic performance.

(4) *Basins for Froude numbers higher than 4.5.*—For basins where the Froude number value of the incoming flow is higher than 4.5, a true hydraulic jump will form. The installation of accessory devices such as blocks, baffles, and sills along the floor of the basin produces a stabilizing effect on the jump, which permits shortening the basin and provides a factor of safety against sweep-out due to inadequate tailwater depth.

The basin shown on figure 9-34, which is designated as a type III basin, can be adopted where incoming velocities do not exceed 50 feet per second. This basin utilizes chute blocks, impact baffle blocks, and an end sill to shorten the jump length and to dissipate the high-velocity flow within the shortened basin length. This basin relies on dissipation of energy by the impact blocks and also on the turbulence of the jump phenomena for its effectiveness. Because of the large impact forces to which the baffles are subjected by the impingement of high incoming velocities and because of the possibility of cavitation along the surfaces of the blocks and floor, the use of this basin must be limited to heads where the velocity does not exceed 50 feet per second.

Cognizance must be taken of the added loads placed upon the structure floor by the dynamic force brought against the upstream face of the baffle blocks. This dynamic force will approximate that of a jet impinging upon a plane normal to the direction of flow. The force, in pounds, may be expressed by the formula:

$$\text{Force} = 2wA(d_1 + h_{v_1}) \qquad (15)$$

where:

$$w = \text{the unit weight of water,}$$

$$A = \text{the area of the upstream face of the block, and}$$

$$(d_1 + h_{v_1}) = \text{the specific energy of the flow entering the basin.}$$

Negative pressure on the back face of the blocks will further increase the total load. However, since the baffle blocks are placed a distance equal to $0.8d_2$ beyond the start of the jump, there will be some cushioning effect by the time the incoming jet reaches the blocks and the force will be less than that indicated by the above equation. If the full force computed by equation (15) is used, the negative pressure force may be neglected.

Where incoming velocities exceed 50 feet per second, or where impact baffle blocks are not employed, the basin designated as type II on figure 9-35 can be adopted. Because the dissipation is accomplished primarily by hydraulic jump action, the basin length will be greater than that indicated for the type III basin. However, the chute blocks and dentated end sill will still be effective in reducing the length from that which would be necessary if they were not used. Because of the reduced margin of safety against sweep-out, the water depth in the basin should be about 5 percent greater than the computed conjugate depth.

(b) *Rectangular Versus Trapezoidal Stilling Basin.*—The utilization of a trapezoidal stilling basin in lieu of a rectangular basin may often be proposed where economy favors sloped side lining over vertical wall construction. Model tests have shown, however, that the hydraulic jump action in a trapezoidal basin is much less complete and less stable than it is in the rectangular basin. In the trapezoidal basin the water in the triangular areas along the sides of the basin adjacent to the jump is not opposed by the incoming high-velocity jet. The jump, which tends to occur vertically, cannot spread sufficiently to occupy the side areas. Consequently, the jump will form only in the central portion of the basin, while areas along the outside will be occupied by upstream-moving flows which ravel off the jump or come from the lower end of the basin.

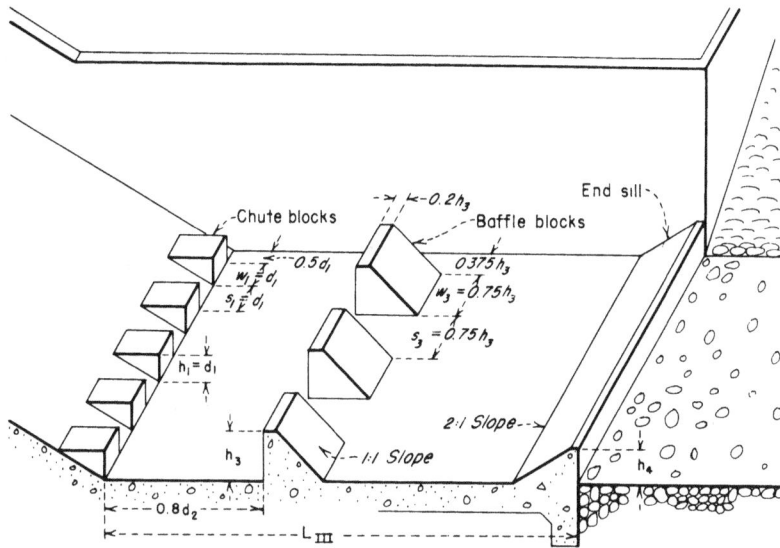

(A) TYPE III BASIN DIMENSIONS

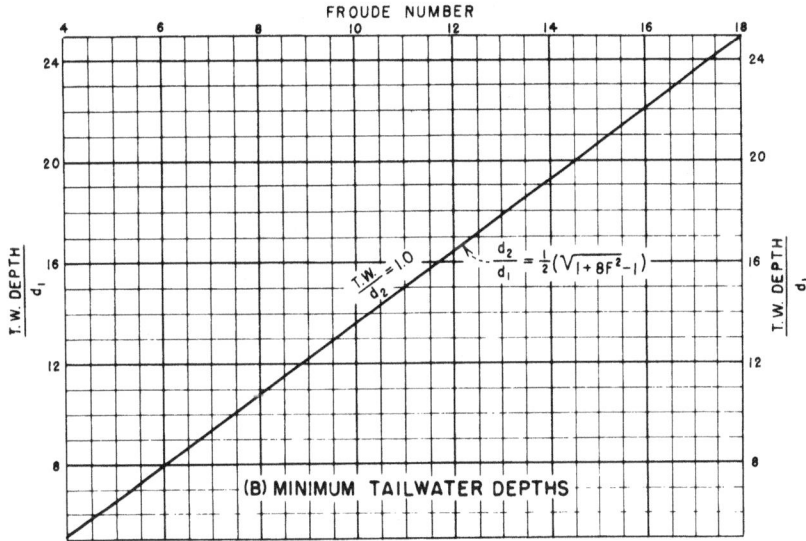

(B) MINIMUM TAILWATER DEPTHS

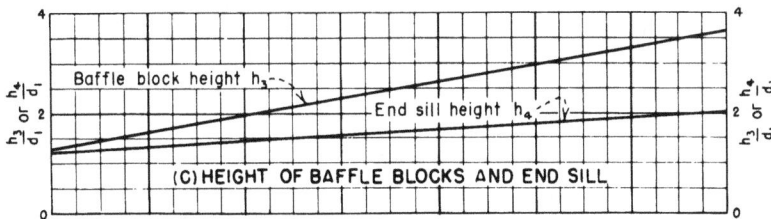

$$\frac{d_2}{d_1} = \frac{1}{2}\left(\sqrt{1+8F^2}-1\right)$$

$$\frac{T.W.}{d_2} = 1.0$$

(C) HEIGHT OF BAFFLE BLOCKS AND END SILL

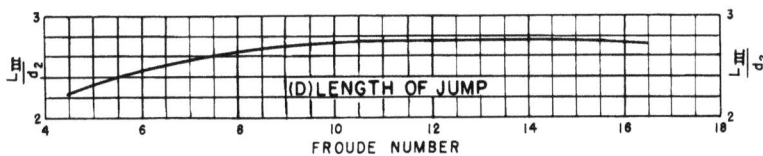

(D) LENGTH OF JUMP

Figure 9-34. Stilling basin characteristics for Froude numbers above 4.5 where incoming velocity does not exceed 50 feet per second.—288-D-3051

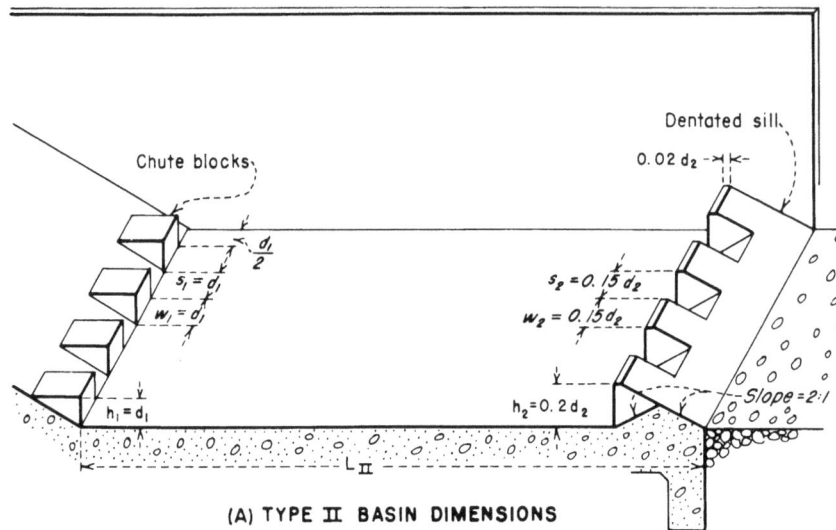

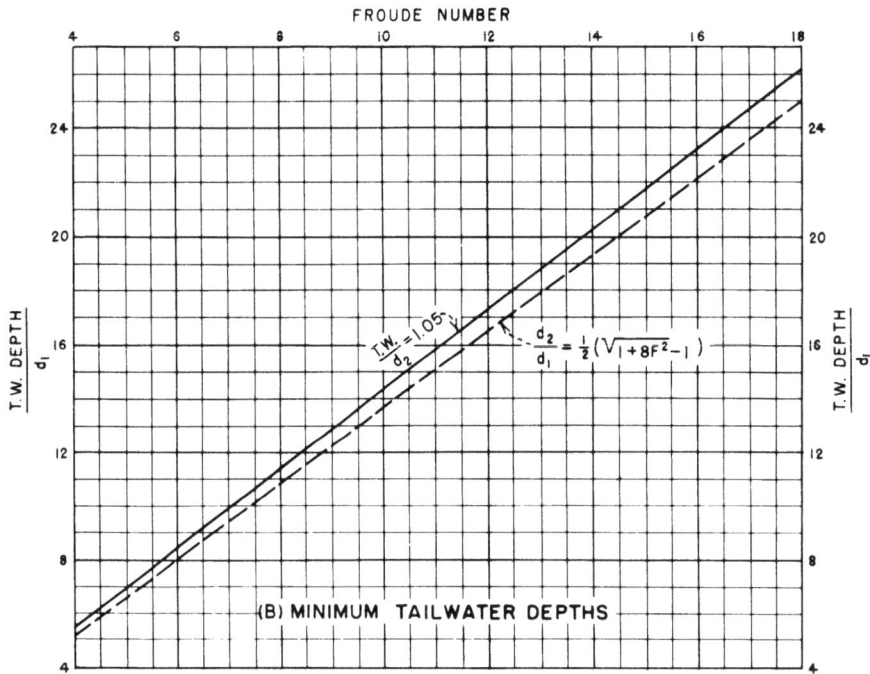

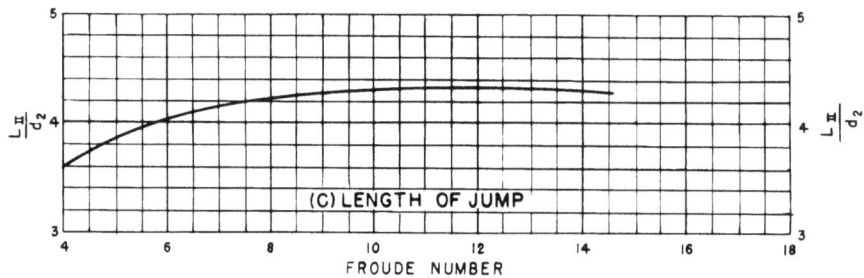

Figure 9-35. Stilling basin characteristics for Froude numbers above 4.5.—288-D-3052

The eddy or horizontal roller action resulting from this phenomenon tends to interfere and interrupt the jump action to the extent that there is incomplete dissipation of the energy and severe scouring can occur beyond the basin. For good hydraulic performance, the sidewalls of a stilling basin should be vertical or as near vertical as is practicable.

(c) *Basin Depths by Approximate Methods.*—The nomograph shown on figure 9-36 will aid in determining approximate basin depths for various basin widths and for various differences between reservoir and tailwater levels. Plottings are shown for the condition of no loss of head to the upstream end of the stilling basin, and for 10, 20, and 30 percent loss. (These plottings are shown on the nomographs as scales A, B, C, and D, respectively.) The required conjugate depths, d_2, will depend on the specific energy available at the entrance of the basin, as determined by the procedure discussed in section 9-17. Where only a rough determination of basin depths is needed, the choice of the loss to be applied for various spillway designs may be generalized as follows:

(1) For a design of an overflow spillway where the basin is directly downstream from the crest, or where the chute is not longer than the hydraulic head, consider no loss of head.

(2) For a design of a channel spillway where the channel length is between one and five times the hydraulic head, consider 10 percent loss of head.

(3) For a design of a spillway where the channel length exceeds five times the hydraulic head, consider 20 percent loss of head.

The nomograph on figure 9-36 gives values of d_2, the conjugate depth for the hydraulic jump. Tailwater depths for the various types of basin described in subsection (a) above should be increased as noted in that subsection.

(d) *Tailwater Considerations.*—The tailwater rating curve, which gives the stage-discharge relationship of the natural stream below the dam, is dependent on the natural conditions along the stream and ordinarily cannot be altered by the spillway design or by the release characteristics. As discussed in section 9-7(d), retrogression or aggradation of the river below the dam, which will affect the ultimate stage-discharge conditions, must be recognized in selecting the tailwater rating curve to be used for stilling basin design. Usually riverflows which approach the maximum design discharges have never occurred, and an estimate of the tailwater rating curve must either be extrapolated from known conditions or computed on the basis of assumed or empirical criteria. Thus, the tailwater rating curve at best is only approximate, and factors of safety to compensate for variations in tailwater must be included in the design.

For a given stilling basin design, the tailwater depth for each discharge seldom corresponds to the conjugate depth needed to form a perfect jump. The basin floor level must therefore be selected to provide tailwater depths which most nearly agree with the conjugate depths. Thus, the relative shapes and relationships of the tailwater curve to the conjugate depth curve will determine the required minimum depth to the basin floor. This is illustrated on figure 9-37. The tailwater rating curve is shown in (A) as curve 1, and a conjugate depth versus discharge curve for a basin of a certain width, W, is represented by curve 3. Since the basin must be made deep enough to provide for conjugate depth (or some greater depth to include a factor of safety) at the maximum spillway design discharge, the curves will intersect at point D. For lesser discharges the tailwater depth will be greater than the conjugate depth, thus providing an excess of tailwater which is conducive to the formation of a so-called drowned jump. (With the drowned jump condition, instead of achieving good jump-type dissipation by the intermingling of the upstream and downstream flows, the incoming jet plunges to the bottom and carries along the entire length of the basin floor at high velocity.) If the basin floor is made higher than indicated by the position of curve 3 on the figure, the depth curve and tailwater rating curve will intersect to the left of point D, thus indicating an excess of tailwater for smaller discharges and a deficiency of tailwater for higher discharges.

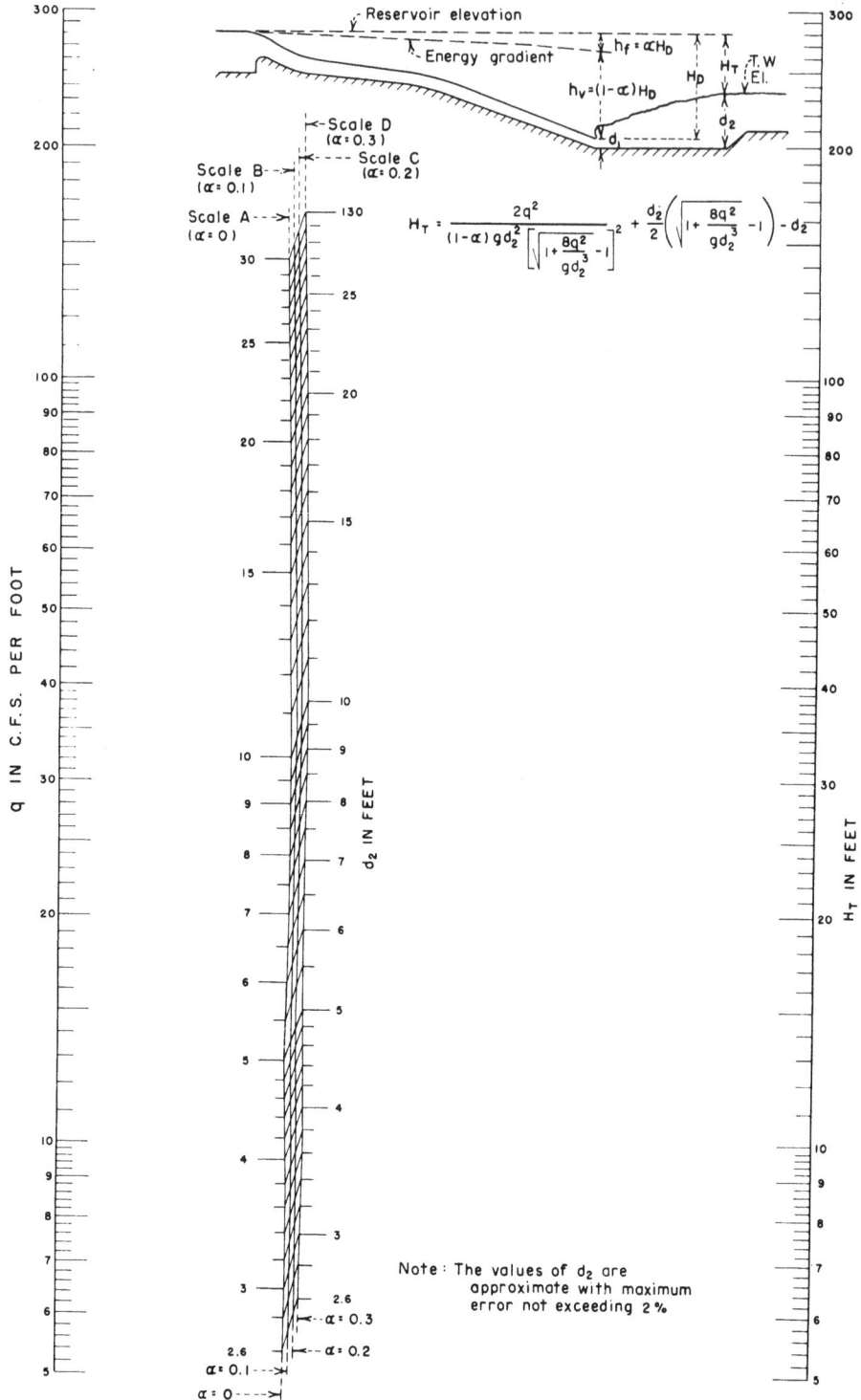

$$H_T = \frac{2q^2}{(1-\alpha)\,g\,d_2^2\left[\sqrt{1+\dfrac{8q^2}{g\,d_2^3}}-1\right]^2} + \frac{d_2}{2}\left(\sqrt{1+\frac{8q^2}{g\,d_2^3}}-1\right) - d_2$$

Note: The values of d_2 are approximate with maximum error not exceeding 2%

Figure 9-36. Stilling basin depths versus hydraulic heads for various channel losses.—288-D-3053

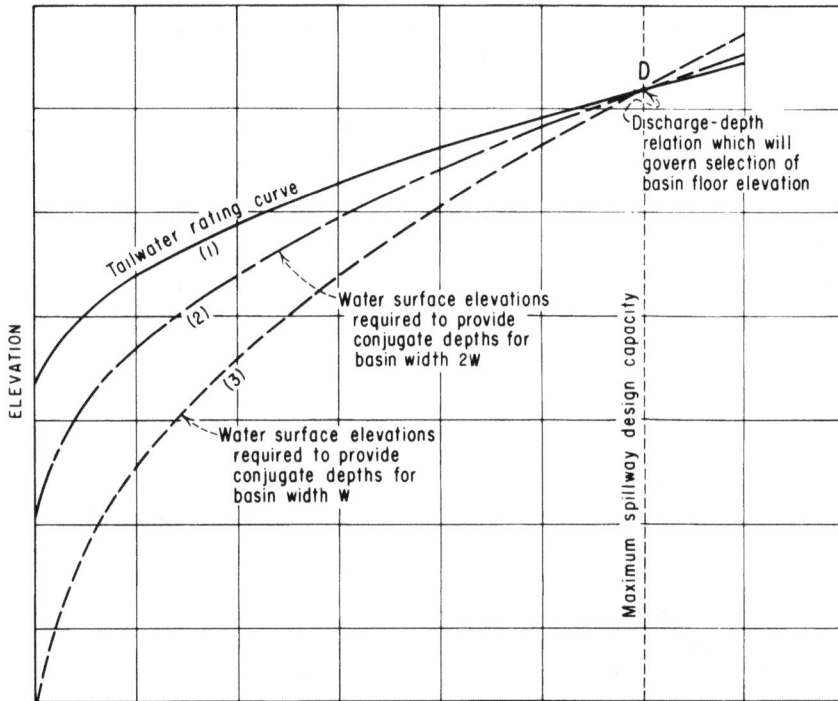

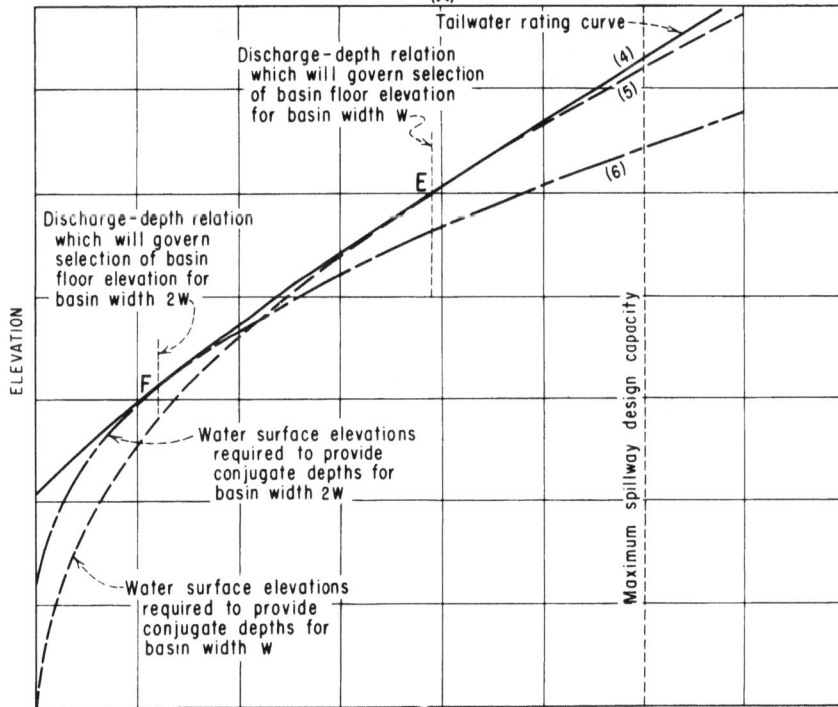

Figure 9-37. Relationships of conjugate depth curves to tailwater rating
curves.—288-D-2439

As an alternative to the selected basin which is represented by curve 3, a wider basin might be considered for which the conjugate depth curve 2 will apply. This design will provide a shallower basin, in which the conjugate depths will more nearly match the tailwater depths for all discharges. The choice of basin widths, of course, involves consideration of economics, as well as hydraulic performance.

Where a tailwater rating curve shaped similar to that represented by curve 4 on figure 9-37(B) is encountered, the level of the stilling basin floor must be determined for some discharge other than the maximum design capacity. If the tailwater rating curve were made to intersect the required water surface elevation at the maximum design capacity, as in figure 9-37(A), there would be insufficient tailwater depth for most smaller discharges. In this case the basin floor elevation is selected so that there will be sufficient tailwater depth for all discharges. For the basin of width W whose required tailwater depth is represented by curve 5, the position of the floor would be selected so that the two curves would coincide at the discharge represented by point E on the figure. For all other discharges the tailwater depth will be in excess of that needed for forming a satisfactory jump. Similarly, if a basin width of $2W$ were considered, the basin floor level would be selected so that curve 6 would intersect the tailwater rating curve at point F. Here also, the selection of basin widths should be based on economic aspects as well as hydraulic performance.

Where exact conjugate depth conditions for forming the jump cannot be attained, the question of the relative desirability of having insufficient tailwater depth as compared to having excessive tailwater depth should be considered. With insufficient tailwater the back pressure will be deficient and sweep-out of the basin will occur. With an excess of tailwater the jump will be formed and energy dissipation within the basin will be quite complete until the drowned jump phenomenon becomes critical. Chute blocks, baffles, and end sills will further assist in energy dissipation, even with a drowned jump.

(e) *Stilling Basin Freeboard*.—A freeboard of 5 to 10 feet is usually provided to allow for surging and wave action in the stilling basin. For smaller, low-head basins, the required freeboard will be nearer the lower value, whereas the higher value will normally be used for larger, high-head spillways. A minimum freeboard may be used if overtopping by the waves will not cause significant damage. Engineering judgment should be used in setting the height of freeboard by comparing the cost of additional wall height against possible damage caused by overflow of the stilling basin walls. Wherever practical, a hydraulic model should be used in determining the amount of freeboard.

9-22. *Deflector Buckets*.—Where the spillway discharge may be safely delivered directly to the river without providing an energy dissipating or stilling device, the jet is often projected beyond the structure by a deflector bucket or lip. Flow from these deflectors leaves the structure as a free-discharging upturned jet and falls into the stream channel some distance from the end of the spillway. The path the jet assumes depends on the energy of flow available at the lip and the angle at which the jet leaves the bucket.

With the origin of the coordinates taken at the end of the lip, the path of the trajectory is given by the equation:

$$y = x \tan \theta - \frac{x^2}{K[4(d + h_v) \cos^2 \theta]} \quad (16)$$

where:

θ = the angle between the curve of the bucket at the lip and the horizontal (or lip angle), and

K = a factor, equal to 1 for the theoretical jet.

To compensate for loss of energy and velocity reduction due to the effect of air resistance, internal turbulences, and disintegration of the jet, a value for K of about 0.85 should be assumed.

The horizontal range of the jet at the level of the lip is obtained by making y in equation (16) equal to zero. Then:

$$x = 4K(d + h_v) \tan \theta \, \cos^2 \theta$$
$$= 2K(d + h_v) \sin 2\theta \qquad (17)$$

The maximum value of x will be equal to $2K(d + h_v)$ when θ is 45°. The lip angle is influenced by the bucket radius and the height of the lip above the bucket invert. It usually varies from 20° to 45°, with 30° being the preferred angle.

The bucket radius should be made long enough to maintain concentric flow as the water moves around the curve. The rate of curvature must be limited similar to that of a vertical curve in a discharge channel (sec. 9-18), so that the floor pressures will not alter the streamline distribution of the flow. The minimum radius of curvature can be determined from equation (13), except that values of p not exceeding 500 pounds per square foot will produce values of the radius which have proved satisfactory in practice. However, the radius should not be less than five times the depth of water. Structurally, the cantilever bucket must be of sufficient strength to withstand this normal dynamic force in addition to the other applied forces.

Figure 9-38 shows the deflector at the end of the spillway tunnel at Hungry Horse Dam in operation.

9-23. Submerged Bucket Energy Dissipators.—When the tailwater depth is too great for the formation of a hydraulic jump, dissipation of the high energy of flow can be effected by the use of a submerged bucket deflector. The hydraulic behavior in this type of dissipator is manifested primarily by the formation of two rollers; one is on the surface moving counterclockwise (if flow is to the right) and is contained within the region above the curved bucket, and the other is a ground roller moving in a clockwise direction and is situated downstream from the bucket. The movements of the rollers, along with the intermingling of the incoming flows, effectively dissipate the high energy of the water and prevent excessive scouring downstream from the bucket.

Two types of roller bucket have been developed and model tested [6]. Their shape and dimensional arrangements are shown on figure 9-39. The general nature of the dissipating action for each type is represented on figure 9-40. Hydraulic action of the two buckets has the same characteristics, but distinctive features of the flow differ to the extent that each has certain limitations. The high-velocity flow leaving the deflector lip of the solid bucket is directed upward. This creates a high boil on the water surface and a violent ground roller moving clockwise downstream from the bucket. This ground roller continuously pulls loose material back towards the lip of the bucket and keeps some of the intermingling material in a constant state of agitation. In the slotted bucket the high-velocity jet leaves the lip at a flatter angle, and only a part of the high-velocity flow finds its way to the surface. Thus, a less violent surface boil occurs and there is a better dispersion of flow in the region above the ground roller which results in less concentration of high-energy flow throughout the bucket and a smoother downstream flow.

Use of a solid bucket dissipator may be objectionable because of the abrasion on the concrete surfaces caused by material which is swept back along the lip of the deflector by the ground roller. In addition, the more turbulent surface roughness induced by the severe surface boil carries farther down the river, causing objectionable eddy currents which contribute to riverbank sloughing. Although the slotted bucket provides better energy dissipation with less severe surface and streambed disturbances, it is more sensitive to sweep-out at lower tailwaters and is conducive to a diving and scouring action at excessive tailwaters. This is not the case with the solid bucket. Thus, the tailwater range which will provide good performance with the slotted bucket is much narrower than that of the solid bucket. A solid bucket dissipator should not be used wherever the tailwater limitations of the slotted bucket can be met. Therefore, only the design of the slotted bucket will be discussed.

Flow characteristics of the slotted bucket are illustrated on figure 9-41. For deficient tailwater depths the incoming jet will sweep the surface roller out of the bucket and will produce a high-velocity flow downstream, both

Figure 9-38. Deflector bucket in operation for the spillway at Hungry Horse Dam in Montana.—P447-105-5924

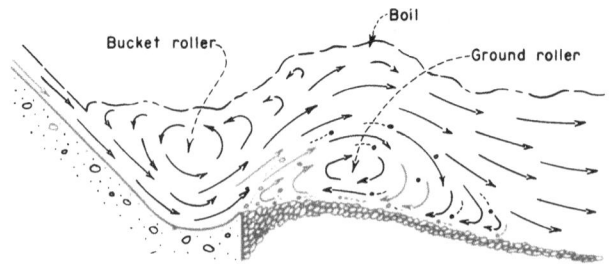

(A) SOLID TYPE BUCKET

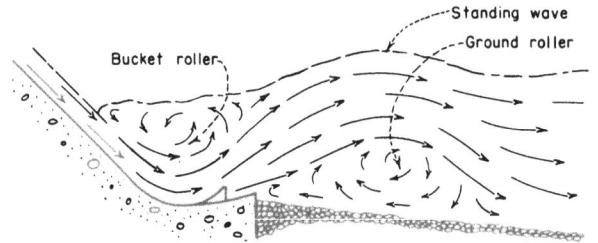

(B) SLOTTED TYPE BUCKET

Figure 9-40. Hydraulic action in solid and slotted buckets.—288-D-2431

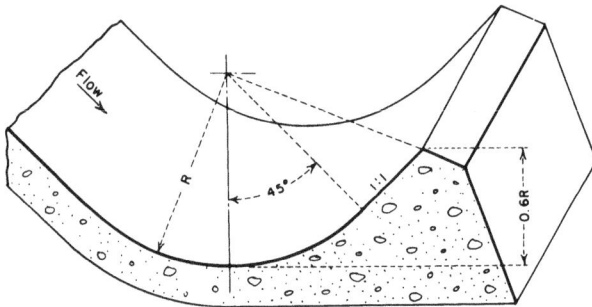

(A) SOLID BUCKET

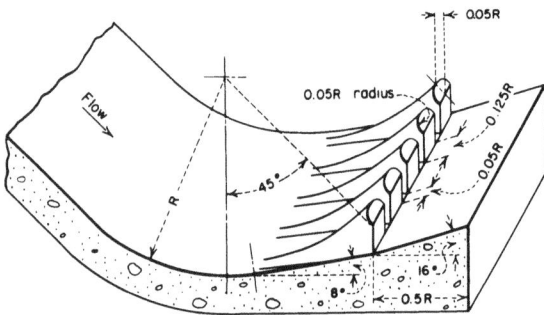

(B) SLOTTED BUCKET

Figure 9-39. Submerged bucket energy dissipators.—288-D-2430

along the water surface and along the riverbed. This action is depicted as stage (A) on figure 9-41. As the tailwater depth is increased, there will be a depth at which instability of flow will occur, where sweep-out and submergence will alternately prevail. To obtain continuous operation at the submerged stage, the minimum tailwater depth must be above this instable state. Flow action within the acceptable operating stage is depicted as stage (B) on figure 9-41.

When the tailwater becomes excessively deep, the phenomenon designated as diving flow will occur. At this stage the jet issuing from the lip of the bucket will no longer rise and continue along the surface but intermittently will become depressed and dive to the riverbed. The position of the downstream roller will change with the change in position of the jet. It will occur at the surface when the jet dives and will form along the river bottom as a ground roller when the jet rides the surface. Scour will occur in the streambed at the point of impingement when the jet dives but will be filled in by the ground

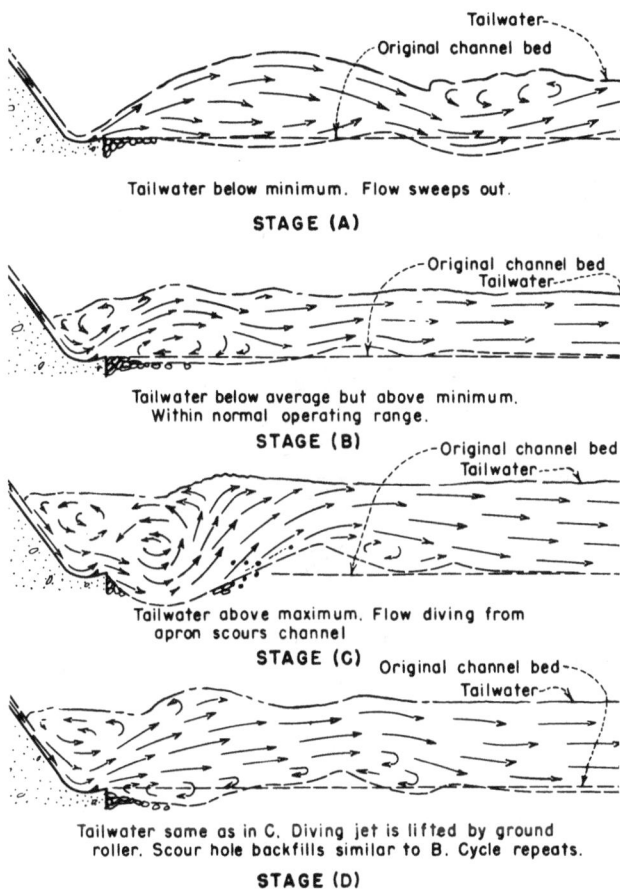

Tailwater below minimum. Flow sweeps out.
STAGE (A)

Tailwater below average but above minimum.
Within normal operating range.
STAGE (B)

Tailwater above maximum. Flow diving from
apron scours channel
STAGE (C)

Tailwater same as in C. Diving jet is lifted by ground
roller. Scour hole backfills similar to B. Cycle repeats.
STAGE (D)

Figure 9-41. Flow characteristics in a slotted
bucket.—288-D-2432

roller when the jet rides. The characteristic flow pattern for the diving stage is depicted in (C) and (D) of figure 9-41. Maximum tailwater depths must be limited to forestall the diving flow phenomenon.

The design of the slotted bucket involves determination of the radius of curvature of the bucket and the allowable range of tailwater depths. These criteria, as determined from experimental results, are plotted on figure 9-42 in relation to the Froude number parameter.

The Froude number values are for flows at the point where the incoming jet enters the bucket. Symbols are defined on figure 9-43.

9-24. *Plunge Pools.*—When a free-falling overflow nappe drops almost vertically into a pool in a riverbed, a plunge pool will be scoured to a depth which is related to the height of the fall, the depth of tailwater, and the concentration of the flow [7]. Depths of scour are influenced initially by the erodibility of the stream material or the bedrock and by the size or the gradation of sizes of any armoring material in the pool. However, the armoring or protective surfaces of the pool will be progressively reduced by the abrading action of the churning material to a size which will be scoured out, and the ultimate scour depth will, for all practical considerations, stabilize at a limiting depth irrespective of the material size. An empirical approximation based on experimental data has been developed by Veronese [8] for limiting scour depths, as follows:

$$d_s = 1.32\, H_T^{0.225}\, q^{0.54} \qquad (18)$$

where:

d_s = the maximum depth of scour
        below tailwater level
        in feet,

H_T = the head from reservoir level to
        tailwater level in feet,
        and

q = the discharge in c.f.s. per
        foot of width.

Three Bureau of Reclamation dams which have plunge pools for energy dissipators have been tested in hydraulic models. Reports of the results of these tests are given in references [9], [10], and [11].

F. HYDRAULICS OF MORNING GLORY (DROP INLET) SPILLWAYS

9-25. *General Characteristics.*—The flow conditions and discharge characteristics of a morning glory spillway are unique in the respect that, in normal operation, the control changes as the head changes. As brought out in the following discussion, at low heads the crest

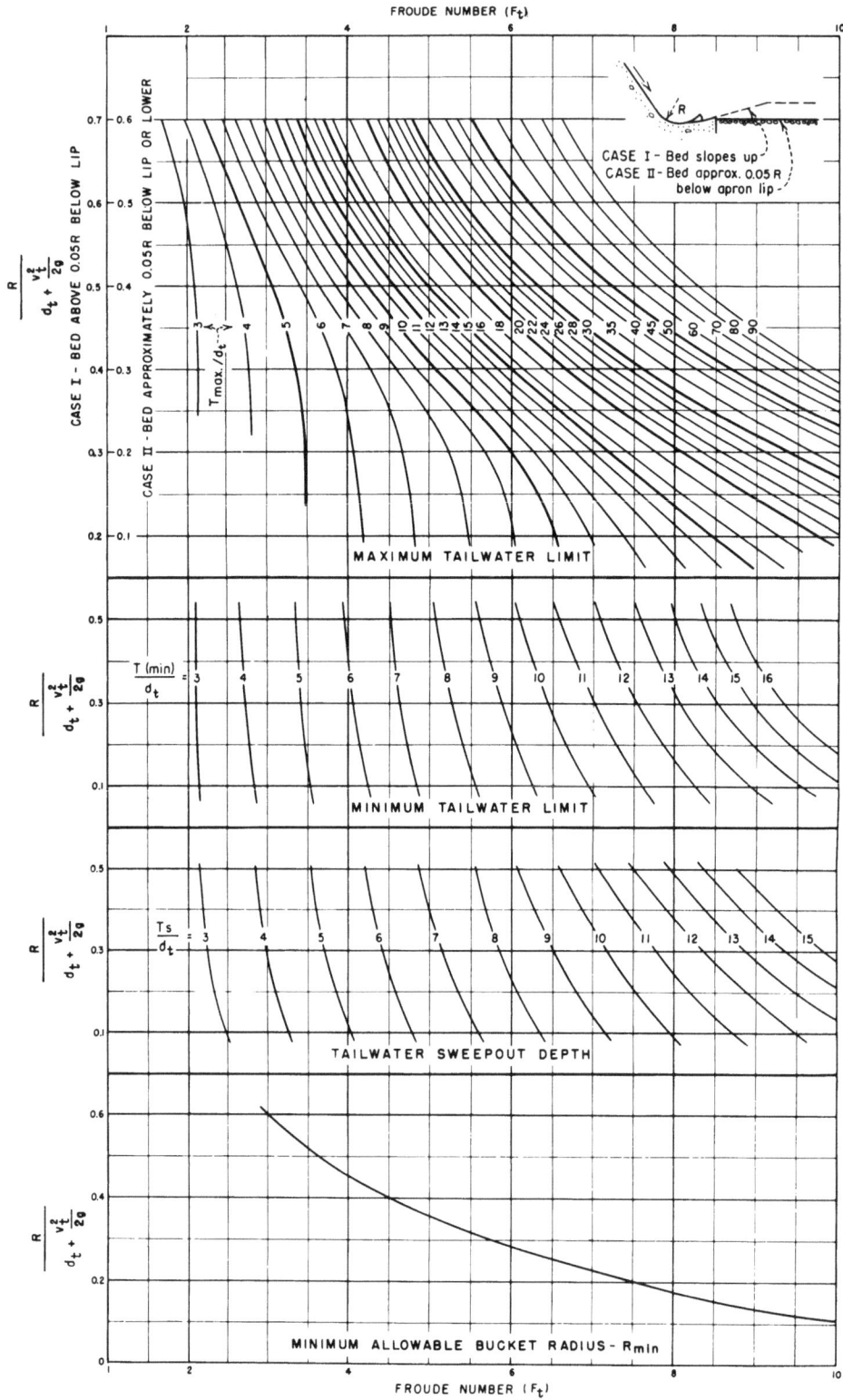

Figure 9-42. Limiting criteria for slotted bucket design.—288-D-2433

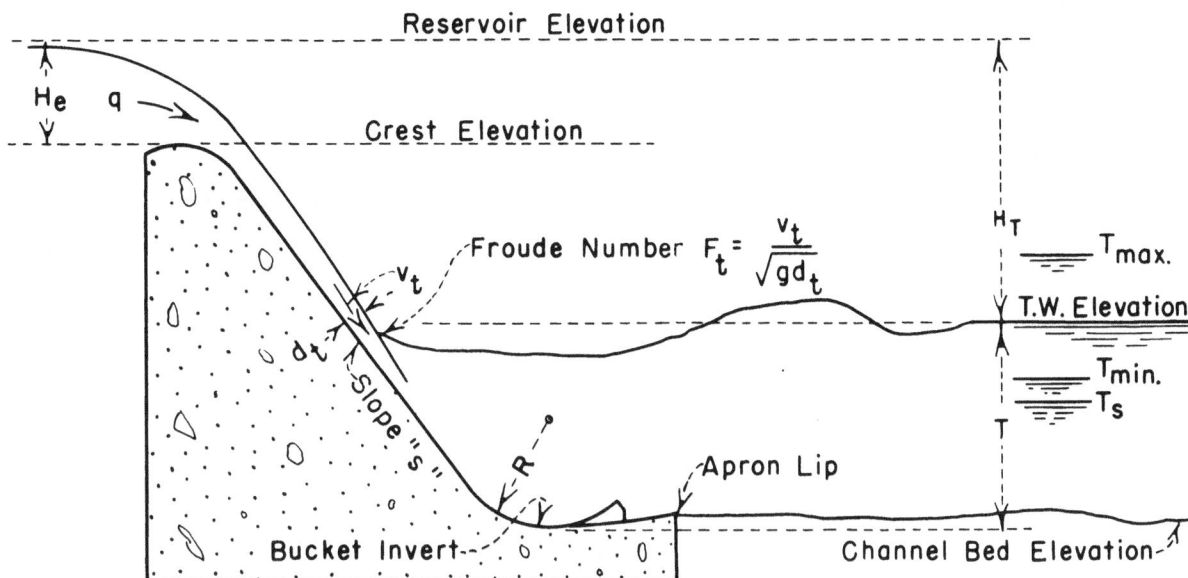

Figure 9-43. Definition of symbols—submerged bucket.—288-D-2434

is the control and the orifice and tunnel serve only as the discharge channel; whereas at progressively higher heads the orifice and then the tunnel serves as the control. Because of this uniqueness the hydraulics of morning glory spillways are discussed separately from other spillway components.

Typical flow conditions and discharge characteristics of a morning glory spillway are represented on figure 9-44. As illustrated on the discharge curve, crest control (condition 1) will prevail for heads between the ordinates of a and g; orifice control (condition 2) will govern for heads between the ordinates of g and h; and the spillway tunnel will flow full for heads above the ordinate of h (represented as condition 3).

Flow characteristics of a morning glory spillway will vary according to the proportional sizes of the different elements. Changing the diameter of the crest will change the curve ab on figure 9-44 so that the ordinate of g on curve cd will be either higher or lower. For a larger diameter crest, greater flows can be discharged over the crest at low heads and orifice control will occur with a lesser head on the crest, tending to fill up the transition above the orifice. Similarly, by altering the size of the orifice, the position of curve cd will shift, changing the head above which orifice control

will prevail. If the orifice is made of sufficient size that curve cd is moved to coincide with or lie to the right of point j, the control will shift directly from the crest to the downstream end of the tunnel. The details of the hydraulic flow characteristics are discussed in following sections.

9-26. *Crest Discharge.*—For small heads, flow over the morning glory spillway is governed by the characteristics of crest discharge. The throat, or orifice, will flow partly full and the flow will cling to the sides of the shaft. As the discharge over the crest increases, the overflowing annular nappe will become thicker, and eventually the nappe flow will converge into a solid vertical jet. The point where the annular nappe joins the solid jet is called the crotch. After the solid jet forms, a "boil" will occupy the region above the crotch: both the crotch and the top of the boil become progressively higher with larger discharges. For high heads the crotch and boil may almost flood out, showing only a slight depression and eddy at the surface.

Until such time as the nappe converges to form a solid jet, free-discharging weir flow prevails. After the crotch and boil form, submergence begins to affect the weir flow and ultimately the crest will drown out. Flow is then governed either by the nature of the

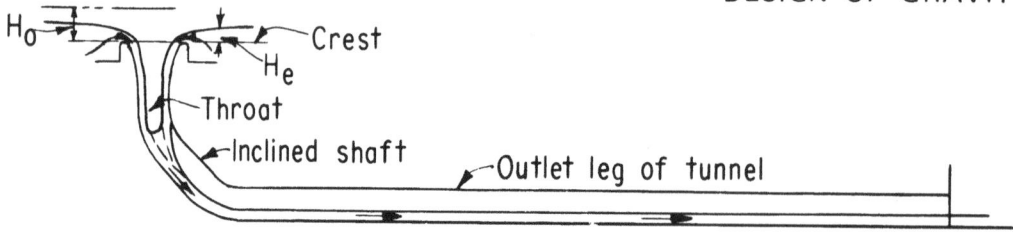

CONDITION 1. CREST CONTROL

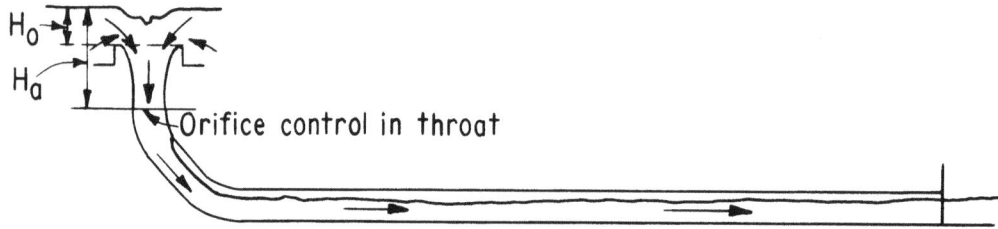

CONDITION 2. ORIFICE CONTROL

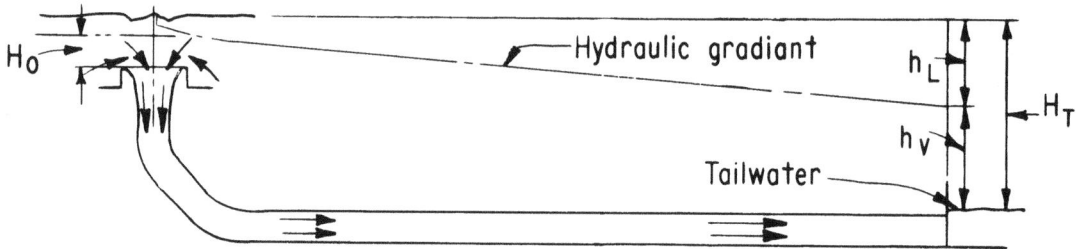

CONDITION 3. TUNNEL CONTROL

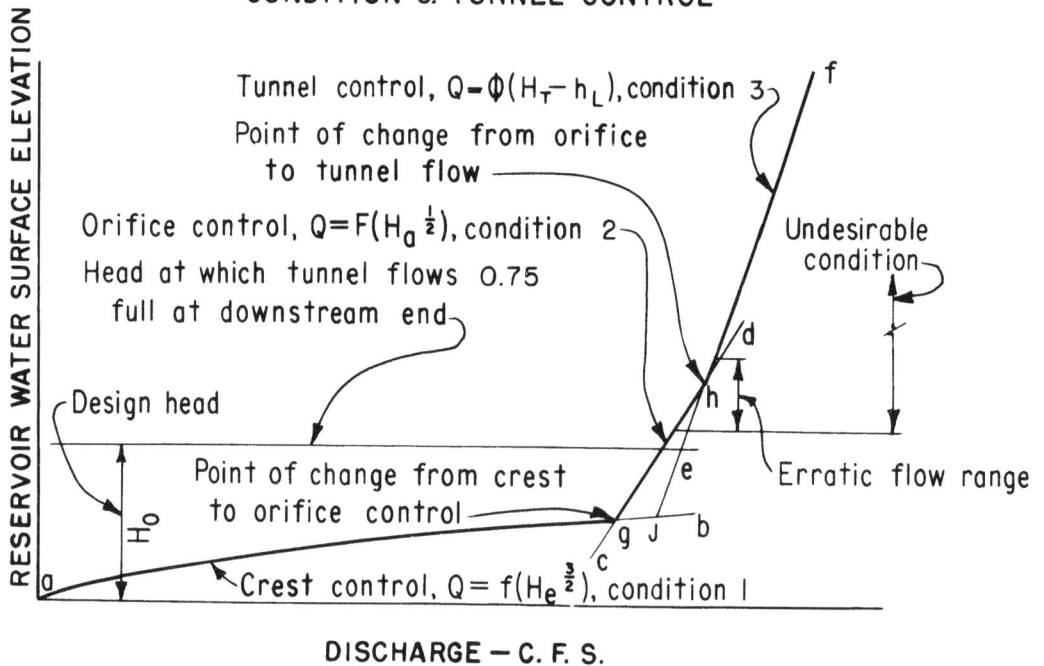

Figure 9-44. Flow and discharge characteristics of a morning glory
spillway.—288-D-3054

contracted jet which is formed by the overflow entrance, or by the shape and size of the throat as determined by the crest profile if it does not conform to the jet shape. Vortex action must be minimized to maintain converging flow into the inlet. Guide piers are often employed along the crest for this purpose [12, 13, 14].

If the crest profile conforms to the shape of the lower nappe of a jet flowing over a sharp-crested circular weir, the discharge characteristics for flow over the crest and through the throat can be expressed as:

$$Q = CLH^{3/2} \qquad (3)$$

where H is the head measured either to the apex of the under nappe of the overflow, to the spring point of the circular sharp-crested weir, or to some other established point on the overflow. Similarly, the choice of the length L is related to some specific point of measurement such as the length of the circle at the apex, along the periphery at the upstream face of the crest, or along some other chosen reference line. The value of C will change with different definitions of L and H. If L is taken at the outside periphery of the overflow crest (the origin of the coordinates in figure 9-45) and if the head is measured to the apex of the overflow shape, equation (3) can be written:

$$Q = C_o(2\pi R_s)H_o^{3/2} \qquad (19)$$

It will be apparent that the coefficient of discharge for a circular crest differs from that

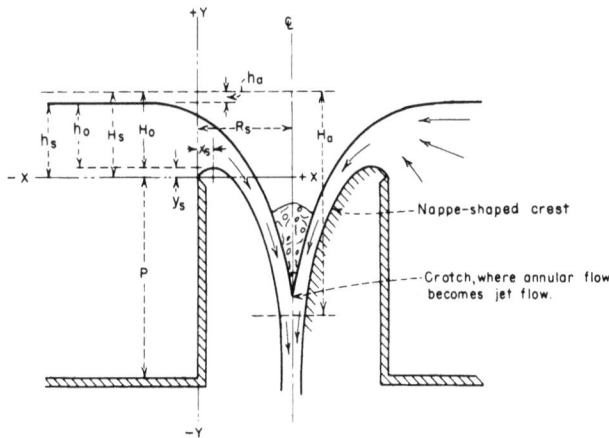

Figure 9-45. Elements of nappe-shaped profile for a circular crest.—288-D-2440

for a straight crest because of the effects of submergence and back pressure incident to the joining of the converging flows. Thus, C_o must be related to both H_o and R_s, and can be expressed in terms of $\dfrac{H_o}{R_s}$. The relationship of C_o, as determined from model tests [15], to values of $\dfrac{H_o}{R_s}$ for three conditions of approach depth is plotted on figure 9-46. These coefficients are valid only if the crest profile conforms to that of the jet flowing over a sharp-crested circular weir at H_o head and if aeration is provided so that subatmospheric pressures do not exist along the lower nappe surface contact.

When the crest profile conforms to the profile of the under nappe shape for an H_o head over the crest, free flow prevails for $\dfrac{H_o}{R_s}$ ratios up to approximately 0.45, and crest control governs. As the $\dfrac{H_o}{R_s}$ ratio increases above 0.45 the crest partly submerges and flow showing characteristics of a submerged crest is the controlling condition. When the $\dfrac{H_o}{R_s}$ ratio approaches 1.0 the water surface above the crest is completely submerged. For this and higher stages of $\dfrac{H_o}{R_s}$ the flow phenomena is that of orifice flow. The weir formula, $Q = CLH^{3/2}$, is used as the measure of flow over the crest regardless of the submergence, by using a coefficient which reflects the flow conditions through the various $\dfrac{H}{R_s}$ ranges. Thus, from figure 9-46 it will be seen that the crest coefficient is only slightly changed from that normally indicated for values of $\dfrac{H_o}{R_s}$ less than 0.45, but reduces rapidly for the higher $\dfrac{H_o}{R_s}$ ratios.

It will be noted that for most conditions of flow over a circular crest the coefficient of discharge increases with a reduction of the

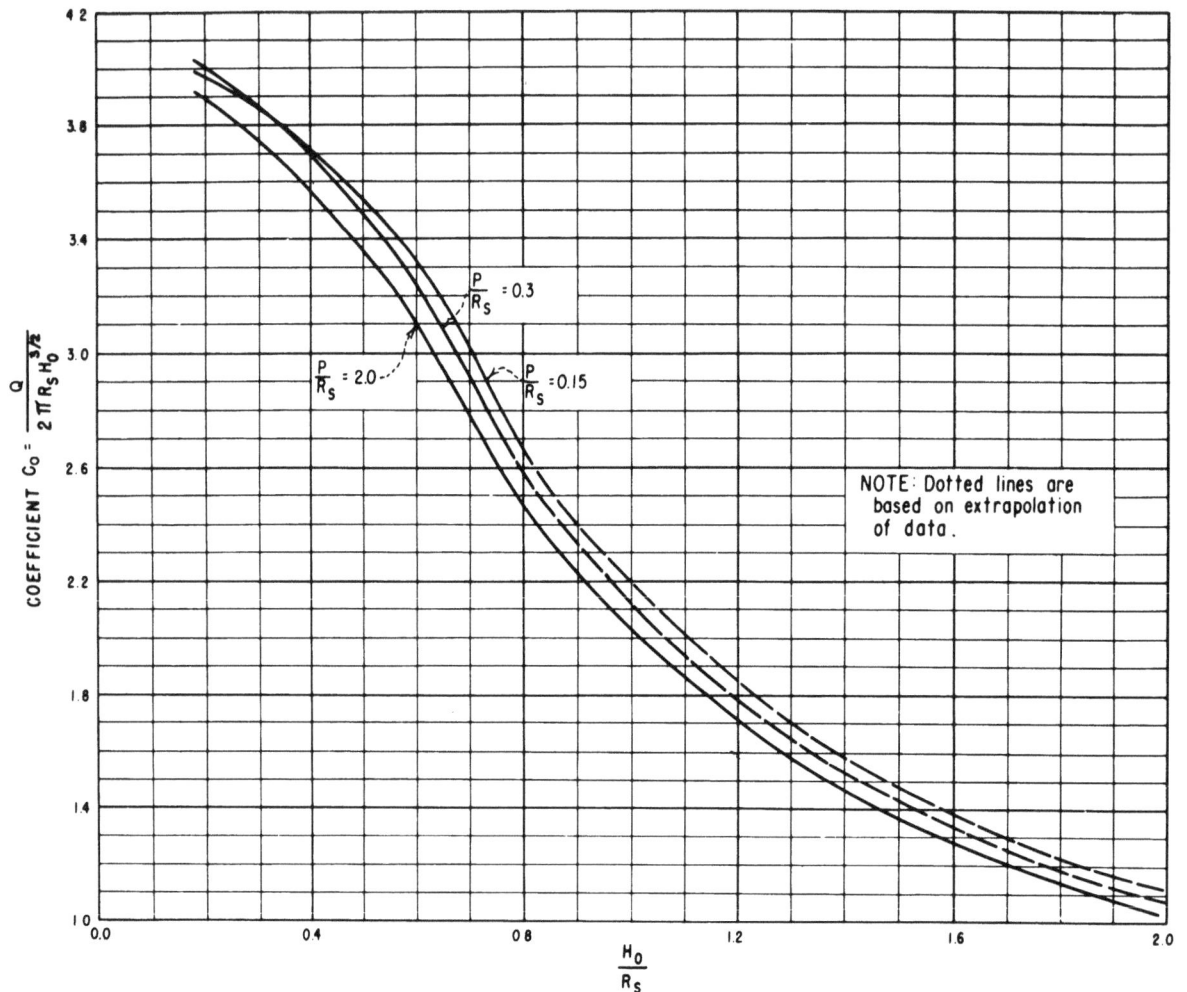

Figure 9-46. Relationship of circular crest coefficient C_O to $\dfrac{H_O}{R_s}$ for different approach depths (aerated nappe).—288-D-2441

approach depth, whereas the opposite is true for a straight crest. For both crests a shallower approach lessens the upward vertical velocity component and consequently suppresses the contraction of the nappe. However, for the circular crest the submergence effect is reduced because of a depressed upper nappe surface, giving the jet a quicker downward impetus, which lowers the position of the crotch and increases the discharge.

Coefficients for partial heads of H_e on the crest can be determined from figure 9-47 to prepare a discharge-head relationship. The designer must be cautious in applying the above criteria, since subatmospheric pressure or submergence effects may alter the flow

conditions differently for variously shaped profiles. These criteria, therefore, should not be applied for flow conditions where $\dfrac{H_e}{R_s}$ exceeds 0.4.

9-27. *Crest Profiles.*—In this discussion, the crest profile is considered to extend from the crest to the orifice control, and forms the transition to the orifice. Values of coordinates to define the shape of the lower surface of a nappe flowing over an aerated sharp-crested circular weir for various conditions of $\dfrac{P}{R_s}$ and $\dfrac{H_s}{R_s}$ are shown in tables 9-2, 9-3, and 9-4. These

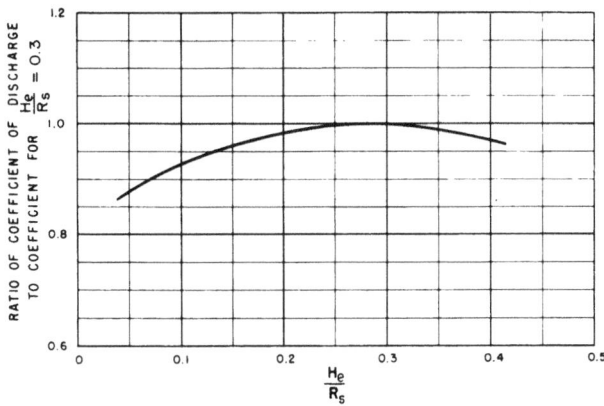

Figure 9-47. Circular crest coefficient of discharge for other than design head.–288-D-2446

data are based on experimental tests [15] conducted by the Bureau of Reclamation. The relationships of H_s to H_o are shown on figure 9-48. Typical upper and lower nappe profiles for various values of $\dfrac{H_s}{R_s}$ are plotted on figure 9-49 in terms of $\dfrac{x}{H_s}$ and $\dfrac{y}{H_s}$ for the condition of

$$\frac{P}{R_s} = 2.0.$$

Illustrated on figure 9-50 are typical lower nappe profiles, plotted for various values of H_s for a given value of R_s. In contrast to the straight crest where the nappe springs farther from the crest as the head increases, it will be seen from figure 9-50 that the lower nappe profile for the circular crest springs farther only in the region of the high point of the profile, and then only for $\dfrac{H_s}{R_s}$ values up to about 0.5. The profiles become increasingly suppressed for larger $\dfrac{H_s}{R_s}$ values. Below the high point of the profile, the paths cross and the shapes for the higher heads fall inside those for the lower heads. Thus, if the crest profile is designed for heads where $\dfrac{H_s}{R_s}$ exceeds about 0.25 to 0.3, it appears that subatmospheric pressure will occur along some portion of the profile when heads are less than the designed

maximum. If subatmospheric pressures are to be avoided along the crest profile, the crest shape should be selected so that it will give support to the overflow nappe for the smaller $\dfrac{H_e}{R_s}$ ratios. Figure 9-51 shows the approximate increase in radius required to minimize subatmospheric pressures on the crest. The crest shape for the enlarged crest radius is then based on a $\dfrac{H'_s}{R'_s}$ ratio of 0.3.

9-28. *Orifice Control.*–The diameter of a jet issuing from a horizontal orifice can be determined for any point below the water surface if it is assumed that the continuity equation, $Q = av$, is valid and if friction and other losses are neglected.

For a circular jet the area is equal to πR^2. The discharge is equal to $av = \pi R^2 \sqrt{2gh_v}$. Solving for R, $R = \dfrac{Q_a^{1/2}}{5H_a^{1/4}}$ where H_a is equal to the difference between the water surface and the elevation under consideration. The diameter of the jet thus decreases indefinitely with the distance of the vertical fall for normal design applications.

If an assumed total loss (to allow for jet contraction losses, friction losses, velocity losses due to direction change, etc.) is taken as $0.1H_a$, the equation for determining the approximate radius of the circular jet can be written:

$$R = 0.204 \frac{Q^{1/2}}{H_a^{1/4}} \qquad (20)$$

Since this equation is for the shape of the jet, its use for determining the theoretical size and shape of a shaft in the area of the orifice would result in the minimum size shaft which would not restrict the flow and would not develop pressures along the sides of the shaft.

A theoretical jet profile or shaft as determined by equation (20) is shown by the dot-dash lines *abc* on figure 9-52. Superimposed on the jet of that figure is an overflow crest with a radius R_s, which serves as an entrance to the shaft. If both the crest and the shaft are designed for the same water

Table 9-2.—Coordinates of lower nappe surface for different values of $\dfrac{H_s}{R}$ when $\dfrac{P}{R} = 2$.

[Negligible approach velocity and aerated nappe]

For portion of the profile above the weir crest — $\dfrac{Y}{H_s}$

$\dfrac{X}{H_s}$ / $\dfrac{H_s}{R}$	0.00	0.10*	0.20	0.25	0.30	0.35	0.40	0.45	0.50	0.60	0.80	1.00	1.20	1.50	2.00
0.000	0.0000	0.0000	0.0000	0.0000	0.0000	0.0000	0.0000	0.0000	0.0000	0.0000	0.0000	0.0000	0.0000	0.0000	0.0000
.010	.0150	.0145	.0133	.0130	.0128	.0125	.0122	.0119	.0116	.0112	.0104	.0095	.0086	.0077	.0070
.020	.0280	.0265	.0250	.0243	.0236	.0231	.0225	.0220	.0213	.0202	.0180	.0159	.0140	.0115	.0090
.030	.0395	.0365	.0350	.0337	.0327	.0317	.0308	.0299	.0289	.0270	.0231	.0198	.0168	.0126	.0085
.040	.0490	.0460	.0435	.0417	.0403	.0389	.0377	.0363	.0351	.0324	.0268	.0220	.0176	.0117	.0050
.050	.0575	.0535	.0506	.0487	.0471	.0454	.0436	.0420	.0402	.0368	.0292	.0226	.0168	.0092	
.060	.0650	.0605	.0570	.0550	.0531	.0510	.0489	.0470	.0448	.0404	.0305	.0220	.0147	.0053	
.070	.0710	.0665	.0627	.0605	.0584	.0560	.0537	.0514	.0487	.0432	.0308	.0201	.0114	.0001	
.080	.0765	.0710	.0677	.0655	.0630	.0603	.0578	.0550	.0521	.0455	.0301	.0172	.0070		
.090	.0820	.0765	.0722	.0696	.0670	.0640	.0613	.0581	.0549	.0471	.0287	.0135	.0018		
.100	.0860	.0810	.0762	.0734	.0705	.0672	.0642	.0606	.0570	.0482	.0264	.0089			
.120	.0940	.0880	.0826	.0790	.0758	.0720	.0683	.0640	.0596	.0483	.0195				
.140	.1000	.0935	.0872	.0829	.0792	.0750	.0 05	.0654	.0599	.0460	.0101				
.160	.1045	.0980	.0905	.0855	.0812	.076?	.0710	.0651	.0585	.0418					
.180	.1080	.1010	.0927	.0872	.0820	.0766	.0705	.0637	.0559	.0361					
.200	.1105	.1025	.0938	.0877	.0819	.0756	.0688	.0611	.0521	.0292					
.250	.1120	.1035	.0926	.0850	.0773	.0683	.0596	.0495	.0380	.0068					
.300	.1105	.1000	.0850	.0764	.0668	.0559	.0446	.0327	.0174						
.350	.1060	.0930	.0750	.0650	.0540	.0410	.0280	.0125							
.400	.0970	.0830	.0620	.0500	.0365	.0220	.0060								
.450	.0845	.0700	.0450	.0310	.0170	.000									
.500	.0700	.0520	.0250	.0100											
.550	.0520	.0320	.0020												
.600	.0320	.0080													
.650	.0090														

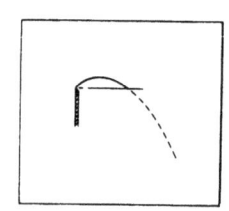

For portion of the profile below the weir crest — $\dfrac{X}{H_s}$

$\dfrac{Y}{H_s}$ / $\dfrac{H_s}{R}$	0.00	0.10	0.20	0.25	0.30	0.35	0.40	0.45	0.50	0.60	0.80	1.00	1.20	1.50	2.00
0.000	0.668	0.615	0.554	0.520	0 487	0.450	0.413	0.376	0 334	0.262	0.158	0.116	0.093	0.070	0.048
−.020	.705	.652	.592	.560	.526	.488	452	.414	369	.293	.185	.145	.120	.096	.074
−.040	.742	.688	.627	.596	.563	.524	487	.448	.400	.320	.212	165	.140	.115	.088
−.060	.777	.720	.660	.630	.596	.557	.519	.478	.428	.342	232	.182	.155	.129	.100
−.080	.808	.752	.692	.662	.628	.589	.549	.506	.454	363	.250	.197	.169	.140	.110
−.100	.838	.784	.722	.692	657	.618	.577	.532	.478	381	.266	.210	.180	.150	.118
−.150	.913	.857	.793	.762	.725	.686	.641	589	.531	.423	.299	.238	.204	.170	.132
−.200	.978	.925	.860	.826	790	.745	.698	.640	.575	.459	.326	.260	.224	.181	.144
−.250	1.040	.985	.919	.883	.847	.801	.750	.683	.613	.490	.348	.280	.239	.196	.153
−.300	1.100	1.043	.976	.941	.900	.852	.797	.722	.648	.518	.368	.296	.251	.206	.160
−.400	1.207	1.150	1.079	1.041	1.000	.944	.880	.791	.706	.562	.400	.322	.271	.220	.168
−.500	1.308	1.246	1.172	1.131	1.087	1.027	.951	.849	.753	.598	.427	.342	.287	.232	.173
−.600	1.397	1.335	1.260	1.215	1.167	1.102	1.012	.898	.793	.627	.449	.359	.300	.240	.179
−.800	1.563	1.500	1.422	1.369	1.312	1.231	1.112	.974	.854	.673	.482	.384	.320	.253	.184
−1.000	1.713	1.646	1.564	1.508	1.440	1.337	1.189	1.030	.899	.710	.508	.402	.332	.260	.188
−1.200	1.846	1.780	1.691	1.635	1.553	1.422	1.248	1.074	.933	.739	.528	.417	.340	.266	
−1.400	1.970	1.903	1.808	1.748	1.653	1.492	1.293	1.108	.963	.760	.542	.423	.344		
−1.600	2.085	2.020	1.918	1.855	1.742	1.548	1.330	1.133	.988	.780	.553	.430			
−1.800	2.196	2.130	2.024	1.957	1.821	1.591	1.358	1.158	1.008	.797	.563	.433			
−2.000	2.302	2.234	2.126	2.053	1.891	1.630	1.381	1.180	1.025	.810	.572				
−2.500	2.557	2.475	2.354	2.266	2.027	1.701	1.430	1.221	1.059	.838	.588				
−3.000	2.778	2.700	2.559	2.428	2.119	1.748	1.468	1.252	1.086	.853					
−3.500	....	2.916	2.749	2.541	2.171	1.777	1.489	1.267	1.102						
−4.000	......	3.114	2.914	2.620	2.201	1.796	1.500	1.280							
−4.500	..	3.306	3.053	2.682	2.220	1.806	1.509								
−5.000	....	3.488	3.178	2.734	2.227	1.811									
−5.500	....	3.653	3.294	2.779	2.229										
−6.000	....	3.820	3.405	2.812	2.232										
$\dfrac{H_s}{R}$	0.00	0.10	0.20	0.25	0.30	0.35	0.40	0.45	0.50	0.60	0.80	1.00	1.20	1.50	2.00

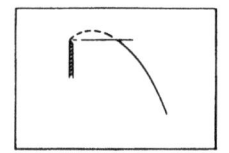

*The tabulation for $\dfrac{H_s}{R}$ =0.10 was obtained by interpolation between $\dfrac{H_s}{R}$ =0 and 0.20.

After Wagner

Table 9-3.—*Coordinates of lower nappe surface for different values of* $\dfrac{H_s}{R}$ *when* $\dfrac{P}{R}$ = 0.30.

$\dfrac{H_s}{R}$	0.20	0.25	0.30	0.35	0.40	0.45	0.50	0.60	0.80
$\dfrac{X}{H_s}$	$\dfrac{Y}{H_s}$ For portion of the profile above the weir crest								
0.000	0.0000	0.0000	0.0000	0.0000	0.0000	0.0000	0.0000	0.0000	0.0000
.010	.0130	.0130	.0130	.0125	.0120	.0120	.0115	.0110	.0100
.020	.0245	.0242	.0240	.0235	.0225	.0210	.0195	.0180	.0170
.030	.0340	.0335	.0330	.0320	.0300	.0290	.0270	.0240	.0210
.040	.0415	.0411	.0390	.0380	.0365	.0350	.0320	.0285	.0240
.050	.0495	.0470	.0455	.0440	.0420	.0395	.0370	.0325	.0245
.060	.0560	.0530	.0505	.0490	.0460	.0440	.0405	.0350	.0250
.070	.0610	.0575	.0550	.0530	.0500	.0470	.0440	.0370	.0245
.080	.0660	.0620	.0590	.0565	.0530	.0500	.0460	.0385	.0235
.090	.0705	.0660	.0625	.0595	.0550	.0520	.0480	.0390	.0215
.100	.0740	.0690	.0660	.0620	.0575	.0540	.0500	.0395	.0190
.120	.0800	.0750	.0705	.0650	.0600	.0560	.0510	.0380	.0120
.140	.0840	.0790	.0735	.0670	.0615	.0560	.0515	.0355	.0020
.160	.0870	.0810	.0750	.0675	.0610	.0550	.0500	.0310	
.180	.0885	.0820	.0755	.0675	.0600	.0535	.0475	.0250	
.200	.0885	.0820	.0745	.0660	.0575	.0505	.0435	.0180	
.250	.0855	.0765	.0685	.0590	.0480	.0390	.0270		
.300	.0780	.0670	.0580	.0460	.0340	.0220	.0050		
.350	.0660	.0540	.0425	.0295	.0150				
.400	.0495	.0370	.0240	.0100					
.450	.0300	.0170	.0025						
.500	.0090	−.0060							
.550									
$\dfrac{Y}{H_s}$	$\dfrac{X}{H_s}$ For portion of the profile below the weir crest								
−0.000	0.519	0.488	0.455	0.422	0.384	0.349	0.310	0.238	0.144
−.020	.560	.528	.495	.462	.423	.387	.345	.272	.174
−.040	.598	.566	.532	.498	.458	.420	.376	.300	.198
−.060	.632	.601	.567	.532	.491	.451	.406	.324	.220
−.080	.664	.634	.600	.564	.522	.480	.432	.348	.238
−.100	.693	.664	.631	.594	.552	.508	.456	.368	.254
−.150	.760	.734	.701	.661	.618	.569	.510	.412	.290
−.200	.831	.799	.763	.723	.677	.622	.558	.451	.317
−.250	.893	.860	.826	.781	.729	.667	.599	.483	.341
−.300	.953	.918	.880	.832	.779	.708	.634	.510	.362
−.400	1.060	1.024	.981	.932	.867	.780	.692	.556	.396
−.500	1.156	1.119	1.072	1.020	.938	.841	.745	.595	.424
−.600	1.242	1.203	1.153	1.098	1.000	.891	.780	.627	.446
−.800	1.403	1.359	1.301	1.227	1.101	.970	.845	.672	.478
−1.000	1.549	1.498	1.430	1.333	1.180	1.028	.892	.707	.504
−1.200	1.680	1.622	1.543	1.419	1.240	1.070	.930	.733	.524
−1.400	1.800	1.739	1.647	1.489	1.287	1.106	.959	.757	.540
−1.600	1.912	1.849	1.740	1.546	1.323	1.131	.983	.778	.551
−1.800	2.018	1.951	1.821	1.590	1.353	1.155	1.005	.797	.560
−2.000	2.120	2.049	1.892	1.627	1.380	1.175	1.022	.810	.569
−2.500	2.351	2.261	2.027	1.697	1.428	1.218	1.059	.837	
−3.000	2.557	2.423	2.113	1.747	1.464	1.247	1.081	.852	
−3.500	2.748	2.536	2.167	1.778	1.489	1.263	1.099		
−4.000	2.911	2.617	2.200	1.796	1.499	1.274			
−4.500	3.052	2.677	2.217	1.805	1.507				
−5.000	3.173	2.731	2.223	1.810					
−5.500	3.290	2.773	2.228						
−6.000	3.400	2.808							
$\dfrac{H_s}{R}$	0.20	0.25	0.30	0.35	0.40	0.45	0.50	0.60	0.80

After Wagner

Table 9-4.—Coordinates of lower nappe surface for different values of $\dfrac{H_s}{R}$ when $\dfrac{P}{R} = 0.15$.

$\dfrac{H_s}{R}$	0.20	0.25	0.30	0.35	0.40	0.45	0.50	0.60	0.80
$\dfrac{X}{H_s}$	$\dfrac{Y}{H_s}$ For portion of the profile above the weir crest								
0.000	0.0000	0.0000	0.0000	0.0000	0.0000	0.0000	0.0000	0.0000	0.0000
.010	.0120	.0120	.0115	.0115	.0110	.0110	.0105	.0100	.0090
.020	.0210	.0200	.0195	.0190	.0185	.0180	.0170	.0160	.0140
.030	.0285	.0270	.0265	.0260	.0250	.0235	.0225	.0200	.0165
.040	.0345	.0335	.0325	.0310	.0300	.0285	.0265	.0230	.0170
.050	.0405	.0385	.0375	.0360	.0345	.0320	.0300	.0250	.0170
.060	.0450	.0430	.0420	.0400	.0380	.0355	.0330	.0265	.0165
.070	.0495	.0470	.0455	.0430	.0410	.0380	.0350	.0270	.0150
.080	.0525	.0500	.0485	.0460	.0435	.0400	.0365	.0270	.0130
.090	.0560	.0530	.0510	.0480	.0455	.0420	.0370	.0265	.0100
.100	.0590	.0560	.0535	.0500	.0465	.0425	.0375	.0255	.0065
.120	.0630	.0600	.0570	.0520	.0480	.0435	.0365	.0220	
.140	.0660	.0620	.0585	.0525	.0475	.0425	.0345	.0175	
.160	.0670	.0635	.0590	.0520	.0460	.0400	.0305	.0110	
.180	.0675	.0635	.0580	.0500	.0435	.0365	.0260	.0040	
.200	.0670	.0625	.0560	.0465	.0395	.0320	.0200		
.250	.0615	.0560	.0470	.0360	.0265	.0160	.0015		
.300	.0520	.0440	.0330	.0210	.0100				
.350	.0380	.0285	.0165	.0030					
.400	.0210	.0090							
.450	.0015								
.500									
.550									
$\dfrac{Y}{H_s}$	$\dfrac{X}{H_s}$ For portion of the profile below the weir crest								
−0.000	0.454	0.422	0.392	0.358	0.325	0.288	0.253	0.189	0.116
−.020	.499	.467	.437	.404	.369	.330	.292	.228	.149
−.040	.540	.509	.478	.444	.407	.368	.328	.259	.174
−.060	.579	.547	.516	.482	.443	.402	.358	.286	.195
−.080	.615	.583	.550	.516	.476	.434	.386	.310	.213
−.100	.650	.616	.584	.547	.506	.462	.412	.331	.228
−.150	.726	.691	.660	.620	.577	.526	.468	.376	.263
−.200	.795	.760	.729	.685	.639	.580	.516	.413	.293
−.250	.862	.827	.790	.743	.692	.627	.557	.445	.319
−.300	.922	.883	.843	.797	.741	.671	.594	.474	.342
−.400	1.029	.988	.947	.893	.828	.749	.656	.523	.381
−.500	1.128	1.086	1.040	.980	.902	.816	.710	.567	.413
−.600	1.220	1.177	1.129	1.061	.967	.869	.753	.601	.439
−.800	1.380	1.337	1.285	1.202	1.080	.953	.827	.655	.473
−1.000	1.525	1.481	1.420	1.317	1.164	1.014	.878	.696	.498
−1.200	1.659	1.610	1.537	1.411	1.228	1.059	.917	.725	.517
−1.400	1.780	1.731	1.639	1.480	1.276	1.096	.949	.750	.531
−1.600	1.897	1.843	1.729	1.533	1.316	1.123	.973	.770	.544
−1.800	2.003	1.947	1.809	1.580	1.347	1.147	.997	.787	.553
−2.000	2.104	2.042	1.879	1.619	1.372	1.167	1.013	.801	.560
−2.500	2.340	2.251	2.017	1.690	1.423	1.210	1.049	.827	
−3.000	2.550	2.414	2.105	1.738	1.457	1.240	1.073	.840	
−3.500	2.740	2.530	2.153	1.768	1.475	1.252	1.088		
−4.000	2.904	2.609	2.180	1.780	1.487	1.263			
−4.500	3.048	2.671	2.198	1.790	1.491				
−5.000	3.169	2.727	2.207	1.793					
−5.500	3.286	2.769	2.210						
−6.000	3.396	2.800							
$\dfrac{H_s}{R}$	0.20	0.25	0.30	0.35	0.40	0.45	0.50	0.60	0.80

After Wagner

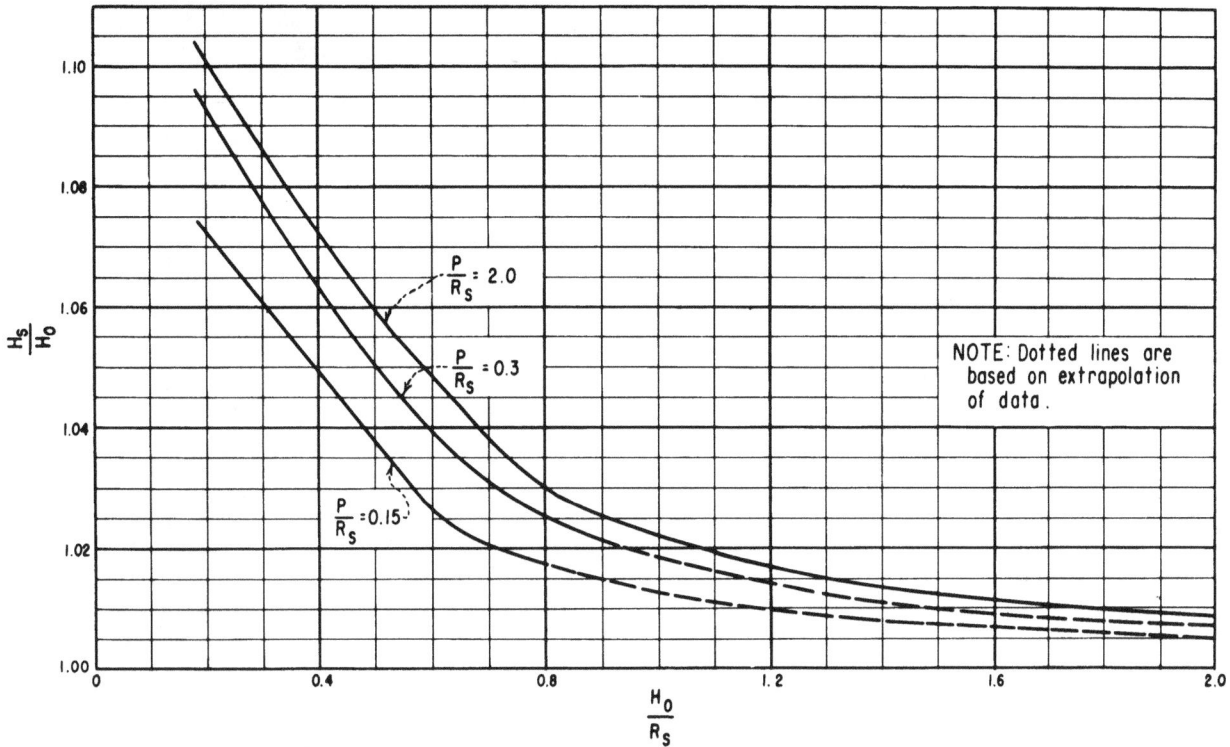

$$\textit{Figure 9-48.} \quad \text{Relationship of } \frac{H_s}{H_o} \text{ to } \frac{H_o}{R_s} \text{ for circular sharp-crested weirs.}-244\text{-}D\text{-}2443$$

surface elevation, the crest profile shape will be the same as the undernappe of the weir discharge, the shaft will flow full at section A-A, and there will be no pressure on the crest or in the shaft for the design head. For higher heads, section A-A will act as an orifice control. The shaft above section A-A will flow full and under pressure. Below section A-A, it will flow full but will not be under pressure. For lower heads, the crest will control and the shaft will flow partially full. Assuming the same losses, equation (20) can be rewritten, as follows, to determine the orifice discharge:

$$Q = 23.90 \, R^2 H_a^{1/2} \qquad (21)$$

If the profile is modified to enlarge the shaft as shown by the solid lines be and aeration is provided, the shaft will not flow full. Neglecting losses, the jet below section A-A will then occupy an equivalent area indicated by the lines bc.

Aeration is usually provided at the orifice

control, either through introduction of air at a sudden enlargement of the shaft or at the installation of a deflector to ensure free flow below the control section A-A. Waterway sizes and slopes must be such that free flow is maintained below the section of control. Failure to provide adequate aeration at the section of control may induce cavitation.

For submerged flow at the crest, the corresponding nappe shape as determined from section 9-27 for a design head H_o will be such that along its lower levels it will closely follow the profile determined from equation (20) if H_e approximates H_o. It must be remembered that on the basis of the losses assumed in equation (20), profile abc will be the minimum shaft size which will accommodate the required flow and that no part of the crest shape should be permitted to project inside this profile. As has been noted in section 9-12, small subatmospheric crest pressures can be tolerated if proper precautions are taken to obtain a smooth surface and if the negative pressure

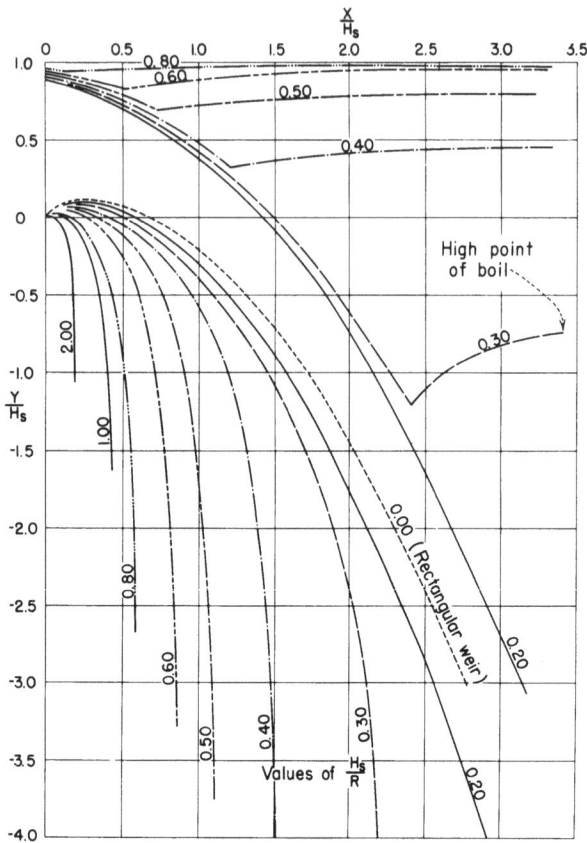

Figure 9-49. Upper and lower nappe profiles for a circular weir (aerated nappe and negligible approach velocity).—288-D-2444

forces are recognized in the structural design. The choice of the minimum crest and orifice control shapes in preference to some wider shape then becomes a matter of economics, structural arrangement, and layout adaptability.

Where the orifice control profile corresponds to the continuation of the crest shape as determined by tables 9-2, 9-3, and 9-4, the discharge can be computed from equation (19) using a coefficient from figure 9-46. Where the orifice control profile differs from the crest shape profile so that a constricted control section is established, the discharge must be determined from equation (20). On figure 9-44 the discharge head relationship curve *ag* will then be computed from the coefficients determined from figure 9-47 while the discharge head relationship curve *gh* will be based on equation (20).

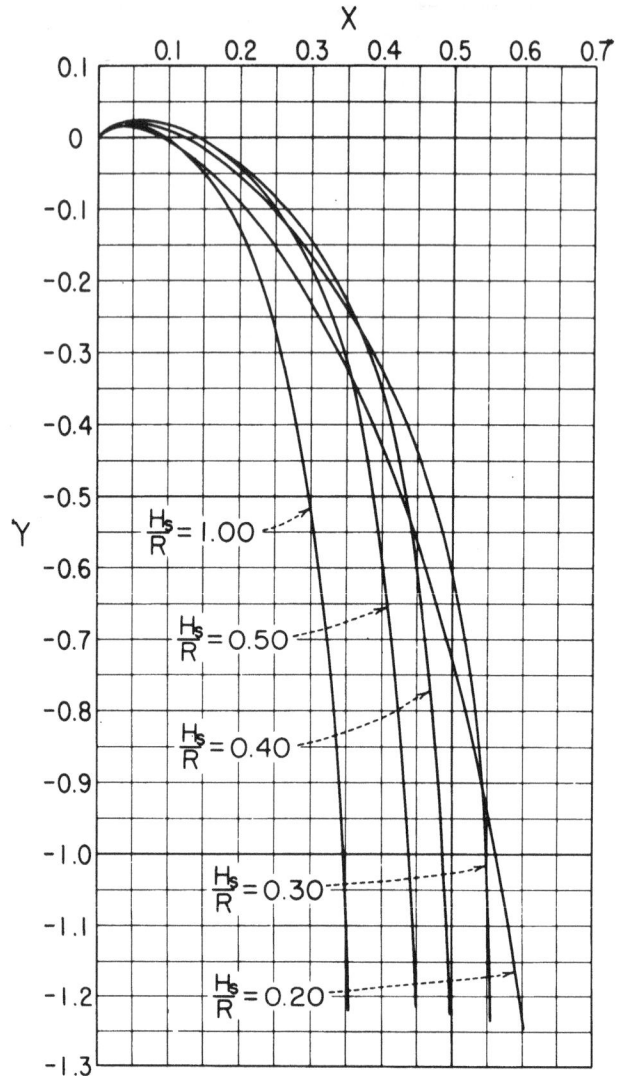

Figure 9-50. Comparison of lower nappe shapes for a circular weir for different heads.—288-D-2445

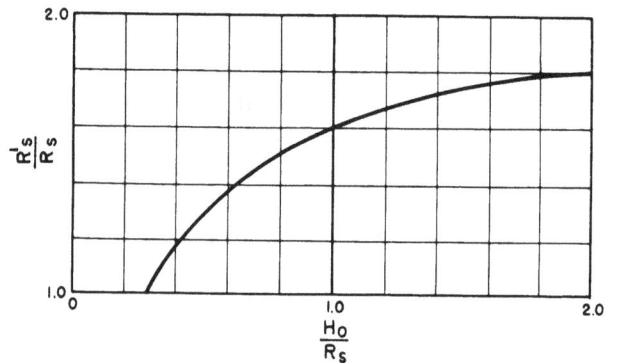

Figure 9-51. Increased circular crest radius needed to minimize subatmospheric pressure along crest.—288-D-2446

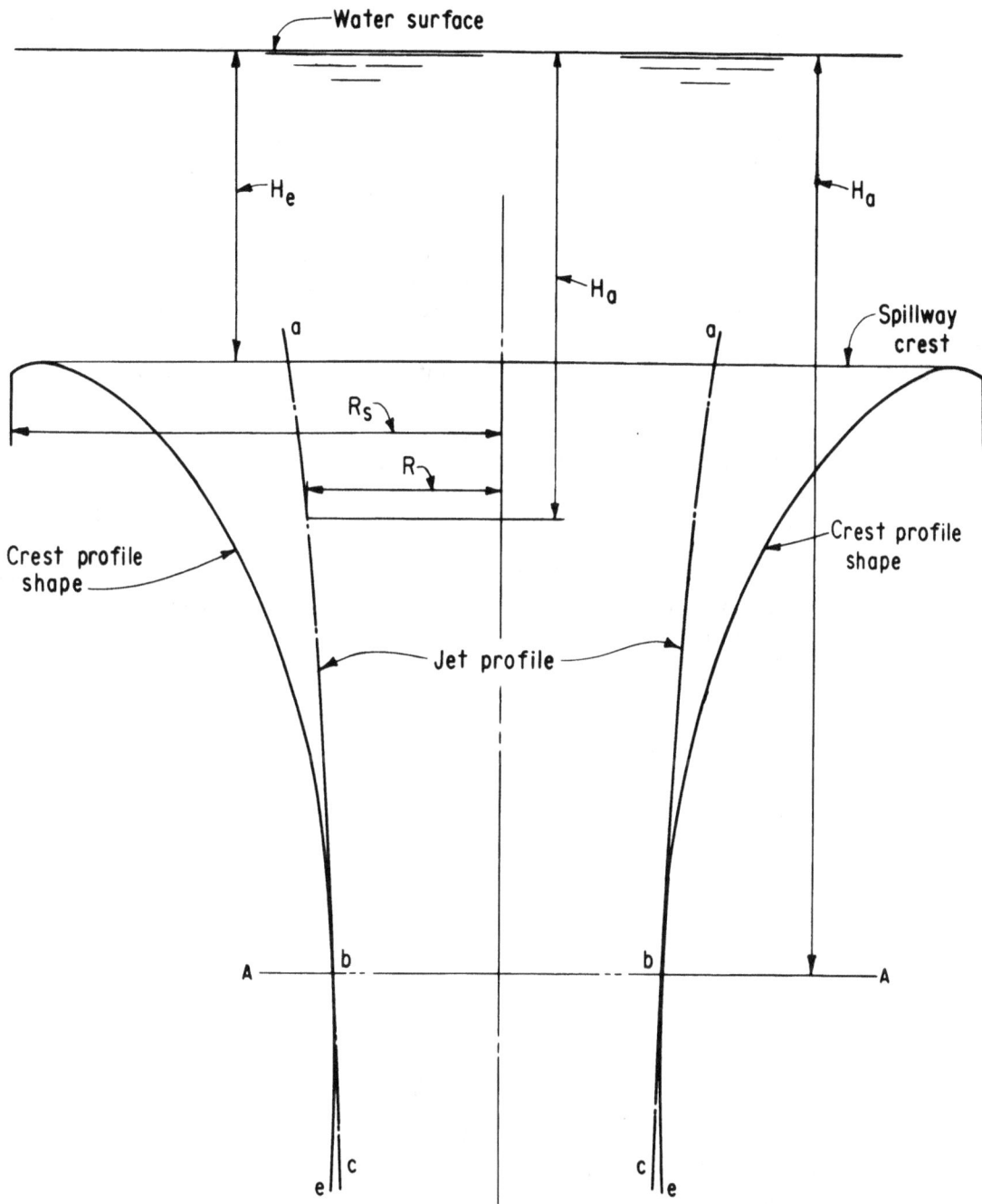

Figure 9-52. Comparison of crest profile shape with theoretical jet profile.—288-D-3058

9-29. *Tunnel Design.* —If, for a designated discharge, the tunnel of a morning glory spillway were to flow full without being under pressure, the required size would vary along its length. So long as the slope of the hydraulic gradient which is dictated by the hydraulic losses is flatter than the slope of the tunnel, the flow will accelerate and the tunnel could decrease in size. When the tunnel slope becomes flatter than the slope of the hydraulic gradient, flow will decelerate and a larger tunnel may be required. All points along the tunnel will act simultaneously to control the rate of flow. For heads in excess of that used to proportion the tunnel, it will flow under pressure with the control at the downstream end; for heads less than that used to determine the size, the tunnel will flow partly full for its entire length and the control will remain upstream. On figure 9-44 the head at which the tunnel just flows full is represented by point h. At heads above point h the tunnel flows full under pressure; at heads less than h the tunnel flows partly full with controlling conditions dictated by the crest or orifice control design.

Because it is impractical to build a tunnel with a varying diameter, it is ordinarily made of a constant diameter. Thus the tunnel from the control point to the downstream end will have an excess of area. If atmospheric pressure can be maintained along the portion of the tunnel flowing partly full, the tunnel will continue to flow at that stage even though the downstream end fills. Progressively greater discharges will not alter the part full flow condition in the upper part of the tunnel, but full flow conditions under pressure will occupy increasing lengths of the downstream end of the tunnel. At the discharge represented by point h on figure 9-44, the full flow condition has moved back to the throat control section and the tunnel will flow full for its entire length.

If the tunnel flows at such a stage that the downstream end flows full, both the inlet and outlet will be sealed. To forestall siphon action by the withdrawal of air from the tunnel would require an adequate venting system. Unless venting is effected over the entire length of tunnel, it may prove inadequate to prevent subatmospheric pressures along some portion of the length because of the possibility of sealing at any point by surging, wave action, or eddy turbulences. Thus, if no venting is provided or if the venting is inadequate, a make-and-break siphon action will attend the flow in the range of discharges approaching full flow conditions. This action is accompanied by erratic discharges, by thumping and vibrations, and by surges at the entrance and outlet of the spillway. This is an undesirable condition and should be avoided.

To avoid the possibility of siphonic flow, the downstream tunnel size for ordinary designs (and especially for those for higher heads) is chosen so that the tunnel will never flow full beyond the throat. To allow for air bulking, surging, etc., the tunnel is selected of such a size that its area will not flow more than 75 percent full at the downstream end at maximum discharges. Under this limitation, air ordinarily will be able to pass up the tunnel from the downstream portal and thus prevent the formation of subatmospheric pressure along the tunnel length. Precautions must be taken, however, in selecting vertical or horizontal curvature of the tunnel profile and alinement to prevent sealing along some portion by surging or wave action.

G. STRUCTURAL DESIGN

9-30. *General.*—The structural design of a spillway and the selection of specific structural details follow the determination of the spillway type and arrangement of components and the completion of the hydraulic design. The design criteria for each component part should be established for any condition which may exist at any time during the life of the structure. Design loads are different for each type of spillway. Each component should be carefully analyzed for loads that can be applied to it.

Structures in or on the dam should be designed for the stresses in the dam due to external loadings and temperatures, as well as the hydraulic load and other loads applied directly to the structure. Slabs, walls, and ogee crests should be designed for dead load and hydraulic pressures plus any other loads such as fill, surcharge, and control or operating equipment. Appurtenant structures not built on the dam which are subject to uplift due to the reservoir water and tailwater should be designed accordingly.

Because of the velocities involved, dynamic

water pressures should be considered in addition to the static water pressures in all cases. Wherever practicable, laboratory model tests should be used to determine hydraulic loads, particularly dynamic loads.

Normal methods of design should be used for walls, slabs, etc. Where special design problems are encountered, the finite element method of analysis (appendix F) can be used to determine the stresses.

H. BIBLIOGRAPHY

9-31. *Bibliography.*

[1] Bureau of Reclamation, "Studies of Crests of Overfall Dams," Bulletin 3, Part VI, Hydraulic Investigations, Boulder Canyon Project Final Reports, 1948.

[2] Bureau of Reclamation, "Discharge Coefficients For Irregular Overfall Spillways," Engineering Monograph No. 9, 1952.

[3] Hinds, Julian, "Side Channel Spillways," Trans. ASCE, vol. 89, 1926, p. 881.

[4] Ball, J. W., "Construction Finishes and High-Velocity Flow," Journal of the Construction Division, ASCE Proceedings, September 1963.

[5] Colgate, D. M., "Hydraulic Model Studies of Aeration Devices For Yellowtail Dam Spillway Tunnel," Pick-Sloan Missouri River Basin Program, Montana, REC-ERC-71-47, 1971.

[6] Bureau of Reclamation, "Hydraulic Design of Stilling Basin and Bucket Energy Dissipators," Engineering Monograph No. 25, 1964.

[7] Doddiah, D., Albertson, M. L., and Thomas, R. A., "Scour From Jets," Proceedings, Minnesota International Association for Hydraulic Research and Hydraulics Division, ASCE, Minneapolis, Minn., August 1953, p. 161.

[8] Scimemi, Ettore, "Discussion of Paper 'Model Study of Brown Canyon Debris Barrier' by Bermeal and Sanks," Trans. ASCE, vol. 112, 1947, p. 1016.

[9] Bureau of Reclamation, "Hydraulic Model Studies of Morrow Point Dam Spillway, Outlet Works and Powerplant Tailrace," Report No. HYD-557, 1966.

[10] Bureau of Reclamation, "Hydraulic Model Studies of the Pueblo Dam Spillway and Plunge Basin," REC-ERC-71-18, 1971.

[11] Bureau of Reclamation, "Hydraulic Model Studies of Crystal Dam Spillway and Outlet Works," REC-ERC-72-01, 1972.

[12] Peterka, A. J., "Morning-Glory Shaft Spillways," Trans. ASCE, vol. 121, 1956, p. 385.

[13] Bradley, J. 'N., "Morning-Glory Shaft Spillways: Prototype Behavior," Trans. ASCE, vol. 121, 1956, p. 312.

[14] Blaisdell, F. W., "Hydraulics of Closed Conduit Spillways—Parts II through VII—Results of Tests on Several Forms of the Spillway," University of Minnesota, Saint Anthony Falls Hydraulic Laboratory, Technical Paper No. 18, series B, March 1958.

[15] Wagner, W. E., "Morning Glory Shaft Spillways: Determination of Pressure-Controlled Profiles," Trans. ASCE, vol. 121, 1956, p. 345.

Chapter X

Outlet Works and Power Outlets

A. INTRODUCTION

10-1. *Types and Purposes.* —An outlet works is a combination of structures and equipment required for the safe operation and control of water released from a reservoir to serve various purposes. Outlet works are usually classified according to their purpose such as river outlets, which serve to regulate flows to the river and control the reservoir elevation; irrigation or municipal water supply outlets, which control the flow of water into a canal, pipeline, or river to satisfy specified needs; or power outlets which provide passage of water to the turbines for power generation. Each damsite has its own requirements as to the type and size of outlet works needed. The outlet works may be designed to satisfy a single requirement or a combination of multipurpose requirements. Typical outlet works installations are shown on figures 10-1 and 10-2.

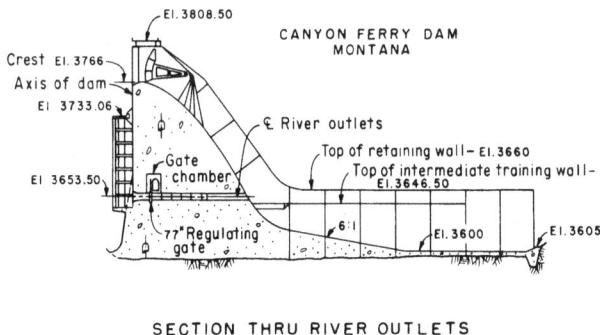

SECTION THRU RIVER OUTLETS

Typical river outlet works which discharges into spillway stilling basin. This outlet consists of a conduit through the dam and a regulating gate controlled from a chamber in the dam. The intake and trashrack are on the upstream face of dam.

Figure 10-1. **Typical river outlet works with stilling basin.—288-D-3060**

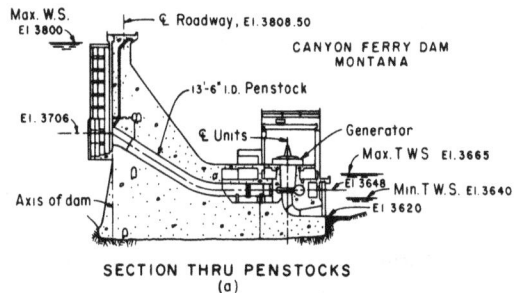

SECTION THRU PENSTOCKS
(a)

(a) Typical power outlet with penstock through dam

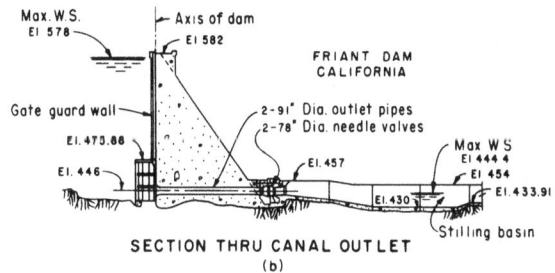

SECTION THRU CANAL OUTLET
(b)

(b) Typical canal outlet works consisting of conduits through the dam with needle valves. A trashrack is on the upstream face and a hydraulic jump stilling basin is utilized to dissipate the energy downstream.

Figure 10-2. **Typical power outlet and canal outlet works. —288-D-3062**

Downstream water requirements, preservation of aquatic life, abatement of stream pollution, and emergency evacuation of the reservoir are some of the factors that influence the design of a river outlet. In certain instances, the river outlet works may be used to increase the flow past the dam in conjunction with the normal spillway discharge. It may also act as a flood control regulator to release waters temporarily stored

217

in flood control storage space or to evacuate storage in anticipation of flood inflows. Further, the river outlet works may serve to empty the reservoir to permit inspection, to make needed repairs, or to maintain the upstream face of the dam or other structures normally inundated.

The general details of operation and design of irrigation or municipal and industrial outlets are similar to those for river outlets. The quantity of irrigation water is determined from project or agricultural needs and is related to the anticipated use and to any special water requirements of the irrigation system. The quality and quantity of water for domestic use is determined from the commercial, industrial, and residential water needs of the area served. The number and size of irrigation and

municipal and industrial outlet works will depend on the capacity requirements with the reservoir at a predetermined elevation, and on the amount of control required as the elevation of the reservoir fluctuates.

Power outlets provide for the passage of water to the powerplant; therefore, they should be designed to minimize hydraulic losses and to obtain the maximum economy in construction and operation. If the powerplant can be located at the toe of the dam, a layout with the penstocks embedded through the dam usually is most economical. Where the powerplant must be located away from the toe of the dam, the penstocks can be located in tunnels or embedded in the dam in the upper portion of their length and run exposed down the abutment to the powerplant.

B. OUTLET WORKS OTHER THAN POWER OUTLETS

10-2. *General.*—An outlet works consists of the equipment and structures which together release the required water for a given purpose or combination of purposes. The flows through river outlets and canal or pipeline outlets vary throughout the year and may involve a wide range of discharges under varying heads. The accuracy and ease of control are major considerations, and a great amount of planning may be justified in determining the type of control devices that can be best utilized.

Ordinarily in a concrete dam, the most economical outlet works consists of an intake structure, a conduit or series of conduits through the dam, discharge flow control devices, and an energy dissipating device where required downstream of the dam. The intake structure includes a trashrack, an entrance transition, and stoplogs or an emergency gate. The control device can be placed (1) at the intake on the upstream face, (2) at some point along the conduit and be regulated from galleries inside the dam, or (3) at the downstream end of the conduit with the operating controls placed in a gatehouse on the downstream face of the dam. When there is a powerplant or other structure near the face of

the dam, the outlet conduits can be extended further downstream to discharge into the river channel beyond these features. In this case, a control valve may be placed in a gate structure at the end of the conduit.

10-3. *Layout.*—The layout of a particular outlet works will be influenced by many conditions relating to the hydraulic requirements, the height and shape of dam, the site adaptability, and the relationship of the outlet works to the construction procedures and to other appurtenances of the development. An outlet works leading to a high-level canal or into a closed pipeline will differ from one emptying into the river. Similarly, a scheme in which the outlet works is used for diversion may vary from one where diversion is effected by other methods. In certain instances, the proximity of the spillway may permit combining some of the outlet works and spillway components into a single structure. As an example, the spillway and outlet works layout might be arranged so that discharges from both structures will empty into a common stilling basin.

The topography and geology of a site will have a great influence on the layout. The downstream location of the channel, the nearby location of any steep cliffs, and the width of the canyon are all factors affecting the selection of the most suitable type and location of outlet works. The river outlets should be located close to the river channel to minimize the downstream excavation. Geology, such as the location, type, and strength of bedrock, is also an important factor to consider when making the layout of an outlet works. An unfavorable foundation such as deep overburden or inferior foundation rock requires special consideration when selecting an impact area; with a weak foundation, a stilling basin may be required to avoid erosion and damage to the channel.

An outlet works may be used for diverting the riverflow or portion thereof during a phase of the construction period, thus avoiding the necessity for supplementary installations for that purpose. The outlet structure size dictated by this use rather than the size indicated for ordinary outlet requirements may determine the final outlet works capacity.

The establishment of the intake level is influenced by several considerations such as maintaining the required discharge at the minimum reservoir operating elevation, establishing a silt retention space, and allowing selective withdrawal to achieve suitable water temperature and/or quality. Dams which will impound waters for irrigation, domestic use, or other conservation purposes must have the outlet works low enough to be able to draw the water down to the bottom of the allocated storage space. Further, if the outlets are to be used to evacuate the reservoir for inspection or repair of the dam, they should be placed as low as practicable. However, it is usual practice to make an allowance in a reservoir for inactive storage for silt deposition, fish and wildlife conservation, and recreation.

Reservoirs become thermally stratified and taste and odor vary between elevations; therefore, the outlet intake should be established at the best elevation to achieve satisfactory water quality for the purpose intended. Downstream fish and wildlife requirements may determine the temperature at which the outlet releases should be made.

Municipal and industrial water use increases the emphasis on water quality and requires the water to be drawn from the reservoir at the elevation which produces the most satisfactory combination of odor, taste, and temperature. Mineral concentrations, algae growth, and temperature are factors which influence the quality of the water and should be taken into consideration when establishing the intake elevation. Water supply releases can be made through separate outlet works at different elevations if the requirements for the individual water uses are not the same and the reservoir is stratified in temperature and quality of water.

Downstream water requirements may change throughout the year and the stratifications of water temperature and quality may fluctuate within the reservoir; therefore, the elevation at which the water should be drawn from the reservoir may vary. Selective withdrawal can be accomplished by a multilevel outlet arrangement in which the stratum of water that is most desirable can be released through the outlet works. Two schemes of multilevel outlet works are common. The first consists of a series of river outlet conduits through the dam at various elevations, and the second consists of a single outlet through the dam with a shutter arrangement on the trashrack structure. The shutters can be adjusted to allow selective withdrawal from the desired reservoir elevation. Figure 15-1 in chapter XV shows an example of a multilevel outlet works consisting of four outlet conduit intakes at different elevations, and figure 15-2 shows a typical example of a shutter arrangement on a trashrack structure.

Another factor to consider in determining a layout for an outlet works is the effect of a particular scheme on construction progress. A scheme which slows down or interferes with the normal construction progress of the concrete dam should be avoided if possible. Usually a horizontal conduit through the dam has the least effect on construction progress; however, sometimes other conditions restrict its use. Generally speaking, the fewer conduit or other outlet works components that must be installed within the mass concrete, the more rapid the rate of construction.

10-4. *Intake Structures.* —In addition to forming the entrance into the outlet works, an intake structure may accommodate control devices. It also supports necessary auxiliary appurtenances (such as trashracks, fish screens, and bypass devices), and it may include temporary diversion openings and provisions for installation of bulkhead or stoplog closure devices.

An intake structure may take one of many forms, depending on the functions it must serve, the range in reservoir head under which it must operate, the discharge it must handle, the frequency of reservoir drawdown, the trash conditions in the reservoir, the reservoir ice conditions, and other considerations.

An intake structure for a concrete dam usually consists of a submerged structure on the upstream face of the dam; however, intake towers in the reservoir have been used in some instances. The most common intake structure consists of a bellmouth intake, a transition between the bellmouth and conduit if required, a trashrack structure on the upstream face of the dam, and guides to be used with a bulkhead gate or stoplogs to seal off the conduit for maintenance and repair. The bulkhead gate or stoplogs are usually installed and removed by use of either a gantry or a mobile crane operating on top of the dam or from a barge in the reservoir.

(a) *Trashrack.* —A trashrack is used to keep trash and other debris from entering the outlet conduit and causing damage or fouling of the control device. Two basic types of trashracks are used for outlet works. One type is a concrete or metal frame structure on which metal trashracks are placed, and the other is an all-concrete structure that consists of relatively large openings formed in the concrete and is without metal racks. The metal trashrack type of structure provides for the screening of small debris when protection is needed to prevent damage to the conduit or control devices. Metal trashracks usually consist of relatively thin, flat steel bars which are placed on edge from 2 to 9 inches apart and assembled in rack sections. The required area of the trashrack is fixed by a limiting velocity through the rack, which in turn depends on the nature of the

trash which must be excluded. Where the trashracks are inaccessible for cleaning, the velocity through the racks ordinarily should not exceed 2 feet per second. A velocity up to approximately 5 feet per second may be tolerated for racks which are accessible for cleaning.

An example of a concrete trashrack structure with metal racks is shown on figure 10-3. The concrete frame structure consists of a base cantilevered from the upstream face of the dam on which the trashrack structure is supported, a series of columns placed in a semicircle around the centerline of the intake, and a series of horizontal ribs spaced along the full height of the structure. The spacing between columns is dependent upon the structural requirements for the head differential that may be applied to the trashracks and the size of the metal rack section that can conveniently be fabricated and shipped. The vertical height of the trashrack structure is divided into a series of bays by arch-shaped ribs that are attached to the face of the dam and give lateral support to the columns. A solid concrete slab is usually constructed as a top for the structure with a slot formed, where required, to allow for placement and removal of the stoplogs or bulkhead gate. Grooves are formed into the vertical columns to hold the metal trashracks which are lowered into position from the top. When the intakes are deeply submerged, it may be desirable to remove and install the metal trashracks from the reservoir water surface. Guides can be supported on a curved concrete wall or "silo" which will facilitate the removal and installation of the trashrack sections.

An all-metal trashrack structure contains horizontal steel arches spaced along the height of the structure with vertical steel supports between the arches. The structure can be constructed so that the racks slide into the metal frame similar to the system used with the concrete frame, or the frame and trashracks can be fabricated into composite units and these arch-trashrack sections assembled to create the final structure. The top of the all-metal structure usually consists of trashrack bars supported as required and containing the slot required for placement and removal of the stoplogs or bulkhead gate.

When small trash is of no consequence and can be washed through the outlet works without damage to the conduit or control device, an all-concrete structure having only formed openings in the concrete can be used. The height and size of this trashrack structure, as well as the size of the formed openings, are dependent upon the desired discharge, the velocity at the intake, and the size and amount of debris in the reservoir. The openings for this type of trashrack usually range from 12 inches to 3 feet. The shape of the trashrack structure in plan can be rectangular, circular, or built in chords for ease of construction as shown on figure 10-4.

The frame used to support metal trashracks requires considerable construction time when formed of concrete; therefore, the use of a metal frame is often desirable because of the shorter construction time required for installation. This type also interferes least with the rapid placement of concrete in a dam.

Where winter reservoir storage is maintained in cold climates, the effect of possible icing conditions on the intake structure must be considered. Where reservoir surface ice can freeze around an intake structure, there is danger to the structure not only from the ice pressure acting laterally, but also from the uplift forces if a filling reservoir lifts the ice mass vertically. These effects should be considered in the design of the trashrack and the inlet structure, and may be a factor in determining the height of the trashrack structure. If practicable, the structure should be submerged at all times. However, if the structure will likely be above the reservoir water surface at times and ice loadings will present a hazard, an air bubbling system can be installed around the structure to circulate the warmer water from lower in the reservoir which will keep the surface area adjacent to the structure free of ice. Such a system will require a constant supply of compressed air and must be operated continuously during the winter months.

(b) *Entrance and Transition.*—The entrance to a conduit should be streamlined and provide smooth, gradual changes in the direction of flow to minimize head losses and to avoid zones where cavitation pressures can develop. Any abrupt change in the cross section of a conduit or any projection into the conduit, such as a gate frame, creates turbulence in the flow which increases in intensity as the velocity increases. These effects can be minimized by shaping the entrance to conform to the shape of a jet issuing from a standard orifice. These bellmouth entrances, as they are called, are discussed in section 10-11. Any time that a change in cross section of the outlet works is required, such as where the outlet changes from the size and shape of the entrance to that of the conduit, a smooth gradual transition should be utilized.

10-5. *Conduits.* —The outlet conduits through a concrete dam are the passageways that carry the water from the reservoir downstream to the river, canal, or pipeline. A conduit may consist of a formed opening through or a steel liner embedded in the mass concrete. The shape may be rectangular or round, or it may transition from one shape to the other depending on the shape of the intake entrance and on the type and location of the control equipment. The outlet works may contain one or more conduits depending on the discharge requirements for a predetermined reservoir water surface elevation. Two smaller conduits are preferable to one larger one, so that one outlet can be operated while the other is shut down for inspection and maintenance.

The design of the conduits required to pass a given discharge through a concrete dam is based upon the head, velocity of flow, type of control, length of conduit, and the associated economic considerations. Generally, the most economical conduit for an outlet is one that is horizontal and passes through the narrowest portion of the dam; however, most outlet works require that the conduit inlet and outlet be at different elevations to meet controlling requirements upstream and downstream. The number of bends required in an outlet conduit should be minimized and all the radii should be made as long as practicable to reduce head loss.

10-6. *Gates and Outlet Controls.* —The discharges from a reservoir outlet works vary throughout the year depending upon

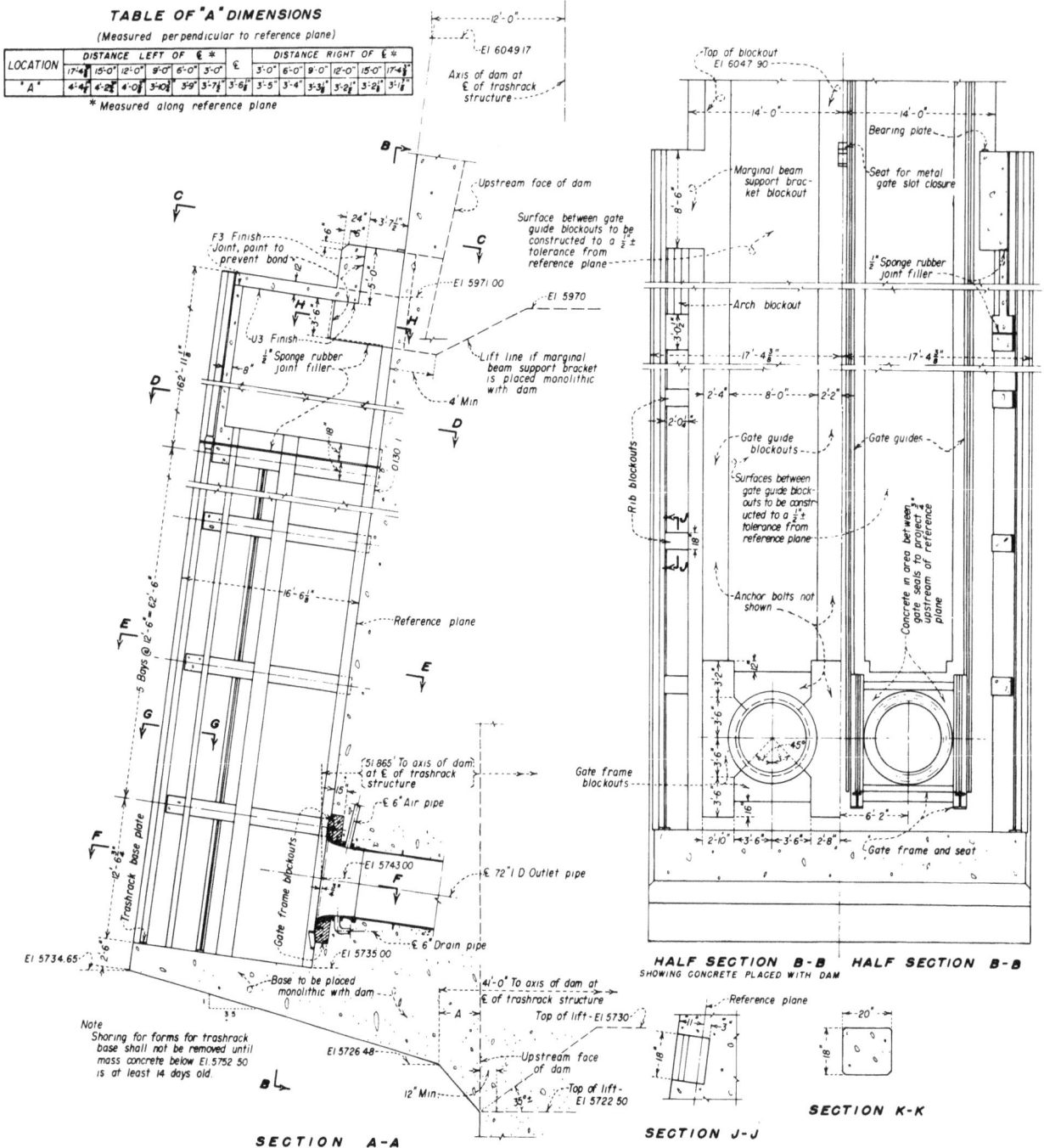

Figure 10-3. River outlet trashrack structure—plans and sections (sheet 1 of 2). —288-D-3063 (1/2)

downstream water needs and reservoir flood control requirements. Therefore, the impounded water must be released at specific regulated rates. To achieve this discharge control, gates or valves must be installed at some point along the conduit.

Control devices for outlet works are categorized according to their function in the

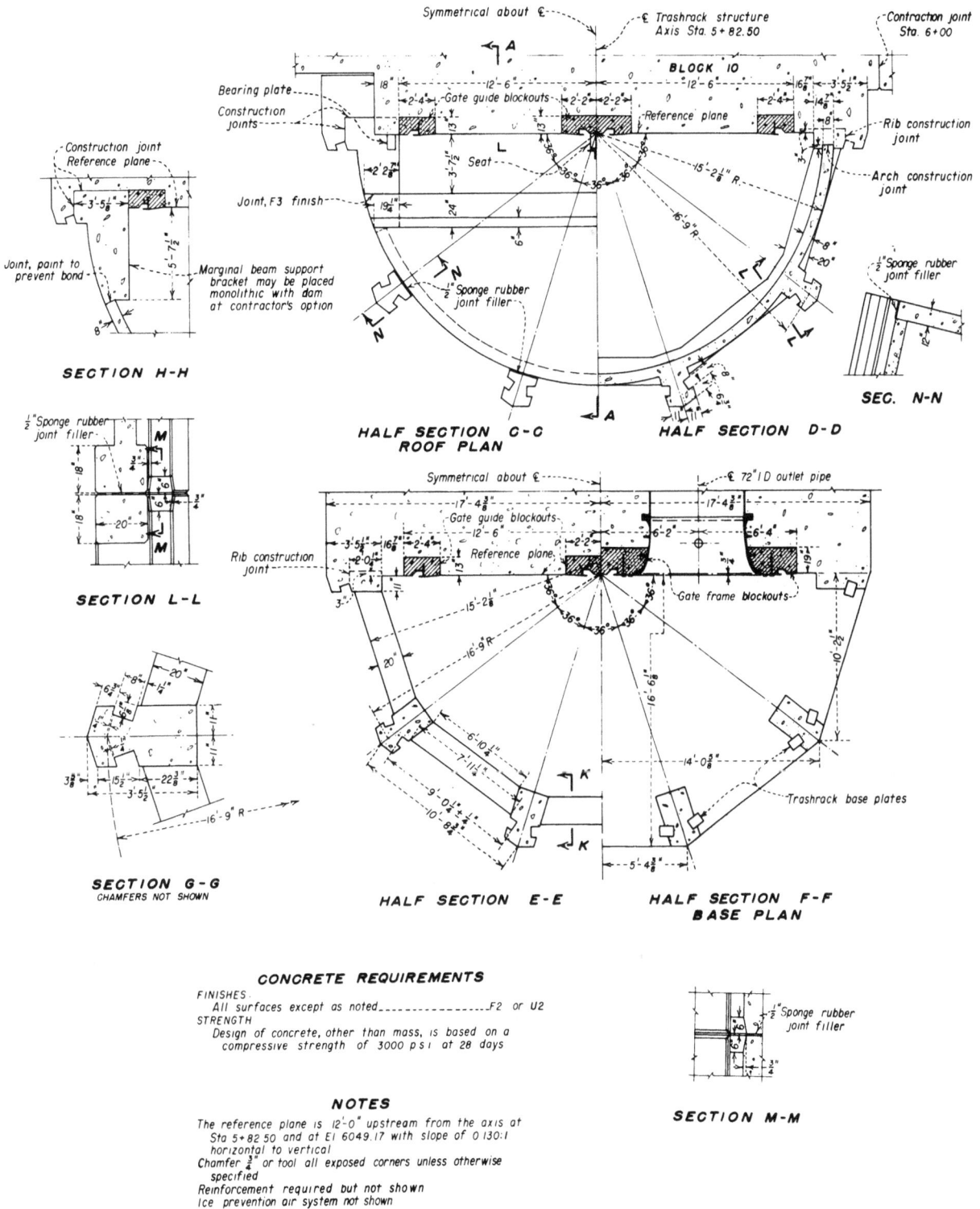

Figure 10-3. River outlet trashrack structure—plans and sections (sheet 2 of 2). −288-D-3063 (2/2)

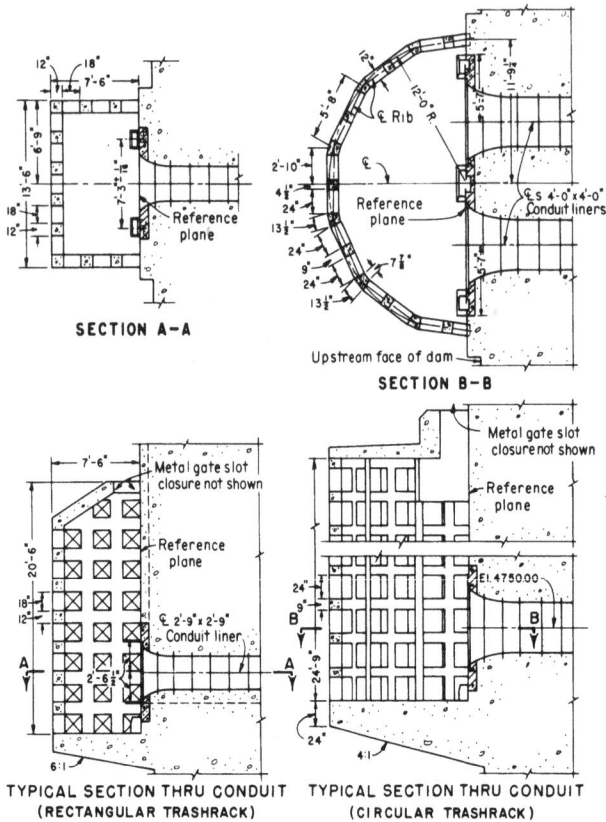

SECTION A-A

SECTION B-B

Upstream face of dam

TYPICAL SECTION THRU CONDUIT
(RECTANGULAR TRASHRACK)

TYPICAL SECTION THRU CONDUIT
(CIRCULAR TRASHRACK)

Figure 10-4. Typical trashrack installations. —288-D-3064

structure. Operating gates and regulating valves are used to control and regulate the outlet works flow and are designed to operate in any position from closed to fully open. Guard or emergency gates are designed to effect closure in the event of failure of the operating gates, or where unwatering is desired either to inspect the conduit below the guard gates or to inspect and repair the operating gates.

Guides may be provided at the conduit entrance to accommodate stoplogs or bulkheads so that the conduit can be closed during an emergency period or for maintenance. For such installations, guard gates may or may not be provided, depending on whether the stoplogs can be readily installed if an emergency arises during normal reservoir operating periods.

Standard commercial gates and valves are available and may be adequate for low-head installations involving relatively small discharges. High-head installations, however,

usually require specially designed equipment. The type of control device should be utilized that least affects flow in the conduit. For example, if possible, control and emergency gates or valves should be used that will not require transitions from one size and shape of conduit to another because these transitions are costly and can contribute greatly to the head loss through the conduit.

(a) *Location of Control Devices.*—The control gate for an outlet works can be placed at the upstream end of the conduit, at an intermediate point along its length, or at the outlet end of the conduit. Where flow from a control gate is released directly into the open as free discharge, only that portion of the conduit upstream from the gate will be under pressure. Where a control gate or valve discharges into a closed pressure pipe, the control will serve only to regulate the releases; full pipe flow will occur in the conduit both upstream and downstream from the control gate. For the pressure-pipe type, the location of the gate or valve will have little influence on the design insofar as internal pressures are concerned. However, where a control discharges into a free-flowing conduit, the location of the control gate becomes an important consideration in the design of the outlet.

Factors that should be considered in locating the control devices to be used on an outlet works include the size of the conduits required, the type of dam, the downstream structures, and the topography. The use of gates at the upstream or downstream face of the dam may be precluded if a satisfactory location for the gate and operating equipment or access is not available due to the layout of the dam or to the surrounding topography. The use of gate chambers within the dam is possible only if the thickness of the dam is great enough to safely contain the required chamber. When the outlet works discharges onto a spillway apron, the control device may, of necessity, have to be located either at a chamber within the dam or at the upstream face of the dam.

The most desirable location for the control device is usually at the downstream end of the conduit. This location permits most of the energy to be dissipated outside of the conduit, removing a possible cause of cavitation and vibration from the conduit. By eliminating gate operation at the entrance and within the conduit, better flow conditions can be maintained throughout the entire conduit length. Also, the size of the intake structure can sometimes be reduced if the control gate is not incorporated into the structure, and this may give the downstream location an additional advantage of economy.

(b) *Types of Gates and Valves.*—Many types of valves and gates are available for the control of outlet works. Each outlet works plan requires gates or valves that are well suited for the operating conditions and the characteristics of that plan. The location of the control device along the conduit, the amount of head applied, and the size and shape of conduit are all factors used in determining the type of control device considered likely to be most serviceable. Some types of gates and valves operate well at any opening, thus can be used as control gates, while others operate satisfactorily only at full open and can be used only as emergency or guard gates.

Where the control device is located at the outlet works intake and is to be operated under low head, the most commonly used device would be a slide gate. If the control is at an intermediate point along the conduit, control devices such as high-pressure slide gates, butterfly valves, or fixed-wheel gates can be used for the discharge control. Control at the downstream end of the outlet conduit may be accomplished by the use of a high-pressure slide gate, a jet-flow gate, or a hollow-jet valve discharging into the channel or stilling device. These are control devices that are commonly used; other types of gates or valves can be utilized if found to be more suitable for a particular situation.

Emergency or guard gates or valves are installed in the outlets upstream from the control device, to provide an emergency means of closing the conduit. These emergency devices may consist of a fixed-wheel gate to close the entrance to the conduit, a duplicate of the control gate or valve in tandem and operated from a chamber or gallery in the dam

or in a control house on the downstream face, or a gate such as a ring-follower gate in tandem with the control gate. A ring-follower gate is well suited to serve as an emergency or guard gate (which operates either fully open or fully closed), since the ring-follower gate when fully open is the same size and shape as the conduit and causes little disturbance to the flow.

Stoplogs or a bulkhead gate on the face of the dam can be used to permit unwatering of the entire waterway and both are usually designed to operate under balanced pressure. Either device is lowered into place over the entrance with the control gate or an emergency gate closed and the conduit is then unwatered. A means of bypassing water from the reservoir into the conduit to balance the pressure on both sides of the stoplogs or bulkhead gate before they are raised must be provided. Adequate air passageways should be provided immediately downstream from the stoplogs or bulkhead gate, to prevent air from being trapped and compressed when the water is admitted to the conduit through the filling bypasses, and to reduce or eliminate negative pressure during unwatering.

10-7. *Energy Dissipating Devices.*—The discharge from an outlet, whether through gates, valves, or free-flow conduits, will emerge at a high velocity, usually in a near horizontal direction. The discharge may be released directly into the channel or riverbed if downstream structures are not endangered by the high-velocity flow and if the geology and topography are such that excessive erosion will not occur. However, if scouring and erosion are likely to be present, some means of dissipating the energy of the flow should be incorporated in the design. This may be accomplished by the construction of a stilling basin or other energy dissipating structure immediately downstream of the outlet.

The two types of energy dissipating devices most commonly used in conjunction with outlet works on concrete dams are hydraulic jump stilling basins and plunge pools. On some dams, it is possible to arrange the outlet works in conjunction with the spillway to utilize the spillway stilling device for dissipating the energy of the water discharging from the river outlets. Energy dissipating devices for free-flow conduit outlet works are essentially the same as those for spillways, discussed in chapter IX. The design of devices to dissipate jet flow is discussed in section 10-12.

1. Hydraulic Design of Outlet Works

10-8. General Considerations.—The hydraulics of outlet works involves either one or both of two conditions of flow—open channel (or free) flow and full conduit (or pressure) flow. Analysis of open channel flow in outlet works, either in an open waterway or in a partly full conduit, is based on the principle of steady nonuniform flow conforming to the law of conservation of energy. Full pipe flow in closed conduits is based on pressure flow, which involves a study of hydraulic losses to determine the total heads needed to produce the required discharges.

Hydraulic jump basins, plunge pools, or other stilling devices can be employed to dissipate the energy of flow at the end of the outlet works if the conditions warrant their use.

10-9. Pressure Flow in Outlet Conduits.—Most outlet works for concrete dams have submerged entrance conditions and flow under pressure with a control device on the downstream end.

For flow in a closed pipe system, as shown on figure 10-5, Bernoulli's equation can be written as follows:

$$H_T = h_L + h_{v_1} \qquad (1)$$

where:

> H_T = the total head needed to overcome the various head losses to produce discharge,
> h_L = the cumulative losses of the system, and
> h_{v_1} = the velocity head exit loss at the outlet.

Equation (1) can be expanded to list each loss, as follows:

$$H_T = h_t + h_e + h_{f(8)} + h_{f(7)} + h_{b(7)}$$
$$+ h_{f(6)} + h_{f(5)} + h_{b(5)} + h_{f(4)}$$
$$+ h_{c(4\text{-}3)} + h_{g(3)} + h_{ex(3\text{-}2)}$$
$$+ h_{f(2)} + h_{b(2)} + h_{c(2\text{-}1)}$$
$$+ h_{g(1)} + h_{v(1)} \qquad (2)$$

where:

> h_t = trashrack losses,
> h_e = entrance losses,
> h_b = bend losses,
> h_c = contraction losses,
> h_{ex} = expansion losses,
> h_g = gate or valve losses, and
> h_f = friction losses.

In equation (2) the number subscripts refer to the various components, transitions, and reaches to which head losses apply.

For a free-discharging outlet, H_T is measured from the reservoir water surface to the center of the outlet gate or the outlet opening. If the outflowing jet is supported on a downstream floor, the head is measured to the top of the emerging jet at the point of greatest contraction; if the outlet portal is submerged the head is measured to the tailwater level.

The various losses are related to the velocity head in the individual components, and equation (2) can be written:

$$H_T = K_t \left(\frac{v_9{}^2}{2g} \right) + K_e \left(\frac{v_8{}^2}{2g} \right) + \frac{fL_8}{D_8} \left(\frac{v_8{}^2}{2g} \right)$$

$$+ \frac{fL_7}{D_7} \left(\frac{v_7{}^2}{2g} \right) + K_{b_7} \left(\frac{v_7{}^2}{2g} \right) + \frac{fL_6}{D_6} \left(\frac{v_6{}^2}{2g} \right)$$

$$+ \frac{fL_5}{D_5} \left(\frac{v_5{}^2}{2g} \right) + K_{b_5} \left(\frac{v_5{}^2}{2g} \right) + \frac{fL_4}{D_4} \left(\frac{v_4{}^2}{2g} \right)$$

(Equation continued on next page.)

$$+ K_{c(4\text{-}3)} \left(\frac{v_3{}^2}{2g} - \frac{v_4{}^2}{2g} \right) + K_{g3} \left(\frac{v_3{}^2}{2g} \right)$$

$$+ K_{ex(3\text{-}2)} \left(\frac{v_3{}^2}{2g} - \frac{v_2{}^2}{2g} \right) + \frac{fL_2}{D_2} \left(\frac{v_2{}^2}{2g} \right)$$

$$+ K_{b2} \left(\frac{v_2{}^2}{2g} \right) + K_{c(2\text{-}1)} \left(\frac{v_1{}^2}{2g} - \frac{v_2{}^2}{2g} \right)$$

$$+ K_{g1} \left(\frac{v_1{}^2}{2g} \right) + K_v \left(\frac{v_1{}^2}{2g} \right) \qquad (3)$$

where:

$D =$ diameter of conduit,
$g =$ acceleration due to force of gravity,
$L =$ length of conduit,
$v =$ velocity,
$K_t =$ trashrack loss coefficient,
$K_e =$ entrance loss coefficient,
$K_b =$ bend loss coefficient,

$f =$ friction factor in the Darcy-Weisbach equation for pipe flow,
$K_{ex} =$ expansion loss coefficient,
$K_c =$ contraction loss coefficient,
$K_g =$ gate loss coefficient, and
$K_v =$ exit velocity head coefficient at the outlet.

Equation (3) can be simplified by expressing the individual losses in terms of an arbitrarily chosen velocity head. This velocity head is usually selected as that in a significant section of the system. If the various velocity heads for the system shown on figure 10-5 are related to that in the downstream conduit, with an area (2), the conversion for any area (x) is found as shown below.

By the principle of continuity,

$$Q = av = a_2 v_2 = a_x v_x$$

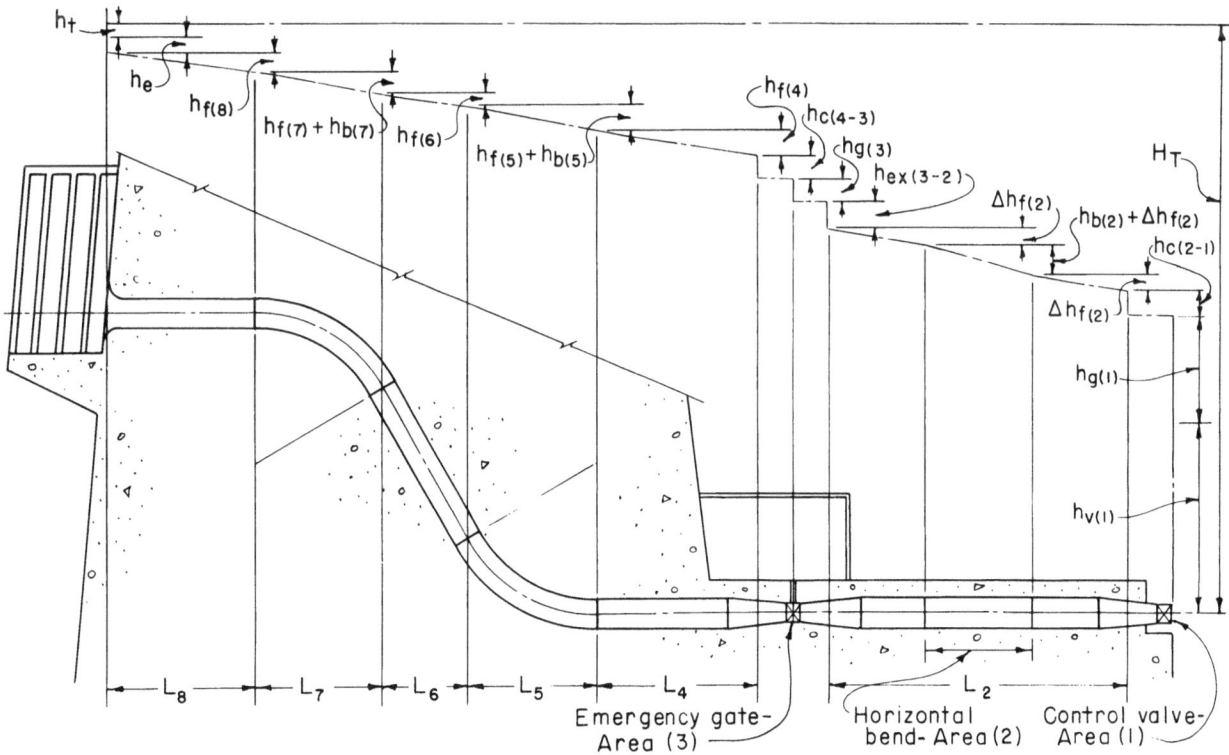

Figure 10-5. Pictorial representation of typical head losses in outlet under pressure. —288-D-3065

where:

> Q = discharge,
> a = cross-sectional area of conduit, and
> v = velocity.

Then:

$$a_2^2 v_2^2 = a_x^2 v_x^2 \text{, and}$$

$$\frac{a_2^2 v_2^2}{2g} = \frac{a_x^2 v_x^2}{2g}$$

from which:

$$\frac{v_x^2}{2g} = \left(\frac{a_2}{a_x}\right)^2 \frac{v_2^2}{2g}$$

Equation (3) then can be written:

$$
\begin{aligned}
H_T = \frac{v_2^2}{2g} \Bigg[& \left(\frac{a_2}{a_9}\right)^2 (K_t) + \left(\frac{a_2}{a_8}\right)^2 \left(K_e + \frac{fL_8}{D_8}\right) \\
& + \left(\frac{a_2}{a_7}\right)^2 \left(\frac{fL_7}{D_7} + K_{b_7}\right) + \left(\frac{a_2}{a_6}\right)^2 \left(\frac{fL_6}{D_6}\right) \\
& + \left(\frac{a_2}{a_5}\right)^2 \cdot \left(\frac{fL_5}{D_5} + K_{b_5}\right) + \left(\frac{a_2}{a_4}\right)^2 \left(\frac{fL_4}{D_4} - K_{c(4\text{-}3)}\right) \\
& + \left(\frac{a_2}{a_3}\right)^2 \left(K_{c(4\text{-}3)} + K_{g_3} + K_{ex(3\text{-}2)}\right) \\
& + \left(\frac{fL_2}{D_2} - K_{ex(3\text{-}2)} + K_{b_2} - K_{c(2\text{-}1)}\right) \\
& + \left(\frac{a_2}{a_1}\right)^2 \left(K_{c(2\text{-}1)} + K_{g_1} + K_v\right) \Bigg]
\end{aligned}
\tag{4}
$$

If the bracketed part of the expression is represented by K_L, the equation can be written:

$$H_T = K_L \frac{v_2^2}{2g} \tag{5}$$

Then:

$$Q = a_2 \sqrt{\frac{2gH_T}{K_L}} \tag{6}$$

10-10. Pressure Flow Losses in Conduits.

—Head losses in outlet works conduits are caused primarily by the frictional resistance to flow along the conduit sidewalls. Additional losses result from trashrack interferences, entrance contractions, contractions and expansions at gate installations, bends, gate and valve constrictions, and other interferences in the conduit. For a conservative design, greater than average loss coefficients should be assumed for computing required conduit and component sizes, and smaller loss coefficients should be used for computing energies of flow at the outlet. The major contributing losses of a conduit or pipe system are discussed in the remainder of this section.

(a) *Friction Losses.*—For flow in large pipes, the Darcy-Weisbach formula is most often employed to determine the energy losses due to frictional resistances of the conduit. The loss of head is stated by the equation:

$$h_f = \frac{fL}{D} \frac{v^2}{2g} \tag{7}$$

where f is the friction loss coefficient and other symbols are as previously defined. This coefficient varies with the conduit surface roughness and with the Reynolds number. The latter is a function of the diameter of the pipe and the velocity, viscosity, and density of the fluid flowing through it. Data and procedures for evaluating the loss coefficient are presented in Engineering Monograph No. 7 [1].[1] Since f is not a fixed value, many engineers are unfamiliar with its variations and would rather use Manning's coefficient of roughness, n, which has been more widely defined. If the influence of the Reynolds number is neglected, and if the roughness factor in relation to the pipe size is assumed constant, the relation of f in the Darcy-Weisbach equation to n in the Manning equation will be:

$$f = \frac{116.5n^2}{r^{1/3}} = \frac{185n^2}{D^{1/3}} \tag{8}$$

---

[1] Numbers in brackets refer to items in the bibliography, sec. 10-26.

where:

r = hydraulic radius, and
D = conduit diameter.

Relationships between the Darcy-Weisbach and Manning's coefficients can be determined graphically from figure 10-6.

Where the conduit cross section is rectangular in shape, the Darcy-Weisbach formula does not apply because it is for circular pipes, and the Manning equation may be used to compute the friction losses. Manning's equation (see sec. F-2(c) in appendix F) as applied to closed conduit flow is:

$$h_f = 29.1 \, n^2 \, \frac{L}{r^{4/3}} \, \frac{v^2}{2g} \qquad (9)$$

Maximum and minimum values of n which may be used to determine the conduit size and the energy of flow are as follows:

Conduit material	Maximum n	Minimum n
Concrete pipe or cast-in-place conduit	0.014	0.008
Steel pipe with welded joints	.012	.008

(b) *Trashrack Losses.*—Trashrack structures which consist of widely spaced structural members without rack bars will cause very little head loss, and trashrack losses in such a case might be neglected in computing conduit losses. When the trash structure consists of racks of bars, the loss will depend on the bar thickness, depth, and spacing. As shown in reference [2], an average approximation can be obtained from the equation:

$$\text{Loss} = K_t \, \frac{v_n^2}{2g} \qquad (10)$$

where:

$$K_t = 1.45 - 0.45 \frac{a_n}{a_g} - \left(\frac{a_n}{a_g}\right)^2$$

In the above:

K_t = the trashrack loss coefficient (empirical),
a_n = the net area through the rack bars,
a_g = the gross area of the racks and supports, and,
v_n = the velocity through the net trashrack area.

Where maximum loss values are desired, assume that 50 percent of the net rack area is clogged. This will result in twice the velocity through the trashrack. For minimum trashrack losses, assume no clogging of the openings when computing the loss coefficient, or neglect the loss entirely.

(c) *Entrance Losses.*—The loss of head at the entrance of a conduit is comparable to the loss in a short tube or in a sluice. If H is the head producing the discharge, C is the coefficient of discharge, and a is the area, the discharge is

$$Q = Ca \, \sqrt{2gH}$$

and the velocity is

$$v = C\sqrt{2gH}$$

or

$$H = \frac{1}{C^2} \, \frac{v^2}{2g} \qquad (11)$$

Since H is the sum of the velocity head h_v and the head loss at the entrance h_e, equation (11) may be written:

$$\frac{v^2}{2g} + h_e = \frac{1}{C^2} \, \frac{v^2}{2g}$$

or

$$h_e = \left(\frac{1}{C^2} - 1\right) \frac{v^2}{2g}$$

Then:

$$K_e = \frac{1}{C^2} - 1 \qquad (12)$$

"f" ————— "n" ————— "D" or "r"
Key

Darcy's equation for
friction loss in circular pipes

$$h_f = f \frac{L}{D} \frac{v^2}{2g}$$

Manning's equation
for friction loss in pipes

$$h_f = \frac{185 \, n^2}{D^{1/3}} \frac{L}{D} \frac{v^2}{2g}$$

Relationship between
Manning's "n" and
Darcy's "f"

$$f = \frac{185 \, n^2}{D^{1/3}} = \frac{116.5 n^2}{r^{1/3}}$$

DARCY'S "f"

MANNING'S "n"

DIAMETER "D"

HYDRAULIC RADIUS "r"

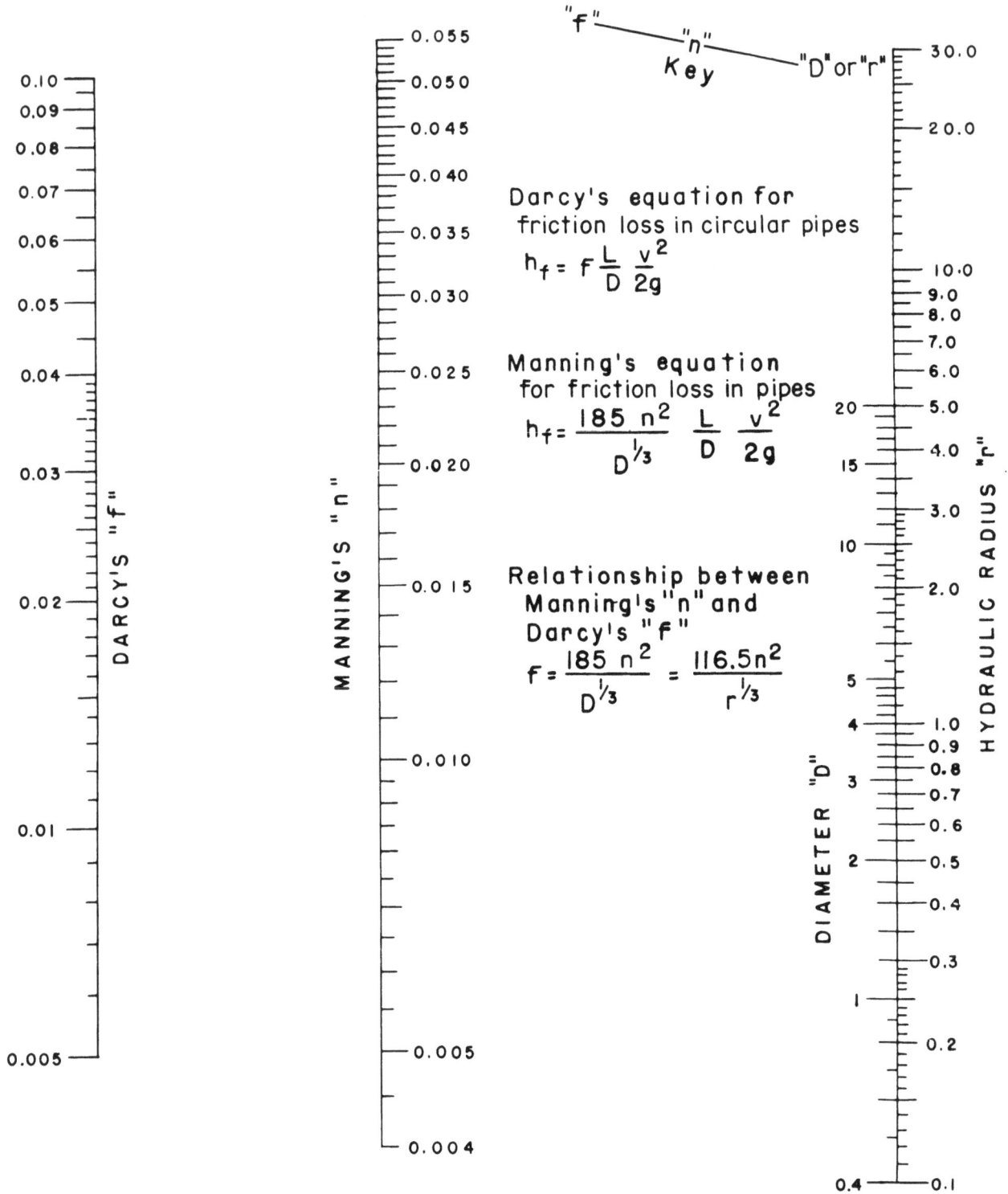

Figure 10-6. Relationship between Darcy's *f* and Manning's *n* for flow in pipes. —288-D-3066

Coefficients of discharge and loss coefficients for typical entrances for conduits, as given in various texts and technical papers, are listed in table 10-1.

(d) *Bend Losses.*—Bend losses in closed conduits in excess of those due to friction loss through the length of the bend are a function of the bend radius, the pipe diameter, and the angle through which the bend turns.

Graphs taken in part from reference [3] giving K_b as a function of these parameters are shown on figure 10-7. Figure 10-7(b) shows the coefficients for 90° bends for various ratios of radius of bend to diameter of pipe. Figure 10-7(c) indicates the coefficients for other than 90° bends. The value of the loss coefficient,

K_b, for various values of $\dfrac{R_b}{D}$ can be applied

directly for circular conduits; for rectangular conduits D is taken as the height of the section in the plane of the bend.

(e) *Transition Losses.*—Head losses in gradual contractions or expansions in a conduit can be considered in relation to the increase or decrease in velocity head, and will vary according to the rate of change of the area and the length of the transition. For contractions the loss of head, h_c, will be approximately

equal to $K_c \left(\dfrac{v_2^2}{2g} - \dfrac{v_1^2}{2g} \right)$, where K_c varies from

0.1 for gradual contractions to 0.5 for abrupt contractions. Where the flare angle does not exceed that indicated in section 10-11, the loss coefficient can be assumed as 0.1. For greater flare angles, the loss coefficient can be assumed to vary in a straight-line relationship to a

maximum of 0.5 for a right angle contraction.

For expansions, the loss of head, h_{ex}, will

be approximately equal to $K_{ex} \left(\dfrac{v_1^2}{2g} - \dfrac{v_2^2}{2g} \right)$

where K_{ex} is as follows:

Flare angle α	2°	5°	10°	12°	15°	20°
K_{ex} [4]	0.03	0.04	0.08	0.10	0.16	0.31
K_{ex} [5]	.02	.12	.16	–	.27	.40

Flare angle α	25°	30°	40°	50°	60°
K_{ex} [4]	0.40	0.49	0.60	0.67	0.72
K_{ex} [5]	.55	.66	.90	1.00	–

(f) *Gate and Valve Losses.*—No gate loss need be assumed where a gate is mounted at the entrance to the conduit so that when wide open it does not interfere with the entrance flow conditions. Also, emergency gates that are of the same size and shape as the conduit, such as ring-follower gates in a circular conduit, do not affect the flow and their associated losses are negligible. Emergency gates such as wheel-mounted or roller-mounted gates, although only operated at full open, have a K_g of not exceeding 0.1 due to the effect of the slot.

For control gates, as with emergency gates, mounted in a conduit so that the floor, sides, and roof, both upstream and downstream, are continuous with the gate opening, only the losses due to the slot will need to be considered, for which a value of K_g not exceeding 0.1 might be assumed. For partly open gates, the coefficient of loss will depend

Table 10-1.—*Coefficients of discharge and loss coefficients for conduit entrances.*

Type of entrance	Coefficient C			Loss coefficient K_e		
	Maximum	Minimum	Average	Maximum	Minimum	Average
(1) Square-cornered	0.85	0.77	0.82	0.70	0.40	0.50
(2) Slightly rounded	.92	.79	.90	.60	.18	.23
(3) Fully rounded	.96	.88	.95	.27	.08	.10
$\dfrac{r}{D} \geq 0.15$						
(4) Circular bellmouth	.98	.95	.98	.10	.04	.05
(5) Square bellmouth	.97	.91	.93	.20	.07	.16
(6) Inward projecting	.80	.72	.75	.93	.56	.80

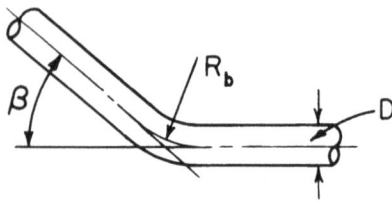

(a) DEFINITION SKETCH

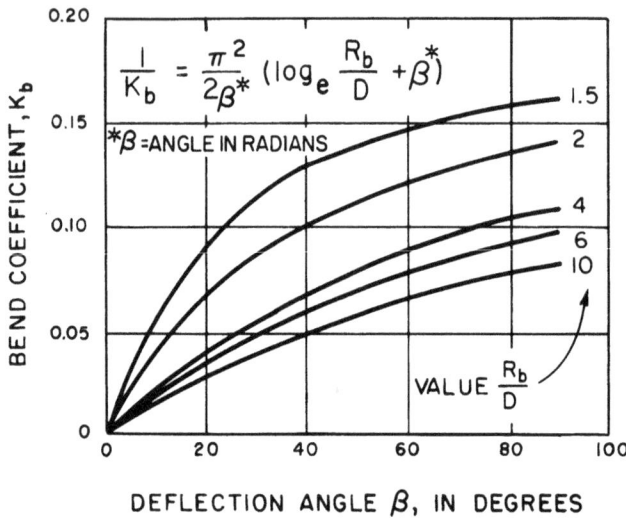

$$\frac{1}{K_b} = \frac{\pi^2}{2\beta^*} (\log_e \frac{R_b}{D} + \beta^*)$$

*β = ANGLE IN RADIANS

VALUE $\frac{R_b}{D}$

BEND COEFFICIENT, K_b

DEFLECTION ANGLE β, IN DEGREES

(c) K_b VS DEFLECTION ANGLE

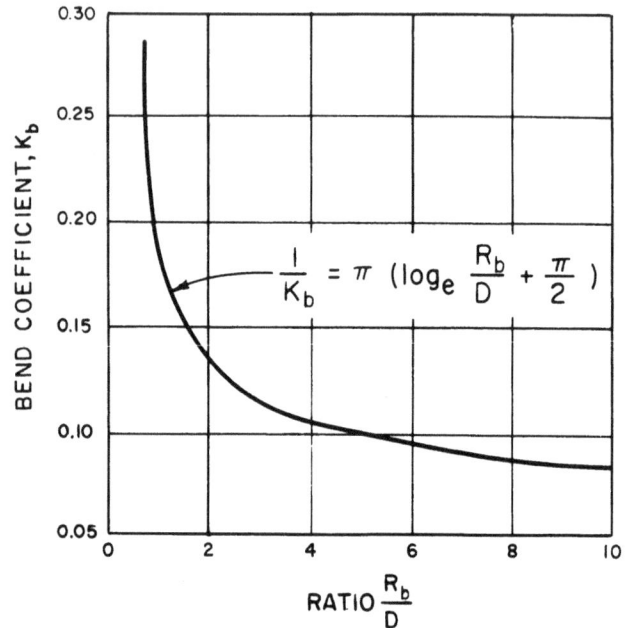

$$\frac{1}{K_b} = \pi (\log_e \frac{R_b}{D} + \frac{\pi}{2})$$

BEND COEFFICIENT, K_b

RATIO $\frac{R_b}{D}$

(b) K_b VS $\frac{R_b}{D}$ FOR 90° BENDS

Figure 10-7. Coefficient for bend losses in a closed conduit. −288-D-3067

on the top contraction.

The loss and discharge coefficients for the individual control gates and valves vary with each type and design; therefore, the actual coefficients used in design should be acquired from the manufacturer or from tests performed in a laboratory. As stated above, the K_g also varies for partial openings of the gate or valve.

(g) *Exit Losses.*—No recovery of velocity head will occur where the release from a pressure conduit discharges freely, or is submerged or supported on a downstream floor. The velocity head loss coefficient, K_v, in these instances is equal to 1.0. When a diverging tube is provided at the end of a conduit, recovery of a portion of the velocity head will be obtained if the tube expands gradually and if the end of the tube is

submerged. The velocity head loss coefficient will then be reduced from the value of 1.0 by the degree of velocity head recovery. If a_1 is the area at the beginning of the diverging tube and a_2 is the area at the end of the tube, K_v is equal to $\left(\frac{a_1}{a_2}\right)^2$.

10-11. *Transition Shapes.*— (a) *Entrances.*—To minimize head losses and to avoid zones where cavitation pressures can develop, the entrance to a pressure conduit should be streamlined to provide smooth, gradual changes in the flow. To obtain the best inlet efficiency, the shape of the entrance should simulate that of a jet discharging into air. As with the nappe-shaped weir, the entrance shape should guide and support the jet with minimum interference until it is contracted to the dimensions of the conduit. If the entrance curve is too sharp or too short, subatmospheric pressure areas may develop which can induce cavitation. A bellmouth entrance which conforms to or slightly encroaches upon the free-jet profile will provide the best entrance shape. For a circular entrance, this shape can be approximated by an elliptical entrance curve represented by the equation:

$$\frac{x^2}{(0.5D)^2} + \frac{y^2}{(0.15D)^2} = 1 \qquad (13)$$

where x and y are coordinates whose x-x axis is parallel to and $0.65D$ from the conduit centerline and whose y-y axis is normal to the conduit centerline and $0.5D$ downstream from the entrance face. The factor D is the diameter of the conduit at the end of the entrance transition.

The jet issuing from a square or rectangular opening is not as easily defined as one issuing from a circular opening; the top and bottom curves may differ from the side curves both in length and curvature. Consequently, it is more difficult to determine a transition for a square or rectangular opening which will eliminate subatmospheric pressures. An elliptical curved entrance which will tend to minimize the negative pressure effects is defined by the equation:

$$\frac{x^2}{D^2} + \frac{y^2}{(0.33D)^2} = 1 \qquad (14)$$

where D is the vertical height of the conduit for defining the top and bottom curves, and is the horizontal width of the conduit for defining the side curves. The major and minor axes are positioned similarly to those indicated for the circular bellmouth.

(b) *Contractions and Expansions.*—To minimize head losses and to avoid cavitation tendencies along the conduit surfaces, contraction and expansion transitions to and from gate control sections in a pressure conduit should be gradual. For contractions, the maximum convergent angle should not exceed that indicated by the relationship:

$$\tan \alpha = \frac{1}{U} \qquad (15)$$

where:

α = the angle of the conduit wall surfaces with respect to its centerline, and
U = an arbitrary parameter = $\dfrac{v}{\sqrt{gD}}$.

The values of v and D are the averages of the velocities and diameters at the beginning and end of the transition.

Expansions should be more gradual than contractions because of the danger of cavitation where sharp changes in the side walls occur. Furthermore, as has been indicated in section 10-10(e), loss coefficients for expansions increase rapidly after the flare angle exceeds about 10°. Expansions should be based on the relationship:

$$\tan \alpha = \frac{1}{2U} \qquad (16)$$

The notations are the same as for equation (15). For usual installations, the flare angle should not exceed about 10°.

The criteria for establishing maximum contraction and expansion angles for conduits flowing partly full are the same as those for open channel flow, as given in section 9-18(b) of chapter IX.

10-12. *Energy Dissipating Devices.*—Whenever practicable, the outlet works should be located so that the spillway energy dissipating structures can also be used to still the flow of the outlet works. Deflector buckets and hydraulic jump basins are commonly designed for stilling both outlet works and spillway flows when the outlet works flow can be directed into the spillway stilling basin. The hydraulic design for free-flow spillways and outlet works is discussed in chapter IX. Plunge pools and hydraulic jump stilling basins designed only for outlet works are discussed below.

(a) *Hydraulic Jump Basins.*—Where the outlet works discharge consists of jet flow, the open-channel flow hydraulic jump stilling basins mentioned above are not applicable. The jet flow either has to be directed onto the transition floor approaching the basin so it will become uniformly distributed, thus establishing open-channel flow conditions at the basin, or a special basin has to be designed.

The design of such a basin that will work well at all discharges is difficult using theoretical calculations, and model tests should be conducted to finalize all designs if practicable. The Bureau of Reclamation hydraulic laboratory has developed generalized designs of several kinds of basins based upon previously run model tests. General design rules are presented so that the necessary dimensions for a particular structure may be easily and quickly determined. One such example is the design of a hydraulic jump basin to still the jet flow from a hollow-jet valve. This basin is about 50 percent shorter than a conventional basin. The stilling basin is designed to take advantage of the hollow-jet shape, so solid jets cannot be used. The general design procedure can be found in Engineering Monograph No. 25 [6].

(b) *Plunge Pools.*—Where the flow of an outlet conduit issues from a downstream control valve or freely discharging pipe, a riprap- or concrete-lined plunge pool might be utilized. Such a pool should be employed only where the jet discharges into the air and then plunges downward into the pool.

When a free-falling overflow nappe drops vertically into a pool in a riverbed, a plunge pool will be scoured to a depth which is related to the height of the fall, the depth of tailwater, and the concentration of the flow [7]. Depths of scour are influenced initially by the erodibility of the stream material or the bedrock and by the size or the gradation of sizes of any armoring material in the pool. However, the armoring or protective surfaces of the pool will be progressively reduced by the abrading action of the churning material to a size which will be scoured out and the ultimate scour depth will, for all practical considerations, stabilize at a limiting depth irrespective of the material size. An empirical approximation based on experimental data has

been developed by Veronese [8] for limiting scour depths, as follows:

$$d_s = 1.32 H_T^{0.225} q^{0.54} \qquad (17)$$

where:

d_s = the maximum depth of scour below tailwater level in feet,
H_T = the head from the reservoir to tailwater levels in feet, and
q = the discharge in cubic feet per second per foot of width. (The width used for a circular valve or discharge pipe should be the diameter.)

Plunge pools used as energy dissipators should be tested in hydraulic models or, if possible, compared with similar designs in use or previously tested in a hydraulic model.

10-13. *Open Channel Flow in Outlet Works.*—If the outlet control gate or valve is at the upstream end or at some point along the conduit, open channel flow may exist downstream of the control; however, upstream of the control the flow is under pressure and the analysis is similar to that discussed in previous sections. The conduit downstream of the control may be enlarged or flared to assure nonpressure conditions, if desired. When open channel flow conditions exist, the design procedures are similar to those for open channel spillway flow discussed in chapter IX. An example of an outlet works with open channel flow downstream of the control gate is shown on figure 10-8.

2. Structural Design of Outlet Works

10-14. *General.*—The structural design of an outlet works is dependent upon the actual characteristics of that feature, the head, where the outlet works are incorporated in the dam, the stresses in the dam due to external loadings, and temperature. The design criteria for each component of the outlet works should be established for the conditions which exist or

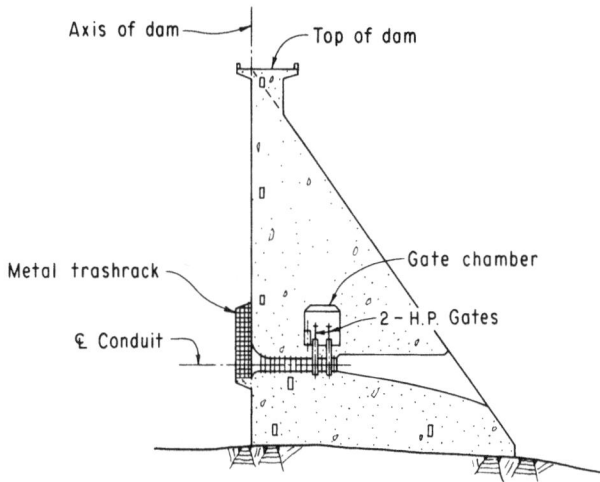

Figure 10-8. **A river outlet works with open channel flow. —288-D-3069**

may be expected to exist at any time during the life of the structure.

10-15. Trashrack.—A trashrack structure, regardless of the type, should be designed for a head differential due to the possible clogging of the rack with trash. This head differential will depend upon the location of the trashrack and its susceptibility to possible clogging, but should be a minimum of 5 feet. Temperature loads during construction should also be investigated in the design. If the trashrack will sometimes be exposed or partially exposed above the reservoir in areas subject to freezing, lateral loads from ice should be considered. In these instances, ice loads due to vertical expansion and the vertical load applied to the structure as ice forms on the members should also be included in the final analysis.

10-16. Conduit.—The outlet works conduit through a concrete dam may either be lined or unlined, but when the conduit is lined it may be assumed that a portion of the stress is being taken by the liner and not all is being transferred to the surrounding concrete. The temperature differential between the relatively cool water passing through the conduit and the relatively warm concrete mass will produce tensile stresses in the concrete in the immediate vicinity of the conduit. Also, the opening through the dam formed by the conduit will alter the distribution of stress in the dam in the vicinity of the conduit, tending to produce tensile stresses in the concrete at the periphery of the conduit. In addition, the bursting effect from hydrostatic pressures will cause tensile stresses at the periphery of the conduit. The above tensile stresses and possible propagation of concrete cracking usually extend only a short distance from the opening of the conduit, so it is common practice to reinforce only the concrete adjacent to the opening. The most useful method for determining the stresses in the concrete surrounding the outlet conduit is the finite element method of analysis, discussed in appendix C and in subchapter E of chapter IV.

10-17. Valve or Gate House.—The design of a control house depends upon the location and size of the structure, the operating and control equipment required, and the conditions of operation. The loadings and temperature conditions used in the design should be established to meet any situation which may be expected to occur during construction or during operation of the outlet works. The basic design approach should be the same as that for any commercial building.

10-18. Energy Dissipating Devices.—The structural design of an energy dissipating device is accomplished by usual methods of analysis for walls, slabs, and other structural members. Because each type of outlet works usually requires a different type of energy dissipator, the design loads depend upon the type of basin used, and have to be determined for the characteristics of the particular outlet works. Because of the dynamic pressures exerted on the structure from the hydraulic stilling process, laboratory tests or other means are usually required to establish the actual design loadings.

C. POWER OUTLETS

10-19. *General*.—Power outlets are outlet works that serve as a passage for water from the reservoir to the turbines within a powerplant. The power outlets consist of: (1) an intake structure which normally includes the emergency gates, a bulkhead gate or stoplog slots and guides, a trashrack structure on the face of the dam, and a bellmouth intake entrance; (2) a transition to the circular shape at the upstream end of the penstock; and (3) a penstock. The penstock acts as a pressure conduit between the turbine scroll case and the intake structure. The power outlets should be as hydraulically efficient as practicable to conserve available head; moreover, the intake structure should be designed to satisfactorily perform all of the tasks for which it was intended.

10-20. *Layout*.—The location and arrangement of the power outlets will be influenced by the size and shape of the concrete dam, the location of the river outlet works and the spillway, the relative location of the dam and powerplant, and the possibility of incorporating the power outlets with a diversion tunnel or the river outlets. For low-head concrete dams, penstocks may be formed in the concrete of the dam; however, a steel lining is desirable to insure watertightness. The penstocks may be completely embedded within the mass concrete of the concrete dam as shown on figure 10-9(a), embedded through the dam while the downstream portions between the dam and powerplant are above ground as shown on figure 10-9(b), or in an abutment tunnel as shown on figure 10-10.

When a powerplant has two or more turbines, the question arises whether to use an individual penstock for each turbine or a single penstock with a header system to serve all units. Considering only the economics of the layout, the single penstock with a header system will usually be less in initial cost; however, the cost of this item alone should not dictate the design. Flexibility of operation should be given consideration, because with a single penstock system the inspection or repair of the penstock will require shutting down the

entire plant. Further, a single penstock with a header system requires complicated branch connections and a valve to isolate each turbine. Also, the bulkhead gates will be larger, requiring heavier handling equipment. In concrete dams, it is desirable to have all openings as small as possible. The decision as to the penstock arrangement must be made considering all factors of operation, design, and overall cost of the entire installation.

Proper location of the penstock intake is important. The intake is usually located on the upstream face of the dam, which facilitates operation and maintenance of the intake gates. However, other structures or topographic conditions may influence the arrangement, and the penstock intake may best be situated in an independent structure located in the reservoir. Regardless of the arrangement, the intake should be placed at an elevation sufficiently below the low reservoir level and above the anticipated silt level to allow an uninterrupted flow of water under all conditions. Each intake opening should be protected against floating trash and debris by means of a trashrack structure.

Bends increase head loss and can cause the development of a partial vacuum during certain operating conditions. Therefore, penstock profiles from intake to turbine should, whenever practicable, be laid on a continuous slope. When vertical or horizontal bends are required in a penstock, their effect should be kept to a minimum by using as long a radius and as small a central angle as practicable.

10-21. *Intake Structures*.—The intake structure consists of several components, each of which is designed to accomplish a specific purpose. A trashrack is incorporated to keep trash from entering the penstocks and causing damage to the turbines; a bellmouth intake is used to establish flow lines at the entrance which minimize the amount of head loss; a transition, from the entrance size and shape to the circular diameter of the penstock, is established to least affect the flow and to minimize head loss. Also, the emergency gates can be incorporated into the intake structure

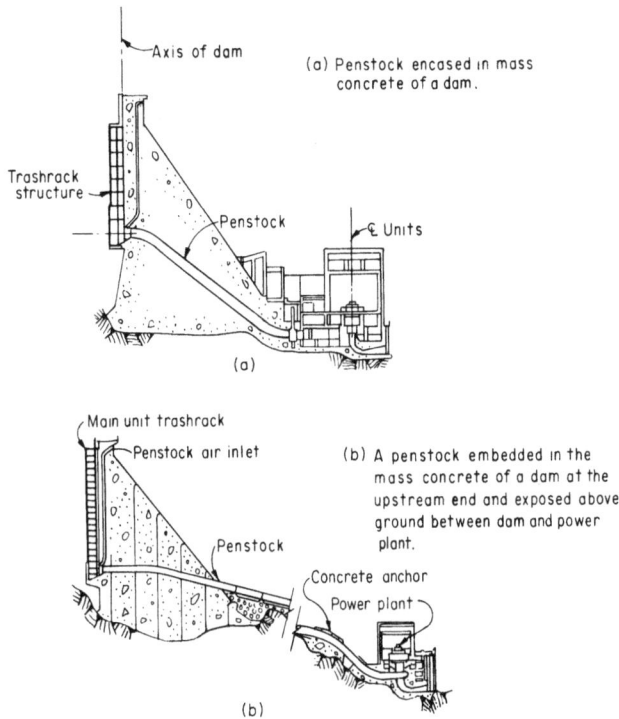

Figure 10-9. **Typical penstock installations.** −288-D-3071

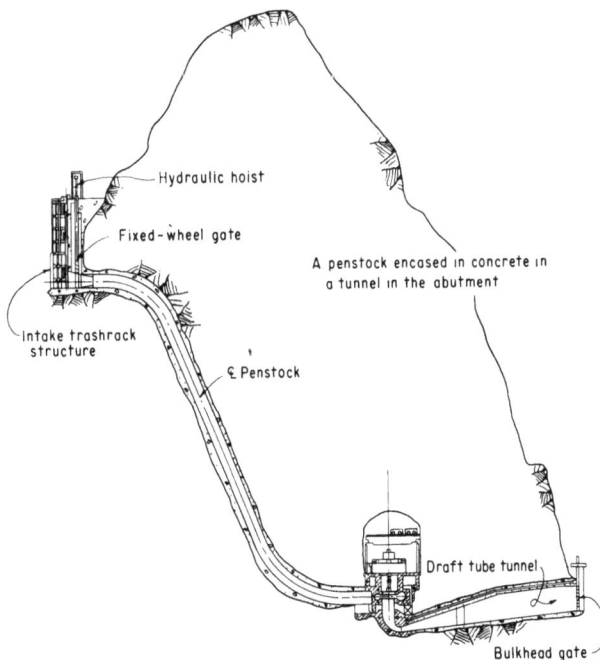

Figure 10-10. **Embedded penstock in abutment tunnel.** − 288-D-3073

to close off the flow through the penstock. Stoplogs are provided upstream of the emergency gates to unwater the entrance area and the emergency gate seats and guides for inspection and maintenance.

The velocity of flow in power intakes is usually much less than that in high-velocity river outlet works. For this reason, a smaller and less costly entrance structure can usually be designed for a power intake than for a river outlet works of equivalent physical size.

(a) *Trashracks.*—The trashrack structures for power intakes are similar to those required for other outlet works. However, because of the possible damage to the turbine and other hydraulic machinery, metal trashracks consisting of closely spaced bars are almost always required on power outlets to prevent the passage of even small trash and debris. With the lower velocity of flow through power outlets, large bellmouth openings at the intakes are not needed, and the length that the trashrack structure is required to span may be less than that for a high-velocity outlet works of equivalent physical size. The structure on which the trashracks are placed may consist of structural steel or of reinforced concrete as shown on figure 10-11. The determination of the type of trashrack structure depends not only upon the comparison of costs between the various structures but also upon the influence on the total time of construction for each scheme. Construction time may be reduced in some instances by using an all-metal or precast concrete structure instead of a cast-in-place structure.

Submerged trashracks should be used, if at all possible, because fully submerged racks normally require less maintenance than those which are alternately wet and dry. Experience has shown that steel will last longer if fully submerged. However, by bolting the all-metal trashrack structure to the concrete with stainless steel bolts, the racks can be replaced by divers if necessary.

When the reservoir surface fluctuates above and below the top of the trashrack structure, trash can accumulate on top of the structure and create a continuous maintenance problem. Normally, in large reservoirs submerged

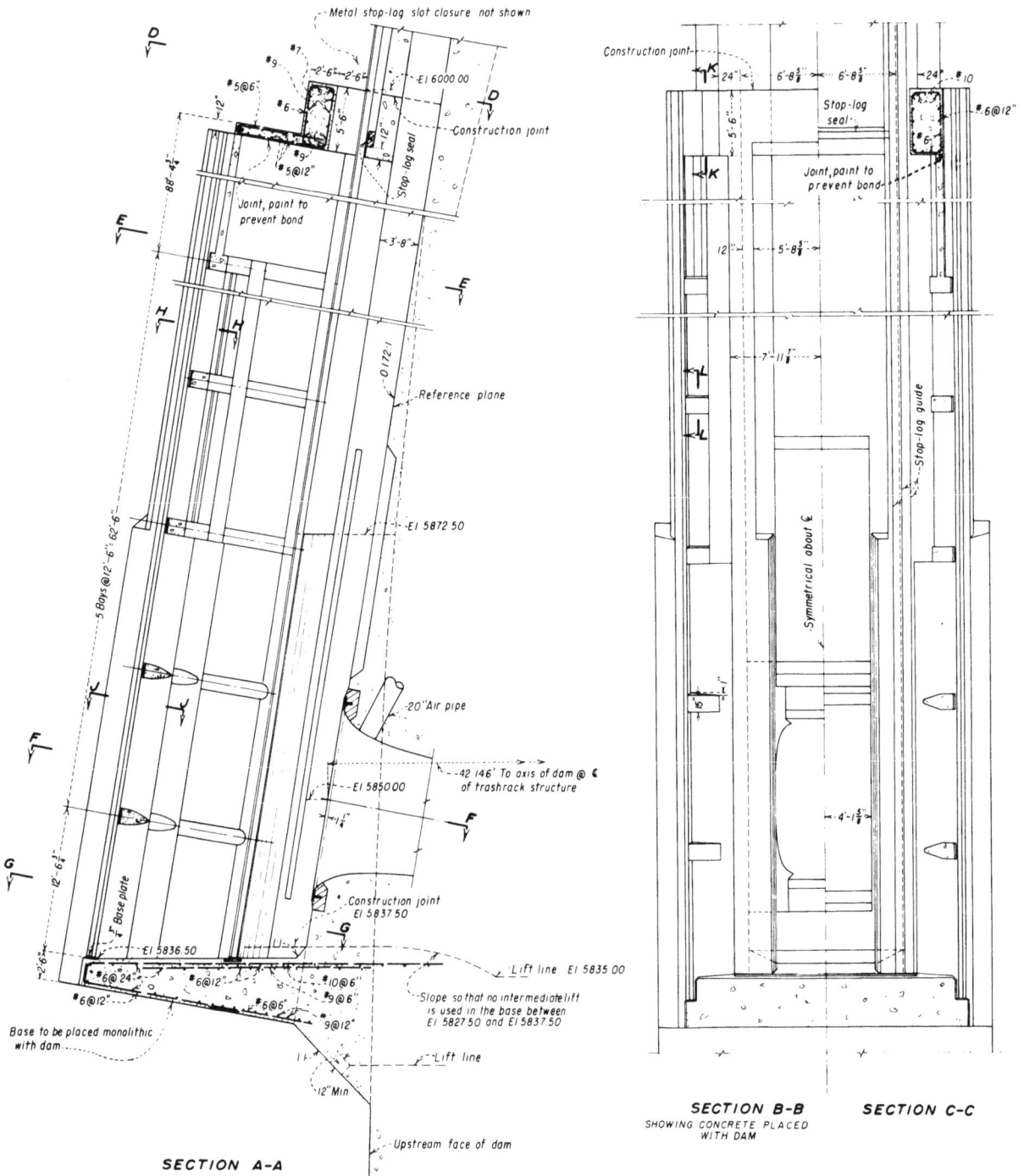

Figure 10-11. **Typical concrete trashrack structure for a penstock (sheet 1 of 2). −288-D-3074 (1/2)**

trashracks do not have to be raked as a result of trash accumulations, except during the initial filling. Ice loads must be considered if the trashrack structure is above the reservoir at times during cold winters. Ice loadings may be prevented by the installation of an air bubbling system around the structure. This system circulates the warmer water from lower in the reservoir around the structure to keep the members ice free.

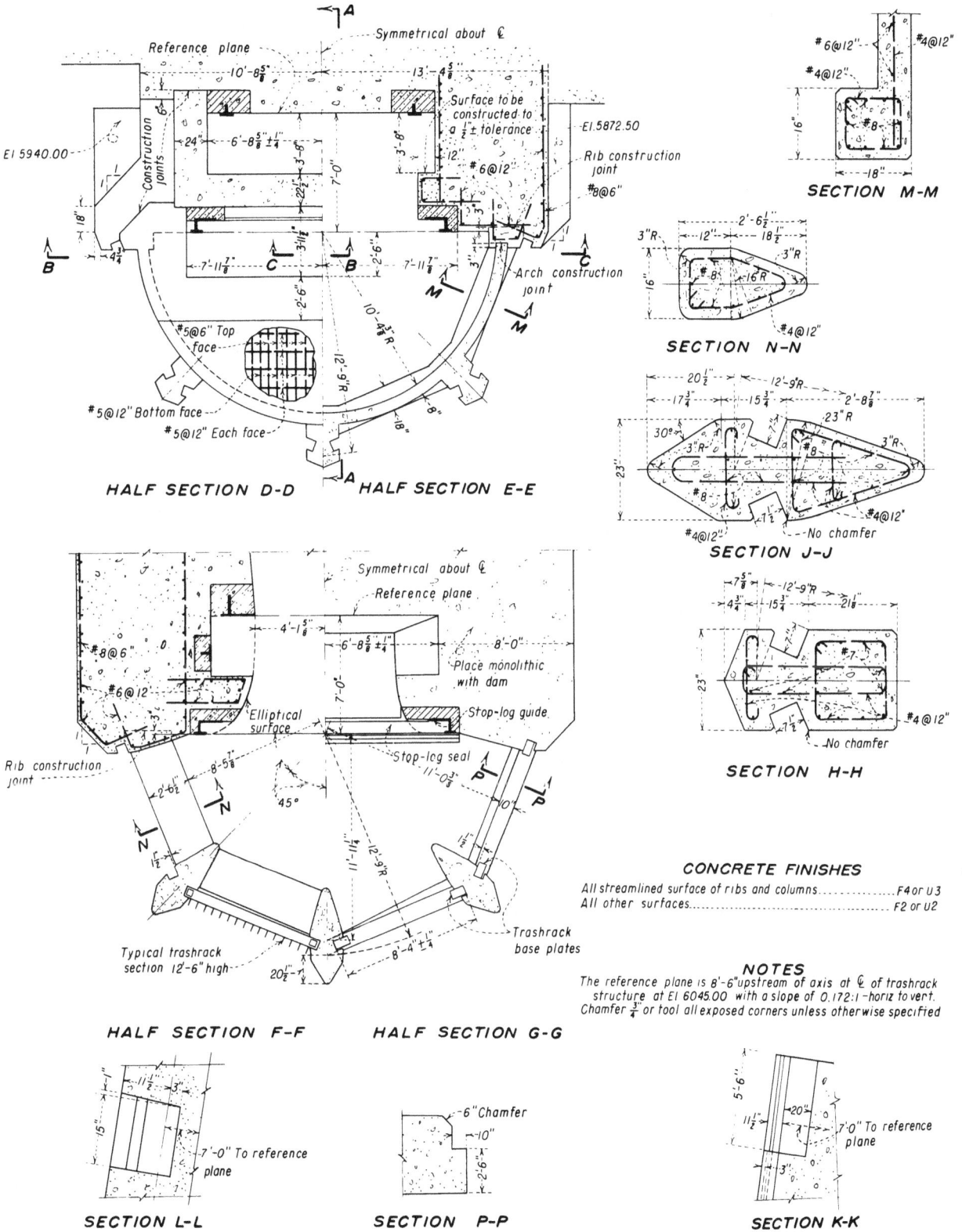

Figure 10-11. Typical concrete trashrack structure for a penstock (sheet 2 of 2). −288-D-3074 (2/2)

The trash bars usually consist of relatively thin, flat steel bars which are placed on edge from 2 to 9 inches apart and assembled in rack sections. The spacing between the bars is related to the size of trash in the reservoir and the size of trash that can safely be passed through the turbines without damage. The required area of the trashrack is fixed by a limiting velocity through the rack, which in turn depends on the nature of the trash which must be excluded. Where the trashracks are inaccessible for cleaning, the velocity should not exceed approximately 2 feet per second; however, a velocity up to approximately 5 feet per second may be tolerated for racks accessible for cleaning.

(b) *Bellmouth Entrance.*—It was brought out in section 10-11 that the entrance to a river outlet should be streamlined to provide smooth, gradual changes in the flow, thus minimizing head losses and avoiding disturbances of the flow in the conduit. This is also true for power outlets; however, because the velocities in penstocks are considerably lower, the bellmouths do not have to be as streamlined as those designed for the high-velocity river outlets. Experience on hydraulic models has shown that relatively simple rounding of corners eliminates most of the entrance losses when velocities are low. With the low velocities, pressure gradients in the bellmouth area are less critical.

(c) *Transition.*—Like the bellmouth entrance, the transition for the power outlets does not need to be as gradual as does the transition for the high-velocity river outlet works. The area throughout the transition can remain approximately the same, changing only from the shape of the gate to that of the penstock, with the gate area nearly equal to that of the penstock.

10-22. *Penstocks.*—The penstock is the pressure conduit which carries the water from the reservoir to the powerplant. The penstock for a low-head concrete dam may be formed in the mass concrete; however, a steel shell or lining is normally used to assure watertightness and prevent leakage into a gallery or chamber or to the downstream face. In large concrete dams under a high-head condition, steel penstock liners are always used to provide the required watertightness in the concrete. Penstocks can be embedded in concrete dams, encased in concrete, or installed in tunnels and backfilled with concrete. The penstocks should be as short as practicable and should be designed hydraulically to keep head loss to a minimum. The size of the penstock is determined from economic and engineering studies that determine the most efficient diameter for overall operation.

10-23. *Gates or Valves.*—Emergency gates or valves are used only to completely shut off the flow in the penstocks for repair, inspection, maintenance, or emergency closure. The wicket gates of the turbines act to throttle the flow in normal operation. The gates or valves, then, need to be designed only for full open operation. Many types of gates or valves can be utilized in the power outlets. Common emergency gates used in a concrete dam are fixed-wheel gates either at the face of the dam and controlled from the top of dam (see fig. 10-12), or in a gate slot in the dam and controlled from a chamber beneath the roadway.

An in-line control device, such as a butterfly valve, can be used anywhere along the length of the penstock and can be controlled from a chamber or control house. Also, in-line controls should be used on each individual penstock if more than one penstock branches off the main power outlet header, to permit the closure of each penstock without interfering with the flow of the others. In addition to butterfly valves, other types of in-line control devices that can be used to close off the flow include gate valves, ring-follower gates, and sphere valves. A determination of the type of valve or gate to be used is influenced by many factors such as the size of penstocks, the location best suited for controls and operators, the operating head, and the general layout of the power outlets. Another factor to consider in determining the control device to be used is the amount of head loss through each alternative type of gate or valve.

10-24. *Hydraulic Design of Power Outlets.*—The hydraulics of power outlets involves pressure flow through a closed

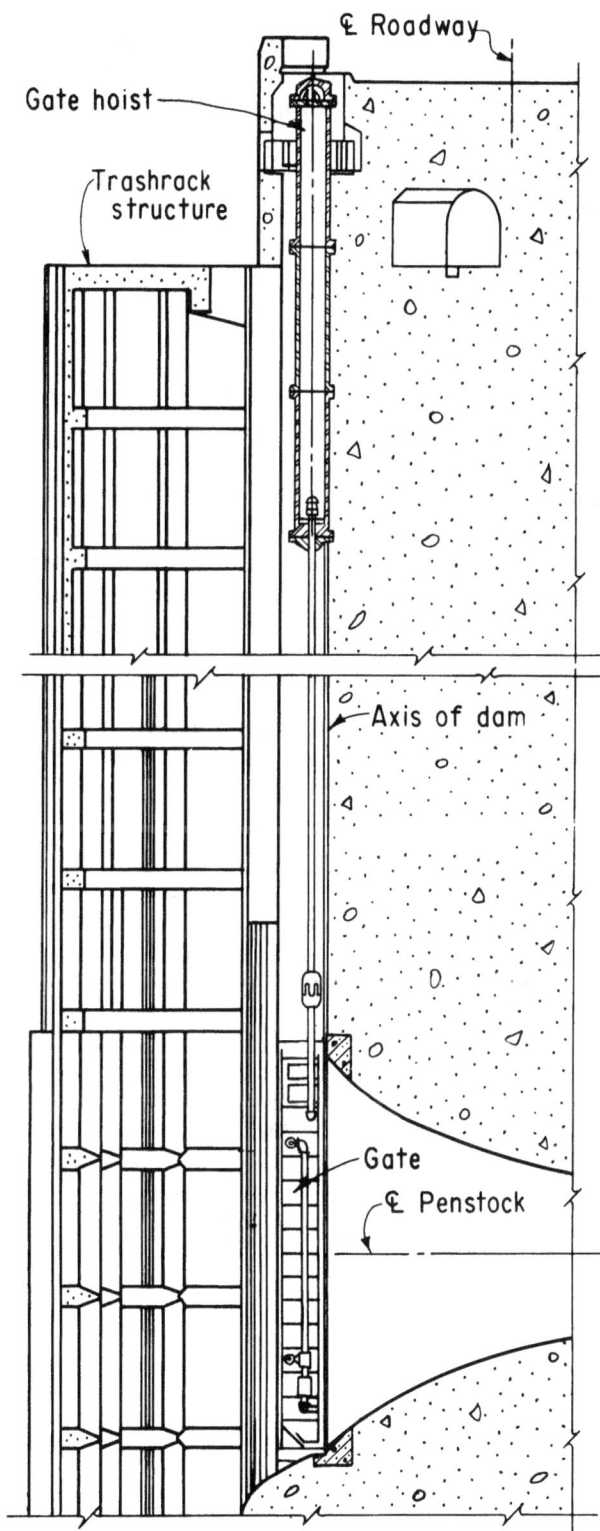

Figure 10-12. Typical fixed-wheel gate installation at upstream face of dam. —288-D-3075

conduit. The methods of hydraulic analysis are similar to those required for other outlet works. A power outlet is designed to carry water to a turbine with the least loss of head consistent with the overall economy of installation. An economic study will size a penstock from a monetary standpoint, but the final diameter should be determined from combined engineering and monetary considerations.

(a) *Size Determination of Penstock.*—A method for determining the economic diameter of a penstock is given in Engineering Monograph No. 3 [9]. All the variables used in this economic study must be obtained from the most reliable sources available, so as to predict as accurately as possible the average variables for the life of the project. The designer must assure himself that all related costs of construction are considered during the economic study.

The head losses used in the economic study for the power outlet are similar to the losses in other outlet works. Because of the lower velocities, these losses are usually small. But over a long period, even a small loss of head can mean a sizable loss of power revenue. The various head losses which occur between reservoir and turbine are as follows:

(1) Trashrack losses.
(2) Entrance losses.
(3) Losses due to pipe friction.
(4) Bend losses.
(5) Contraction losses (if applicable).
(6) Losses in gate or valve.

Engineering Monograph No. 3 gives a complete discussion of these losses and how they should be used in the determination of the economic size of a penstock.

(b) *Intake Structure.*—As stated in earlier sections, the lower velocity through a power outlet requires less streamlining of the intake structure to achieve economically acceptable hydraulic head losses. The gate can be made smaller, the bellmouths can be designed with sharper curvature, and the transition need not be made as gradual as for a high-velocity river outlet works. The design of the trashrack structure is similar to that for the river outlet works, discussed in section 10-4(a).

10-25. *Structural Design of Power Outlets.*—The structural design of a power outlet is dependent upon the actual characteristics of the power outlet works; the head; and where applicable, the stresses within the dam, due to temperature, gravity, and external loads. The design criteria for power outlet works should be established for the conditions which exist or may be expected to exist at any time during the operation or life of the structure.

(a) *Trashrack.*—The design of a trashrack structure for a power outlet should be based on a head differential of 5 feet due to partial clogging of trash. This small head differential minimizes power loss and is sufficient for the low velocities at which the power outlets operate. Ice loads should be applied in cold climates if the trashrack is exposed or partially exposed above the reservoir. Temperature loads during construction should also be investigated in the design procedures.

(b) *Penstocks.*—The penstocks through a concrete dam are usually lined with a steel shell; however, for low heads, penstocks may be simply a formed opening through the dam. Most penstock linings begin downstream from the transition. Therefore, when designing reinforcement around the penstocks, two conditions, lined and unlined, are usually present. In the area through which the penstocks are lined, a reasonable portion of the stresses may be assumed to be taken by the liner and not transferred to the surrounding concrete. Hydrostatic bursting pressures, concentration of the stresses within the dam, and temperature differentials between the water in the penstock and the mass concrete all may create tensile stresses in the concrete at the periphery of the penstock. Reinforcement is therefore placed around the penstock within the areas of possible tensile stress. A common method of analysis to determine the stresses in the concrete is a finite element study using a computer for the computations. Bursting pressures, dam loadings, and temperature variations can all be incorporated into this analysis to design the required reinforcement.

D. BIBLIOGRAPHY

10-26. *Bibliography*.

[1] Bradley, J. N., and Thompson, L. R., "Friction Factors for Large Conduits Flowing Full," Engineering Monograph No. 7, Bureau of Reclamation, March 1951.

[2] Creager, W. P., and Justin, J. D., "Hydroelectric Handbook," second edition, John Wiley & Sons, Inc., New York, N. Y., 1954.

[3] "Hydraulic Design Criteria, Sheet 228-1, Bend Loss Coefficients," Waterways Experiment Station, U.S. Army Engineers, Vicksburg, Miss.

[4] King, W. H., "Handbook of Hydraulics," fourth edition, McGraw Hill Book Co., Inc., New York, N. Y., 1954.

[5] Rouse, Hunter, "Engineering Hydraulics," John Wiley & Sons, Inc., New York, N. Y., 1950.

[6] "Hydraulic Design of Stilling Basins and Energy Dissipators," Engineering Monograph No. 25, Bureau of Reclamation, 1964.

[7] Doddiah, D., Albertson, M. L., and Thomas, R. A., "Scour From Jets," Proceedings, Minnesota International Hydraulics Convention (Joint Meeting of International Association for Hydraulic Research and Hydraulics Division, ASCE), Minneapolis, Minn., August 1953, p. 161.

[8] Scimemi, Ettore, "Discussion of Paper 'Model Study of Brown Canyon Debris Barrier' by Bermeal and Sanks," Trans. ASCE, vol. 112, 1947, p. 1016.

[9] "Welded Steel Penstocks," Engineering Monograph No. 3, Bureau of Reclamation, 1967.

Chapter XI

Galleries and Adits

11-1. *General.*—A gallery is an opening within the dam that provides access into or through the dam. Galleries may run either transversely or longitudinally and may be either horizontal or on a slope. Where used as a connecting passageway between other galleries or to other features such as powerplants, elevators, and pump chambers, the gallery is usually called an adit. Where a gallery is enlarged to permit the installation of equipment, it is called a chamber or vault.

11-2. *Purpose.*—The need for galleries varies from dam to dam. Some of the more common uses or purposes of galleries are:

(1) To provide a drainageway for water percolating through the upstream face or seeping through the foundation.

(2) To provide space for drilling and grouting the foundation.

(3) To provide space for headers and equipment used in artificially cooling the concrete blocks and grouting contraction joints.

(4) To provide access to the interior of the structure for observing its behavior after completion.

(5) To provide access to, and room for, mechanical and electrical equipment such as that used for the operation of gates in the spillways and outlet works.

(6) To provide access through the dam for control cables and/or power cables.

(7) To provide access routes for visitors.

Other galleries may be required in a particular dam to fulfill a special requirement.

Galleries are named to be descriptive of their location or use in the dam; for example, the foundation gallery is the gallery that follows the foundation of the dam, and the gate gallery is the gallery for servicing the gates. A typical gallery layout is shown on figures 11-1 and 11-2.

11-3. *Location and Size.*—The location and size of a gallery will depend upon its intended use or purpose. Some of the more common types of galleries are:

(a) *Foundation Gallery.*—The foundation gallery generally extends the length of the dam near the foundation rock surface, conforming in elevation to the transverse profile of the canyon; in plan it is near the upstream face and approximately parallel to the axis of the dam. It is from this gallery that the holes for the main grout curtain are drilled and grouted and from which the foundation drain holes are drilled. Its size, normally 5 feet wide by 7½ feet high, is sufficient to accommodate a drill rig. There should be a minimum of 5 feet of concrete between the floor of the gallery and the foundation rock.

(b) *Drainage Gallery.*—In high dams a supplementary drainage gallery is sometimes located further downstream, about two-thirds of the base width from the upstream face, for the purpose of draining the downstream portion of the foundation. This gallery usually extends only through the deepest portion of the dam. Drainage holes may be drilled from this gallery, so the 5- by 7½-foot size is usually adopted.

(c) *Gate Galleries and Chambers.*—Gate galleries and chambers are placed in dams to provide access to, and room for, the mechanical and electrical equipment required

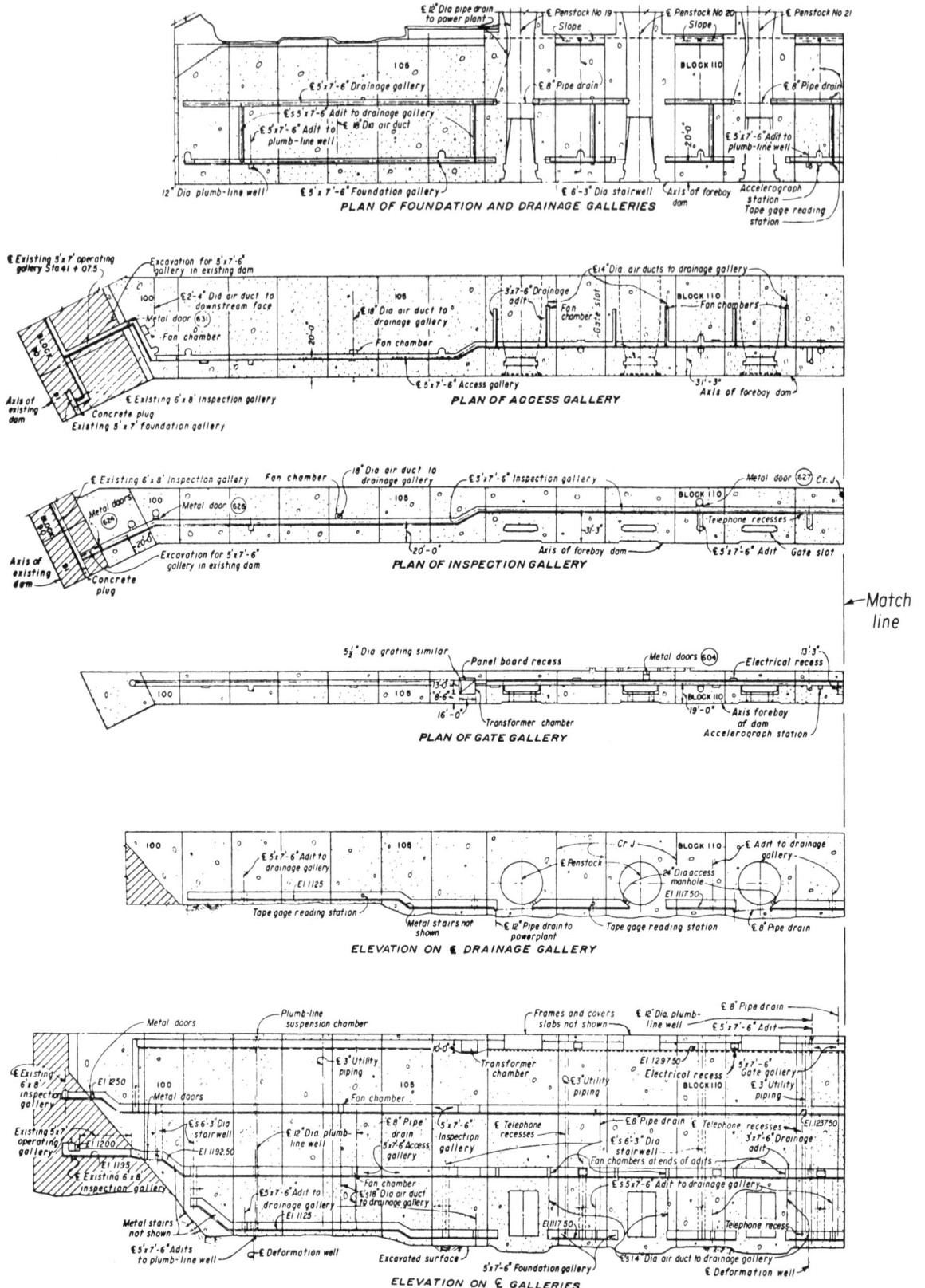

Figure 11-1. Galleries and shafts in Grand Coulee Forebay Dam—plans, elevations, section (sheet 1 of 2). −288-D-3077 (1/2)

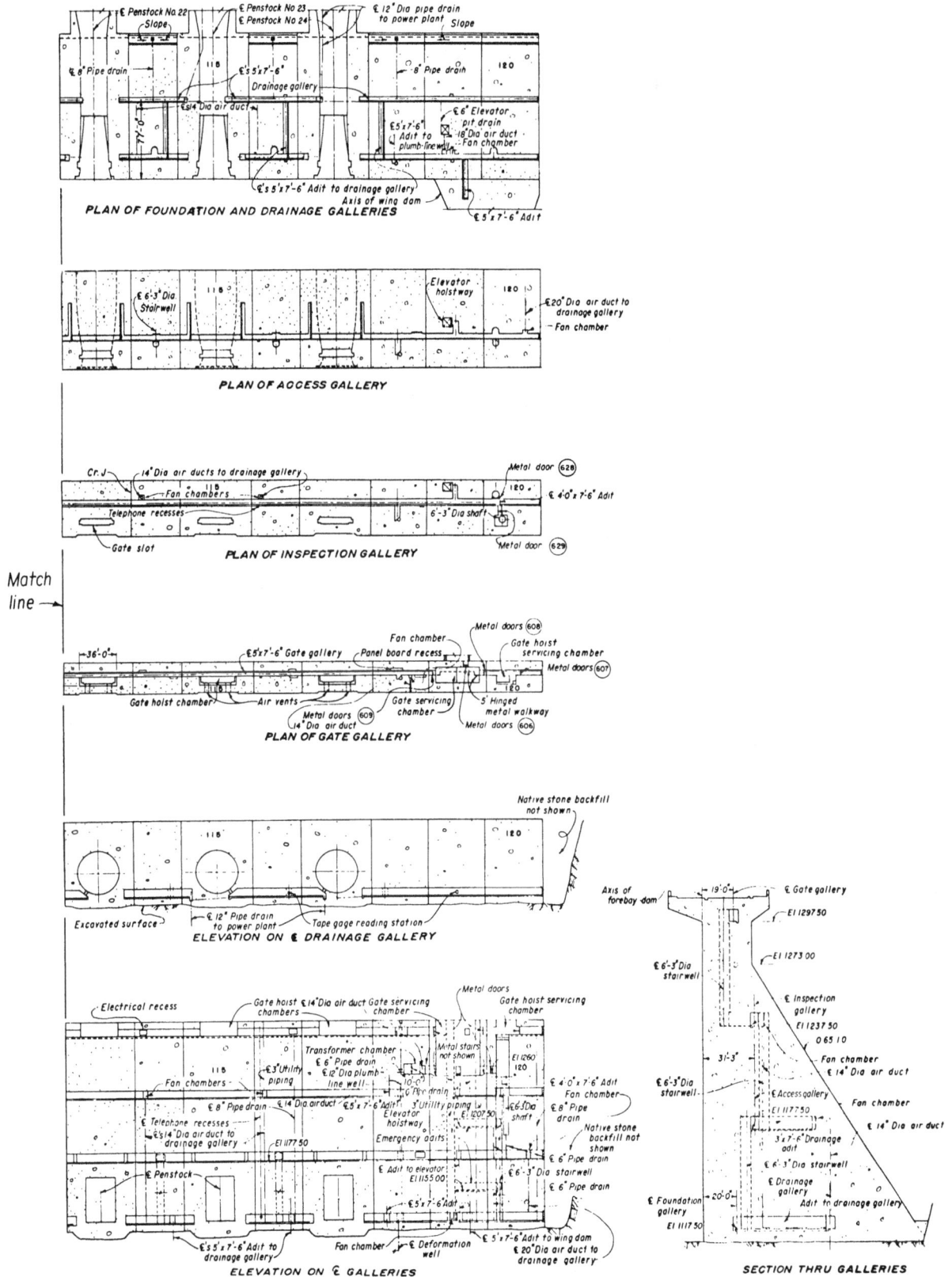

PLAN OF FOUNDATION AND DRAINAGE GALLERIES

PLAN OF ACCESS GALLERY

PLAN OF INSPECTION GALLERY

PLAN OF GATE GALLERY

ELEVATION ON ℄ DRAINAGE GALLERY

ELEVATION ON ℄ GALLERIES

SECTION THRU GALLERIES

Figure 11-1. Galleries and shafts in Grand Coulee Forebay Dam—plans, elevations, section (sheet 2 of 2). −288-D-3077 (2/2)

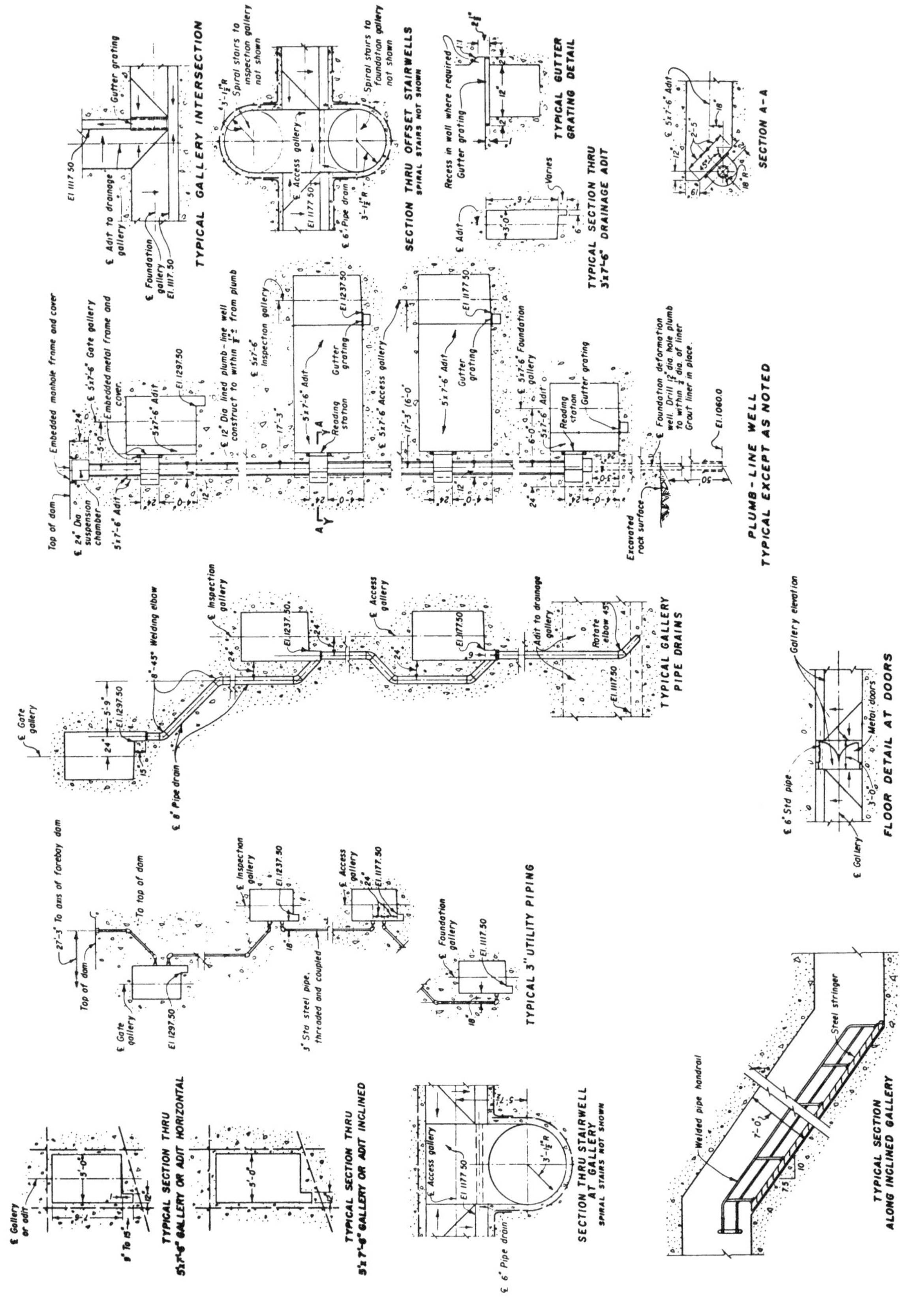

Figure 11-2. Galleries and shafts in Grand Coulee Forebay Dam—sections. —288-D-3079

for the operation of gates for outlets, power penstocks, or the spillway. Their size will depend on the size of the gates to be served.

(d) *Grouting Galleries.*—If it is impracticable to grout contraction joints from the face of the dam, the grout-piping system should be arranged so as to locate the supply, return, and vent headers in galleries placed near the top of each grout lift. The piping system for artificial cooling of the blocks may also be arranged to terminate in these galleries.

Transverse galleries or adits may be required for foundation consolidation grouting.

(e) *Visitors' Galleries.*—Visitors' galleries are provided to allow visitors into points of interest or as part of a tour route between visitors' facilities and the powerplant. The size would depend upon the anticipated number of visitors.

(f) *Cable Galleries.*—Galleries may be utilized, in conjunction with tunnels, cut and cover sections or overhead lines, as a means to carry control cables or power cables from the powerplant to the switchyard or spreader yard. The size of the gallery will depend upon the number of cables, the space required for each cable, and the space required for related equipment.

(g) *Inspection Galleries.*—Inspection galleries are located in a dam to provide access to the interior of the mass in order to inspect the structure and take measurements which are used to monitor the structural behavior of the dam after completion. All the galleries discussed above, which are located primarily for other specific purposes, also serve as inspection galleries.

As mentioned previously, galleries are usually made rectangular and 5 feet wide by 7½ feet high with a 12-inch-wide gutter along the upstream face for drainage. The 4-foot width is a comfortable width for walking and the 7½-foot height corresponds with the 7½-foot placement lift in mass concrete. Experience has shown that this size of gallery will provide adequate work area and access for equipment for normal maintenance except where special equipment is required such as at gate chambers. Galleries as narrow as 2 feet

have been used; however, a minimum of 3 feet is recommended.

11-4. *Drainage Gutter.*—All galleries should have gutters to carry away any seepage which gets into the gallery. On horizontal runs, the depth of gutter may vary from 9 to 15 inches to provide a drainage slope. Pipes should collect the water at low points in the gutter and take it to lower elevations where it will eventually go to the pump sump or drain directly to the downstream face by gravity.

11-5. *Formed Drains.*—Five-inch-diameter drains are formed in the mass concrete to intercept water which may be seeping into the dam along joints or through the concrete. By intercepting the water, the drains minimize the hydrostatic pressure which could develop within the dam. They also minimize the amount of water that could leak through the dam to the downstream face where it would create an unsightly appearance.

The drains are usually located about 10 feet from the upstream face and are parallel to it. They are spaced at approximately 10-foot centers along the axis of the dam. The lower ends of the drains extend to the gallery, or are connected to the downstream face near the fillet through a horizontal drainpipe or header system if there are no galleries. The tops of the drains are usually located in the crest of the dam to facilitate cleaning when required. Where the top of the dam is thin, the drains may be terminated at about the level of the normal reservoir water surface. A 1½-inch pipe then connects the top of the drain with the crest of the dam and can be used to flush the drains.

11-6. *Reinforcement.*—Reinforcement is usually required around galleries in a dam only where high tensile stresses are produced, such as around large openings, openings whose configuration produces high tensile stress concentrations, and openings which are located in areas where the surrounding concrete is in tension due to loads on the dam or temperature or shrinkage. Reinforcement should also be utilized where conditions are such that a crack could begin at the gallery and propagate through the dam to the reservoir.

Stresses around openings can be determined using the finite element method for various loading assumptions such as dam stresses, temperature, and shrinkage loads. Reinforcement is usually not required if the tensile stresses in the concrete around the opening are less than 5 percent of the compressive strength of the concrete. If tensile stresses are higher than 5 percent of the compressive strength, reinforcement should be placed in these areas to limit cracking. Each gallery should be studied individually using the appropriate dam section and loads.

In areas of high stress or where the stresses are such that a crack once started could propagate, reinforcement should be used. If unreinforced, such a crack could propagate to the surface where it would be unsightly and/or admit water to the gallery. It could also threaten the structure safety. The stresses determined by the finite element analysis can be used to determine the amount of reinforcement required around the opening to control the cracking.

In some cases, reshaping or relocating the gallery can reduce or eliminate the tensile stresses.

11-7. *Services and Utilities.*—Service lines, such as air and water lines, can be installed in the gallery to facilitate operation and maintenance after the dam has been completed. To supply these lines, utility pipe should be embedded vertically between the galleries and from the top gallery to the top of the dam. This will enable the pipe at the top of the dam to be connected with an air compressor, for example, and deliver compressed air to any gallery. The number and size of the utility piping would depend upon anticipated usage.

Galleries should have adequate lighting and ventilation so as not to present a safety hazard to persons working in the galleries. The ventilation system should be designed to prevent pockets of stale air from accumulating.

Telephones should be installed at appropriate locations in the gallery for use in an emergency and for use of operations and maintenance personnel.

The temperature of the air in the gallery should be about the same as that of the surrounding mass concrete to minimize temperature stresses. This may require heating of incoming fresh air, particularly in colder climates. Galleries used for high-voltage power cables may require cooling since the cables give off considerable heat.

11-8. *Miscellaneous Details.*—Horizontal runs of galleries, where practicable, should be set with the floor at the top of a placement lift in the dam for ease of construction. Galleries on a slope should provide a comfortable slope for walking on stairs. A 7½ to 10 slope is reasonable for stairs, yet is steep enough to follow most abutments. A slope of 7½ to 9 has been used on steeper abutments. Ramp slopes may be used where small or gradual changes in elevation are required. Ramp slopes should be less than 10° but can be up to 15° if special nonslip surfaces and handrails are provided.

Spiral stairs in a vertical shaft are used where the abutments are steeper than can be followed by sloping galleries. These shafts are usually made 6 feet 3 inches in diameter to accommodate commercially available metal stairs.

To minimize the possibility of a crack developing between the upstream face of the dam and a gallery which would leak water, galleries are usually located a minimum distance of 5 percent of the reservoir head on the gallery from the upstream face. A minimum of 5 feet clear distance should be used between galleries and the faces of the dam and contraction joints, to allow room for placement of mass concrete and to minimize stress concentrations.

Chapter XII

Miscellaneous Appurtenances

12-1. *Elevator Tower and Shaft.*—Elevators are placed in concrete dams to provide access between the top of the dam and the gallery system, equipment and control chambers, and powerplant. The elevators can also be used by the visiting public for tours through the dam. The elevator structure consists of an elevator shaft that is formed within the mass concrete, and a tower at the crest of the dam. The shaft should have connecting adits which provide access into the gallery system and into operation and maintenance chambers. These adits should be located to provide access to the various galleries and to all locations at which monitoring and inspection of the dam or maintenance and control of equipment may be required. Stairways and/or emergency adits to the gallery system should be incorporated between elevator stops to provide an emergency exit.

The tower provides a sheltered entrance at the top of the dam and houses the elevator operating machinery and equipment. Moreover, the tower may be designed to provide space for utilities, storage, and offices. Tourist concession and information space may also be provided in the tower at the top of the dam, if the project is expected to have a large tourist volume. The height of the tower above the roadway is dependent upon the number of floors needed to fulfill the space requirements of the various functions for which the tower is intended. On large dams more than one elevator may be incorporated into the design to make access more available. Moreover, separate elevators may be constructed for visitors other than the elevators provided for operation and maintenance. Since the towers provide the entrance to the interior of the structure and are used by most visitors, they are a focal point of interest and their architectural considerations should be an important factor in their design and arrangement. The architectural objective should be simplicity and effectiveness blending with the massiveness of the dam to present a pleasing and finished appearance to the structure.

The machinery and equipment areas should include sufficient space for the required equipment and adequate additional space for maintenance and operation activities. Electrical, telephone, water, air, and any other services which may be required should be provided to the appropriate areas. Restrooms for visitors as well as those for maintenance personnel may also be included in the layout of the tower. Stairways, either concrete or metal, are usually included for access to machinery and equipment floors to facilitate maintenance and repair. Stairways can also be provided as emergency access between levels. An example of the layout of a typical elevator shaft and tower can be seen on figures 12-1 and 12-2.

(a) *Design of Shaft.*—The design of reinforcement around a shaft can be accomplished by the use of finite element studies, with the appropriate loads applied to the structure. The stresses within the dam near the shaft and any appropriate temperature loads should be analyzed to determine if tension can develop at the shaft and be of such magnitude that reinforcement would be required. A nominal amount of reinforcement should be placed around the shaft if it is near any waterway or the upstream face of the dam to minimize any chance of leakage through any

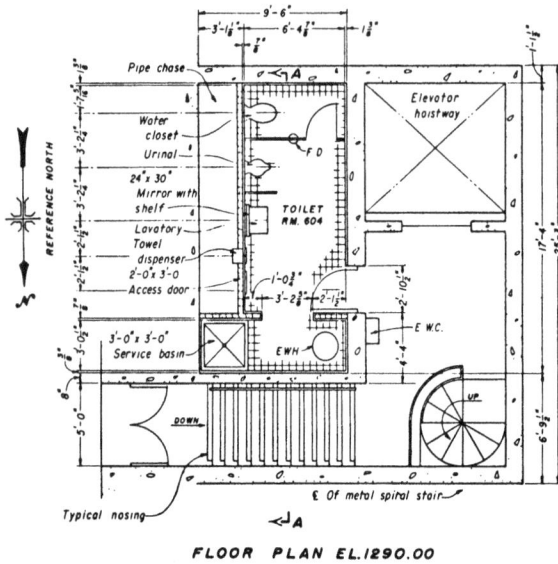

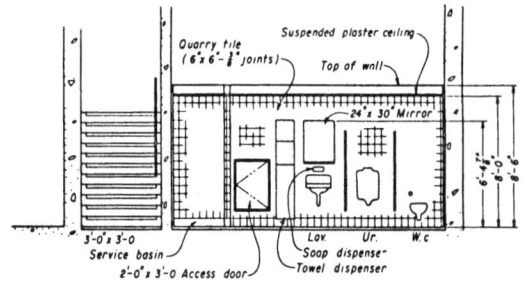

FLOOR PLAN EL.1290.00

SECTION A-A

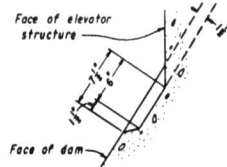

DETAIL A
MODIFICATION OF ARCHITECTURAL
GROOVE TYPE G-4

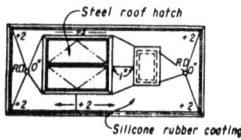

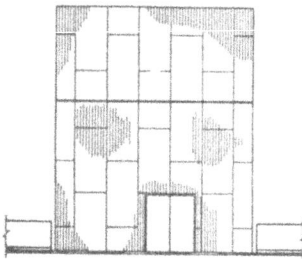

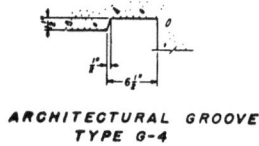

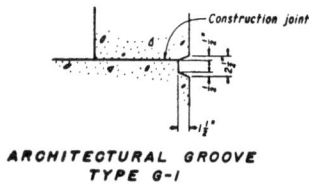

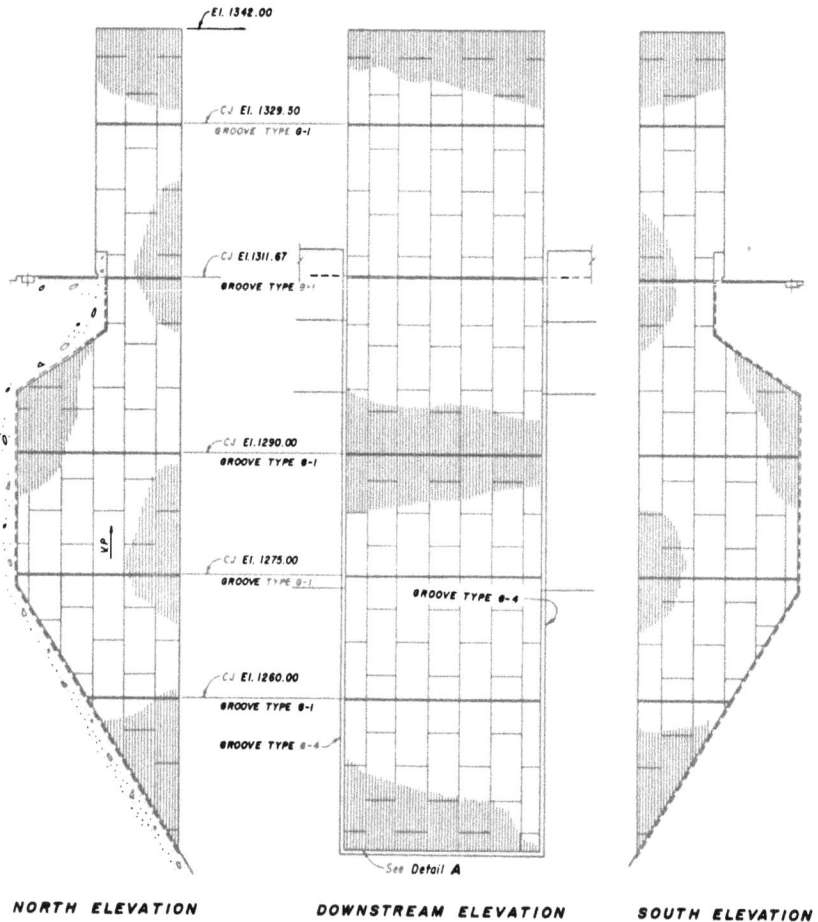

ROOF PLAN

REFERENCE NORTH

UPSTREAM ELEVATION

ARCHITECTURAL GROOVE
TYPE G-4

ARCHITECTURAL GROOVE
TYPE G-1

NORTH ELEVATION

DOWNSTREAM ELEVATION

SOUTH ELEVATION

Figure 12-1. Architectural layout of elevator tower in Grand Coulee Forebay Dam. –288-D-3082

cracks which may open. Reinforcement should also be placed around the periphery of the shaft as it approaches the downstream face of the dam, where tensile stresses due to temperature loadings become more likely to occur.

(b) *Design of Tower.*—The structural design of the elevator tower above mass concrete should be accomplished by using standard design procedures and the appropriate loads that can be associated with the structure. Live loads, dead loads, temperature loads, wind loads, and earthquake loads should all be included in the design criteria. The magnitude of earthquake load on the tower (see (2)b below) may be increased substantially by the resonance within the structure and must be determined by actual studies. Reinforcement to be placed in the structure at all the various components should be designed with respect to the characteristics of the structure and the requirements of the reinforced concrete code.

Dead loads and live loads usually used in the design of an elevator tower are as follows:

(1) *Dead loads:*

Reinforced concrete—150 pounds per cubic foot

Roofing—varies with type of material

(2) *Live loads:*

a. Uniformly distributed floor loads, pounds per square foot.

Lobby 150
Office space 100
Roof (includes snow) 50
Toilets 100
Stairways 100
Elevator—machinery floor *250
Storage space—heavy 250
Storage space—light 125

*Concentrated loads from the elevator machinery may control the design instead of the uniform load given.

b. Other loads:

Wind loads . . . . 30 pounds per square
                                   foot on vertical
                                   projection

Earthquake loads:

Horizontal . . . . 0.1 gravity
Vertical . . . . 0.05 gravity

12-2. *Bridges.* —Bridges may be required on the top of the dam to carry a highway over the spillway or to provide roadway access to the top of the dam at some point other than at the end of the dam. A bridge may also be provided over a spillway when bulkhead gates for river outlets or spillway crest gates require the use of a traveling crane for their operation or maintenance. Where there is no highway across the dam and no crane operations are required, a spillway bridge designed only to facilitate operation and maintenance may be constructed. When a bridge is to be used for a highway or to act as a visitors' access route, architectural treatment should be undertaken to give the structure a pleasing appearance. This architectural treatment should be based on the size of dam, the size and type of other appurtenant structures, local topography, and a type of bridge structure which blends pleasingly with the entire feature.

Design criteria for highway bridges usually conform to the standard specifications adopted by the American Association of State Highway Officials, modified to satisfy local conditions and any particular requirement of the project. The width of roadway for two-way traffic should be a minimum of 24 feet curb to curb plus sidewalk widths as required. However, with new highway regulations requiring greater widths, both Federal and local codes should be consulted to establish a final width. The structural members can consist of reinforced concrete, structural steel, or a combination of both types of materials. The bridge structure can be one of many types such as barrel-arch, slab and girder, or slab, depending on the required architecture, loads, and span. The structure should be designed to carry the class of traffic which is to use the bridge; however, the traffic design load used should generally not be less than the HS-20 classification.

Special heavy loads during the construction period, such as powerplant equipment hauled on specially constructed trailers, may produce stresses far in excess of those produced by the normal highway traffic and these should be considered in the design criteria. If the bridge deck is to be used for servicing gates or other mechanical equipment, the loading imposed by the weight of the crane, the force necessary to

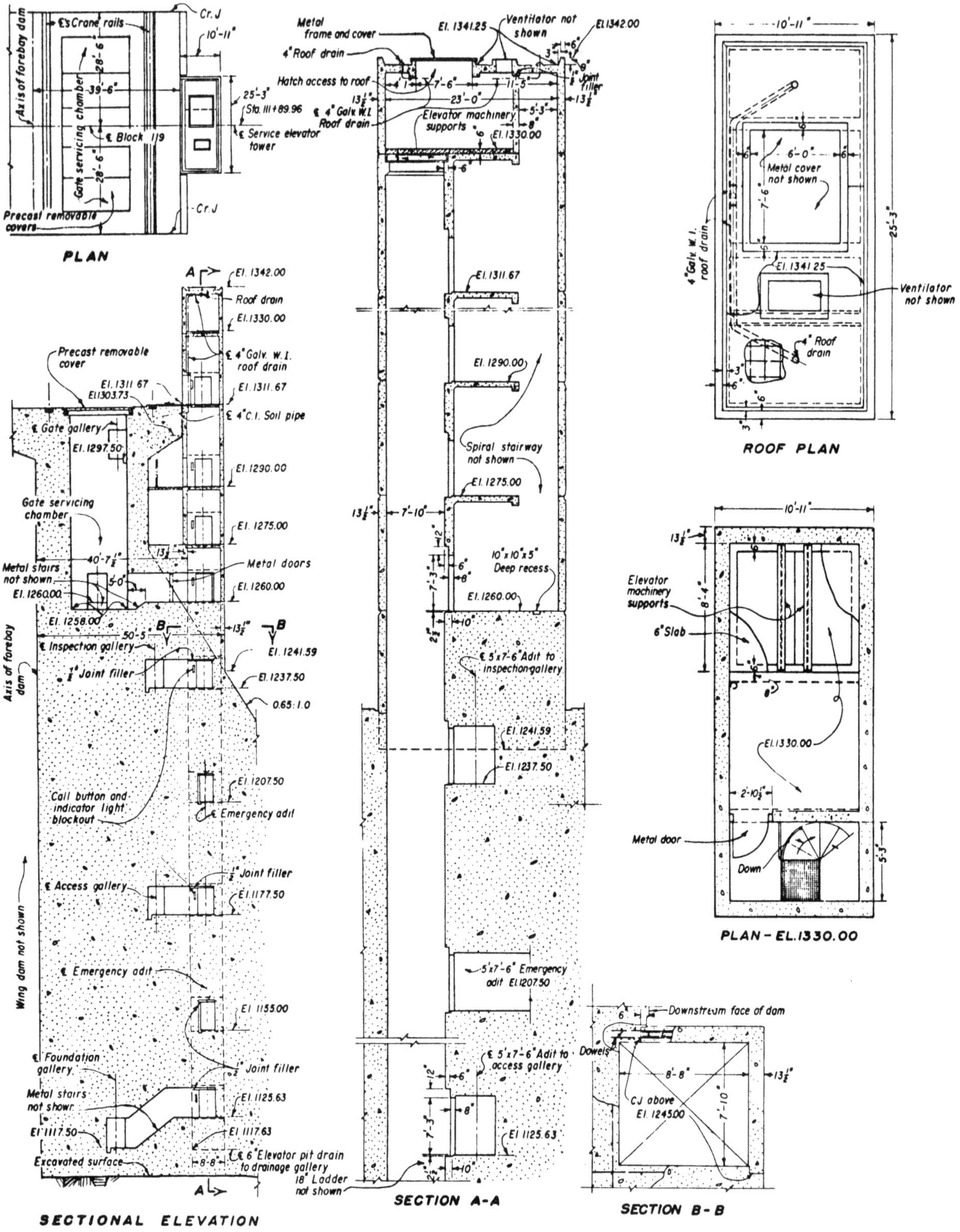

Figure 12-2. Structural layout of elevator shaft and tower in Grand Coulee Forebay Dam (sheet 1 of 2).—
288-D-3084 (1/2)

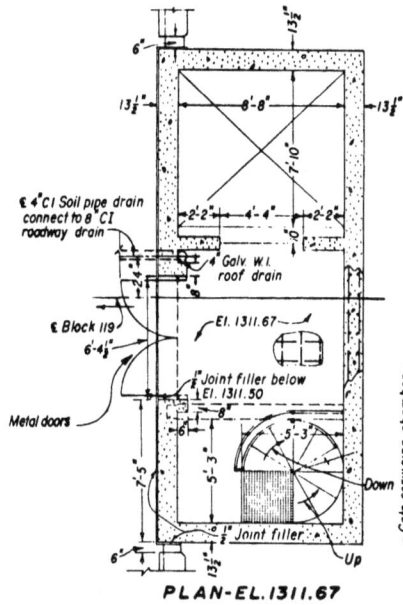

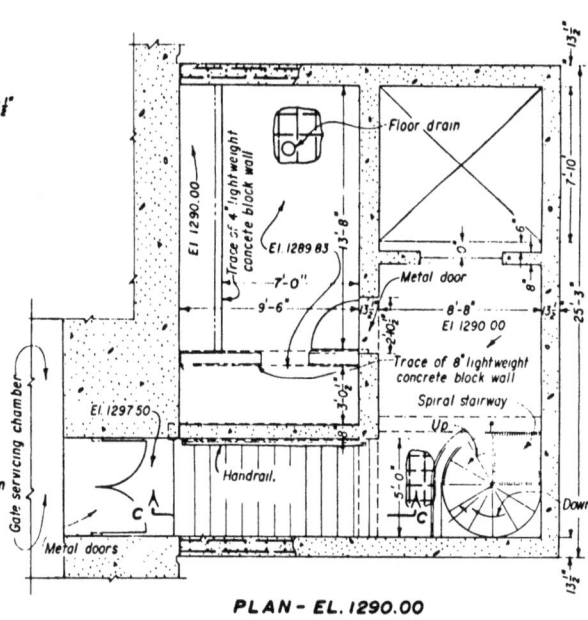

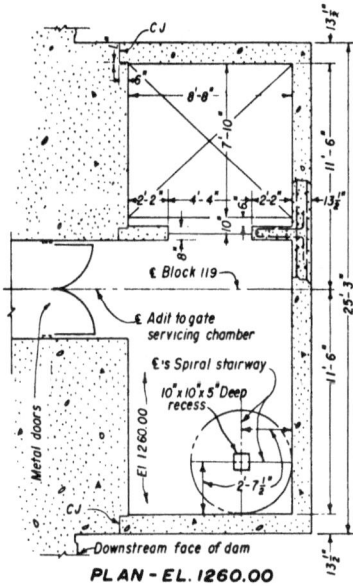

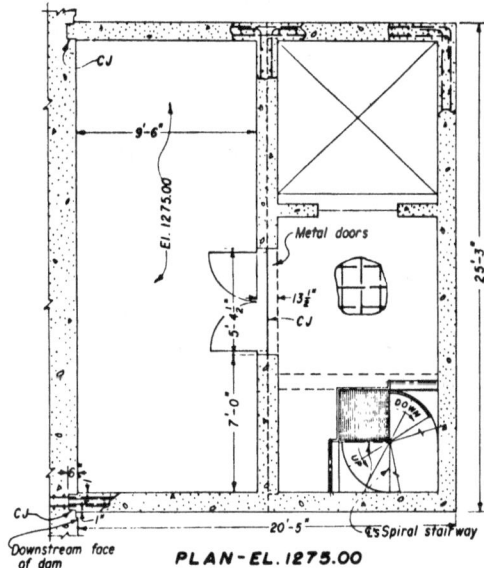

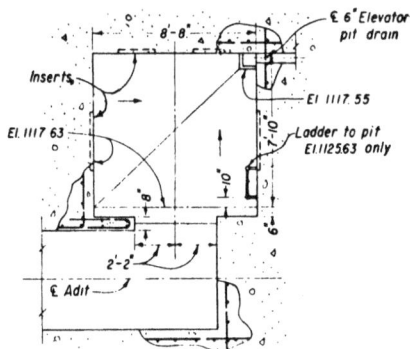

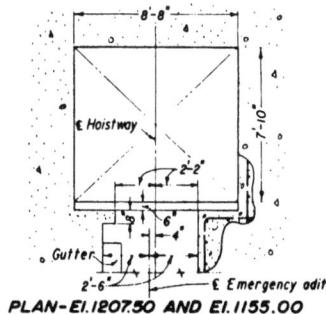

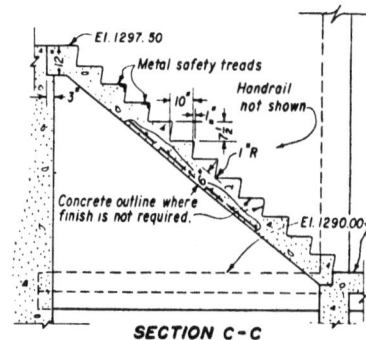

Figure 12-2. Structural layout of elevator shaft and tower in Grand Coulee Forebay Dam (sheet 2 of 2).—
288-D-3084 (2/2)

lift the gate or equipment, as well as the normal traffic loads should all be included in the design. Sidewalk and pedestrian bridge design loads should be a minimum live load of 85 pounds per square foot. Other considerations which should be covered in the design are camber, crown of roadway slab, storm drainage, and roadway lighting.

12-3. *Top of Dam.*—The top of the dam may contain a highway, maintenance road, or walkway depending upon the requirements at the site. If a roadway is to be built across the dam, the normal top of the dam can be widened by the use of cantilevers from the upstream and downstream faces of the dam. Operation and maintenance areas, and where conditions warrant visitors' parking, may also be provided on the top of the dam by further enlarging the cantilevers to the required size. The width of the roadway on the top of the dam is dependent upon the type and size of roadway, sidewalks, and maintenance and operation spaces that are needed to accomplish the tasks required. The minimum width for a two-lane roadway is 24 feet between curbs; however, the actual width should be established by the class of roadway crossing the dam. For highways, the roadway between curbs should be made the width required by the American Association of State Highway Officials or stipulated by local considerations. The sidewalks should be a minimum of 18 inches wide; however, the actual width should be determined by the proposed usage and the overall layout and space required for operation and maintenance. The top of Grand Coulee Forebay Dam, which contains a two-lane roadway, can be seen in figure 12-3.

When a highway is not to be taken across the dam, the top width should be established to meet the requirement for operation and maintenance. A width can be established which allows a truck to be taken out on the dam if operation requires it, or a walkway may be all that is needed for normal operation and maintenance. If only a walkway is required, the minimum width should be no less than the actual top width minus the width required for handrails and/or parapets. Widened areas for service decks can be constructed, where

required, to facilitate operation of outlet works, power outlets, and spillways.

Parapets or handrails are required both upstream and downstream on the top of the dam and should be designed not only to meet the safety requirements but also to blend into the architectural scheme. On dams where a large tourist traffic is expected, extreme care should be taken to assure the safety of the public. Therefore, the parapets should be of a height sufficient to keep anyone from falling over the side. The minimum height of parapet above the sidewalk should be 3 feet 6 inches; however, the minimum height may be more on some dams because of local conditions. When a handrail is used, chain-link fabric may be used to prevent a child falling or crawling between the rails. A solid upstream parapet may be used to increase the freeboard above the top of dam if additional height is needed.

Adequate drainage and lighting should be provided along the top of the dam. Service lines such as electricity, water, and air should also be provided as required. Crane rails may be embedded in the top of the dam if a gantry crane will be used for operation and maintenance (see fig. 12-3).

The design of the reinforcement for the top of the dam involves determining the amount of reinforcement required for the live and dead loadings on the roadway cantilevers and any temperature stresses which may develop. If a highway is to cross the dam, the cantilevers should be designed for a minimum AASHO loading of HS-20; however, special heavy loads which could occur during the construction period should also be investigated. Crane loads should also be included in the design criteria if a crane is to be used for operation and maintenance. A sidewalk live load of 85 pounds per square foot should be used in the design. Concrete parapets should be designed for a transverse force of 10,000 pounds spread over a longitudinal length of 5 feet; moreover, the parapets should be designed to withstand the appropriate waterload if the parapet is expected to create additional freeboard.

The temperature reinforcement requirement at the top of the dam is dependent upon the configuration and size of the area and the temperature conditions which may occur at the site. Many dams have a gallery or chamber below the roadway, which complicates the analysis and increases the amount of reinforcement needed to resist stresses caused by variations between the outside air temperature and the temperature within the opening in the dam. All temperature studies should be based on historic temperature data from that area and the temperatures occurring in galleries or chambers within the dam. After the temperature distributions are determined by studies, the temperature stresses that occur can be analyzed by the use of finite element methods.

12-4. *Fishways*.—The magnitude of the fishing industry in various localities has resulted in Federal, State, and local regulations controlling construction activities which interfere with the upstream migration and natural spawning of anadromous fish. All dams constructed on rivers subject to fish runs must be equipped with facilities enabling the adult fish to pass the obstruction on their way upstream, or other methods of fish conservation must be substituted. Since it is required that all facilities for fish protection designed by Federal agencies be approved by the U.S. Fish and Wildlife Service, this agency and similar State or local agencies should be consulted prior to the final design stage.

Low dams offer little difficulty in providing adequate means for handling fish. High dams, however, create difficulties not only in providing passage for adult fish on their way upstream, but also in providing safe passage for the young fish on their journey downstream. Fish ladders for high dams may require such length and size as to become impracticable. Large reservoirs created by high dams may cause flooding of the spawning areas. The velocity and turbulence of the flow over the spillway or the sudden change in pressure in passing through the outlet works may result in heavy mortality for the young fish. These difficulties often necessitate the substitution of artificial propagation of fish in lieu of installation of fishways.

Several types of fishways have been developed, the most common of which is the fish ladder. In its simplest form, it consists of an inclined flume in which vertical baffles are constructed to form a series of weirs and pools. The slope of the flume is usually 10 horizontal to 1 vertical. The difference in elevation of successive pools and the depth of water flowing over the weirs are made such that the fish are induced to swim rather than leap from pool to pool, thereby insuring that the fish will stay in the ladder for its entire length. The size of the structure is influenced by the size of the river, height of dam, size of fish, and magnitude of the run.

Another type of fishway in common use is the fish lock. This structure consists of a vertical water chamber, gate-controlled entrance and exit, and a system of valves for alternately filling and draining the chamber. Fish locks are usually provided with a horizontal screen which can be elevated, thereby forcing the fish to rise in the chamber to the exit elevation.

12-5. *Restrooms*.—Restrooms should be placed throughout a dam and its appurtenant works at convenient locations. The number required depends on the size of dam, ease of access from all locations, and the estimated amount of usage. At least one restroom should be provided at all dams for the use of operation and maintenance personnel. Separate restrooms should be provided for tourists at dams which may attract visitors. In larger dams, restrooms should be placed at convenient locations throughout the gallery system as well as in appurtenant structures such as elevator towers and gate houses.

12-6. *Service Installations*.—Various utilities, equipment, and services are required for the operation and maintenance of mechanical and electrical features of the dam, outlet works, spillway, and other appurtenant structures. Other utilities and services are required for the convenience of operating personnel and visitors. The amount and type of services to be provided will vary with the requirements imposed by the size, complexity, and function of the various appurtenant structures. The elaborateness of installations for personal convenience will depend on the size of the operating forces and the number of

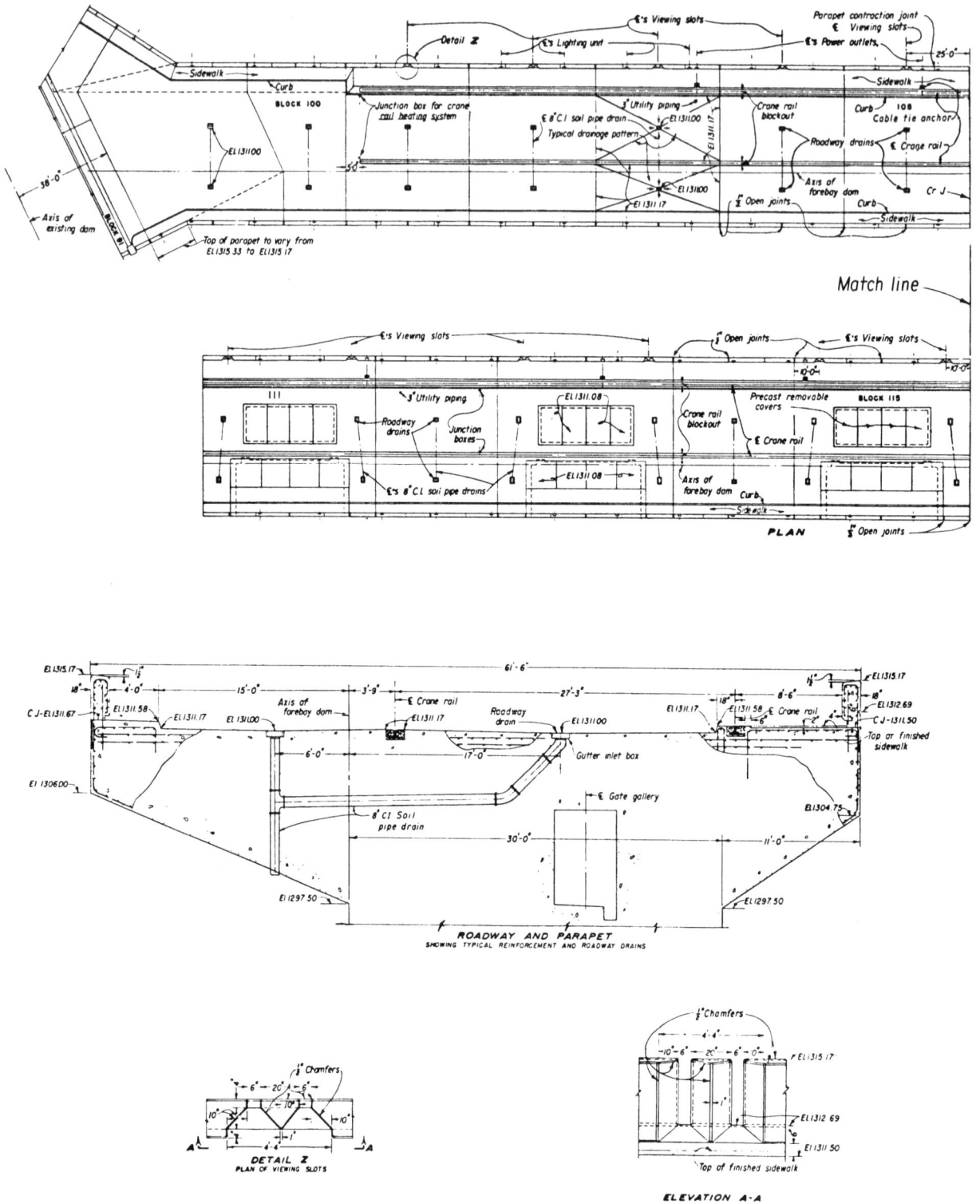

Figure 12-3. Typical arrangement at top of a gravity dam (Grand Coulee Forebay Dam) (sheet 1 of 2).—
285-D-3085 (1/2)

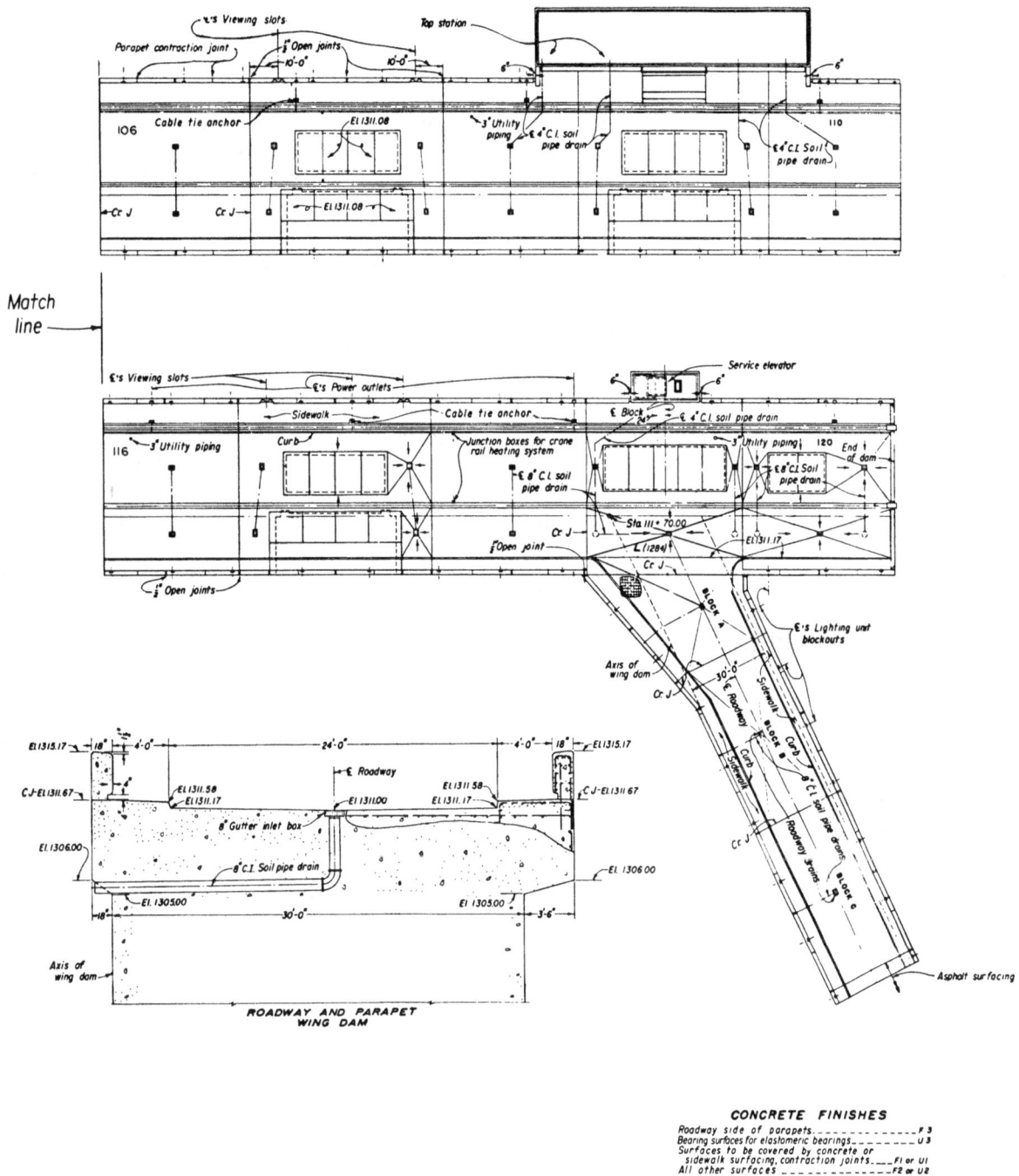

Figure 12-3. Typical arrangement at top of a gravity dam (Grand Coulee Forebay Dam) (sheet 2 of 2).—
285-D-3085 (2/2)

tourists attracted to the project.

(a) *Electrical Services.*—Electrical services to be installed include such features as the power supply lines to gate operating equipment, drainage pumps, elevators, crane hoists, and all lighting systems. Adequate lighting should be installed along the top of the dam, at all service and maintenance yards, and internally in the galleries, tunnels, and appurtenant structures. Power outlet receptacles should be provided at the top of the dam, in all appurtenant structures, throughout the gallery system and at any location which may require a power source.

(b) *Mechanical Services.*—Mechanical installations and equipment that may be required include such features as overhead traveling cranes in gate or valve houses, gantry cranes on top of the dam for gate operation and trashrack servicing, hoisting equipment for accessories located inside the dam, and the elevator equipment. Compressed air lines should be run into the gallery system, into service and maintenance chambers, into appurtenant structures, and anywhere else where compressed air could be utilized.

(c) *Other Service Installations.*—Chambers or recesses in the dam may be provided for the storage of bulkhead gates when these are not in use. Adequate storage areas should be provided throughout the dam such as in the gallery system, elevator towers, gate or valve house, and other appurtenant structures for maintenance and operation supplies and equipment. If gantry cranes are to be installed at a dam, recesses in the canyon walls may be provided for housing them when they are not in use. The gallery system and all appurtenant structures should be supplied with a heating and ventilating system where required.

A telephone or other communication system should be established at most concrete dams for use in emergency and for normal operation and maintenance communication. The complexity of the system will depend on the size of the dam, the size of the operating force, and the amount of mechanical control equipment. Telephones are usually placed throughout the gallery system for ease of access and safety in case of an emergency such as flooding or power failure. Telephones are also placed near mechanical equipment such as in gate or valve houses, elevator towers, machinery rooms, and other areas in which maintenance may be required. Telephones should also be placed at convenient locations along the top of the dam.

Water lines should be installed to provide a water source throughout the dam and the appurtenant structures. Water for operation and maintenance should be taken into the gallery system at the various levels of the galleries and into the appurtenant structures where required. The water for operation and maintenance can come from the river or reservoir but water for restrooms and drinking fountains requires a potable water source. Drinking fountains should be placed at convenient locations that are readily accessible to both maintenance personnel and tourist traffic.

Chapter XIII

Structural Behavior Measurements

13-1. *Scope and Purpose.*—Knowledge of the behavior of a concrete gravity dam and its foundation may be gained by studying the service action of the dam and the foundation, using measurements of an external and an internal nature. Of primary importance is the information by which a continuing assurance of the structural safety of the dam can be gaged. Of secondary importance is information on structural behavior and the properties of concrete that may be used to give added criteria for use in the design of future concrete gravity dams.

In order to determine the manner in which a dam and its foundation behave during the periods of construction, reservoir filling, and service operations, measurements are made on the structure and on the foundation to obtain actual values of behavior criteria in terms of strain, temperature, stress, deflection, and deformation of the foundation. Properties of the concrete from which the dam is constructed, such as temperature coefficient, modulus of elasticity, Poisson's ratio, and creep, are determined in the laboratory.

(a) *Development of Methods.*—The investigations of the behavior of concrete dams began at least 50 years ago, and have included scale model and prototype structures. Reports on the investigations are available in references [1], [2], and [3].[1] Along with the development of instruments [4] to use for measurements, and the instrumentation programs, there was the development of a suitable method for converting strain, as determined in the concrete which creeps under

load [5], to stresses that are caused by the measured deformation [6]. The basic method, which departs from simple Hooke's law relationships obtained for elastic materials, has been presented in reference [7] with later refinements presented in other publications [8, 9]. As analyses of the behavioral data from dams were completed, reports on the results of the investigations became available [10, 11].

Similarly, reports on the results of investigations of foundation behavior have become available [12, 13].

(b) *Two General Methods.*—At a major concrete dam, two general methods of measurement are used to gain the essential behavioral information, each method having a separate function in the overall program.

The first method of measurement involves several types of instruments that are embedded in the mass concrete of the structure and on features of the dam and appurtenances to the dam. Certain types of instruments are installed at the rock-concrete interfaces on the abutments and at the base of the dam for measuring deformation of the foundation. Others are installed on the steel liners of penstocks for measuring deformation from which stress is determined, and at the outer surface of the penstocks for measuring hydrostatic head near the conduit. This type of instrumentation may also be used with rock bolts in walls of underground openings such as a powerplant or tunnel and in reinforcement steel around penstocks and spillway openings to measure deformation from which stress is determined.

The second method involves several types of precise surveying measurements which are

---

[1] Numbers in brackets refer to items in the bibliography sec. 13-11.

made using targets on the downstream face of a dam, through galleries and vertical wells in a dam, in tunnels, on the abutments, and with targets on the top of a dam.

13-2. Planning.—From the modest programs for measurements provided at the earliest dams, there have evolved the extensive programs which are presently in operation in recently constructed Bureau of Reclamation dams. The formulation of programs for the installation of structural behavior instruments and measurement systems in dams has required careful and logical planning and coordination with the various phases of design and of construction.

Plans for a measurement program for a dam should be initiated at the time the feasibility plans are prepared for the structure. The layout should include both the embedded instrument system and systems for external measurements. Appropriate details must be included with those layouts to provide sufficient information for preparing a cost estimate of items needed for the program.

The information which a behavioral measurement system is to furnish is usually somewhat evident from the analytical design investigations which have been made for the dam and from a study of past experience with behavioral measurements at other dams. This information includes temperature, strain, stress, hydrostatic pressure, contraction joint behavior, deformation of foundation, and deformation of the structure, all as influenced by the loading which is imposed on the structure with respect to time.

The cost of a program is contingent on the size of structure, the number of segments which make up the program, the types of instruments to be used, and the number of instruments of the various types needed to obtain the desired information.

13-3. Measurement Systems.—Measurement systems, their layouts, and the locations and use of the various devices embedded in the mass concrete of dams for determining volumetric changes are discussed in the following sections. Measurement systems which employ surveying methods for determining

deformation changes in a dam are discussed separately.

The locations of the instruments to be installed in a gravity dam are shown on the plan, elevation, and section of figures 13-1 and 13-2.

(a) *Embedded Instrument Measurements.*— Embedded instruments in a concrete gravity dam usually consist of those which measure length change (strain), stress, contraction joint opening, temperature, concrete pore pressure, and foundation deformation. Instruments to measure stress may be installed at locations in reinforcing steel such as around a spillway opening or other opening in the dam and on the steel liners of penstocks. All instruments are connected through electrical cables to terminal boards located at appropriate reading stations in the gallery system of the dam. At those stations readings from the instruments are obtained by portable readout units. Mechanical-type deformation gages which utilize invar-type tapes, and a micrometer-type reading head may be installed vertically in cased wells which extend from the foundation gallery into the foundation to any desired depth. They may also be installed horizontally in tunnels in the abutments.

In a gravity dam such as shown on figures 13-1 and 13-2, the logical section for instrumentation is the maximum section where the greater stresses and deformations may be expected to occur. For investigation of the dam's behavior, instrumentation to determine temperature, stress, and deformation is required. Stress is investigated by clusters of strain measuring instruments in three-dimensional configuration, located at several positions on a horizontal gageline streamwise on the centerline of the maximum block near the base of the dam. For a structure of unusual size, similar installations are made along horizontal gagelines streamwise at intermediate elevations between the base and the top of the dam and at that same elevation in blocks near each abutment.

The instruments are installed at several locations along each gageline in clusters of 12 instruments each, designated as groups, for

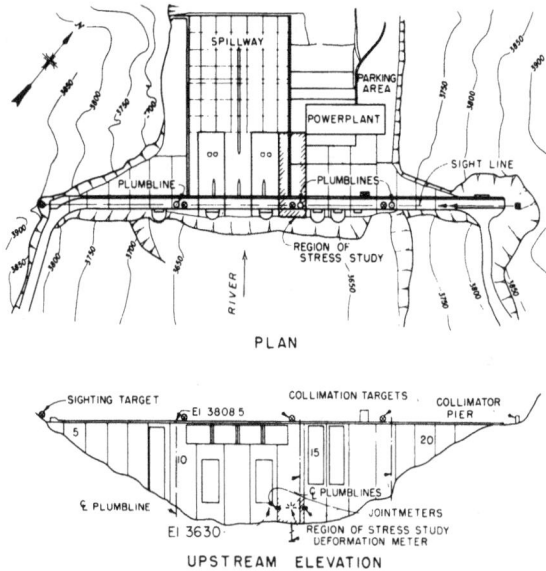

Figure 13-1. Locations of instrumentation installed in a gravity dam—plan and elevation.—288-D-3087

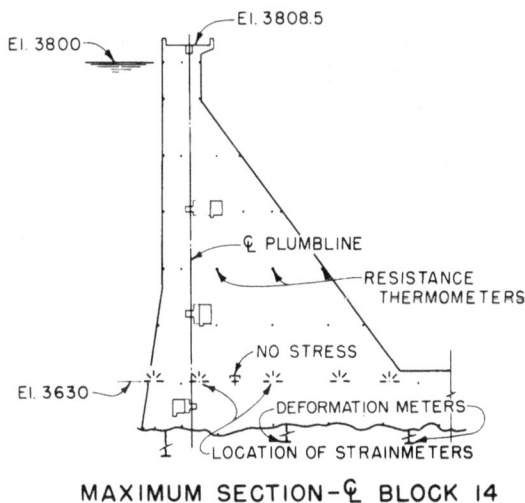

Figure 13-2. Locations of instrumentation installed in a gravity dam—maximum section.—288-D-3089

determination of multidimensional stress at the cluster locations. From these configurations, stress distribution normal to vertical and to horizontal planes at the gagelines may be determined as well as shear stresses and principal stresses. Duplicate instruments are installed on the three major orthogonal axes in each cluster. Eleven instruments of each cluster are supported by a holding device or spider. The twelfth instrument is placed vertically beside the cluster. Clusters are located along each gageline near each face and at midpoint between. An additional cluster usually is located between the interior cluster and the one near each face.

A pair of instruments, one vertical and one horizontal, placed in the concrete under a supported surface, is usually installed near the centrally located cluster of instruments. This pair of instruments is needed to determine stress-free behavior of the mass concrete. Instruments in various arrays may be installed near the faces or near contraction joints to determine conditions of special interest in the concrete, or in structural elements. Data are obtained from all instruments at frequent intervals so that time lapse variations of stress will be available for study during the entire period of observation, usually several years.

Instruments are installed across the contraction joints bounding the blocks containing the instrument clusters. These instruments provide a means of monitoring the behavior of the joints to determine the beginning and extent of joint opening due to cooling of the mass concrete. They serve as indicators of maximum joint opening to indicate when grouting should be performed. The instruments also give an indication of the effectiveness of grouting and show whether any movement in the joint occurs after grouting.

Several deformation measuring instruments are installed at selected locations in the foundation below the concrete of the maximum section and other sections of a gravity dam.

A pattern of temperature-sensing devices is included in the maximum section of the dam. In a structure of unusual size, similar installations could be made in additional sections when deemed desirable to determine the manner in which heat of hydration from the mass concrete is generated and dissipated. These instruments should be located on gagelines at several elevations in a section. They are not located near the instrument clusters, as the stress instrumentation also senses temperature.

An installation of instruments, when required for investigation of stress in the steel liner of a penstock, consists of instruments attached circumferentially to the penstock by supporting brackets. Instruments to detect pore pressure are placed on the outer surface of the penstock when it is embedded in concrete or when it extends through rock. The instruments are connected by electrical cables to terminal boards at appropriately located reading stations.

Instruments for measuring stress are sometimes installed in reinforcing steel which surrounds openings through a dam such as penstocks, spillway openings, or galleries. Similarly, instruments may be installed in rock bolts used to stabilize rock masses. Temperature-sensing devices installed on a grid pattern in the maximum section or in several sections of a dam have been used to determine the distribution of temperature. This is of great importance because the volume change caused by temperature fluctuation is one of the factors which contributes significantly to stress and deflection. Temperature-sensing devices are also used for control in the cooling operations. Another extensive use of these devices has been the development of concrete temperature histories to study the heat of hydration which is generated and dissipated, and to evaluate conditions which contribute to or accompany the formation of thermal cracking in mass concrete.

Concrete surface temperatures of dams are obtained by temperature-sensing devices embedded at various random locations on the downstream faces and embedded at uniform vertical intervals between the base and crest on the upstream faces. The latter installations furnish information on temperature due to the thermal variations in the reservoir.

Measurements of strain obtained by extensometers used with appropriate gage point anchors have been made on the faces of dams and on gallery walls. These furnish records of change in strain due to change in surface stress. Similar measurements which have been made across contraction joints and across cracks in concrete have furnished records of the joint or crack opening or closing as variations occur with time.

(b) *Deformation Measurements.*—The deformation measuring systems for a dam usually contain provisions for determining horizontal structural deformation between its base and top elevation. An additional system is needed to determine horizontal deformations with respect to references located on the abutments. Both systems employ methods of surveying to obtain the required information. Their locations are shown on figure 13-1.

Plumblines are installed in gravity dams to determine horizontal deformation of the structure which occurs between its top and the base. They are located in vertical wells usually formed in the maximum section and in sections about midway between the abutments and the maximum section. Each plumbline consists of a wire with a weight hung on it at the lowest accessible elevation, or the wire is anchored at the bottom of the well and suspended by a float in a tank of liquid at the top. Access to the plumblines for measurements is from stations at the several elevations where galleries are located in the dam. Figure 13-3 shows the layout of a typical plumbline well with reading stations at several elevations.

Horizontal deformation of the structure which occurs at its top elevation with respect to off-dam reference stations is determined by collimation measurements normal to the axis at several locations. These measurements are made between the stations at the top of the dam and sight lines between the off-dam reference stations. The measurements are made using a movable reference on the dam, the on-line position being indicated by an operator with a sighting instrument at one off-dam reference station. The horizontal deformation is obtained from differences between successive measurements.

To determine vertical deformations of the structure, a line of leveling across the top of the dam is used. Stations for measurements are located on several blocks. The leveling should begin and end at locations sufficiently distant from the dam to avoid locations which would be materially affected by vertical displacement of the dam.

Similarly, leveling measurements are made in other locations such as in powerplants and on gate structures to detect settlement or tipping of large machine units and appurtenant

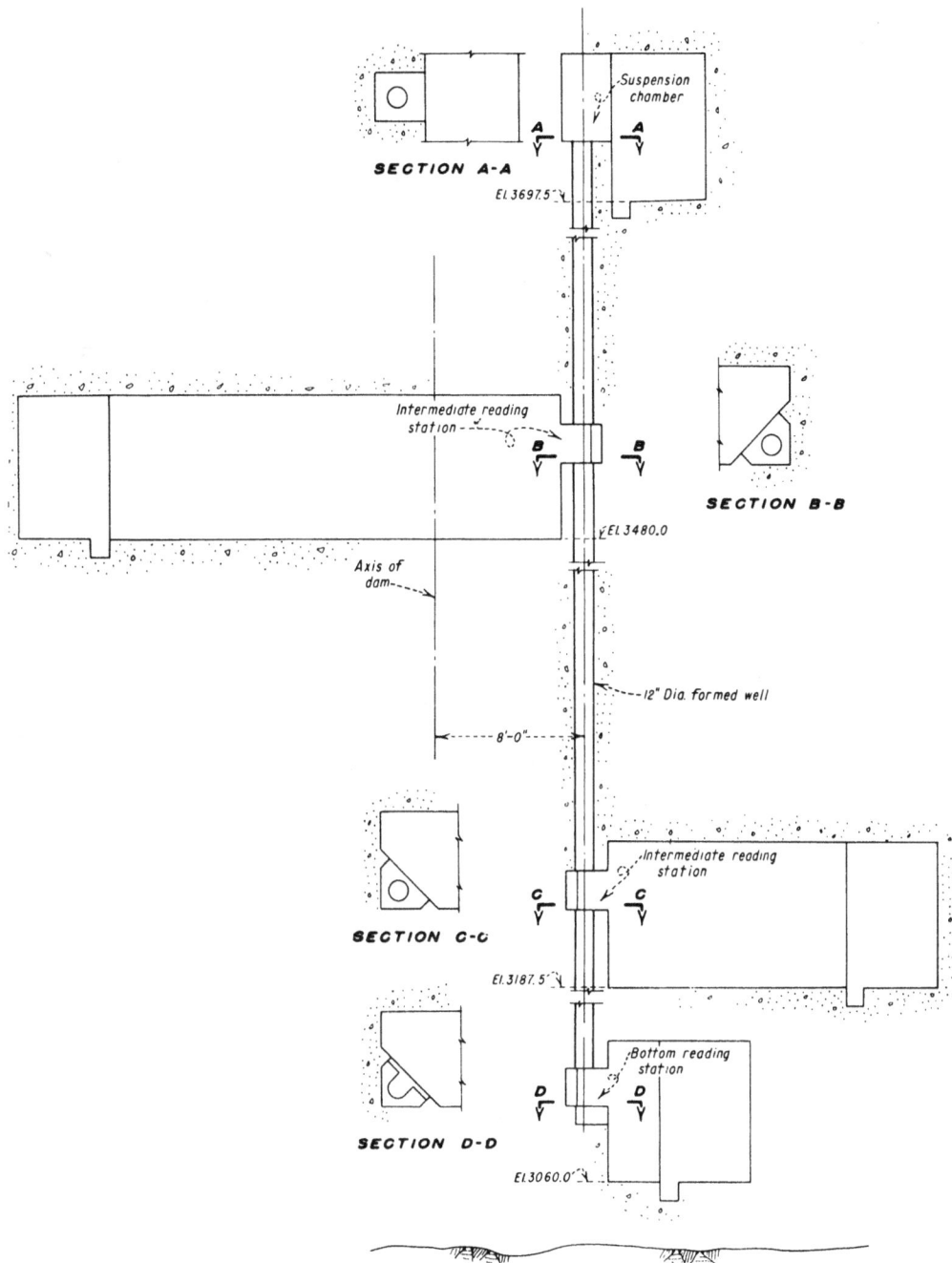

Figure 13-3. Typical plumbline well in a concrete dam with reading stations at several elevations.—288-D-3090

features of a dam.

13-4. *Embedded Instrumentation.*—The instruments to be used for the embedded measurements in a concrete dam may be selected from several types presently available on the commercial market. One type, the Carlson elastic wire instrument [14], is available in patterns suited to most purposes. They are dual-purpose instruments and measure temperature as well as the function for which designed. These instruments have proved reliable and stable for measurements which extend over long periods of time. Installations have been made in many Bureau of Reclamation dams and experience with the instruments covers a period of many years. The description of the instruments, their operation, and the manner in which they have been installed appear in other publications [4, 15, 16]. Foreign made instruments have been used occasionally, as they were more applicable to particular installations than the Carlson-type instruments. Satisfactory results have been furnished by those instruments.

For locations where only temperature measurements which are a part of the behavior program are desired, resistance thermometers are used. Temperature measurements of a special nature and of short duration such as for concrete cooling operations are made with thermocouples.

The instruments which are used for determining stress in a gravity or other type concrete dam are strain meters in groups of 12. Eleven strain meters are supported by a framework, or "spider," and installed in a cluster as shown on figure 13-4. The twelfth strain meter is placed vertically adjacent to the cluster.

Stress meters as shown on figure 13-5 are used for some special applications such as determining vertical stress at the base of the maximum section for comparison and to check

Figure 13-5. A stress meter partially embedded in concrete.—PX-D-74011

results from strain meters. Contraction joint openings are measured by joint meters as shown on figure 13-6. Temperatures are measured by resistance thermometers, and foundation deformation is measured by a special joint meter which has a range of movement greater than that of the joint meter used on a contraction joint. Investigation of hydrostatic pressure is made by means of pore pressure meters.

The meters are terminated through electrical cables which connect the instruments to terminal boards as shown on figure 13-7, located at appropriate reading stations in the system of galleries throughout the dam. At each station, readings from the instruments are obtained with special type portable wheatstone bridge test sets shown on figure 13-8.

Mechanical deformation gages which utilize an invar tape and a micrometer reading head may be installed vertically in each of several cased wells which extend from the foundation gallery to distances of 30, 60, 90 feet, or more below the base of the dam, usually at locations in the maximum section. The locations of the instruments are shown on figures 13-1 and 13-2.

Groups of strain meters in multidimensional configuration as shown on figure 13-4 are embedded in the mass concrete on the gagelines through the dam as shown on figure 13-1 to measure volume changes from which the stresses can be computed. The strain meters also measure temperature. The gagelines of strain meter groups usually are identical to the

Figure 13-4. A cluster of strain meters supported on a "spider" and ready for embedment in concrete.—P557-420-05870

Figure 13-6. A joint meter in position at a contraction joint.—P591-421-3321

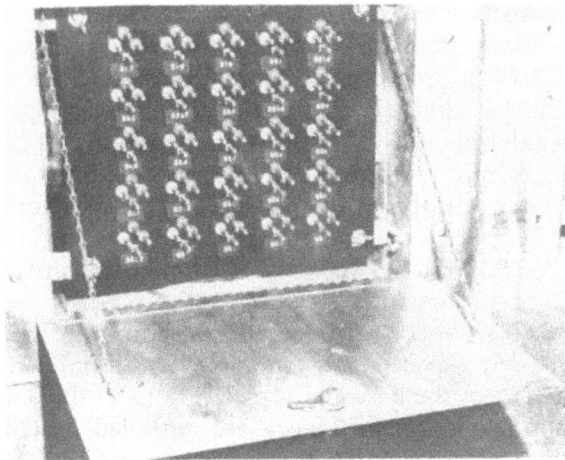

Figure 13-7. An instrument terminal board and cover box.—PX-D-74012

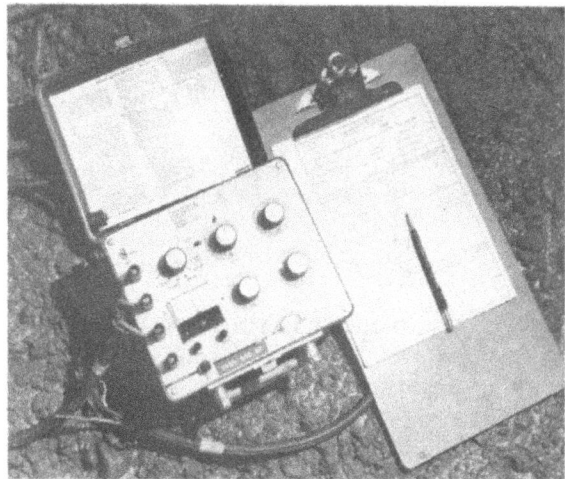

Figure 13-8. A special portable wheatstone bridge test set for reading strain meters.—C-8343-2

centerlines of the construction blocks. In the maximum section of a dam, gagelines of meter groups in addition to the gageline at the block centerline may be installed near each contraction joint at the elevation of the meter groups on the block centerline. The three gagelines of meter groups permit more extensive determination of stress distributions within the block than those resulting from a single gageline.

Vertical and horizontal stresses are determined at the base of the maximum section where maximum cantilever stresses may be expected. Vertical and horizontal stresses are also determined at other locations in the dam.

Data regarding the volume changes in the concrete that take place in the absence of loading are required for analysis of stress. "No-stress" strain meters as shown on figure 13-9(a) are installed to supply this information. A pair of "no-stress" strain meters are installed near each gageline of strain meter groups on the block centerline. These strain meters are installed in a truncated cone of mass concrete as shown on figure 13-9(b) under a free surface at the interior of the dam so that the instruments are not affected by vertical or horizontal loads.

In some instrumentation layouts, stress meters may be installed companion to each strain meter of a selected strain meter group as shown on figure 13-10. Strain meters in groups indicate length changes which are used to compute structural stresses. The stress meters indicate stress conditions from which stresses are obtained with only a minimum of computation. These serve as a check on results from the strain meters.

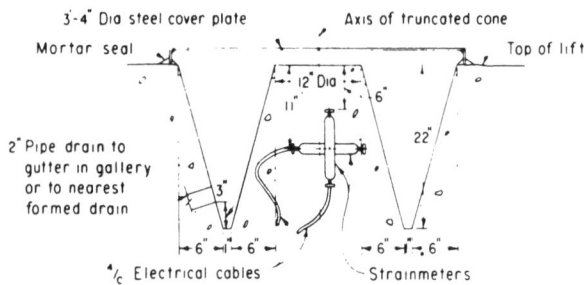

(a) STRAIN METER LAYOUT.—288-D-3091

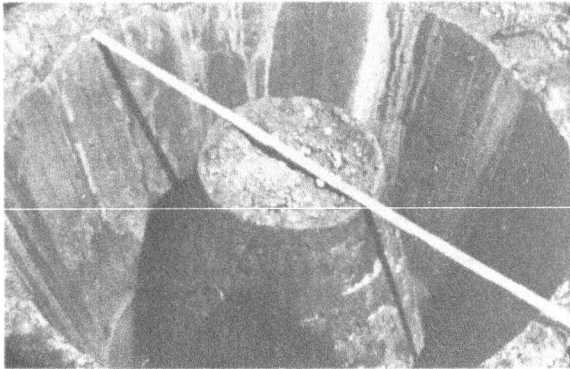

(b) TRUNCATED CONE OF MASS CONCRETE CONTAINING STRAIN METERS.—P622-427-3434NA

Figure 13-9. "No-stress" strain meter installation.

Trios of mutually perpendicular strain meters are sometimes installed as shown on figure 13-11 near the upstream and the downstream face of a dam to determine strain gradients near the surfaces. The trios of strain meters are located on the gagelines of strain meter groups, and are installed at distances of 2, 4, and 6 feet from the upstream and downstream faces of a dam.

Gagelines of strain meter groups may be installed near large openings which extend through a dam such as a spillway. Each gageline usually contains two meter groups from which the stress distribution near those openings is determined.

In conjunction with the installations of strain meter groups and stress meter arrays at the various locations throughout a dam, joint meters as shown on figure 13-6 are placed on the radial contraction joints at the same elevations as the groups of strain meters and the stress meters.

In designs where stress in reinforcement steel is to be investigated, reinforcement meters as shown on figure 13-12 are installed in the reinforcement placed around a penstock, spillway, or other opening. The instruments are placed on at least one bar of each row of reinforcement at selected locations around the opening to measure deformation in the reinforcement from which stress is determined.

Where stress in the steel liners of penstocks is to be investigated, strain meters are attached to the outer surface of a penstock by supporting brackets also shown on figure 13-12. The instruments are installed at each of three equally spaced circumferential locations and at two or more elevations on a steel liner.

At each location of a penstock strain meter installation, pore pressure meters shown on figure 13-13 are installed at the outer surface of the steel liner to measure possible hydrostatic pressure which may develop between the liner and the surrounding concrete. The pore pressure meters are particularly useful in cases where backfill concrete is placed around a penstock in a tunnel.

Pore pressure meters as shown on figure 13-14 are sometimes placed at several locations at the same elevation in the concrete on the centerline of a block near the base of the maximum section to measure hydrostatic pressure in pores of the concrete if it develops. The meters usually are spaced 1, 3, 6, 10, 15, 20, 30, and 40 to 50 feet from the upstream face of the dam.

Resistance thermometers as shown on figure 13-15, spaced at equal vertical intervals between the base and top elevations of the dam at the upstream face, are installed in the maximum section to record reservoir water temperature at various depths. Resistance thermometers usually are installed at two elevations at the downstream face of a dam in the maximum section to record temperature of the concrete caused by solar heat.

At several general locations between the foundation and the mass concrete at the base of a dam, deformation meters as shown on figure 13-16 are installed. These meters employ a joint meter as the measuring device and are

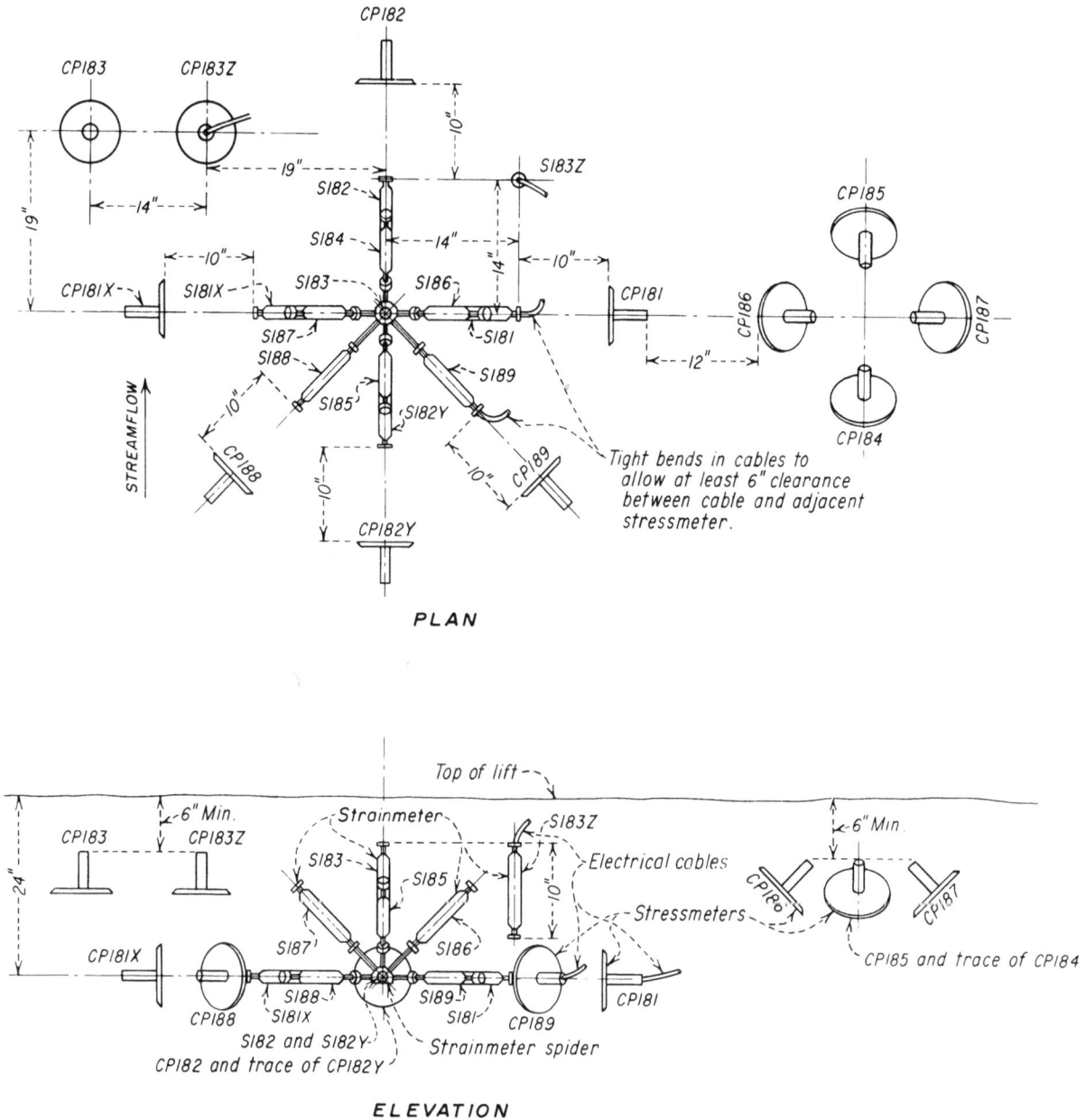

Figure 13-10. Meter group comprising strain meters and stress meters.—288-D-3092

installed in cased holes to detect deformation of the foundation rock, usually over depths of 30 to 90 feet below the rock-concrete contact surface.[2]

Ordinarily, two deformation meters are installed between the upstream and downstream boundaries of the blocks. In areas such as beside foundation and other galleries in the base of the dam, where access is available at a blockout on a gallery wall or floor location, a mechanical-type deformation gage is installed in place of a deformation meter.

The deformation gages, which utilize invar-type tapes and micrometer-type reading heads as shown on figure 13-17, are installed vertically in cased wells in the base of the dam

---

[2] Depths of 200 feet are planned for the deformation meters to be installed at Crystal Dam, currently under construction in Colorado (1973).

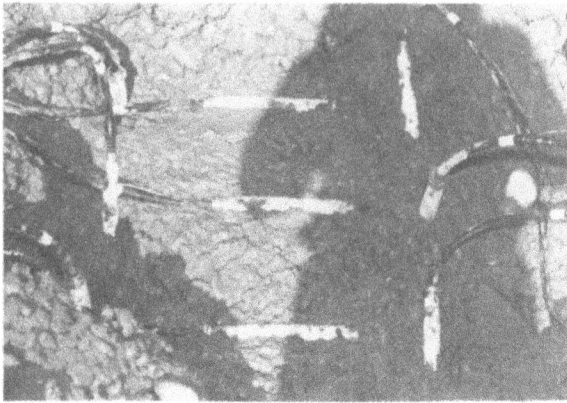

Figure 13-11. Trios of mutually perpendicular strain meters installed near face of dam.—P557-420-7933

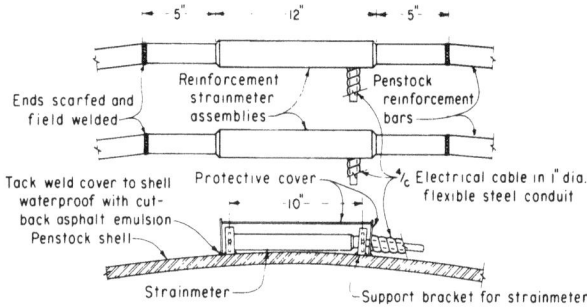

Figure 13-12. Penstock and reinforcement strain meters.—288-D-3093

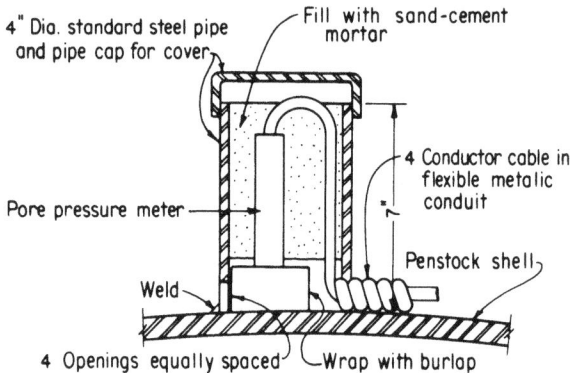

Figure 13-14. Pore pressure meters installed in mass concrete.—HH2653

Figure 13-13. Pore pressure meter installed on a penstock.—288-D-3094

Figure 13-15. Resistance thermometer installed at upstream face of a dam.—3PXI 3/10/71-3

in the maximum section. These gages extend 30 to 90 feet below the surface of contact between the rock and the concrete from appropriate reading stations in the foundation gallery. The gages show length change over their depths into the rock in the same manner as the deformation meter shows the amount of vertical compressive deformation caused by the

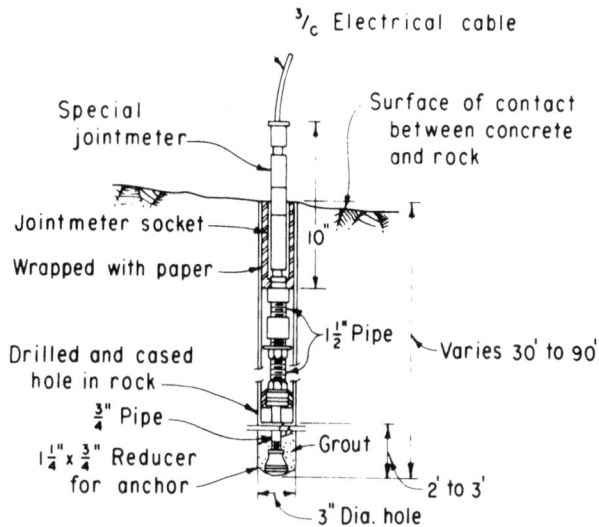

Figure 13-16. Deformation meter installed in cased well under dam to measure deformation of foundation rock.—288-D-3095

weight of the dam and by the loading on the dam.

Similar deformation gages may be installed horizontally in tunnels which have been excavated into the abutment formations of a dam. Figure 13-18 shows the micrometer-type reading head of one portion of a horizontal installation which is comprised of several 100-foot sections.

The strain, stress, pore pressure, foundation deformation, and reinforcement meters, and the resistance thermometers embedded in the mass concrete of a dam will furnish data over a long period of time for determining the stress behavior of the structure and conditions of stress which develop in features that have been instrumented. The joint meters detect the amount of contraction joint opening for information during joint grouting.

All of the above-mentioned instruments except the deformation gages employ a wheatstone bridge measuring circuit, and the same portable resistance bridge as shown on figure 13-8 can be used in common with all instruments. Also, the same bridge is used for obtaining temperature from resistance thermometers.

Data supplied by the strain meters, stress meters, joint meters, pore pressure meters, reinforcement meters, and deformation meters

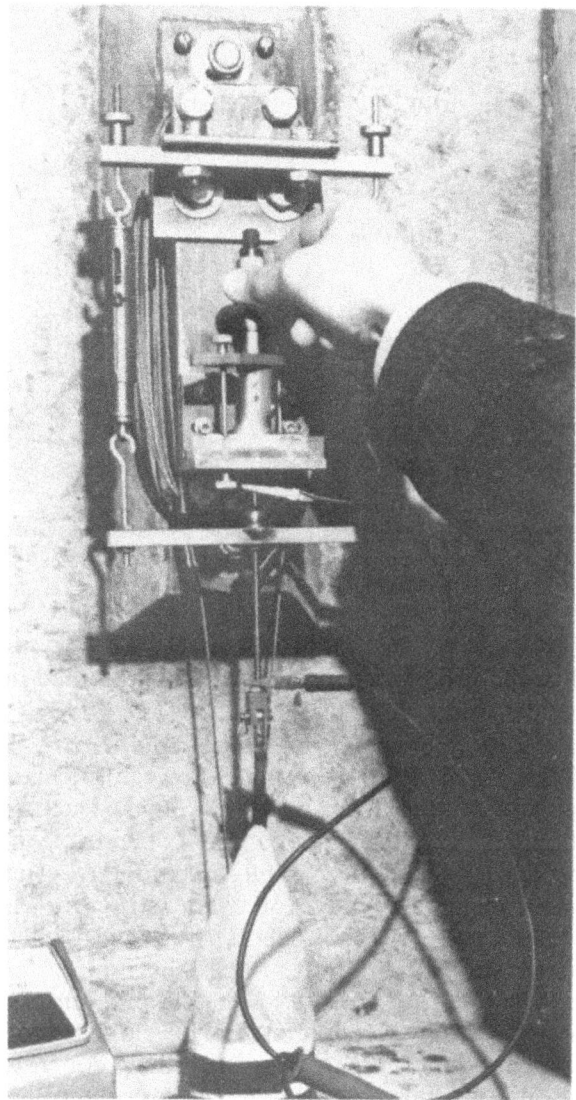

Figure 13-17. Micrometer-type reading head for use with foundation deformation gage.—P622B-427-3916NA

are in terms of total ohmic resistance and in terms of the ratio of the resistance of the two coils contained in the meter. Data supplied by the resistance thermometers are in terms of ohmic resistance of the coil of the thermometer. All data are recorded on appropriate data sheets. Computations of stress, temperature, hydrostatic pressure, joint opening, and foundation deformation are made from the field data by computer. Results from the computations are plotted as functions of time by an electronic plotter. Distributions of stress and temperature on gagelines of the

Figure 13-18. Micrometer reading head and invar tape used with horizontal tape gage in abutment tunnel.—P557-420-9328NA

Figure 13-19. Creep tests in progress on 18- by 36-inch mass concrete cylinders.—P557-D-34369

instruments are then prepared for various loading conditions on the dam and presented in report form.

13-5. *Supplementary Laboratory Tests.*—The determination of stress in the mass concrete of a dam or other large structure requires a knowledge of the concrete from which the structure is built. Accordingly, after the concrete mix for the structure is determined in the laboratory, and when practicable, prior to the beginning of construction at the site, a testing program for that specific concrete is developed and expedited. The results of the concrete properties and creep tests are an important part of a behavior program, as that information is needed for the solution of stress from the clusters of strain meters which are embedded in a dam.

The program includes creep tests, test cylinders for which are shown on figure 13-19, as well as the usual concrete strength tests and tests for determining elastic modulus, Poisson's ratio, thermal coefficient, autogeneous growth and drying shrinkage. All these tests are made on specimens which are fabricated in the laboratory and utilize materials from which the structure will be built. The materials, which are shipped to the laboratory from the damsite, are mixed in the same proportions as the mix for the structure, and cast into appropriate

cylinders. The cylinders are stored and tested under controlled environmental conditions. Reports are available on the methods of testing and on the creep tests (see references [17], [18], [19], [20], and [21]).

13-6. *Deformation Instrumentation.*—Of equal importance to the measurements made by embedded instruments are the measurements which are made with surveying instruments and by mechanical devices using precise surveying methods. These measurements involve plumblines, tangent line collimation, precise leveling, and triangulation deflection targets on the face of a dam. Over a period of several years, results from those measurements show the range of deformation of a structure during the cyclic loading conditions of temperature and water to which a dam is subjected.

Plumblines provide a convenient and relatively simple way to measure the manner in which a dam deforms due to the waterload and temperature change. In early Bureau of Reclamation dams where elevator shafts were provided in the structures, plumblines were sometimes contained in these shafts. This proved generally to be unsatisfactory and, at present, plumblines are suspended in vertically formed wells which extend from the top of the dam to near the foundation at three or more locations in the dam. Wherever feasible, reading stations are located at intermediate elevations, as well as at the lowest possible elevation to measure the deflected position of the section over the full height of the structure. A typical well with reading stations is shown on figure 13-3.

The wells are usually 1 foot in diameter and maintained to within one-half inch of plumb as the dam increases in elevation. In some dams, pipe or casing has been used and left in place for forming the well, while in other dams the wells have been formed with slip-forms. The reading stations on a plumbline are located at galleries in the dam. A doorframe is set in the concrete of the gallery wall at each reading station, and doors seating against sponge rubber seals are provided as closures. The doors of the reading stations are kept locked except when readings are being made, to prevent the plumbline being disturbed. Reading stations are oriented so that measurements may be made in the directions of anticipated movements, hence avoiding the need for trigonometric resolution. In the older dams orientation of the reading stations requires that measurements be made at 45° to the directions of dam movements, thus necessitating computation. Measurements of deformation are made with a micrometer slide device having either a peep sight or a microscope for viewing. The measured movements indicate deformation of the structure with respect to the plumbline.

Plumbline installations of two types have been used. These are the weighted plumbline and the float-suspended plumbline. For the weight-supporting plumbline the installation consists of a weight near the base of the dam suspended by a wire from near the top of the dam. The suspension is located in a manhole at the roadway or when practicable in a utility gallery near the top. The components of equipment for the installation are shown on figure 13-20. Recent plumbline installations are float suspended, using antifreeze in a tank at the top of the dam with a float holding the wire. Figure 13-21 shows one type of float and tank. When the lower end of the plumbline is at a gallery reading station, the wire is fixed at the bottom location. In other cases, where a pipe well is extended below the foundation, the wire is attached to a weight which is lowered into that well from the lowest reading

Figure 13-20. Components of equipment for weighted plumbline installation.—PX-D-74010

Figure 13-21. Tank and float for use with float-suspended plumbline.—C-8163-1NA

station. Figure 13-22 shows the weight and weight support. The support is attached to the plate which closes the lower end of the pipe well prior to lowering the pipe well into the hole in the foundation. Figure 13-23 shows a typical reading station and reading devices.

In conjunction with the plumbline installations in dams, reference monuments have been established in cased wells below the foundation near the base elevation of the plumblines. These are used to determine whether horizontal movement of the dam

(a) SUPPORT AND WEIGHT.

(b) WEIGHT RESTING ON SUPPORT, AND OTHER WEIGHTS.

Figure 13-22. Anchorage for float-suspended plumbline.–C-8170-2NA, C-8170-1NA

Figure 13-23. Typical plumbline reading station and reading devices.–P459-640-3593NA

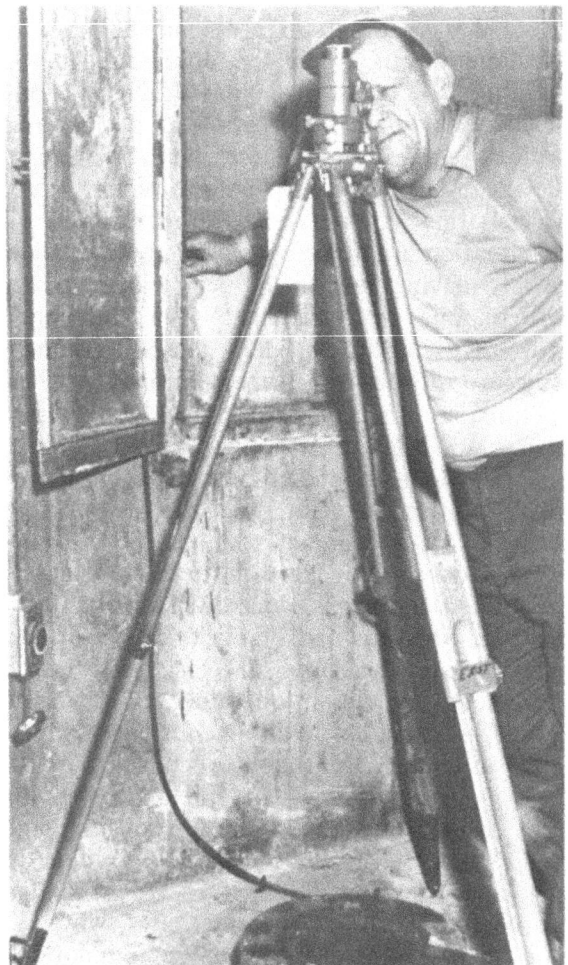

Figure 13-24. Foundation deformation well, optical plummet, and reference grid.–P459-640-4221

occurs. The locations of these monuments are periodically determined with respect to the top elevation of the well to determine whether movement at the elevation of the measurement location has occurred. Figure 13-24 shows the optical plummet and the reference grid that are used for measurements at the gallery elevation of a well which extends into the foundation of a dam.

Tangent line, or collimation measurements, are a useful means for determining the deformation of a dam at its top elevation with respect to off-dam references. This method has been used for measuring the deflection at the top of some Bureau of Reclamation gravity and arch dams. It is also used at major structures and is convenient for measurement at small dams that have no assigned survey personnel, since the survey can usually be conducted by two persons.

Collimation measurements are made with a theodolite or jig-transit. An instrument pier as shown on figure 13-25 is constructed on one reservoir bank on the axis and at a higher elevation than the dam. A reference target as shown on figure 13-26 is installed on the opposite reservoir bank on the axis and at about the same elevation as the pier. The target and pier locations are selected so that a sight line between them will be approximately on the axis or parallel to it at the location of a

Figure 13-26. A reference sighting target for use in obtaining collimation measurements.—P526-400-7852

movable measuring target as shown on figure 13-27 on the top of the dam. Progressive differences in the position of the movable target from the sight line indicate the deformation change in fractions of an inch at the measurement station. Usually three to four stations on a dam are sufficient to obtain the desired information. The results are correlated with plumbline measurements to provide sufficient data for charting the deformation behavior of the structure. A typical layout for a collimation system and locations of the items of equipment are shown on figure 13-28.

A more elaborate installation, and one that requires experienced and trained personnel, is that of triangulation targets on the face of a dam from off-dam references. Although the installation is better suited for an arch dam than for a gravity dam, a minimal installation is entirely satisfactory for gravity dam deformation measurements. The layout of targets on the face of the dam is made compatible with the layout of the embedded instruments. Targets are located on the gagelines of instruments, and on the locations of plumbline reading stations projected radially from the plane of the axis.

This system requires a net of instrument piers and a baseline downstream from the dam. The configuration is laid out to provide the greatest strength of the geometrical figures [22] and to afford sight lines to each target from as many instrument piers as is feasible. The nature of the terrain and the topography

Figure 13-25. An instrument pier for use with collimation or triangulation systems.—P526-400-7877

Figure 13-27. A movable collimation target at a measuring station on top of a dam.—PX-D-74009

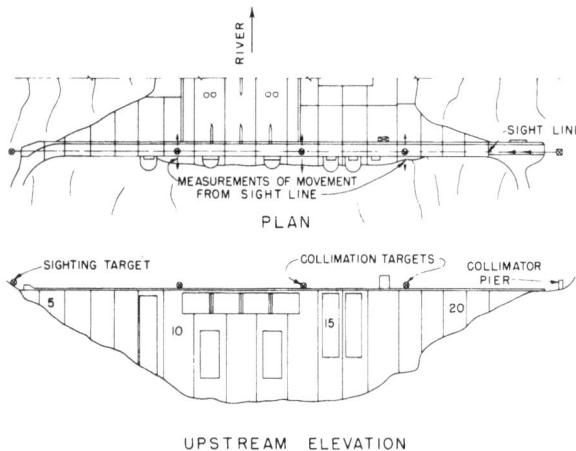

Figure 13-28. A collimation system layout for a gravity dam.—288-D-3097

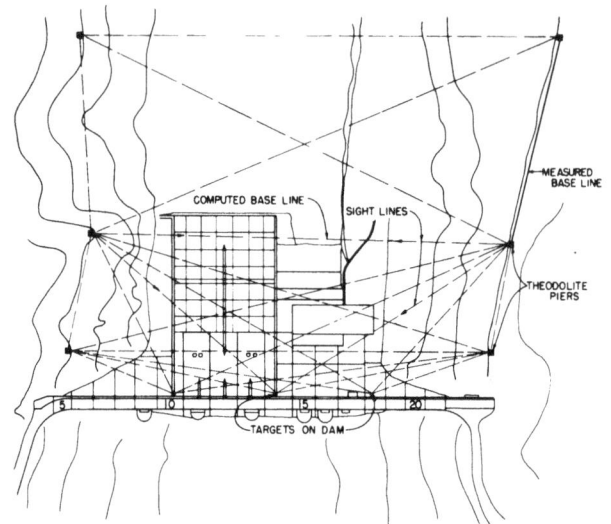

Figure 13-29. A triangulation system layout for a gravity dam.—288-D-3099

of the area are governing factors in the size of the net layout. The measurements are made using first-order equipment, methods, and procedures insofar as feasible. The results from these measurements show deformation of a dam with respect to off-dam references and deformation of the canyon downstream from a dam in the streamwise and cross-stream directions. The layout of a system and locations of items of equipment are shown on figure 13-29. Figure 13-25 shows a pier suitable

for theodolite stations. Figure 13-30 shows the tensioning device used with the tape for precise baseline measurements, and figure 13-31 shows targets used on the face of a dam and on the theodolite piers.

Leveling measurements are used to determine vertical displacements of a structure with respect to off-dam references. These measurements employ first-order equipment and procedures [23]. Base references for the measurements should be far enough from the dam to assure that they are unaffected by vertical displacement caused by the dam and reservoir.

Combinations of the several precise surveying-method measurements are included in the behavior measurement layouts for new dams. Except for plumbline and deformation well measurements, all are readily adaptable to older dams, should monitoring of behavior become desirable.

13-7. *Other Measurements.*—Under this general category may be included types of measurements which are related to and have an influence on the structural behavior measurements, but which are not included as a part of the program for those measurements. The measurements of primary interest are those of air temperature as recorded at an official weather station, air temperature as may be recorded at certain locations on a project,

Figure 13-30. A tensioning device used with a tape for precise baseline measurements.—P557-D-58714

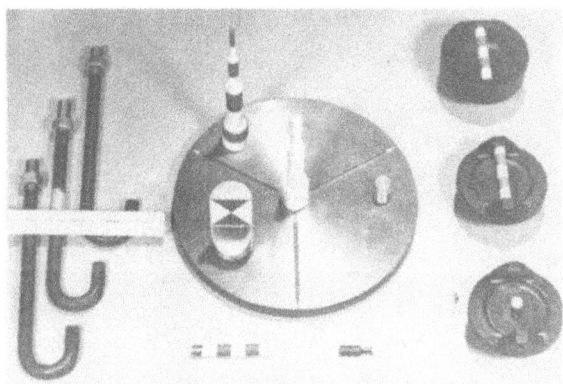

Figure 13-31. A pier plate, pier targets, and dam deformation targets.—P557-D-58717

river water temperature, concrete temperature during the construction of a dam, reservoir water elevation, uplift pressure under a dam, and flow of water from drainage systems. The latter two items are discussed more fully.

(a) *Uplift Pressure Measurements.*—A system of piping is installed in several blocks at the contact between the foundation rock and the concrete of a gravity dam as shown on figure 13-32. The piping is installed to determine whether any hydraulic underpressures may be present at the base of the dam due to percolation or seepage of water along underlying foundation seams or joint systems after filling the reservoir. Measured values of uplift pressure also may indicate the effectiveness of foundation grouting and of

drainage. The uplift pressure gradient through the section of a dam used for design is an assumed variation between the upstream and downstream faces of a dam as shown in reference [24] and in section 3-9 of this manual.

Uplift pressures are determined by pressure gages or by soundings. When a pipe is under pressure, the pressure is measured by a Bourdon-type pressure gage calibrated in feet, attached through a gage cock to the pipe. When zero pressure is indicated at a pipe, the water level in the pipe is determined by sounding.

Another system for measuring uplift pressure at the base of a structure where no galleries are included near the foundation into which a piping system may be routed is to install pore pressure cells at the locations to be investigated. Electrical cables may be routed from the cell locations to appropriate reading stations on the downstream face or top of the structure where measurements can be obtained. The installation of pore pressure cells is particularly applicable to installations beneath concrete apron slabs downstream from an overflow section of a dam, spillway training walls, and powerhouse structures. A typical pressure cell installation is shown on figure 13-33. Details on figure 13-34 show the manner in which contraction joints can be crossed by electrical instrument cable which is encased in electrical conduit.

(b) *Drainage Flow Measurements.*—A system of foundation drains is installed during the construction of a gravity dam. The drainage system usually consists of 3- or 4-inch-diameter pipes placed on approximately 10-foot centers in the axis direction, in the floors of the foundation gallery and foundation tunnels. Periodic measurements of flow from the individual drains should be made and recorded. When flows from drains are minimal, measurements may be made using any suitable container of known volume and noting the number of containers filled per minute. When flows are too great to measure by that method, measuring weirs may be installed as needed in the drainage gutters of the galleries and adits of a dam. Weirs should be located as required to measure flows from specific zones in a dam.

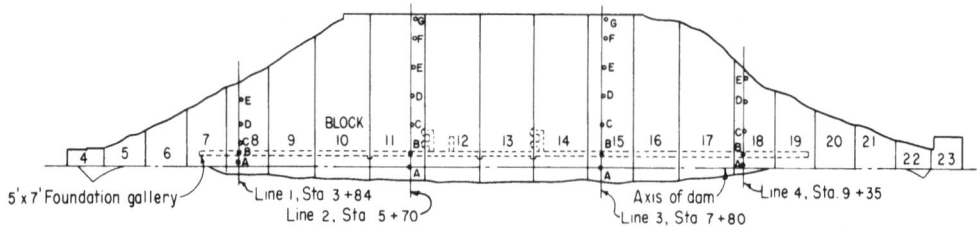

PLAN AT FOUNDATION GALLERY

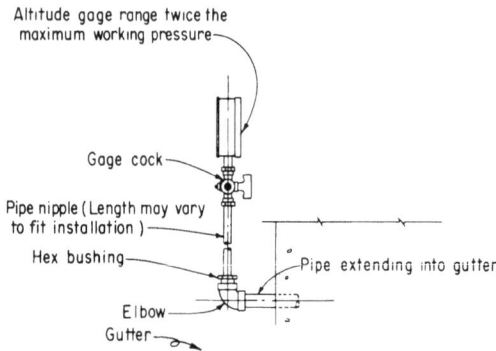

5'x7' Foundation gallery
Line I, Sta 3+84
Line 2, Sta 5+70
Line 3, Sta 7+80
Axis of dam
Line 4, Sta 9+35
BLOCK

Altitude gage range twice the maximum working pressure

Gage cock

Pipe nipple (Length may vary to fit installation)

Hex bushing

Pipe extending into gutter

Elbow

Gutter

TYPICAL ALTITUDE GAGE INSTALLATION

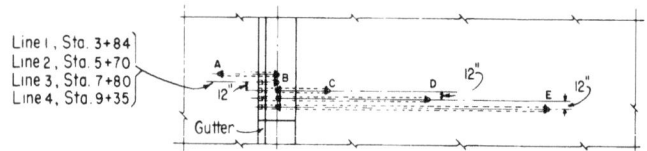

Line I, Sta. 3+84
Line 2, Sta. 5+70
Line 3, Sta. 7+80
Line 4, Sta. 9+35
Gutter

SECTION A-A
LINE I
LINE 2,3 AND 4 SIMILAR (EXCEPT FOR NUMBER OF PIPES AND SPACING)

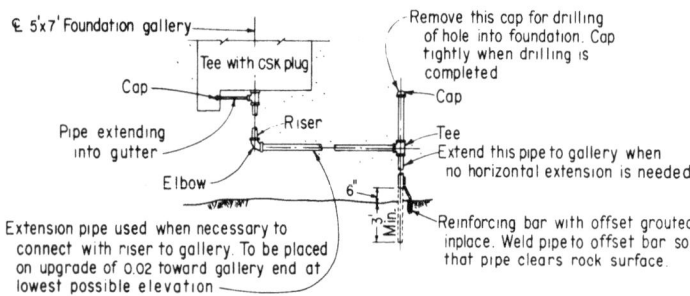

₵ 5'x7' Foundation gallery

Tee with CSK plug

Cap

Remove this cap for drilling of hole into foundation. Cap tightly when drilling is completed

Cap

Riser

Pipe extending into gutter

Tee

Elbow

Extend this pipe to gallery when no horizontal extension is needed

Extension pipe used when necessary to connect with riser to gallery. To be placed on upgrade of 0.02 toward gallery end at lowest possible elevation

Reinforcing bar with offset grouted in place. Weld pipe to offset bar so that pipe clears rock surface.

6"
3" Min.

TYPICAL PIPE DETAILS
ALL PIPES ARE 2½" STANDARD STEEL EXCEPT FOR PIPES TO GUTTER WHICH ARE ½". ALL FITTINGS ARE STANDARD MALLEABLE IRON

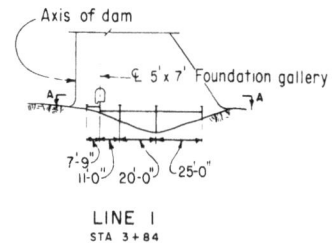

Axis of dam

₵ 5'x7' Foundation gallery

7'-9"
11'-0"
20'-0"
25'-0"

LINE I
STA 3+84

Figure 13-32. An uplift pressure measurement system for a gravity dam.—288-D-3101

When flows of drainage water are sufficient to be measured by weirs, the measurements are usually made on a monthly schedule and records maintained on appropriate data sheets. Any sudden increase or decrease in drainage should be noted and correlated with the reservoir water surface elevation and any change in the percolating conditions of the drains. All drains should be protected against obstructions and should be kept free-flowing.

13-8. *Measurement Program Management*.—The overall planning, execution, and control of a measurement program must be under the supervision of the central design office to expedite the various phases of the entire program. Control of the program starts with the installation of the various instruments and measurement systems during the construction period.

Cooperation between the central design office, the project construction office, the contractor's organization, and later with the operations and maintenance organization is important and necessary in obtaining reliable installations of instruments and reliable information from the various phases of the measurement program.

A schedule for installation of

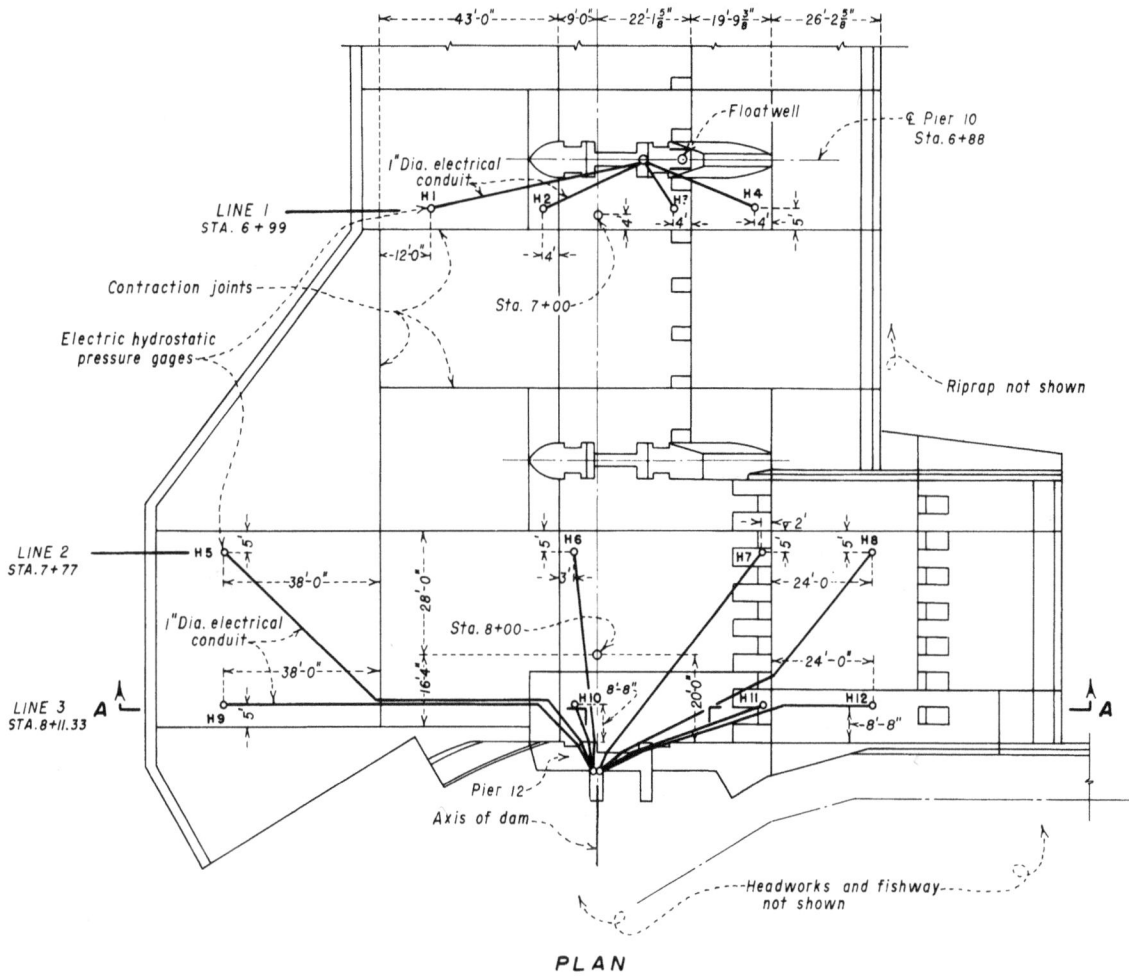

PLAN

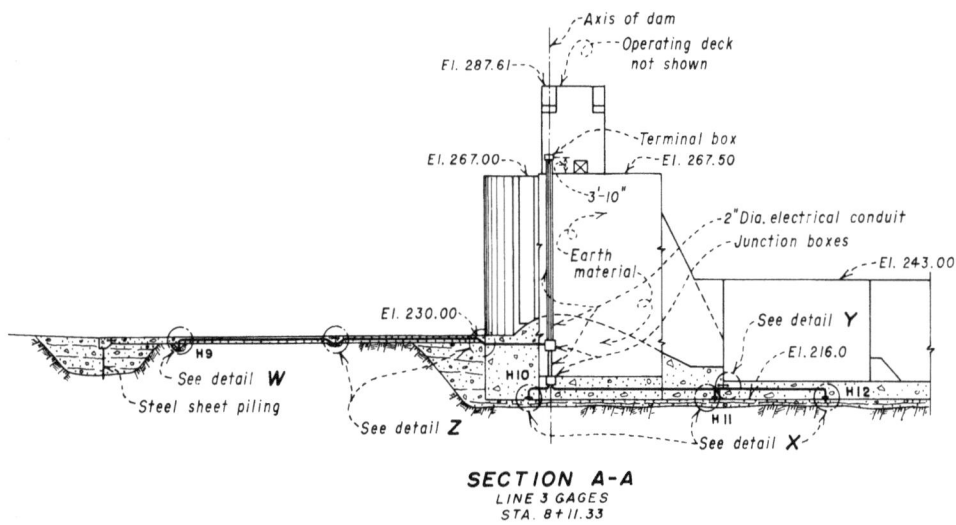

SECTION A-A
LINE 3 GAGES
STA. 8+11.33

Figure 13-33. A pore pressure meter installation for determining uplift pressure.—288-D-3102

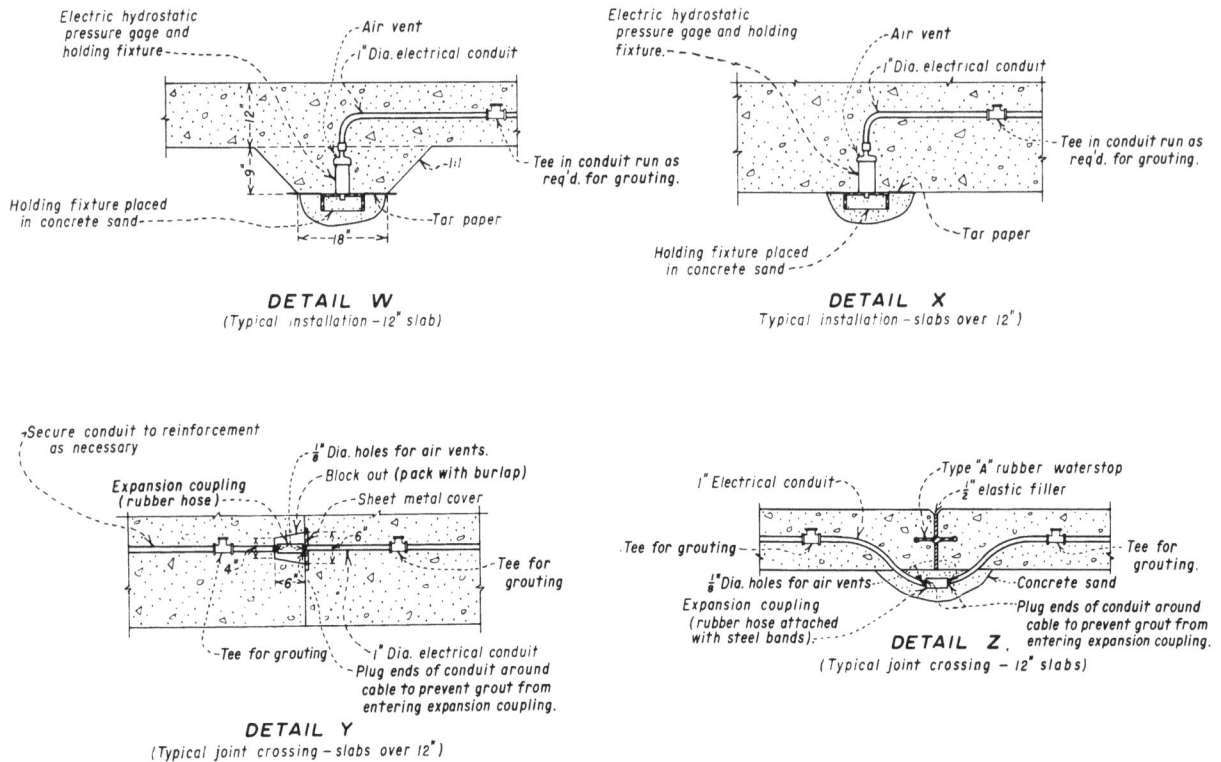

Figure 13-34. Details of pore pressure meter installation illustrated on figure 13-33.—288-D-3103

instrumentation and for obtaining readings at a new dam begins almost with the placement of the first bucket of concrete, continues through the construction period, and then extends into the operating stage, possibly for an undetermined period of time.

The information which is obtained is forwarded to the design office in the form of a report prepared at monthly intervals as explained in reference [25]. It includes all tabulations of instrument readings obtained during the prior month and other pertinent information, such as daily records of air and water temperature, reservoir and tailwater elevations when the operating period is reached, any other data which may have an effect on the structural action of a dam, and comments concerning the operation of instruments or measurement devices. Photographs and sketches should be used freely to convey information.

The schedules for obtaining data from structural behavior installations are somewhat varied. Embedded instrument readings are required more frequently immediately after embedment than at later periods. The reading frequency is usually weekly or every 10 days during construction and semimonthly after construction.

In some cases, instrument readings at monthly intervals can be allowed. Although the wider spread of intervals is not desirable for strain meters, it is satisfactory for stress meters, reinforcement meters, joint meters, pressure cells, and thermometers. During periods of reservoir filling or rapid drawdown, readings at more frequent intervals are preferred. In this case, schedules for readings may be accelerated for the periods of time involved.

Data from deflection measuring devices such as plumblines and collimation are preferred weekly. During events of special interest, such as a rapidly rising or falling reservoir, readings at closer intervals may be desired.

Data from uplift pressure measurement systems may be obtained monthly except during the initial filling of a reservoir when data are obtained at weekly or 10-day intervals.

Pore pressure gages may be read monthly. At dams where drain flow is of a sufficient quantity to be measure, these data should be obtained at monthly intervals.

The target deflection and pier net triangulation measurements should be conducted at least semiannually during the periods of minimum and maximum air temperature so as to obtain the furthest downstream and upstream deformed positions of a dam. During the early stages of reservoir filling and operation, additional measurements are desirable and are made approximately midway between those of minimum and maximum air temperature conditions. These latter data are useful in noting deformation trends and for correlating collimation and plumbline information.

Periodic leveling should be conducted in the vicinity of and across the top of a dam to detect possible vertical displacement of the structure.

The planned program for measurements should cover a time period which will include a full reservoir plus two cycles of reservoir operation, after which a major portion of the measurements are suspended. After the suspension of a major portion of readings, some types of measurements, such as those from plumblines, collimation, foundation deformation meters and gages, and from certain clusters of embedded meters which are considered essential for long-time structural surveillance, are continued indefinitely. For these measurements, the intervals between successive readings may be lengthened.

13-9. *Data Processing*.—The installations of instruments and measurement systems in dams and the associated gathering of quantities of data require that a program for processing be planned in advance. This requires definitely established schedules and adherence to the processing plan. Otherwise, seemingly endless masses of data can accumulate from behavior instrumentation and become overwhelming with no apparent end point in sight. Careful planning with provisions for the execution of such a program, possibly during a period of several years, cannot be too strongly emphasized.

For some measurements, computations and plots can be made and used to advantage by construction or operating personnel at a damsite. Under this category are data from resistance thermometers, joint meters, extensometers, Bourdon pressure gages, and the less complex systems for measuring deformation, such as collimation and plumblines. X-Y hand plotting of these data can be maintained with relative ease, as required.

For measurements from the other instruments such as strain meters, stress meters, and reinforcement meters, the obtaining of results is complex and time consuming.

The methods and details for computing which are used to reduce the instrument data to temperature, stress, and deflection are completely described in separate reports (see references [6], [8], [26], and [27]). The results of the Bureau's laboratory creep test program, which covers a period of more than 20 years, are described in references [5], [18], [28], and [29].

The processing of large masses of raw data is efficiently and economically handled by computer methods. The instrument and deformation data are processed in the Bureau's E&R Center in Denver. Processing of the majority of these data is presently done by punched cards, magnetic tape, and electronic computers, using programs of reference [26] for the computing which have been devised for the specific purposes. Plotted results are from output material which is fed into an electronic X-Y plotter. Reports are prepared from these results.

13-10. *Results*.—The interpretation of data and compiled results includes the careful examination of the measurements portion of the program as well as examination of other influencing effects, such as reservoir operation, air temperature, precipitation, drain flow and leakage around a structure, contraction joint grouting, concrete placement schedule, seasonal shutdown during construction, concrete testing data, and periodic instrument evaluations. All of these effects must be reviewed and, when applicable, fitted into the interpretation. The presentation of results, both tabular and graphical, must be simple, forceful, and readily understood.

The interpretation of the measurement results, as shown in references [27], [30], and [31], progresses along with the processing of the gathered data. Progress reports usually cover the findings which are noted during the periods of construction and initial reservoir filling stages for a dam. The resume of findings, as a final report, is usually not forthcoming until several years after completion of the structure, since the factors of a full reservoir, its seasonal operating cycle, the seasonal range of concrete temperature, and local effects of temperature on concrete are all time-governed.

13-11. *Bibliography*.

[1] "Arch Dam Investigation," vol. I, Engineering Foundation Committee on Arch Dam Investigation, ASCE, 1927.

[2] "Arch Dam Investigation," vol. II, Committee on Arch Dam Investigation, The Engineering Foundation, 1934.

[3] "Arch Dam Investigation," vol. III, Committee on Arch Dam Investigation, The Engineering Foundation, 1933.

[4] Raphael, J. M., and Carlson, R. H., "Measurement of Structural Action in Dams," James J. Gillick and Co., Berkeley, Calif., 1956.

[5] McHenry, Douglas, "A New Aspect of Creep in Concrete and its Application to Design," Proc. ASTM, vol. 43, pp. 1069-1087, 1943.

[6] Jones, Keith, "Calculations of Stress from Strain in Concrete," Engineering Monograph No. 25, Bureau of Reclamation, October 29, 1961.

[7] Raphael, J. M., "Determination of Stress from Measurements in Concrete Dams," Question No. 9, Report 54, Third Congress on Large Dams, ICOLD, Stockholm, Sweden, 1948.

[8] Roehm, L. H., and Jones, Keith, "Structural Behavior Analysis of Monticello Dam for the Period September 1955 to September 1963," Technical Memorandum No. 622, with Appendixes I and II, Bureau of Reclamation, September 1964.

[9] Carlson, R. W., "Manual for the Use of Stress Meters, Strain Meters, and Joint Meters in Mass Concrete," second edition, 1958, R. W. Carlson, 55 Maryland Avenue, Berkeley, Calif.

[10] Raphael, J. M., "The Development of Stress in Shasta Dam," Trans. ASCE, vol. 118, pp. 289-321, 1953.

[11] Copen, M. D., and Richardson, J. T., "Comparison of the Measured and the Computed Behavior of Monticello (Arch) Dam," Question No. 29, Report 5, 8th Congress on Large Dams, ICOLD, Edinburgh, Scotland, 1964.

[12] Rice, O. L., "In Situ Testing of Foundation and Abutment Rock on Large Dams," Question No. 28, Report 5, 8th Congress on Large Dams, ICOLD, Edinburgh, Scotland, 1964.

[13] Rouse, G. C., Richardson, J. T., and Misterek, D. L.,

"Measurement of Rock Deformations in Foundations on Mass Concrete Dams," ASTM Symposium, Instrumentation and Apparatus for Soil and Rock, 68th Annual Meeting, Purdue University, 1965.

[14] Technical Bulletin Series, Bulletins 16 through 23, Terrametrics Division of Earth Sciences, Teledyne Co., Golden, Colo., 1972.

[15] Technical Record of Design and Construction, "Glen Canyon Dam and Powerplant," Bureau of Reclamation, pp. 117-138 and 449-453, and p. 464, December 1970.

[16] Technical Record of Design and Construction, "Flaming Gorge Dam and Powerplant," Bureau of Reclamation, 1968.

[17] Hickey, K. B., "Effect of Stress Level on Creep and Creep Recovery of Lean Mass Concrete," Report REC-OCE-69-6, Bureau of Reclamation, December 1969.

[18] "A Loading System for Compressive Creep Studies on Concrete Cylinders," Concrete Laboratory Report No. C-1033, Bureau of Reclamation, June 1962.

[19] Best, C. H., Pirtz, D., and Polivka, M., "A Loading System for Creep Studies of Concrete," ASTM Bulletin No. 224, pp. 44-47, September 1957.

[20] "A 10-Year Study of Creep Properties of Concrete," Concrete Laboratory Report No. SP-38, Bureau of Reclamation, July 1953.

[21] Hickey, K. B., "Stress Studies of Carlson Stress Meters in Concrete," Report REC-ERC-71-19, Bureau of Reclamation, April 1971.

[22] "Manual of Geodetic Triangulation," Special Publication No. 247, Coast and Geodetic Survey, Department of Commerce, Washington, D.C., 1950.

[23] "Manual of Geodetic Leveling," Special Publication No. 239, Coast and Geodetic Survey, Department of Commerce, Washington, D.C., 1948.

[24] Design Criteria for Concrete Gravity and Arch Dams," Engineering Monograph No. 19, Bureau of Reclamation, p. 3, December 1960.

[25] Reclamation Instructions, Part 175, Reports of Construction and Structural Behavior (L-21 Report) Bureau of Reclamation, 1972.

[26] "Calculations of Deflections Obtained by Plumblines," Electronic Computer Description No. C-114, Bureau of Reclamation, 1961.

[27] Roehm, L. H., "Investigation of Temperature Stresses and Deflections in Flaming Gorge Dam," Technical Memorandum 667, Bureau of Reclamation, 1967.

[28] "Twenty-Year Creep Test Results on Shasta Dam Concrete," Laboratory Report No. C-805A, Bureau of Reclamation, February 1962.

[29] "Properties of Mass Concrete in Bureau of Reclamation Dams," Laboratory Report No. C-1009, Bureau of Reclamation, December 1961.

[30] Roehm, L. H., "Deformation Measurements of Flaming Gorge Dam," Proc. ASCE, Journal of the Surveying and Mapping Division, vol. 94, No. SU1, pp. 37-48, January 1968.

[31] Richardson, J. T., "Measured Deformation Behavior of Glen Canyon Dam," Proc. ASCE, Journal of the Surveying and Mapping Division, vol. 94, No. SU2, pp. 149-168, September 1968.

Chapter XIV

Concrete Construction

14-1. *General.*—Concrete control and concrete construction operations are of vital concern to the designer of a concrete structure. The ideal situation would be to have the engineer responsible for the design of a structure go to the site and personally supervise the construction to assure its intended performance. Since this is not practicable, it falls on the construction engineer and his inspection staff, the design engineer's closest contact with the work, to assure that the concrete meets the requirements of the design.

The safety of any structure is related to certain design criteria which include factors of safety. Only when all concrete control and construction operations are of high quality will the factors of safety be valid for the completed structure. Whereas steel used for structures can be tested for material requirements and structural properties, with the full knowledge that another piece of that same steel will react in the same manner, concrete is mixed and placed under varying conditions. Concrete is placed in the structure knowing what it has done in the past under similar circumstances. From experience, we know what concrete *can do*. Time alone will tell if it *will do this*. A high assurance that it *will* can be obtained by the concrete inspector by making certain that the concrete is mixed and placed, and the structure is completed, in full compliance with the specifications.

Appendix H covers those specifications paragraphs relating to concrete that are normally required for construction of concrete dams.

14-2. *Design Requirements.*—Basically, a concrete structure must be capable of performing its intended use for what may be an unknown but usually long period of time. To serve its purpose, the concrete in the structure must be of such strength and have such physical properties as are necessary to carry the design loads in a safe and efficient manner. The concrete throughout the structure must be of uniform quality because a structure is only as strong as its weakest part. The concrete must be durable and resistant to weathering, chemical attack, and erosion. The structure must be relatively free of surface and structural cracks. Because of increasing environmental demands, the final completed structure must be pleasing in appearance. And, last but not least, the construction processes and procedures should reflect an economical design and use of materials, manpower, and construction effort.

A number of the above design requirements are the responsibility of the designer. These include the determination of the configuration and dimensions of the structure, the sizes and positioning of reinforcing bars, and the finishes necessary to minimize erosion and cavitation on the surfaces of the structure. Additional design requirements are determined from field investigations of the site conditions, including such items as the type and condition of the foundation for the structure, and the availability of sand and coarse aggregates. Other design requirements may be obtained from concrete laboratory investigations on the concrete mix, from hydraulic laboratory model studies, and from environmental studies on the desired appearance of the structure. The fulfillment of all design requirements is dependent upon actual construction processes and procedures. A continuing effort must therefore be made by all inspection personnel to assure the satisfactory construction of the desired structure.

Aggregates for use in concrete should be of good quality and reasonably well graded. Usually, an aggregate source has been selected and tested during preconstruction investigations. Also, in some cases, concrete mix design studies have been made as part of the preconstruction investigations using the aggregates from the deposit concerned. When good quality natural sand and coarse aggregate are available, use of crushed sand and/or coarse aggregate is generally limited to that needed to make up deficiencies in the natural materials, as crushing generally increases the cost of the aggregates and resulting concrete. In these instances, crushing is usually restricted to crushing of oversize materials and/or the excess of any of the individual sizes of coarse aggregate. Where little or no natural coarse aggregate is available in a deposit, it may be necessary to use crushed coarse aggregate from a good quality quarry rock.

14-3. *Composition of Concrete.*—The concrete for a specific concrete structure is proportioned to obtain a given strength and durability. Concrete with a higher strength than required could be designed by adding more cement, and perhaps admixtures, but this higher strength concrete is not desirable from the standpoint of economy of design. In cold climates, where frequent cycles of freezing and thawing often occur, it may be advantageous to use a special mix for the face concrete of the dam to assure adequate durability. A higher cement content and lower water to portland cement ratio, or when the mix contains pozzolan a lower water to portland cement plus pozzolan ratio, is often used in these outer portions of the dam. On the larger and more important Bureau of Reclamation structures, trial mixes are made in the laboratories at the Engineering and Research Center not only to obtain an economical and workable mix but to assure that the required strength and durability can be obtained with the cement and aggregates proposed for the construction.

Adjustments in the field are sometimes necessary to obtain a workable mix. These may be occasioned by variations in aggregate characteristics within the deposit being worked, or by a change in the characteristics of the cement being used.

The amount of cement to be used per cubic yard of concrete is determined by mix investigations which are primarily directed toward obtaining the desired strength and durability of the concrete. The type of cement, however, may be determined by other design considerations.

Considerable bad experience has been encountered where alkali reactive aggregates are used in concrete. Where field and laboratory investigations of aggregate sources indicate that alkali reactive aggregates will be encountered, a low-alkali cement is normally required to protect against disruptive expansion of the concrete which may occur due to alkali-aggregate reaction (a chemical reaction between alkalies in the cement and the reactive aggregates). Another means of controlling alkali-reactive aggregates is by use of a suitable fly ash or natural pozzolan. If a highly reactive aggregate is to be used, it may be necessary to use both low-alkali cement and a pozzolan.

Another design consideration is the type of cement to be used. Type II cement is normally used by the Bureau of Reclamation in mass concrete dams. Limitations on the heat of hydration of this cement are specified when determined necessary to minimize cracking in the concrete structure. Use of a type II cement will generally reduce the heat of hydration to an acceptable level, particularly since type II cement is usually used in conjunction with other methods of heat reduction. These include use of lower cement contents, inclusion of a pozzolan as part of the cementitious material, use of a pipe cooling system, and use of a specified maximum placing temperature of the concrete, which may be as low as 50° F. Use of all or some of these methods will usually reduce or eliminate the need for stringent limitations on the heat of hydration of the cement. However, a limitation of 58 percent on the tricalcium aluminate plus tricalcium silicate ($C_3A + C_3S$) content of the type II cement may be required where heat of hydration of cement must be kept low. Further limitation on the heat of hydration, if more stringent control of heat is needed, can be obtained with a type II cement by providing a maximum limitation on the cement of 70 calories per gram at 7 days or 80 calories per gram at 28 days, or both.

If the above measures are insufficient, use of type IV cement, an extremely low heat of hydration cement, may be specified. This type of cement, referred to as low-heat cement, was developed many years ago for mass concrete when thick, very massive, high-cement-content concrete dams were being built. Maximum limitations on heat of hydration of type IV cement are 60 calories per gram at 7 days and 70 calories per gram at 28 days. The amount and type of cement used must be compatible with strength, durability, and temperature requirements.

Admixtures are incorporated into the mix design as needed to obtain economy, workability, or certain other desired objectives such as permitting placement over extended periods of time. Admixtures have varying effects on concretes, and should be employed only after a thorough evaluation of their effects. Most commonly used admixtures are accelerators; air-entraining agents; water-reducing, set-controlling admixtures (WRA); and pozzolans. Calcium chloride should not be used as an accelerator where aluminum or galvanized metalwork is embedded. When accelerators are used, added care will be necessary to prevent cold joints during concrete placing operations. Air-entraining agents should be used to increase the durability of the concrete, especially if the structure will be exposed to cycles of freezing and thawing. Use of a WRA will expedite the placing of concrete under difficult conditions, such as for large concrete placements in hot weather. Also, use of a WRA will aid in achieving economy by producing higher strengths with a given cement content.

Good quality pozzolans can be used as a replacement for cement in the concrete without sacrificing later-age strength. Pozzolan is generally less expensive than cement and will, as previously indicated, aid in reducing heat of hydration. Since the properties of pozzolans vary widely, if a pozzolan is to be used in a concrete dam it is necessary to obtain one that will not introduce adverse qualities into the concrete. Pozzolan, if used in face concrete of the dam, must provide adequate durability to the exposed surfaces. Concrete containing pozzolan requires thorough curing to assure good resistance to freezing and thawing.

The water used in the concrete mix should be reasonably free of silt, organic matter, alkali, salts, and other impurities. Water containing objectionable amounts of chlorides or sulfates is particularly undesirable, because these salts prevent the full development of the desired strength.

14-4. *Batching and Mixing*.—Inherently, concrete is not a homogeneous material. An approach to a "homogeneous" concrete is made by careful and constant control of batching and mixing operations which will result in a concrete of uniform quality throughout the structure. Because of its effect on strength, the amount of water in the mix must be carefully controlled. This control should start in the stockpiles of aggregate where an effort must be made to obtain a uniform and stable moisture content. Water should be added to the mix by some method which will assure that the correct amount of water is added to each batch.

Close control of the mixing operation is required to obtain the desired uniform mix. Sand, rock, and cement pockets will result in a structure weaker in some sections than in others. A nonuniform concrete mix will also result in stress concentrations which cause a redistribution of stresses within the structure. These redistributed stresses may or may not be detrimental depending on where the stresses occur.

Segregation of sand and coarse aggregates can also result in surface defects such as rock pockets, surface scaling and crazing, and sand streaks. These are not only unsightly but are the beginning of surface deterioration in structures subjected to severe weathering.

14-5. *Preparations for Placing*.—The integrity of a concrete structure is dependent to a large extent on the proper preparation of construction joints before placing fresh concrete upon the construction joint surfaces. Bond is desired between the old and new concretes and every effort must be directed toward obtaining this bond. All laitance and inferior surface concrete must be removed from the old surface with air and water jets and wet sandblasting as necessary. All surfaces should be washed thoroughly prior to placing the new concrete, but should be surface dry at the time they are covered with the fresh concrete. Rock surfaces to be covered with concrete must be sound and free of loose material and should also be saturated, but surface dry, when covered with fresh concrete or mortar. Mortar should be placed only on those rock surfaces which are highly porous or are horizontal or nearly horizontal absorptive surfaces.

14-6. *Placing.* —Mass concrete placement can result in a nonuniform concrete when the concrete is dropped too great a distance or in the wrong manner. The same effect will occur when vibrators are used to move the concrete into its final position. All discharge and succeeding handling methods should therefore be carefully watched to see that the uniformity obtained in mixing will not be destroyed by separation.

Thorough vibration and revibration is necessary to obtain the dense concrete desired for structures. Mass concrete is usually placed in 5- or 7½-foot lifts and each of these lifts is made up of 18- to 20-inch layers. Each successive layer must be placed while the next lower layer is still plastic. The vibrators must penetrate through each layer and revibrate the concrete in the upper portion of the underlying layer to obtain a dense monolithic concrete throughout the lift. Such a procedure will also prevent cold joints within the placement lift.

14-7. *Curing and Protection.* —One of the major causes of variation in attained concrete strength is the lack of proper curing. Laboratory tests show that strength of poorly cured concrete can be as much as one-third less than that of well-cured concrete. This variance is more for some cements than for others. Curing of concrete is therefore important if

high quality is to be obtained. The full effectiveness of water curing requires that it be a continuous, not intermittent, operation. Curing compounds, if used, must be applied as soon as the forms are stripped, and must be applied to completely cover all exposed surfaces.

Poor curing often results in the formation of surface cracking. These cracks affect the durability of the structure by permitting weathering and freeze-thaw actions to cause deterioration of the surface. The larger structural cracks often begin with the cracks caused by poor curing.

Protection of the newly placed concrete against freezing is important to the designer, since inadequate protective measures will be reflected by lower strength and durability of the concrete. Protective measures include addition of calcium chloride to the mix and maintaining a minimum 40° F. placement temperature. Although calcium chloride in a quantity of not to exceed 1 percent, by weight of cement, is normally required when weather conditions in the area of the work will permit a drop in temperature to freezing, its use should not preclude the application of more positive means to assure that early age concrete will not freeze. When freezing temperatures may occur, enclosures and surface insulation should also be required. One of the most important factors associated with protection of concrete is advance preparation for the placement of concrete in cold weather. Arrangements for covering, insulating, or otherwise protecting newly placed concrete must be made in advance of placement and should be adequate to maintain the temperature and moisture conditions recommended for good curing.

14-8. *Finishes and Finishing.* —Suitable finishing of concrete surfaces is of particular concern to the designer. Some surfaces of concrete, because of their intended function, can be rough and of varying texture and evenness; whereas, others in varying degree must be smooth and uniform, some necessitating stringent allowable irregularity limits. The Bureau of Reclamation uses a letter-number system to differentiate between the different types of finishes, using F1, F2, F3, and F4 for formed surfaces and U1, U2, and U3 for unformed surfaces. Each finish is defined as to allowable abrupt and gradual irregularities. For formed surfaces, the particular forming materials permitted are also related to the letter-number system.

Finish F1 applies to formed surfaces that will be covered by fill material or concrete, which includes vertical construction and contraction joints, and upstream faces of mass concrete dams that are below the minimum water pool. Finish F2 applies to formed surfaces that will be permanently exposed to view but which do not require any special architectural appearance or treatment, or which do not involve surfaces that are subject to high-velocity waterflow. Finish F3 is used for formed surfaces for which, because of prominent exposure to public view, an aesthetic appearance from an architectural standpoint is considered desirable. Finish F4 is for formed flow surfaces of hydraulic structures where accurate alinement and evenness of surfaces are required to eliminate destructive effects of high-velocity water.

Finish U1 applies to unformed surfaces that will be covered by fill material or concrete. This is a screeded surface where considerable roughness can be tolerated. Screeding of an unformed surface is preliminary to the application of a U2 finish. The U2 finish is a wood-floated finish. This finish applies to all exposed unformed surfaces, and is a preliminary to applying a U3 finish, which requires steel troweling. A U3 finish is required on high-velocity flow surfaces of spillway tunnels and elsewhere where a steel-troweled surface is considered desirable.

When finishing the surfaces of newly placed concrete, overtroweling is to be avoided in all instances. Surfaces which are overtroweled are susceptible to weathering and erosion and usually result in a requirement for early repair measures on the structure.

14-9. *Tolerances.*—The prescribed tolerances on all structures should be maintained at all times. Some of these tolerances are placed in the specifications to control the overall construction and are necessary if the structure is to be completed as designed. Some tolerances are for appearances and others are to minimize future maintenance of the structure. For example, near-horizontal surfaces with very slight slopes are hard to finish without leaving depressions in the surface. These depressions collect moisture and usually begin weathering at an early age.

14-10. *Repair of Concrete.*—Repair of concrete covers not only the patching of holes remaining after construction operations but also the repair of cracks and damaged concrete. Repair of concrete in Bureau of Reclamation structures is required to conform to the Bureau's "Standard Specifications for Repair of Concrete." These specifications generally provide for concrete to be repaired with concrete, dry pack or portland cement mortar, or, at the option of the contractor, with epoxy-bonded concrete or epoxy-bonded epoxy mortar, where and as permitted by the specifications for the particular repair to be made. Repairs to high-velocity flow surfaces of concrete in hydraulic structures are required to be made with concrete, epoxy-bonded concrete, or epoxy-bonded epoxy mortar. Concrete is used for areas of extensive repair which exceed 6 inches in depth, while epoxy-bonded concrete is used for areas having depths of 1½ to 6 inches. Epoxy-bonded epoxy mortar is used for shallow surface repairs for depths ranging from 1½ inches to featheredges.

Before making any repair, all deteriorated and defective concrete must be removed. Unsound or questionable concrete may negate the successful repair of any concrete. Removal of the defective concrete should be followed by a thorough washing of the surfaces. A surface-dry condition should exist at the time replacement concrete is placed.

Cracks should not be repaired until all evidences indicate that the crack has stablized. The cause of the crack should also be investigated and, if possible, corrective measures initiated so that the crack will not reopen. All repairs should be thoroughly cured to minimize drying shrinkage in the repair concrete. Except where repairs are made with epoxy-bonded epoxy mortar, featheredges should be avoided in all repair of concrete.

Chapter XV

Ecological and Environmental Considerations

A. INTRODUCTION

15-1. *General Considerations*.—The rapid increase in world population and the increasing demands this population has made on the planet's natural resources have called into question the long-term effect of man upon his environment. The realization that man is an integral part of nature, and that his interaction with the fragile ecological systems which surround him is of paramount importance to his continued survival, is prompting a reevaluation of the functional relationships that exist between the environment, its ecology, and man.

Of increasing concern is the effect which man's structures have upon the ecosystems in which they are placed, and especially on the fish, wildlife, and human inhabitants adjacent to these structures. The need to store water for use through periods of drought, to supply industry and agriculture with water for material goods and foodstuffs, to provide recreational opportunity in ever-increasing amounts, and to meet the skyrocketing electric power demand has required the development of water resources projects involving the construction and use of dams and other related structures. These structures help man and yet at the same time cause problems in the environment and in the ecosystems into which they are placed. Many of these problems are exceedingly complex, and few answers which encompass the total effect of a structure on its environment are readily available.

Included in the answer to these problems must be the development and protection of a quality environment which serves both the demands of nature for ecological balance and the demands of man for social and psychological balance. The present challenge is to develop and implement new methods of design and construction which minimize environmental disturbances, while also creating aesthetic and culturally pleasing conditions under which man can develop his most desirable potentialities. This challenge can only be answered by the reasoned, pragmatic approach of sensitive, knowledgeable human beings.

The purpose of this chapter is to provide practical solutions to some of the environmental and ecological problems which confront the designer. This discussion is not exhaustive and it is hoped that the reader will consult the references at the end of this chapter (and numerous others available on this topic) for a more extensive coverage. The amount of scientific data concerning the environment and man's relation to it is expanding rapidly, and new design methods should become available soon. The practical information presented here can provide a useful introduction to the designer and a basis for maximizing the project's benefits and minimizing its negative environmental and ecological effects.

Recognizing the importance of man's environment, the 91st Congress passed the National Environmental Policy Act of 1969. This act established a three-member Council on Environmental Quality in the Executive Office of the President. Before beginning construction of a project, an Environmental Impact

Statement must be prepared by the agency having jurisdiction over project planning and submitted through proper channels to appropriate governmental agencies and interested private entities for review and comments.

The term environment is meant here to include the earth resources of land, water, air, and vegetation and manmade structures which surround or are directly related to the proposed structure. The term ecology is meant to encompass the pattern of relationships that exist between organisms (plant, animal, and human) and their environment.

15-2. *Planning Operations.*—One of the most important aspects of dealing correctly and completely with the ecological and environmental impact of any structure is proper planning. If possible, an environmental team should be formed consisting of representatives from groups who will be affected by the structure and experts from various scientific fields who can contribute their ideas and experience. The team approach will help assure that environmental considerations are placed in proper perspective with other vital issues such as reliability, cost, and safety, and that the relative advantages and disadvantages of each proposal are carefully weighed. It should also assure that the project is compatible with the natural environment. A suggested list of participants is given below:

(1) Concerned local and community officials.

(2) Design personnel.

(3) Environment and ecology experts.

(4) Fish biologists and wildlife experts.

(5) Building architects.

(6) Landscape architects.

(7) Recreational consultants.

This team should be responsible for the submittal of an ecological and environmental report to the designers with a list of criteria which the designs should encompass. Some of the topics which should be discussed in the report are covered briefly in this chapter. Since each site will present unique problems, only a general outline of the most important considerations is provided herein.

B. FISH AND WILDLIFE CONSIDERATIONS

15-3. *General.*—The placement of a dam and its reservoir within the environment should be done with due consideration to the effects on the fish and wildlife populations of the specific area. These considerations often involve complex problems of feeding patterns and mobility, and where possible an expert in this field should be consulted. The Fish and Wildlife Service of the Department of the Interior, the Forest Service of the Department of Agriculture, and appropriate State agencies can supply considerable expertise on the environmental impact of a proposed structure. It should be remembered that dams and reservoirs can be highly advantageous in that they provide a year-round supply of drinking water for wildlife, breeding grounds for waterfowl, and spawning areas for fish. At the time of design, as many benefits as practicable to fish, wildlife, and waterfowl populations should be included and provisions should be made for the future protection of these populations. The following sections discuss some of the items which affect fish and wildlife and outline what can be done to aid them.

15-4. *Ecological and Environmental Considerations for Fish.*—Critically important to the survival of fish population are three items: (1) water quality, (2) water temperature, and (3) mobility. Water quality is obviously important to the survival of fish, and an effort should be made to see that the quantity of pollutants which enter the stream during construction and the reservoir after completion is kept to the minimum. Strict regulations concerning pollutants should be instituted and enforced. Quantities of degradable, soluble, or toxic pollutants should not be left within the reservoir area after construction. Heavy pesticide runoffs can cause fish kills, and some means such as a holding pond or contour ditches should be used to reduce their presence and steps taken to eventually eliminate them. Substances that can cloud or darken the water interfere with the ability of sight-feeding fish to forage, and should not be allowed to enter the water.

In mining areas where heavy erosion often occurs, careful consideration must be given to the effects of siltation which may rapidly reduce the reservoir capacity. Consideration must also be given to the acidic character of the water since it can cause fish kills. Control dams may be the solution, and in one case the Bureau of Reclamation has constructed an off-reservoir dam to reduce the rapid sedimentation of the main reservoir and to limit the amount of acidic inflow to an acceptable level.

The temperature of the water controls timing of migration, breeding, and hatching and affects the appetite, growth, rate of heartbeat, and oxygen requirements of all fish. Each species of fish has an optimum temperature range within which it can survive, and consideration must be given to the temperature range which will exist both in the reservoir and in the stream below the dam due to the reservoir releases. For example, if a low dam is constructed in a mountainous area, the cool water entering the shallow reservoir can be warmed by the sun during the summer months to an undesirable extent. The warm temperatures of the shallow water within the lake and also of the downstream releases could then prevent the spawning of cold water species of fish such as trout.

To remedy this problem, care must be taken to provide sufficiently deep reservoir areas where cold water will remain, and to use an outlet works which is capable of selectively withdrawing the colder water from the lower reservoir depths. Federal and State fish and wildlife agencies should be consulted as to the correct depth for the outlets in a specific area. Figure 15-1 shows the selective withdrawal outlet works to be used at Pueblo Dam in Colorado. Figure 15-2 shows a selective withdrawal outlet works used at Folsom Dam in California. Movable shutters were placed

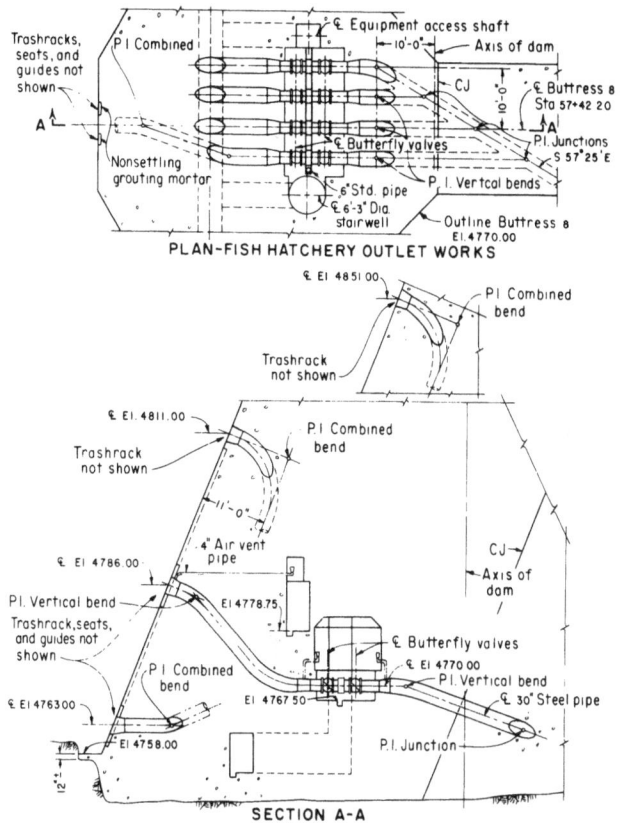

Figure 15-1. Selective withdrawal outlet at Pueblo Dam in Colorado. Water can be withdrawn at any of four levels.—288-D-3104

upstream from the trashracks. By manipulation of the shutters water may be drawn from the desired level. The Bureau of Reclamation has used selective withdrawal outlet works at several locations to create favorable temperatures for fish spawning downstream of the dam. Further information concerning these structures is available in references [1] and [2].[1] Another reason for providing adequate reservoir depth, in addition to creating favorable conditions for spawning, is to prevent fish kill in the winter due to extreme cold. However, shallow reservoir areas are sometimes required to develop a warm water fishery or for waterfowl habitats.

Although salmon are commonly thought of as the only migrators, other species of fish such

[1]Numbers in brackets refer to items in the bibliography, section 15-12.

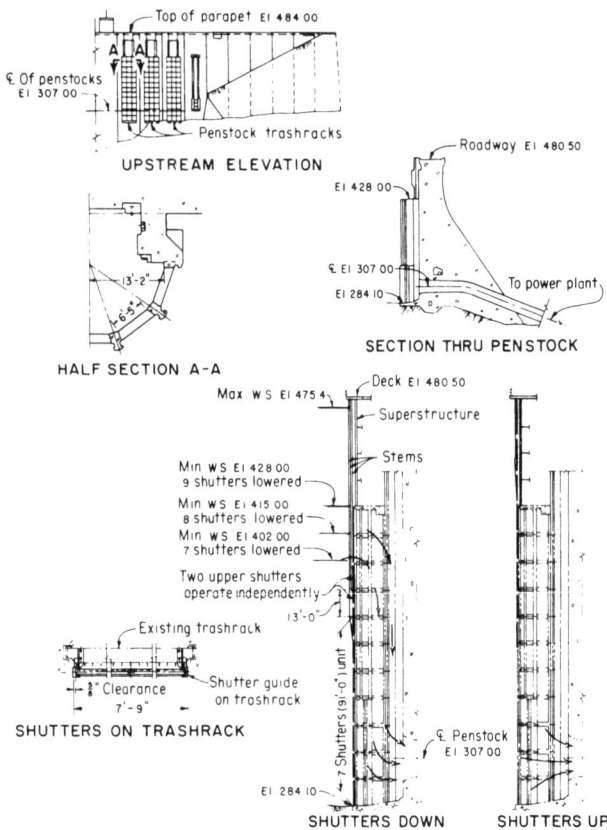

Figure 15-2. Selective withdrawal outlet at Folsom Dam in California. This outlet makes use of an adjustable shutter arrangement.–288-D-3105

Figure 15-3. Fish ladder used on the left abutment of R e d B l u f f D i v e r s i o n D a m i n California.–P602-200-4543 NA

as shad, steelhead trout, and other trout also require mobility considerations. The most common method for allowing fish to pass by a dam is use of the fish ladder. Figure 15-3 shows the fish ladder used by the Bureau of Reclamation on the Red Bluff Diversion Dam in California. Specific design requirements for fish ladders may be obtained from the Forest Service, the Fish and Wildlife Service, or appropriate State agencies. Where practicable, fish should be prohibited from entering spillways, outlet pipes, penstocks, and other restricted areas by use of fish screens.

Where fish populations are concerned, care should be taken to avoid the destruction of vegetation in the reservoir area since this becomes a food source after the reservoir is filled. Certain amounts of standing trees or tree debris left in the reservoir area can provide a habitat for several species of fish as can brush

piles, which are staked down to prevent them from being washed away. Figure 15-4 shows a reservoir in which trees have been left standing to benefit the fish population.

Although certain aquatic plants are desirable for water birds, such as ducks, coots, and wading birds, they can be detrimental to fish production and should be controlled when necessary. Shallow shorelines in the inlet portions of the reservoir can be deepened to eliminate the growth of any plant life found not useful.

In newly constructed reservoirs, arrangements should be made for stocking with the appropriate type or types of fish. Consultation with a fisheries expert is recommended to determine the correct type of fish and the proper time for stocking them.

The oxygen content of some reservoirs can decrease with time and an examination of available reservoir reaeration devices may prove helpful. The oxygen content of the reservoir water may be increased during release from the reservoir by the use of reaeration devices such as the U-tube [3]. Reaeration is also aided by increasing the contact of the water released in the spillways and outlet works with air. A

Figure 15-4. An aerial view of a small reservoir with trees left at the water's edge to provide a fish habitat.—288-D-2869

bibliography of reaeration devices compiled by the Bureau of Reclamation is contained in reference [4].

In some cases, fish hatcheries can be built in conjunction with the dam. Figure 15-5 shows the hatchery below Nimbus Dam in California. Canals also may provide spawning areas for fish, although considerable cost and special equipment may be required. Figure 15-6 shows an artist's conception of the "gravel cleaner" which will be provided at a salmon spawning area in the Tehama-Colusa Canal in California. Special gravel and special gravel sizes were required in the canal bottom to facilitate spawning.

15-5. *Environmental Considerations for Wildlife.* —Three common detrimental effects of reservoirs on wildlife involve: (1) removal of feeding areas, (2) loss of habitat, and (3) limitation of mobility. The severity of each of these effects can be significantly reduced.

When reservoirs inundate wildlife feeding areas, new areas should be planted to lessen the impact and, if possible, new types of grasses which are suitable and which provide more food per unit area should be planted. In addition, the new feeding areas can sometimes be irrigated with reservoir water to cause rapid, heavy growth. If the reservoir water is not immediately needed for irrigation, the water level can be left below the normal water surface to allow sufficient time for the feeding areas which are to be flooded to be replaced by areas of new growth.

Where flooding of the homes of a large number of smaller animals such as muskrat and

Figure 15-5. Fish hatchery at Nimbus Dam in California.—AR2964-CV

Figure 15-6. An artist's conception of the gravel cleaner to be used at a salmon spawning area on the Tehama-Colusa Canal in California.—P602-D-54534-520

beaver will occur, consideration should be given to adjusting the required excavation, reducing the reservoir water levels, or relocating the dam so that the number of animals affected will be minimized. It may also be possible to provide special dikes and drainage conditions which can lessen the effect. Problems in this area are difficult to solve and the advice of a specialist should be sought.

Provisions for ducks, geese, and other waterfowl at reservoirs can be made by planting vegetation beneficial to nesting and by leaving areas of dense grass and weeds at the water's edge. If suitable areas already exist at the damsite, an effort should be made to selectively excavate to leave the habitat in place. Assistance concerning the appropriate reservoir treatment can be obtained from the Fish and Wildlife Service of the Department of the Interior, the Forest Service of the Department of Agriculture, and appropriate State agencies.

C. RECREATIONAL CONSIDERATIONS

15-6. *General.*—The nation's increase in population, the decrease in working hours, and the great mobility of large numbers of people have caused a significant increase in the use of reservoirs for recreational activities. These activities include fishing, boating, water skiing, swimming, scuba diving, camping, picnicking, and just simply enjoying the outdoor experience of the reservoir setting. Many of the reservoirs constructed in past years have become the recreation centers of the present, and this will undoubtedly be repeated in the future. Provisions should be made to obtain the maximum recreational benefits from the completed reservoir, and a future development plan should provide for area modifications as the recreation use increases.

15-7. *Recreational Development.*— Considerations for recreational development should start when project planning is begun and should be integrated into the total site plan. Areas of significant natural beauty should be left intact if possible, and recreational facilities should be developed around them. Boat ramps and boat docking facilities are beneficial to most reservoir areas and should be constructed at the same time as the dam. Figure 15-7 shows the docking facilities at the Bureau of Reclamation's Canyon Ferry Reservoir in Montana. Camping facilities for truck campers, trailers, and tenters, and picnicking areas can often be provided at reasonably low costs.

Trash facilities should be provided at convenient locations to help in litter control, and the excessive use of signs and billboards near the reservoir area should be prohibited. The signs which are used should be blended with the surroundings. Toilet facilities should be available at all camping grounds and proper sewage disposal facilities should be installed, especially where the possibility of reservoir pollution exists.

If the reservoir is near a population center it may prove advantageous to provide bicycle paths, equestrian paths, and foot paths for public use. At the damsite or nearby, a reservoir viewing location and possibly a visitors' center should be built. Exhibits showing the history of the project, local history, or other appropriate exhibits can enhance the visitors' enjoyment of the reservoir. These centers should be aesthetically designed to fit the location. Figure 15-8 shows a viewing area at Glen Canyon Dam.

Buildings adjacent to the reservoir should be of low profile and blend with the reservoir surroundings; however, in some cases it may be desirable to contrast the buildings with their surroundings.

Fishing benefits can be maximized by stocking the reservoir with several types of fish and by replenishing these stocks yearly.

Proper maintenance requirements for the recreation areas should be instituted after completion of the dam and reservoir complex, and should include repair of broken and damaged equipment, repainting, and rebuilding to meet expanded facility demands. Trash should be removed from the campgrounds and adjacent recreation facilities at regular intervals, and the possibility of recycling aluminum and other metal products should be explored. Recreational areas which are overused should be rotated to prevent their deterioration, and single areas which receive exceptionally heavy use should be fenced off completely for short intervals to prevent their ruination. Protection of the reservoir banks from sloughing may be required for steep slopes, and excessive erosion at any part of the site should be prevented.

D. DESIGN CONSIDERATIONS

15-8. *General.*—Design requirements should be devoted to the accomplishment of three goals: (1) keeping the natural beauty of the surrounding area intact, (2) creating

Figure 15-7. Boat docking facilities at Canyon Ferry Reservoir in Montana.—P296-600-949

aesthetically satisfying structures and landscapes, and (3) causing minimal disturbance to the area ecology. Designers should try to accomplish these goals in the most economical way. The following paragraphs discuss some items to be considered during design and will provide some practical suggestions for designers. Many of the items discussed here should be considered during the project planning stages and the critical decisions made at that time.

If it is necessary to excavate rock abutments above the crest of the dam, consideration should be given to the use of presplitting techniques since they leave clean, aesthetic surfaces. As discussed in sections 15-6 and 15-7, a scenic overlook should be provided for viewing the dam and reservoir. The overlook should have adequate parking and, if practicable, a visitors' center.

The diversion schemes (see ch. V), should be

such that excessive silt created during construction will not find its way into the downstream water. Materials from excavations should be placed in the reservoir area upstream of the dam to prevent unsightly waste areas in the downstream approaches. In some cases, boat ramps, picnic areas, or view locations can be constructed with excavated material. Where foundation conditions permit it, spillway structures of a type which minimizes the required surface excavations on the dam abutments should be used. (See ch. IX.)

If a section of canal is constructed in connection with a dam, spoil piles should be shaped to the natural landscape slopes along the canal length; this material can also be used to construct recreation areas where appropriate. Pipelines should be buried as should electrical wiring; where this is not possible, the pipelines and electrical apparatus should be painted to blend with their

Figure 15-8. Viewing area at Glen Canyon Dam in Arizona.—P557-400-1133

background. Protective railings used on the dam crest and on bridges near the site should be low enough so that the reservoir may be seen from a passenger car.

15-9. Landscape Considerations.—As much natural vegetation as possible should be left in place. If significant areas of natural beauty exist near the project, every effort should be made to utilize or preserve them, and additional right-of-way should be obtained to include any such adjacent areas.

Access roads to the damsite and roads used by the contractor during construction should be kept to a minimum, and those roads not planned for use after completion of the dam should be obliterated and replanted with grass or other natural vegetation. Access roads subject to excessive erosion should have protective surfacing. Roads which are required for maintenance of the dam or appurtenant works should be protectively fenced if excessive visitor usage will cause erosion.

Erosion control should be started at the beginning of the job. Roads and cut slopes should be provided with terraces, berms, or other check structures if excessive erosion is likely. Exploratory trenches which are adjacent to the damsite should be refilled and reseeded.

Quarry operations and rock excavations should be performed with care. The minimum amount of material should be removed, correct blasting techniques should be used, unsightly waste areas should not be left, and final rock slopes for the completed excavation should be shaped to have a pleasing appearance. Presplitting and/or controlled blasting should be considered for final slope cutting to permit a clean, pleasing view.

Road relocations near the dam can often eliminate deep cuts in hillsides, allow scenic alinements, and provide reservoir viewing locations. Adequate road drainage should be used and slopes should be cut so that reseeding operations will be convenient. For projects where power transmission lines will be required the publications "Environmental Criteria for Electric Transmission Systems" [5] and "Environmental Considerations in Design of Transmission Lines" [6] will provide many helpful suggestions which will lessen their environmental impact.

15-10. *Protective Considerations.*—At locations where accidents are most likely to occur, protective devices and warning systems should be installed. The most dangerous locations at a damsite are near the spillway (especially if it is a chute type), the outlet works intake tower, and the stilling basins of both the spillway and outlet works. Any portion of the spillway and outlet works stilling basins which might prove hazardous should be fenced off and marked by warning signs.

Canals with steep side slopes which prevent a person or an animal from climbing out are extremely dangerous, as also are siphons. Considerable information concerning canal safety is contained in the Bureau's publication "Reducing Hazards to People and Animals on Reclamation Canals"[7].

When the project encompasses the generation of electricity, the problems of providing adequate safety precautions are considerably increased and the advice of an expert in that field should be sought.

15-11. *Construction Considerations.*—The environmental and ecological design requirements presented in the specifications are converted from an abstraction into a reality by the builder. The contractor and his personnel should be informed that this is a most important step in the planning, design, and construction sequence. In this regard, a preconstruction conference may be invaluable in assuring an understanding of the job requirements by the builder and in enlisting his cooperation. The owner should insure compliance by having competent inspectors and by having specifications which clearly spell out the construction requirements. Excessive air and water pollution during construction should be prevented, and specifications covering these items are included in appendix I; they should provide a framework for the inclusion of other important environmental provisions. The builder should also institute safety precautions during construction, and the publication "Safety and Health Regulations for Construction" [8] will provide helpful information. The builder should be encouraged to bring forward any obvious defects in the environmental considerations which he encounters during construction and to suggest improvements.

Construction campsites should be placed within the reservoir area below normal water level. All trees, shrubs, and grassland areas which are to be protected should be staked or roped off. Any operations which would affect a large wildlife population should be moved to a different location if at all possible. Large volumes of water should not be taken from the stream if there are prior downstream commitments, and the water going downstream should be muddied as little as possible and kept pollution free; siltation ponds may be needed in extreme cases. The builder should be required to remove or bury all trash and debris collected during the construction period and to remove all temporary buildings. Every opportunity should be taken to use the timber in the reservoir area for commercial operations. The burning of trees cleared within the reservoir area should be prevented if excessive air pollution will result or if State laws prevent it. At Pueblo Dam in Colorado, the Bureau required that all brush and timber smaller than 7 inches in diameter be chipped into mulch and stockpiled for future use on the reseeded areas. The chipping operation at Pueblo Dam is shown in figure 15-9. Slightly larger timber can be cut into firewood for use at camping and recreation areas, and still larger timber can be channeled into some commercial use such as production of lumber, wallboard, or boxes.

A temporary viewing site for the project, having signs which show the completed project and explain its purpose, is helpful in promoting good community relations.

Figure 15-9. **Chipping operations at Pueblo Dam in Colorado. All brush and timber smaller than 7 inches in diameter are chipped and stored for use as a mulch on reseeded areas.**—P382-700-790 NA

E. BIBLIOGRAPHY

15-12. *Bibliography*.

[1] "Register of Selective Withdrawal Works in United States," Task Committee on Outlet Works, Committee on Hydraulic Structures, Journal of the Hydraulics Division, ASCE, vol. 96, No. HY9, September 1970, pp. 1841-1872.

[2] Austin, G. H., Gray, D. A., and Swain, D. G., "Multilevel Outlet Works at Four Existing Reservoirs," Journal of the Hydraulics Division, ASCE, vol. 95, No. HY6, November 1970, pp. 1793-1808.

[3] Speece, R. E., and Orosco, R., "Design of U-Tube Aeration Systems," Journal of the Sanitary Engineering Division, ASCE, vol. 96, No. SA3, June 1970, pp. 715-725.

[4] King, D. L., "Reaeration of Streams and Reservoirs—Analysis and Bibliography," REC-OCE-70-55, Bureau of Reclamation, December 1970.

[5] U.S. Department of the Interior and U.S. Department of Agriculture, "Environmental Criteria for Electric Transmission Systems," Government Printing Office, Washington, D.C., 1970.

[6] Brenman, H., and Covington, D. A., "Environmental Considerations in Design of Transmission Lines," ASCE National Meeting on Transportation Engineering, Washington, D.C., July 1969.

[7] Latham, H. S., and Verzuh, J. M., "Reducing Hazards to People and Animals on Reclamation Canals," REC-OCE-70-2, Bureau of Reclamation, January 1970.

[8] Bureau of Reclamation, "Safety and Health Regulations for Construction," latest edition.

* _____, "Environmental Quality—Preservation and Enhancement," Reclamation Instructions, Series 350, Part 376, 1969.

U. S. Department of the Interior, "Man—An Endangered Species," Government Printing Office, Washington, D.C., 1968.

_____, "River of Life, Water: The Environmental Challenge," Government Printing Office, Washington, D.C., 1970.

_____, "The Population Challenge—What It Means to America," Government Printing Office, Washington, D.C., 1966.

_____, "The Third Wave," Government Printing Office, Washington, D.C., 1967.

Benson, N. G. (editor), "A Century of Fisheries in North America," American Fisheries Society, Washington, D.C., 1970.

Clawson, M., and Knetsch, J. L., "Economics of Outdoor Recreation," The John Hopkins Press, Baltimore, Md., 1966.

Dasmann, R. E., "Environmental Conservation," John Wiley & Sons, Inc., New York, N.Y., 1968.

Dober, R. P., "Environmental Design," Van Nostrand Reinhold Co., New York, N.Y., 1969.

"Environmental Quality," First annual report of the Council on Environmental Quality, Government Printing Office, Washington, D.C., August 1970.

*References without numbers are not mentioned in text.

Flawn, P. T., "Environmental Geology: Conservation, Land-Use Planning, and Resources Management," Harper & Row, New York, N.Y., 1970.

McCullough, C. A., and Nicklen, R. R., "Control of Water Pollution During Dam Construction," Journal of the Sanitary Engineering Division, ASCE, vol. 97, No. SA1, February 1971, pp. 81-89.

Prokopovich, N. P., "Siltation and Pollution Problems in Spring Creek, Shasta County, California," Journal of American Water Works Association, vol. 57, No. 8, August 1965, pp. 986-995.

Reid, G. K., "Ecology of Inland Water and Estuaries," Van Nostrand Reinhold Co., New York, N.Y., 1961.

"Report of the Committee on Water Quality Criteria," U.S. Department of the Interior, Federal Water Pollution Control Administration, April 1, 1968.

Seaman, E. A., "Small Fish Pond Problem-Management Chart," Technical Publication No. 2, West Virginia Conservation Commission, Charleston, W. Va.

Smith, G. (editor), "Conservation of Natural Resources," third edition, John Wiley & Sons, Inc., New York, N.Y., 1965.

"Transactions," American Fisheries Society, Washington, D.C.

U.S. Department of the Interior, "Quest for Quality," Government Printing Office, Washington, D.C., 1965.

Vernberg, J. F., and Vernberg, W. B., "The Animal and the Environment," Holt, Reinhart & Winston, Inc., New York, N.Y. 1970.

Watt, K. E., "Ecology and Resources Management; A Qualitative Approach," McGraw-Hill, New York, N.Y., 1968.

Wing, L. W., "Practice of Wildlife Conservation," John Wiley & Sons, New York, N.Y., 1951.

The Gravity Method of Stress and Stability Analysis

A-1. *Example of Gravity Analysis—Friant Dam.* —The example presented in this appendix was taken from the gravity analysis of the revised Friant Dam. Friant Dam was constructed during the period 1939 to 1942 and is located in the Central Valley of California. A plan, elevation, and sections of the dam are shown on figure A-1.

The assumptions and constants used for the analysis are given below:

(1) Unit weight of water = 62.5 pounds per cubic foot.

(2) Unit weight of concrete = 150 pounds per cubic foot.

(3) Unit shear resistance of both concrete and rock = 450 pounds per square inch.

(4) Coefficient of internal friction of concrete, or of concrete on rock = 0.65.

(5) Weight of 18-foot drumgate = 5,000 pounds per linear foot.

(6) Top of nonoverflow section, elevation 582.

(7) Crest of spillway section, elevation 560.

(8) Normal reservoir water surface, elevation 578.

(9) Tailwater surface, elevation 305.

(10) Horizontal component of assumed earthquake has an acceleration of 0.1 gravity, a period of vibration of 1 second, and a direction which is at right angles to axis of dam.

Note. Figure A-2 is a graph showing values of the coefficient K_E, which was used to determine hydrodynamic effects for the example given. However, this procedure is not consistent with current practice. A discussion of the coefficient C_m, which is presently used to determine hydrodynamic pressures, is given in section 4-34.

(11) Vertical component of assumed earthquake shock has an acceleration of 0.1 gravity and a period of 1 second.

(12) For combined effects, horizontal and vertical accelerations are assumed to occur simultaneously.

(13) Uplift pressure on the base or on any horizontal section varies from full-reservoir pressure at the upstream face to zero, or tailwater pressure, at the downstream face, and is considered to act over two-thirds the area of the section. Uplift is assumed to be unaffected by earthquake shock, and to have no effect on stresses in the interior of the dam.

Note. This uplift assumption is no longer used by the Bureau of Reclamation. See section 3-9 for uplift assumptions now in use.

A-2. *List of Conditions Studied.* —A list of conditions studied for Friant Dam for both the nonoverflow and the overflow section is tabulated below:

(1) Reservoir empty.

(2) Reservoir full.

(3) Reservoir empty plus earthquake.

(4) Reservoir full plus earthquake.

Loads for reservoir empty are dead loads consisting of the weight of the dam and gates. Loads for full-reservoir operation include, in addition to dead loads, the vertical and horizontal components of normal waterloads on the faces of the dam.

Loads for earthquake effects with reservoir empty include inertia forces caused by acceleration of the mass of dead loads. Loads for earthquake effects with reservoir full include, in addition to the above, the inertia force of the mass of water and the hydrodynamic force caused by the movement

299

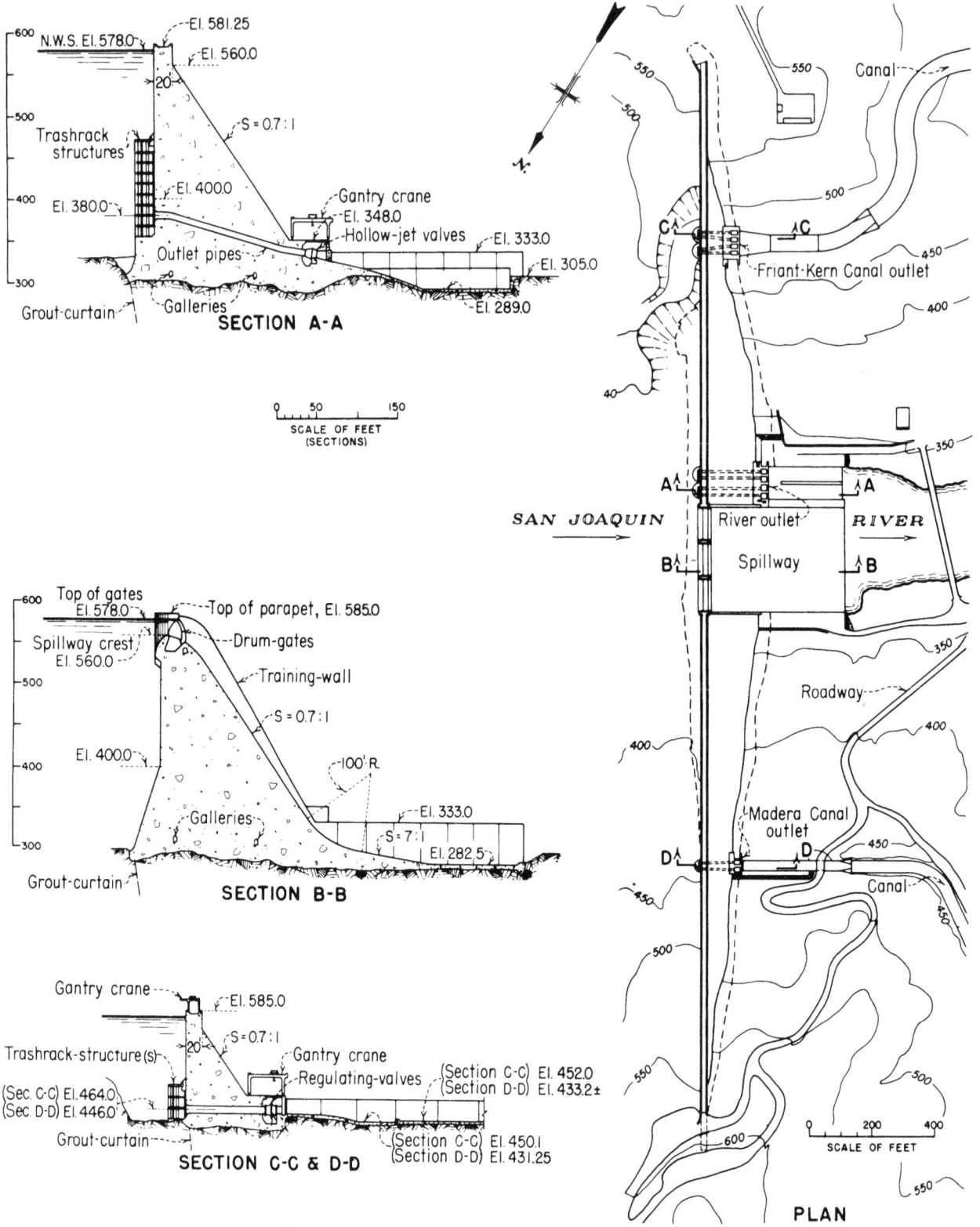

Figure A-1. Friant Dam—plan and sections. —288-D-3156

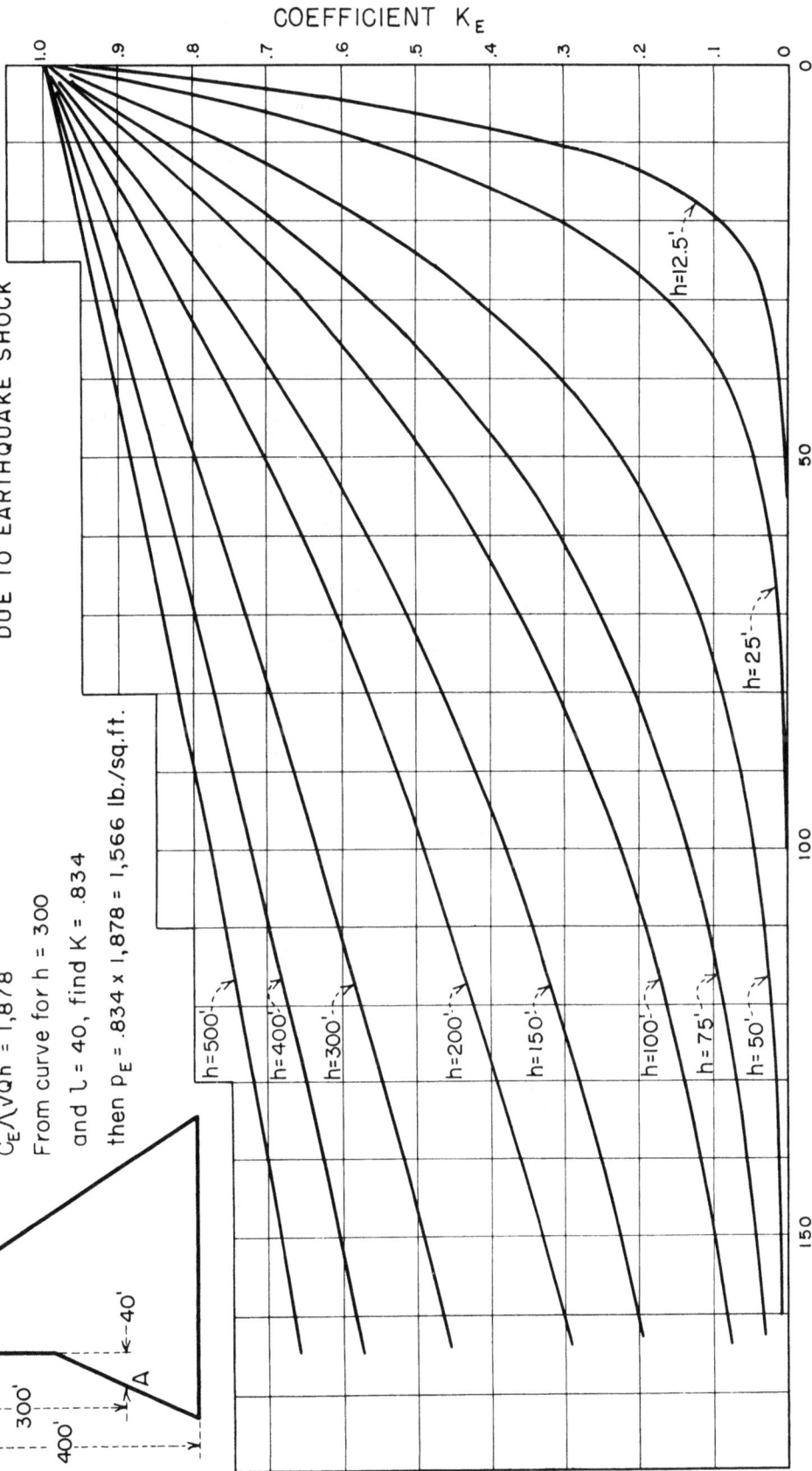

COEFFICIENT K_E

CURVES FOR COEFFICIENT K_E

FOR COMPUTING CHANGE IN PRESSURE

ON INCLINED FACES OF DAMS

DUE TO EARTHQUAKE SHOCK

Example: To find pressure change at A

General Formula: $p_E = K_E C_E \lambda \sqrt{Qh}$

$Q = 400'$ $h = 300$ $\lambda = .1$

$C_E \lambda \sqrt{Qh} = 1,878$

From curve for $h = 300$

and $L = 40$, find $K = .834$

then $p_E = .834 \times 1,878 = 1,566$ lb./sq.ft.

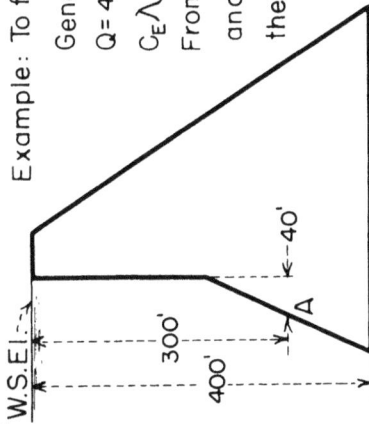

L = HORIZONTAL DISTANCE IN FEET TO POINT ON FACE FROM INTERSECTION OF WATER SURFACE AND FACE OF DAM

Figure A-2. Curves for coefficient K_E for computing change in pressure due to earthquake shock. —288-D-3157

of the dam against the water of the reservoir. Uplift forces are assumed to be unaffected by earthquake shocks.

The effects of earthquake were studied for each of the following directions of the acceleration:

(1) Horizontal upstream.

(2) Horizontal downstream.

(3) Vertical upward.

(4) Vertical downward.

(5) Horizontal upstream plus vertical upward.

(6) Horizontal upstream plus vertical downward.

(7) Horizontal downstream plus vertical upward.

(8) Horizontal downstream plus vertical downward.

A-3. *Computations and Forms.* —Computations for the gravity analysis of the nonoverflow section of Friant Dam are shown as figures A-3 to A-9, inclusive. These are for reservoir-full conditions with earthquake accelerations upstream and upward. Equations used are shown at the top of the forms. Standard forms are used.

A-4. *Final Results.* —Final results are given on figures A-10 to A-18, inclusive, which show normal and shear stresses, stability factors, and principal stresses for each loading condition on the overflow and nonoverflow sections.

A-5. *Summary and Conclusions.* —Following is a summary of results and conclusions obtained from the gravity analysis of Friant Dam. These are presented for the purpose of showing the type of information usually obtained from such an analysis.

(1) The analyses of the maximum nonoverflow and spillway sections of Friant Dam indicate stresses and stability factors within safe limits for all loading conditions.

(2) The maximum compressive stress, maximum horizontal shear stress, and minimum shear-friction factor all occur for normal full-reservoir operation during earthquake accelerations "horizontal upstream" and "vertical upward."

(3) The maximum tensile stress occurs for reservoir-empty conditions combined with earthquake acceleration "horizontal downstream" acting alone or in conjunction with earthquake acceleration "vertical upward."

(4) The maximum sliding factor occurs for normal full-reservoir conditions combined with earthquake accelerations "horizontal upstream" and "vertical downward."

(5) Points of application of resultant forces on the bases and horizontal sections of the nonoverflow and spillway sections are well within the middle-third for most loading conditions.

(6) Maximum stresses occur at the downstream face of the maximum nonoverflow section; the maximum compressive and shear stresses occur at the base elevation, and the maximum tensile stress occurs at elevation 400. Maximum direct stresses all act parallel to the face.

(7) The maximum sliding factor occurs at elevation 400 and the minimum shear-friction factor occurs at the base elevation of the nonoverflow section.

(8) The maximum compressive stress is 409 pounds per square inch and the maximum tensile stress is 46 pounds per square inch.

(9) The maximum horizontal shear stress is 192 pounds per square inch. The maximum sliding factor is 0.999, and the minimum shear-friction factor is 5.45.

(10) Since tensile stresses occur at points not subjected to water pressure, the possibility of uplift forces acting in tension cracks is eliminated.

Complete results for nonoverflow and spillway sections are tabulated in table A-1.

FRIANT DAM   NONOVERFLOW SECTION   STUDY No. 3

GRAVITY STRESS ANALYSIS OF MAXIMUM PARALLEL-SIDE CANTILEVER

(NOTE: Origin at downstream face)   VALUES AND POWERS OF **y**   By J.T.R.   Date 1-30-40

y, y^2, and y^3 (Feet)

ELEV.		U.S.	6	5	4	3	2	1	D.S.				
						VERTICAL PLANE							
	y												
	y^2												
	y^3												
	y	27.3		7.3					0				
550	y^2	745.29		53.29									
	y^3	20,346,417		389.017									
	y	62.3		42.3	2.3				0				
500	y^2	3,881.29		1,789.29	5.29								
	y^3	241,804.37		75,686.97	12.167								
	y	97.3		77.3	37.3				0				
450	y^2	9,467.29		5,975.29	1,391.29								
	y^3	921,167.32		461,889.92	51,895.117								
	y	132.3		112.3	72.3	32.3			0				
400	y^2	17,503.29		12,611.29	5,227.29	1,043.29							
	y^3	2,315,685.27		1,416,247.87	377,933.07	33,698.267							
	y	182.3		147.3	107.3	67.3	27.3		0				
350	y^2	33,233.29		21,697.29	11,513.29	4,529.29	745.29						
	y^3	6,058,428.77		3,196,010.82	1,235,376.02	304,821.22	20,346.417						
	y	217.3	211.8	171.8	131.8	91.8	51.8	11.8	0				
315	y^2	47,219.29	44,859.24	29,515.24	17,371.24	8,427.24	2,683.24	139.24					
	y^3	10,260,752	9,501,187	5,070,718.2	2,289,529.4	773,620.63	138,991.83	1,643.032					
	y												
	y^2												
	y^3												
	y												
	y^2												
	y^3												
	y												
	y^2												
	y^3												
	y												
	y^2												
	y^3												
	y												
	y^2												
	y^3												
	y												
	y^2												
	y^3												

Figure A-3. Friant Dam study—values and powers of y. —DS2-2(6)

FRIANT DAM NONOVERFLOW SECTION. RESERVOIR W.S. EL. 578. TAILWATER EL. NONE. STUDY No. 3

GRAVITY STRESS ANALYSIS OF MAXIMUM PARALLEL-SIDE CANTILEVER

INCLUDING EFFECTS OF TAILWATER AND HORIZONTAL EARTHQUAKE

NORMAL STRESS ON HORIZONTAL PLANES

By J.T.R. Date 2-5-40

EARTHQUAKE ←

NOTE: $\omega_c = 150$
$\omega = 434.03$

$$a = \sigma_{ZD} = \tfrac{1}{T}(\Sigma W) - \tfrac{6}{T^2}(\Sigma M) \qquad b = \tfrac{12}{T^3}(\Sigma M) \qquad \sigma_{ZU} = \tfrac{1}{T}(\Sigma W) + \tfrac{6}{T^2}(\Sigma M)$$

Check: (for $y = T$), $\sigma_z = a + by$

σ_z Pounds per Square Foot

ELEV.	T	$\tfrac{1}{T}$	$\tfrac{6}{T^2}$	$\tfrac{12}{T^3}$	ΣW	ΣM	b	U.S.	6	5	4	3	2	1	D.S.	σ_{zu}
					Reservoir Full											
550		.036,630,036	.008,05055,2	0,589,784,4	111,880	-170,700	100676,20	2,708.4		4,733.5					5438	2,723.94
500		.016,051,364	.001,5458,77,7	0,049,626,693	481,490	-2,947,200	146260,38	3,122.56		6,042.8	11,983.2				12,285	3,172.56
450		.010,277,492	.000,633,761,04	0,013,026,948	1,139,800	-4,328,000	186650,11	2,618.75		6,344.8	13,896.8				20,144	2,633.76
400		.007,258,579	.000,342,792,70	0,005,820,51,3	2,086,900	-9,701,000	205732,62	2,115.78		6,244.4	14,101.7	22,758 0			29,204	2,164.78
350		.005,485,463,5	.000,180,541,86	0,001,980,711,6	3,594,000	-86,900,000	172,123,84	4,028.66		10,070.0	16,118.0	23,165 0	30,213,9		35,246	4,025.67
315		.004,601,932,8	.000,127,066,71	0,001,169,450	4,925,000	-132,780,000	155,286,87	5,740.60	6,646.7	12,889.2	19,132.6	25,176.1	31,219	37,704.1	39,275	5,792.60

Figure A-4. Friant Dam study—normal stresses on horizontal planes. —DS2-2(7)

FRIANT DAM. NONOVERFLOW SECTION. RESERVOIR W.S. EL. 57.8. TAILWATER EL. NONE. STUDY No. 3

GRAVITY STRESS ANALYSIS OF MAXIMUM PARALLEL-SIDE CANTILEVER
INCLUDING EFFECTS OF TAILWATER AND HORIZONTAL EARTHQUAKE
SHEAR STRESS ON HORIZONTAL AND VERTICAL PLANES $\tau_{zy} = \tau_{yz} = a_1 + b_1 y + c_1 y^2$

EARTHQUAKE ←   By J.T.R. Date 2-6-40.

$$\tau_{zyu} = -(\sigma_{zu} - p \pm p_E) \tan \phi_u \qquad a_1 = \tau_{zyD} = (\sigma_{zD} - p' \pm p_E') \tan \phi_D$$

$$b_1 = -\left[\frac{6}{T^2}(\Sigma V) + \frac{2}{T}(\tau_{zyu}) + \frac{4}{T}(\tau_{zyD})\right] \qquad c_1 = \frac{6}{T^3}(\Sigma V) + \frac{3}{T^2}(\tau_{zyu}) + \frac{3}{T^2}(\tau_{zyD})$$

(‡ Use (+) sign if horizontal earthquake acceleration is upstream.)   (‡ Use (–) sign if horizontal earthquake acceleration is upstream.)

$$C = \sqrt{\frac{51.0 \text{ lbs./cu.ft.}}{1-0.72\left(\frac{Q \cdot \text{Sec}}{1000 \cdot \text{N·ft}}\right)^2}} = \dots \qquad Q = \dots \qquad Q' = \dots$$

$$p_E = C K \lambda \sqrt{Q h} \qquad C' = \sqrt{\frac{51.0 \text{ lbs./cu.ft.}}{1-0.72\left(\frac{Q' \cdot \text{Sec}}{1000 \cdot \text{N·ft}}\right)^2}} = \dots \qquad p_E' = C' K' \lambda \sqrt{Q' h}$$

ELEV.	2/T	4/T	3/T²	6/T²	6/T³	ΣV	p=ωh	p_E	K	tan φ_J / p'=ωh	tan φ_D	b₁	c₁	U.S.	6	5	4	3	2	1	D.S.	τ_zyu
						Reservoir Full																
550	0.07326 0027	0.140,520,14	0.004,025,786	0.008,058,372	0.0294,892,7	-45,502	1,925			0	.70	-194,955,32	+201,369,3	0		1714.2					0	0
500	0.03210 2728	0.064,245,456	0.001,722,958	0.100,545,878	0.0024,813,9	-29,1880	5,362.5			0	.70	-100,905,42	-595,884,6	0		3,264.7	8,364.0					0
450	0.020,534,39	0.041,109,968	0.03,168,9952	0.06,337,610,4	0.03,106,573,7	-748,720	8,800			0	.70	-123,902,34	-264,138,16	0		3,400.4	9,562.3					0
400	0.05,117,158	0.030,234,316	0.171,396,30	0.03,342,793	0.002,591,0	-1,413,200	12,238			0	.70	-137,432,33	-136,314,55	0		3,415.5	9,919.3	15,111				0
350	0.00,970 927	0.021,941,854	0.09,027,093	0.0,990,356,7	0.990,356,7	-2,289,400	15,675			50	.70	-168,786,83	+285,314,75	3,424.80		6,110.9	9,969.8	14,102	20,142		24,172	3,494.8
315	0.00,920 956	0.018,407,731	0.06,353,357	0.0,127,067	0.584,752,5	-3,031,500	18,081			30	.70	-158,170,88	+219,857,91	3,626.51 4,238.6		6,949.9	10,675	15,194	20,139	25,179	27,192	3,666.5

$\tau_{zy} = \tau_{yz}$ Pounds per Square Foot   Vertical Plane

Note: Values for p_E and K were computed separately and are not shown.

Figure A-5. Friant Dam study—shear stresses on horizontal and vertical planes.—DS2-2(8)

FRIANT DAM NONOVERFLOW SECTION. RESERVOIR W.S. EL. 578 TAILWATER EL. NONE STUDY No. 3

GRAVITY STRESS ANALYSIS OF MAXIMUM PARALLEL-SIDE CANTILEVER

INCLUDING EFFECTS OF TAILWATER AND HORIZONTAL EARTHQUAKE

By H.P.W. Date 2-19-40

$\omega_c = .150$
$\mathfrak{Z} = 434.03$

EARTHQUAKE ← (ACCEL)

PARTIAL DERIVATIVES FOR OBTAINING σ_y

$K_1 = \frac{4}{T}p - \frac{4}{T^2}\Sigma W + \frac{\ddagger 4}{T}p_E - \frac{12}{T^3}\Sigma M$ $K_2 = \frac{2}{T^2}\Sigma W + \frac{\ddagger 2}{T}p_E - \frac{2}{T}p' - \frac{12}{T^3}\Sigma M$ $K_3 = \frac{12}{T^3}\Sigma M + \frac{2}{T^2}\Sigma W - \frac{2}{T}p' - \frac{12}{T^2}\Sigma V$ $K_4 = \frac{12}{T^3}\Sigma M + \frac{2}{T^2}\Sigma W + \frac{4}{T}p' - \frac{\ddagger 4}{T}p_E$ $\frac{\partial\sigma_{zD}}{\partial z} = \omega_c + K_1\tan\phi_U + K_2\tan\phi_U + K_3\tan\phi_U + K_4\tan\phi_D + \frac{6}{T^2}\Sigma'$

(‡Use(+)sign if horizontal earthquake acceleration is upstream) (⊗Use(−)sign if horizontal earthquake acceleration is upstream) (⊗ω−omitted if water on face is absent)

$\frac{\partial\tau_{zyU}}{\partial z} = \tan\phi_U\left(\omega^\otimes - \frac{\partial\sigma_{zU}}{\partial z} + \frac{\ddagger}{}\frac{\partial p_E}{\partial z}\right) + \frac{\partial\tan\phi_U}{\partial z}(p - \sigma_{zU} + \frac{\ddagger}{}p_E)$

$\frac{\partial\tau_{zyD}}{\partial z} = \tan\phi_D\left(\frac{\partial\sigma_{zD}}{\partial z} - \omega^\otimes + \frac{\ddagger}{}\frac{\partial p_E}{\partial z}\right) + \frac{\partial\tan\phi_D}{\partial z}(\sigma_{zD} - p' + \frac{\ddagger}{}p_E')$

$\frac{\partial\Sigma V}{\partial z} = -(p - p' + \lambda\omega_c T + \frac{\ddagger}{}p_E + p_E')$ $\frac{\partial T}{\partial z} = \tan\phi_U + \tan\phi_D$

ELEV.	$\frac{\partial\tan\phi_U}{\partial z}=\frac{\Delta\tan\phi_U}{\Delta z}$ $\frac{2}{T}$	$\frac{\partial\tan\phi_D}{\Delta z}=\frac{\Delta\tan\phi_D}{\Delta z}$ $\frac{4}{T}$	$\frac{2}{T^2}$	$\frac{6}{T^2}$	$\frac{12}{T^3}$	$\frac{\partial p_E}{\partial z}=\frac{18}{T^4}$	K_1	K_2	$K_3=\frac{\Delta p_E'}{\Delta z}$	K_4	$\frac{\partial\sigma_{zD}}{\partial z}$	$\frac{\partial p_E}{\partial z}$	$\frac{\partial\tan\phi_U}{\partial z}\frac{\partial\tau_{zyU}}{\partial z}$	$\frac{\partial p_E'}{\partial z}$	$\frac{\partial\tan\phi_D}{\partial z}\frac{\partial\tau_{zyD}}{\partial z}$	$\frac{\partial T}{\partial z}$	$\lambda\omega_c T$	$\frac{\partial\Sigma V}{\partial z}$
550				Reservoir Full			—	—	—	⁻701,140,432	5,518,153	0	0	0	⁻25,887,293	.7	15 T	⁻2,783.47
500		Note: K_1, K_2 and K_3 not required above El. 400 because U.S. face is vertical and multiplier is zero.					—	—	—	⁺642,476851	51,476,99	0	0	0	⁺62,283,893	.7	15 T	⁻7,046.36
450							—	—	—	⁺668,224,001	56,752,77	0	0	0	⁺65,976,935	.7	15 T	⁻11,219.45
400							—	—	—	⁻682,648,68	56,580,57	0	0	0	⁺65,856,399	.7	15 T	⁻15,354.50
350							83,484,033	88,412,99	33,601,79	127,803,97	604,702,13	101,699,85	0	8,669,4630	0	⁻27,439,895	1.0	15 T	⁻19,690.70
315							70,914,684	363,888,06	40,793,158	113,103,54	572,489251	100,529,19	0	6,512,052,6	0	⁺26,620,433	1.0	15 T	⁻22,716.50

Figure A-6. Friant Dam study—partial derivatives for obtaining σ_y. —DS2-2(9)

WITH EARTHQUAKE

INTERMEDIATE COMPUTATIONS FOR OBTAINING STRESSES — GRAVITY ANALYSIS OF FRIANT DAM

FRIANT DAM   STUDY NO. 3 - NONOVERFLOW   RES. W.S. EL. 578   T.W.S. EL. NONE   BY H.P.W.   DATE

	$\frac{dt}{dz}$	$\frac{12}{T^3}\Sigma V$	$\frac{4}{T^2}t_{zyu}$	$\frac{2}{T^2}t_{zyu}$	$\frac{6}{T^2}\frac{\partial \Sigma V}{\partial z}$	$\frac{2}{T}\frac{\partial t_{zyu}}{\partial z}$	$\frac{4}{T}\frac{dt_{zyD}}{dz}$	$\frac{\partial b}{\partial z}=$	$-\frac{dt}{dz}$	$\frac{18}{T^4}\Sigma V$	$\frac{6}{T^3}t_{zyu}$	$\frac{6}{T^3}t_{zyD}$	$\frac{6}{T^3}\frac{\partial \Sigma V}{\partial z}$	$\frac{3}{T^2}\frac{dt_{zyu}}{dz}$	$\frac{3}{T^2}\frac{dt_{zyD}}{dz}$	$\frac{\partial c}{\partial z}=$
		Normal		Full	Reservoir	Operation										
550	.7	26.836370	0	2055941	22.408484	0	3.7930098	21.807619	-.7	7.4945258	0	1.1296377	8.2082359	0	7.0420357	6.8360555
500	.7	14.495098	0	8862218,6	10.892.811	0	3.9989658	2.9578296	-.7	3.4875836	0	2.13.37603	1.7484447	0	5.4814164	3.93520
450	.7	9.7533565	0	6.150173,57	11.04503	0	2.712,30991	1.8757863	-.7	1.50.36284	0	0.9481.2538	0.73077596	0	0.20.90681	0.3.28558
400	.7	7.323.2749	0	4.700430,75	2.63410,5	0	1.9911,23.2	1.436,2964	-.7	0.83.03027	0	0.532426	0.3.978.3904	0	0.11.287540	0.0.680,3
350	1.0	4.534.6411	210.319,23	2982875,3	3.5549956	0.9511204,6	6.02.082,17	1.516.354,8	-1.0	0.37.3119.13	0.0346.105	0.2454.369	0.09500.799	0.07826089	0.0247.024	0.0693404
315	1.0	3.545.3544	156.144,66	2.344.424,1	2.8865.109	0.5933.657	4.9002.177	1.29.767,4	-1.0	0.24.47322	0.02.155.202	0.16.183.323	0.3.283530	0.4.37.3256	0.09.295.500	0.0504431

Figure A-7. Friant Dam study—intermediate computations for obtaining stresses. —DS2-2(10)

FRIANT DAM NONOVERFLOW SECTION. RESERVOIR W.S. EL. 578 TAILWATER EL. NONE STUDY No. 3

GRAVITY STRESS ANALYSIS OF MAXIMUM PARALLEL-SIDE CANTILEVER

INCLUDING EFFECTS OF TAILWATER AND HORIZONTAL EARTHQUAKE

EARTHQUAKE ←

NORMAL STRESS ON VERTICAL PLANES

$\sigma_y = a_2 + b_2 y + c_2 y^2 + d_2 y^3$

By H.P.W. Date 4-2-40

$\frac{\partial a_1}{\partial z} = \frac{\partial \tau_{zyD}}{\partial z}$

$\frac{\partial b_1}{\partial z} = \frac{\partial T}{\partial z}\left[\frac{12}{T^3}(\Sigma V) + \frac{2}{T^2}(\tau_{zyu}) + \frac{4}{T^2}(\tau_{zyD})\right] - \frac{6}{T^2}\left(\frac{\partial \Sigma V}{\partial z}\right) - \frac{2}{T}\left(\frac{\partial \tau_{zyu}}{\partial z}\right) - \frac{4}{T}\left(\frac{\partial \tau_{zyD}}{\partial z}\right)$

$\frac{\partial c_1}{\partial z} = -\frac{\partial T}{\partial z}\left[\frac{18}{T^4}(\Sigma V) + \frac{6}{T^3}(\tau_{zyu}) + \frac{6}{T^3}(\tau_{zyD})\right] + \frac{6}{T^3}\left(\frac{\partial \Sigma V}{\partial z}\right) + \frac{3}{T^2}\left(\frac{\partial \tau_{zyu}}{\partial z}\right) + \frac{3}{T^2}\left(\frac{\partial \tau_{zyD}}{\partial z}\right)$

(‡ Use (+) sign if horizontal earthquake acceleration is upstream) (‡ Use (−) sign if horizontal earthquake acceleration is upstream)

$a_2 = \sigma_{yD} = a_1 \tan\phi_D + p' \overset{\ddagger}{\pm} p'_E$ $b_2 = b_1 \tan\phi_D + \frac{\partial a_1}{\partial z} \overset{\ddagger}{\pm} \lambda\omega_c$ $c_2 = c_1 \tan\phi_D + \frac{1}{2}\frac{\partial b_1}{\partial z}$ $d_2 = \frac{1}{3}\frac{\partial c_1}{\partial z}$ Check for $y = T$; $\sigma_{yu} = (p \overset{\ddagger}{\pm} p_E) - \tau_{zyu}\tan\phi_U$

ELEV.	$\frac{\partial a_1}{\partial z}$	$\frac{\partial b_1}{\partial z}$	$\frac{\partial c_1}{\partial z}$	b_2	c_2	d_2	σ_y Pounds per Square Foot — VERTICAL PLANE: U.S.	6	5	4	3	2	1	D.S.	σ_{yu}
			Reservoir	Full											
550	−25.887,293	+21.807,619	−683,605,49	−177.356,02	+12.304,768	−227,868,5	2,373.97		1914.9					2,619.5	2,373.9
500	+62.283,893	+2.957,829,6	−.031,935,70	23.349,901	+861,795,3	−010,645,07	6,142 86		6143.9	5,941.2				6,042.5	6,111.9
450	+65.976,939	+1.875,786,3	−013,285,579	−35.754,699	+.752,996,4	−004,428,53	9,759 95		9869.5	9,667.6				10,171.5	9,760.0
400	+65.856,399	+1.436,296,4	−007,680,134	45.346,232	+.627,728,0	−002,560,05	13,393.0		13,594.1	13,493.9	13,494.5			14,100.8	13,370.0
350	+27.439,895	+.516,354,8	−006,934,037	−105.710,89	+.957,897,7	−002,311,35	15,110 75		15,105.4	14,988.3	13,896.6	15,105.9		17,120.9	15,907.8
315	+26.620,433	+1.219,767,4	−005,044,313	99.099,183	+.799,784,2	−001,681,44	18,127.0	18,127.7	17,121.4	16,114.13	15,109.7	16,112.8	18,127.1	19,135.9	18,351.0

Figure A-8. **Friant Dam study—normal stresses on vertical planes. —DS2-2(11)**

FRIANT DAM. NONOVERFLOW SECTION. RES. W.S. EL. 578 TAILWATER EL NONE STUDY No. 3

GRAVITY STRESS ANALYSIS OF MAXIMUM PARALLEL-SIDE CANTILEVER

RESERVOIR FULL WITH EARTHQUAKE   PRINCIPAL STRESSES   By H.P.W. Date 4-6-40

$$\sigma_{PI} = \frac{\sigma_z + \sigma_y}{2} \pm \sqrt{\left(\frac{\sigma_z - \sigma_y}{2}\right)^2 + (t_{zy})^2}$$

If $(\sigma_z - \sigma_y) > 0$, use (+); If $(\sigma_z - \sigma_y) < 0$, use (−). Alternate sign gives σ_{P2} which is perpendicular to σ_{PI}.

$$\phi_{PI} = \tfrac{1}{2}\arctan\left(-\frac{t_{zy}}{\frac{\sigma_z - \sigma_y}{2}}\right)$$

ELEV.		VERTICAL PLANE								Stress in pounds per sq. inch							
		U.S.	6	5	4	3	2	1	D.S.	U.S.	6	5	4	3	2	1	D.S.
550	$\tfrac{1}{2}(\sigma_z+\sigma_y)$	2,548.96		3,345.7					4,076.9								
	$\tfrac{1}{2}(\sigma_z-\sigma_y)$	174.99		1,391.8					1,395.5								
	σ_{PI}	2,723.95		6,219.4					8,153.8	19		43					57
	σ_{P2}	2,373.97		472.0					0	16		3					0
	$\tan 2\phi_{PI}$	0		1.806,438					2.745,023								
	ϕ_{PI}	0		−30°31'					−35°00'								
500	$\tfrac{1}{2}(\sigma_z+\sigma_y)$	4,642.21		6,111.9	8,959.7				9,152.0								
	$\tfrac{1}{2}(\sigma_z-\sigma_y)$	7,469.65		−14.05	2,988.5				3,132.6								
	σ_{PI}	3,172.56		2,847.2	17,841.6				18,304	22		20	124				127
	σ_{P2}	6,111.86		9,376.6	77.8				0	42		65	1				0
	$\tan 2\phi_{PI}$	0		−232.362	2.798,728				2.745,071								
	ϕ_{PI}	0		+44°53'	−35°10'				−35°00'								
450	$\tfrac{1}{2}(\sigma_z+\sigma_y)$	6,196.85		8,123.2	11,753.2				15,492.1								
	$\tfrac{1}{2}(\sigma_z-\sigma_y)$	3,563.10		1,756.4	2,079.6				5,302.8								
	σ_{PI}	2,633.75		4,296.0	21,543.9				30,984.2	18		30	150				215
	σ_{P2}	9,789.95		11,950.4	1,962.5				0	68		83	14				0
	$\tan 2\phi_{PI}$	0		1.936,005	4600,548				2.745,035								
	ϕ_{PI}	0		31°20'	−38°52'				−35°00'								
400	$\tfrac{1}{2}(\sigma_z+\sigma_y)$	7,767.39		9,906.3	13,957.8	18,117.3			21,890.5								
	$\tfrac{1}{2}(\sigma_z-\sigma_y)$	5,602.61		3,626.9	550.90	4,620.7			7,492.7								
	σ_{PI}	2,164.78		4,924.3	23,892.4	34,758.7			43,781.0	15		34	166	241			304
	σ_{P2}	13,370.0		14,888.3	4,023.2	1,475.9			0	93		103	28	10			0
	$\tan 2\phi_{PI}$	0		.941,713	18005,627	3.459,865			2745,105								
	ϕ_{PI}	0		21°38'	−43°25'	−36°56'			−35°00'								
350	$\tfrac{1}{2}(\sigma_z+\sigma_y)$	9,966.71		12,611.7	15,556.6	18,843.8	22,916.9		26,375.9								
	$\tfrac{1}{2}(\sigma_z-\sigma_y)$	5,941.05		2,561.7	1,378.4	4,976.2	7,788.0		9,028.0								
	σ_{PI}	3,073.98		5,985.6	25,608.4	34,378.0	44,741.3		52,751.8	21		42	178	239	311		366
	σ_{P2}	16,859.44		19,237.8	5,504.8	3,309.6	1,092.5		0	117		134	38	23	8		0
	$\tan 2\phi_{PI}$	.588,246		2.385,486	7.223,447	2.957,196	2.617,809		2.745,092								
	ϕ_{PI}	15°14'		33°38'	−41°04'	−38°39'	−34°33'		−35°00'								
315	$\tfrac{1}{2}(\sigma_z+\sigma_y)$	12,071.82	12,466.20	15,142.8	17,712.4	20,497.9	23,822.2	28,008.1	29,454.7								
	$\tfrac{1}{2}(\sigma_z-\sigma_y)$	6,279.22	5,819.50	2,284.6	1,357.2	4,783.2	7,670.4	9,696.0	10,081.8								
	σ_{PI}	4,790.41	5,383.20	7,788.1	28,446.3	36,249.9	45,310.1	55,607.1	58,909.8	33	37	54	198	252	315	386	409
	σ_{P2}	19,353.23	19,549.20	22,497.5	6,978.5	4,745.9	2,334.3	409.1	0	134	136	156	48	33	16	3	0
	$\tan 2\phi_{PI}$	.587,097	.693,805	3.060,011	7.845,417	3.137,690	2.616,839	2.664,986	2.745,096								
	ϕ_{PI}	15°14'	17°23'	35°57'	−41°22'	−36°10'	−34°33'	−34°43'	−35°00'								
	$\tfrac{1}{2}(\sigma_z+\sigma_y)$																
	$\tfrac{1}{2}(\sigma_z-\sigma_y)$																
	σ_{PI}																
	σ_{P2}																
	$\tan 2\phi_{PI}$																
	ϕ_{PI}																

Figure A-9. Friant Dam study—principal stresses. —DS2-2(12)

RESERVOIR EMPTY

VERTICAL STRESS	STRESS PARALLEL TO FACE	HORIZ. SHEAR STRESS
Pounds Per Square Inch		
43	43	0
99	99	0
149	149	0
201	201	0
201	219	-60
213	232	-64
69	69	0
130	130	0
188	188	0
197	215	-59
219	239	-66

NORMAL RES. W.S. EL. 578

WATER PRESSURE	VERTICAL STRESS	STRESS PARALLEL TO FACE	HORIZ. SHEAR STRESS
Pounds Per Square Inch			
12	30	30	0
34	46	46	0
56	53	53	0
77	61	61	0
99	74	71	8
114	87	84	8
34	38	38	0
56	58	58	0
77	73	73	0
99	85	84	4
121	107	105	4

UPSTREAM FACE

NORMAL RES. W.S. EL. 578

SLIDING FACTOR	SHEAR FRICTION FACTOR	MAX.HORIZ. SHEAR STRESS	WATER PRESSURE	VERTICAL STRESS	STRESS PARALLEL TO FACE	HORIZ. SHEAR STRESS
				Pounds Per Square Inch		
.286	74.48	15	0	22	33	15
.565	22.38	36	0	52	78	36
.659	13.30	66	0	95	141	66
.704	9.58	97	0	138	206	97
.677	8.23	123	0	175	261	123
.658	7.50	140	0	200	297	140
.581	26.10	33	0	46	69	33
.611	14.77	61	0	87	129	61
.643	10.39	89	0	127	190	89
.629	8.74	116	0	166	248	116
.613	7.59	139	2	200	297	139

RESERVOIR EMPTY

VERTICAL STRESS	STRESS PARALLEL TO FACE	HORIZ. SHEAR STRESS
9	14	6
-1	-1	-0.7
-2	-2	-1
-1	-2	-1
33	50	23
51	75	35
10	15	7
11	17	8
10	15	7
40	60	28
63	94	44

DOWNSTREAM FACE

HORIZONTAL SECTION

MAXIMUM NON-OVERFLOW SECTION

TOP OF DAM EL.582
RES. W.S. EL.578
EL. 550
EL. 500
EL. 450
EL. 400
EL. 350
EL. 315
SLOPE = 0.70
SLOPE = 0.30
AXIS OF DAM

MAXIMUM SPILLWAY SECTION

CREST EL. 560
RES. W.S. EL.578
EL. 500
EL. 450
EL. 400
EL. 350
EL. 300
TAILWATER SURFACE EL.305
SLOPE = 0.70
SLOPE = 0.30
AXIS OF DAM

↓ Resultant—concrete weight only. ↘ Resultant—water pressure and weight. ↗ Resultant—water pressure, weight and uplift.

Sliding factor = Horizontal Force / (Weight−Uplift)

Shear−friction factor = [(Weight−Uplift) x Coefficient of Internal Friction + Horizontal Area x Unit Shear Resistance] / Horizontal Force

Unit weight of concrete = 150 pounds per cubic foot. Unit shear resistance = 450 pounds per square inch.
Coefficient of internal friction = 0.65.
Uplift pressure varies as a straight line from reservoir water pressure at upstream face to zero or tailwater pressure at downstream face, acting over two-thirds the area of the horizontal section.
Total load carried by vertical cantilever.
All normal stresses are compressive except those preceded by a negative sign, which are tensile.
Positive shear stresses are caused by shear forces acting thus ⇒. Negative shear stresses are caused by shear forces acting thus ⇐.
Weight of gate included in analysis of spillway section.

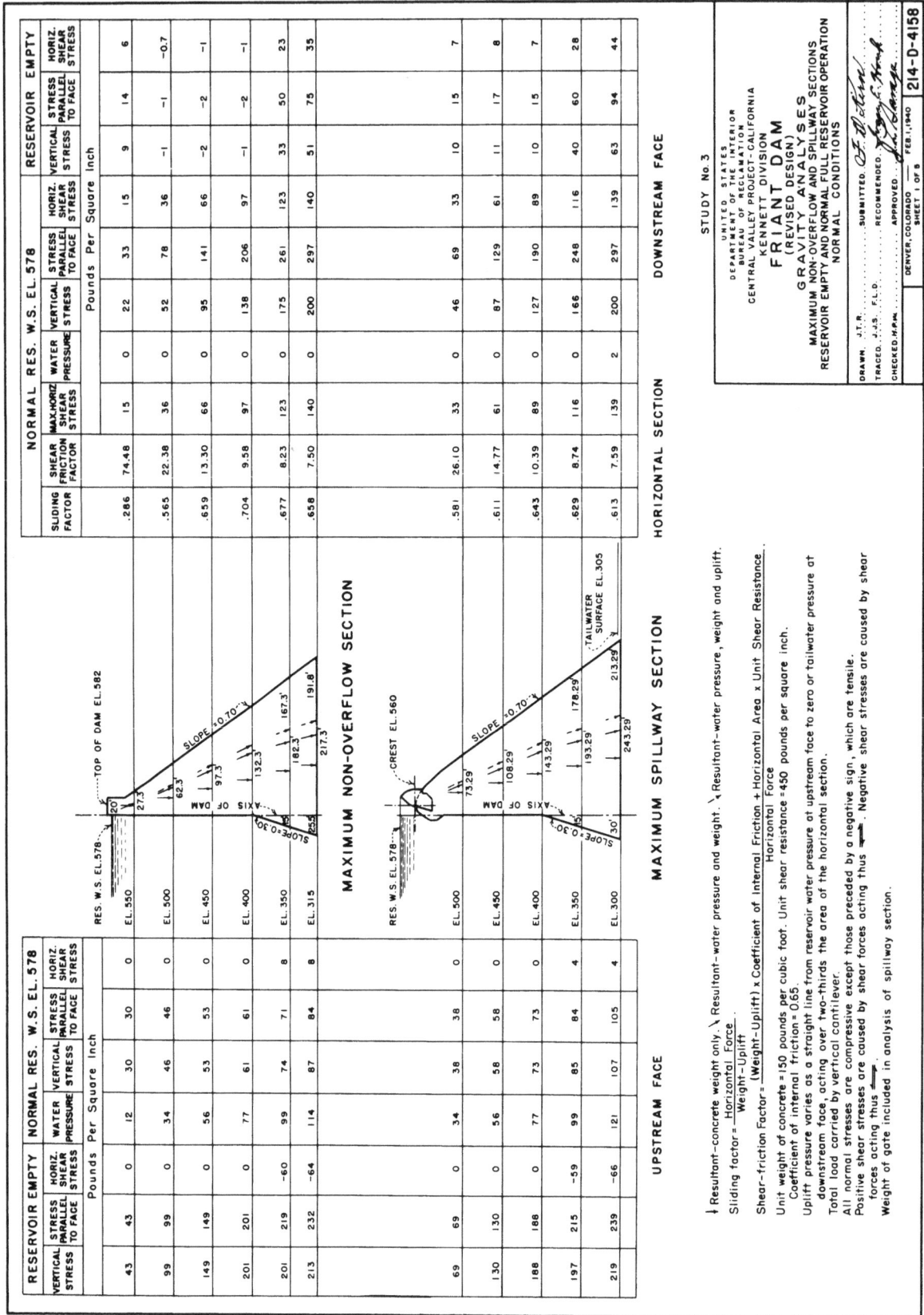

STUDY No 3

UNITED STATES
DEPARTMENT OF THE INTERIOR
BUREAU OF RECLAMATION
CENTRAL VALLEY PROJECT-CALIFORNIA
KENNETT DIVISION
FRIANT DAM
(REVISED DESIGN)
GRAVITY ANALYSES
MAXIMUM NON-OVERFLOW AND SPILLWAY SECTIONS
RESERVOIR EMPTY AND NORMAL FULL RESERVOIR OPERATION
NORMAL CONDITIONS

DRAWN. J.T.R. .... SUBMITTED.
TRACED. J.J.S. F.L.D. .... RECOMMENDED.
CHECKED.M.P.M. .... APPROVED.
DENVER, COLORADO .... FEB.1,1940   SHEET 1 OF 8   214-D-4158

Figure A-10. Friant Dam study—gravity analyses for normal conditions.

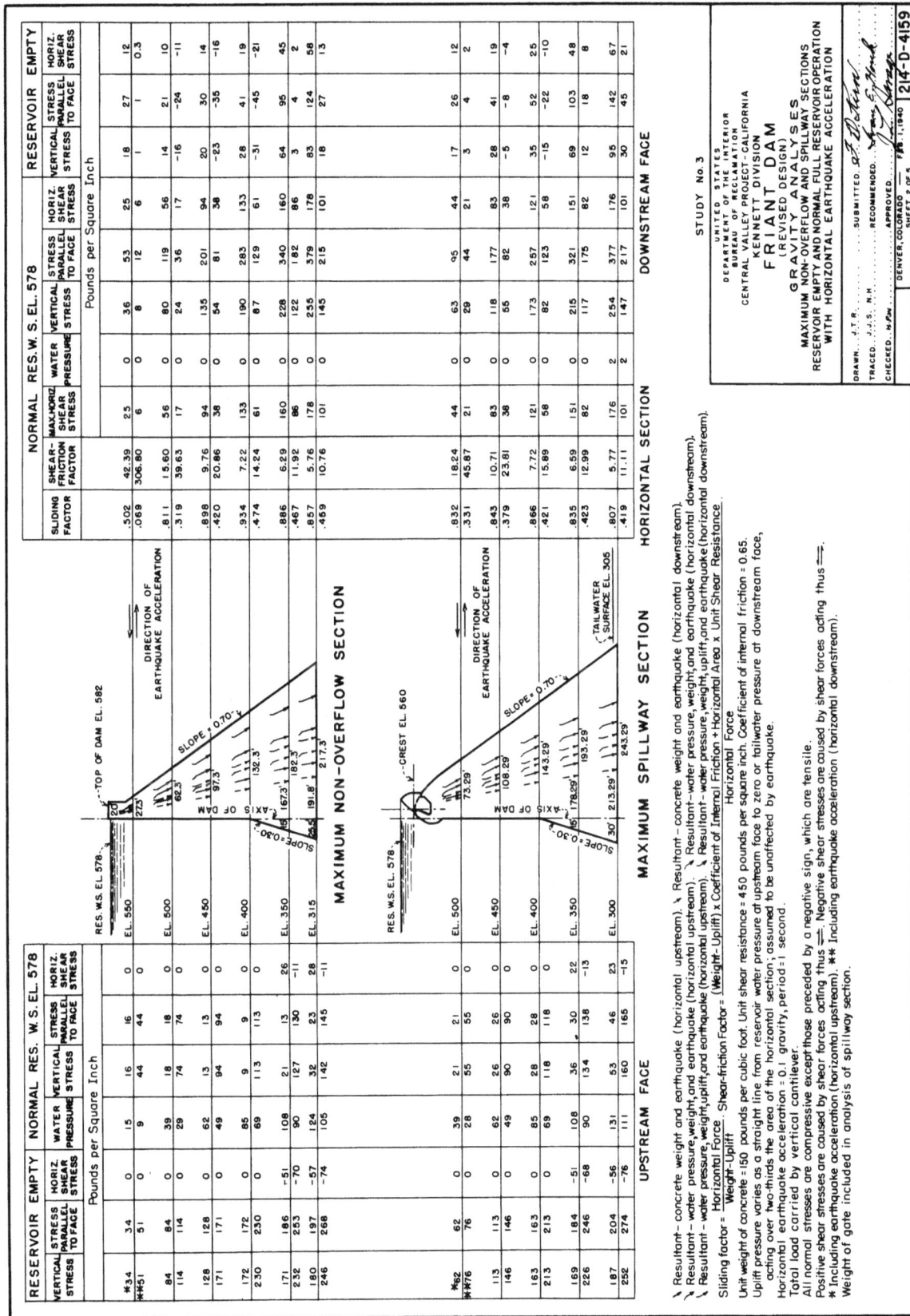

Figure A-11. Friant Dam study—gravity analyses with horizontal earthquake acceleration.

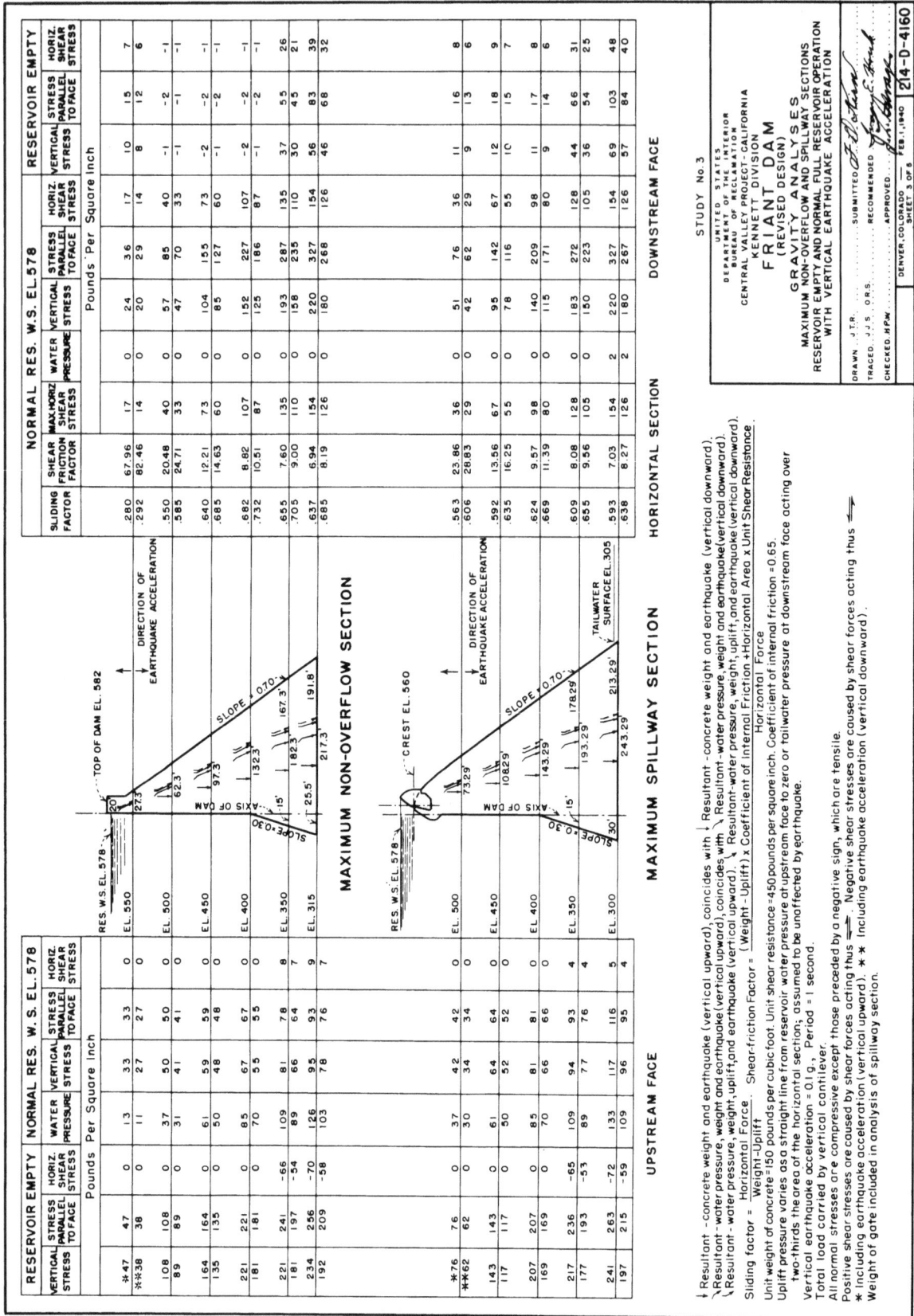

Figure A-12. Friant Dam study—gravity analyses with vertical earthquake acceleration.

DOWNSTREAM FACE (NORMAL RES. W.S. EL.578 and RESERVOIR EMPTY)

SLIDING FACTOR	SHEAR FRICTION FACTOR	MAX.HORIZ. SHEAR STRESS	WATER PRESSURE	VERTICAL STRESS	STRESS PARALLEL TO FACE	HORIZ. SHEAR STRESS	VERTICAL STRESS	STRESS PARALLEL TO FACE	HORIZ. SHEAR STRESS
			NORMAL RES. W.S. EL.578				RESERVOIR EMPTY		
.474	40.25	27	0	38	57	27	19	28	13
.088	218.08	7	0	10	15	7	1	2	0.9
.768	14.68	60	0	85	127	60	14	21	10
.332	33.89	21	0	29	44	20	-16	-24	-11
.851	9.19	102	0	144	215	102	20	30	14
.429	18.21	45	0	64	95	45	-24	-35	-16
.885	6.81	143	0	204	304	143	27	41	19
.479	12.56	70	0	100	150	70	-31	-46	-21
.839	5.93	172	0	246	366	172	67	100	47
.471	10.58	98	0	140	209	98	6	9	4
.812	5.45	192	0	275	409	192	88	132	62
.462	9.58	110	0	157	235	110	23	34	16
.783	17.15	48	0	68	102	48	18	27	13
.342	39.25	24	0	34	51	24	4	6	3
.797	10.08	89	0	127	189	89	29	43	20
.388	20.70	45	0	64	95	45	-4	-6	-3
.820	7.28	130	0	185	276	130	36	54	25
.428	13.96	67	0	95	142	67	-14	-21	-10
.790	6.23	162	0	232	345	162	73	109	51
.428	11.49	94	0	134	199	94	16	24	11
.764	5.46	190	2	274	407	190	101	151	71
.422	9.88	115	3	167	247	115	37	25	26

HORIZONTAL SECTION   DOWNSTREAM FACE — Pounds Per Square Inch

UPSTREAM FACE (RESERVOIR EMPTY and NORMAL RES. W.S. EL.578)

VERTICAL STRESS	STRESS PARALLEL TO FACE	HORIZ. SHEAR STRESS	WATER PRESSURE	VERTICAL STRESS	STRESS PARALLEL TO FACE	HORIZ. SHEAR STRESS
RESERVOIR EMPTY			NORMAL RES. W.S. EL.578			
*38	38	0	16	19	19	0
**56	56	0	10	47	47	0
93	93	0	42	22	22	0
123	123	0	32	78	78	0
143	143	0	68	18	18	0
186	186	0	54	99	99	0
192	192	0	93	15	15	0
250	250	0	77	119	119	0
208	191	-57	118	28	20	27
275	252	-76	100	134	137	-10
220	202	-61	135	40	32	28
291	267	-80	116	150	153	-10
*69	69	0	43	25	25	0
**83	83	0	32	59	59	0
126	126	0	68	32	32	0
159	159	0	54	95	95	0
182	182	0	93	35	35	0
232	232	0	77	126	126	0
245	205	-56	118	45	38	22
267	228	-74	100	143	147	-13
209	228	-63	143	64	57	24
274	298	-82	123	171	175	-14

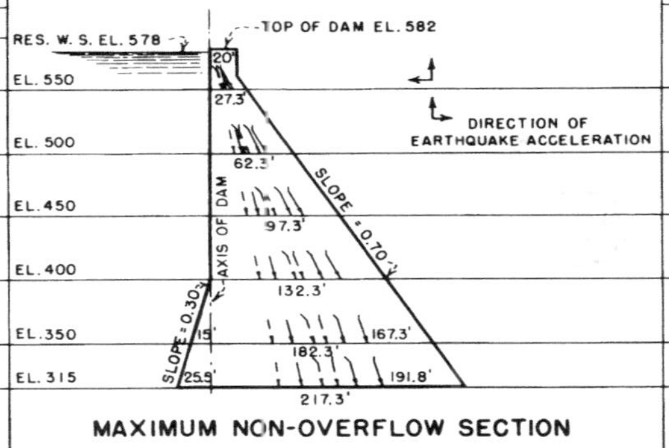

UPSTREAM FACE — Pounds Per Square Inch

MAXIMUM NON-OVERFLOW SECTION
DIRECTION OF EARTHQUAKE ACCELERATION
TOP OF DAM EL.582 — AXIS OF DAM — RES. W.S. EL.578
SLOPE = 0.70 — SLOPE = 0.30
EL.550, EL.500, EL.450, EL.400, EL.350, EL.315
97.3', 132.3', 167.3', 182.3', 191.8', 217.3', 62.3', 20', 27.3'

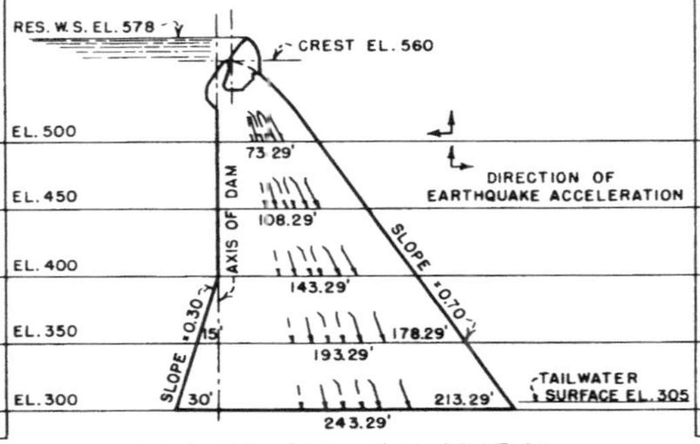

MAXIMUM SPILLWAY SECTION
DIRECTION OF EARTHQUAKE ACCELERATION
CREST EL.560 — AXIS OF DAM — RES. W.S. EL.578
SLOPE = 0.70 — SLOPE = 0.30
EL.500, EL.450, EL.400, EL.350, EL.300
73.29', 108.29', 143.29', 178.29', 193.29', 213.29', 243.29'
TAILWATER SURFACE EL.305

ALL RESULTANT FORCES INCLUDE VERTICAL EARTHQUAKE ACCELERATION UPWARD.
Resultant–concrete weight and earthquake (horizontal upstream).
Resultant–water pressure, weight, and earthquake (horizontal downstream).
Resultant–water pressure, weight, uplift, and earthquake (horizontal upstream).
Resultant–water pressure, weight, uplift, and earthquake (horizontal downstream).

$$\text{Sliding factor} = \frac{\text{Horizontal Force}}{\text{Weight} - \text{Uplift}}$$

$$\text{Shear-friction Factor} = \frac{(\text{Weight} - \text{Uplift}) \times \text{Coefficient of Internal Friction} + \text{Horizontal Area} \times \text{Unit Shear Resistance}}{\text{Horizontal Force}}$$

Unit weight of concrete = 150 pounds per cubic foot. Unit shear resistance = 450 pounds per square inch. Coefficient of internal friction = 0.65.
Uplift pressure varies as a straight line from reservoir water pressure at upstream face to zero or tailwater pressure at downstream face, acting over two-thirds the area of the horizontal section; assumed to be unaffected by earthquake.
Vertical earthquake acceleration and horizontal earthquake acceleration = 0.1 gravity, period = 1 second.
Total load carried by vertical cantilever.
All normal stresses are compressive except those preceded by a negative sign, which are tensile.
Positive shear stresses are caused by shear forces acting thus ⟶ . Negative shear stresses are caused by shear forces acting thus ⟵ .
* Including earthquake acceleration (horizontal upstream and vertical upward). ** Including earthquake acceleration (horizontal downstream and vertical upward).
Weight of gate included in analysis of spillway section.

STUDY No.3

UNITED STATES
DEPARTMENT OF THE INTERIOR
BUREAU OF RECLAMATION
CENTRAL VALLEY PROJECT–CALIFORNIA
KENNETT DIVISION
FRIANT DAM
(REVISED DESIGN)
GRAVITY ANALYSES
MAXIMUM NON-OVERFLOW AND SPILLWAY SECTIONS
RESERVOIR EMPTY AND NORMAL FULL RESERVOIR OPERATION
WITH HORIZONTAL AND VERTICAL EARTHQUAKE EFFECTS
VERTICAL ACCELERATION UPWARD

DRAWN. J.T.R.   SUBMITTED.
TRACED. J.E.S. D.W.S.   RECOMMENDED.
CHECKED. H.F.W.   APPROVED.
DENVER, COLORADO   FEB. 1, 1940   SHEET 4 OF 5
214-D-4161

Figure A-13. Friant Dam study—gravity analyses with horizontal and vertical earthquake effects, vertical acceleration upward.

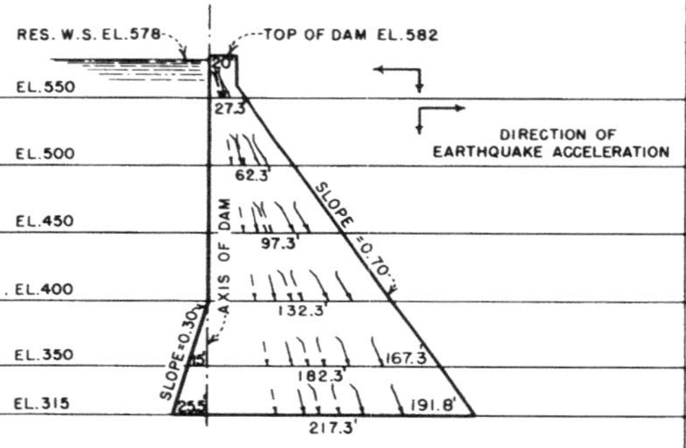

MAXIMUM NON-OVERFLOW SECTION

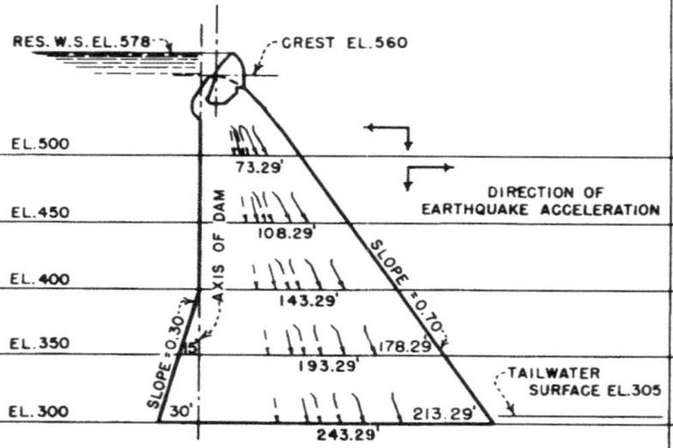

MAXIMUM SPILLWAY SECTION

UPSTREAM FACE

DOWNSTREAM FACE

HORIZONTAL SECTION

DIRECTION OF EARTHQUAKE ACCELERATION

ALL RESULTANT FORCES INCLUDE VERTICAL EARTHQUAKE ACCELERATION DOWNWARD.

Resultant–concrete weight and earthquake (horizontal upstream).
Resultant–concrete weight and earthquake (horizontal downstream).
Resultant–water pressure, weight, and earthquake (horizontal upstream).
Resultant–water pressure, weight, and earthquake (horizontal downstream).
Resultant–water pressure, weight, uplift, and earthquake (horizontal upstream).
Resultant–water pressure, weight, uplift, and earthquake (horizontal downstream).

Sliding factor = $\dfrac{\text{Horizontal Force}}{\text{Weight} - \text{Uplift}}$.

Shear-friction Factor = $\dfrac{(\text{Weight} - \text{Uplift}) \times \text{Coefficient of Internal Friction} + \text{Horizontal Area} \times \text{Unit Shear Resistance}}{\text{Horizontal Force}}$

Unit weight of concrete = 150 pounds per cubic foot. Unit shear resistance = 450 pounds per square inch. Coefficient of internal friction = 0.65.
Uplift pressure varies as a straight line from reservoir water pressure at upstream face to zero or tailwater pressure at downstream face, acting over two-thirds the area of the horizontal section; assumed to be unaffected by earthquake.
Total load carried by vertical cantilever.
Vertical earthquake acceleration and horizontal earthquake acceleration = 0.1 gravity, period = 1 second.
All normal stresses are compressive except those preceded by a negative sign, which are tensile.
Positive shear stresses are caused by shear forces acting thus →. Negative shear stresses are caused by shear forces acting thus ←.
*Including earthquake acceleration (horizontal upstream and vertical downward). **Including earthquake acceleration (horizontal downstream and vertical downward).
Weight of gate included in analysis of spillway section.

STUDY No.3

UNITED STATES
DEPARTMENT OF THE INTERIOR
BUREAU OF RECLAMATION
CENTRAL VALLEY PROJECT–CALIFORNIA
KENNETT DIVISION
FRIANT DAM
(REVISED DESIGN)
GRAVITY ANALYSES
MAXIMUM NON-OVERFLOW AND SPILLWAY SECTIONS
RESERVOIR EMPTY AND NORMAL FULL RESERVOIR OPERATION
WITH HORIZONTAL AND VERTICAL EARTHQUAKE EFFECTS
VERTICAL ACCELERATION DOWNWARD

DRAWN. J.T.R.   SUBMITTED.
TRACED. J.J.S.P.L.D.   RECOMMENDED.
CHECKED. M.P.K.   APPROVED.
DENVER, COLORADO   FEB. 1, 1940   SHEET 3 OF 5   214-D-4162

Figure A-14. Friant Dam study—gravity analyses with horizontal and vertical earthquake effects, vertical acceleration downward.

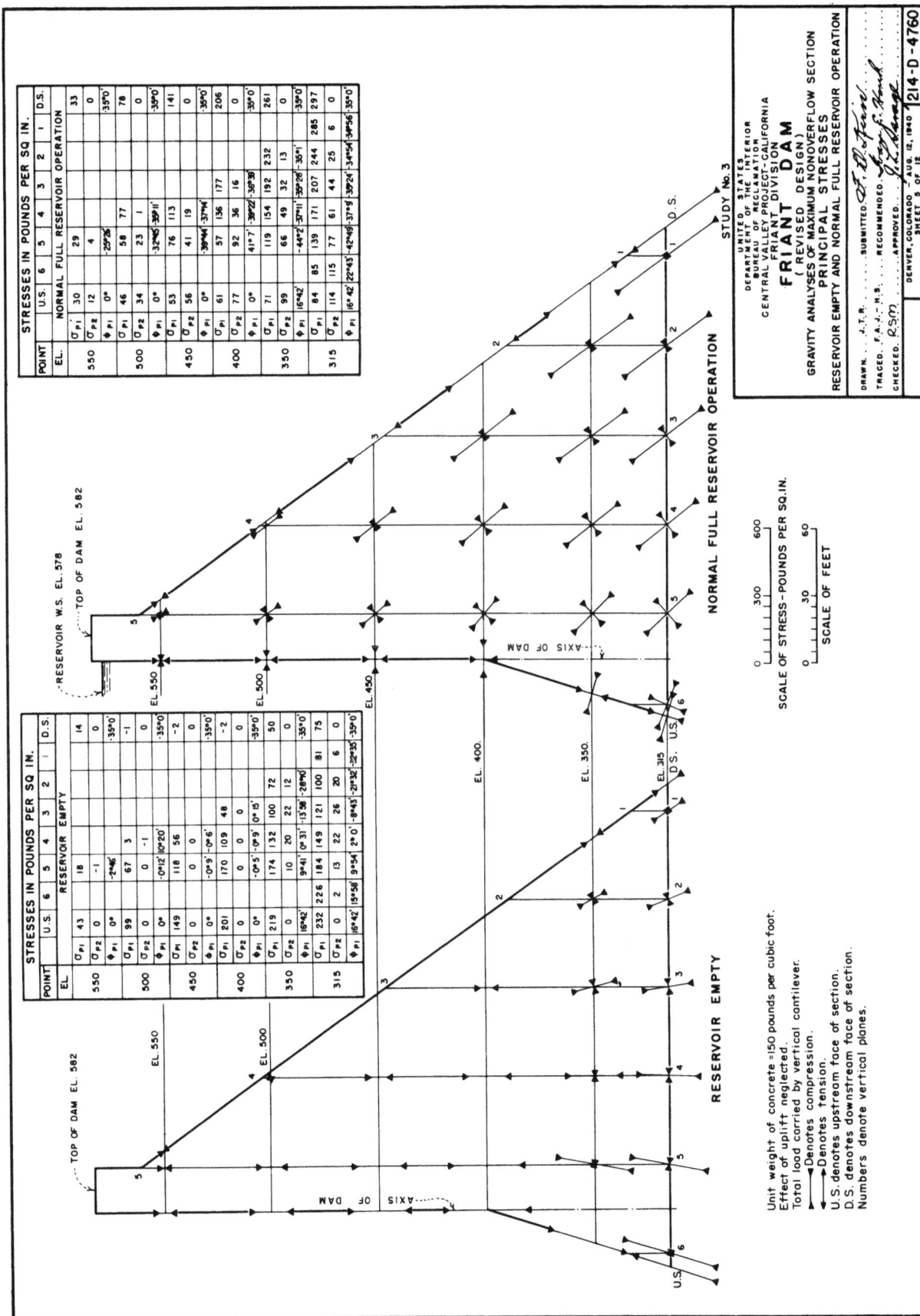

Figure A-15. Friant Dam study—principal stresses on the maximum nonoverflow section, normal conditions.

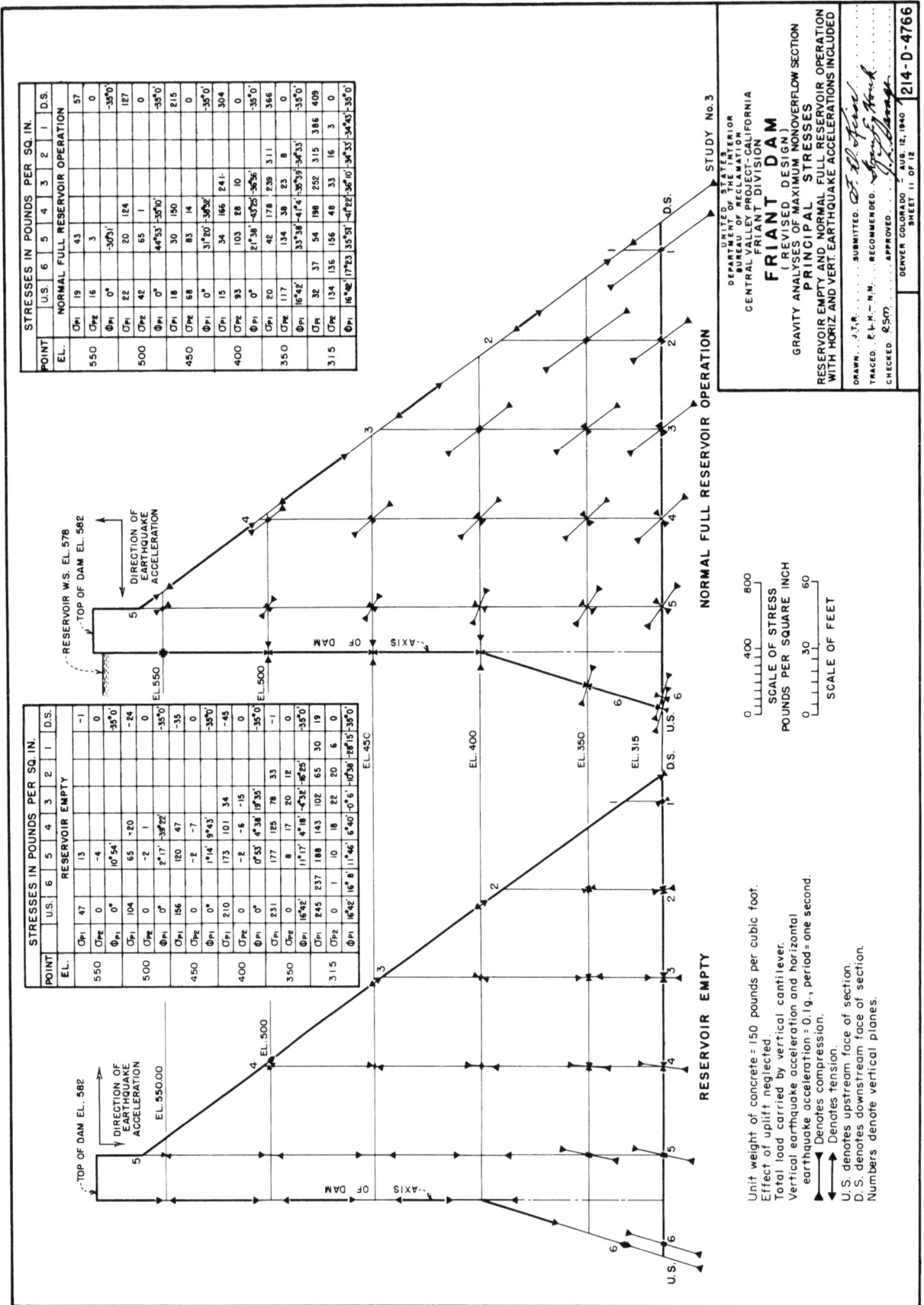

Figure A-16. Friant Dam study—principal stresses on the maximum nonoverflow section, horizontal and vertical earthquake accelerations included.

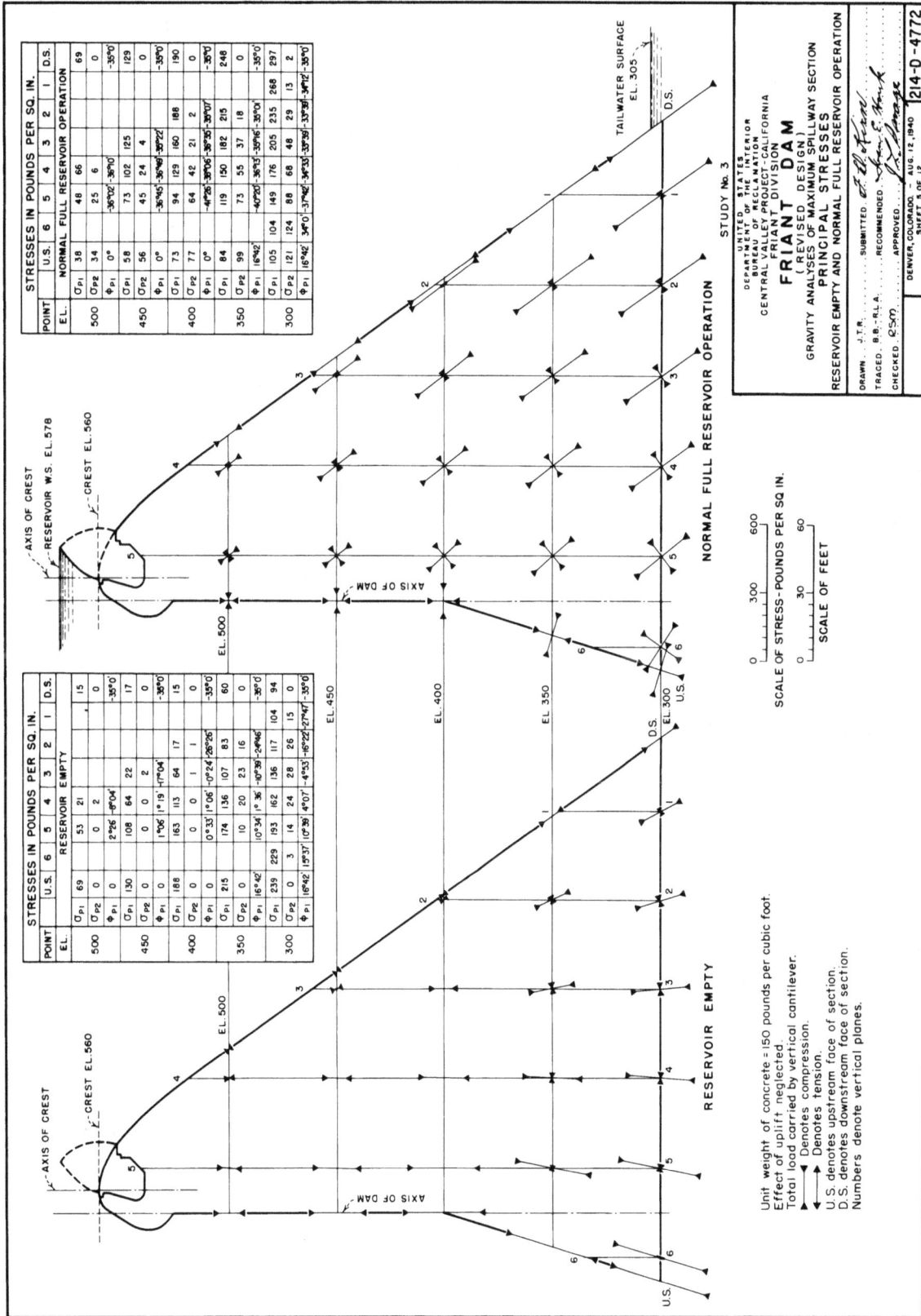

Figure A-17. Friant Dam study—principal stresses on the spillway section for normal conditions.

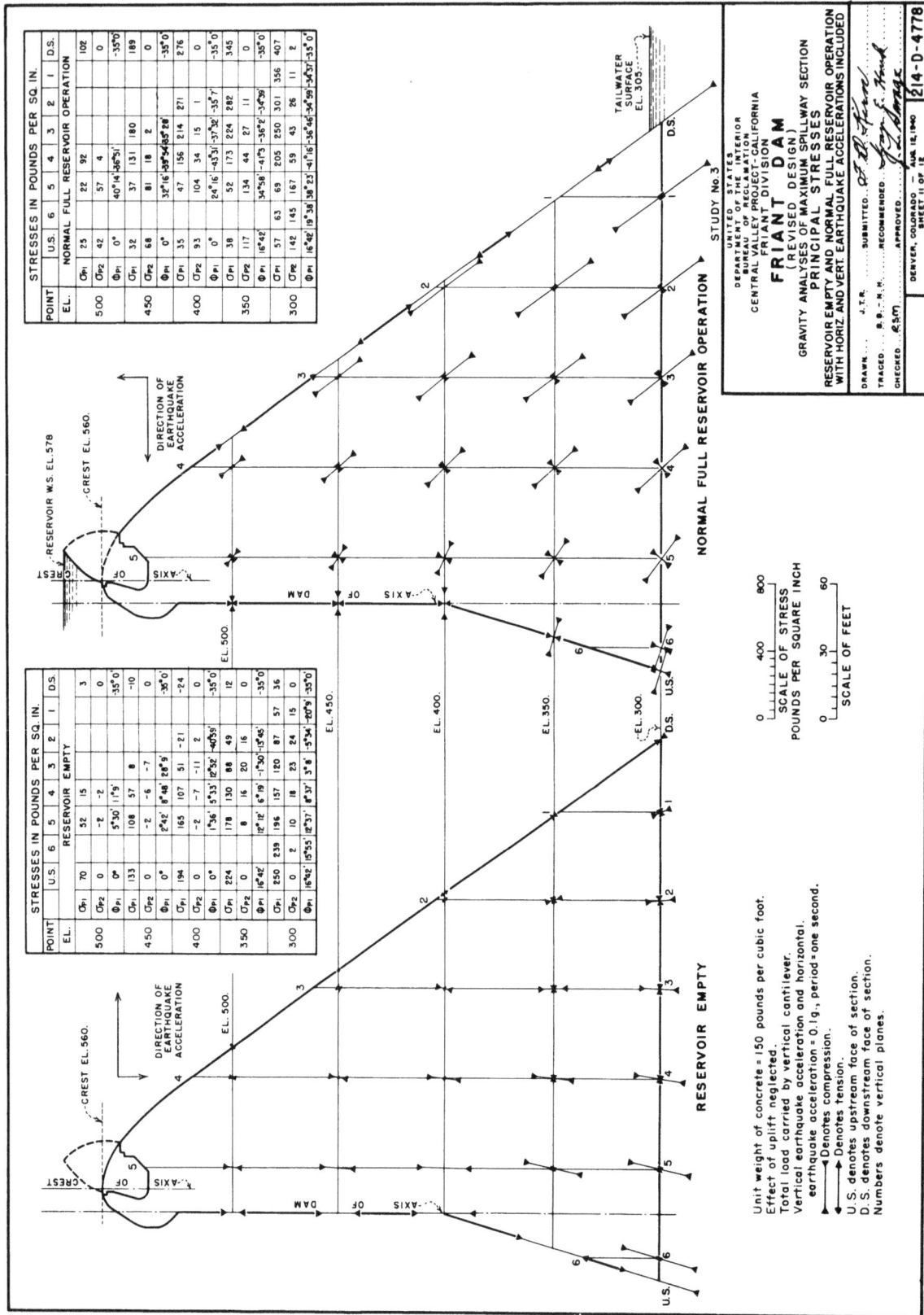

Figure A-18. Friant Dam study—principal stresses on the spillway section, horizontal and vertical earthquake accelerations included.

Table A-1.—Friant Dam, nonoverflow and spillway sections (revised design)—maximum stresses, sliding factors, and minimum shear-friction factors. DS2-2/22)

Loading conditions	Nonoverflow section					Spillway section				
	Stress, lbs. per sq. in.			Max. sliding factor	Min. shear-friction factor	Stress, lbs. per sq. in.			Max. sliding factor	Min. shear-friction factor
	Direct		Max. shear			Direct		Max. shear		
	Compr.	Tens.				Compr.	Tens.			
A. Normal conditions:										
1. Reservoir empty	232	2	64	—	—	239	none	66	—	—
2. Normal full reservoir operation	297	none	140	0.704	7.50	297	none	139	0.643	7.59
B. Including earthquake effect:										
1. Reservoir empty	291	46	80	—	—	298	24	82	—	—
2. Normal full reservoir operation	409	none	192	0.999	5.45	407	none	190	0.926	5.46

Trial-Load Twist Analysis—Joints Grouted

B-1. *Example of Twist Analysis, Joints Grouted—Canyon Ferry Dam.*—Illustrations from a trial-load twist analysis, joints grouted, of a gravity dam are given on the following pages. The dam selected is Canyon Ferry Dam, and the plan, elevation, and selected elements are shown on figure B-1.

B-2. *Design Data.*—The following design data and assumption are presented for Canyon Ferry Dam:

(1) Elevation top of dam, 3808.5.

(2) Elevation of spillway crest, 3766.0.

(3) Maximum and normal reservoir water surface, elevation 3800.0.

(4) Minimum tailwater surface with gates closed, elevation 3633.0.

(5) Concentrated ice load of 7 tons per linear foot at elevation 3798.75. Provision is to be made so that no ice will form against the radial gates.

(6) Sustained modulus of elasticity of concrete in tension and compression, 3,000,000 pounds per square inch.

(7) Sustained modulus of elasticity of foundation and abutment rock, 3,000,000 pounds per square inch.

(8) Maximum horizontal earthquake assumed to have an acceleration of 0.1 gravity, a period of vibration of 1 second, and a direction of vibration normal to the axis of the dam.

(9) Maximum vertical earthquake assumed to have an acceleration of 0.1 gravity, a period of vibration of 1 second, and a direction that gives maximum stress conditions in the dam.

Note. Figure A-2 is a graph showing values of the coefficient K_E, which was used to determine hydrodynamic effects for the example given. However, this procedure is not consistent with current practice. A discussion of the coefficient C_m, which is presently used to determine hydrodynamic pressures, is given in section 4-34.

(10) Poisson's ratio for concrete and foundation rock, 0.20.

(11) Unit weight of water, 62.5 pounds per cubic foot.

(12) Unit weight of concrete, 150 pounds per cubic foot.

(13) Weight of radial gates, 3,000 pounds per linear foot.

(14) Weight of bridge, 5,500 pounds per linear foot.

(15) Unit shear resistance of concrete or concrete on rock, 400 pounds per square inch.

(16) Coefficient of internal friction of concrete on rock, 0.65.

(17) Uplift pressure on the base or horizontal sections above the base varies from full-reservoir water pressure at the upstream face to zero or tailwater pressure at the downstream face and acts over two-thirds the area of the base or horizontal sections.

Note. This uplift assumption is no longer used by the Bureau of Reclamation. See section 3-9 for uplift assumptions now in use.

(18) Effects of spillway bucket are included in the analyses.

(19) Effects of increased horizontal thickness of beams in spillway section are included.

B-3. *Abutment Constants.*—The method of determining abutment constants for elements of a concrete dam is shown in section 4-14.

B-4. *Deflections and Slopes Due to Unit Loads.*—Certain data pertaining to unit loads

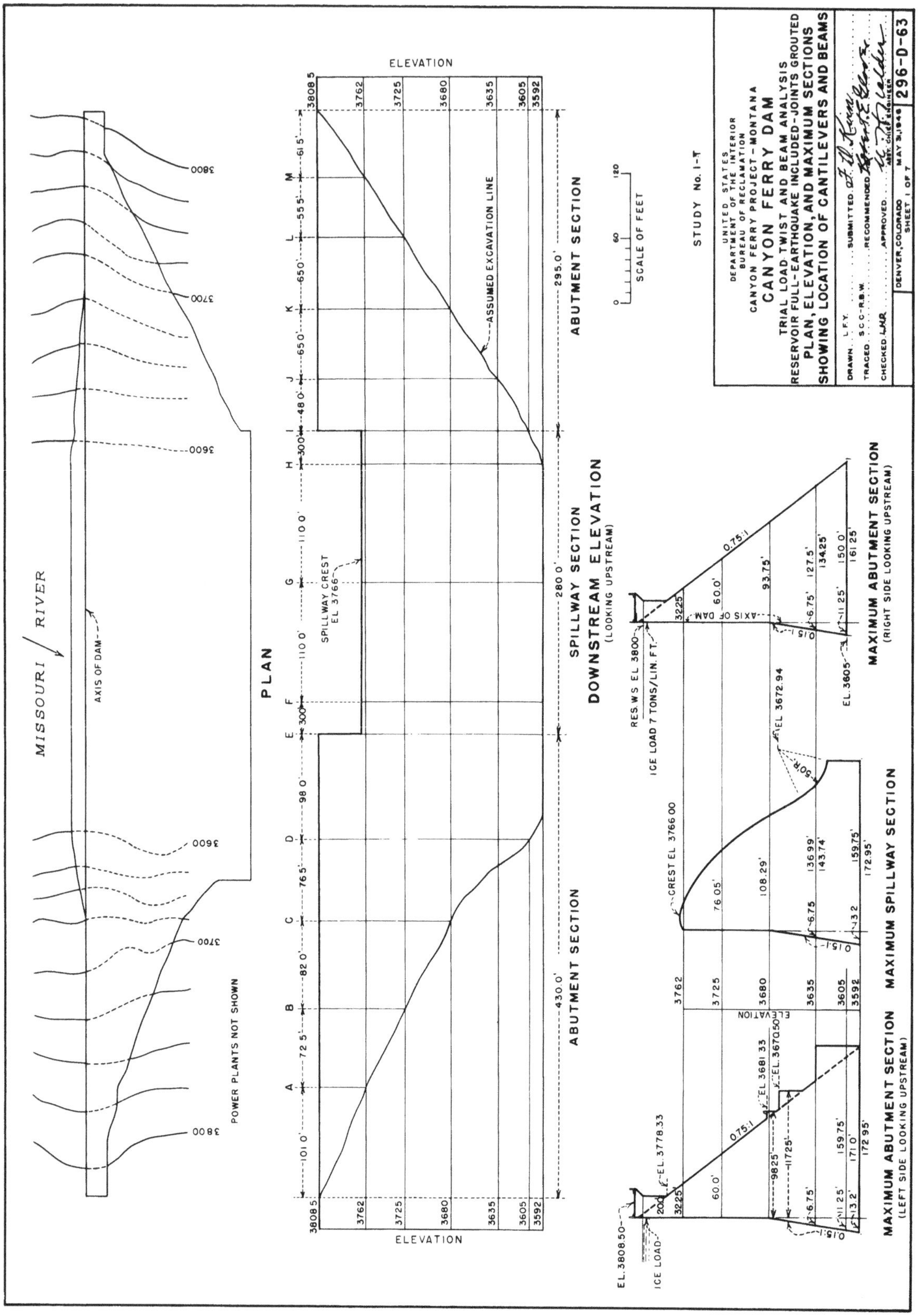

Figure B-1. Canyon Ferry Dam study—plan, elevation, and maximum sections.

are required prior to starting an adjustment. These include beam deflections for each unit triangular load, uniform load, and concentrated load and moment at the dividing plane; the slope of the beam at the abutment and at the dividing plane, due to unit loads; shears and twisted-structure deflections due to unit triangular, uniform, and concentrated shear loads on horizontal elements of the twisted structure; deflections of the vertical elements of the twisted structure due to unit triangular loads; cantilever deflections due to unit triangular normal loads; and shears and rotations of vertical elements of the twisted structure due to unit loads. Typical tabulations of these values are shown on figures B-2 through B-7. Calculations were by equations given in sections 4-29, 4-17, and 4-19. For identification of the cantilevers and beams in these drawings, see figure B-1. In the beam symbols, L means the left portion of the beam and R the right. A ΔG load is a triangular load with a value of 1,000 pounds per square foot at the abutment and zero at G, and so on for other loads. Cantilever loads are designated by the elevation at which the load is peaked.

B-5. Deflections of Cantilevers due to Initial Loads. —Cantilever deflections due to initial loads must be calculated prior to making a deflection adjustment. These deflections represent the position from which deflections of the cantilevers are measured when subjected to trial loads. Figure B-8 shows a tabulation of deflections due to initial loads on the cantilevers. These were computed by means of equation (17) in section 4-17. The initial loads are not shown but include loads of the type discussed in the latter part of section 4-16.

B-6. Trial-Load Distribution. —The total horizontal waterload is divided by trial between the three structures. However, it must be remembered that the twisted-structure load is split in half (see sec. 4-25), one-half to be placed on the horizontal elements and one-half on the vertical elements. In order to accomplish the trial-load distribution, the horizontal load ordinates must be determined at locations of the vertical elements, as illustrated on figure B-9. By multiplying these ordinates by loads on the horizontal elements,

the equivalent loads on the vertical elements are obtained. The first trial-load distribution on elements of the left half of the dam is given on figure B-10, and the sixth and final trial-load distribution for these elements is shown on figure B-11.

The total waterload at any point must equal the cantilever load plus the loads on the horizontal and vertical twisted elements (or twice the load on the horizontal twisted element) plus the beam load. Accordingly, at elevation 3680 for cantilever G, the total waterload in kips is equal to 7.269 plus (1.9 x 2 x 0) plus (0.8 x 2) plus 0.2, or 9.069.

The values for P and M for beam loads are required to provide slope and deflection agreement at the dividing plane. These may be established by trial, or more easily by calculation by assuming approximate values of deflection components from previous trials, and computing the P and M necessary to give the same slope (not equal to zero) and deflection of left and right portions of the beam at the crown. Two equations involving V_c and M_c are obtained from the conditions that the slope and deflection of the two halves of the beam must be in agreement at the dividing plane. The simultaneous solution of these two equations gives the amount of shear V_c (or P) and moment M_c necessary at the crown of the beam to restore continuity in the beam structure.

B-7. Cantilever Deflections. —Cantilever deflections due to final trial loads are shown on figure B-12 for the left half of the dam. On the upper half of the sheet are deflections due to normal loads. These are obtained by multiplying loads given in the upper right-hand section of figure B-11 by corresponding deflections for unit normal loads. On the lower half of the figure are deflections due to shear loads on vertical elements of the twisted structure. These loads are given in the lower right of figure B-11. The loads are multiplied by cantilever deflections due to unit shear loads (see fig. B-4) to obtain the values shown. At the bottom of figure B-12 are inserted the values for abutment movements due to beam and twisted-structure elements which have common abutments with the cantilever

CANYON FERRY DAM. .... SECTION. .... STUDY NO. I
PARALLEL-SIDE CANTILEVER--STRESS ANALYSIS—TRIAL-LOAD TWIST
DEFLECTION OF BEAM DUE TO UNIT NORMAL LOADS — LEFT SIDE
BEAM 3725L

By L.R.S.   Date 3-2-46

POINT LOAD	Δy						φ	
	B	C	D	E	F	G	B	G
ΔG	⁻.0³562,79	⁻.015,269	⁻.039,130	⁻.075,211	⁻.086,608	⁻.128,414	⁻.0³048,408	⁻.0³379,78
ΔF	⁻.0³380,22	⁻.007,927,0	⁻.018,741	⁻.033,597	⁻.038,074	⁻.054,736	⁻.0³025,474	⁻.0³51,48
ΔE	⁻.0³333,95	⁻.006,335,5	⁻.014,511	⁻.025,413	⁻.028,741	⁻.040,941	⁻.0²020,486	⁻.0³110,91
ΔD	⁻.0³193,33	⁻.002,368,3	⁻.004,647,4	⁻.007,519,9	⁻.008,399,3	⁻.011,624	⁻.0³007,975,0	⁻.0³029,311
ΔC	⁻.0³094,75	⁻.0³601,47	⁻.0³998,39	⁻.001,506,9	⁻.001,662,5	⁻.002,233,3	⁻.0⁵002,234,1	⁻.0³005,188,6
Unif.	⁻.001,257.5	⁻.046,000	⁻.126,530	⁻.258,534	⁻.301,761	⁻.461,466	⁻.0³144,23	⁻.001,450,34
Conc. P	⁻.0³004,694	⁻.0³232,51	⁻.0³692,17	⁻.001,534,8	⁻.001,828,4	⁻.002,945,3	⁻.0⁶722,46	⁻.0³010,190
Conc. M	⁻.0⁶005,033,9	⁻.0⁶585,76	⁻.0³001,907,2	⁻.0³004,699,6	⁻.0³005,771,9	⁻.0³010,190	⁻.0⁶001,809,4	⁻.0⁶043,637

Figure B-2. Canyon Ferry Dam study—deflection of a beam due to unit normal loads. —DS2-2(32)

CANYON FERRY DAM. .... SECTION. .... STUDY NO. 1.
PARALLEL-SIDE CANTILEVER--STRESS ANALYSIS—TRIAL-LOAD TWIST
DEFLECTION OF HORIZONTAL ELEMENT* DUE TO UNIT SHEAR LOADS — LEFT SIDE
ELEMENT 3725 L

By L.R.S. Date 3-2-46

POINT LOAD	Δy					
	B	C	D	E	F	G
ΔG	$-.0_3430,89$	$.001,950,1$	$-.002,807,7$	$-.003,330,0$	$-.003,384,3$	$-.003,435,4$
ΔF	$-.0_3311,35$	$-.001,318,9$	$-.001,753,5$	$-.001,892,9$	$-.001,894,3$	$-.001,894,3$
ΔE	$-.0_2278,75$	$-.001,148,3$	$-.001,477,1$	$.001,547,1 \rightarrow$	$.001,547,9$	$.001,547,9$
ΔD	$-.0_3172,25$	$-.0_3602,37$	$-.0_3656,86$	$\rightarrow$	$\rightarrow$	$-.0_3656,86$
ΔC	$-.0_3089,11$	$-.0_3218,82$		$\rightarrow$	$\rightarrow$	$-.0_3218,82$
Unif.	$-.0_3861,78$	$-.004,235,7$	$-.006,753,7$	$-.008,825,5$	$-.009,167,9$	$-.009,720,4$
Conc. P	$-.0_2002,173,5$	$-.0_5011,664$	$-.0_3020,518$	$-.0_3031,860$	$-.0_3034,600$	$-.0_3044,644\cdot$

* Beam or twisted-structure element

Figure B-3. Canyon Ferry Dam study—deflection of a horizontal element due to unit shear loads.—DS2-2(33)

CANYON FERRY DAM.   SECTION.   STUDY NO. 1
PARALLEL-SIDE CANTILEVER-STRESS ANALYSIS — TRIAL-LOAD TWIST
DEFLECTION OF CANTILEVER DUE TO UNIT SHEAR LOADS
CANTILEVER D

By C.W.J. Date 3-2-46

ELEV/LOAD	3808.5	3762	3725	3680	3635	3605
3808.5	$-.0^3549,3$	$-.0^3432,9$	$-.0^3290,5$	$-.0^3191,2$	$-.0^3125,4$	$-.0^3093,2$
3762	$-.0^3820,1$	$-.0^3703,7$	$-.0^3521,7$	$-.0^3343,4$	$-.0^3225,2$	$-.0^3167,4$
3725	$-.0^3493,3$	$-.0^3493,3$	$-.0^3453,8$	$-.0^3337,2$	$-.0^3221,2$	$-.0^3164,4$
3680	$-.0^3370,1$	$-.0^3370,1$	$-.0^3370,1$	$-.0^3332,6$	$-.0^3242,7$	$-.0^3180,4$
3635	$-.0^3216,8$	$-.0^3216,8$	$-.0^3216,8$	$-.0^3216,8$	$-.0^3190,6$	$-.0^3150,3$
3605	$-.0^3069,3$	$-.0^3069,3$	$-.0^3069,3$	$-.0^3069,3$	$-.0^3069,3$	$-.0^3060,1$

Figure B-4. Canyon Ferry Dam study—deflection of a cantilever due to unit shear loads. —DS2-2(34)

CANYON FERRY DAM...........SECTION...........STUDY NO. 1.........
......PARALLEL-SIDE CANTILEVER—STRESS ANALYSIS—TRIAL-LOAD TWIST......
........DEFLECTION OF CANTILEVER DUE TO UNIT NORMAL LOADS........
CANTILEVER D
By C.W.J. Date 3-2-46

ELEV \ LOAD	3808.5	3762	3725	3680	3635	3605
3808.5	-.00$3$552,6	-.001,853,5	-.0^{8}982,3	-.0^{3}477,6	-.0^{3}220,3	-.0^{3}119,5
3762	-.003,904,7	-.002,444,7	.001,497,9	-.0^{3}773,3	-.0^{3}370,9	-.0^{3}207,6
3725	-.001,716,5	-.001,352,0	.001,032,5	-.0^{3}632,1	-.0^{3}327,2	-.0^{1}193,3
3680	-.0^{3}967,9	-.0^{3}818,7	-.0^{3}700,0	-.0^{3}523,9	-.0^{3}318,7	-.0^{3}200,7
3635	-.0^{3}416,0	-.0^{3}370,7	-.0^{3}334,6	-.0^{3}290,7	-.0^{3}222,6	-.0^{3}158,2
3605	-.0^{3}102,4	-.0^{3}095,0	-.0^{3}089,1	-.0^{3}082,0	+.0^{3}074,8	-.0^{3}061,0

Figure B-5. Canyon Ferry Dam study—deflection of a cantilever due to unit normal loads. —DS2-2(35)

CANYON FERRY DAM .... SECTION. STUDY NO. I

PARALLEL-SIDE CANTILEVER-STRESS ANALYSIS — TRIAL-LOAD TWIST

SHEARS IN TWISTED STRUCTURE (V_T) DUE TO UNIT LOADS — LEFT SIDE

By L.R.S. Date 2-4-46

POINT LOAD	B	C	3635L	D	E	F	G	L'
Element 3725 L								
ΔG	-198,250	-124,729	-97,108	-71,430	-24,716	-15,259	0	396.5
ΔF	-143,250	-72,985	-48,964	-28,593	-1,571	0		286.5
ΔE	-128,250	-59,357	-36,854	-18,721	0			256.5
ΔD	-79,250	-18,461	-4,922	0				158.5
ΔC	-41,000	0						82.0
Unif.	-396,500	-314,500	-277,500	-238,000	-140,000	-110,000	0	
Conc.P	-1.000						-1,000	
Element 3680 L								
ΔG		-157,250	-122,426	-90,054	-31,161	-19,237	0	314.5
ΔF		-102,250	-68,597	-40,059	-2,200	0		204.5
ΔE		-87,250	-54,173	-27,519	0			174.5
ΔD		-38,250	-10,198	0				76.5
Unif.		-314,500	-277,500	-238,000	-140,000	-110,000	0	
Conc.P		-1.000					-1,000	
Element 3635 L								
ΔG			-138,750	-102,061	-35,315	-21,802	0	277.5
ΔF			-83,750	-48,907	-2,687	0		167.5
ΔE			-68,750	-34,924	0			137.5
ΔD			-19,750	0				39.5
Unif.			-277,500	-238,000	-140,000	-110,000	0	
Conc.P			-1.000				-1,000	

Figure B-6. Canyon Ferry Dam study—shears in twisted structure due to unit loads.—DS2-2(36)

CANYON FERRY DAM. .... SECTION. .... STUDY NO. I.

PARALLEL-SIDE CANTILEVER—STRESS ANALYSIS — TRIAL-LOAD TWIST

ROTATIONS OF VERTICAL TWISTED-STRUCTURE ELEMENTS DUE TO UNIT COUPLE LOADS.

CANTILEVERS C, D, AND 3635 L

By C.W.J.   Date 3-2-46

Cantilever D

ELEV/LOAD	3808.5	3762	3725	3680	3635	3605
3808.5	$-0{,}^{7}001{,}177{,}9$	$-0{,}^{6}640{,}72$	$-0{,}^{6}146{,}90$	$-0{,}^{6}045{,}008$	$-0{,}^{6}016{,}639$	$-0{,}^{6}009{,}510$
3762	$-0{,}^{3}001{,}347{,}6$	$-0{,}^{6}810{,}43$	$-0{,}^{6}263{,}79$	$0{,}^{6}080{,}822$	$0{,}^{6}029{,}879$	$0{,}^{6}017{,}077$
3725	$-0{,}^{6}233{,}74$	$-0{,}^{6}233{,}74$	$-0{,}^{6}180{,}92$	$-0{,}^{6}079{,}369$	$-0{,}^{6}029{,}342$	$-0{,}^{6}016{,}770$
3680	$-0{,}^{6}087{,}113$	$0{,}^{6}087{,}113$	$0{,}^{6}087{,}113$	$0{,}^{6}066{,}633$	$0{,}^{6}032{,}205$	$0{,}^{6}018{,}406$
3635	$0{,}^{6}030{,}712$	$-0{,}^{6}030{,}712$		$-0{,}^{6}030{,}712$	$0{,}^{6}023{,}738$	$0{,}^{6}015{,}339$
3605	$0{,}^{6}007{,}635$	$-0{,}^{6}007{,}635$		$-0{,}^{6}007{,}635$	$0{,}^{6}007{,}635$	$0{,}^{6}006{,}135$

Cantilever C

ELEV/LOAD	3808.5	3762	3725	3680
3808.5	$-0{,}^{6}001{,}161{,}6$	$-0{,}^{6}624{,}37$	$-0{,}^{6}130{,}55$	$-0{,}^{6}028{,}664$
3762	$-0{,}^{6}001{,}318{,}3$	$-0{,}^{6}781{,}08$	$-0{,}^{6}234{,}44$	$0{,}^{6}051{,}473$
3725	$-0{,}^{6}204{,}92$	$-0{,}^{6}204{,}92$	$-0{,}^{6}152{,}10$	$0{,}^{6}050{,}548$
3680	$-0{,}^{6}048{,}220$	$0{,}^{6}048{,}220$	$0{,}^{6}048{,}220$	$0{,}^{6}027{,}740$

Base of Cantilever 3635 L

ELEV/LOAD	3635
3808.5	$-0{,}^{6}010{,}331$
3762	$-0{,}^{6}018{,}551$
3725	$-0{,}^{6}018{,}218$
3680	$-0{,}^{6}019{,}995$
3635	$-0{,}^{6}009{,}997{,}7$

Figure B-7. Canyon Ferry Dam study—rotations of vertical twisted-structure elements due to unit couple loads. — DS2-2(37)

CANYON FERRY DAM.   SECTION.   STUDY NO. 1.
PARALLEL-SIDE CANTILEVER–STRESS ANALYSIS—TRIAL-LOAD TWIST
DEFLECTION OF CANTILEVERS DUE TO INITIAL LOADS

By L.R.S.   Date 3-5-46

Elev.	A	B	C	D	E	F	G	H	I	J	K	L	M
						Cantilevers							
3808.5	$\overline{0}{}^3,675,3$	$\overline{0}01,746,3$	$\overline{0}03,056,8$	$\overline{0}04,850,5$	$\overline{0}05,762,9$	←— Spillway		→	$\overline{0}05,002,1$	$\overline{0}04,167,2$	$\overline{0}03,086,7$	$\overline{0}01,698,1$	$\overline{0}{}^3,658,8$
3762	$\overline{0}{}^3,096,8$	$\overline{0}{}^3,738,9$	$\overline{0}01,774,7$	$\overline{0}03,492,5$	$\overline{0}04,311,3$	$\overline{0}06,625,6$		→	$\overline{0}03,602,5$	$\overline{0}02,804,0$	$\overline{0}01,798,2$	$\overline{0}{}^3,713,9$	$\overline{0}{}^3,086,1$
3725		$\overline{0}{}^3,205,7$	$\overline{0}01,015,0$	$\overline{0}02,672,2$	$\overline{0}03,416,7$	$\overline{0}04,850,6$		→	$\overline{0}02,749,1$	$\overline{0}01,979,7$	$\overline{0}01,033,3$	$\overline{0}{}^3,191,1$	
3680			$\overline{0}{}^3,388,1$	$\overline{0}01,972,0$	$\overline{0}02,626,4$	$\overline{0}03,665,9$		→	$\overline{0}02,008,4$	$\overline{0}01,274,3$	$\overline{0}{}^3,400,2$		
3635				$\overline{0}01,394,4$	$\overline{0}01,958,8$	$\overline{0}02,627,7$		→	$\overline{0}01,392,7$	$\overline{0}{}^3,693,8$			
3605				$\overline{0}01,024,6$				→	$\overline{0}{}^3,997,7$				
3592		$\overline{0}{}^3,517,1$			$\overline{0}01,353,8$	$\overline{0}01,703,5$							

Base of 3635L

Figure B-8. Canyon Ferry Dam study–deflections of cantilevers due to initial loads. –DS2-2(38)

CANYON FERRY DAM.   SECTION.   STUDY NO. 1.
PARALLEL-SIDE CANTILEVER--STRESS ANALYSIS—TRIAL-LOAD TWIST
LOAD ORDINATES AT CANTILEVER POINTS.— LEFT SIDE
BEAM OR TWISTED-STRUCTURE LOADS
By L.R.S. Date 3-5-46

Beam	Load	Abt.	A	B	C	3635L	D	E	F	G
							Cantilevers			
3808.5	ΔE	1.0	.765,12	.596,51	.405,81	.319,77	.227,91	0		
	ΔD	1.0	.695,78	.477,41	.230,42	.118,98	0			
	ΔC	1.0	.604,70	.320,94	0					
	ΔB	1.0	.417,87	0						
3762	ΔG		1.0	.845,42	.670,58	.591,68	.507,46	.298,51	.234,54	0
	ΔF		1.0	.798,05	.569,64	.466,57	.356,55	.083,57	0	
	ΔE		1.0	.779,63	.530,40	.417,93	.297,87	0		
	ΔD		1.0	.686,15	.331,17	.171,00	0			
	ΔC		1.0	.530,74	0					
	ΔB		1.0	0						
3725	ΔG			1.0	.793,19	.599,87	.600,25	.353,09	.277,43	0
	ΔF			1.0	.713,79	.584,64	.446,77	.104,71	0	
	ΔE			1.0	.680,31	.536,06	.382,07	0		
	ΔD			1.0	.482,65	.249,21	0			
	ΔC			1.0	0					
3680	ΔG				1.0	.882,35	.756,76	.445,15	.349,76	0
	ΔF				1.0	.819,07	.625,92	.146,70	0	
	ΔE				1.0	.787,97	.561,60	0		
	ΔD				1.0	.516,34	0			
3635	ΔG					1.0	.857,66	.504,50	.396,40	0
	ΔF					1.0	.764,18	.179,10	0	
	ΔE					1.0	.712,73	0		
	ΔD					1.0	0			

Figure B-9. Canyon Ferry Dam study—load ordinates at cantilever points. —DS2-2(39)

CANYON FERRY DAM.   SECTION.   STUDY NO. 1.
PARALLEL-SIDE CANTILEVER--STRESS ANALYSIS—TRIAL-LOAD TWIST
TRIAL-LOAD DISTRIBUTION—LEFT SIDE
(TRIAL NO. 1)

By L.R.S.   Date 3-8-46

Normal cantilever loads

Elev.	A	B	C	3635L	D	E	F	G
38085							—	—
3762	+.973	1.397	1.877	2.093	2.324	2.898	3.073	3.073
3725		+.203	1.806	2.529	3.301	5.217	5.803	5.803
3680			+.069	1.697	+3.436	7.749	9.069	9.069
3635				+.251	+3.081	10.102	12.251	12.251
3605					+3.0			
3592						15.288	12.469	12.469

Horizontal twisted-structure loads

Elev.	Unif.	ΔB	ΔC	ΔD	ΔE	ΔF	ΔG	Conc.
38085								
3762						+1.0		
3725						+2.4		
3680						+3.5		
3635						+4.0		
3605								
3592								

Vertical twisted-structure or cantilever shear loads

Elev.	A	B	C	3635L	D	E	F	G	Conc.M	Conc.P
38085							0	0	—	—
3762	+1.0	+.798	+.570	+.467	+.357	+.084			+565	
3725		+2.4	+1.713	+1.403	+1.072	+.251			-3,645	
3680			+3.5	+2.867	+2.191	+.513			-6,060	
3635				+4.0	+3.057	+.716			-9,478	
3605					+3.5					
3592						0				

Beam loads

Elev.	Unif.	ΔB	ΔC	ΔD	ΔE	ΔF	ΔG	Conc.P	Conc.M
38085	-.001				+.015				
3762						+.1			
3725						+.8			
3680						+2.0			
3635						+4.0			
3605									
3592									

Figure B-10. Canyon Ferry Dam study—trial-load distribution (trial No. 1).—DS2-2(40)

CANYON FERRY DAM.   SECTION.   STUDY NO. 1.
PARALLEL-SIDE CANTILEVER--STRESS ANALYSIS— TRIAL-LOAD TWIST.
TRIAL-LOAD DISTRIBUTION—LEFT SIDE
(FINAL)

By L.R.S. Date 4-6-46.

Horizontal twisted-structure loads / Normal Cantilever loads

	ΔG	ΔF	ΔE	ΔD	ΔC	ΔB	Unif.	Conc.	A	B	C	3635L	D	E	F	G
3808.5	—	—		+.15			—	—	-.161	-.135	-.105	-.066	-.024	+.04	—	—
3762	—	+4.0	-4.5	+.2	+.25		+0.05	-5.	3.033	3.027	3.020	2.897	2.766	2.296	2.973	2.973
3725	+.3	+3.0	-2.0	-2.5	+1.0		+.45	-30.		5.053	5.060	4.161	3.200	3.910	4.595	+4.803
3680	+1.9	+3.5	-3.0	-.5			+.8	-50.			+1.469	+1.789	+2.130	+4.257	+5.940	+7.269
3635		+4.0	-2.5				+1.7	-110.				+1.151	+2.662	+6.412	+8.651	+8.651
3605													+3.000			
3592												*Estimated*	*Estimated*	+15.288	+12.469	+12.469

Beam loads / Vertical twisted-structure or cantilever shear loads

	ΔG	ΔF	ΔE	ΔD	ΔC	ΔB	Unif.	Conc.P	Conc.M	A	B	C	3635L	D	E	F	G
3808.5	—	—	+.28	-.163	-.179		-.04	—	—	+.104	+.072	+.035	+.018	0	0	—	—
3762		+.10		-.06				+.596	-570.8	0	+.004	+.008	+.070	+.136	+.384	+.050	+.050
3725	+.15						+.1	-.136	-5,324.3		+.250	+.262	+.719	+1.206	+.870	+.533	+.450
3680		+2.0					+.2	+2.474	-12,793			+2.700	+2.721	+2.744	+2.159	+1.465	+.800
3635		+4.5					+.2	0	-17,953				+3.200	+2.975	+2.416	+1.700	+1.700
3605														+3.500			
3592													*Estimated*		0	0	0

Figure B-11. Canyon Ferry Dam study—trial-load distribution (final)—DS2-2(41)

CANYON FERRY DAM. .... SECTION. .... STUDY NO. 1
PARALLEL–SIDE CANTILEVER––STRESS ANALYSIS — TRIAL–LOAD TWIST
CANTILEVER DEFLECTION COMPONENTS – LEFT SIDE
(FINAL)

By.............. Date..............

Cantilever Δy due to normal loads

	A	B	C	3635L	D	E	F	G
38085	‾001,686	·006,309	‾013,045		‾019,684	·029,482	—	—
3762	‾$0^3$312	‾003,106	‾008,112		‾014,059	‾023,258	‾021,650	‾023,238
3725		‾$0^3$993	‾004,634		‾010,073	‾018,391	‾018,326	‾019,678
3680			·001,531		‾006,286	‾013,292	‾014,028	‾015,051
3635				‾001,263	‾003,563	‾009,038	‾009,815	‾010,477
3605					‾002,222			
3592						·005,471	‾005,978	·006,345

Cantilever Δy due to shear loads

	A	B	C	3635L	D	E	F	G
38085	·$0^3$022	‾$0^3$056,2	‾$0^3$427,1		‾002,609,5	·002,431,9	—	—
3762	‾$0^3$010	‾$0^3$047,3	‾$0^3$422,1		‾002,593,7	‾002,387,2	‾001,369,2	‾001,056,5
3725		‾$0^3$026,4	‾$0^3$405,3		‾002,521,3	‾002,282,9	‾001,351,0	·001,040,9
3680			·$0^3$268,6		·002,253,6	·002,032,0	·001,248,4	‾$0^5$967,9
3635				‾$0^5$647,8	·001,772,9	‾001,628,6	·001,030,0	‾$0^3$812,3
3605					·001,373,5			
3592						·001,181,2	‾$0^3$756,9	‾$0^3$6026

Abutment Δy due to beam and twisted–structure loads.

	A	B	C	3635L	D
	‾$0^4$119,61	‾$0^3$669,7	‾002,423,2	·003,569,6	·002,382

Note: These deflections are due to load on vertical twisted–structure element.

Figure B-12. Canyon Ferry Dam study—cantilever deflection components (final). —DS2-2(42)

structure. The three component deflections given on figure B-12 represent the deflections due to trial loads on the structure which must be added algebraically to the deflections due to initial loads (see fig. B-8) to obtain the total deflection of the cantilever structure. These values are shown on Figure B-13. It should be noted at this point that the abutment movements of each structure are equal.

B-8. *Twisted-Structure Deflections.*—Shears due to loads on the horizontal elements of the twisted structure and angular rotations of vertical elements due to these shears are shown on figure B-14. Loads on horizontal twisted elements in the upper left of figure B-11 operate on unit shears to give the shear at each point in the horizontal twisted-structure element. The shear is divided by negative 1,000 to get units of twist load to operate on the unit rotations given on figure B-7 because the maximum ordinate for a unit twist load was assumed to be minus 1,000 foot-pounds per square foot. At each point where the vertical element and beam have a common base and abutment, it is desirable to note the value of abutment rotation of the vertical element due to load on the beam. These values are obtained for each element from figure B-16 and are indicated by asterisks (*) on figure B-14. At the base of element *D* there is no beam and a value is estimated.

In the upper half of figure B-15, rotations of vertical elements are integrated from the abutment to the crown using values calculated in figure B-14. Here the abutment rotations of the beams have been included. These are deflections of the horizontal elements due to rotation of vertical elements and abutment rotation of the beams. In the lower half of the figure are given the shear detrusions of horizontal elements due to loads on the beams (see the lower left-hand section of figure B-11). Detrusions are obtained by using deflections due to unit shear loads on horizontal elements as shown on figure B-3.

The lower half of figure B-16 shows values of shear detrusions due to twisted-structure loads. These are calculated by using deflections due to unit shear loads on horizontal elements, from figure B-3. Not only are these values

components of the twisted-structure deflections, but they are also components of deflections of the beam structure, as will be shown later.

At the base of the deflection columns for cantilevers *A* to *D*, inclusive, on the lower half of figure B-16, the abutment movements of the cantilever and of the beam due to moment only, $M\alpha_2$, are entered for inclusion in the total twisted-structure deflection. Thus, the abutment movement at the base of cantilever *A* is equal to $-.0^3,023$ (fig. B-15), plus $-.0^3,086$, plus $-.0^3,430$ (fig. B-16), or equal to $-.0^3,539$. This is equal to the abutment movement at the base of the cantilever structure at *A* (see fig. B-13). Final twisted-structure deflections are given on figure B-13. These are compared with beam and cantilever deflections given on this same sheet.

B-9. *Beam-Structure Deflections.*—Deflections of beams due to bending are calculated in the upper half of figure B-16. These are determined by means of beam loads given on figure B-11 and unit deflections given on figure B-2. Slopes at the abutment and at the crown are also shown. Slopes at the crown include rotation of the common abutment due to twist loads on the vertical elements, but the slope shown at the abutment is only the rotation due to beam loads. Immediately above each deflection due to bending, the deflection of the beam due to rotation of the vertical element at the abutment is entered. Deflections are calculated by multiplying the slope at the abutment by the horizontal distance to each cantilever. At the abutment of each beam there are also additional movements due to initial, trial normal, and trial shear loads on the cantilevers which are entered at the bottom of figure B-16. Another component of the total beam deflection is due to shear detrusion for twisted-structure loads on horizontal elements. These values were previously calculated for the twisted structure and are shown in the lower half of figure B-16. Total deflections of beams may now be calculated by adding deflections due to bending, rotation, shear detrusion, and abutment movement. For example, the total deflection at the abutment of beam 3762, which coincides with the base of cantilever *A*,

CANYON FERRY DAM. ... SECTION. STUDY NO. 1
PARALLEL-SIDE CANTILEVER--STRESS ANALYSIS--TRIAL-LOAD TWIST
TOTAL DEFLECTIONS--LEFT SIDE
(FINAL)

By G.B.   Date 4-12-46

TOTAL DEFLECTIONS — LEFT SIDE (FINAL)

		Beam deflection								Cantilever deflection			
	Abt.	A	B	C	D	E	F	G	A	B	C	3635L
3808.5	-.0³034	-.002,949	-.010,225	-.020,639	-.029,166	-.037,843	—	—	-.002,503	-.008,781	-.018,952	—
3762		-.0³539	-.005,198	-.013,444	-.021,169	-.028,365	-.029,707	-.031,175	-.0³539	-.004,562	-.012,732	—
3725			-.001,895	-.009,121	-.016,821	-.023,437	-.024,414	-.025,152		-.001,895	-.008,477	—
3680				-.004,611	-.012,097	-.017,796	-.019,123	-.019,594			-.004,611	
3635	-.005,998				-.008,789	-.013,173	-.013,799	-.014,110				-.005,998

Twisted-structure deflection / Cantilever deflection

		Twisted-structure deflection								Cantilever deflection			
	Abt.	A	B	C	D	E	F	G.	D	E	F	G
3808.5	-.0³034	-.003,049	-.008,411	-.017,789	-.027,889	-.035,712	—	—	-.029,526	-.037,677	—	—
3762		-.0³539	-.004,549	-.012,732	-.022,003	-.029,494	-.030,064	-.030,076	-.022,527	-.029,956	-.029,645	-.030,920
3725			-.001,895	-.009,084	-.017,499	-.024,421	-.025,143	-.025,147	-.017,640	-.024,091	-.024,528	-.025,570
3680				-.004,546	-.012,201	-.018,356	-.019,038	-.019,162	-.012,893	-.017,950	-.018,942	-.019,685
3635	-.005,998				-.008,873	-.013,141	-.013,616	-.013,563	-.009,112	-.012,626	-.013,473	-.013,917
3605									-.007,002	-.008,006	-.008,438	-.008,651
3592												

Estimated

Figure B-13. Canyon Ferry Dam study—total deflections (final).—DS2-2(43)

CANYON FERRY DAM. .... SECTION.   STUDY NO. 1

PARALLEL-SIDE CANTILEVER—STRESS ANALYSIS — TRIAL-LOAD TWIST

SHEARS (V) IN HORIZONTAL ELEMENTS AND ROTATIONS (ΔΦ) OF VERTICAL ELEMENTS

DUE TO TWISTED-STRUCTURE LOAD — LEFT SIDE .... (FINAL) By G.F.B. Date 4-9-46

El.	C — V	C — ΔΦ	3635L — V	3635L — ΔΦ	D — V	D — ΔΦ	E — V	E — ΔΦ	F — V	F — ΔΦ	G — V	G — ΔΦ
3808.5	−1,322	−0⁴121,07	−352.5	—	0	−0⁴126,09	0	−0³030,18	← Spillway		←	
3762	−37,992	−0³099,95	−36,556	—	−32,494	−0³108,63	−7,012	−0³030,18	−500.	−0³012,36	+5,000.	+0³013,09
3725	−203,032	−0³067,802	−184,886	—	−146,866	−0³083,11	−45,128	−0³027,43	−24,078	−0³011,72	+30,000	+0³012,18
3680	−577,375	−0³028,273	−477,081	—	−369,152	−0³054,69	−128,906	−0³019,64	−74,550	−0³009,37	+50,000	+0³009,63
3635			−524,875	−0³018,84	−402,918	−0³030,17	−138,748	−0³011,82	−77,000	−0³006,25	+110,000	+0³006,66
3605					−450,000	—						
3592		*−0³012,407		*−0³008,8773	Not required *−0³004		Estimated					

El.	A — V	A — ΔΦ	B — V	B — ΔΦ
3808.5L				
3808.5	−12,054	−0³041,44	−5,675	−0³079,407
3762	−38,613	−0³014,22	−38,479	−0³055,688
3725		−0³007,800	−224,025	.0⁵020,019
	*+0⁶25			*−0³011,952

Notes: At any point in structure −1,000 pounds shear represents one unit of twist load.

* Rotation of abutment due to beam loads.

Figure B-14. Canyon Ferry Dam study—shears in horizontal elements and rotations of vertical elements due to twisted-structure load (final).—DS2-2(44)

CANYON FERRY DAM. SECTION. STUDY NO. 1.
PARALLEL-SIDE CANTILEVER--STRESS ANALYSIS — TRIAL-LOAD TWIST

TWISTED-STRUCTURE DEFLECTION DUE TO ROTATIONS OF VERTICAL ELEMENT—LEFT SIDE.
TWISTED-STRUCTURE DEFLECTION DUE TO BEAM LOADS. (FINAL) By G.F.B. Date 4-10-46

*Note: $\Delta\phi$ includes rotations due to beam abutment forces.

Elev.	3808.5		3762		3725		3680		3635		
$\Delta\frac{x}{2}$	$\Delta\phi$*	$\int\Delta\phi$	$\Delta\phi$*	$\int\Delta\phi$	$\Delta\phi$*	$\int\Delta\phi$	$\Delta\phi$*	$\int\Delta\phi$	$\Delta\phi$*	$\int\Delta\phi$	$\Delta\frac{x}{2}$
Abt.	$+.0^6,25$	0		0		0						
A 50.5	$-.0^3,049,240$	$-.002,474$	$-.0^3,022,02$								
36.25											
B 41.0	$-.0^3,091,359$	$-.007,571$	$-.0^3,067,640$	$.003,250$	$-.0^3,031,971$						
38.25											
C	$-.0^3,133,48$	$-.016,789$	$-.0^3,112,36$	$.010,630$	$-.0^3,080,209$	$.004,599$	$-.0^3,040,68$	0	$-.0^3,027,72$	0	ABT 19.75
D 49.0	$-.0^3,130,09$	$-.026,870$	$-.0^3,112,63$	$.019,236$	$-.0^3,087,11$	$.010,999$	$-.0^3,058,69$	$.003,801$	$-.0^3,034,17$	$.001,222$	49.0
E 15.0	$-.0^3,030,18$	$-.034,724$	$-.0^3,030,18$	$.026,234$	$-.0^3,027,43$	$.016,612$	$-.0^3,019,64$	$.007,639,6$	$-.0^3,011,82$	$.003,476$	15.0
F 55.0			$-.0^3,012,36$	$.026,871$	$-.0^3,011,72$	$.017,199$	$-.0^3,009,37$	$.008,074$	$-.0^3,006,25$	$.003,747$	55.0
G			$+.0^3,013,09$	$.026,832$	$+.0^3,012,18$	$.017,174$	$+.0^3,009,63$	$.008,060$	$-.0^3,006,66$	$.003,724$	

Twisted-structure Δy for beam loads

	Abt	A	B	C	D	E	F	G
3808.5	$+.0^3,013$	$+.0^3,108$	$+.0^3,061$	$-.0^3,007$	$-.0^3,014$	$+.0^3,017$	—	—
3762		$+.0^3,023$	$.0^3,182$	$-.0^3,309$	$-.0^3,380$	$-.0^3,413$	$-.0^3,417$	$-.0^3,431$
3725			$-.0^3,151$	$-.0^3,714$	$-.001,094$	$-.001,378$	$-.001,420$	$-.001,481$
3680				$-.0^3,752$	$-.001,858$	$-.002,401$	$-.002,456$	$-.002,551$
3635				$-.001,577$	$-.002,285$	$-.002,957$	$-.002,999$	$-.003,057$

Figure B-15. Canyon Ferry Dam study—twisted-structure deflection due to rotations of vertical element, and twisted-structure deflection due to beam loads (final). —DS2-2(45)

CANYON FERRY DAM. SECTION. STUDY NO. I.
PARALLEL-SIDE CANTILEVER-STRESS ANALYSIS—TRIAL-LOAD TWIST
BEAM DEFLECTION DUE TO BEAM LOADS AND ABUTMENT ROTATIONS.—LEFT SIDE. (FINAL).
DEFLECTION OF HORIZONTAL ELEMENTS DUE TO TWISTED-STRUCTURE LOAD. By G.F.B. Date 4-10-46.

Beam Δy due to bending and shear.
Beam Δy due to rotation of abutment caused by twist on vertical elements.

Elev.	Rotation due to twist	Abt.	A	B	C	D	E	F	G	Rotation due to beam loads φ Abt.	φ Cr. (Total slope of beam at crown)
3808.5	0	$+0,^3014$	-.002,216	.009,324	-.019,646	-.028,161	-.036,838	-.005,105	-.006,669	$+0,^625$	$-0,^5085$
φxL			0	.001,031	-.002,197	-.003,285	-.004,678				
3762	$-0,^3014$		$-0,^3034$	.003,061	-.009,465	-.015,508	-.020,851	-.021,747	-.021,708	$-0,^3008$	$+0,^3011$
φxL				0	.001,642	-.003,173	-.005,134	-.005,735	-.007,938		
3725	$-0,^5020$			$-0,^3183$	-.003,740	-.008,274	-.011,904	-.012,187	-.010,754	$-0,^5012$	$+0,^3012$
φxL					0	.002,163	-.004,934	-.005,782	-.008,892		
3680	$-0,^5028$				$-0,^5815$	-.003,455	-.004,609	-.004,896	-.002,214	$-0,^3012$	—
φxL					ABT.	$-0,^5744$	-.002,591	-.003,156	-.005,228		
3635	$-0,^3019$				.001,655	-.002,756	-.003,943	-.003,850	-.002,177	$-0,^3009$	—

Beam Δy, also twisted-structure Δy (shear detrusion) due to twisted-structure loads.

Elev.	Abt.	A	B	C	D	E	F	G
3808.5	$-0,^3048$	$-0,^5683$	$-0,^5901$	$-0,^5993$	-.001,005	-.001,005		
3762		$-0,^3086$	$-0,^5687$	.001,363	-.001,957	-.002,417	-.002,436	-.002,383
3725			$-0,^5487$	.002,514	-.004,149	-.005,174	-.005,267	-.005,235
3680				.001,609	.004,292	.006,066	-.066,258	.006,301
3635				.001,915	.002,861	.004,211	.004,365	.004,277

Abutment movements of beam due to loads on other elements.
(3725) (3762) (3680) (3635)
$-0,^3419$ .001,225 -.002,187 -.002,428

Abutment movements of twisted structure due to loads on loads on other elements.
$-0,^3430$ -.001,257 .002,250 .002,505

Figure B-16. Canyon Ferry Dam study—beam deflection due to beam loads and abutment rotations, and deflection of horizontal elements due to twisted-structure loads (final).—DS2-2(46) is equal to $-.0^3,033,974$, plus $-.0^3,419$, plus $-.0^3,085,64$ (fig. B-16), or $-.0^3,539$. Inspection of figure B-13 shows that this agrees with the cantilever and twisted-structure deflections at the same point.

B-10. *Total Deflections.*—Total deflections for the right side of the dam are given on figure B-17. Note that at the crown point, *G*, the deflections agree closely with those computed for *G* for the left side of the dam (see fig. B-13).

B-11. *Moment and Shear due to Trial Loads on Beams.*—Total bending moments for each beam are calculated by multiplying final beam loads by bending moments in beams due to unit loads. The total shear is obtained by adding the beam load and the twisted-structure load on the horizontal element, and multiplying the result by the shear due to unit load. These moments and shears are tabulated for the left side of the dam on figure B-18.

B-12. *Beam Stresses.*—Stresses at the faces of beams due to pure bending are calculated from the well-known formula, $\sigma_x = \pm Mc/I$. No weight is carried by the beams, since it has been assumed that weight is assigned to the cantilevers. Beam stresses are calculated in pounds per square foot, but are tabulated in pounds per square inch. These calculations are not shown due to their simplicity.

B-13. *Cantilever Stresses.*—Vertical cantilever stresses at the faces are calculated by means of the usual formula, $W/A \pm Mc/I$. The inclined cantilever stress parallel to either face of the dam at any point is calculated by dividing the corresponding vertical cantilever stress by the square of the cosine of the angle, ϕ, between the face and a vertical line, and subtracting from this quotient the product of the net normal water pressure and the square of the tangent of the angle ϕ. (See the lower part of figure 4-2 for equation and method of allowing for earthquake effect.)

In the example given here, an upward vertical earthquake acceleration was assumed. Consequently, the effective weight of the dam is found by multiplying by 1.1. The total moment is found by adding algebraically the moments due to weight, horizontal earthquake, vertical waterload, vertical earthquake, ice

load, and trial load on the cantilever. Stresses at the faces are then calculated, using the formulas mentioned in the preceding paragraph. Principal stresses are calculated by means of equations given on figure 4-3.

Stability factors on horizontal planes are computed by formulas previously given in section 4-10. In computing the stability factors on inclined abutment planes, the equivalent horizontal force is the total shearing force due to the sum of the shears from the cantilever element and the abutting horizontal element of the twisted structure.

Assuming a unit area on the sloping surface, the total inclined abutment shear is computed by the equation,

$$\Sigma V = V_c \sin \psi + V_T \cos \psi$$

where:

ΣV = total inclined abutment shear on unit area,
V_c = shear in horizontal plane at base of cantilever,
V_T = shear in vertical plane at abutment of horizontal element, and
ψ = angle between vertical and inclined plane of contact.

The total force normal to the inclined abutment plane is equal to the resultant of the total vertical force and horizontal thrust transferred from the vertical cantilever and horizontal element, respectively. This force (see fig. B-19) is equal to

$$F_N = \frac{(W + U) \sin \psi}{\sin \psi} = W + U$$

where:

F_N = total force normal to inclined abutment plane,
U = uplift force, and
ψ = angle between the vertical and the inclined abutment plane.

After the above values have been obtained, the sliding factor is computed by dividing the total inclined abutment shear by the normal

CANYON FERRY DAM.   SECTION.   STUDY NO. 1.
PARALLEL-SIDE CANTILEVER-STRESS ANALYSIS—TRIAL-LOAD TWIST
TOTAL BEAM AND TWISTED-STRUCTURE DEFLECTIONS—RIGHT SIDE
(FINAL)

By L.R.S.   Date 4-10-46

Beam deflection / Cantilever deflection

Abt.		M	L	K	J	I	H	G		J	K	L	M
				Beam deflection							Cantilever deflection		
3808.5	$-.0_3006$	-.002,408	-.008,162	-.017,065	-.025,711	-.031,664	—	—		-.024,270	-.016,625	-.009,264	-.002,915
3762		$-.0_3649$	-.004,945	-.012,494	-.020,041	-.024,672	-.026,983	-.031,495		-.018,237	-.011,121	-.004,820	$-.0_3649$
3725			-.002,168	-.009,215	-.016,377	-.020,372	-.022,163	-.025,882		-.013,896	-.007,379	-.002,168	
3680				-.004,018	-.010,172	-.014,003	-.015,831	-.019,603		-.009,547	-.004,018		
3635					-.006,033	-.009,327	-.010,912	-.014,019		-.006,033			

Twisted-structure deflection

Abt.		M	L	K	J	I	H	G		I	H	G
3808.5	$-.0_3006$	-.001,993	-.007,178	-.016,576	-.026,167	-.031,239	—	—		-.031,092	—	—
3762		$-.0_3649$	-.004,670	-.012,365	-.020,548	-.025,360	-.027,278	-.030,503				
3725			-.002,168	-.008,698	-.015,580	-.020,008	-.021,963	-.025,401		-.024,027	-.026,447	-.030,919
3680				-.004,018	-.009,820	-.013,695	-.015,571	-.018,667		-.018,920	-.021,739	-.025,570
3635					-.006,033	-.009,644	-.011,393	-.014,419		-.013,642	-.016,673	-.019,685
3605										-.009,255	-.011,856	-.013,917
3592										-.006,758	-.007,434	-.008,652

Figure B-17. Canyon Ferry Dam study—total beam and twisted-structure deflections (final). —DS2-2(47)

CANYON FERRY DAM........................SECTION.   STUDY NO. 1
PARALLEL-SIDE CANTILEVER-STRESS ANALYSIS — TRIAL-LOAD TWIST
BENDING MOMENTS (ΣM) IN BEAM DUE TO TRIAL LOADS — LEFT SIDE   (FINAL)
TOTAL SHEAR (ΣV) IN HORIZONTAL ELEMENTS DUE TO TRIAL LOADS   By G.F.B. Date 4-20-46

Beam		Abt.	A	B	C	3635L	D	E	F	G
3808.5	ΣM	+11,268	-260,706	-125,422	+68,972		+89,935		Spillway	
3808.5	ΣV	-17,975	-12,675	-8,314	-2,819		+ 793			
3762	ΣM		-1,323,134	-584,898	+5,699		+331,591	+486,106	+505,240	+570,800
3762	ΣV		-50,229	-47,244	-43,652		-35,372	-7,733	-1,096	+4,404
3725	ΣM		-6,412,700		-1,539,820		+1,674,450	+4,190,326	+4,650,338	+5,324,300
3725	ΣV		-293,276		-253,056		-181,244	-62,699	-37,231	+30,136
3680	ΣM				+1,816,170		+3,121,450	+10,442,630	+11,310,860	+12,793,000
3680	ΣV				-847,249		-499,344	-163,780	-99,024	+47,526
3635	ΣM						-0,789,800	+15,872,100	+16,743,000	+17,953,000
3635	ΣV						-957,250	-178,840	-99,000	+110,000
	$\Sigma_t M$						+2,898,360			
	ΣV						-670,600			

Note:
ΣM = Beam trial loads times unit moment (M'_B) in beam due to unit loads.
ΣV = Beam plus twisted-structure trial loads times unit shear (V_T) in
horizontal element due to unit loads.

Figure B-18. Canyon Ferry Dam study—bending moments in beam due to trial loads (final), and total shear in horizontal elements due to trial loads (final). —DS2-2(48)

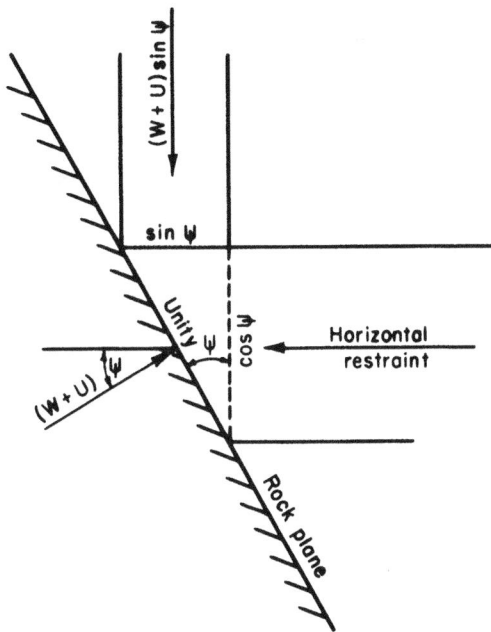

Figure B-19. **Force normal to an inclined abutment plane. —DS2-2(49)**

resisting force. The shear-friction factor is also computed. See section 4-10 for equations and a discussion of these factors. If the computed factors are not within the allowable values, the dam must be reproportioned to correct this condition.

B-14. *Final Results.*—Final results of the trial-load twist analysis of Canyon Ferry Dam are given on figures B-20 to B-25, inclusive. These show load distribution and adjustment on horizontal and vertical elements; stresses in horizontal beams and cantilevers; principal stresses at the faces of the dam; and stability factors for both the twist analysis and the gravity analysis.

The following conclusions were made from the twist analysis:

(1) Results determined from the trial-load twist analysis show that the maximum compressive principal stress is 263 pounds per square inch and occurs at elevation 3680 at the downstream face of cantilever C.

(2) The maximum tensile principal stress occurs at the upstream face of the right abutment of the beam at elevation 3725 and amounts to 146 pounds per square inch.

(3) The maximum rock-plane shearing stress occurs at the base of cantilever G, elevation 3592, and also at the left abutment of the beam at elevation 3635, and amounts to 101 pounds per square inch.

(4) The maximum sliding factor on horizontal planes is 0.812 and occurs at elevation 3725 in cantilever G. The maximum sliding factor on inclined abutment planes occurs at the base of cantilever L and is 1.197.

(5) The minimum shear-friction factor of safety on horizontal planes is 6.78 and occurs at the base elevation of cantilever G. The minimum shear-friction factor on inclined abutment planes is 6.32 and occurs at the base of cantilever C.

(6) Tensile principal stresses which occur at the left and right abutments of the dam at practically all elevations at the upstream face indicate that some diagonal cracking may occur in the concrete in these regions.

(7) In order to reduce the extent of diagonal cracking, it is recommended that the concrete in the dam be subcooled 8° F. or more, if possible, below mean annual temperature prior to grouting the contraction joints.

(8) Maximum compressive stresses in the beams and cantilevers, principal compressive stresses, and rock-plane shear stresses are conservative and well within allowable design limits for good concrete.

(9) The maximum sliding factor of 1.197 that occurs at the inclined base of cantilever L indicates that somewhat unsatisfactory stability conditions may be considered to exist at higher elevations along the abutments of the dam if sliding factors are used as the criterion for judging whether or not the dam is safe against failure by sliding. However, if shear-friction factors are used as the criterion instead of the sliding factors, stability conditions in the dam can be considered as being satisfactory. The minimum value for the shear-friction factor calculated from the trial-load twist analysis was 6.32.

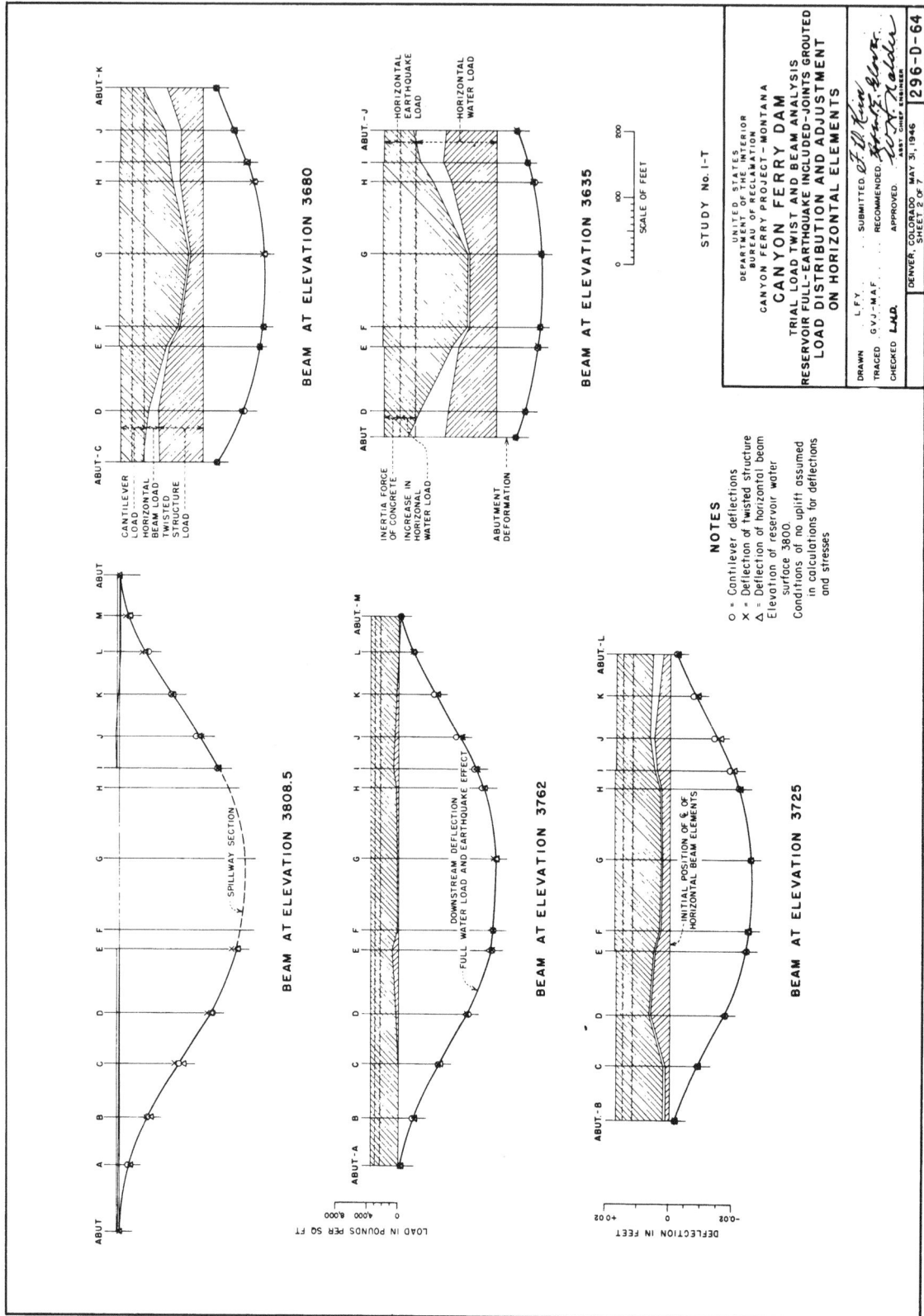

Figure B-20. Canyon Ferry Dam study—load distribution and adjustment on horizontal elements.

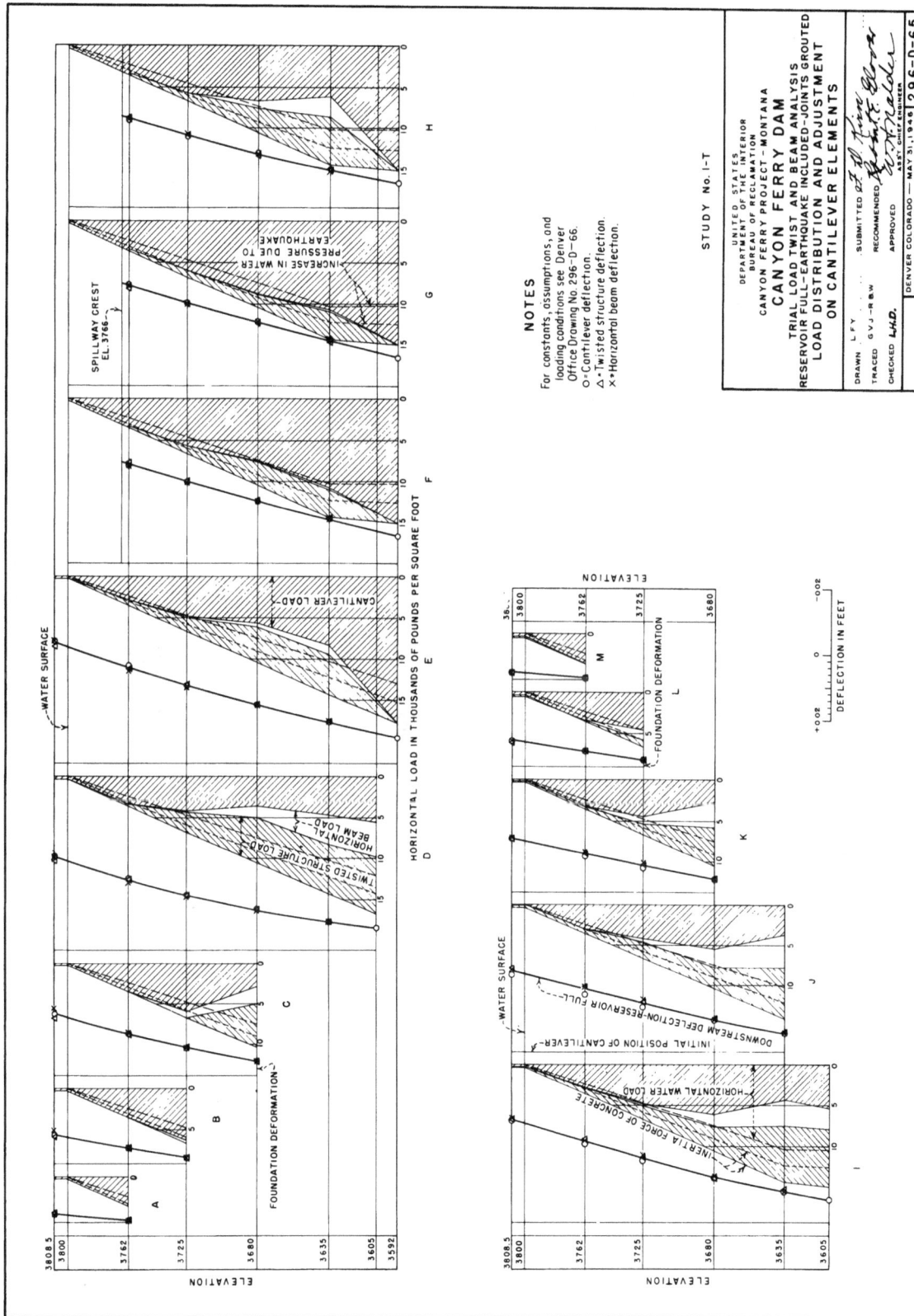

Figure B-21. Canyon Ferry Dam study—load distribution and adjustment on cantilever elements.

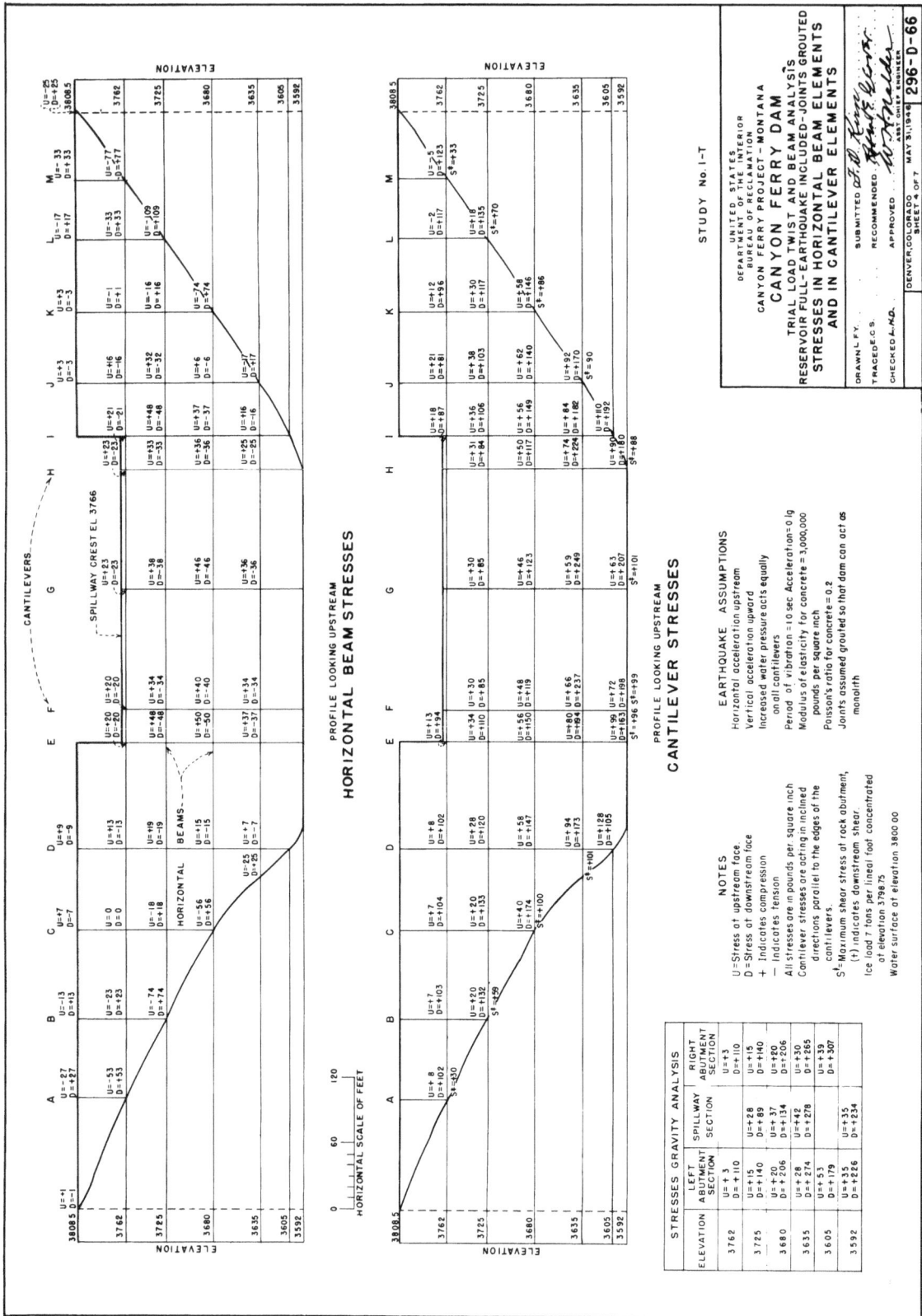

Figure B-22. Canyon Ferry Dam study—stresses in horizontal beam elements and in cantilever elements.

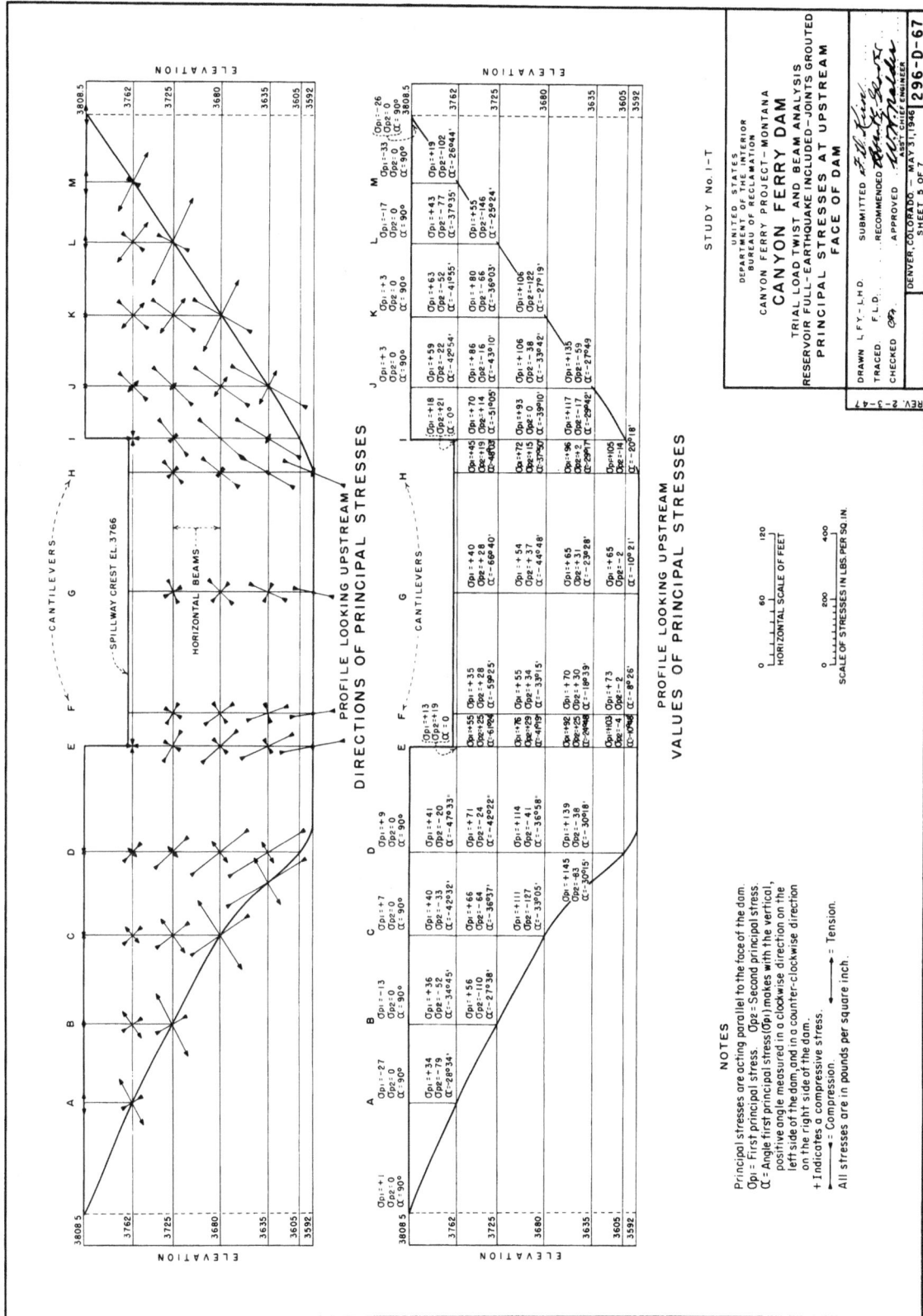

Figure B-23. Canyon Ferry Dam study—principal stresses at upstream face of dam.

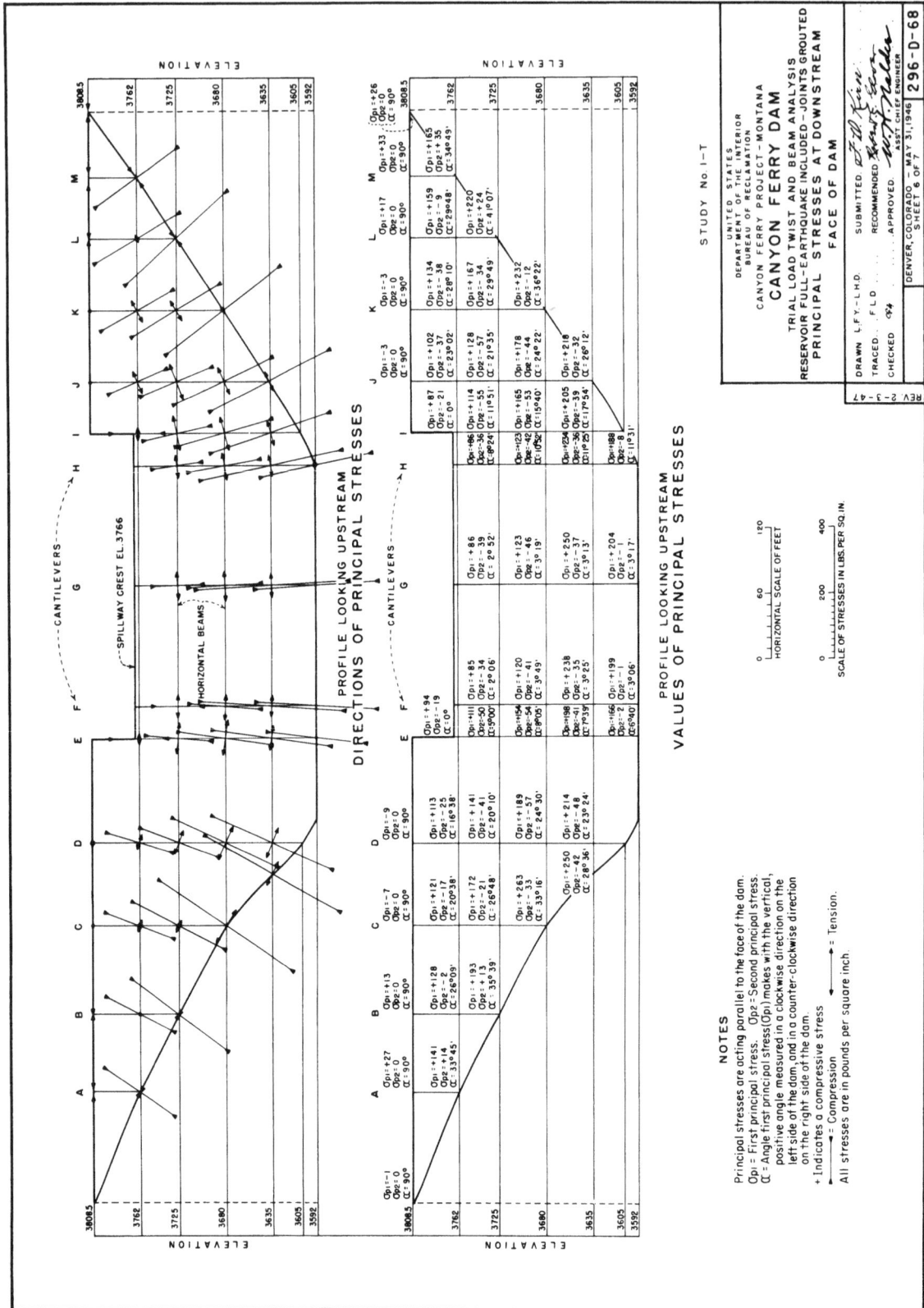

Figure B-24. Canyon Ferry Dam study—principal stresses at downstream face of dam.

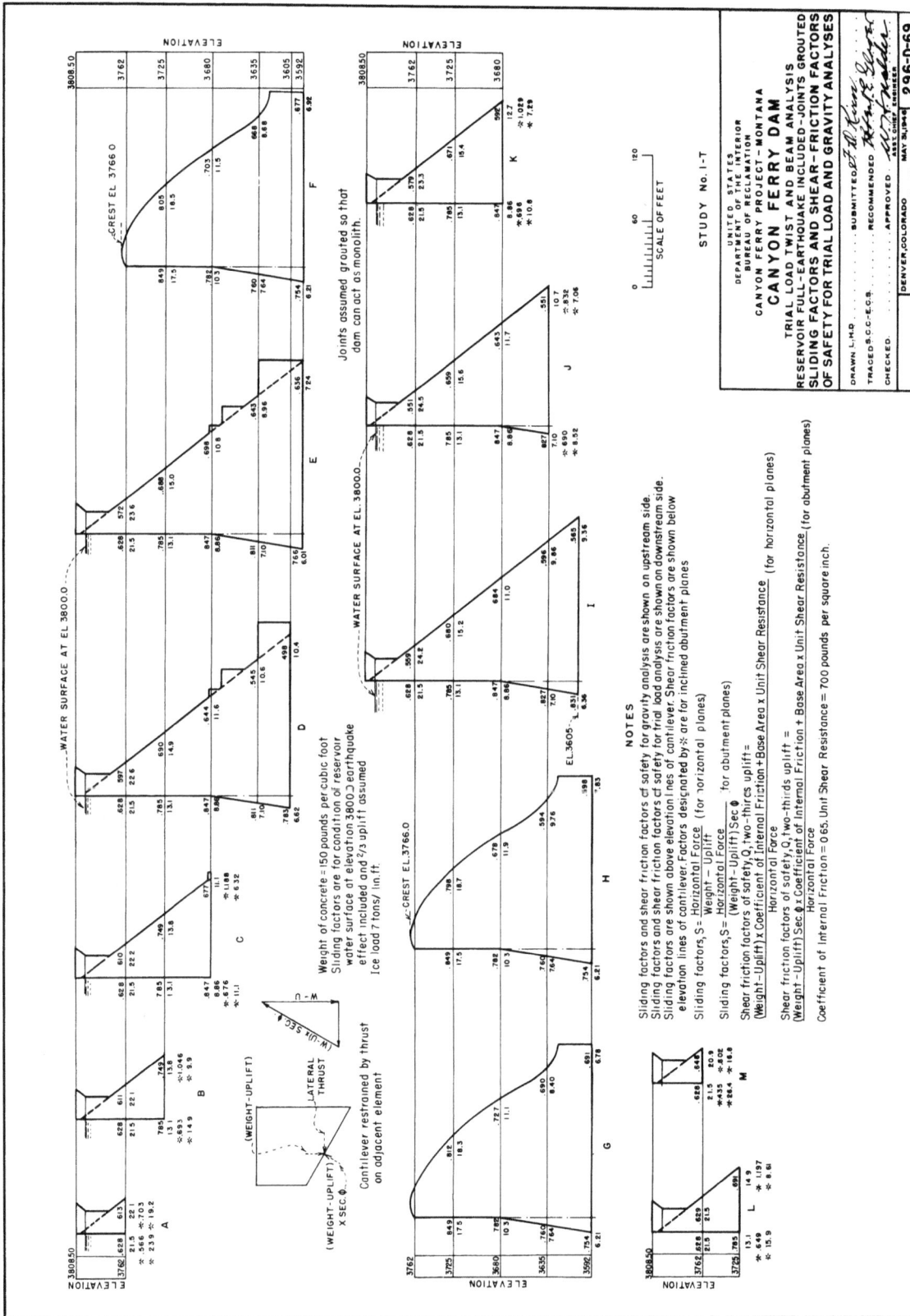

Figure B-25. Canyon Ferry Dam study—sliding factors and shear-friction factors of safety for trial-load and gravity analyses.

Finite Element Method of Analysis

A. TWO-DIMENSIONAL FINITE ELEMENT ANALYSIS

C-1. *Introduction.*—The two-dimensional finite element analysis, discussed in sections 4-36 through 4-44, is illustrated by the following foundation study of the Grand Coulee Forebay Dam. Figure C-1 shows a partial grid of section DG through the dam, reservoir, and foundation.

C-2. *Description of Problem.*—Foundation rock under Grand Coulee Forebay Dam and reservoir has a wide range of deformation moduli, with several faults or planes of weakness. One fault area, because of its low modulus, causes the concrete in the dam immediately above it to bridge over the fault causing horizontal tensions. By treating this fault (replacing part of the low-modulus fault material with concrete) these stresses in the dam will be minimized. This study was made to determine the depth of treatment necessary to obtain satisfactory stress.

C-3. *Grid and Numbering System.*—Figure C-1 shows a portion of the grid used in this study. The nodes are numbered starting in the upper right corner and from left to right at each elevation. The entire grid has 551 nodal points. The elements are designated by a number in a circle. The numbering starts in the upper right corner and proceeds from left to right in horizontal rows. The entire grid has 517 elements. Numbers in squares designate the material numbers. The boundaries for each material are defined by elements. There are 23 materials assumed in this study.

C-4. *Input.*—Printouts of portions of the input are shown on figures C-2, C-3, and C-4.

Figure C-2 shows the number of nodal points, the number of elements, and the number of different materials as indicated above. An acceleration of −1.0 in the Y-direction is a means of including the weight of the materials. Each material is defined for mass density, moduli of elasticity in compression and tension, and Poisson's ratio. Figure C-3 is a listing of the nodal points showing type of restraint (if any), X and Y coordinates, load or displacement in the X or Y direction, and temperature. As an example node 19 is free to move in either direction; it is 653.0 feet to the right of the X reference line and 799.0 feet upward from the Y reference line; and a horizontal load of 27.0 kips is acting on the node in a direction to the left. There is no load in the Y direction and no temperature change.

Figure C-4 is a listing of the nodes enclosing an element and the element material. As an example, element 45 is bounded by nodes 53, 52, 63, 64 and is composed of material number 6.

C-5. *Output.*—The results of an analysis are given as the displacements of the nodes in the X and Y directions and the stresses in the elements.

A printout of displacements for nodes 51 through 100 for the condition of no treatment of the foundation is shown on figure C-5. A similar printout for a loading condition where the foundation is treated for 25 feet is shown on figure C-6. Without treatment, node 69 is displaced 0.007,05 foot in the X direction to the left and 0.037,6 foot downward. After the

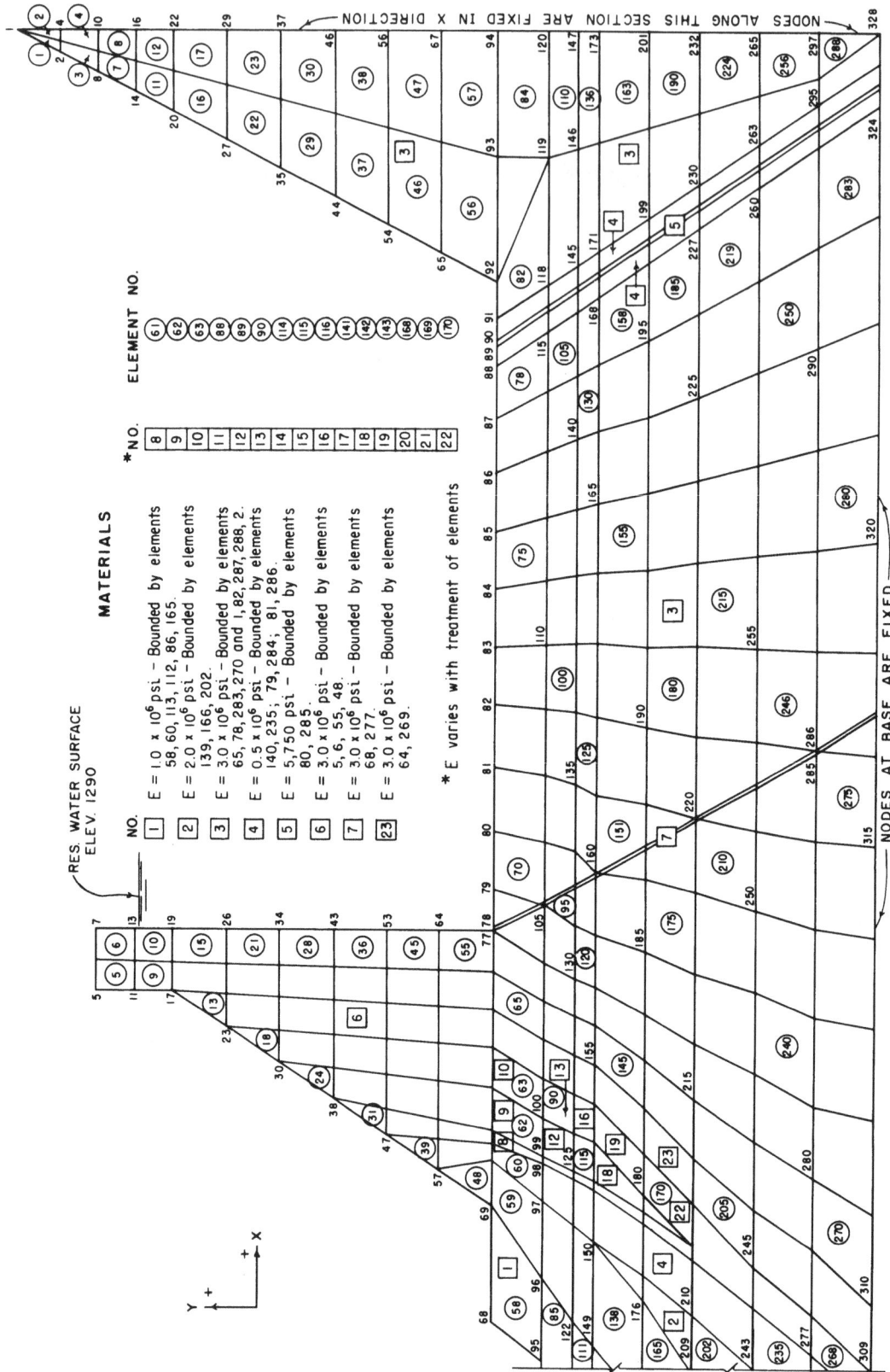

Figure C-1. Grid layout for section *DG* of Grand Coulee Forebay Dam, including excavated cut slope along canyon wall at right. —288-D-3159

ANALYSIS OF PLANE PROBLEMS

COULEE 3RD **FOUNDATION** SEC. DG, GRID 9, HYDRO LOAD, NO TREATMENT

********* INPUT DATA *********   NOTE-- INPUT UNITS MATCH OUTPUT UNITS UNLESS SPECIFIED

                     DATA PREPARED BY----

                     DATA CHECKED BY----

COULEE 3RD **FOUNDATION** SEC. DG, GRID 9, HYDRO LOAD, NO TREATMENT

NUMBER OF NODAL POINTS------551

NUMBER OF ELEMENTS---------517

NUMBER OF DIFF. MATERIALS--- 23

NUMBER OF PRESSURE CARDS---- -0

X-ACCELERATION-------------- -0.0000+000

Y-ACCELERATION-------------- -1.0000+000

REFERENCE TEMPERATURE------- -0.0000+000

NUMBER OF APPROXIMATIONS---- 1

MATERIAL NUMBER = 1,   NUMBER OF TEMPERATURE CARDS = 1,   MASS DENSITY = -0.0000+000

TEMPERATURE	E(C)	NU	E(T)	G/H2	ALPHA	X-STRESS	Y-STRESS
-0.000	144000.0000000	0.1300000	144000.0000000	-0.0000000	-0.0000000	-0.0000000	-0.0000000

MATERIAL NUMBER = 2,   NUMBER OF TEMPERATURE CARDS = 1,   MASS DENSITY = -0.0000+000

TEMPERATURE	E(C)	NU	E(T)	G/H2	ALPHA	X-STRESS	Y-STRESS
-0.000	288000.0000000	0.1300000	288000.0000000	-0.0000000	-0.0000000	-0.0000000	-0.0000000

MATERIAL NUMBER = 3,   NUMBER OF TEMPERATURE CARDS = 1,   MASS DENSITY = -0.0000+000

TEMPERATURE	E(C)	NU	E(T)	G/H2	ALPHA	X-STRESS	Y-STRESS
-0.000	432000.0000000	0.1300000	432000.0000000	-0.0000000	-0.0000000	-0.0000000	-0.0000000

MATERIAL NUMBER = 4,   NUMBER OF TEMPERATURE CARDS = 1,   MASS DENSITY = -0.0000+000

TEMPERATURE	E(C)	NU	E(T)	G/H2	ALPHA	X-STRESS	Y-STRESS
-0.000	72000.0000000	0.1300000	72000.0000000	-0.0000000	-0.0000000	-0.0000000	-0.0000000

MATERIAL NUMBER = 5,   NUMBER OF TEMPERATURE CARDS = 1,   MASS DENSITY = -0.0000+000

TEMPERATURE	E(C)	NU	E(T)	G/H2	ALPHA	X-STRESS	Y-STRESS
-0.000	828.0000000	0.2500000	828.0000000	-0.0000000	-0.0000000	-0.0000000	-0.0000000

Figure C-2. Two-dimensional input data—control data and material properties.—288-D-3160

ANALYSIS OF PLANE PROBLEMS

COULEE 3RD **FOUNDATION** SEC. DG, GRID 9, HYDRO LOAD, NO TREATMENT

PAGE NUMBER 4
DATE 05/27/70

NODAL POINT	TYPE	X-ORDINATE (FT)	Y-ORDINATE (FT)	X LOAD OR DISPLACEMENT (KIPS)	(FT)	Y LOAD OR DISPLACEMENT (KIPS)	(FT)	TEMPERATURE (DEG F)
1	1.00	1081.000	880.000	-0.0000000+000	-0.0000000+000	-0.0000000+000	-0.0000000+000	-0.000
2	0.00	1072.000	859.000	-0.0000000+000	-0.0000000+000	-0.0000000+000	-0.0000000+000	-0.000
3	0.00	1076.500	859.000	0.0000000+000	0.0000000+000	0.0000000+000	0.0000000+000	0.000
4	1.00	1081.000	859.000	-0.0000000+000	-0.0000000+000	-0.0000000+000	-0.0000000+000	-0.000
5	0.00	623.000	836.000	-0.0000000+000	-0.0000000+000	-0.0000000+000	-0.0000000+000	-0.000
6	0.00	637.500	836.500	0.0000000+000	0.0000000+000	0.0000000+000	0.0000000+000	0.000
7	0.00	652.000	837.000	-0.0000000+000	-0.0000000+000	-0.0000000+000	-0.0000000+000	-0.000
8	0.00	1060.000	837.000	-0.0000000+000	-0.0000000+000	-0.0000000+000	-0.0000000+000	-0.000
9	0.00	1070.000	837.000	-0.0000000+000	-0.0000000+000	-0.0000000+000	-0.0000000+000	-0.000
10	1.00	1080.000	837.000	-0.0000000+000	-0.0000000+000	-0.0000000+000	-0.0000000+000	-0.000
11	0.00	623.000	819.000	-0.0000000+000	-0.0000000+000	-0.0000000+000	-0.0000000+000	-0.000
12	0.00	638.000	819.000	-0.0000000+000	-0.0000000+000	-0.0000000+000	-0.0000000+000	-0.000
13	0.00	653.000	819.000	-1.1250000+000	-1.1250000+000	-0.0000000+000	-0.0000000+000	-0.000
14	0.00	1051.000	819.000	1.1250000+000	1.1250000+000	-5.0600000-001	-5.0600000-001	-0.000
15	0.00	1065.500	819.000	0.0000000+000	0.0000000+000	0.0000000+000	0.0000000+000	0.000
16	1.00	1080.000	819.000	-0.0000000+000	-0.0000000+000	-0.0000000+000	-0.0000000+000	-0.000
17	0.00	623.000	799.000	-0.0000000+000	-0.0000000+000	-0.0000000+000	-0.0000000+000	-0.000
18	0.00	638.000	799.000	0.0000000+000	0.0000000+000	0.0000000+000	0.0000000+000	0.000
19	0.00	653.000	799.000	-2.7000000+001	-2.7000000+001	-0.0000000+000	-0.0000000+000	-0.000
20	0.00	1042.000	799.000	2.7000000+001	2.7000000+001	-1.2938000+001	-1.2938000+001	-0.000
21	0.00	1061.500	799.000	0.0000000+000	0.0000000+000	0.0000000+000	0.0000000+000	0.000
22	1.00	1081.000	799.000	-0.0000000+000	-0.0000000+000	-0.0000000+000	-0.0000000+000	-0.000
23	0.00	606.000	771.000	-0.0000000+000	-0.0000000+000	-0.0000000+000	-0.0000000+000	-0.000
24	0.00	621.667	771.000	-0.0000000+000	-0.0000000+000	-0.0000000+000	-0.0000000+000	-0.000
25	0.00	637.333	771.000	0.0000000+000	0.0000000+000	0.0000000+000	0.0000000+000	0.000
26	0.00	653.000	771.000	-7.7000000+001	-7.7000000+001	-0.0000000+000	-0.0000000+000	-0.000
27	0.00	1028.000	771.000	7.7000000+001	7.7000000+001	-3.8500000+001	-3.8500000+001	-0.000
28	0.00	1054.500	771.000	0.0000000+000	0.0000000+000	0.0000000+000	0.0000000+000	0.000
29	1.00	1081.000	771.000	-0.0000000+000	-0.0000000+000	-0.0000000+000	-0.0000000+000	-0.000
30	0.00	587.000	743.000	-0.0000000+000	-0.0000000+000	-0.0000000+000	-0.0000000+000	-0.000
31	0.00	603.500	743.000	0.0000000+000	0.0000000+000	0.0000000+000	0.0000000+000	0.000
32	0.00	620.000	743.000	0.0000000+000	0.0000000+000	0.0000000+000	0.0000000+000	0.000
33	0.00	636.500	743.000	0.0000000+000	0.0000000+000	0.0000000+000	0.0000000+000	0.000
34	1.00	653.000	743.000	-1.2867800+002	-1.2867800+002	-0.0000000+000	-0.0000000+000	-0.000
35	0.00	1014.000	743.000	1.2867800+002	1.2867800+002	-6.3210000+001	-6.3210000+001	-0.000
36	0.00	1047.500	743.000	0.0000000+000	0.0000000+000	0.0000000+000	0.0000000+000	0.000
37	1.00	1081.000	743.000	-0.0000000+000	-0.0000000+000	-0.0000000+000	-0.0000000+000	-0.000
38	0.00	570.000	714.000	-0.0000000+000	-0.0000000+000	-0.0000000+000	-0.0000000+000	-0.000
39	0.00	586.600	714.000	0.0000000+000	0.0000000+000	0.0000000+000	0.0000000+000	0.000
40	0.00	603.200	714.000	0.0000000+000	0.0000000+000	0.0000000+000	0.0000000+000	0.000
41	0.00	619.800	714.000	0.0000000+000	0.0000000+000	0.0000000+000	0.0000000+000	0.000
42	0.00	636.400	714.000	0.0000000+000	0.0000000+000	0.0000000+000	0.0000000+000	0.000
43	0.00	653.000	714.000	-1.7586800+002	-1.7586800+002	-0.0000000+000	-0.0000000+000	-0.000
44	0.00	1000.000	714.000	1.7909000+002	1.7909000+002	-8.7934000+001	-8.7934000+001	-0.000
45	0.00	1040.500	714.000	0.0000000+000	0.0000000+000	0.0000000+000	0.0000000+000	0.000
46	1.00	1081.000	714.000	-0.0000000+000	-0.0000000+000	-0.0000000+000	-0.0000000+000	-0.000
47	0.00	552.000	687.000	-0.0000000+000	-0.0000000+000	-0.0000000+000	-0.0000000+000	-0.000
48	0.00	565.000	687.000	-0.0000000+000	-0.0000000+000	-0.0000000+000	-0.0000000+000	-0.000
49	0.00	582.600	687.000	0.0000000+000	0.0000000+000	0.0000000+000	0.0000000+000	0.000
50	0.00	600.200	687.000	0.0000000+000	0.0000000+000	0.0000000+000	0.0000000+000	0.000

Figure C-3. Two-dimensional input data—loading and description of section by nodal points.—288-D-3161

ANALYSIS OF PLANE PROBLEMS

PAGE NUMBER 16
DATE 05/27/70

COULEE 3RD **FOUNDATION** SEC. DG, GRID 9, HYDRO LOAD, NO TREATMENT

ELEMENT NO.	I	J	K	L	MATERIAL
1	1	2	3	3	3
2	1	3	4	4	3
3	3	2	8	9	3
4	4	3	9	10	3
5	6	5	11	12	6
6	7	6	12	13	6
7	9	8	14	15	3
8	10	9	15	16	3
9	12	11	17	18	6
10	13	12	18	19	6
11	15	14	20	21	3
12	16	15	21	22	3
13	17	23	24	24	6
14	18	17	24	25	6
15	19	18	25	26	6
16	21	20	27	28	3
17	22	21	28	29	3
18	23	30	31	31	6
19	24	23	31	32	6
20	25	24	32	33	6
21	26	25	33	34	6
22	28	27	35	36	3
23	29	28	36	37	3
24	30	38	39	39	6
25	31	30	39	40	6
26	32	31	40	41	6
27	33	32	41	42	6
28	34	33	42	43	6
29	36	35	44	45	3
30	37	36	45	46	3
31	38	47	48	48	6
32	39	38	48	49	6
33	40	39	49	50	6
34	41	40	50	51	6
35	42	41	51	52	6
36	43	42	52	53	6
37	45	44	54	55	3
38	46	45	55	56	3
39	47	57	58	58	6
40	48	47	58	59	6
41	49	48	59	60	6
42	50	49	60	61	6
43	51	50	61	62	6
44	52	51	62	63	6
45	53	52	63	64	6
46	55	54	65	66	3
47	56	55	66	67	3
48	57	69	70	70	6
49	58	57	70	71	6
50	59	58	71	72	6

Figure C-4. Two-dimensional input data—elements defined by nodal points with material. –288-D-3162

COULEE 3RD **FOUNDATION** SEC. DG, GRID 9, HYDRO LOAD, NO TREATMENT

NODAL POINT	DISPLACEMENT--X (FT)	DISPLACEMENT--Y (FT)
51	-1.9811064-002	-2.0667878-002
52	-1.9941570-002	-1.6607736-002
53	-2.0141816-002	-1.2279052-002
54	1.2091843-003	-2.4071188-002
55	5.5973499-004	-2.3599376-002
56	0.0000000+000	-2.3429267-002
57	-1.2884864-002	-3.7561138-002
58	-1.3239801-002	-3.5288567-002
59	-1.3559575-002	-3.2812637-002
60	-1.3979795-002	-2.8402294-002
61	-1.3880246-002	-2.4046112-002
62	-1.3516598-002	-1.9974260-002
63	-1.3437772-002	-1.6478346-002
64	-1.3871664-002	-1.2207339-002
65	1.2985815-003	-2.4044677-002
66	5.8831362-004	-2.3523886-002
67	0.0000000+000	-2.3303277-002
68	-2.9159215-003	-3.2254766-002
69	-7.0522162-003	-3.7597583-002
70	-7.6117066-003	-3.5823494-002
71	-7.5138281-003	-3.4742080-002
72	-7.3896957-003	-3.3570870-002
73	-7.0165442-003	-2.8818410-002
74	-6.7566923-003	-2.1328360-002
75	-6.9195435-003	-1.8176320-002
76	-6.8467254-003	-1.5494703-002
77	-6.2642486-003	-1.3235448-002
78	-6.1773772-003	-1.3248088-002
79	-5.0546562-003	-1.3404921-002
80	-3.9716033-003	-1.3389674-002
81	-3.5272857-003	-1.3428153-002
82	-3.2858108-003	-1.3391417-002
83	-3.1882931-003	-1.3314550-002
84	-3.1574305-003	-1.3196625-002
85	-3.1735731-003	-1.2980730-002
86	-3.2264351-003	-1.2680855-002
87	-3.2859588-003	-1.2212955-002
88	-3.2372784-003	-1.1813686-002
89	-2.8409063-003	-1.2637024-002
90	1.1806029-003	-2.5426777-002
91	1.0295768-003	-2.4358245-002
92	1.0241282-003	-2.4091306-002
93	5.8904893-004	-2.3338998-002
94	0.0000000+000	-2.3113335-002
95	-4.0600306-003	-3.1606605-002
96	-4.5308027-003	-3.3783533-002
97	-5.0386242-003	-3.5799620-002
98	-5.0921594-003	-3.5052235-002
99	-3.9714688-003	-3.2638101-002
100	-3.9967861-003	-3.1880739-002

Figure C-5. Nodal point displacements (no treatment). −288-D-3163

ANALYSIS OF PLANE PROBLEMS

COULEE 3RD **FOUNDATION** SEC. DG, GRID 9, HYDRO LOAD, 25 FT TREATMENT

NODAL POINT	DISPLACEMENT--X (FT)	DISPLACEMENT--Y (FT)
51	-1.8724543-002	-2.0415804-002
52	-1.8937366-002	-1.7466395-002
53	-1.9147395-002	-1.4287073-002
54	1.2031721-003	-2.4049716-002
55	5.5581550-004	-2.3582274-002
56	0.0000000+000	-2.3412686-002
57	-1.3924929-002	-3.1812965-002
58	-1.4099223-002	-3.0061762-002
59	-1.4272226-002	-2.8128268-002
60	-1.4290854-002	-2.5340594-002
61	-1.4190080-002	-2.2586042-002
62	-1.4060031-002	-1.9924903-002
63	-1.4032529-002	-1.7248892-002
64	-1.4441423-002	-1.3770922-002
65	1.2863362-003	-2.4022650-002
66	5.8064917-004	-2.3506172-002
67	0.0000000+000	-2.3285657-002
68	-4.8971975-003	-2.9753268-002
69	-8.6758414-003	-3.2517384-002
70	-9.3259331-003	-3.0508820-002
71	-9.5034153-003	-2.9319722-002
72	-9.5773006-003	-2.8390606-002
73	-9.7625006-003	-2.4257549-002
74	-9.2909087-003	-2.1070602-002
75	-8.8710655-003	-1.8504728-002
76	-8.4077281-003	-1.6204246-002
77	-7.6451459-003	-1.4086363-002
78	-7.5582207-003	-1.4071215-002
79	-6.2934332-003	-1.3983721-002
80	-5.0034774-003	-1.3774928-002
81	-4.4302627-003	-1.3719672-002
82	-4.0935786-003	-1.3606782-002
83	-3.9344774-003	-1.3477032-002
84	-3.8588070-003	-1.3316513-002
85	-3.8335854-003	-1.3053523-002
86	-3.8570314-003	-1.2710341-002
87	-3.8942547-003	-1.2192559-002
88	-3.8333803-003	-1.1747112-002
89	-3.4139745-003	-1.2555023-002
90	1.1148291-003	-2.5411830-002
91	9.9502980-004	-2.4338743-002
92	9.9658980-004	-2.4071306-002
93	5.7744601-004	-2.3319313-002
94	0.0000000+000	-2.3094178-002
95	-5.1226583-003	-2.9567080-002
96	-5.5237841-003	-3.0422893-002
97	-5.9118189-003	-3.1153132-002
98	-5.6404379-003	-3.0135814-002
99	-5.5543256-003	-2.9776455-002
100	-5.1357266-003	-2.6788933-002

Figure C-6. Nodal point displacements (25-foot treatment). —288-D-3164

25-foot treatment, node 69 is displaced 0.008,67 foot to the left and 0.032,5 foot downward.

Printouts of stresses for the analysis with the no-treatment condition are shown on figure C-7. This listing gives the element number, the location of the stresses in X and Y ordinates, stresses in the X and Y planes, the shear stress in the XY plane, and the principal stresses with the angle from the horizontal to the maximum stress. In this case, a shear stress along a specified plane and a stress normal to that plane were found. A similar printout for the condition with the foundation treated for 25 feet is shown on figure C-8. Stresses in element 51 are the key to this foundation problem. By treating the foundation, the compressive stresses in element 51 are increased from 8 to 33 pounds per square inch in the horizontal direction, and from 26 to 120 pounds per square inch in the vertical direction.

Microfilm plots of the grid and stresses are also provided by the computer as part of the regular output. Principal stresses in the dam for the no-treatment condition are shown on figure C-9. Principal stresses shown on figure C-10 are for the condition where the foundation is treated for 25 feet. These latter principal stresses are derived from the vertical stresses shown on figure C-11, the horizontal stresses shown on figure C-12, and the shear stresses shown on figure C-13.

Occasionally the finite element mesh is so fine that sufficient detail cannot be portrayed on the microfilm. In order to gain greater detail of a particular area and its stresses, the area can be plotted to an enlarged scale and more accurate stresses thus obtained.

B. THREE-DIMENSIONAL FINITE ELEMENT ANALYSIS

C-6. *Introduction.*—The analysis of the Grand Coulee Forebay Dam demonstrates the capabilities of the three-dimensional finite element system of stress analysis, discussed in sections 4-45 through 4-48. Distribution of stresses around the penstock is of special interest because of the large size of the opening in relation to the size of the block.

C-7. *Layout and Numbering System.*—A three-dimensional drawing of half of a block with the opening for a penstock is shown on figure C-14. To clarify the penstock area, vertical sections normal to the penstock are also shown. Although no foundation is shown, a treated foundation was assumed in the analysis. The block is divided into hexahedron elements. Nodal points are numbered consecutively from left to right starting at the top. There are 588 nodes in the example problem. The elements are numbered starting at the top and follow the general pattern set up for the nodes. There are 374 elements in this example.

C-8. *Input.*—Examples of the required input data are shown on the printouts in figures C-15 through C-18. Figure C-15 shows the numbers of elements, nodes, boundary nodes, loaded nodes, and different materials. Also shown is the maximum band width expected. Data given for each of the materials are modulus of elasticity, Poisson's ratio, and the mass density. The nodal points are described using ordinates in the X, Y, and Z directions as shown on figure C-16. For example, node 45 is 14.0 feet from the centerline of the block in the X direction, 19.58 feet from the upstream face in the Y direction, and at 273.0 (elevation 1273) in the Z direction. The nodal points that enclose the elements, the element material, and the integration rule are shown on figure C-17. Element 41 is bounded by nodal points 49, 55, 103, 97, 50, 56, 104, and 98. It contains material number 1 and is to be integrated by rule 2.

Forces or loads are applied at the nodal points. In this problem the loads are due to weight of the concrete, the hydraulic pressure on the upstream face, the uplift pressure at the base of the dam, and the internal pressure in the penstock and gate shaft. An example of

ANALYSIS OF PLANE PROBLEMS

COULEE 3RD **FOUNDATION** SEC. DG, GRID 9, HYDRO LOAD, NO TREATMENT

EL.NO.	X (FT)	Y (FT)	X-STRESS (PSI)	Y-STRESS (PSI)	XY-STRESS (PSI)	MAX-STRESS (PSI)	MIN-STRESS (PSI)	ANGLE (DEG)	SHEAR-PLANE (PSI)	NORMAL-PLANE (PSI)
51	569.25	649.50	-82.9413-001	-26.5010+000	-10.1157-001	-37.8874-001	-31.0064+000	-24.01	-13.3122+000	-20.2235+000
52	588.25	649.25	-18.2965+000	-18.0532+001	81.2799+000	15.4183+000	-21.4246+001	22.53	29.8317+000	-21.0304+001
53	606.75	649.00	-22.4601+000	-25.5836+001	-65.2115+000	-54.7449-001	-27.2822+001	-14.60	-11.4819+001	-20.7597+001
54	625.25	649.00	-92.6967-001	-14.1428+001	-89.3897+000	35.8151+001	-18.6511+001	-26.76	-11.0425+001	-87.8800+000
55	643.75	649.00	16.2583+000	33.5292+000	-10.0265+001	12.5530+001	-75.7425+000	-47.46	-82.5144+000	82.5048+000
56	995.13	648.75	-33.7595+000	-26.2658+000	27.2580+000	-24.9834-001	-57.5269+000	48.91	25.4795+000	-40.3968+000
57	1051.38	648.25	-34.9132+000	-27.1232+000	68.5384-001	-23.1349+000	-38.9015+000	59.80	78.8312-001	-31.0719+000
58	480.00	623.00	-47.9419+000	-14.3210+000	-24.0107+000	-18.2099+000	-60.4419+000	-62.50	-12.3886+000	-45.6785-001
59	519.00	623.00	-29.7315+000	-72.1687+000	-32.0523+000	-12.5108+000	-89.3894+000	-28.25	-38.3674+000	-53.2998+000
60	540.00	623.00	-35.8968+001	-33.0459+000	-93.5017-001	-87.2364+000	-35.7632+000	-16.20	-15.4615+000	-26.3976+000
61	548.75	623.00	56.7466-002	-86.0317-002	54.4996-002	75.1718-002	-14.4457-001	18.68	11.5035-001	-10.3717-001
62	560.90	622.75	28.7067+000	-14.7544+000	-14.5284+000	30.8445-001	-14.9683+000	6.25	-27.1509-001	-14.5503+000
63	580.20	622.50	-48.5870-002	-21.8737-001	10.5970-001	22.3277-003	-26.9556-001	25.62	49.2252-002	-26.0324-001
64	599.00	622.50	-76.1447+000	-43.0465+001	-65.8382+000	-64.3064+000	-44.2303+001	-10.19	-14.5598+001	-37.3811+001
65	617.80	622.50	-30.8871+000	-21.0267+001	-11.7020+000	-30.1269+000	-21.1027+001	-3.72	-54.9792+000	-19.2400+001
66	636.60	622.50	21.3056+000	-85.4769+000	-59.5000-001	21.6351+000	-85.8074+000	-3.18	-31.8485+000	-75.3488+000
67	653.50	622.50	14.2442+000	37.3463+000	-44.3987+000	71.6720+000	-20.0815+000	-52.29	-32.6749+000	57.9981+000
68	661.00	622.50	11.4532+001	-18.7765+000	-73.9562+000	14.7738+001	-51.6827+000	-23.99	-97.3751+001	27.1116+000
69	665.67	626.67	17.2429+000	-37.2502+000	-77.6679+000	19.8064+001	-62.8853+000	-18.27	-11.9682+001	15.6296+000
70	685.78	622.50	66.0723+000	-63.1979+000	-33.8962+000	74.4211+000	-71.5467+000	-13.84	-61.6725+000	-37.5903+000
71	715.08	622.50	32.9545+000	-72.6528+000	-12.5610+000	34.4280+000	-74.1262+000	-6.69	-37.2800+000	-59.2979+000
72	744.64	622.50	13.9267+000	-76.8051+000	-66.6803-001	14.4141+000	-77.2925+000	-4.18	-28.4576+000	-69.3932+000
73	773.69	622.50	26.2043-001	-77.4449+000	-42.0971-001	28.4116-001	-77.6657+000	-3.00	-23.6621+000	-69.9767+000
74	801.25	622.50	-43.5231-001	-77.9217+000	-28.6859-001	-42.4063-001	-78.0334+000	-2.23	-20.8766+000	-71.5592+000
75	829.56	622.50	-93.4967+000	-17.2536+000	-19.7601-001	-92.9265+000	-77.8196+000	-1.65	-18.8145+000	-72.1917+000
76	858.86	622.50	-12.3082+000	-77.7409+000	-68.1714-002	-12.3011+000	-77.7480+000	-0.60	-16.9486+000	-73.0169+000
77	888.67	622.50	-12.6495+000	-75.7173+000	-12.2243-001	-12.6258+000	-75.7409+000	1.11	-14.7083+000	-72.1037+000
78	917.72	622.50	-90.0447-001	-82.2540+000	82.2306-001	-80.9269-001	-83.1658+000	6.33	-11.1910+000	-81.4587+000
79	936.00	622.50	54.4767-001	-42.5930+000	-81.2811-001	67.8562-001	-43.9310+000	-9.35	-19.0493+000	-35.3109+000
80	941.75	622.50	46.3713-001	-17.2536+000	-93.3747-001	80.7889-001	-20.6953+000	-20.23	-13.5592+000	-11.1185+000
81	947.75	622.50	-63.9990-001	-20.3811+000	81.5241-001	-26.5131-001	-68.1297+000	24.69	35.6488-001	-23.5208+000
82	971.00	626.67	-23.7352+000	-56.2670+000	23.2021-001	-11.6653+000	-68.3369+000	27.48	11.9607+000	-65.6888+000
83	999.50	622.50	-26.3419+000	-36.6750+000	22.7138-001	-82.1449-001	-54.8025+000	38.59	17.0874+000	-47.3397+000
84	1049.88	622.50	-30.7179+000	-33.7637+000	-23.9640-001	-24.7872+000	-39.6944+000	39.11	55.5741-001	-37.2079+000
85	458.50	602.50	-14.0753+000	-15.5933+000	-23.8834+000	90.6109-001	-38.7298+000	-44.09	-21.0631+000	-35.4997-001
86	497.50	602.50	-17.5592+000	-44.1041+000	-30.1450-001	21.0587-001	-63.7692+000	-33.12	-32.7426+000	-27.2534+000
87	528.00	602.50	-67.8666-001	-41.5736+000	-15.1161+000	-11.3608-001	-47.2241+000	-20.50	-21.7875+000	-31.6852+000
88	539.75	602.50	88.7687-002	-23.1891-001	-15.6864-001	15.2742-001	-29.5864-001	22.19	55.6837-001	-28.8843-001
89	550.80	602.50	-13.1075-001	-14.0082+000	11.1145-001	-13.0978-001	-14.0092+000	0.50	-30.7811-001	-13.2132+000
90	569.40	602.50	-49.3957-002	-26.5319-001	18.4298-001	56.2345-002	-37.0949-001	29.82	10.5626-001	-34.3004-001
91	588.00	602.50	-50.9808+000	-27.7548+001	-10.2420+001	-11.5456+000	-31.6983+001	-21.06	-14.5340+000	-21.1161+001
92	606.60	602.50	-41.9202+000	-27.4602+001	-71.6463-001	-41.6998+000	-27.4788+001	-1.76	-64.3666+000	-25.5401+001
93	625.20	602.50	-41.7499+000	-14.6202+001	18.9042+000	-38.4338+000	-14.9518+001	9.95	-97.4159-001	-14.8657+001
94	648.25	602.50	-20.0766+000	-54.3948+000	-63.2339-001	-18.9485+000	-55.5229+000	-10.11	-14.0558+000	-48.9342+000
95	666.67	600.00	-16.8373+000	87.1260-002	-27.6822+000	21.0808+000	-37.0468+000	-53.87	-19.5464+000	13.5261+000
96	672.00	602.50	-78.4602-001	-14.0913+000	-38.3291+000	27.4874+000	-49.4247+000	-42.67	-34.7553+000	54.9163-001
97	684.28	602.50	21.0208+000	-44.9724+000	-39.9128+000	39.8104+000	-63.7620+000	-25.21	-51.0638+000	20.5553+000
98	711.52	602.50	19.3361+000	-62.8762+000	-26.3743+000	27.0697+000	-70.6098+000	-16.34	-43.3940+000	-44.1818+000
99	742.45	602.50	11.4645+000	-73.2695+000	-15.3108+000	14.1461+000	-75.9512+000	-9.93	-34.4430+000	-59.9380+000
100	773.38	602.50	35.1875-001	-75.7009+000	-96.2477-001	46.7134-001	-76.8535+000	-6.83	-28.1402+000	-65.5818+000

Figure C-7. Stresses in elements (no treatment).—288-D-3165

ANALYSIS OF PLANE PROBLEMS

COULEE 3RD **FOUNDATION** SEC. DG, GRID 9, HYDRO LOAD, 25 FT TREATMENT

EL.NO.	X (FT)	Y (FT)	X-STRESS (PSI)	Y-STRESS (PSI)	XY-STRESS (PSI)	MAX-STRESS (PSI)	MIN-STRESS (PSI)	ANGLE (DEG)	SHEAR-PLANE (PSI)	NORMAL-PLANE (PSI)
51	569.25	649.50	-32.7145+000	-12.0034+001	94.2840-001	-31.7080+000	-12.1041+001	6.09	-13.6647+000	-11.8899+001
52	588.25	649.25	74.9617-001	-16.6625+001	-16.8468+000	91.1118-001	-16.8240+001	-5.48	-58.1201+000	-14.6538+001
53	606.75	649.00	20.2207+000	-16.1694+001	-51.0114+000	33.5486+000	-17.5022+001	-14.64	-89.6558+000	-12.4002+001
54	625.25	649.00	21.3625+000	-12.1475+001	-76.0907+000	54.3011+000	-15.4413+001	-23.41	-10.1606+001	-73.8609+000
55	643.75	649.00	26.5266+000	-10.3843+000	-91.3180+000	10.1235+001	-85.0932+000	-39.29	-88.3115+000	37.7473+000
56	995.13	648.75	-33.1891+000	-26.1342+000	27.5627+000	-18.7407-001	-57.4492+000	48.65	25.6338+000	-40.3881+000
57	1051.38	648.25	-34.4154+000	-27.2529+000	69.1634-001	-23.0456+000	-38.6227+000	58.69	77.0035-001	-31.1909+000
58	480.00	623.00	-43.1646+000	-12.0455+000	-21.7198+000	-88.7067-002	-54.3230+000	-62.81	-11.0301+000	-32.7016-001
59	519.00	623.00	-26.6169+000	-52.2818+000	-34.0554+000	-30.5646-001	-75.8422+000	-34.68	-35.9091+000	-33.5348+000
60	540.00	623.00	-39.8110-002	-17.9328+000	-22.0893+000	14.6001+000	-32.9311+000	-34.18	-23.5136+000	-57.1357-001
61	548.75	623.00	-21.4494-001	-45.9332+000	-13.3666+000	16.1279-001	-49.6909+000	-15.70	-22.5228+000	-36.3166+000
62	560.90	623.00	11.5484+001	-46.6719+000	12.8563+000	14.2610+000	-49.3844+000	11.91	-34.2120-001	-49.2000+000
63	580.20	622.75	15.2437+000	-24.6529+001	57.3429+000	27.2540+000	-25.8539+001	11.83	-15.7828+000	-25.7665+001
64	599.00	622.50	-20.0826+001	-24.6460+001	-42.8305+000	-12.2501+000	-25.4292+001	-10.36	-93.6866+000	-20.9880+001
65	617.80	622.50	-65.3651-002	-16.5341+001	-48.8801+000	12.7676+000	-17.8675+001	-15.35	-83.4816+000	-12.9788+001
66	636.60	622.50	43.4393+000	-85.5618+000	-29.8111+000	49.9952+000	-92.1177+000	-12.40	-58.0674+000	-62.0148+000
67	653.50	622.50	30.4139+000	15.7429+000	-54.1453+000	77.7183+000	-31.5615+000	-41.14	-50.5590+000	43.7983+000
68	661.00	622.50	12.4017+001	-21.8999+000	-70.6748+000	15.2636+001	-50.5182+000	-22.04	-97.6854+000	23.2121+000
69	665.67	626.67	19.4144+001	-42.7663+000	-76.2355+000	21.5455+001	-65.1781+000	-16.38	-12.5249+001	11.2214+000
70	685.78	622.50	81.0329+000	-63.9932+000	-37.0082+000	89.9309+000	-72.8911+000	-13.52	-68.3065+000	-35.7742+000
71	715.08	622.50	44.1654+000	-71.8777+000	-14.5382+000	45.9591+000	-73.6713+000	-7.03	-41.6012+000	-56.8352+000
72	744.64	622.50	22.3087+000	-76.6165+000	-78.5996-001	22.9293+000	-77.2372+000	-4.51	-31.5382+000	-66.0598+000
73	773.69	622.50	89.9168+000	-77.3023+000	-50.2472-001	92.8327+000	-77.5939+000	-3.32	-25.9250+000	-69.0093+000
74	801.25	622.50	63.2754-002	-77.8593+000	-34.8766-001	78.7417-002	-78.0140+000	-2.54	-22.6434+000	-70.8575+000
75	829.56	622.50	-54.5643-001	-77.7359+000	-24.6500-001	-53.7246-001	-77.8199+000	-1.95	-20.2046+000	-71.6616+000
76	858.86	622.50	-93.7148-001	-77.7381+000	-10.8770-001	-93.5418-001	-77.7554+000	-0.91	-18.0336+000	-72.6145+000
77	888.67	622.50	-10.5079+000	-75.6816+000	88.8498-002	-10.4958+000	-75.6937+000	0.78	-15.5240+000	-71.7601+000
78	917.72	622.50	-74.7314+000	-82.0137+000	80.6889-001	-66.0970-001	-82.8772+000	6.11	-11.6473+000	-81.0549+000
79	936.00	622.50	66.1293-001	-42.4652+000	-79.2355-001	78.6046-001	-43.7128+000	-8.95	-19.1315+000	-35.2159+000
80	941.75	622.50	58.1203-001	-17.0752+000	-91.3237-001	90.0933-001	-20.2725+000	-19.30	-13.6307+000	-10.9759+000
81	947.75	622.50	-51.6067-001	-20.2284+000	84.1768-001	-13.9780-001	-23.9913+000	24.09	35.2298-001	-23.4279+000
82	971.00	622.50	-22.8807+000	-56.1159+000	23.3454+000	-10.7926+000	-68.1687+000	27.23	11.8811+000	-65.5928+000
83	999.50	626.67	-25.5592+000	-36.7323+000	22.8377+000	-76.3471-001	-54.6568+000	38.13	16.9847+000	-47.4027+000
84	1049.88	602.50	-30.1020+000	-33.9169+000	73.6312-001	-24.4033+000	-39.6156+000	37.74	54.2294-001	-37.3429+000
85	458.50	602.50	-12.4712+001	-15.1605+000	-21.4961+000	77.2229-001	-35.5540+000	-43.21	-19.2285+000	-42.3231-001
86	497.50	602.50	-13.6288+000	-35.8360+000	-25.7387+000	32.9915-001	-52.7640+000	-33.33	-27.8421+000	-21.4791+000
87	528.00	602.50	29.1846-001	-30.7301+000	-10.9243+000	61.5399-001	-33.9657+000	-16.50	-17.8728+000	-23.0140+000
88	539.25	602.50	68.9677-002	-75.2481-002	31.5427-002	75.5649-002	-81.8453-002	11.81	-87.3719-003	-81.3588-002
89	550.80	602.50	45.0945-001	-13.2119+000	17.1196-001	46.7330-001	-13.3777+000	5.47	-29.4823-001	-12.8826+000
90	569.40	602.50	-22.6987-002	-15.5969-001	77.7242-002	13.0444-002	-19.1712-001	24.70	33.9937-002	-18.5903-001
91	588.00	602.50	-87.3317+000	-47.8188+001	-37.7765+000	-83.7141+000	-48.1806+001	-5.47	-13.0429+001	-43.3117+001
92	606.60	602.50	-69.8602+000	-19.5830+001	-10.5696+000	-68.9795+000	-19.6711+001	-4.76	-40.6460+000	-18.2107+001
93	625.20	602.50	-41.3323+000	-12.2190+001	-98.5612-001	-40.1506+000	-12.3372+001	-6.84	-28.7414+000	-11.1850+001
94	648.25	602.50	-83.1921-001	-54.0970+000	-21.2528+000	26.1804-003	-62.4424+000	-21.44	-29.8499+000	-40.4041+000
95	666.67	600.00	-85.5344-001	-41.6698-001	-33.9891+000	27.6996+000	-40.4200+000	-46.85	-28.3388+000	12.5337+000
96	672.00	602.50	31.5872-001	-17.2564+000	-42.5064+000	36.6660+000	-50.7637+000	-38.25	-41.9154+000	53.6433-001
97	684.28	602.50	28.6650+000	-46.0280+000	-43.3399+000	48.5297+000	-65.8926+000	-24.62	-56.2067+000	-19.3545+000
98	711.52	602.50	26.3567+000	-61.4328+000	-29.6761+000	35.4470+000	-70.5231+000	-17.03	-47.6476+000	-40.7140+000
99	742.45	602.50	17.7289+000	-72.5470+000	-17.8019+000	21.1125+000	-75.9306+000	-10.76	-37.9859+000	-57.5988+000
100	773.38	602.50	86.6945-001	-75.3200+000	-11.1828+000	10.1848-001	-76.8353+000	-7.58	-30.8551+000	-64.0024+000

Figure C-8. Stresses in elements (25-foot treatment).—288-D-3166

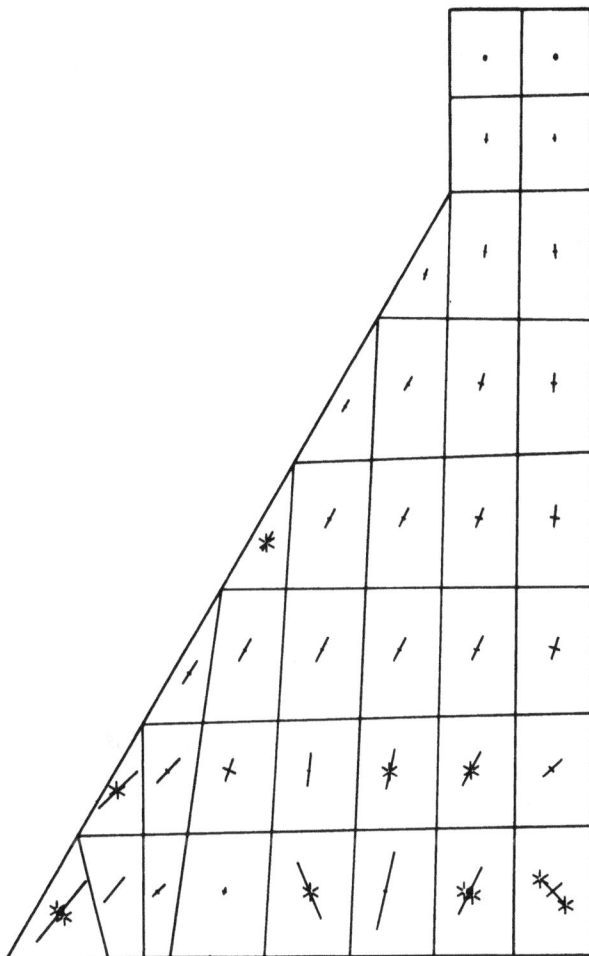

PRINCIPAL STRESSES
(✳) Indicates tension
——————— 1000 P.S.I.
Scale ——————— 50 Feet

Figure C-9. Grand Coulee Forebay Dam foundation
study—microfilm printout showing principal stresses
(no treatment). —288-D-3167

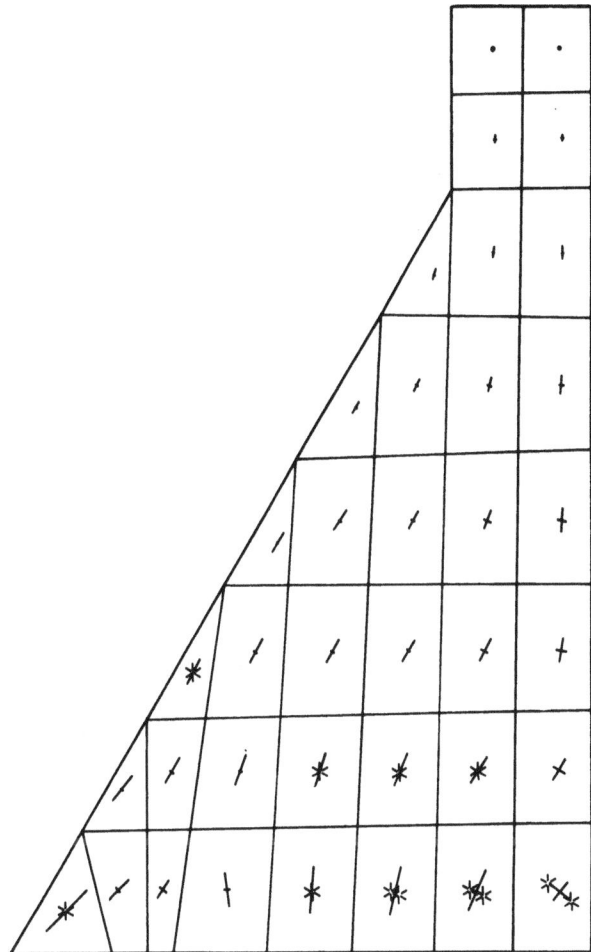

PRINCIPAL STRESSES
(✳) Indicates tension
——————— 1000 P.S.I.
Scale ——————— 50 Feet

Figure C-10. Grand Coulee Forebay Dam
study—microfilm printout showing principal stresses
(25-foot treatment). —288-D-3168

load vectors is shown on figure C-18. Nodal
point 10 has a load of 4,105 pounds in the
positive X direction, 2,711 pounds in the
positive Y direction, and 143,590 pounds in
the negative Z direction.

C-9. Output.—Displacements of the nodes
are given in X, Y, and Z directions. Shear
stresses and stresses normal to each of the three

planes are computed at each node.

Some of the stresses of interest at the base
of the dam and around the penstock are shown
on figure C-19. The maximum compressive
stress is about 255 pounds per square inch and
the maximum tensile stress, 98 pounds per
square inch.

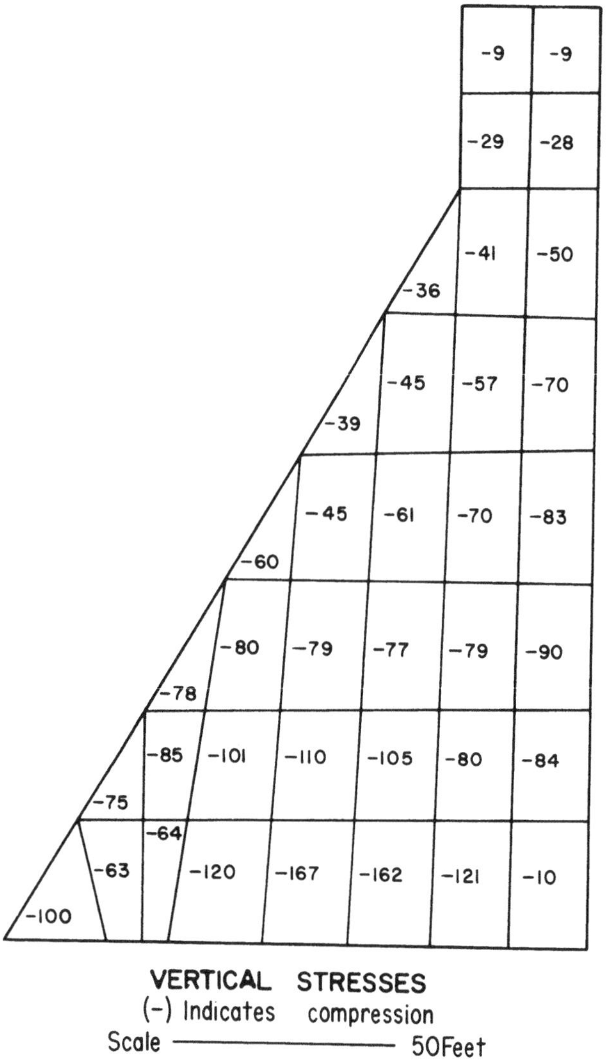

VERTICAL STRESSES
(−) Indicates compression
Scale ——————— 50 Feet

Figure C-11. Grand Coulee Forebay Dam study—microfilm printout showing vertical stresses (25-foot treatment). −288-D-3169

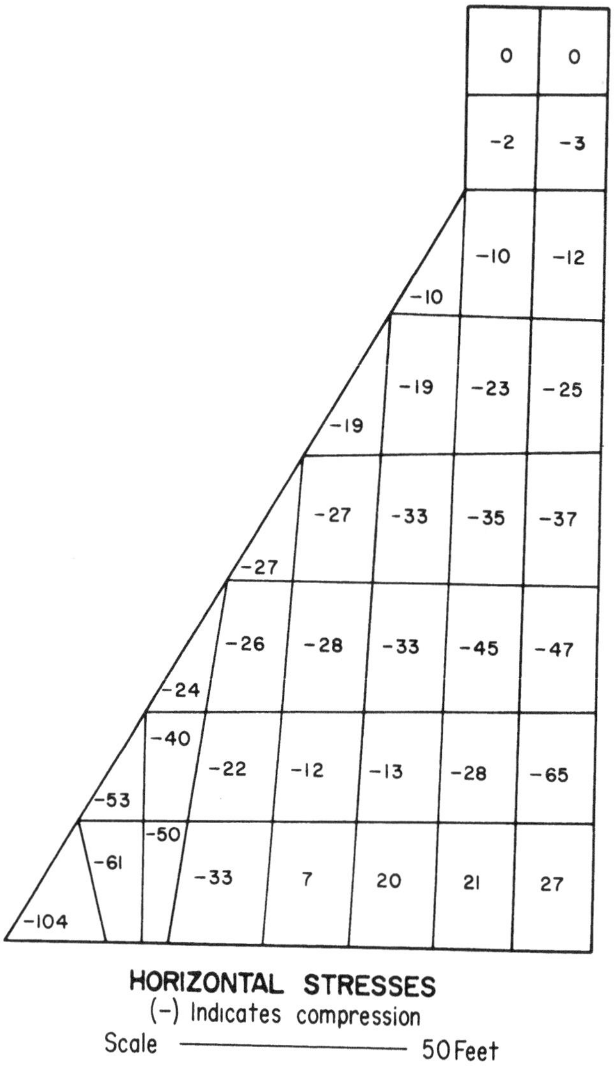

HORIZONTAL STRESSES
(−) Indicates compression
Scale ——————— 50 Feet

Figure C-12. Grand Coulee Forebay Dam study—microfilm printout showing horizontal stresses (25-foot treatment). −288-D-3170

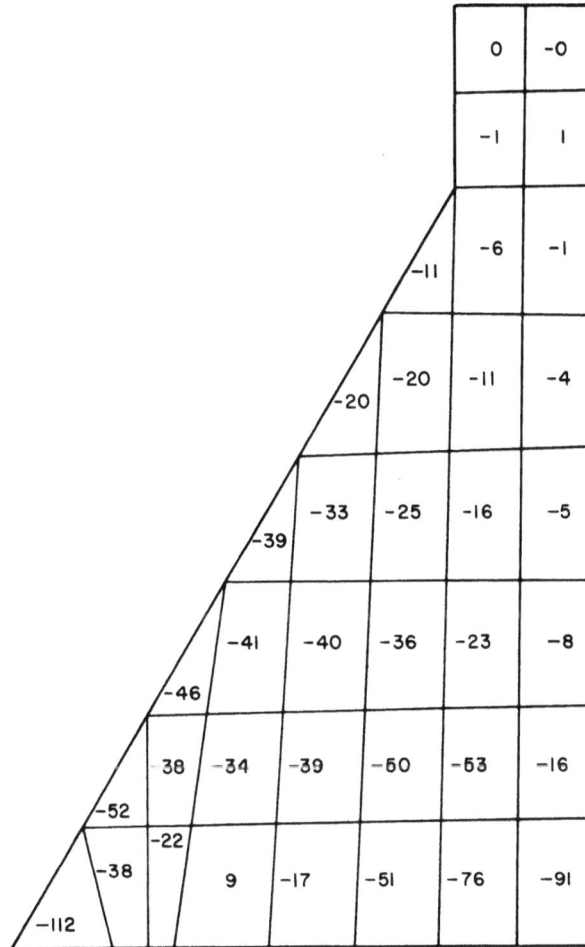

SHEAR STRESSES

Scale —————————— 50 Feet

Figure C-13. Grand Coulee Forebay Dam
study—microfilm printout showing shear stresses
(25-foot treatment). −288-D-3171

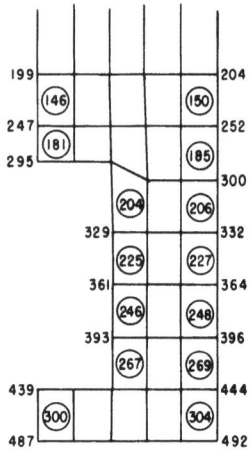

SECTION A-A

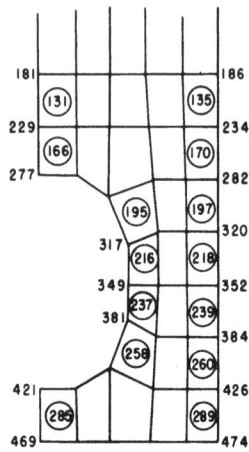

SECTION B-B

NUMBER CODE:

148   Node Number

(45)   Element Number

[2]   Material Number

NOTE:

All volume shown is Material Number 1, except
Material Number 2, water, which rose in
the gate slot as shown.

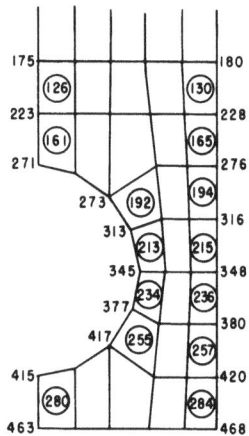

SECTION C-C

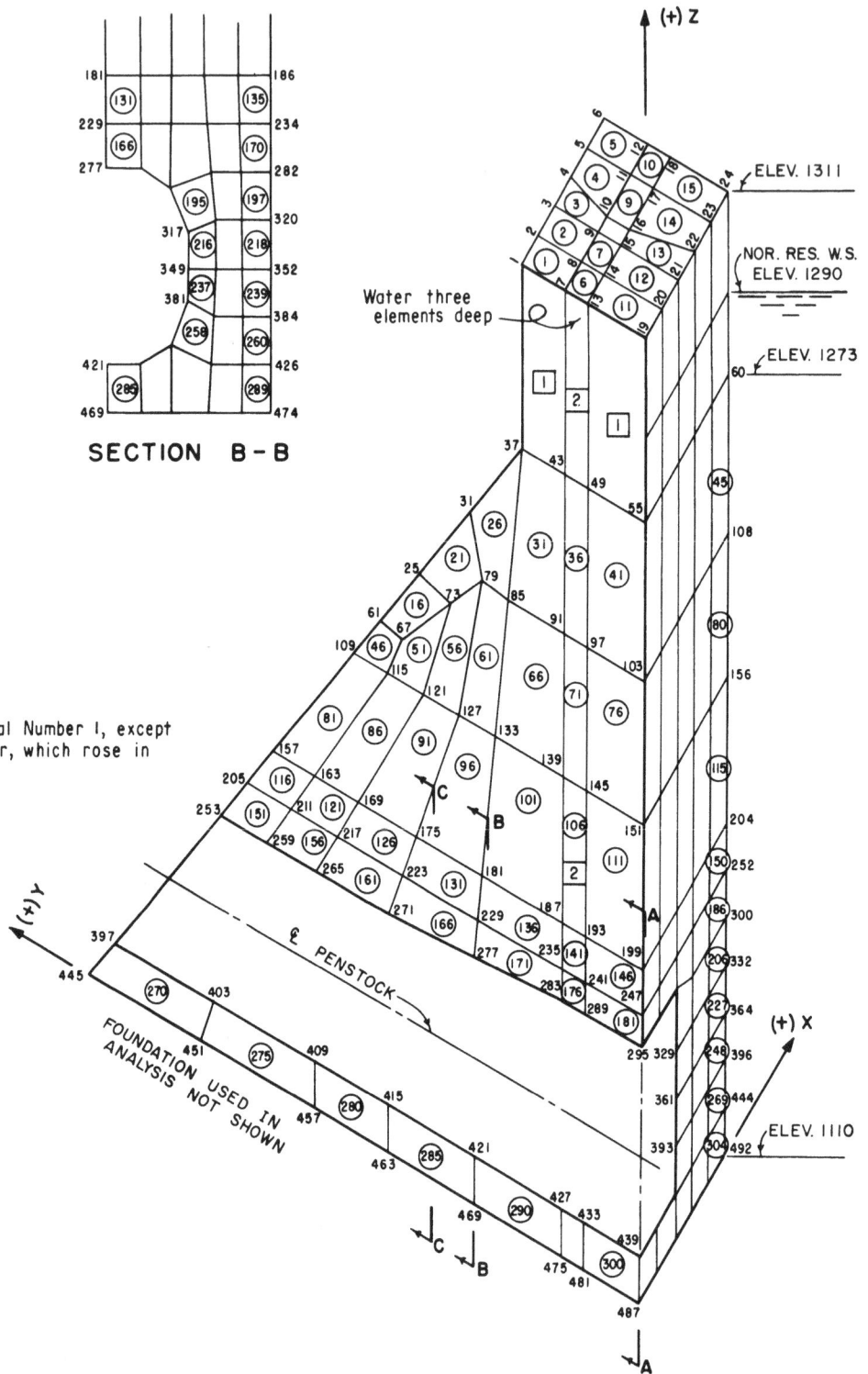

HALF-BLOCK, THREE-DIMENSIONAL VIEW

Figure C-14. Grand Coulee Forebay Dam study—three-dimensional finite element grid. −288-D-3172

GRAND COULEE FOREBAY--FAULT U/S OF HEEL--LOADS DUE TO GRAVITY, HYDROSTATIC, UPLIFT

```
        NUMBER OF ELEMENTS----------- 374
        NUMBER OF NODES-------------- 588
        NUMBER OF BOUNDARY NODES----- 194
        MAXIMUM BAND WIDTH----------- 168
        NUMBER OF MATERIALS----------   5
        NUMBER OF LOADED NODES-------  -0
```

MATERIAL NUMBER	MODULUS	POISSON	DENSITY
1	4.320+008	0.15	150.00
2	0.000+000	0.00	0.00
3	3.880+008	0.13	0.00
4	2.880+008	0.13	0.00
5	1.728+008	0.13	0.00

Figure C-15. Three-dimensional input data—control data and material properties. —288-D-3173

NODE	XORD	YORD	ZORD
1	0.0000	30.0000	311.0000
2	7.0000	30.0000	311.0000
3	14.0000	30.0000	311.0000
4	27.0000	30.0000	311.0000
5	28.0000	30.0000	311.0000
6	35.0000	30.0000	311.0000
7	0.0000	19.5800	311.0000
8	7.0000	19.5800	311.0000
9	14.0000	19.5800	311.0000
10	17.5000	19.5800	311.0000
11	28.0000	19.5800	311.0000
12	35.0000	19.5800	311.0000
13	0.0000	14.2800	311.0000
14	7.0000	14.2800	311.0000
15	14.0000	14.2800	311.0000
16	17.5000	14.2800	311.0000
17	28.0000	14.2800	311.0000
18	35.0000	14.2800	311.0000
19	0.0000	0.0000	311.0000
20	7.0000	0.0000	311.0000
21	14.0000	0.0000	311.0000
22	21.0000	0.0000	311.0000
23	28.0000	0.0000	311.0000
24	35.0000	0.0000	311.0000
25	0.0000	57.9500	230.0000
26	7.0000	57.9500	230.0000
27	14.0000	57.9500	230.0000
28	21.0000	57.9500	230.0000
29	28.0000	57.9500	230.0000
30	35.0000	57.9500	230.0000
31	0.0000	44.9500	250.0000
32	7.0000	44.9500	250.0000
33	14.0000	44.9500	250.0000
34	21.0000	44.9500	250.0000
35	28.0000	44.9500	250.0000
36	35.0000	44.9500	250.0000
37	0.0000	30.0000	273.0000
38	7.0000	30.0000	273.0000
39	14.0000	30.0000	273.0000
40	21.0000	30.0000	273.0000
41	28.0000	30.0000	273.0000
42	35.0000	30.0000	273.0000
43	0.0000	19.5800	273.0000
44	7.0000	19.5800	273.0000
45	14.0000	19.5800	273.0000
46	17.5000	19.5800	273.0000
47	28.0000	19.5800	273.0000
48	35.0000	19.5800	273.0000
49	0.0000	14.2800	273.0000
50	7.0000	14.2800	273.0000
51	14.0000	14.2800	273.0000
52	17.5000	14.2800	273.0000
53	28.0000	14.2800	273.0000
54	35.0000	14.2800	273.0000
55	0.0000	0.0000	273.0000
56	7.0000	0.0000	273.0000
57	14.0000	0.0000	273.0000

Figure C-16. Three-dimensional input data—description
of section by nodal points. —288-D-3174

ELEMENT			CONNECTED	NODES				MATERIAL	INT. RULE
1	1	7	43	37	2	8	44	38	1	2
2	2	8	44	38	3	9	45	39	1	2
3	3	9	45	39	4	10	46	40	1	3
4	4	10	46	40	5	11	47	41	1	3
5	5	11	47	41	6	12	48	42	1	2
6	7	13	49	43	8	14	50	44	2	2
7	8	14	50	44	9	15	51	45	2	2
8	9	15	51	45	10	16	52	46	2	2
9	10	16	52	46	11	17	53	47	1	2
10	11	17	53	47	12	18	54	48	1	2
11	13	19	55	49	14	20	56	50	1	2
12	14	20	56	50	15	21	57	51	1	2
13	15	21	57	51	16	22	58	52	1	3
14	16	22	58	52	17	23	59	53	1	3
15	17	23	59	53	18	24	60	54	1	3
16	61	25	73	67	62	26	74	68	1	3
17	62	26	74	68	63	27	75	69	1	3
18	63	27	75	69	64	28	76	70	1	3
19	64	28	76	70	65	29	77	71	1	3
20	65	29	77	71	66	30	78	72	1	3
21	25	31	79	73	26	32	80	74	1	4
22	26	32	80	74	27	33	81	75	1	4
23	27	33	81	75	28	34	82	76	1	4
24	28	34	82	76	29	35	83	77	1	4
25	29	35	83	77	30	36	84	78	1	4
26	31	37	85	79	32	38	86	80	1	4
27	32	38	86	80	33	39	87	81	1	4
28	33	39	87	81	34	40	88	82	1	4
29	34	40	88	82	35	41	89	83	1	4
30	35	41	89	83	36	42	90	84	1	4
31	37	43	91	85	38	44	92	86	1	3
32	38	44	92	86	39	45	93	87	1	3
33	39	45	93	87	40	46	94	88	1	3
34	40	46	94	88	41	47	95	89	1	3
35	41	47	95	89	42	48	96	90	1	3
36	43	49	97	91	44	50	98	92	2	2
37	44	50	98	92	45	51	99	93	2	2
38	45	51	99	93	46	52	100	94	2	2
39	46	52	100	94	47	53	101	95	1	2
40	47	53	101	95	48	54	102	96	1	2
41	49	55	103	97	50	56	104	98	1	2
42	50	56	104	98	51	57	105	99	1	2
43	51	57	105	99	52	58	106	100	1	3
44	52	58	106	100	53	59	107	101	1	3
45	53	59	107	101	54	60	108	102	1	2
46	61	67	115	109	62	68	116	110	1	3
47	62	68	116	110	63	69	117	111	1	3
48	63	69	117	111	64	70	118	112	1	3
49	64	70	118	112	65	71	119	113	1	3
50	65	71	119	113	66	72	120	114	1	3
51	67	73	121	115	68	74	122	116	1	4
52	68	74	122	116	69	75	123	117	1	4
53	69	75	123	117	70	76	124	118	1	4
54	70	76	124	118	71	77	125	119	1	4
55	71	77	125	119	72	78	126	120	1	4
56	73	79	127	121	74	80	128	122	1	4
57	74	80	128	122	75	81	129	123	1	4

Figure C-17. Three-dimensional input data—elements defined by nodal points with material.—
288-D-3175

LOAD VECTOR

NODE	X-LOAD	Y-LOAD	Z-LOAD
1	0.0000+000	0.0000+000	-5.1970+004
2	0.0000+000	0.0000+000	-1.0394+005
3	0.0000+000	0.0000+000	-1.1508+005
4	0.0000+000	0.0000+000	-1.0394+005
5	0.0000+000	0.0000+000	-9.2803+004
6	0.0000+000	0.0000+000	-5.1970+004
7	-2.6494-007	5.4220+003	-5.1970+004
8	-5.2988-007	1.0844+004	-1.0394+005
9	-5.2988-007	8.1331+003	-9.6515+004
10	4.1053+003	2.7110+003	-1.4359+005
11	0.0000+000	0.0000+000	-1.7745+005
12	0.0000+000	0.0000+000	-7.8403+004
13	0.0000+000	-5.4220+003	-7.1221+004
14	0.0000+000	-1.0844+004	-1.4244+005
15	0.0000+000	-8.1331+003	-1.1870+005
16	4.1053+003	-2.7110+003	-1.8209+005
17	0.0000+000	0.0000+000	-2.3227+005
18	0.0000+000	0.0000+000	-9.7655+004
19	0.0000+000	5.4220+003	-7.1221+004
20	0.0000+000	1.0844+004	-1.4244+005
21	0.0000+000	1.0844+004	-1.3057+005
22	0.0000+000	1.0844+004	-1.4244+005
23	0.0000+000	1.0844+004	-1.5431+005
24	0.0000+000	5.4220+003	-7.1222+004
25	0.0000+000	0.0000+000	-3.5017+004
26	0.0000+000	0.0000+000	-7.0035+004
27	0.0000+000	0.0000+000	-7.0035+004
28	0.0000+000	0.0000+000	-7.0035+004
29	0.0000+000	0.0000+000	-7.0035+004
30	0.0000+000	0.0000+000	-3.5017+004
31	0.0000+000	0.0000+000	-5.6622+004
32	0.0000+000	0.0000+000	-1.1324+005
33	0.0000+000	0.0000+000	-1.1324+005
34	0.0000+000	0.0000+000	-1.1324+005
35	0.0000+000	0.0000+000	-1.1324+005
36	0.0000+000	0.0000+000	-5.6622+004
37	0.0000+000	0.0000+000	-1.4149+005
38	0.0000+000	0.0000+000	-2.8299+005
39	0.0000+000	0.0000+000	-2.7593+005
40	0.0000+000	0.0000+000	-2.8299+005
41	0.0000+000	0.0000+000	-2.9005+005
42	0.0000+000	0.0000+000	-1.4149+005
43	-4.5776-006	1.2828+005	-1.0176+005
44	-9.1553-006	2.5656+005	-2.0353+005
45	-9.1553-006	1.9242+005	-1.7456+005
46	9.7124+004	6.4139+004	-2.7761+005
47	0.0000+000	0.0000+000	-3.5597+005
48	0.0000+000	0.0000+000	-1.5115+005
49	-9.3561-008	-1.2828+005	-1.3307+005
50	-1.8712-007	-2.5656+005	-2.6614+005
51	-1.8712-007	-1.9242+005	-2.2179+005
52	9.7124+004	-6.4139+004	-3.4023+005
53	0.0000+000	0.0000+000	-4.3397+005
54	0.0000+000	0.0000+000	-1.8246+005
55	0.0000+000	1.2828+005	-1.3307+005

Figure C-18. Three-dimensional input data—load vectors. —
288-D-3176

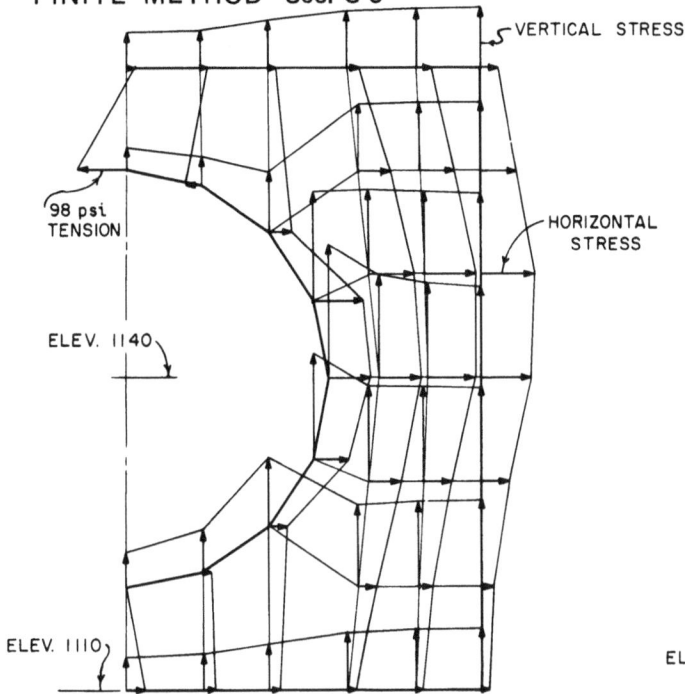

HORIZONTAL AND VERTICAL
STRESSES
AT DOWNSTREAM FACE

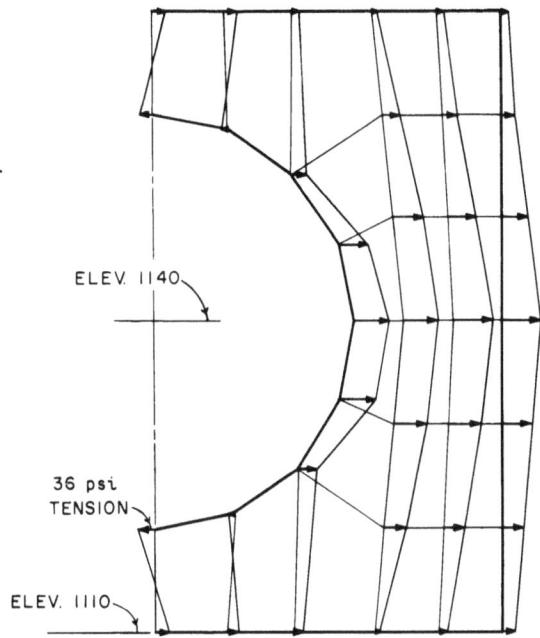

HORIZONTAL STRESSES
AT VERTICAL SECTION
62 FEET FROM UPSTREAM FACE

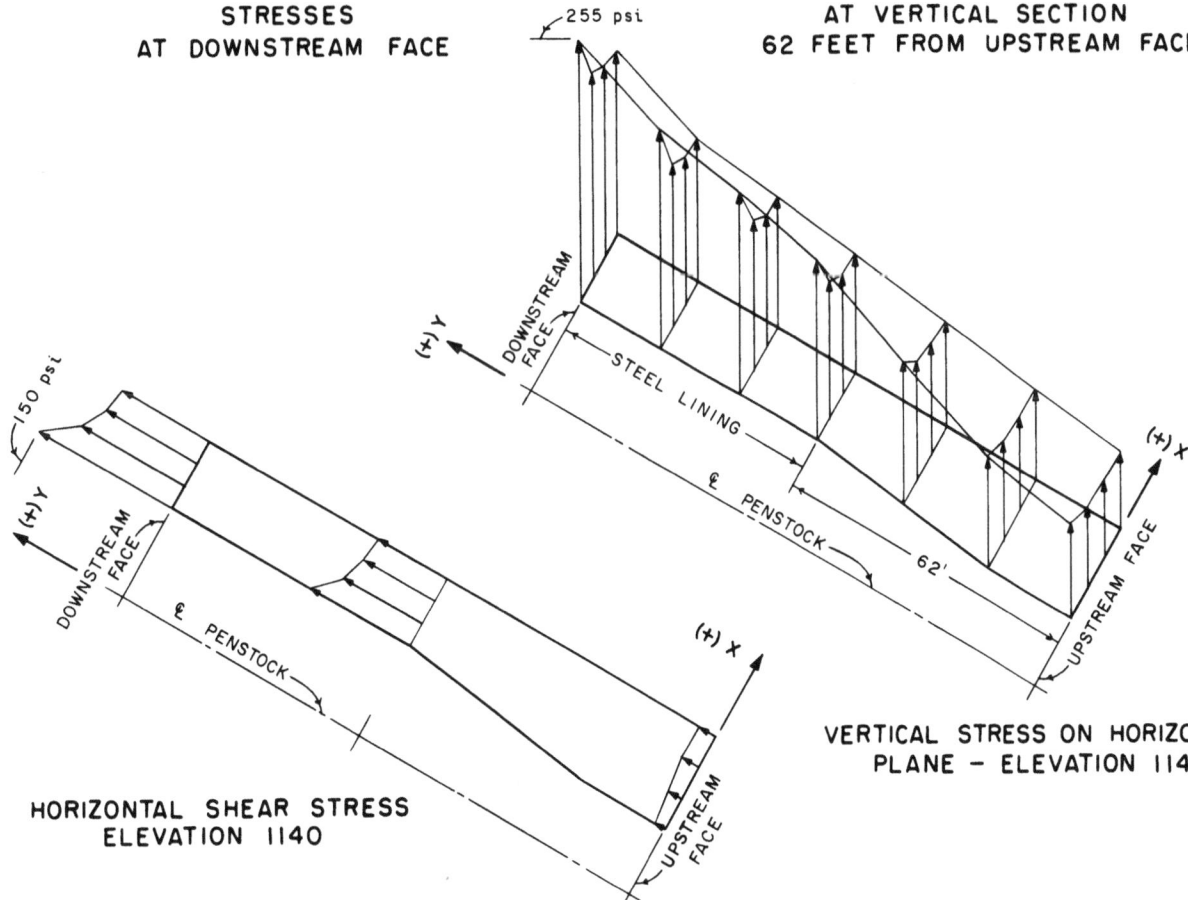

VERTICAL STRESS ON HORIZONTAL
PLANE — ELEVATION 1140

HORIZONTAL SHEAR STRESS
ELEVATION 1140

Figure C-19. Grand Coulee Forebay Dam study—stresses at nodal points. —288-D-3177

Special Methods of Nonlinear Stress Analysis

D-1. *Introduction*.—The systems for determining nonlinear stresses presented here are the "Slab Analogy Method," "Lattice Analogy Method," "Experimental Models," and "Photoelastic Models." None of these methods, except photoelastic models, are used presently in the Bureau of Reclamation because of their complexity and the time consumed in performing the analyses. These methods are included in the discussion since they were used in some of the examples shown in this manual.

Modern two-dimensional and three-dimensional finite element methods provide more sophisticated and more economical analyses for the determination of nonlinear stress distribution than the methods mentioned above. The finite element methods are discussed in subchapter E of chapter IV and in appendix C.

D-2. *Slab Analogy Method*.—Although the exact law of nonuniform stress distribution is unknown, an approach towards a determination of true stresses can be made by means of the theory of elasticity. The "Slab Analogy Method" was developed as a result of a suggestion by H. M. Westergaard in 1930, in connection with the design of Hoover (formerly Boulder) Dam. This method is described in detail in one of the Boulder Canyon Project Final Reports.[1] Consequently, the method will be only briefly described here.

---

[1] "Stress Studies for Boulder Dam," Bulletin 4 of Part V, Boulder Canyon Project Final Reports, Bureau of Reclamation, 1939.

It is a lengthy, laborious method and is justified only for unusually high and important dams. The analysis is based upon an analogy between an Airy's surface, which defines the stresses in a two-dimensional elastic structure, and the deflections of an unloaded slab bent by forces and couples applied around its edges. The slab has the same shape as a cantilever section including a large block of the foundation. The edges of the slab are bent into a form which corresponds to the stresses at the surface of the structure. The analysis is made by dividing the analogous slab into horizontal and vertical beams which are brought into slope and deflection agreement by trial loads. The curvatures in the slab are then proportional to the shears in the structure, and consequently the moments in the horizontal and vertical beams are proportional to the stresses in the vertical and horizontal directions, respectively.

Nonlinear stress investigations by the slab analogy method have been made for three large dams: Hoover, Grand Coulee, and Shasta. Conclusions drawn from several studies of maximum cantilever sections are that stresses in the vicinity of the upstream and downstream edges of the base are greater than those found by the gravity method and warrant special consideration in design. These studies also indicated that nonlinear effects are important within approximately one-third the height of the cantilever, and reach a maximum at the base.

The maximum nonlinear effects which were found in the vicinity of the bases of Hoover,

Grand Coulee, and Shasta Dams are shown in table D-1. The table also shows a comparison between stresses based on linear and nonlinear distribution for the vertical, horizontal, and shear stresses in the regions of the upstream and downstream toes. Since the nonlinear (slab analogy) method bears out the proof by the theory of elasticity that the theoretical maximum shear stresses are infinite at the reentrant corners of the base, the values given are for the maximum computed shear stresses at conjugate beam points nearest the corners. The vertical stresses were the ones which showed the greatest changes when computed by the nonlinear method. The maximum increase in vertical upstream stress was 18 percent, and occurred for Hoover Dam; while the maximum increase in vertical stress at the downstream toe was 64 percent and occurred for the maximum nonoverflow section of Grand Coulee Dam.

The studies of Shasta Dam showed the least departure of stresses from the linear law of any nonlinear studies completed to date. The upstream vertical stresses were decreased by approximately 12 percent and the downstream stresses were increased by corresponding amounts. This close agreement of linear and nonlinear stresses was believed to be due to the fact that the batter of the upstream face at the base of the cantilever was 0.5 to 1, which allowed for a better introduction of stresses from the dam into the foundation than would a sharper reentrant.

Table D-1 shows that horizontal stresses as computed by the nonlinear method may be over twice the values computed by the ordinary linear assumption. This is an important consideration in the design of gallery and drainage systems, outlet works, power penstocks, elevator shafts, and other openings in the dam. The studies show that shear stresses computed by the nonlinear method follow rather closely the parabolic distribution obtained by an ordinary gravity analysis, except of course, at the reentrant corners.

D-3. Lattice Analogy Method.—Many of the two-dimensional problems encountered in engineering are difficult or impossible of solution when treated mathematically.

Necessity has fostered the approximate "Lattice Analogy Method" of dealing with such problems. This section will describe the method and some of its applications rather than the derivation of formulas involved in its use. As far as practical engineering problems are concerned, the field of application is restricted only by two limitations: (1) The shape of the section must be such that it can be built up, exactly or to a satisfactory approximation, from a limited number of square elements; and (2) the value of Poisson's ratio must be equal to one-third. The limitation upon Poisson's ratio is usually unimportant. In many cases, stress distribution is independent of the values of the elastic constants, and in cases where these constants affect the results, the value of one-third will ordinarily be close enough to the true value that only small differences will exist in stresses or displacements.

As in the usual treatment of two-dimensional problems in elasticity, a section of the structure to be analyzed is considered as though it were a slice or plate of unit thickness, in accordance with the generalized theory of plane stress. The plate is simulated in size and shape by a lattice network composed of interconnected elemental square frames, each diagonally connected at the corners. When the plate has irregular boundaries, its outline may be approximated to any desired degree of accuracy depending on the number of frames chosen. As the number is increased, however, the solution becomes more involved so that for any problem a practical decision must be made as to the refinement desired. The validity of the simulation may be shown by demonstrating that in the limit, as the dimensions of the square frames approach zero, the differential equations of equilibrium and compatibility become identical for the lattice and the plate, and the boundary conditions become expressible in the same form. Thus the two solutions become identical and for obvious reasons the lattice is referred to as analogous to the plate.

(a) *Conditions to be Satisfied.*—In the analogy between the lattice and the plate, three

TABLE D-1.—*Maximum nonlinear stress effects in sections of various dams.*—DS2-2(T1)

Dam	BOULDER	GRAND COULEE		SHASTA		
Cantilever Section	Crown	Maximum Non-Overflow	Maximum Spillway With Bucket	Maximum Non-Overflow	Maximum Spillway	Maximum Non-Overflow
Loading Condition	Dead Load plus Trial Load† Water Load	Dead Load plus Full Water Load	Dead Load plus Full Water Load	Dead Load plus Full Water Load	Dead Load plus Full Water Load	Dead Load plus Full Water Load plus Earthquake
Region near Upstream Edge of Base						
Maximum Change, vertical normal stress	555 to 654	261 to 302	245 to 260	227 to 204	239 to 172	155 to 111
Maximum Change, horizontal normal stress	230 to 405	221 to 72	198 to 194	200 to 120	219 to 48	198 to 54
Maximum Change, shear stress*	68 to 32	0 to 160	5 to 95	-15 to -48	-9 to -72	36 to 73
Region near Downstream Edge of Base						
Maximum Change, vertical normal stress	271 to 377	332 to 546	289 to 196	248 to 282	339 to 371	356 to 397
Maximum Change, horizontal normal stress	139 to 299	226 to 406	184 to 369	179 to 256	199 to 310	222 to 317
Maximum Change, shear stress*	140 to 120	190 to 216	195 to 240	213 to 297	240 to 109	271 to 318

Notes:　† Based on trial-load arch dam analysis.

* Theoretical maximum shear is infinite at reentrant corner; therefore value given is the maximum computed stress in vicinity of corner.

Figures to left based on gravity stress analysis; figures to right based on nonlinear stress analysis. (Slab Analogy Method).

fundamental conditions must be satisfied in order that an assemblage of elemental lattices may constitute a plate. These conditions are:

(1) The normal and tangential stresses must be distributed throughout the plate in such a way that the forces acting upon each element are in equilibrium with respect to translation and rotation of the element.

(2) The extensions and detrusions of the elements resulting from these stresses constitute a single-valued system of displacements.

(3) Any special conditions of stress or displacement which are specified at the boundaries must be satisfied.

(b) *Solution.*—Having replaced the plate prototype by a lattice framework, a solution may be devised for the lattice and applied to the plate. The essential concept involved in this solution is a systematic relaxation of restraints at the joints. A description of a relaxing process to aid in an understanding of the adjustment will be given subsequently. After the adjustment of the lattice to remove restraint has been completed, the strains are deduced from relative displacements between successive joints and from these the stresses may be computed.

The fundamental device employed in the lattice analogy is the elemental square frame which is composed of six elastic members, two of which are horizontal, two vertical, and two diagonal. The length of the sides is considered unity in the derivation of the lattice formulas. The six members are assumed to be connected at the corners by frictionless joints. The elastic properties of the frame members are so determined that the behavior of the frame under given boundary conditions will correspond exactly to the square element of the plate section with respect to axial elongation, lateral contraction, and shear detrusion. In a lattice network composed of many frames, the amount of work involved in a conventional solution would be tremendous. However, by using a relaxation method, one may deal with a small region in which equilibrium is easily established and the method can consequently be applied to intricate lattice systems. Adjustment of

equilibrium in a second region disturbs the first, but still leaves it approaching final equilibrium. The operations of adjustment are easily applied when each step is confined to a single joint.

To illustrate the method, consider the simplest case where the boundary conditions are given in terms of displacements. The procedure of adjustment may be visualized as follows: Consider a lattice actually constructed to a given scale, with elastic members coming together at the corners to form frictionless hinged joints. Lay this lattice out on a horizontal board, and before applying any displacements, completely restrain all joints by nailing them to the board. Next, displace and secure again, the various boundary joints through distances corresponding to their assigned displacements. Then, working in a line of joints adjacent to a displaced boundary, free one joint and allow it to move to a new position of equilibrium and resecure it. Repeat the process at successive joints until all have been adjusted (keeping only the boundary joints fixed in conformity with the given displacements) as many times as is required to give a satisfactory approach to the condition of complete transfer of forces from the interior nails to the members of the lattice. Simple relationships then exist between displacements and stresses.

(c) *Equations.*—Lattice equations for the displacement of an interior joint, an exterior corner joint, a reentrant corner joint, and a boundary joint have been developed in terms of loads at the joint and in terms of displacement of the surrounding joints. These equations are shown on figure D-1.

(d) *Boundary Conditions.*—Boundary conditions for the problem can be given either in terms of loadings or displacements. For the design of structures, estimated or computed loads would probably comprise the boundary conditions, but for structures already built it is more likely that boundary conditions would be given in terms of measured displacements. In either case, the loads or displacements for the plate must be expressed in terms of loads or displacements for the boundary joints of the lattice. However, the adjustment of the lattice

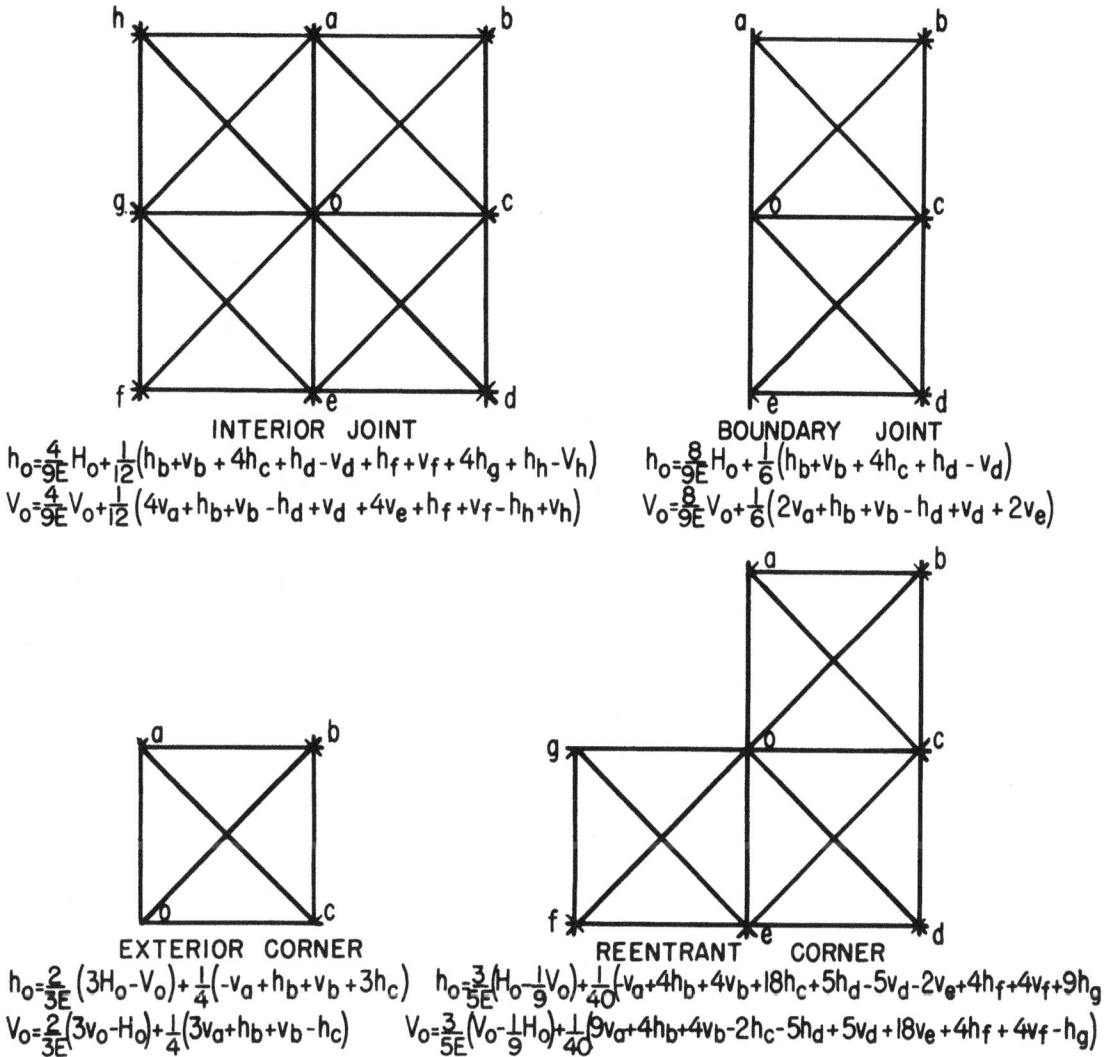

INTERIOR JOINT

$$h_o = \frac{4}{9E}H_o + \frac{1}{12}(h_b + v_b + 4h_c + h_d - v_d + h_f + v_f + 4h_g + h_h - v_h)$$

$$V_o = \frac{4}{9E}V_o + \frac{1}{12}(4v_a + h_b + v_b - h_d + v_d + 4v_e + h_f + v_f - h_h + v_h)$$

BOUNDARY JOINT

$$h_o = \frac{8}{9E}H_o + \frac{1}{6}(h_b + v_b + 4h_c + h_d - v_d)$$

$$V_o = \frac{8}{9E}V_o + \frac{1}{6}(2v_a + h_b + v_b - h_d + v_d + 2v_e)$$

EXTERIOR CORNER

$$h_o = \frac{2}{3E}(3H_o - V_o) + \frac{1}{4}(-v_a + h_b + v_b + 3h_c)$$

$$V_o = \frac{2}{3E}(3V_o - H_o) + \frac{1}{4}(3v_a + h_b + v_b - h_c)$$

REENTRANT CORNER

$$h_o = \frac{3}{5E}(H_o - \frac{1}{9}V_o) + \frac{1}{40}(-v_a + 4h_b + 4v_b + 18h_c + 5h_d - 5v_d - 2v_e + 4h_f + 4v_f + 9h_g)$$

$$V_o = \frac{3}{5E}(V_o - \frac{1}{9}H_o) + \frac{1}{40}(9v_a + 4h_b + 4v_b - 2h_c - 5h_d + 5v_d + 18v_e + 4h_f + 4v_f - h_g)$$

h,v indicate displacements of joint O.

H,V indicate forces per unit thickness at O representing body forces at O in plane of lattice.

E is elastic modulus of prototype material.

Figure D-1. Lattice analogy—equations for displacement of joint O.—103-D-274

is always made by adjusting displacements at the interior joints to remove restraints.

(e) *Stresses.*—Normally, the purpose of computing a lattice would be to determine stresses in the prototype. The adjustment of

the lattice to remove restraint having been completed, the resulting displacements may be applied to the plate. The difference in displacements between successive lattice joints will yield strains, and stresses may be computed from the conventional stress-strain relationship.

(1) *Restraining forces.*—At any time during adjustment of a lattice, the restraining forces at the joints may be computed. For an exact solution, these forces will reduce to zero, and they are, therefore, a measure of the accuracy of the adjustment at any stage. Ordinarily, the computation of the restraining forces involves considerable work so that other methods are used to judge the end of an adjustment. The easiest way is to overadjust the displacements so that reversal occurs in their direction because of passing the end point.

(2) *Body forces.*—The equations previously mentioned concerning displacement at certain joints due to loads at these joints, will apply to the body forces of the structure. Such loads can be introduced into the lattice adjustment by computing the horizontal and vertical components, computing the displacements, and adding these displacements to those produced by displacements of the surrounding joints. Certain limited types of body forces, including gravity forces, may also be handled by Biot's method of applying fictitious boundary pressures.

(3) *Thermal stresses.*—A system has been devised in which displacements due to temperature change are computed by the application of fictitious body and boundary forces. The determination of the fictitious forces is somewhat involved and will not be given here, and the application of body and boundary forces to a lattice system has already been discussed.

(f) *Applications and Limitations.*—The lattice analogy method is used for solving two-dimensional nonlinear stress problems in engineering and has many applications that are involved in the design of masonry dams. The method is adaptable to the computation of stresses in a gravity dam. A section from a gravity dam is normally computed of unit thickness and its outline could be approximated by a lattice network made up of squares. As has been pointed out, boundary forces (waterloads), body forces (dead loads), and thermal forces cause no particular difficulty in adjusting lattice displacements.

The principal limitations placed on application to gravity dam design or other purposes are the time and labor involved in the calculations. The method has been found useful in determining the stress distribution in a body composed of two or more different materials. This represents a problem of great practical importance, especially in the design of reinforced concrete structures. Another problem which is fundamental in the study of concrete structures is that of uniform shrinkage of a two-dimensional section on a rigid foundation. This problem has been analyzed successfully, using the lattice analogy.

D-4. *Experimental Models.*—The use of models is a very valuable addition to the analytical methods used in the design of dams and similar structures. Models are necessary for any careful design development and can be used for checking of theory. All models come under one of two major classifications: (1) similar models, or those that resemble the prototype; and (2) dissimilar models. Principal among the former group are the two-dimensional and three-dimensional types of elastic displacement models; photoelastic models; and models used in studies employing the slab analogy. In the dissimilar group are those employing such analogies as the membrane, electric, and sand-heap analogies. These last-mentioned types, while of considerable value to stress studies of special problems, do not concern us here, and it is only those model types which have proved adaptable to experimental studies of masonry dam structures that will be discussed.

(a) *Three-Dimensional Models.*—Three-dimensional displacement models are those constructed of elastic materials to proportionate size and loading of the prototype so that deformation, structural action, and stress conditions of the latter can be predicted by measurement of displacements of the model.

The following conditions must be fulfilled, in order to obtain similarity between a model and its prototype, while at the same time satisfying theoretical considerations and the requirements of practical laboratory procedure:

(1) The model must be a true scalar representation of the prototype.

(2) The loading of the model must be proportional to the loading of the prototype.

(3) Upon application of load, resulting strains and deflections must be susceptible of measurement with available laboratory equipment. Because of reduced scale this condition ordinarily requires a higher specific gravity and greatly reduced stiffness in the model compared with the prototype.

(4) Because of influence of volume strains on the stress distribution, Poisson's ratio should be the same for both model and prototype.

(5) The model material must be homogeneous, isotropic, and obey Hooke's law within the working-stress limits, since these conditions are assumed to exist in a monolithic structure such as a concrete dam.

(6) Foundations and abutments must be sufficiently extensive to allow freedom for the model to deform in a manner similar to the prototype.

(7) If effects of both live load and gravity forces are to be investigated, the ratio of dead weight to live load should be the same in both model and prototype. If the effects of live load only are to be investigated, the results are not affected by the specific gravity of the model, providing Hooke's law is obeyed and no cracking occurs.

If all requirements of similarity are fulfilled, the relations between model and prototype may be expressed in simple mathematical terms of ratios. Overall compliance with this restriction is not always possible in model tests of masonry dams, but since the purpose of many tests made on dam models, such as the Hoover Dam model tests, is to obtain data for verifying analytical methods, some variation from true similarity does not detract greatly from the value of the test. Complete details of model tests for Hoover Dam are given in the Boulder Canyon Project Final Reports.[2]

(b) *Two-Dimensional Displacement Models.*—Two-dimensional displacement models are often referred to as cross-sectional or slab models. Acting under two-dimensional stress such a model can be compared directly only to a similar slice through the prototype acting as a separate stressed member, since in the actual structure all interior points are under three-dimensional stress. The model slab, having no forces applied normal to the section, is considered to be in a state of plane stress. A cross-sectional element or cantilever acting as an integral part of a masonry dam is stressed by a more complex system of forces, and is under neither plane stress nor plane strain. A state of plane strain is closely approached, however, in the central cantilever element in a long, straight gravity dam and also in a vertical slice through the foundation under the crown cantilever of an arch dam. Assuming a state of plane strain can be realized, similarity of stress and strain can be had if Poisson's ratio is the same for model and prototype. For fairly reliable results in the evaluation of stress distribution in the cantilever section of a dam, the usefulness of the two-dimensional model is limited to the straight gravity type of dam, and then only when applied to the central cantilever element. This usefulness is further limited in its application to arch dam cantilevers, to the immediate neighborhood of the base of the crown cantilever, and to that part of the foundation slab contiguous with it. Two-dimensional arch models, while usually failing to give stress and deformation values which can be taken as representative of those occurring in the prototype, have furnished valuable information in connection with the evaluation of abutment rotation and deformation for use in analytical studies.

D-5. *Photoelastic Models.*—Photoelastic models are used extensively by the Bureau for

[2] Bulletin 2, "Slab Analogy Experiments," Bulletin 3, "Model Tests of Boulder Dam," and Bulletin 6, "Model Tests of Arch and Cantilever Elements," Part V, Technical Investigations, Boulder Canyon Project Final Reports, 1938-40.

design and analysis of localized portions of masonry dams and their appurtenant works. Stresses in photoelastic models are determined by means of the visible optical effects which are produced by passing polarized light through the model while it is under load. The model material must be elastic, transparent, isotropic, and free from initial or residual stresses. Bakelite, celluloid, gelatin, and glass have been successfully used. Studies employing photoelastic models are usually limited to conditions of plane stress or strain, and may be said to have their most important application in the determination of regions of stress concentrations.

Effects of stress in a photoelastic model are made visible by means of an optical instrument known as the photoelastic polariscope. Through a system of polaroids, the polariscope directs a beam of light through the model so polarized than when the material of the model becomes doubly refractive under stress, the familiar photoelastic pattern is projected to the observer on a screen or photographic plate. The alternate color bands of the pattern, or fringes as they are called, furnish a means of measuring the stress quantity, by a known relation between principal stresses and their retardative effect on polarized light-waves passing through the stressed model. This "unit" of measure, called *material fringe value*, is readily evaluated in the laboratory. For bakelite, the most extensively used material, the value is 87 pounds per square inch per inch of thickness, and represents the stress corresponding to one fringe. Values for any number of fringes, or fringe order, are found by direct proportion; and by applying a suitable factor of proportionality, corresponding values of the stress quantity in the prototype structure may be determined. This stress quantity is the difference in principal stresses at any point (twice the maximum shear stress), and has particular significance along free boundaries, where one of the principal stresses is zero.

Where it is desirable to know the magnitude and direction of the individual principal stresses acting at a point within the model, optical instruments such as the photoelastic interferometer or the Babinet compensator are used. The determination of stress from photoelastic models and the techniques used in this type of investigation are subjects too complex to properly come within the scope of this appendix.

Much valuable information has been obtained through photoelastic studies in connection with stress distribution and magnitude in dam and foundation structures. The photoelastic studies made on Shasta Dam furnish a good example of the application of the method. These studies were made to determine what effects would be produced on dam and foundation stresses by several weak-rock conditions which had been exposed in the foundation during the excavation for construction. A 5-foot clay-filled fault seam was discovered lying in a direction making an approximate 60° angle with the proposed axis of the dam. It was desired to determine the depth, if any, to which the seam in question should be excavated and backfilled with concrete in order to keep stresses within allowable limits. Because of the direction of the seam with respect to the dam, three possible locations of the seam were assumed for the tests. Photographs of the photoelastic stress patterns were taken of models constructed and loaded to represent the critical cantilever section under full reservoir load with the fault seam at the three alternate locations and at varying depths under the cantilever. These stress patterns were studied with regard to the effect of the various depths of seam repair on the stress at the downstream-toe fillet, where the most critical stress condition existed. Figure D-2 shows two photographs of the photoelastic model under stress. Figure D-3 gives curves showing the relation between the values of downstream-toe fillet stresses obtained from the photoelastic stress patterns, and the depth of the 5-foot clay-filled fault seam.

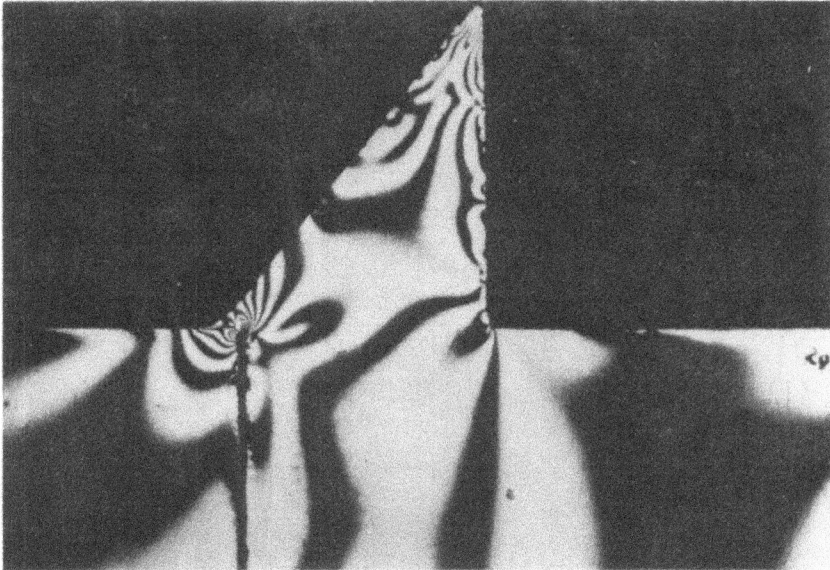

(a) FAULT SEAM UNDISTURBED

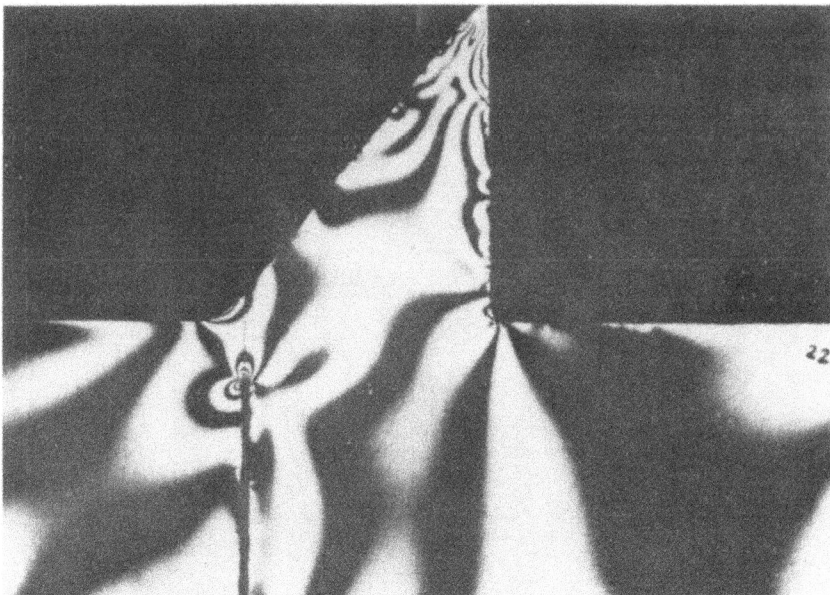

(b) FAULT SEAM EXCAVATED, BACKFILLED 52 FEET

Figure D-2. Photoelastic study of foundation fault seam near downstream face of Shasta Dam—reservoir full.—PX-D-74424

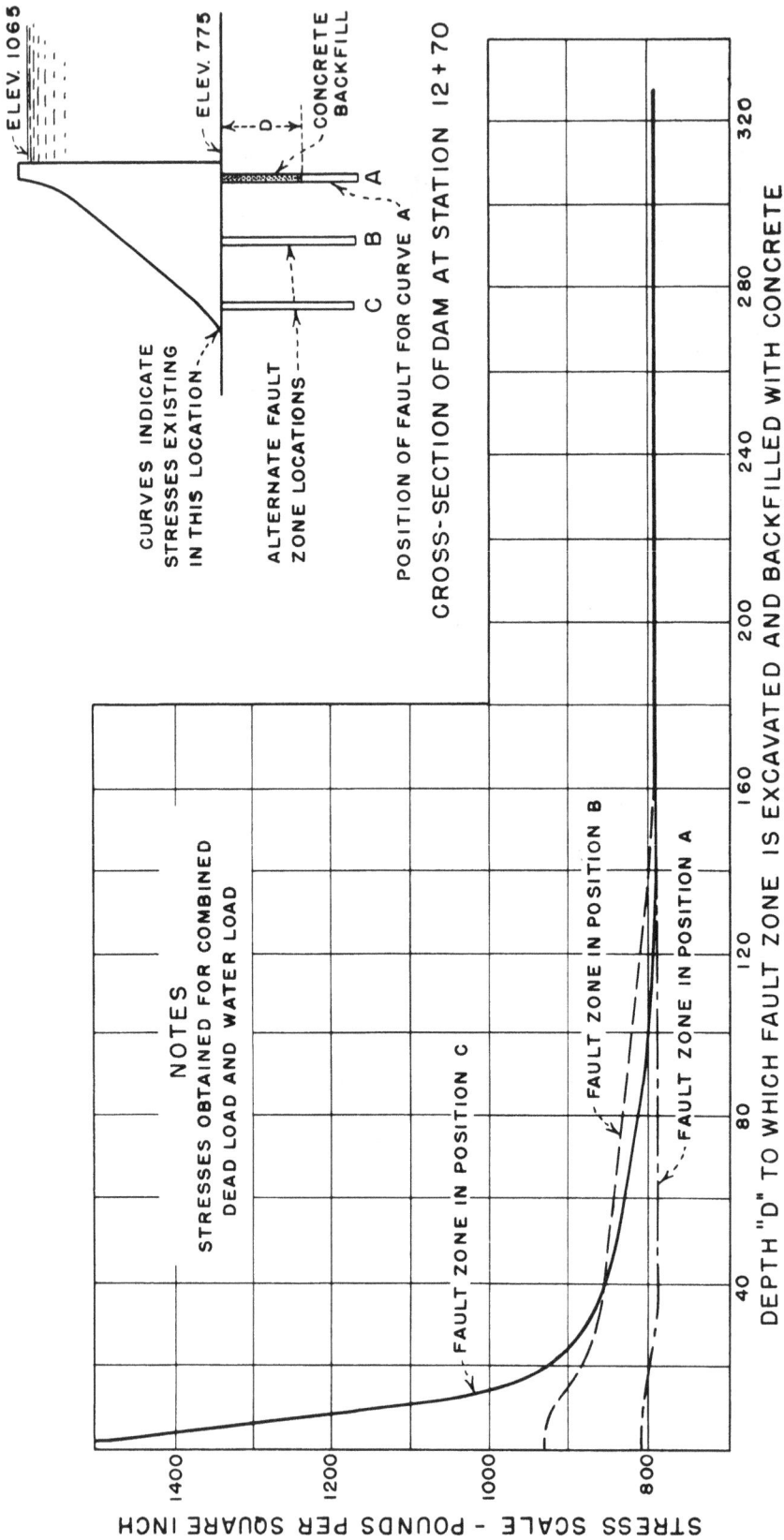

SHASTA DAM

PHOTOELASTIC STUDY OF FOUNDATION

Figure D-3. Relation of stress at toe of dam to depth and location of fault zone.—DS2-2(58)

Comparison of Results by Gravity and Trial-Load Methods

E-1. *Stresses and Stability Factors.* —Stresses and stability factors for normal and maximum loading conditions for 12 gravity dams are given on figures E-1 to E-29, inclusive. All of these dams were analyzed by the "Gravity Method," and, in addition, three were analyzed by the "Trial-Load Twist Method" and one by the "Trial-Load Arch and Cantilever Method." These are Grand Coulee, Kortes, and Angostura Dams; and East Park Dam, respectively. For these four dams, stresses obtained by the gravity analysis are shown on the same sheet with stresses obtained by the trial-load analysis. The same arrangement is used for showing stability factors. This facilitates comparison of results obtained by the two methods.

E-2. *Structural Characteristics of Dams and Maximum Stresses Calculated by the Gravity and Trial-Load Methods.* —A tabulation of structural characteristics, maximum stresses, and maximum stability factors for the 12 gravity dams mentioned in the preceding section is shown in table E-1. The 12 dams are divided into four groups in accordance with their relative heights. Structural characteristics are given in the upper half of the sheet. The ratios of crest-length to height, base to top width, and base to height of the crown cantilever define the relative characteristics of each dam. The cantilever profiles are shown for which the maximum stresses are tabulated in the lower half of the figure. The cross-canyon profile is shown for those dams for which a trial-load analysis was made.

In the lower half of table E-1 is shown the critical stress at the upstream face of each dam.

This critical stress is considered to be that stress at the upstream face which is less than water pressure at the same point. In most cases this stress occurs at the base of the crown cantilever. These critical stresses are tabulated for normal loading conditions and maximum loading conditions. The water pressure at the same point is also shown. Examination of critical cantilever stresses at the upstream face for maximum loading conditions reveals that in all cases the water pressure exceeds the stress shown for the designated loading. Tensile stresses are indicated at the upstream face for three dams; namely, Black Canyon, East Park, and Keswick. However, it is felt that this is an exceptional condition with little likelihood of occurrence. The criterion to be used, therefore, is the normal loading condition, for which in no case is the stress at the upstream face below a value of about 40 percent of the water pressure at the same point.

Maximum stresses parallel to the downstream face for normal operating reservoir load and for maximum loading are also shown in table E-1. Maximum stresses computed by the trial-load analysis are given for comparison with gravity stresses. Generally, the two methods show very little stress disagreement in the central section of the dam, but usually show significant differences in stress and stability factors in the region of the abutments.

Maximum sliding factors and minimum shear-friction factors are also shown in table E-1 for the 12 dams as computed by the gravity and trial-load analyses. These factors are for maximum loading conditions. For

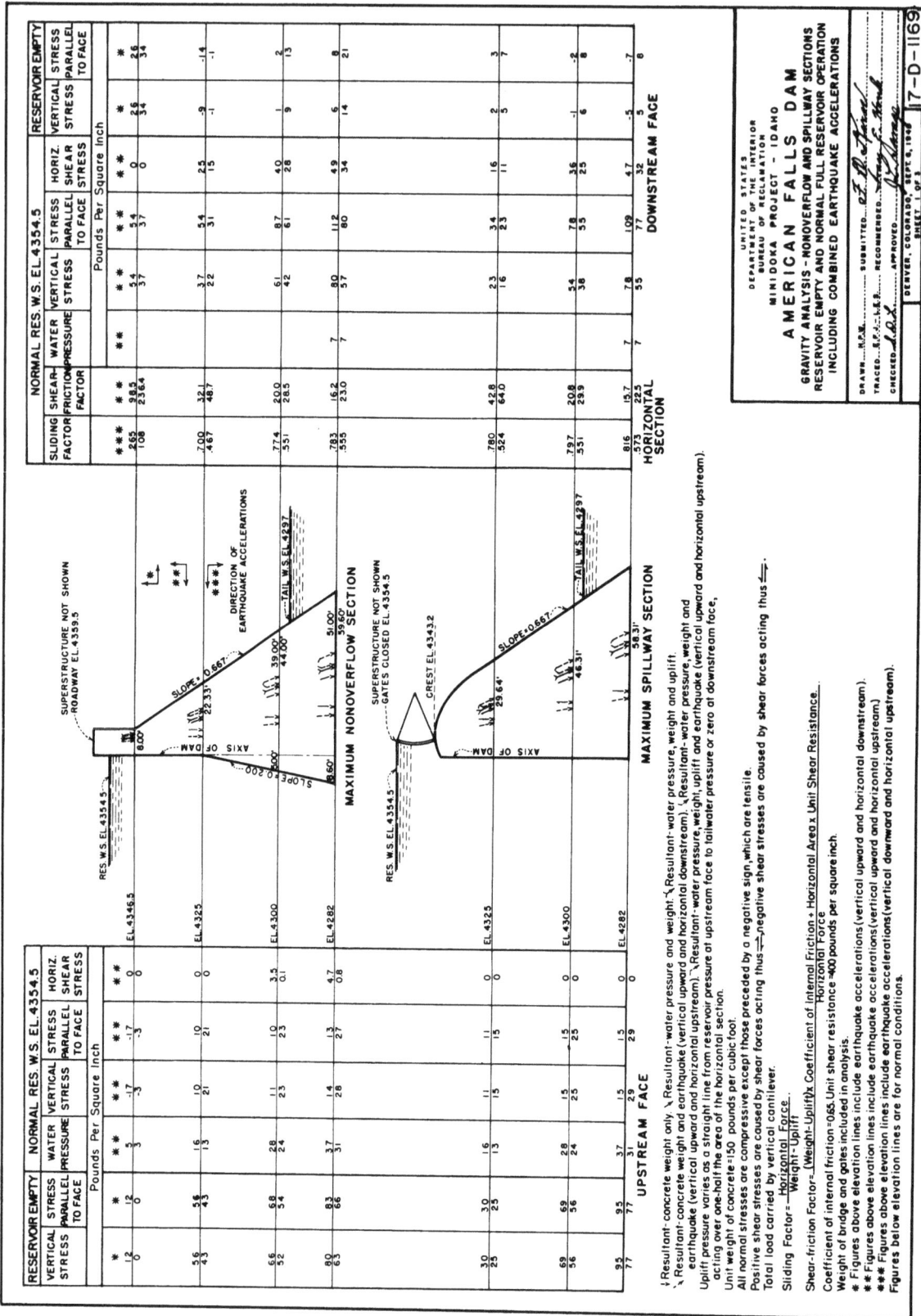

Figure E-1. American Falls Dam—gravity analyses of nonoverflow and spillway sections including effects of earthquake accelerations.

Figure E-2. American Falls Dam—gravity analyses of nonoverflow and spillway sections, normal conditions with ice load.

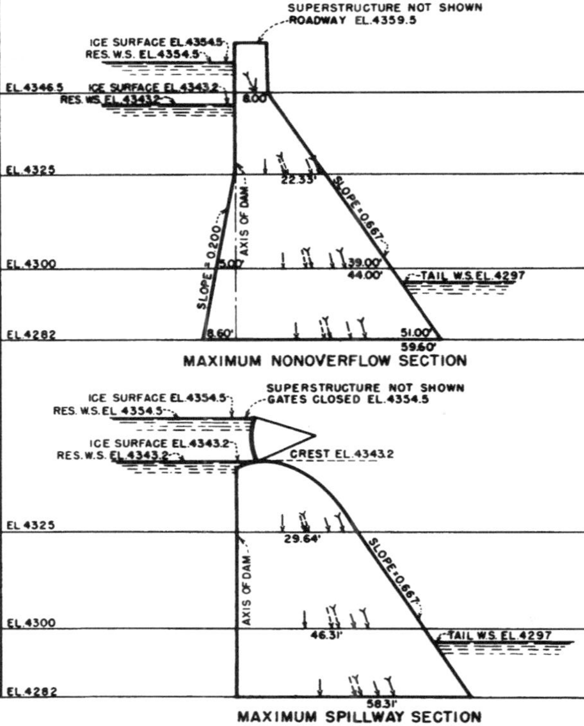

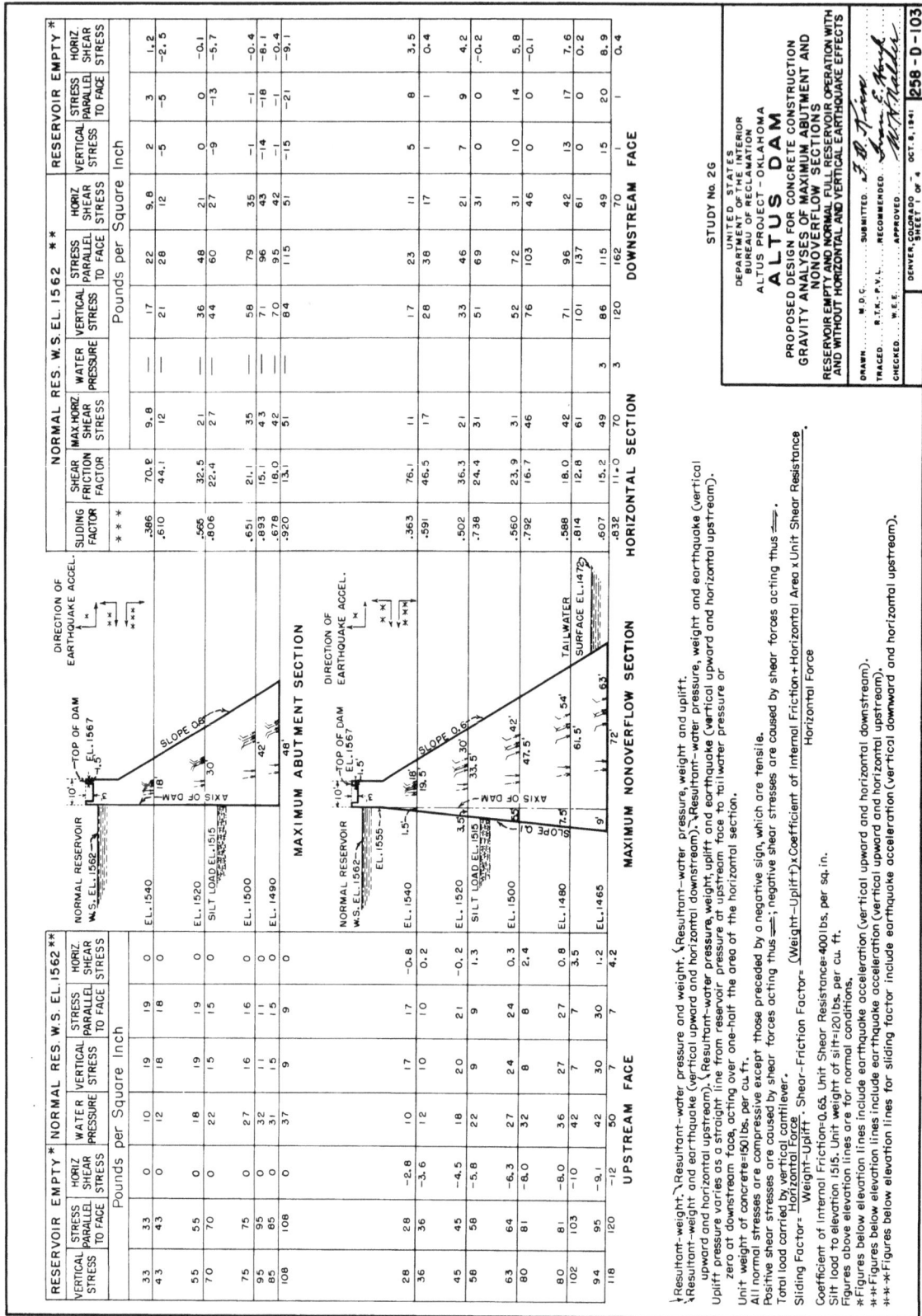

Figure E-3. Altus Dam—gravity analyses of maximum abutment and nonoverflow sections.

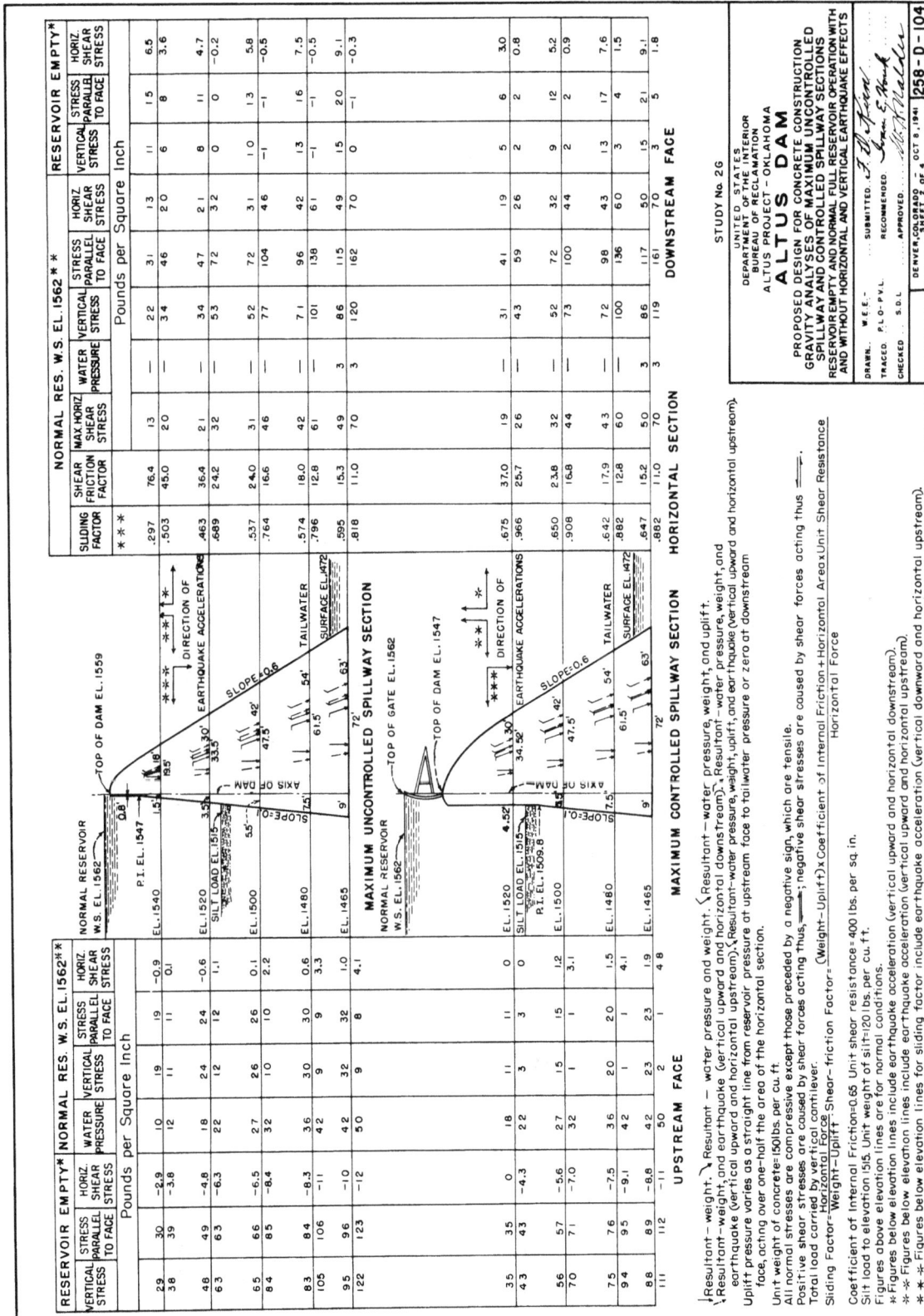

Figure E-4. Altus Dam—gravity analyses of spillway sections.

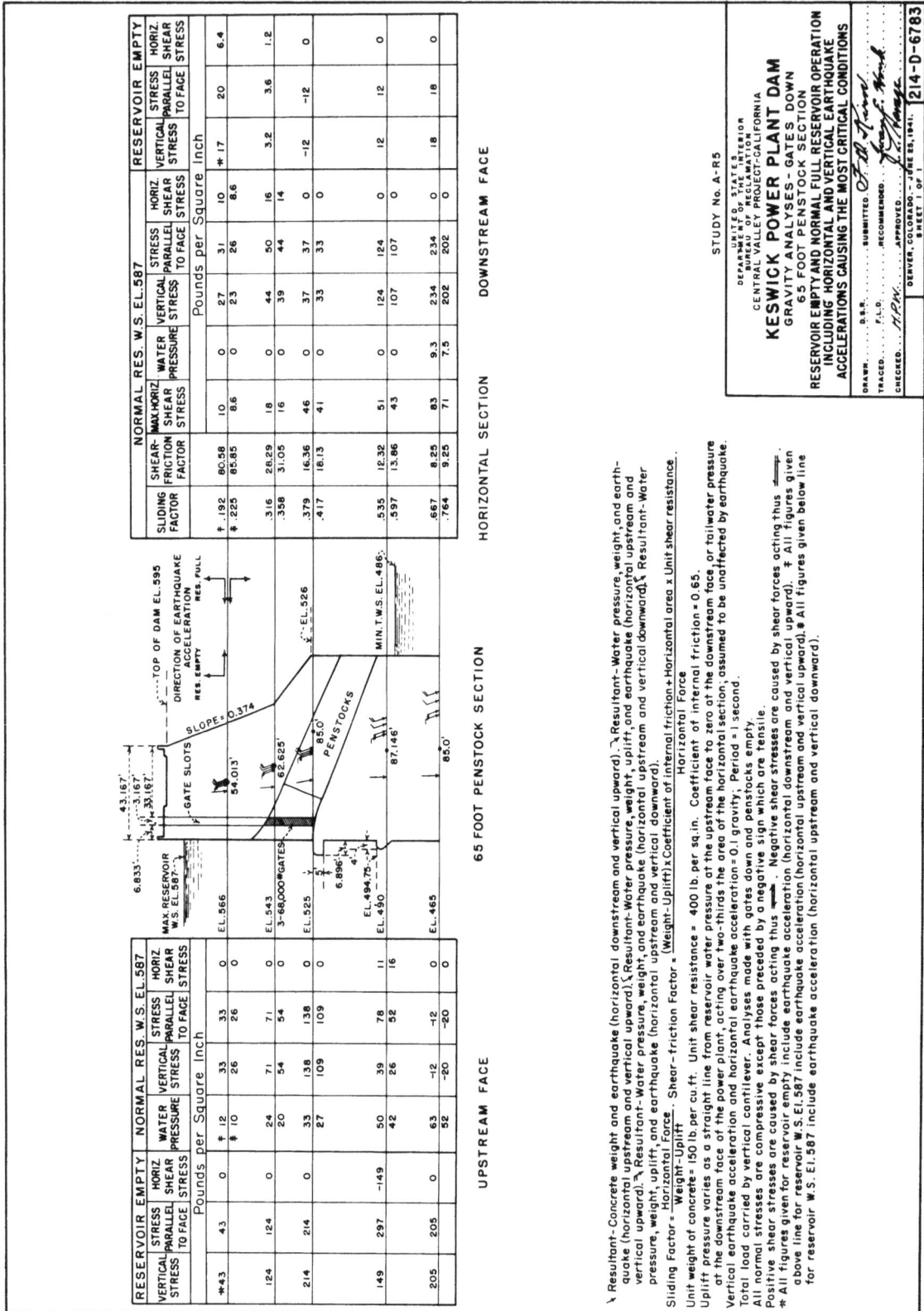

Figure E-5. Keswick Powerplant Dam—gravity analyses of penstock section including effects of earthquake accelerations.

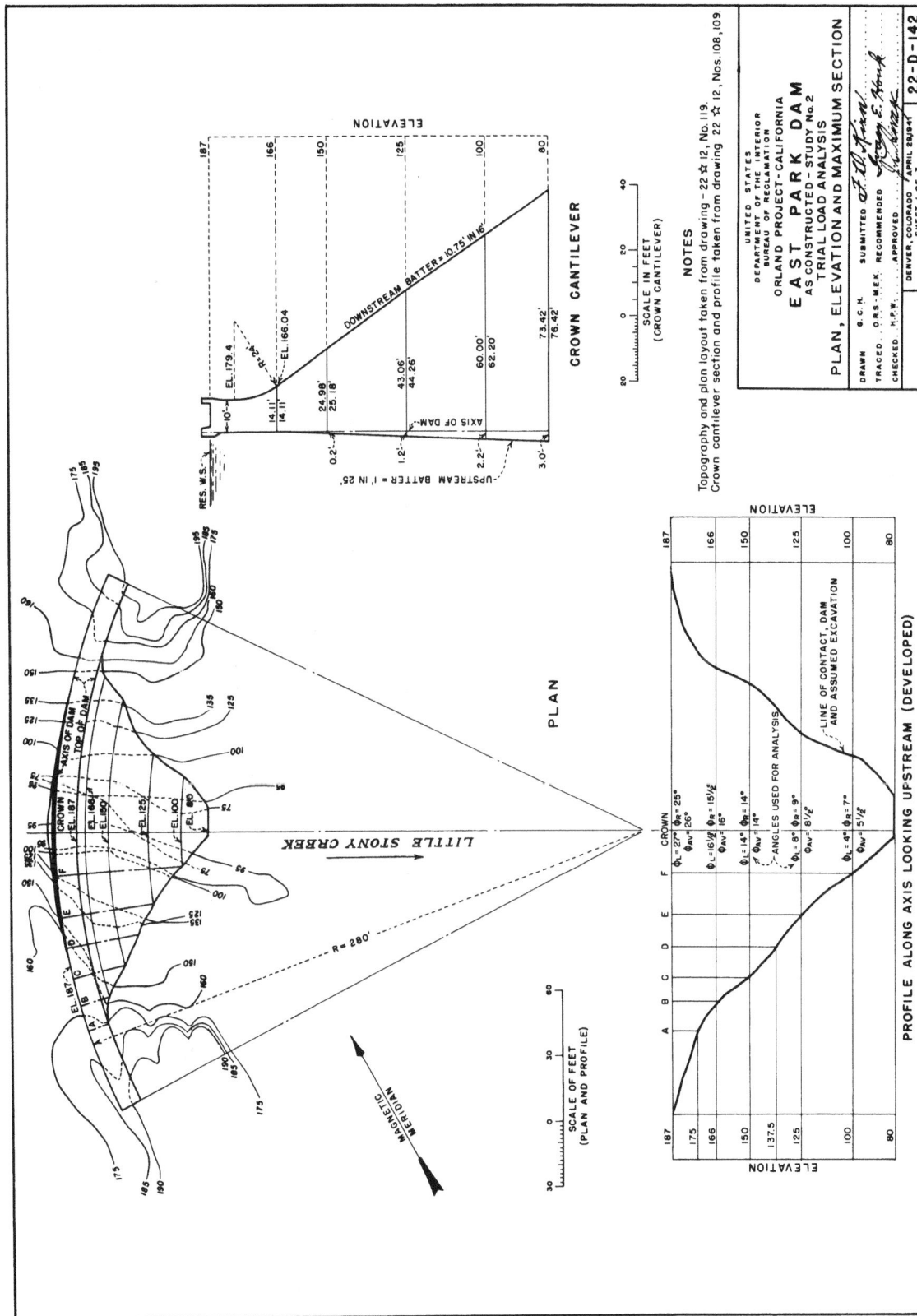

Figure E-6.   East Park Dam—plan, elevation, and maximum section.

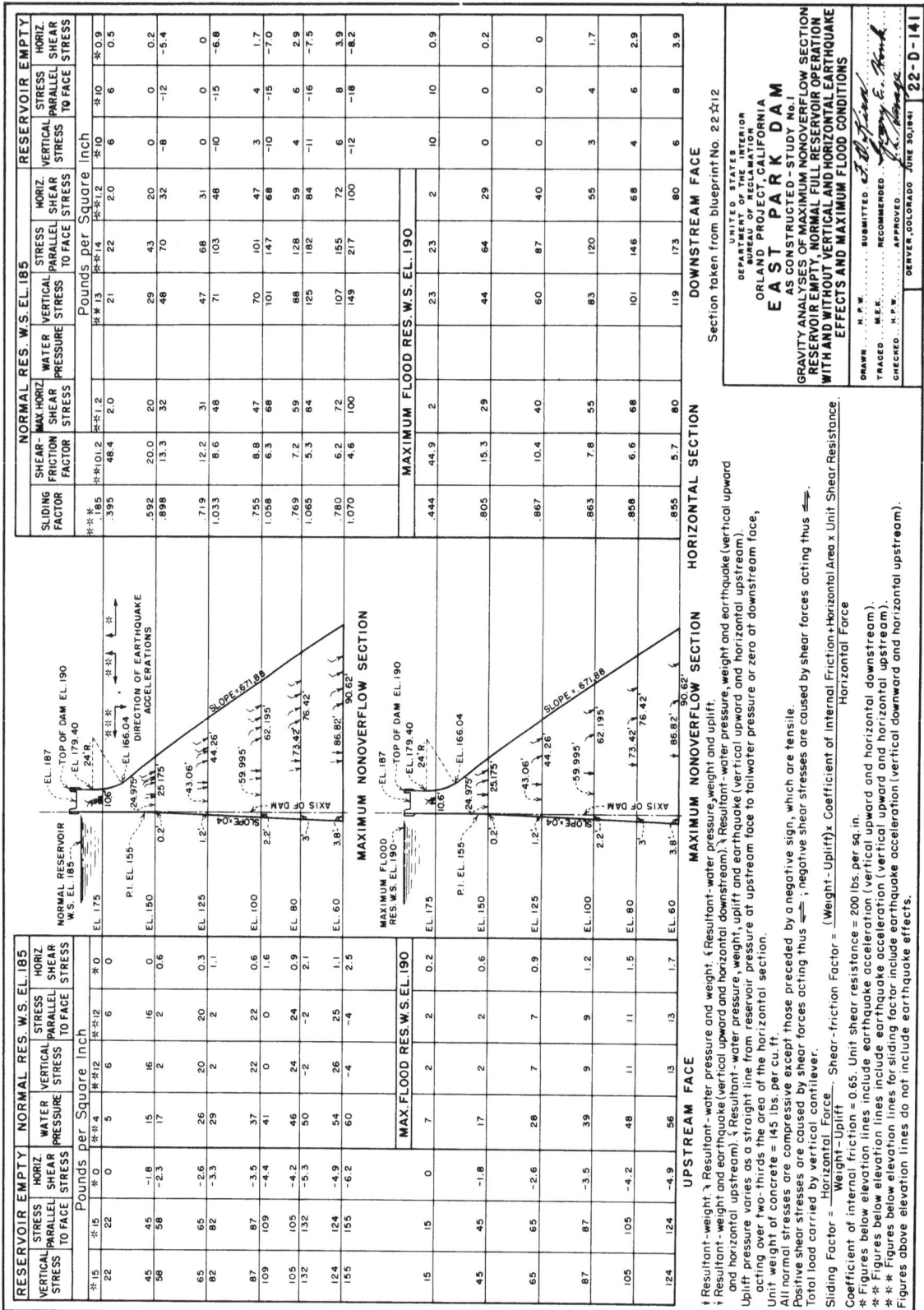

Figure E-7. East Park Dam—gravity analyses of maximum nonoverflow section.

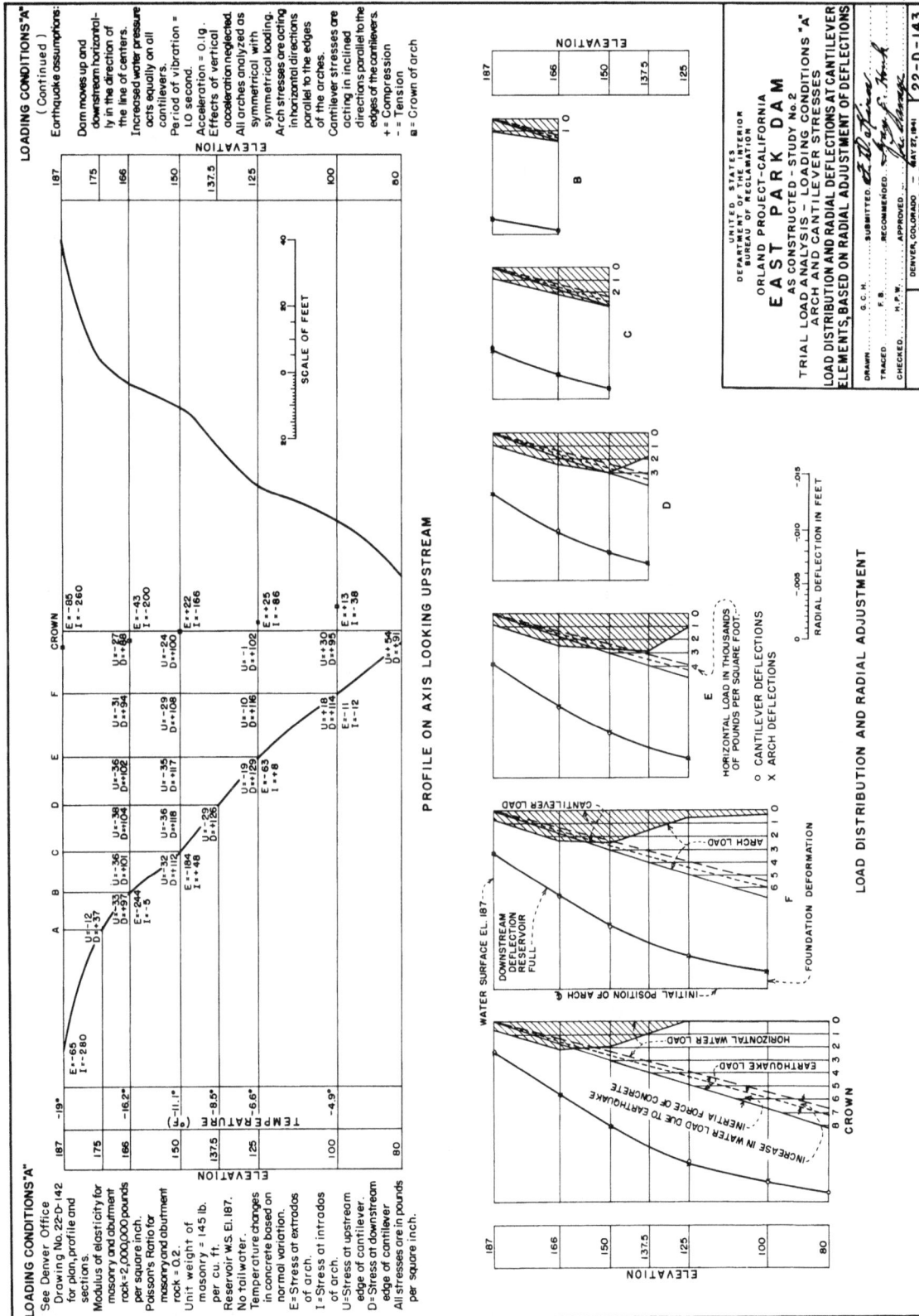

Figure E-8. East Park Dam—stresses, load distribution, and radial deflections from trial-load analysis.

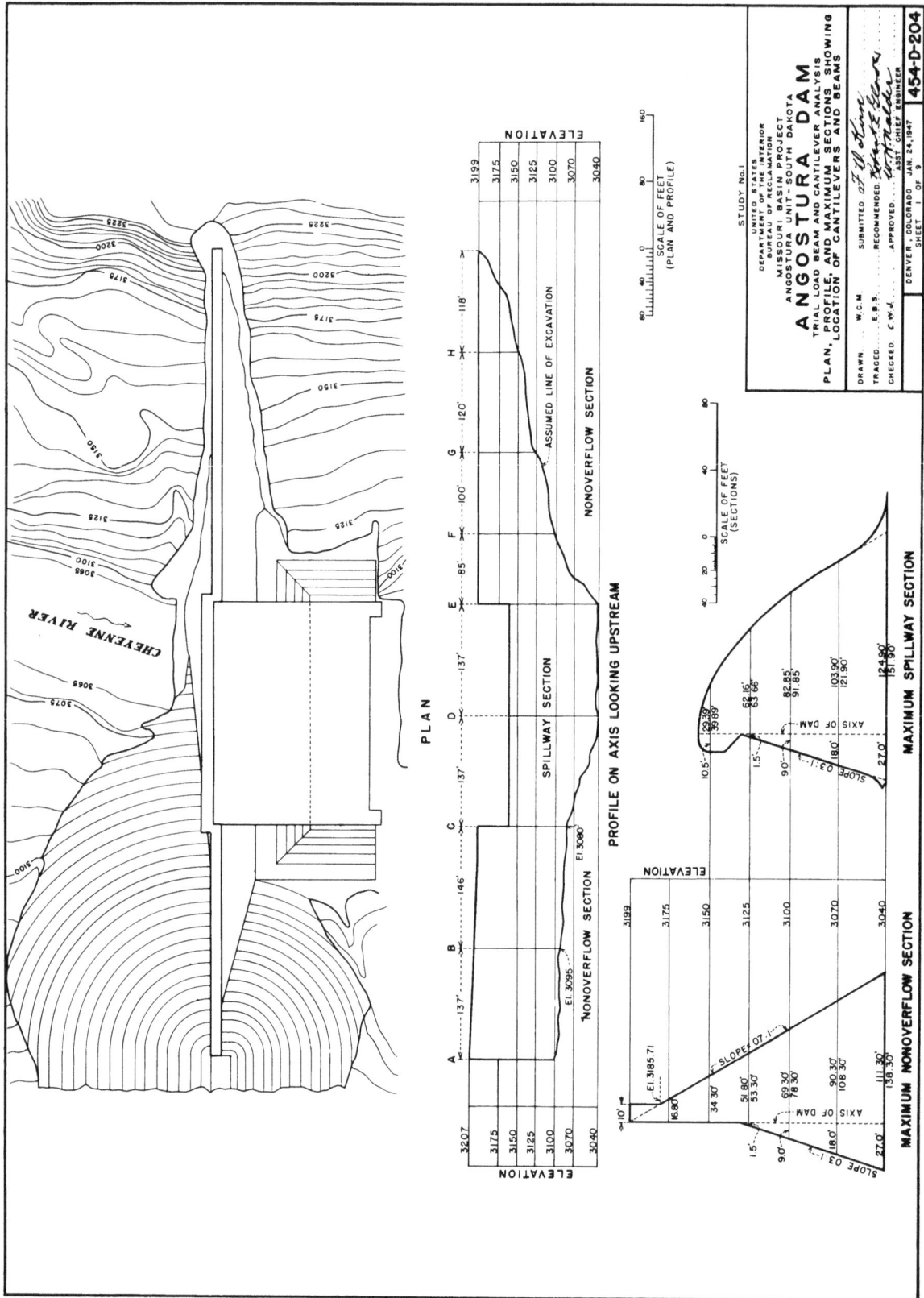

Figure E-9. Angostura Dam—plan, profile, and maximum section.

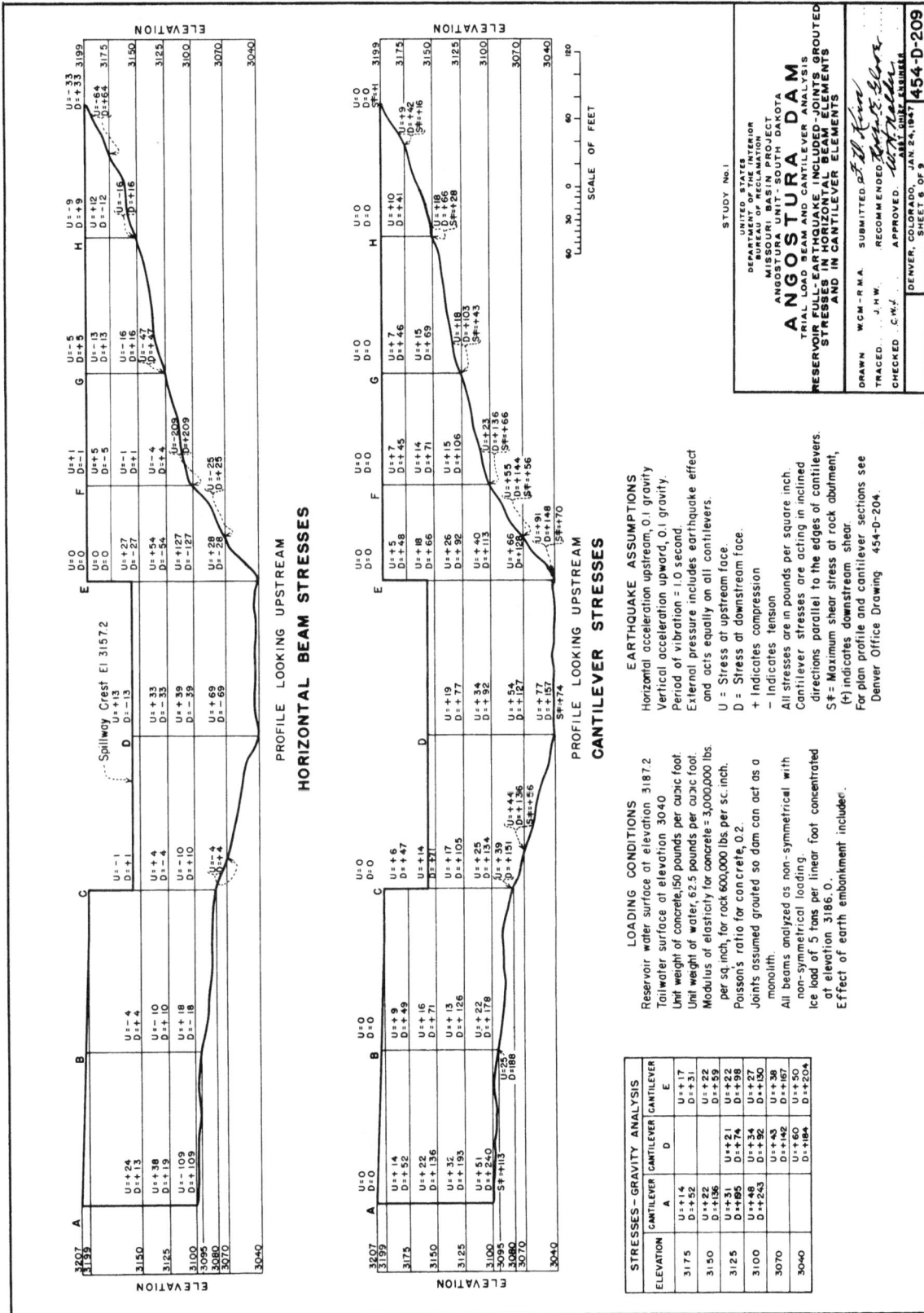

Figure E-10. Angostura Dam—stresses from trial-load beam and cantilever analysis.

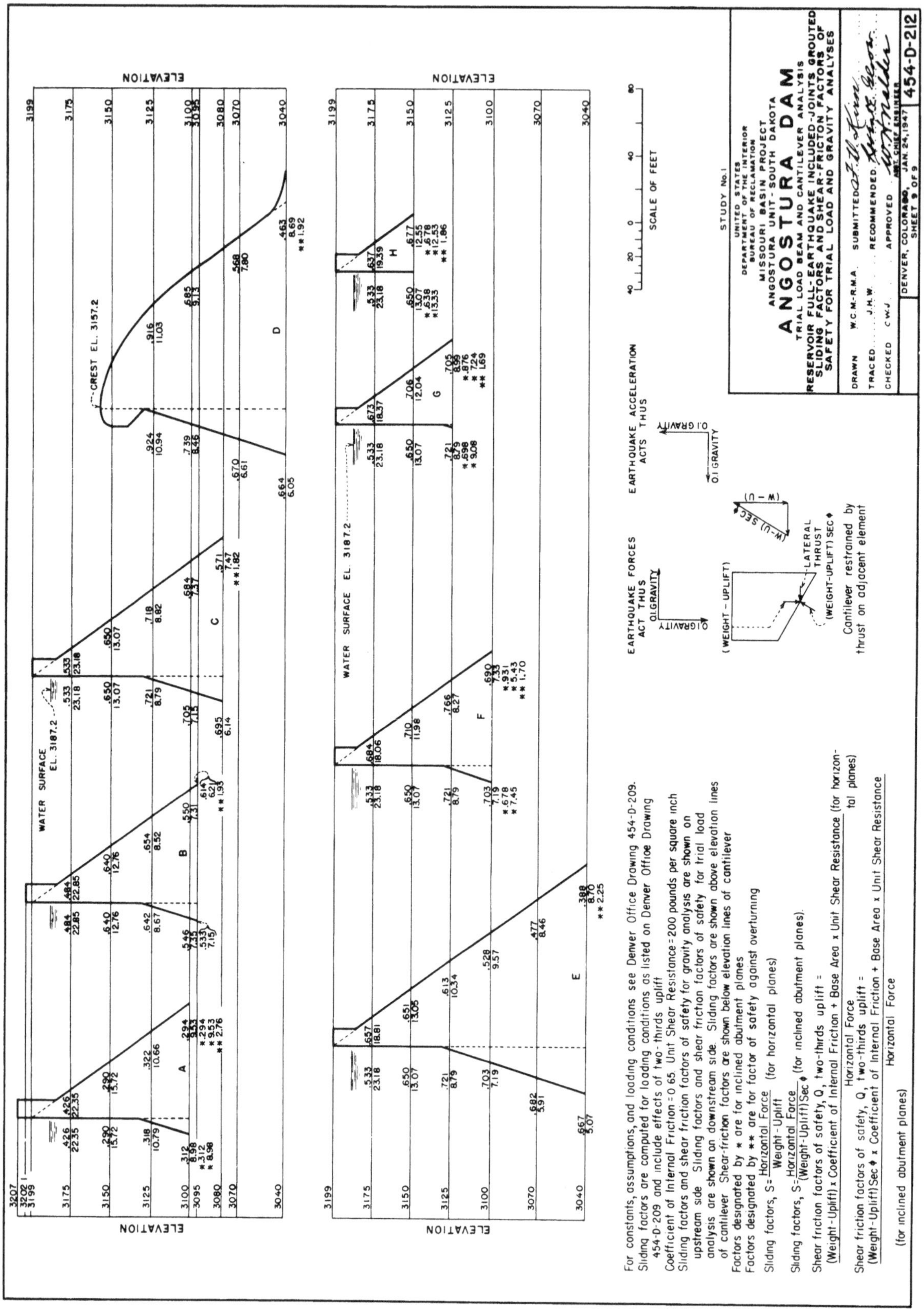

Figure E-11. Angostura Dam—stability factors from trial-load beam and cantilever analysis.

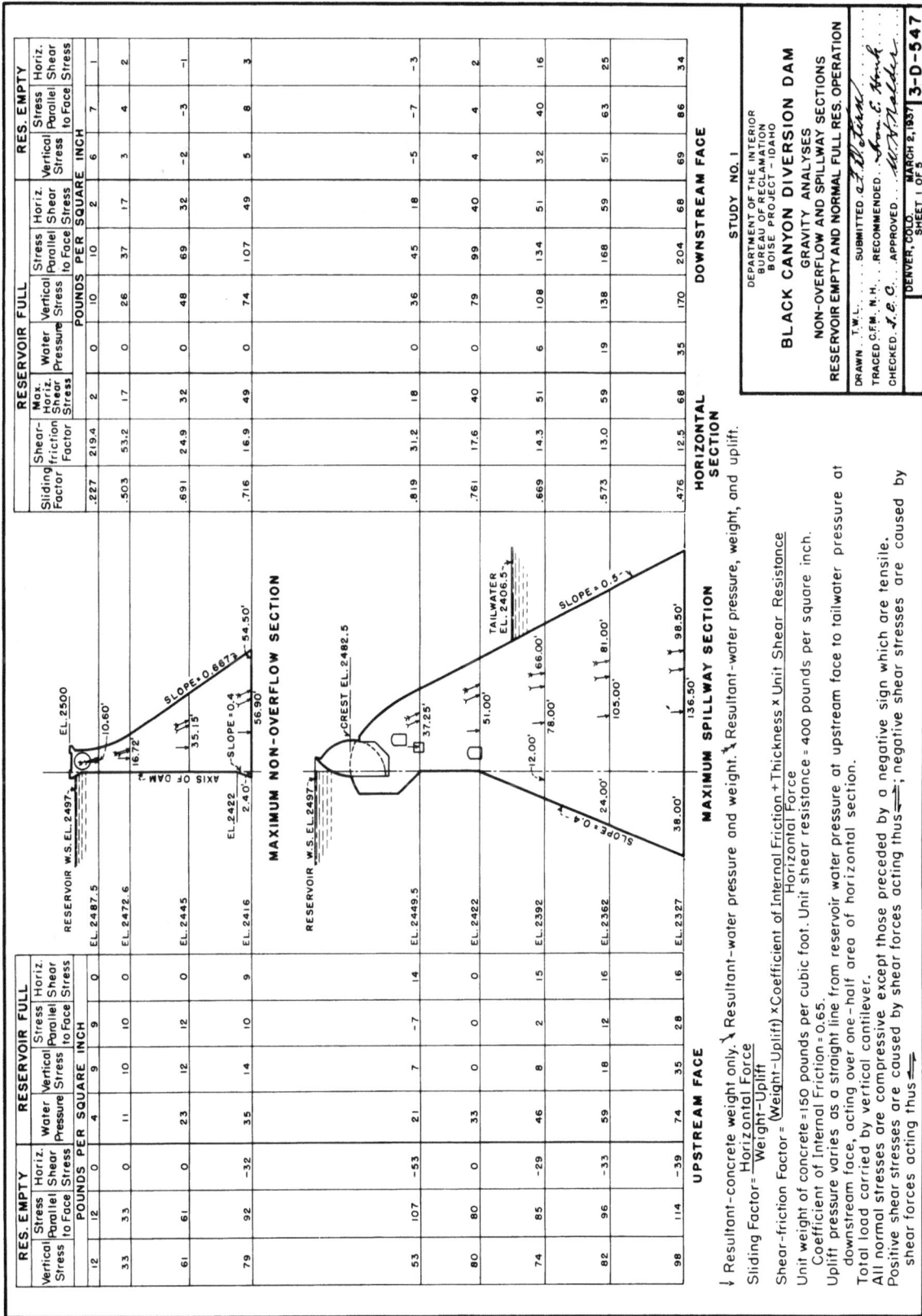

Figure E-12. Black Canyon Diversion Dam—stresses for normal conditions from gravity analyses.

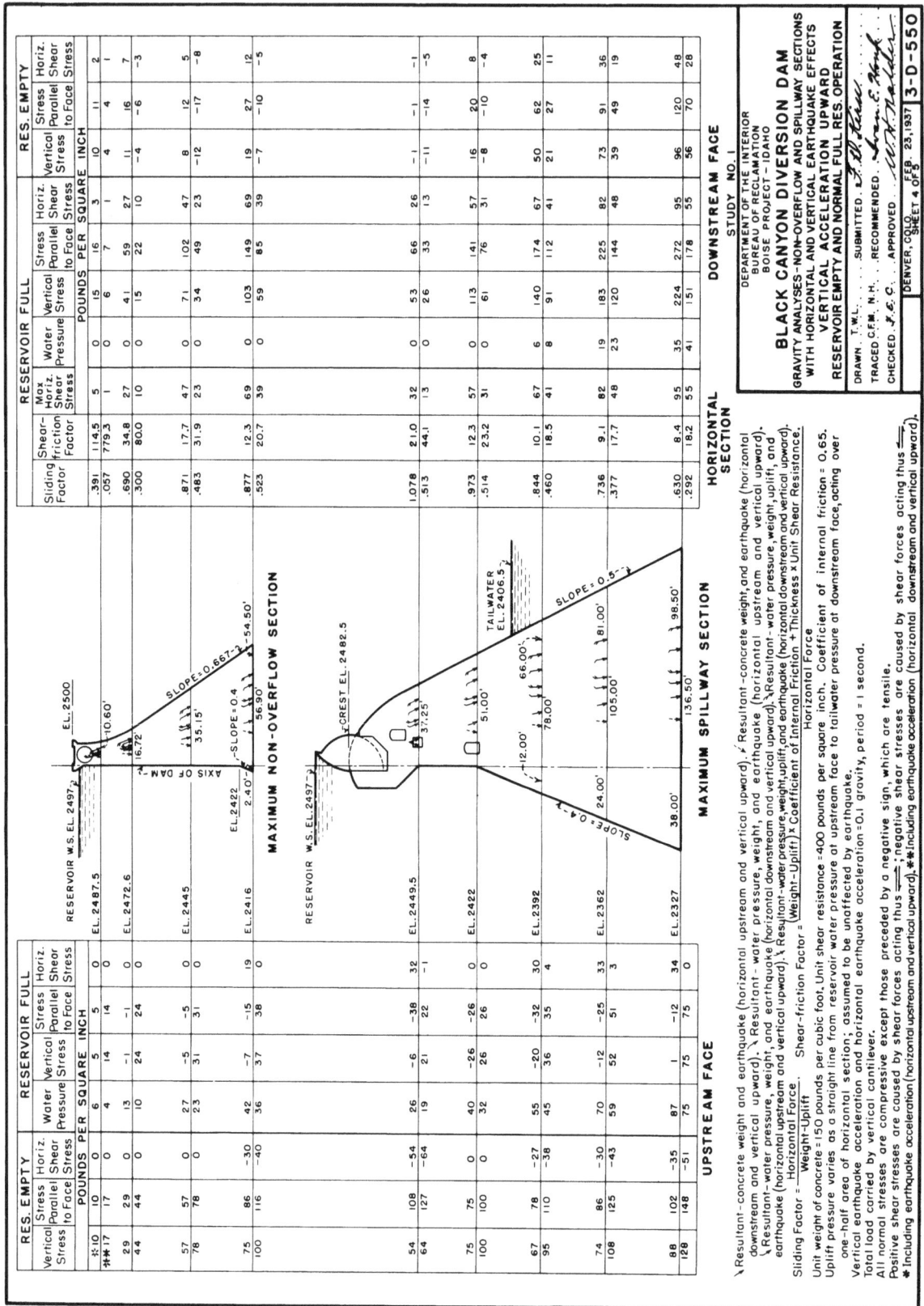

Figure E-13. Black Canyon Diversion Dam—gravity analyses including effects of earthquake, vertical acceleration upward.

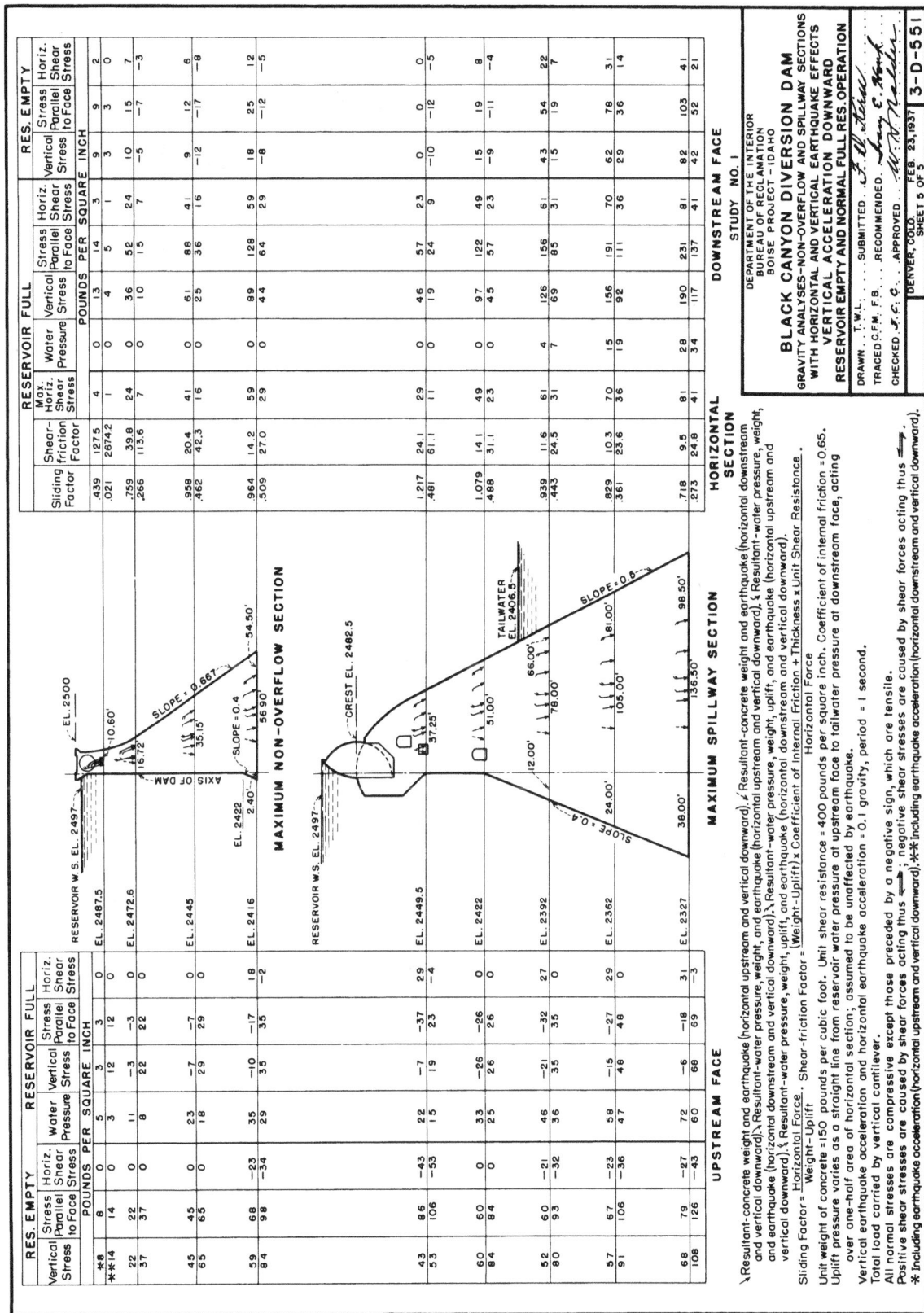

Figure E-14. Black Canyon Diversion Dam—gravity analyses including effects of earthquake, vertical acceleration downward.

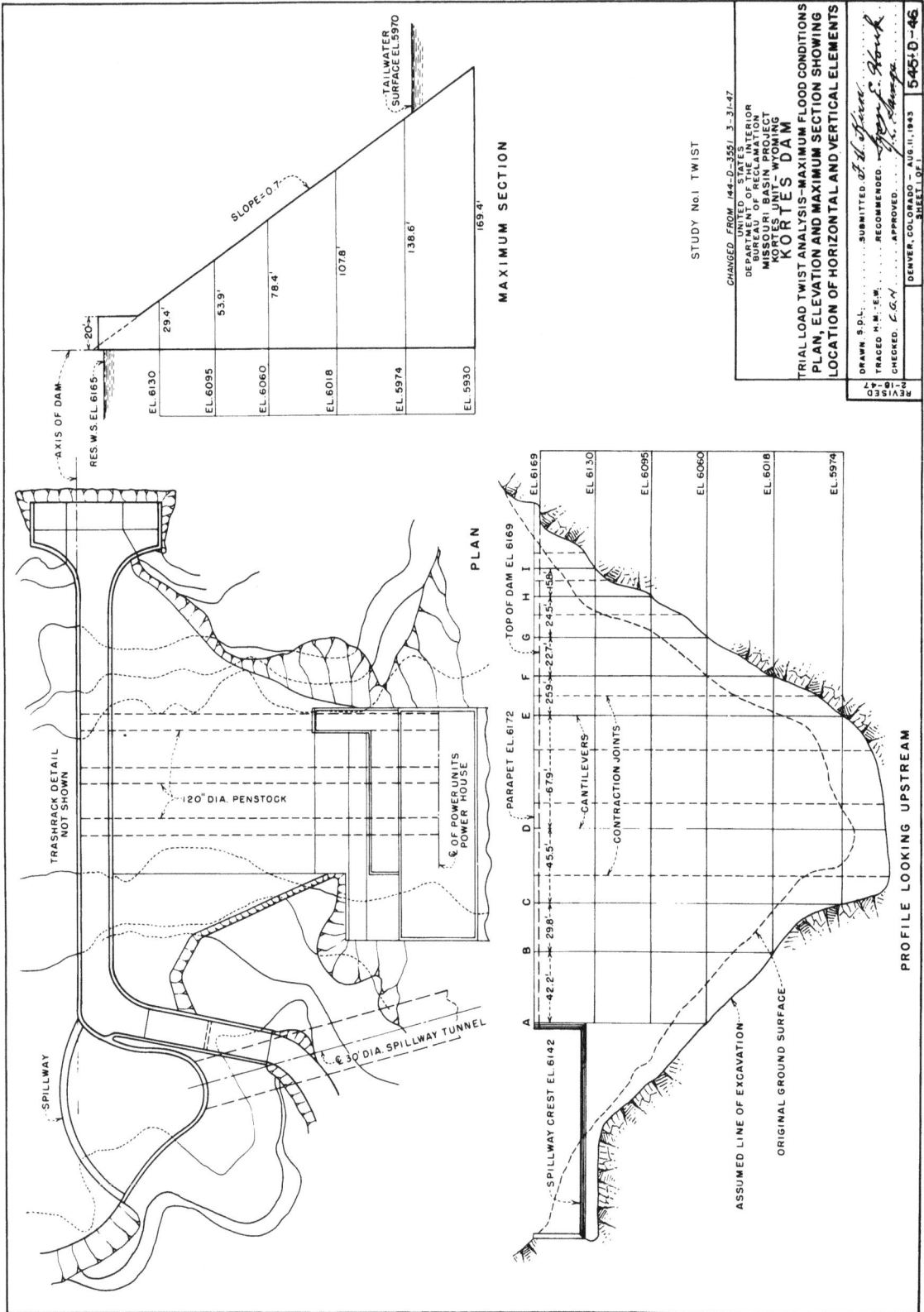

Figure E-15. Kortes Dam–plan, elevation, and maximum section.

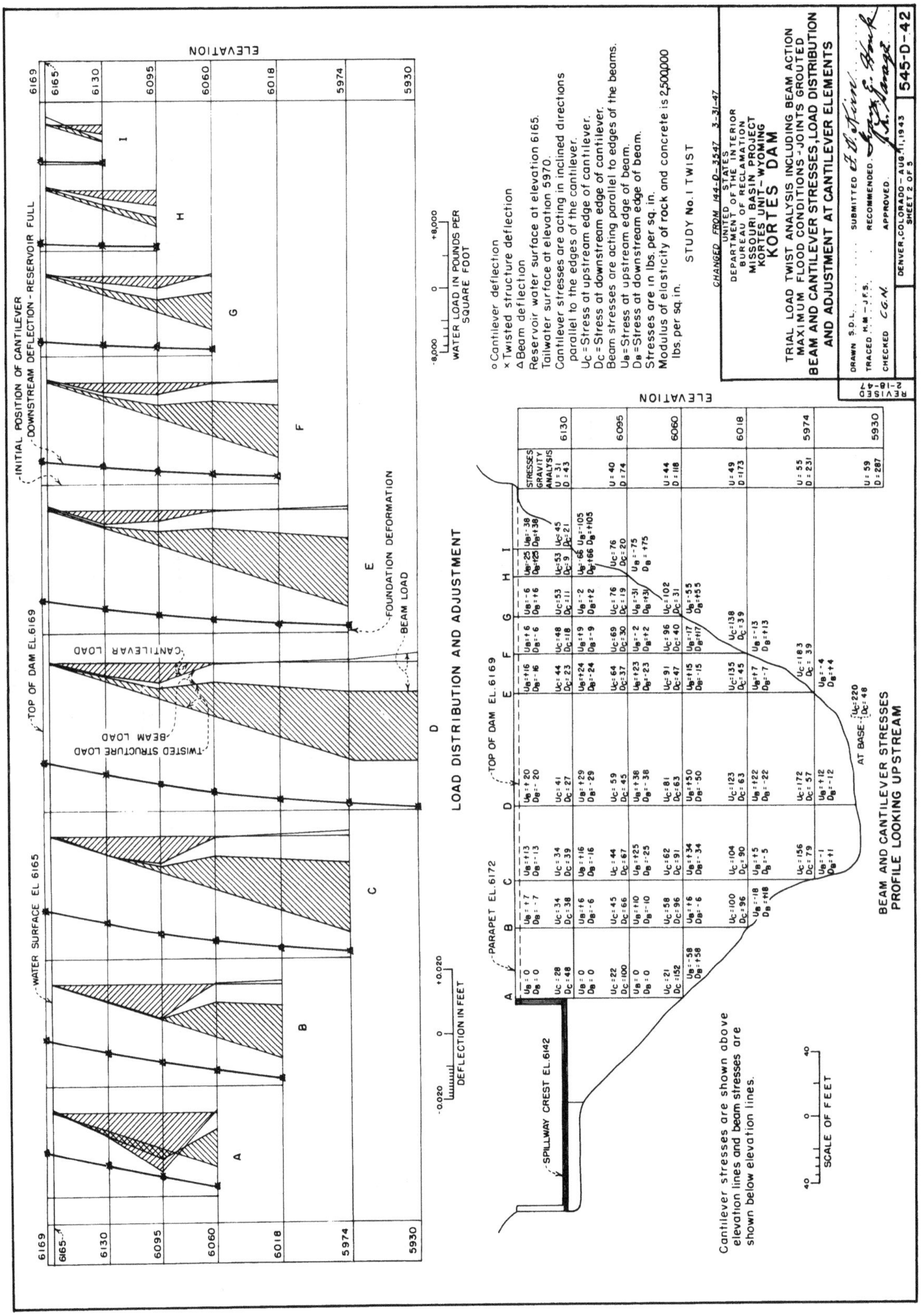

Figure E-16. Kortes Dam—stresses and load distribution from trial-load twist analysis.

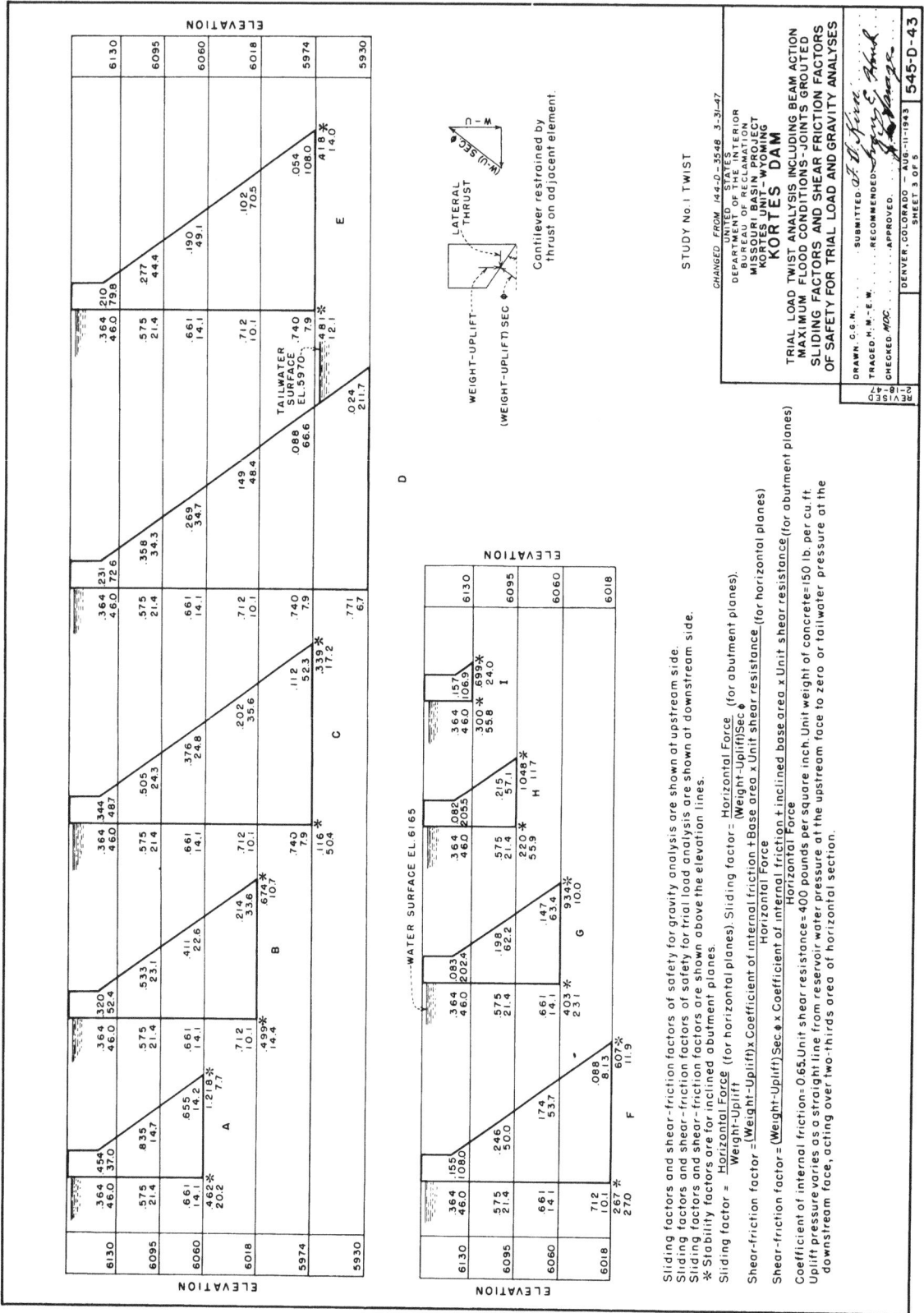

Figure E-17. Kortes Dam—stability factors from trial-load twist analysis.

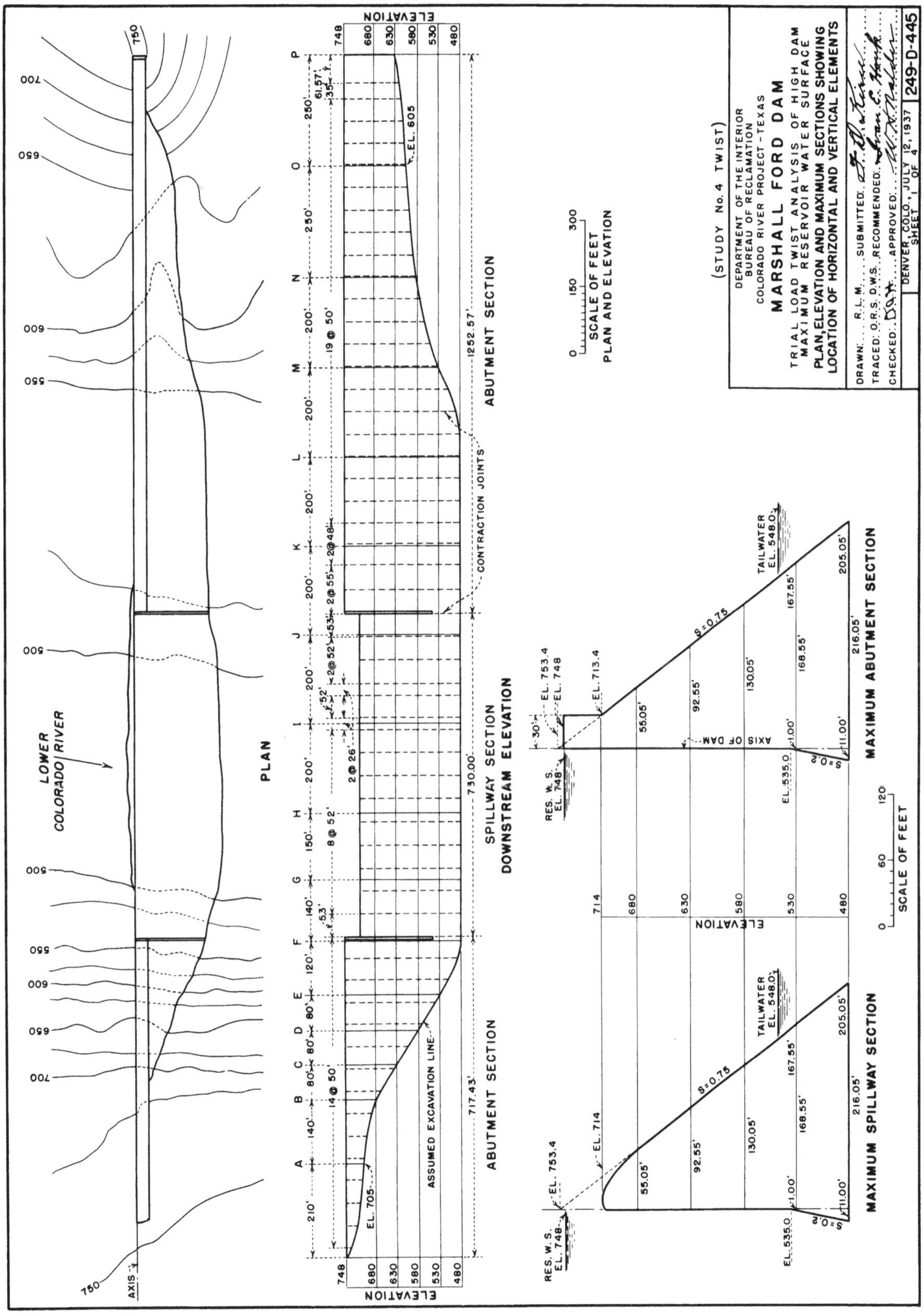

Figure E-18. Marshall Ford Dam—plan, elevation, and maximum sections.

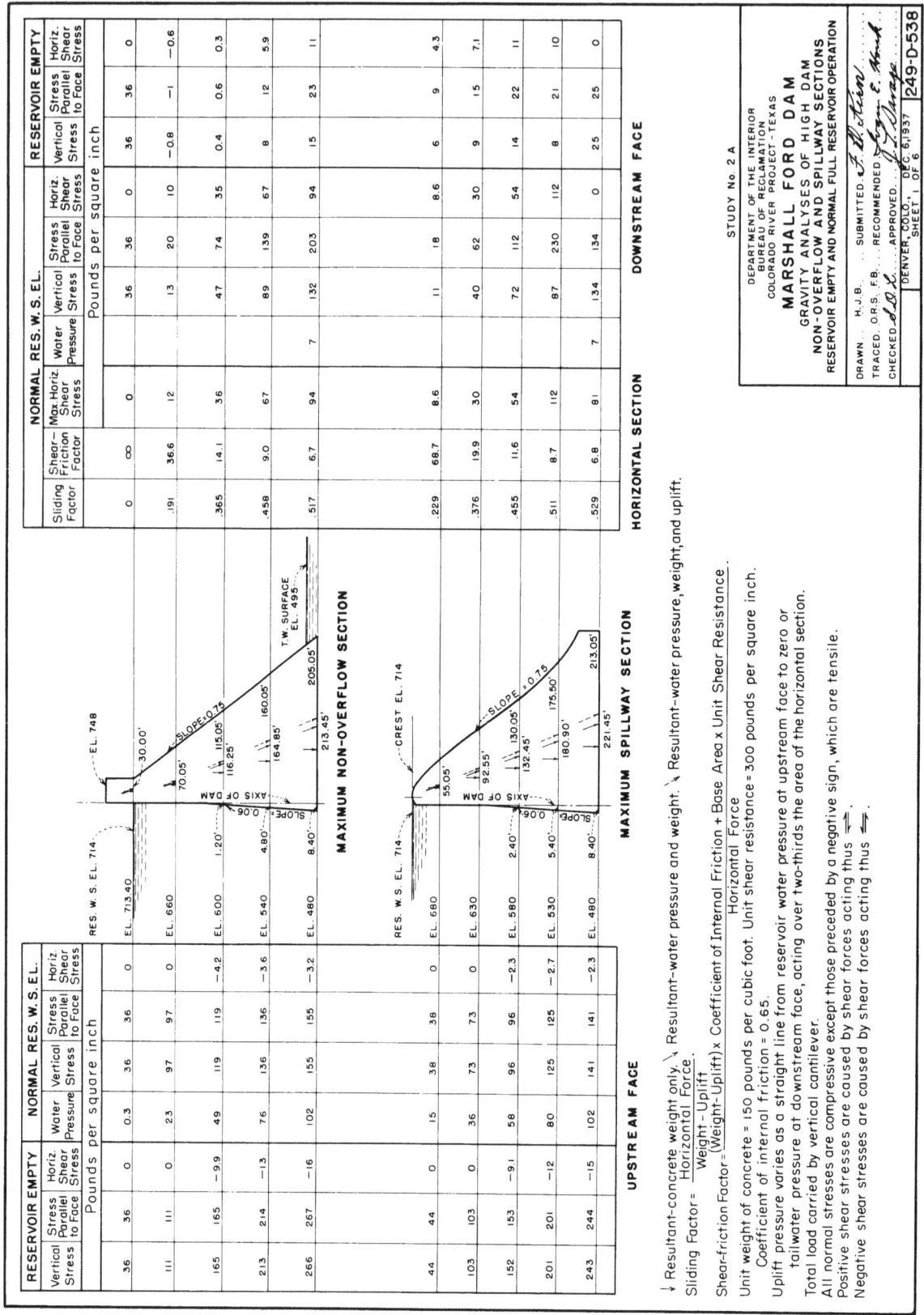

Figure E-19. Marshall Ford Dam—gravity analyses for normal conditions.

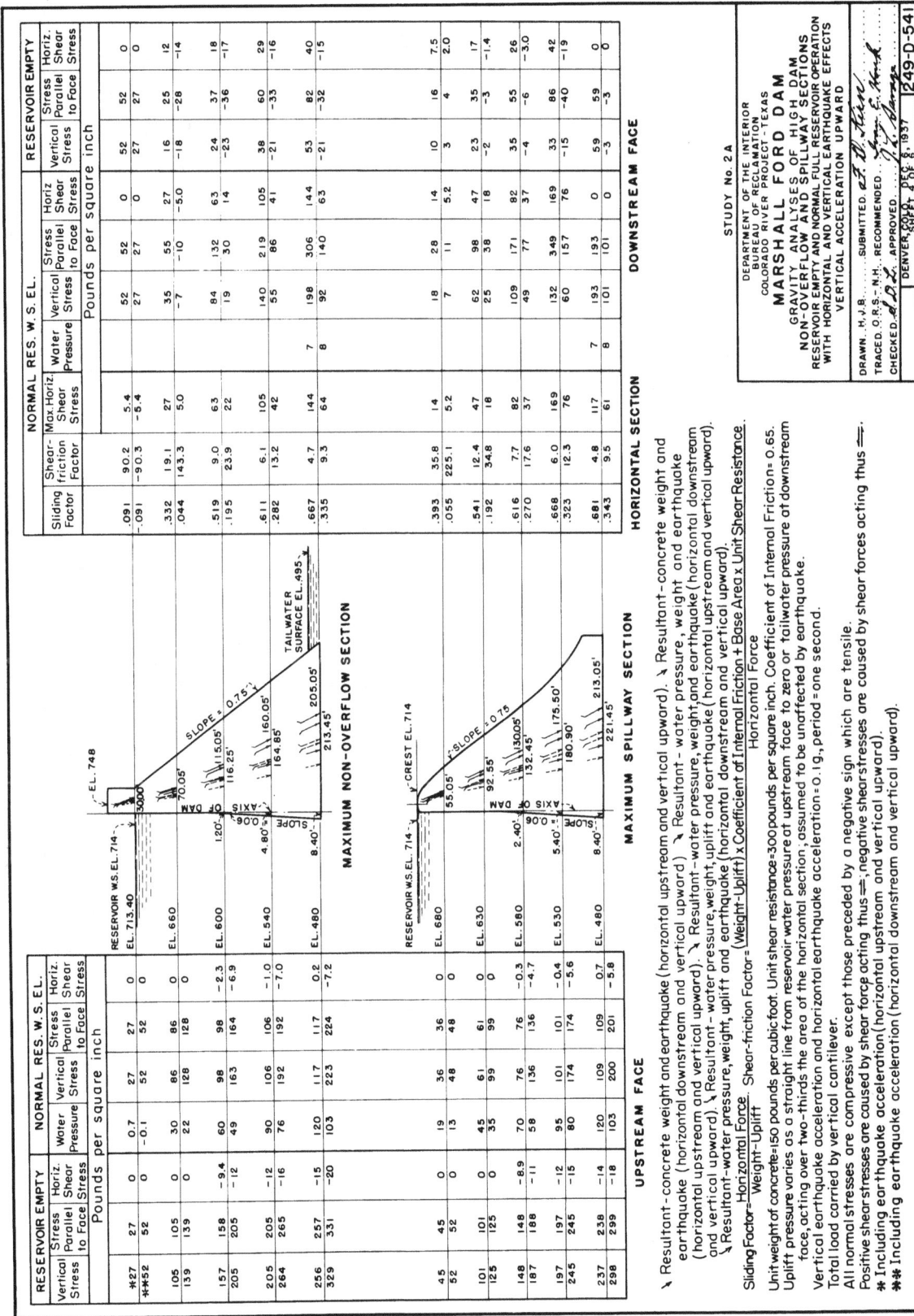

Figure E-20. Marshall Ford Dam—gravity analyses including effects of earthquake, vertical acceleration upward.

UPSTREAM FACE

RESERVOIR EMPTY			NORMAL RES. W. S. EL.			
Vertical Stress	Stress Parallel to Face	Horiz. Shear Stress	Water Pressure	Vertical Stress	Stress Parallel to Face	Horiz. Shear Stress
Pounds per square inch						
*20	20	0	0.7	20	20	0
**45	45	0	-0.2	45	45	0
83	83	0	25	66	66	0
117	117	0	17	109	109	0
125	125	-7.5	50	74	74	-1.4
172	172	-10	39	140	140	-6.1
162	163	-9.7	75	79	79	-0.2
222	222	-13	61	165	165	-6.3
203	204	-12	100	86	86	0.8
276	277	-17	83	192	193	-6.6
36	36	0	17	29	29	0
43	43	0	10	40	40	0
80	80	0	38	47	47	0
105	105	0	28	84	84	0
117	118	-7.0	59	56	56	0.1
156	157	-9.4	46	117	117	-4.2
157	157	-9.4	79	76	76	0.2
204	205	-12	64	149	149	-5.1
188	189	-11	100	81	81	1.2
250	251	-15	83	172	173	-5.4

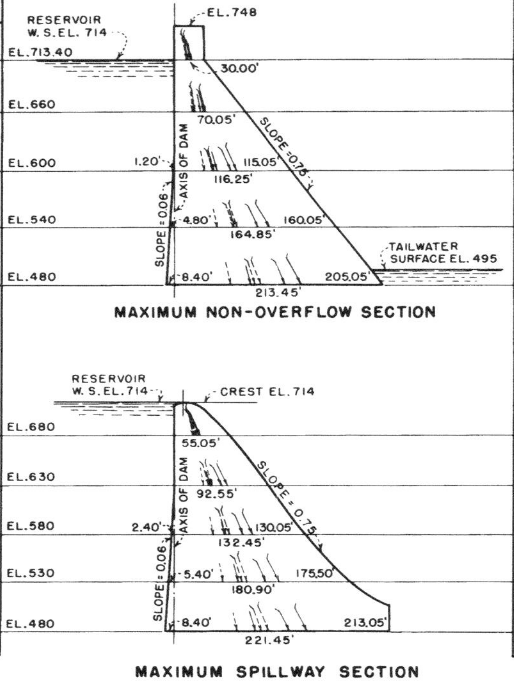

HORIZONTAL SECTION / **DOWNSTREAM FACE**

NORMAL RES. W. S. EL.							RESERVOIR EMPTY		
Sliding Factor	Shear-Friction Factor	Max. Horiz. Shear Stress	Water Pressure	Vertical Stress	Stress Parallel to Face	Horiz. Shear Stress	Vertical Stress	Stress Parallel to Face	Horiz. Shear Stress
		Pounds per square inch							
.112	88.9	5.4		45	45	0	45	45	0
-.112	-89.0	-5.4		20	20	0	20	20	0
.376	20.8	25		33	51	25	16	26	12
.013	621.0	-7.0		-9.3	-15	-7.0	-18	-28	-13
.584	10.0	56		75	117	56	24	37	18
.167	34.7	15		10	15	7.3	-23	-36	-17
.687	6.7	92		123	192	92	37	58	28
.260	17.7	30		37	58	28	-23	-36	-17
.754	5.1	125	5	172	265	125	50	78	37
.319	12.1	45	6	66	99	45	-23	-37	-18
.452	39.5	12		16	25	12	9	14	6.7
.019	967.3	3.5		5	7	3.5	2	2	1.1
.614	13.8	41		55	85	41	21	33	16
.162	52.2	12		17	26	12	-4	-6	-2.8
.695	8.6	71		95	148	71	32	50	24
.246	24.3	26		35	54	26	-7	-11	-5.1
.756	6.6	147		115	303	147	31	82	40
.305	16.4	54		42	111	54	-17	-44	-21
.772	5.3	101	5	166	166	0	54	54	0
.329	12.4	44	6	75	75	0	-8	-8	0

↘ Resultant-concrete weight and earthquake (horizontal upstream and vertical downward). ↘ Resultant-concrete weight and earthquake (horizontal downstream and vertical downward). ↘ Resultant-water pressure, weight and earthquake (horizontal upstream and vertical downward). ↘ Resultant-water pressure, weight and earthquake (horizontal downstream and vertical downward). ↘ Resultant-water pressure, weight, uplift and earthquake (horizontal upstream and vertical downward). ↘ Resultant-water pressure, weight, uplift and earthquake (horizontal downstream and vertical downward).

Sliding Factor = $\frac{\text{Horizontal Force}}{\text{Weight} - \text{Uplift}}$. Shear-friction Factor = $\frac{(\text{Weight} - \text{Uplift}) \times \text{Coefficient of Internal Friction} + \text{Base Area} \times \text{Unit Shear Resistance}}{\text{Horizontal Force}}$.

Unit weight of concrete = 150 pounds per cubic foot. Unit Shear Resistance = 300 pounds per square inch. Coefficient of Internal Friction = 0.65.
Uplift pressure varies as a straight line from reservoir water pressure at upstream face to zero or tailwater pressure at downstream face, acting over two-thirds the area of the horizontal section; assumed to be unaffected by earthquake.
Vertical earthquake acceleration and horizontal earthquake acceleration = 0.1 g., period = one second.
Total load carried by vertical cantilever.
All normal stresses are compressive except those preceded by a negative sign, which are tensile.
Positive shear stresses are caused by shear forces acting thus ⇌; negative shear stresses are caused by shear forces acting thus ⇌.
* Including earthquake acceleration (horizontal upstream and vertical downward).
** Including earthquake acceleration (horizontal downstream and vertical downward).

STUDY No. 2A

DEPARTMENT OF THE INTERIOR
BUREAU OF RECLAMATION
COLORADO RIVER PROJECT - TEXAS
MARSHALL FORD DAM
GRAVITY ANALYSES OF HIGH DAM
NON-OVERFLOW AND SPILLWAY SECTIONS
RESERVOIR EMPTY AND NORMAL FULL RESERVOIR OPERATION
WITH HORIZONTAL AND VERTICAL EARTHQUAKE EFFECTS
VERTICAL ACCELERATION DOWNWARD

DRAWN H. J. B. SUBMITTED F. W. Kun
TRACED O. R. S. - D. W. S. RECOMMENDED [signature]
CHECKED S. D. L. APPROVED J. L. Savage
DENVER, COLO. DEC. 6, 1937 SHEET 5 OF 6 249-D-542

Figure E-21. **Marshall Ford Dam—gravity analyses including effects of earthquake, vertical acceleration downward.**

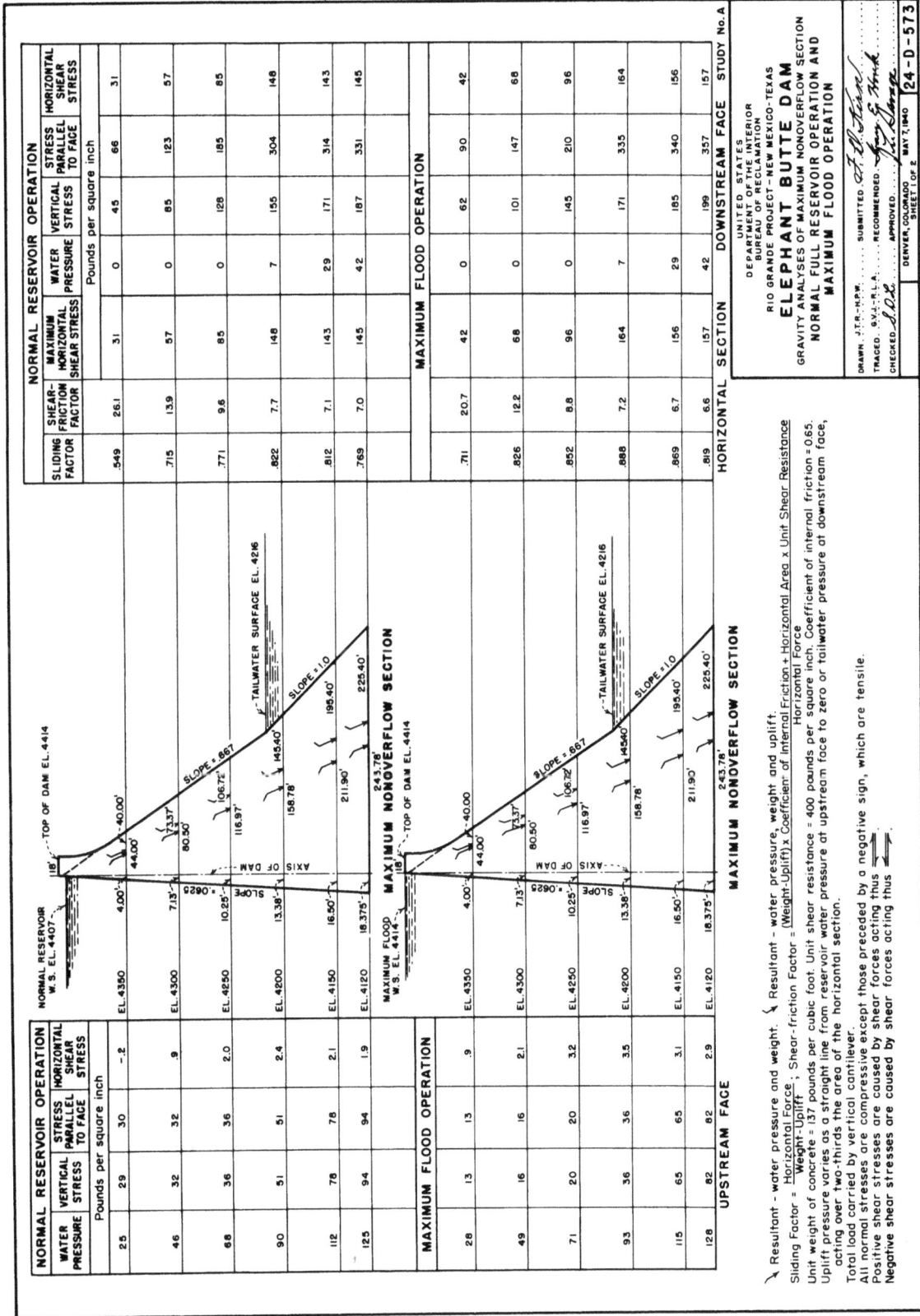

Figure E-22. Elephant Butte Dam—gravity analyses for maximum flood condition.

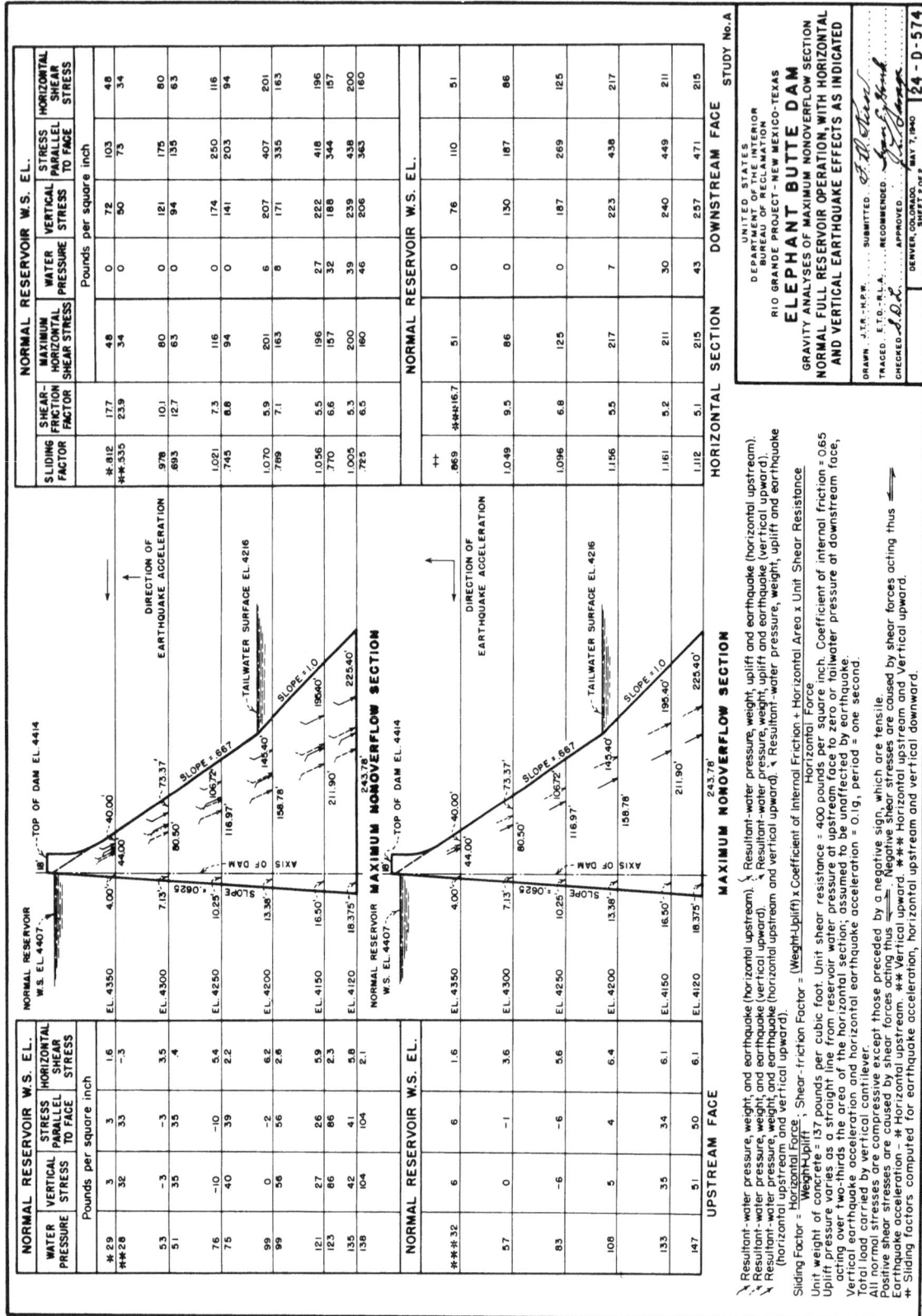

Figure E-23. Elephant Butte Dam—gravity analyses including effects of earthquake accelerations.

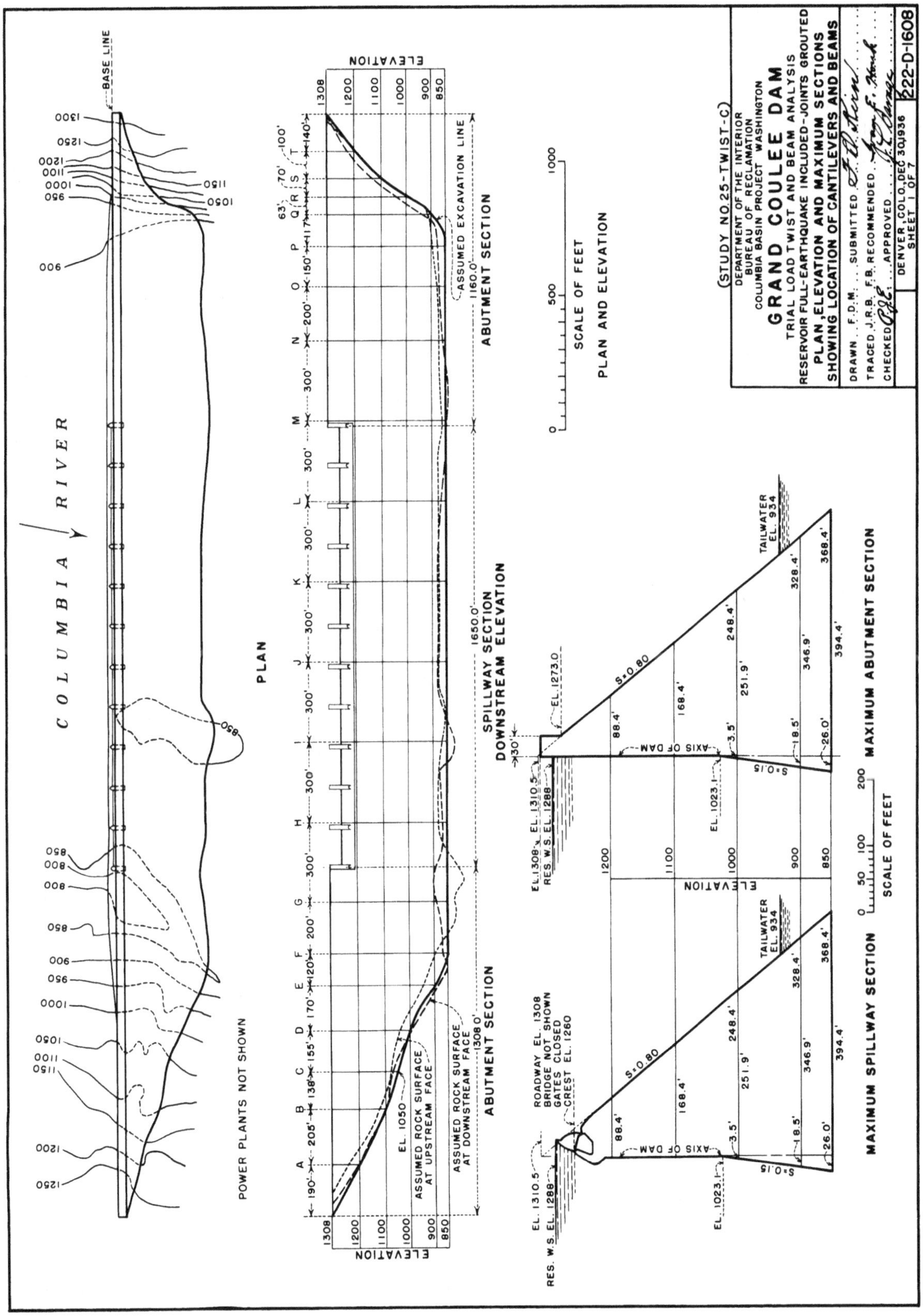

Figure E-24. Grand Coulee Dam—plan, elevation, and maximum sections.

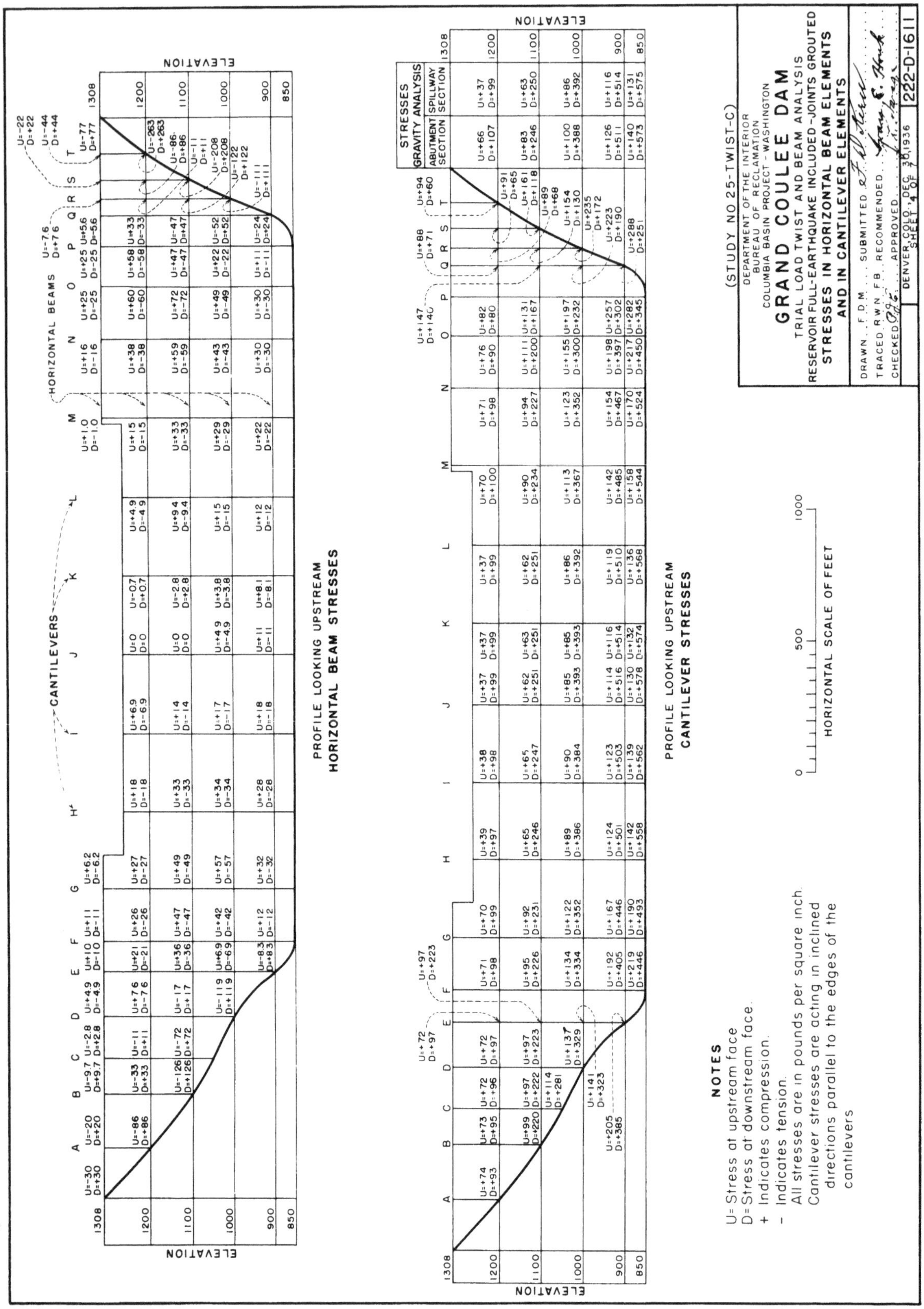

Figure E-25. Grand Coulee Dam—stresses from trial-load twist and beam analysis.

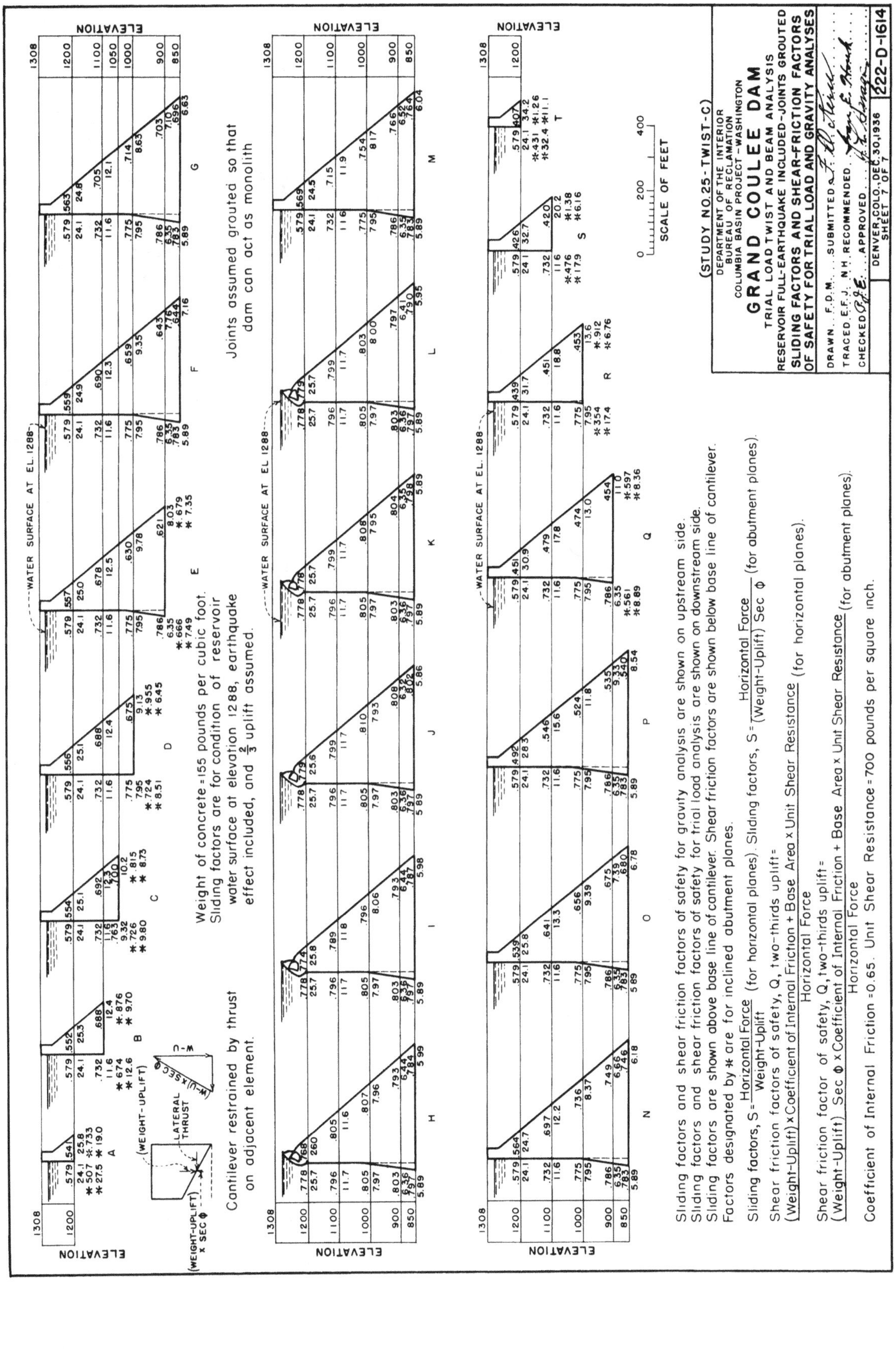

Figure E-26. Grand Coulee Dam—stability factors from trial-load twist and beam analysis.

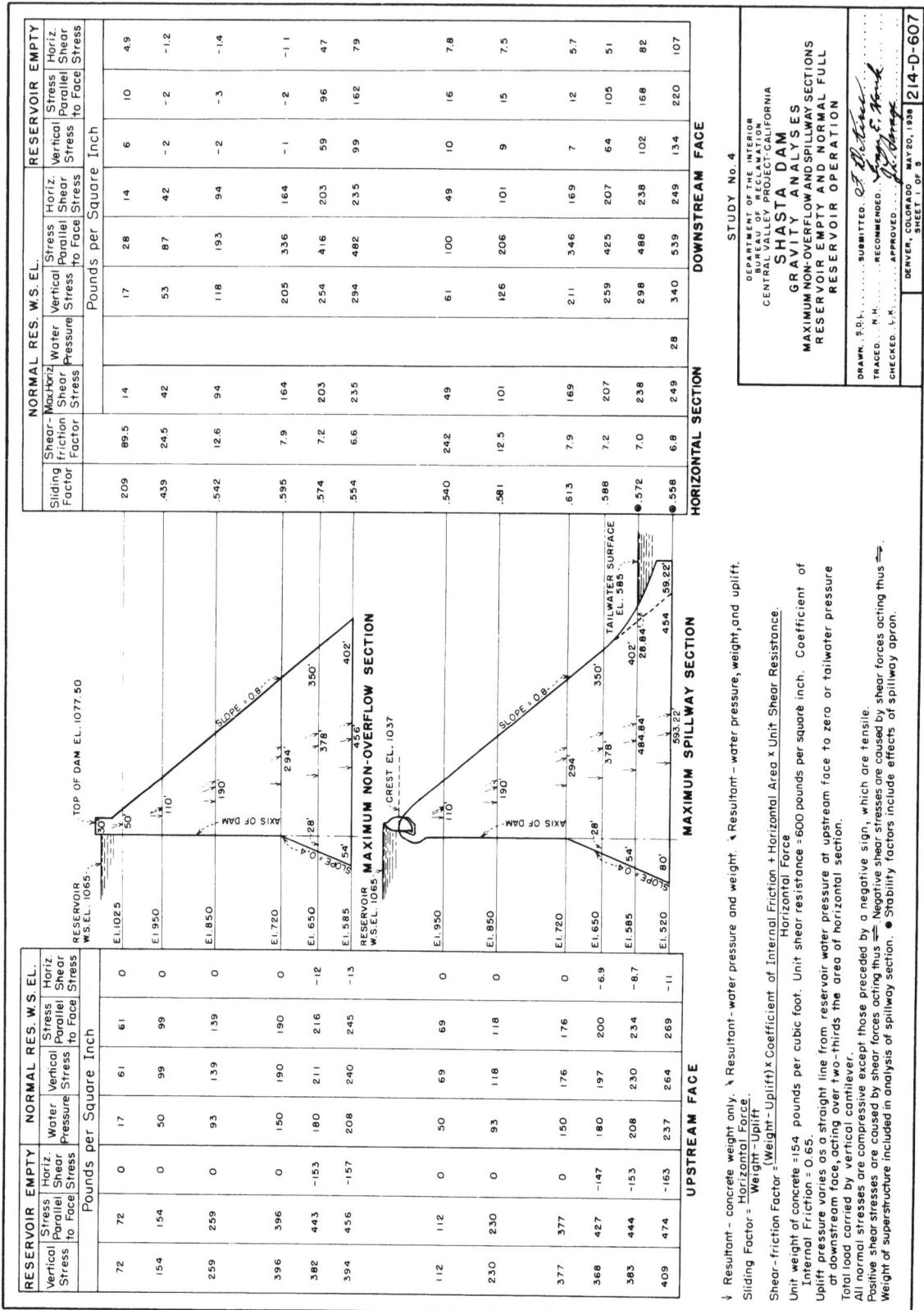

Figure E-27. Shasta Dam—gravity analyses for normal conditions.

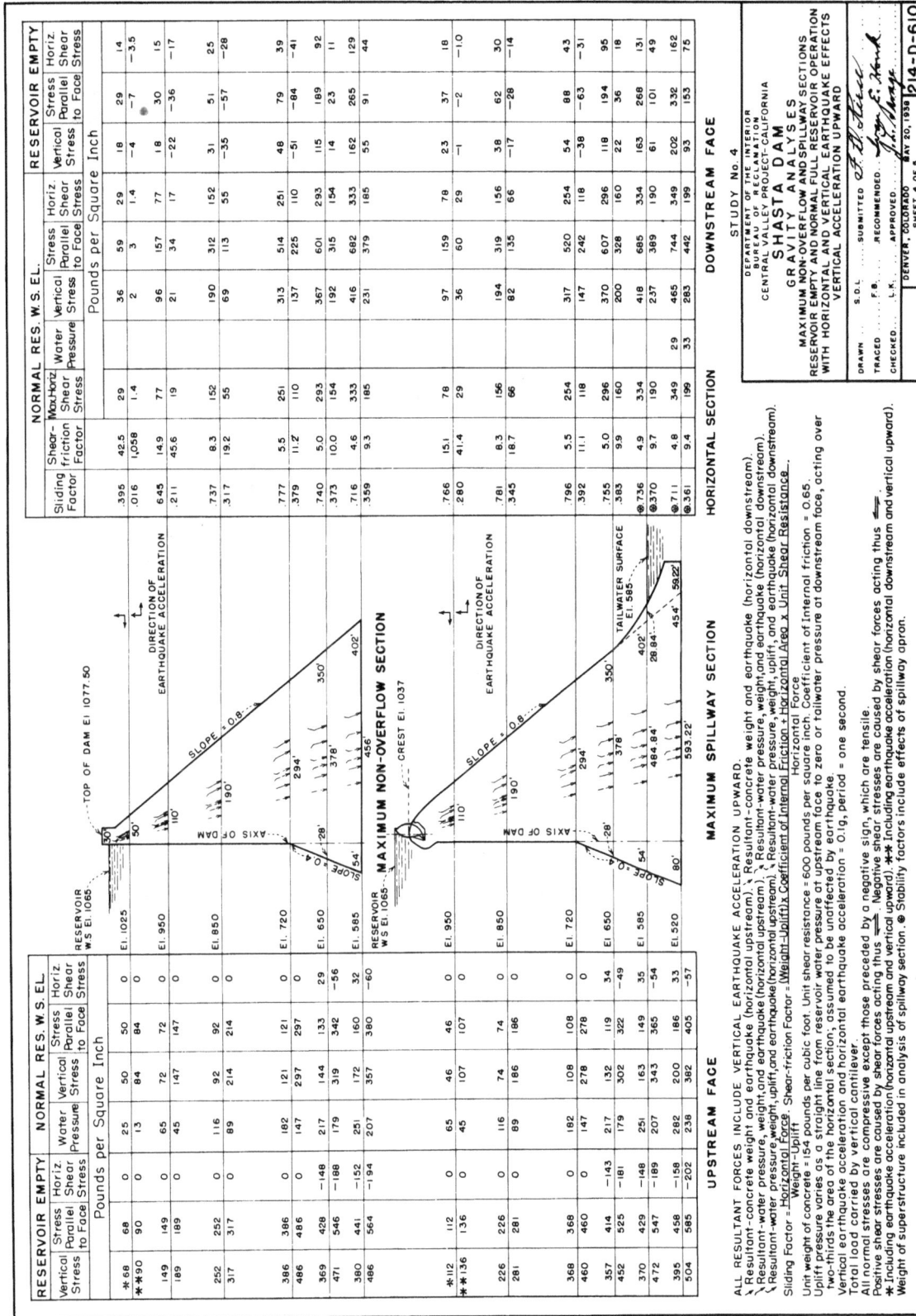

Figure E-28. Shasta Dam—gravity analyses including effects of earthquake, vertical acceleration upward.

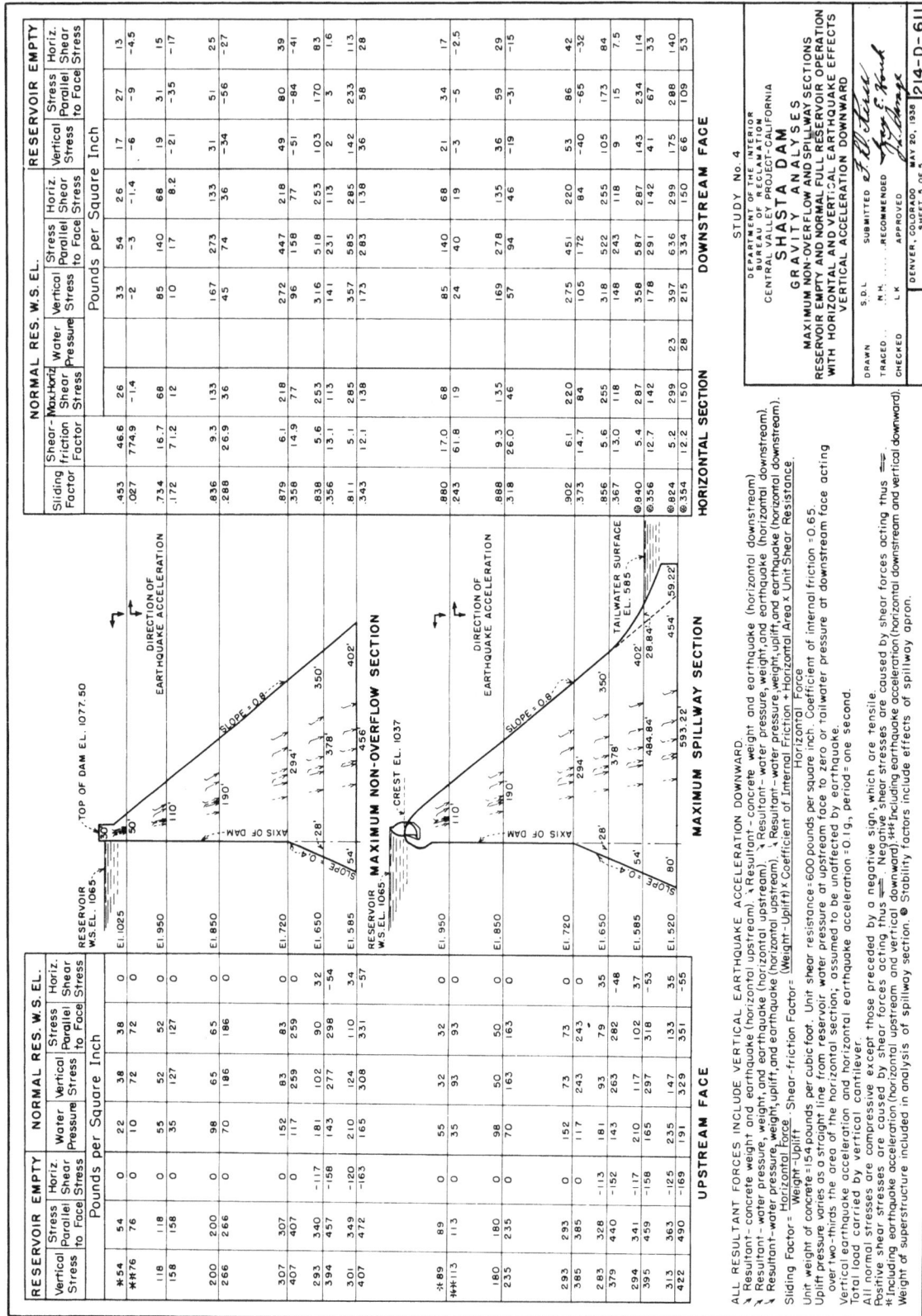

Figure E-29. Shasta Dam—gravity analyses including effects of earthquake, vertical acceleration downward.

TABLE E-1.—*Comparison of stresses and stability factors for 12 dams.* —DS2-2(T2)

		GROUP I				GROUP II			GROUP III			GROUP IV	
		AMERICAN FALLS DAM — SNAKE RIVER, IDAHO	ALTUS DAM — NORTH FORK, RED RIVER, OKLAHOMA	KESWICK — SACRAMENTO RIVER, CALIFORNIA	EAST PARK — LITTLE STONY CREEK, CALIFORNIA	ANGOSTURA — CHEYENNE RIVER, SOUTH DAKOTA	BLACK CANYON — PAYETTE RIVER, IDAHO	KORTES — NORTH PLATTE RIVER, WYO.	FRIANT — SAN JOAQUIN RIVER, CALIFORNIA	MARSHALL FORD — COLORADO RIVER, TEXAS	ELEPHANT BUTTE — RIO GRANDE RIVER, NEW MEXICO	GRAND COULEE — COLUMBIA RIVER, WASHINGTON	SHASTA — SACRAMENTO RIVER, CALIFORNIA
TYPE OF DAM		Straight Gravity	Straight Gravity	Straight Gravity	Curved Gravity	Straight Gravity	Bent Gravity	Straight Gravity	Straight Gravity	Straight Gravity	Straight Gravity	Straight Gravity	Curved Gravity
YEAR COMPLETED		1927	1945	U. C.	1910	U. C.	1925	U C.	1942	1942	1916	1941	1944
MAXIMUM HEIGHT, FT. (ANALYZED SECTION)		77.5	102	130	130	159	170	239	267	268	294	460.5	492.5
CREST LENGTH, FT.		5,227	1,112	1,046	250	980	1,039	440	3,430	5,128	1,674	4,173	3,500
LENGTH-TO-HEIGHT RATIO		67.4	10.9	8.1	1.9	6.2	6.1	1.8	12.8	19.1	5.7	9.1	7.1
THICKNESS AT CROWN, FT.	TOP	8	10	43.2	10.6	10	10.6	20	20	30	18	30	30
	BASE	59.6	72	85	90.6	138.3	136.5	169.4	217.3	221.45	243.78	394.4	593.22
BASE-TO-TOP RATIO		7.5	7.2	2.0	8.5	13.8	12.9	8.5	10.9	7.4	13.5	13.1	19.8
BASE-TO-HEIGHT RATIO		0.8	0.7	0.7	0.7	0.9	0.8	0.7	0.8	0.8	0.8	1.2	1.2
VOLUME OF DAM (CU. YARDS OF CONCRETE)		166,000	70,000	192,000	12,200	224,000	79,000	130,000	2,030,000	1,770,000	605,000	9,790,000	6,440,000
CANTILEVER PROFILE		△	△	◁	△	△	△	◁	◁	◁	◁	◁	◁
CROSS CANYON PROFILE (TRIAL-LOAD ANALYSES)		—	⌣	—	·	◡	—	◠	◡	◡	◡	◡	◡
⊛CRITICAL CANTILEVER STRESS, UPSTREAM FACE	NORMAL LOADING	27 U / 31 P / 0.8 S	30 U / 42 P / 1.2 S		25 U / 54 P / 1.1 S		28 U / 74 P / 16 S		84 U / 114 P / 8 S	141 U / 102 P / -2.3 S	94 U / 125 P / 1.9 S		269 U / 237 P / 1.1 S
	MAXIMUM LOADING	13 U / 37 P / 47 S	7 U / 50 P / 4.2 S	-20 U / 52 P / 0 S	-4 U / 60 U / ‡-19 U / 27 P	50 U / ‡ 64 P / 51 U / 38 P	-18 U / 72 P / 31 S	59 U / 102 P / ‡220 U / 102 P	32 U / 135 P / 28 S	81 U / 100 P / -7.2 S	104 U / 138 P / -10 U / 76 P / 64 S	-131 U / ‡-130 U / 190 P	186 U / 282 P / -57 S
MAXIMUM CANTILEVER STRESS, DOWNSTREAM FACE	NORMAL LOADING	80 D / 34 S	115 D / 49 S		155 D / 72 S		204 D / 68 S		297 D / 140 S	230 D / 112 S	331 D / 149 S		539 D / 249 S
	MAXIMUM LOADING	112 D / 49 S	162 D / 70 S	234 D / 83 S	217 D / 100 S / ‡ 129 D / ‡-200 I	243 D / ‡240 D	272 D / 95 S	287 D / ‡155 D	409 D / 192 S	349 D / 169 S	471 D / 215 S	575 D / ‡578 D	744 D / 349 S
MAXIMUM SLIDING FACTOR	GRAVITY ANALYSIS	0.783	0.832	0.764	1.070	0.924	1.217	0.771	0.999	0.772	1.161	0.805	0.902
	TRIAL-LOAD ANALYSIS	—	—	—	—	†0.931	—	†1.218	—	—	—	†1.38	—
MINIMUM SHEAR-FRICTION FACTOR	GRAVITY ANALYSIS	16.2	11.0	8.25	4.6	5.07	8.4	6.7	5.45	4.8	5.1	5.89	4.8
	TRIAL-LOAD ANALYSIS	—	—	—	—	5.43	—	†7.7	—	—	—	5.86	—
LOADING CONDITIONS, GRAV. ANAL. U.S. FACE	NORMAL LOADING	Res full + T W	Res. full + silt + T.W.	—	Res full w/o T W	—	Res full + T.W.	—	Res full w/o T.W.	Res. full w/o T.W	Res. full + T.W.	—	Res full + T. W.
	MAXIMUM LOADING	Normal + E ↕ or Ice ↗	Normal + E ↗	Normal + E ↕	Normal + E ↕	Normal + E ↗	Normal + E ↓	Max Flood	Normal + E ↕	Normal + E ↓	Normal + E ↕	Normal + E w/o TW	Normal + E ↕
LOADING CONDITIONS, GRAV. ANAL. D.S. FACE	NORMAL LOADING	Res full + T W	Res. full + silt + T W	—	Res. full w/o T W	—	Res full + T. W.	—	Res. full w/o T. W.	Res. full w/o T. W.	Res. full + T. W.	—	Res full + T W
	MAXIMUM LOADING	Normal + E ↕ or Ice ↗	Normal + E ↗	Normal + E ↕	Normal + E ↕	Normal + E ↗	Normal + E ↓	Max Flood + TW	Normal + E ↗	Normal + E ↕	Normal + E ↕	Normal + E w/o TW	Normal + E ↕
MAXIMUM LOADING CONDITIONS, TRIAL-LOAD ANALYSIS		—	—	—	Normal + E + Temp w/o T W	Normal + E + Ice + Earth	—	Max. Flood + T W	—	—	—	Normal + E	—
REFERENCES		Unnumbered Memo. Oct 29, 1940	Unnumbered Memo. Dec 26, 1941	Unnumbered Memo. July 28, 1941	Unnumbered Memo. Aug 25, 1941	Unnumbered Memo. Feb 28, 1947	Tech Memo. 549. Apr 8, 1937	Unnumbered Memo. Sept. 3, 1943	Tech Memo. 612. Sept 21, 1940	Tech Memo 573. May 15, 1938	Unnumbered Memo. June 19, 1940	Tech Memo. 546. Feb 25, 1937	Tech Memo. 575. May 15, 1938

⊛ That stress which is lowest percentage of water pressure at the same point. Maximum compressive and tensile stresses parallel to the face are shown as well as water pressure at the point, if water pressure exceeds stress at face for any given loading condition.

※ Results by Trial-Load Arch and Cantilever Analysis.

‡ Results by Trial-Load Beam and Cantilever Twist Analysis

† Near Abutment

P = Water Pressure

S = Horizontal Shear Stress

I = Intrados Arch Stress

D = Downstream Face

U = Upstream Face

E = Earthquake

T.W. = Tailwater

w/o = Without

U.C. = Under Construction

normal loading conditions sliding factors are considerably smaller and shear-friction factors larger (see figs. E-1 through E-29, and also figs. A-10 through A-14 of app. A). The average maximum sliding factor for the gravity analyses for 12 dams is equal to 0.917 and the minimum shear-friction factor is equal to 7.19.

The maximum effects of twist action in seven gravity dams are shown in table E-2. The most noteworthy effects of twist action on stresses and stability factors obtained by trial-load analysis, as compared with those quantities obtained by gravity analysis, may be summarized briefly as follows:

(1) An increase in sliding factors along the steeper inclined rock planes which form the bases of the cantilevers in the abutment sections.

(2) A decrease in sliding factors in the longer cantilevers whose bases are located in the lower regions of the abutment slopes.

(3) A decrease in shear-friction factor of safety along the steeper inclined rock planes at the abutment cantilevers.

(4) An increase in shear-friction factor of safety at the high cantilevers near the lower ends of the abutment slopes.

(5) Relatively small changes in stresses and stability factors in the longer cantilevers near the central section of the dam where most of the external load is carried by the cantilevers.

(6) A decrease in inclined cantilever compressive stresses along the base of the dam at the downstream edges of the abutment sections and as far toward the center of the structure as appreciable portions of external load may be carried by twist action.

(7) An increase in inclined cantilever compressive stresses along the base of the dam at the upstream edges of the abutment sections and as far toward the center of the structure as appreciable portions of external load may be carried by twist action.

(8) The development of appreciable horizontal compressive stresses at and parallel to the downstream face, decreasing in magnitude from the abutment slopes toward the center of the dam.

(9) The development of appreciable horizontal tensile stresses at and parallel to the upstream face of the dam, with possible resultant cracking, decreasing in magnitude and effect from the abutment slopes toward the center of the dam.

(10) Wherever the deflection curves of the horizontal elements may indicate the possible existence of relatively high tensile stresses, diagonal cracking may occur. This condition may exist especially near the points of contraflexure of horizontal elements in the upper portions of the dam.

It is seen from the above summary that both beneficial and detrimental effects on loads, stresses, and stability factors for straight gravity dams may accrue by twist action. The lateral transfer of load to the abutments causes some reduction in load on the high cantilevers at the lower ends of the abutment slopes. However, the beneficial results of such reductions are usually of minor importance in comparison with the detrimental effects of load increases on the shorter end cantilevers. In some cases, sliding factors at the bases of these shorter cantilevers are increased to more than unity; hence the sections theoretically would move downstream if they were not held in place by the shear resistance and weight of the mass of the dam. Fortunately, shear-friction factors of safety at the bases of gravity sections increase as the heights of the sections decrease. Consequently, the shear resistance at the bases of the shorter end cantilevers is usually great enough to prevent failure even though the sliding factor in these regions may be greater than unity.

Theoretically, it may sometimes be possible to save concrete by reducing slightly the thickness of the cross section at regions where twist action is indicated to be beneficial. In practice, however, it is usually desirable to keep the slopes of the faces constant throughout the length of the dam for economy of construction. Another reason for not making reductions in cross section to allow for

TABLE E-2.—*Maximum effects of twist action in some gravity dams with principal dimensions of twisted structure.*—DS2-2(T3)

GENERAL DIMENSIONS AND DATA

ITEM		Madden	Norris	O'Shaughnessy	Grand Coulee	Friant	Marshall Ford	Davis
Location		Panama Canal Zone	Tennessee	California	Washington	California	Texas	Ariz.-Nevada
River		Chagres	Clinch	Tuolumne	Columbia	San Joaquin	Lower-Colo.	Lower-Colo.
Maximum height of twisted section		210	260	382	458	267	268	153
Length of twisted section		950	1580	850	4118	3390	2700	402
Width at top of dam		22	20	27.5	30	20	30	32
Width at base of maximum section		176	210	308	394	220	216	110
Upstream projection at base		12	15	16	26	17	11	0
Downstream projection at base		144	175	264	338	183	175	78
Loading condition analyzed		Full Reservoir	Full Reservoir	Full Reservoir +Earthquake	Full Reservoir +Earthquake	Full Reservoir +Earthquake	Full Reservoir +Earthquake	Full Reservoir +Earthquake
SLIDING FACTORS	Maximum increase at any position	0.40-0.75	0.44-0.69	0.43-1.49	0.35-0.48 / 0.42-1.19	0.80-0.84 / 0.84-0.93	0.73-0.74 / 0.39-0.68	0.07-3.84
	Maximum decrease at any position	0.90-0.62	0.64-0.37	0.87-0.46	0.79-0.41 / 0.79-0.55	0.88-0.82 / 0.88-0.79	0.66-0.48 / 0.66-0.49	0.50-0.29
SHEAR-FRICTION FACTOR	Maximum increase at any position			22.7-37.4	32.4-54.2 / 24.3-31.5	5.9-6.3 / 7.2-7.9	310-341 / 5.9-8.0	30.1-77.6
	Maximum decrease at any position			37.7-10.9	174-128 / 76.5-52.0	18.3-17.8 / 10.5-9.2	12.7-12.5 / 92.1-72.9	129.4-8.7
CANTILEVER STRESSES COMPRESSION	Maximum increase at upstream face	66-94	81-123	1-126 / 1-228	126-217 / 130-278	31-37 / 22-53	98-124 / 101-161	-11-140
	Maximum decrease at downstream face	198-156	224-163	508-279 / 508-93	511-366 / 520-282	382-374 / 318-274	220-180 / 286-197	292-75
Remarks:				Designed as Gravity Dam Radius 700 ft				Concrete Gravity Penstock Section

Notes: Figures above line - Joints ungrouted. Figures below line - Joints grouted. Dimensions in feet, Stresses in p.s.i., Stresses act parallel to face.

the effects of beneficial twist action is that effects of nonlinear distribution of stress throughout the sections would probably overshadow the beneficial effects of twist action.

Hydraulic Data and Tables

F-1. *Lists of Symbols and Conversion Factors.*—The following list includes symbols used in hydraulic formulas given in chapters IX and X and in this appendix. Standard mathematical notations and symbols having only very limited applications have been omitted.

Symbol	Description
A, a	An area; area of a surface; cross-sectional area of flow in an open channel; cross-sectional area of a closed conduit
a_g	Gross area of a trashrack
a_n	Net area of a trashrack
b	Bottom width of a channel
C	A coefficient; coefficient of discharge
C_d	Coefficient of discharge through an orifice
C_i	Coefficient of discharge for an ogee crest with inclined upstream face
C_o	Coefficient of discharge for a nappe-shaped ogee crest designed for an H_o head
C_s	Coefficient of discharge for a partly submerged crest
D	Diameter; conduit diameter; height of a rectangular conduit or passageway; height of a square or rectangular orifice
d	Depth of flow in an open channel; height of an orifice or gate opening
d_c	Critical depth
d_H	Depth for high (subcritical) flow stage (alternate to d_L)
d_j	Height of a hydraulic jump (difference in the conjugate depths)
d_L	Depth for low (supercritical) flow stage (alternate to d_H)
d_m	Mean depth of flow
d_{m_c}	Critical mean depth
d_n	Depth of flow measured normal to channel bottom

Symbol	Description
d_s	Depth of scour below tailwater in a plunge pool
d_t	Depth of flow in a chute at tailwater level
E	Energy
E_m	Energy of a particle of mass
F	Froude number parameter for defining flow conditions in a channel, $F = \dfrac{v}{\sqrt{gd}}$
F_t	Froude number parameter for flow in a chute at the tailwater level
f	Friction loss coefficient in the Darcy-Weisbach formula $h_f = \dfrac{fL}{D}\dfrac{v^2}{2g}$
g	Acceleration due to the force of gravity
H	Head over a crest; head on center of an orifice opening; head difference at a gate (between the upstream and downstream water surface levels)
H_A	Absolute head above a datum plane, in channel flow
H_a	Head above a section in the transition of a drop inlet spillway
H_1	Head measured to bottom of an orifice opening
H_2	Head measured to top of an orifice opening
h	Head; height of baffle block; height of end sill
h_a	Approach velocity head
h_b	Head loss due to bend
h_c	Head loss due to contraction
H_D	Head from reservoir water surface to water surface at a given point in the downstream channel
h_d	Difference in water surface level, measured from reservoir water surface to the downstream channel water surface
H_E	Specific energy head
H_{E_C}	Specific energy head at critical flow
H_e	Total head on a crest, including velocity of approach

Symbol	Description
h_e	Head loss due to entrance
h_{ex}	Head loss due to expansion
h_f	Head loss due to friction
Δh_f	Incremental head loss due to friction
h_g	Head loss due to gates or valves
h_L	Head losses from all causes
Σh_{L_u}	Sum of head losses upstream from a section
Δh_L	Incremental head loss from all causes
$\Sigma(\Delta h_L)$	Sum of incremental head losses from all causes
H_o	Design head over ogee crest
h_o	Head measured from the crest of an ogee to the reservoir surface immediately upstream, not including the velocity of approach (crest shaped for design head H_o)
H_s	Total head over a sharp-crested weir
h_s	Head over a sharp-crested weir, not including velocity of approach
H_T	Total head from reservoir water surface to tailwater, or to center of outlet of a free-discharging pipe
h_t	Head loss due to trashrack
h_v	Velocity head; head loss due to exit
h_{v_c}	Critical velocity head
K	A constant factor for various equations; a .coefficient
k	A constant
K_a	Abutment contraction coefficient
K_b	Bend loss coefficient
K_c	Contraction loss coefficient
K_e	Entrance loss coefficient
K_{ex}	Expansion loss coefficient
K_g	Gate or valve loss coefficient
K_L	A summary loss coefficient for losses due to all causes
K_p	Pier contraction coefficient
K_t	Trashrack loss coefficient
K_v	Velocity head loss coefficient
L	Length; length of a channel or a pipe; effective length of a crest; length of a hydraulic jump; length of a stilling basin; length of a transition
ΔL	Incremental length; incremental channel length
L_I, L_{II}, L_{III}	Stilling basin lengths for different hydraulic jump stilling basins
L'	Net length of a crest
M	Momentum
M_d	Momentum in a downstream section
M_u	Momentum in an upstream section
ΔM	Difference in momentum between successive sections

Symbol	Description
m	Mass
N	Number of piers on an overflow crest; number of slots in a slotted grating dissipator
n	Exponential constant used in equation for defining crest shapes; coefficient of roughness in the Manning equation
P	Approach height of an ogee weir, hydrostatic pressure of a water prism cross section
p	Unit pressure intensity; unit dynamic pressure on a spillway floor; wetted perimeter of a channel or conduit cross section
Q	Discharge; volume rate of flow
ΔQ	Incremental change in rate of discharge
q	Unit discharge
Q_c	Critical discharge
q_c	Critical discharge per unit of width
Q_i	Average rate of inflow
Q_o	Average rate of outflow
R	Radius; radius of a cross section; crest profile radius; vertical radius of curvature of the channel floor profile; radius of a terminal bucket profile
r	Hydraulic radius; radius of abutment rounding
R_b	Radius of a bend in a channel or pipe
R_s	Radius of a circular sharp-crested weir
S	Storage
ΔS	Incremental storage
s	Friction slope in the Manning equation; spacing
s_b	Slope of the channel floor, in profile
s_{ws}	Slope of the water surface
T	Tailwater depth; width at the water surface in a cross section of an open channel
T_{max}	Limiting maximum tailwater depth
T_{min}	Limiting minimum tailwater depth
t	Time
Δt	Increment of time
T_s	Tailwater sweep-out depth
$T.W.$	Tailwater; tailwater depth
U	A parameter for defining flow conditions in a closed waterway, $U = \dfrac{v}{\sqrt{gD}}$
v	Velocity
Δv	Incremental change in velocity
v_a	Velocity of approach
v_c	Critical velocity
v_t	Velocity of flow in a channel or chute, at tailwater depth
W	Weight of a mass; width of a stilling basin
w	Unit weight of water; width of chute and baffle blocks in a stilling basin

Symbol	Description
x	A coordinate for defining a crest profile; a coordinate for defining a channel profile; a coordinate for defining a conduit entrance
Δx	Increment of length
x_c	Horizontal distance from the break point, on the upstream face of an ogee crest, to the apex of the crest
x_s	Horizontal distance from the vertical upstream face of a circular sharp-crested weir to the apex of the undernappe of the overflow sheet
Y	Drop distance measured from the crest of the overflow to the basin floor, for a free overfall spillway
y	A coordinate for defining a crest profile; a coordinate for defining a channel profile; a coordinate for defining a conduit entrance
$\bar{y}$	Depth from water surface to the center of gravity of a water prism cross section
Δy	Difference in elevation of the water surface profile between successive sections in a side channel trough
y_c	Vertical distance from the break point, on the upstream face of an ogee crest, to the apex of the crest
y_s	Vertical distance from the crest of a circular sharp-crested weir to the apex of the undernappe of the overflow sheet
Z	Elevation above a datum plane
ΔZ	Elevation difference of the bottom profile between successive sections in an open channel
z	Ratio, horizontal to vertical, of the slope of the sides of a channel cross section
α	A coefficient; angular variation of the side wall with respect to the structure centerline
β	Deflection angle of bend in a conduit
θ	Angle from the horizontal; angle from vertical of the position of an orifice; angle from the horizontal of the edge of the lip of a deflector bucket

Table F-1 presents conversion factors most frequently used by the designer of concrete dams to convert from one set of units to another—for example, to convert from cubic feet per second to acre-feet. Also included are some basic conversion formulas such as the ones for converting flow for a given time to volume.

F-2. Flow in Open Channels.—(a) *Energy and Head.*—If it is assumed that streamlines of flow in an open channel are parallel and that velocities at all points in a cross section are equal to the mean velocity v, the energy possessed by the water is made up of two parts: kinetic (or motive) energy and potential (or latent) energy. Referring to figure F-1, if W is the weight of a mass m, the mass possesses Wh_2 foot-pounds of energy with reference to the datum. Also, it possesses Wh_1 foot-pounds of energy because of the pressure exerted by the water above it. Thus, the potential energy of the mass m is $W(h_1 + h_2)$. This value is the same for each particle of mass in the cross section. Assuming uniform velocity, the kinetic energy of m is $W\left(\dfrac{v^2}{2g}\right)$.

Thus, the total energy of each mass particle is

$$E_m = W\left(h_1 + h_2 + \frac{v^2}{2g}\right) \qquad (1)$$

Applying the above relationship to the whole discharge Q of the cross section in terms of the unit weight of water w,

$$E = Qw\left(d + Z + \frac{v^2}{2g}\right) \qquad (2)$$

where E is total energy per second at the cross section.

The portion of equation (2) in the parentheses is termed the absolute head, and is written:

$$H_A = d + Z + \frac{v^2}{2g} \qquad (3)$$

Equation (3) is called the Bernoulli equation.

The energy in the cross section, referred to the bottom of the channel, is termed the specific energy. The corresponding head is referred to as the specific energy head and is expressed as:

$$H_E = d + \frac{v^2}{2g} \qquad (4)$$

Where $Q = av$, equation (4) can be stated:

TABLE F-1.—*Conversion factors and formulas.*—288-D-3199(1/2)

> To reduce units in column 1 to units in column 4, multiply column 1 by column 2
> To reduce units in column 4 to units in column 1, multiply column 4 by column 3

CONVERSION FACTORS				CONVERSION FACTORS			
Column 1	Column 2	Column 3	Column 4	Column 1	Column 2	Column 3	Column 4
LENGTH				**FLOW**			
In............	2.54	0.3937	Cm.	Cu. ft./sec. (c.f.s.) (second-feet) (sec.-ft.).	60.0	0.016667	Cu. ft./min.
	0.0254	39.37	M.		86,400.0	.11574×10⁻⁴	Cu. ft./day.
Ft............	0.3048	3.2808	M.		31.536×10⁶	.31709×10⁻⁷	Cu. ft./yr.
Miles...........	1.609	0.621	Km.		448.83	.2228×10⁻²	Gal./min.
					646,317.0	.15472×10⁻⁵	Gal./day.
AREA					1.98347	.50417	Acre-ft./day.
Sq. in...........	6.4516	0.1550	Sq. cm.		723.98	.13813×10⁻²	Acre-ft./365 days.
Sq. m...........	10.764	.0929	Sq. ft.		725.78	.13778×10⁻²	Acre-ft./366 days.
Sq. miles.........	27.8784×10⁶	0.3587×10⁻⁷	Sq. ft.		55.54	.018005	Acre-ft./28 days.
	640.0	.15625×10⁻²	Acres (1 section).		57.52	.017385	Acre-ft./29 days.
	30.976×10⁵	.3228×10⁻⁶	Sq. yd.		59.50	.016806	Acre-ft./30 days.
	2.59	.386	Sq. km.		61.49	.016262	Acre-ft./31 days.
Acre............	43,560.0	0.22957×10⁻⁴	Sq. ft.		50.0	.020	Miner's inch in Idaho, Kans., Nebr., N. Mex., N. Dak., S. Dak., and Utah.
	4,046.9	.2471×10⁻³	Sq. m.		40.0	.025	Miner's inch in Ariz., Calif., Mont., Nev., and Oreg.
	4,840.0	.2066×10⁻³	Sq. yd.		38.4	.026042	Miner's inch in Colo.
VOLUME					35.7	.028011	Miner's inch in British Columbia.
Cu. ft...........	1,728.0	0.5787×10⁻³	Cu. in.		0.028317	35.31	Cu. m./sec.
	7.4805	.13368	Gal.		1.699	.5886	Cu. m./min.
	6.2321	.16046	Imperial gal.		0.99173	1.0083	Acre-in./hr.
Cu. m...........	35.3145	0.028317	Cu. ft.	Cu. ft./min...........	7.4805	0.13368	Gal./min.
	1.3079	.76456	Cu. yd.		10,772.0	.92834×10⁻⁴	Gal./day.
Gal............	231.0	0.4329×10⁻²	Cu. in.	10⁶ gal./day...........	1.5472	0.64632	C.f.s.
	3.7854	.26417	Liters.		694.44	.1440×10⁻²	Gal./min.
Million gal.......	133,681.0	0.74805×10⁻⁵	Cu. ft.		3.0689	.32585	Acre-ft./day.
	3.0689	.32585	Acre-ft.	In. depth/hr........	645.33	0.15496×10⁻²	C.f.s./sq. mile.
Imperial gal......	1.2003	0.83311	Gal.	In. depth/day.......	26.889	0.03719	C.f.s./sq. mile.
Acre-in...........	3,630.0	.27548×10⁻³	Cu. ft.		53.33	.01878	Acre-ft./sq. mile.
Acre-ft...........	1,233.5	0.81071×10⁻³	Cu. m.	C.f.s./sq. mile......	1.0413	0.96032	In. depth/28 days.
	43,560.0	.22957×10⁻⁴	Cu. ft.		1.0785	.92720	In. depth/29 days.
In. on 1 sq. mile..	232.32×10⁴	0.43044×10⁻⁶	Cu. ft.		1.1157	.89630	In. depth/30 days.
	53.33	.01875	Acre-ft.		1.1529	.86738	In. depth/31 days.
Ft. on 1 sq. mile..	278.784×10⁵	0.3587×10⁻⁷	Cu. ft.		13.574	.073668	In. depth/365 days.
	640.0	.15625×10⁻²	Acre-ft.		13.612	.073467	In. depth/366 days.
VELOCITY AND GRADE				Acre-ft./day........	226.24	0.442×10⁻²	Gal./min.
Miles/hr..........	1.4667	0.68182	Ft./sec.		20.17	.0496	Miner's inch in Calif.
M./sec...........	3.2808	.3048	Ft./sec.		19.36	.0517	Miner's inch in Colo.
	2.2369	.44704	Miles/hr.	Gal./sec............	5.347	0.187	Miner's inch in Calif.
Fall in ft./mile....	189.39×10⁻⁶	5.28×10³	Fall/ft.		5.128	.195	Miner's inch in Colo.
				PERMEABILITY			
				Meinzer (gal./day through 1 sq. ft. under unit gradient).	48.8	0.02049	Bureau of Reclamation (cu. ft./yr. through 1 sq. ft. under unit gradient).

TABLE F-1.—*Conversion factors and formulas.*—Continued.—288-D-3199(2/2)

CONVERSION FACTORS				FORMULAS
Column 1	Column 2	Column 3	Column 4	VOLUME

POWER AND ENERGY

	Column 2	Column 3	Column 4
Hp	555.0	0.18182×10^{-2}	Ft.-lb./sec.
	0.746	1.3405	Kw.
	6,535.	0.15303×10^{-3}	Kw.-hr./yr.
	42.4	.0236	B.t.u./min.
	1.0	1.0	C.f.s. falling 8.8 ft.
Hp.-hr	0.746	1.3405	Kw.-hr.
	198.0×10^{4}	0.505×10^{-6}	Ft.-lb.
	2.545.0	$.393 \times 10^{-3}$	B.t.u.
Kw	8,760.0	0.11416×10^{-3}	Kw.-hr./yr.
	737.56	$.1354 \times 10^{-2}$	Ft.-lb./sec.
	11.8	.0846	C.f.s. falling 1 ft.
	3,412.0	$.29308 \times 10^{-3}$	B.t.u./hr.
Kw.-hr	0.975	1.025	Acre-ft. falling 1 ft.
B.t.u	778.0	0.1285×10^{-2}	Ft.-lb.
	0.1×10^{-3}	10,000	Lb. of coal.
	to	to	
	$.834 \times 10^{-4}$	12,000	

PRESSURE

	Column 2	Column 3	Column 4
Ft. water at max. density	62.425	0.01602	Lb./sq./ft.
	0.4335	2.3087	Lb./sq. in.
	.0295	33.93	Atm.
	.8826	1.133	In. Hg at 30° F.
	773.3	0.1293×10^{-2}	Ft. air at 32° F. and atm. pressure.
Ft. avg. sea water	1.026	0.9746	Ft. pure water.
Atm., sea level, 32° F	14.697	.06804	Lb./sq. in.
Millibars	295.299×10^{-4}	33.863	In. Hg.
	75.008×10^{-2}	1.3331	Mm. Hg.
Atm	29.92	33.48×10^{-3}	In. Hg.

WEIGHT

	Column 2	Column 3	Column 4
P.p.m.	0.00136	735.29	Tons/acre-ft.
	.0584	17.123	Gr./gal.
	8.345	0.1198	Lb./10⁶ gal.
Lb.	7.0×10^{3}	0.14286×10^{-3}	Gr.
Gm.	15.432	.064799	Gr.
Kg.	2.2046	.45359	Lb.
Lb. water at 39.1° F	27.6812	0.03612	Cu. in.
	0.11983	8.345	Gal.
	.09983	10.016	Imperial gal.
	.453617	2.204	Liters.
	.01602	62.425	Cu. ft. pure water.
	.01560	64.048	Cu. ft. sea water.
Lb. water at 62° F	0.01604	62.355	Cu. ft. pure water.
	.01563	63.976	Cu. ft. sea water.

FORMULAS

VOLUME

Average depth in inches, or acre-inch per acre

$$= \frac{(\text{c.f.s.}) \ (\text{hr.})}{\text{acres}}$$

$$= \frac{(\text{gal./min.}) \ (\text{hr.})}{450 \ (\text{acres})}$$

$$= \frac{(\text{miner's in.}) \ (\text{hr.})}{(40^*) \ (\text{acres})}$$

*Where 1 miner's in. = 1/40 c.f.s.

Use 50 where 1 miner's in. = 1/50 c.f.s.

Conversion of inches depth on area to c.f.s.

$$\text{c.f.s.} = \frac{(645) \ (\text{sq. miles}) \ (\text{in. on area})}{(\text{time in hr.})}$$

POWER AND ENERGY

$$\text{hp.} = \frac{(\text{c.f.s.}) \ (\text{head in ft.})}{8.8}$$

$$= \frac{(\text{c.f.s.}) \ (\text{pressure in lb./sq. in.})}{3.8}$$

$$= \frac{(\text{gal./min.}) \ (\text{head in ft.})}{3,960}$$

$$= \frac{(\text{gal./min.}) \ (\text{pressure in lb./sq. in.})}{1,714}$$

$$\text{b. hp.} = \frac{\text{water hp.}}{\text{pump efficiency}}$$

kw.-hr./1,000 gal. pumped/hr.

$$= \frac{(\text{head in ft.}) \ (0.00315)}{(\text{pump efficiency}) \ (\text{motor efficiency})}$$

Kw.-hr. = (plant efficiency) (1.025) (head in ft.) (water in acre-ft.)

$$\text{Load factor} = \frac{(\text{kw.-hr. in time } t)}{(\text{kw. peak load}) \ (\text{time } t \text{ in hr.})}$$

SEDIMENTATION

Tons/acre-ft. = (unit weight/cu. ft.) (21.78)

Tons/day = (c.f.s.) (p.p.m.) (0.0027)

TEMPERATURE

$$^\circ C. = \frac{5}{9} (^\circ F. - 32^\circ) \qquad ^\circ F. = \frac{9}{5} \ ^\circ C. + 32^\circ$$

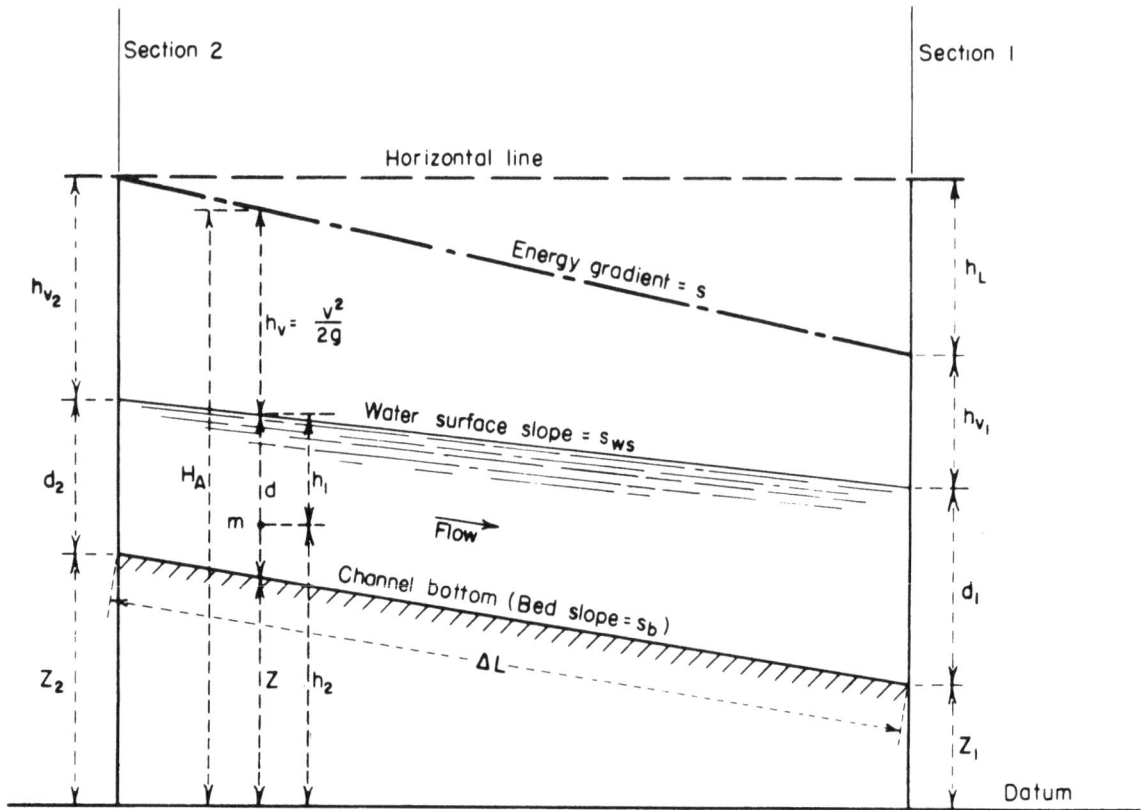

Figure F-1. **Characteristics of open-channel flow.—288-D-2550**

$$H_E = d + \frac{Q^2}{2ga^2} \qquad (5)$$

For a trapezoidal channel where b is the bottom width and z defines the side slope, if q is expressed as $\frac{Q}{b}$ and a is expressed $d(b + zd)$, equation (5) becomes:

$$H_E = d + \frac{q^2}{2gd^2 \left(1 + \dfrac{zd}{b}\right)^2} \qquad (6)$$

Equation (5) is represented in diagrammatic form on figure F-2 to show the relationships between discharge, energy, and depth of flow in an open channel. The diagram is drawn for several values of unit discharge in a rectangular channel.

It can be seen that there are two values of d, d_H, and d_L for each value of H_E, except at the point where H_E is minimum, where only a

single value exists. The depth at energy $H_{E_{min}}$ is called the critical depth, and the depths for other values of H_E are called alternate depths. Those depths lying above the trace through the locus of minimum depths are in the subcritical flow range and are termed subcritical depths, while those lying below the trace are in the supercritical flow range and are termed supercritical depths.

Figure F-3 plots the relationships of d to H_E as stated in equation (6), for various values of unit discharge q and side slope z. The curves can be used to quickly determine alternate depths of flow in open channel spillways.

(b) *Critical Flow.*—Critical flow is the term used to describe open channel flow when certain relationships exist between specific energy and discharge and between specific energy and depth. As indicated in section F-2(a) and as demonstrated on figure F-2, critical flow terms can be defined as follows:

(1) *Critical discharge.*—The maximum

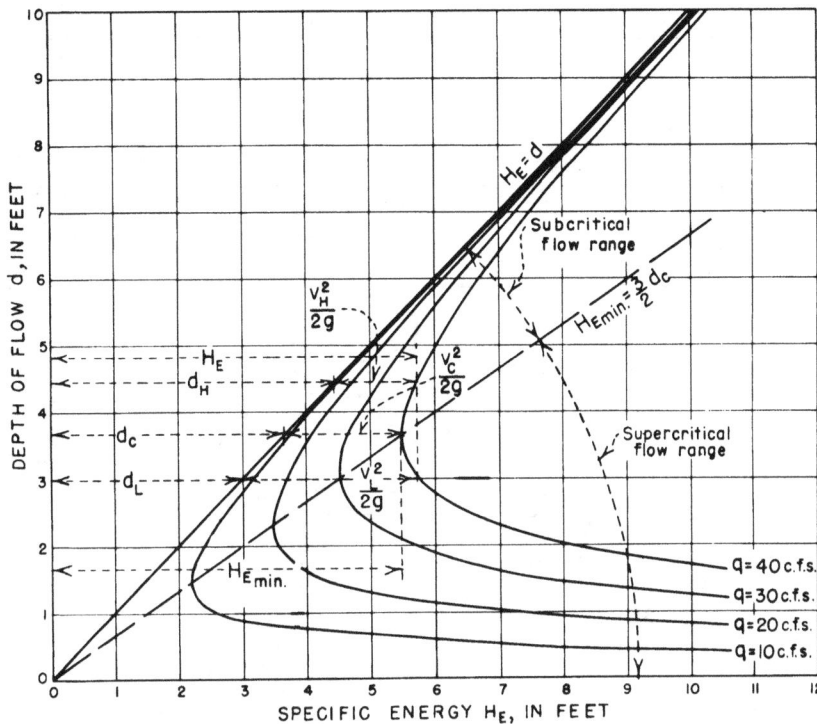

$H_E = d + \dfrac{v^2}{2g} = d + \dfrac{q^2}{2gd^2}$ where q = discharge per unit width.

$d_c = \left(\dfrac{q_c}{\sqrt{g}}\right)^{\frac{2}{3}} = \dfrac{2}{3} H_{E_{min.}}$ where d_c = critical depth

q_c = critical discharge per unit width

$H_{E_{min.}}$ = minimum energy content.

Figure F-2. **Depth of flow and specific energy for rectangular section in open channel.—288-D-2551**

discharge for a given specific energy, or the discharge which will occur with minimum specific energy.

(2) *Critical depth.*—The depth of flow at which the discharge is maximum for a given specific energy, or the depth at which a given discharge occurs with minimum specific energy.

(3) *Critical velocity.*—The mean velocity when the discharge is critical.

(4) *Critical slope.*—That slope which will sustain a given discharge at uniform critical depth in a given channel.

(5) *Subcritical flow.*—Those conditions of flow for which the depths are greater than critical and the velocities are less than critical.

(6) *Supercritical flow.*—Those conditions of flow for which the depths are less than critical and the velocities are greater than critical.

More complete discussions of the critical flow theory in relationship to specific energy are given in most hydraulic textbooks [1, 2, 3, 4, 5].[1] The relationship between cross section and discharge which must exist in order that flow may occur at the critical stage is:

$$\frac{Q^2}{g} = \frac{a^3}{T} \qquad (7)$$

where:

a = cross-sectional area in square feet, and
T = water surface width in feet.

---
[1]Numbers in brackets refer to items in the bibliography, sec. F-5.

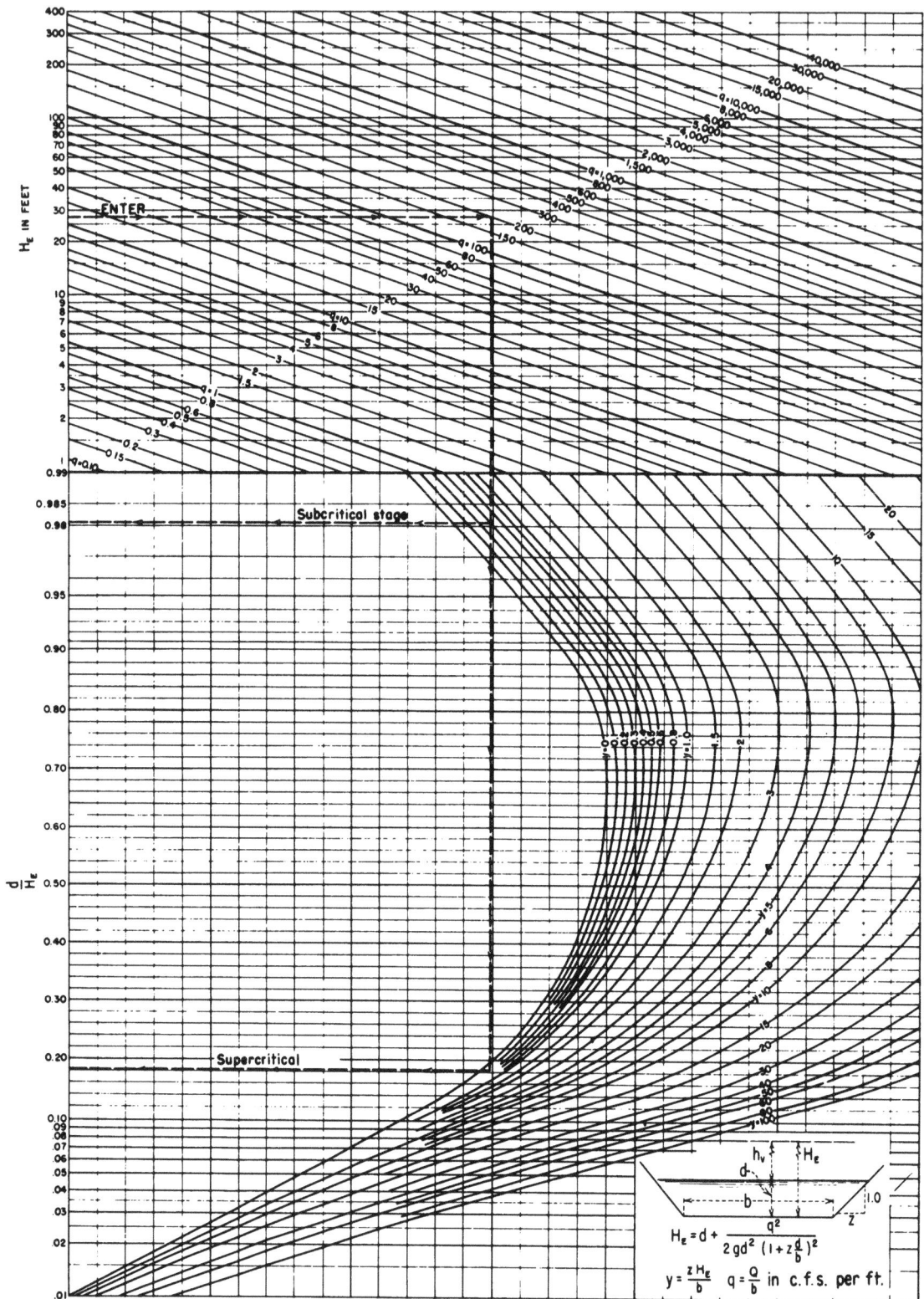

Figure F-3. Energy-depth curves for rectangular and trapezoidal channels.—288-D-3193

Since $Q^2 = a^2 v^2$, equation (7) can be written:

$$\frac{v_c^2}{2g} = \frac{a}{2T} \tag{8}$$

Also, since $a = d_m T$, where d_m is the mean depth of flow at the section, and $\dfrac{v_c^2}{2g} = h_{v_c}$, equation (8) can be rewritten:

$$h_{v_c} = \frac{d_{m_c}}{2} \tag{9}$$

Then equation (4) can be stated

$$H_E = d_c + \frac{d_{m_c}}{2} \tag{10}$$

From the foregoing, the following additional relations can be stated:

$$d_{m_c} = \frac{v_c^2}{g} \tag{11}$$

$$d_{m_c} = \frac{Q_c^2}{a^2 g} \tag{12}$$

$$v_c = \sqrt{g d_{m_c}} \tag{13}$$

$$v_c = \sqrt{\frac{ag}{T}} = 5.67\sqrt{\frac{a}{T}} \tag{14}$$

$$Q_c = a\sqrt{g d_{m_c}} \tag{15}$$

For rectangular sections, if q is the discharge per foot width of channel, the various critical flow formulae are:

$$H_{E_c} = \frac{3}{2} d_c \tag{16}$$

$$d_c = \frac{2}{3} H_{E_c} \tag{17}$$

$$d_c = \frac{v_c^2}{g} \tag{18}$$

$$d_c = \sqrt[3]{\frac{q_c^2}{g}} \tag{19}$$

$$d_c = \sqrt[3]{\frac{Q_c^2}{b^2 g}} \tag{20}$$

$$v_c = \sqrt{g d_c} \tag{21}$$

$$v_c = \sqrt[3]{g q_c} \tag{22}$$

$$v_c = \sqrt[3]{\frac{g Q_c}{b}} \tag{23}$$

$$q_c = d_c^{3/2} \sqrt{g} \tag{24}$$

$$Q_c = 5.67 b d_c^{3/2} \tag{25}$$

$$Q_c = 3.087 b H_{E_c}^{3/2} \tag{26}$$

The critical depth for trapezoidal sections is given by the equation:

$$d_c = \frac{v_c^2}{g} - \frac{b}{2z} + \sqrt{\frac{v_c^4}{g^2} + \frac{b^2}{4z^2}} \tag{27}$$

where z = the ratio, horizontal to vertical, of the slope of the sides of the channel.

Similarly, for the trapezoidal section,

$$v_c = \sqrt{\left(\frac{b + z d_c}{b + 2 z d_c}\right) d_c g} \tag{28}$$

and

$$Q_c = d_c^{3/2} \sqrt{\frac{g(b + z d_c)^3}{b + 2 z d_c}} \tag{29}$$

The solutions of equations (25) and (29) are simplified by use of figure F-4.

(c) *Manning Formula.*—The formula developed by Manning for flow in open channels is used in most of the hydraulic analyses discussed in this text. It is a special form of Chezy's formula; the complete development is contained in most textbooks on elementary fluid mechanics. The formula is written as follows:

Example No. 1
Q_C = 900 c.f.s.
Bottom width "b" = 12'

Side slope	Critical depth "d_C" (feet)
2·1	4.4
Vertical	5.6

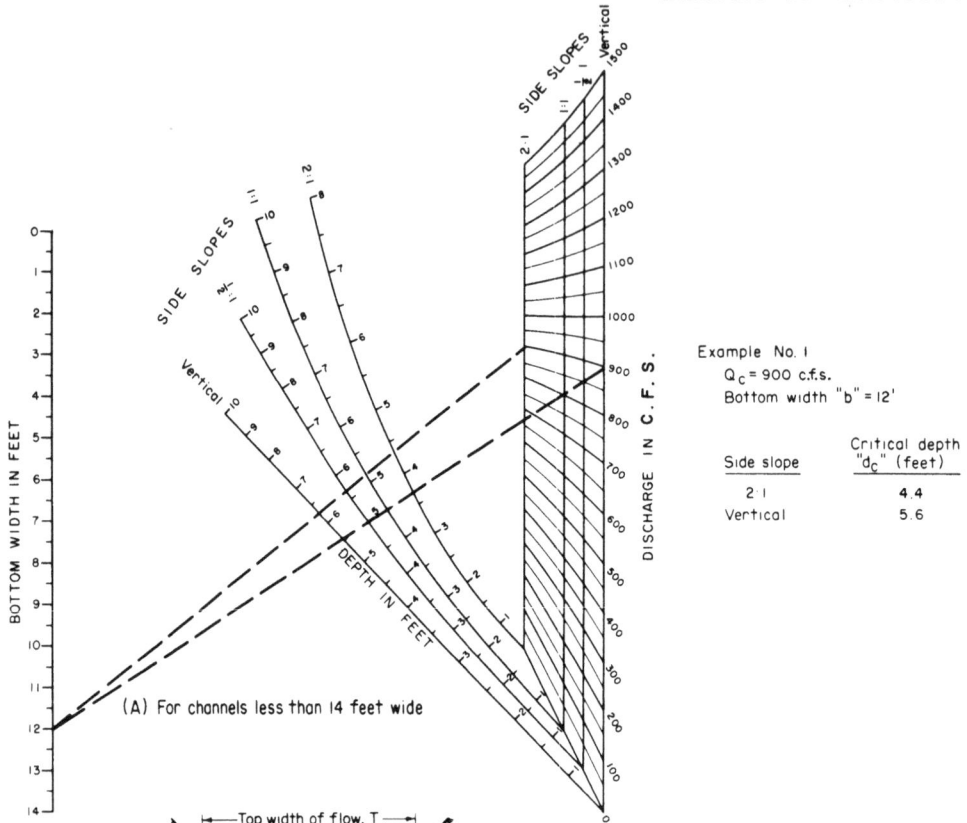

(A) For channels less than 14 feet wide

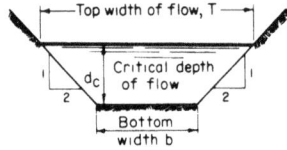

Top width of flow, T

Critical depth
d_C of flow

Bottom
width b

Chart gives values of d_C for known values of Q_C in the relationship $Q_C = (\frac{A^3 g}{T})^{\frac{1}{2}}$ Single solution line gives relationship between Q_C, b, z, and d_C as shown

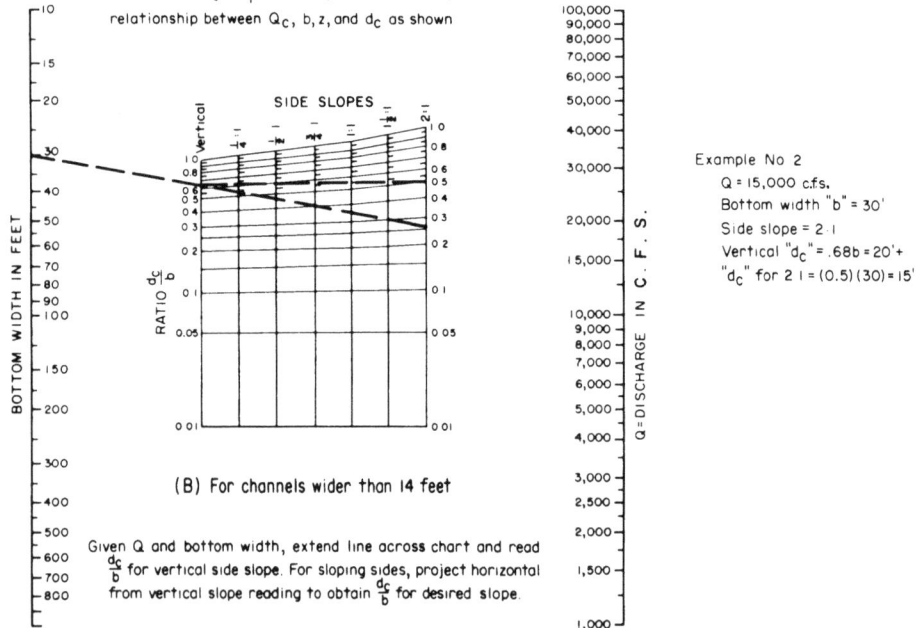

SIDE SLOPES

Example No 2
Q = 15,000 c.f.s.
Bottom width "b" = 30'
Side slope = 2·1
Vertical "d_C" = .68b = 20'+
"d_C" for 2·1 = (0.5)(30) = 15'

(B) For channels wider than 14 feet

Given Q and bottom width, extend line across chart and read $\frac{d_C}{b}$ for vertical side slope. For sloping sides, project horizontal from vertical slope reading to obtain $\frac{d_C}{b}$ for desired slope.

Figure F-4. Critical depth in trapezoidal section.—288-D-3194

$$v = \frac{1.486}{n} r^{2/3} s^{1/2} \qquad (30)$$

or

$$Q = \frac{1.486}{n} a r^{2/3} s^{1/2} \qquad (31)$$

where:

Q = discharge in cubic feet per second (c.f.s.),

a = the cross section of flow area in square feet,

v = the velocity in feet per second,

n = a roughness coefficient,

r = the hydraulic radius

$$= \frac{\text{area } (a)}{\text{wetted perimeter } (p)}, \text{ and}$$

s = the slope of the energy gradient.

The value of the roughness coefficient, n, varies according to the physical roughness of the sides and bottom of the channel and is influenced by such factors as channel curvature, size and shape of cross section, alinement, and type and condition of the material forming the wetted perimeter.

Values of n commonly used in the design of artificial channels are as follows:

Description of channel	Values of n		
	Minimum	Maximum	Average
Earth channels, straight and uniform	0.017	0.025	0.0225
Dredged earth channels	.025	.033	.0275
Rock channels, straight and uniform	.025	.035	.033
Rock channels, jagged and irregular	.035	.045	.045
Concrete lined	.012	.018	.014
Neat cement lined . .	.010	.013	. . . . . .
Grouted rubble paving	.017	.030	. . . . . .
Corrugated metal . .	.023	.025	.024

(d) *Bernoulli Theorem.*—The Bernoulli theorem, which is the principle of conservation of energy applied to open channel flow, may be stated: The absolute head at any section is equal to the absolute head at a section downstream plus intervening losses of head. Referring to figure F-1, the energy equation (3) can be written:

$$Z_2 + d_2 + h_{v_2} = Z_1 + d_1 + h_{v_1} + h_L \qquad (32)$$

where h_L represents all losses in head between section 2 (subscript 2) and section 1 (subscript 1). Such head losses will consist largely of friction loss, but may include minor other losses such as those due to eddy, transition, obstruction, impact, etc.

When the discharge at a given cross section of a channel is constant with respect to time, the flow is steady. If steady flow occurs at all sections in a reach, the flow is continuous and

$$Q = a_1 v_1 = a_2 v_2 \qquad (33)$$

Equation (33) is termed the equation of continuity. Equations (32) and (33), solved simultaneously, are the basic formulas used in solving problems of flow in open channels.

(e) *Hydraulic and Energy Gradients.*—The hydraulic gradient in open channel flow is the water surface. The energy gradient is above the hydraulic gradient a distance equal to the velocity head. The fall of the energy gradient for a given length of channel represents the loss of energy, either from friction or from friction and other influences. The relationship of the energy gradient to the hydraulic gradient reflects not only the loss of energy, but also the conversion between potential and kinetic energy. For uniform flow the gradients are parallel and the slope of the water surface represents the friction loss gradient. In accelerated flow the hydraulic gradient is steeper than the energy gradient, indicating a progressive conversion from potential to kinetic energy. In retarded flow the energy gradient is steeper than the hydraulic gradient, indicating a conversion from kinetic to potential energy. The Bernoulli theorem defines the progressive relationships of these energy gradients.

For a given reach of channel ΔL, the average slope of the energy gradient is $\frac{\Delta h_L}{\Delta L}$, where Δh_L is the cumulative losses through the reach. If these losses are solely from friction, Δh_L will become Δh_f and

$$\Delta h_f = \left(\frac{s_2 + s_1}{2}\right)\Delta L \qquad (34)$$

Expressed in terms of the hydraulic properties at each end of the reach and of the roughness coefficient,

$$\Delta h_f = \frac{n^2}{4.41}\left[\left(\frac{v_2}{r_2^{2/3}}\right)^2 + \left(\frac{v_1}{r_1^{2/3}}\right)^2\right]\Delta L \qquad (35)$$

If the average friction slope, s_f, is equal to $\frac{s_2 + s_1}{2} = \frac{\Delta h_f}{\Delta L}$, and s_b is the slope of the channel floor, by substituting $s_b \Delta L$ for $Z_2 - Z_1$, and H_E for $(d + h_v)$, equation (32) may be written:

$$\Delta L = \frac{H_{E_1} - H_{E_2}}{s_b - s_f} \qquad (36)$$

(f) *Chart for Approximating Friction Losses in Chutes.*—Figure 9-26 is a nomograph from which approximate friction losses in a channel can be evaluated. To generalize the chart so that it can be applied for differing channel conditions, several approximations are made. First, the depth of flow in the channel is assumed equal to the hydraulic radius; the results will therefore be most applicable to wide, shallow channels. Furthermore, the increase in velocity head is assumed to vary proportionally along the length of the channel. Thus, the data given in the chart are not exact and are intended to serve only as a guide in estimating channel losses.

The chart plots the solution of the equation $s = \frac{dh_f}{dx}$, integrated between the limits from zero to L, or

$$h_f = \int_o^L s\, dx,$$

where, from the Manning equation,

$$s = \frac{v^2}{\left(\dfrac{1.486}{n}\right)^2 r^{4/3}}$$

F-3. *Flow in Closed Conduits.*—(a) *Partly Full Flow in Conduits.*—The hydraulics of partly full flow in closed conduits is similar to that in open channels, and open channel flow formulas are applicable. Hydraulic properties for different flow depths in circular and horseshoe conduits are tabulated in tables F-2 through F-5 to facilitate hydraulic computations for these sections.

Tables F-2 and F-4 give data for determining critical depths, critical velocities, and hydrostatic pressures of the water prism cross section for various discharges and conduit diameters. If the area at critical flow, a_c, is represented as $k_1 D^2$ and the top width of the water prism, T, for critical flow is equal to $k_2 D$, equation (7) can be written:

$$\frac{Q_c^2}{g} = \frac{(k_1 D^2)^3}{k_2 D}, \text{ or } Q_c = k_3 D^{5/2} \qquad (37)$$

Values of k_3, for various flow depths, are tabulated in column 3. The hydrostatic pressure, P, of the water prism cross section is $wa\bar{y}$, where $\bar{y}$ is the depth from the water surface to the center of gravity of the cross section. If $a_c = k_1 D^2$ and $\bar{y} = k_4 D$, then

$$P = k_5 D^3 \qquad (38)$$

Values of k_5, for various flow depths, are tabulated in column 4. Column 2 gives the values of h_{v_c} in relation to the conduit diameter, for various flow depths.

Tables F-3 and F-5 give areas and hydraulic radii for partly full conduits and coefficients which can be applied in the solution of the Manning equation. If $A = k_6 \dfrac{\pi D^2}{4}$ and $r = k_7 D$, Manning's equation can be written:

$$Q = \frac{1.486}{n}\left(k_6\frac{\pi D^2}{4}\right)(k_7 D)^{2/3} s^{1/2},$$

or

TABLE F-2.—*Velocity head and discharge at critical depths and static pressures in circular conduits partly full.*—288-D-3195

D = Diameter of pipe.
d = Depth of flow.
h_{v_c} = Velocity head for a critical depth of d.
Q_c = Discharge when the critical depth is d.
P = Pressure on cross section of water prism in cubic units of water. To get P in pounds, when d and D are in feet, multiply by 62.5.

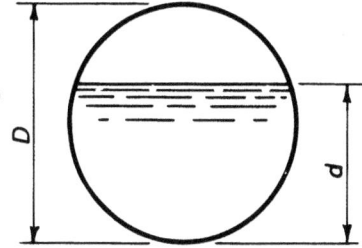

$\dfrac{d}{D}$	$\dfrac{h_{v_c}}{D}$	$\dfrac{Q_c}{D^{5/2}}$	$\dfrac{P}{D^3}$	$\dfrac{d}{D}$	$\dfrac{h_{v_c}}{D}$	$\dfrac{Q_c}{D^{5/2}}$	$\dfrac{P}{D^3}$	$\dfrac{d}{D}$	$\dfrac{h_{v_c}}{D}$	$\dfrac{Q_c}{D^{5/2}}$	$\dfrac{P}{D^3}$
1	2	3	4	1	2	3	4	1	2	3	4
0.01	0.0033	0.0006	0.0000	0.34	0.1243	0.6657	0.0332	0.67	0.2974	2.4464	0.1644
.02	.0067	.0025	.0000	.35	.1284	.7040	.0356	.68	.3048	2.5182	.1700
.03	.0101	.0055	.0001	.36	.1326	.7433	.0381	.69	.3125	2.5912	.1758
.04	.0134	.0098	.0002	.37	.1368	.7836	.0407	.70	.3204	2.6656	.1816
.05	.0168	.0153	.0003	.38	.1411	.8249	.0434	.71	.3286	2.7414	.1875
.06	.0203	.0220	.0005	.39	.1454	.8671	.0462	.72	.3371	2.8188	.1935
.07	.0237	.0298	.0007	.40	.1497	.9103	.0491	.73	.3459	2.8977	.1996
.08	.0271	.0389	.0010	.41	.1541	.9545	.0520	.74	.3552	2.9783	.2058
.09	.0306	.0491	.0013	.42	.1586	.9996	.0551	.75	.3648	3.0607	.2121
.10	.0341	.0605	.0017	.43	.1631	1.0458	.0583	.76	.3749	3.1450	.2185
.11	.0376	.0731	.0021	.44	.1676	1.0929	.0616	.77	.3855	3.2314	.2249
.12	.0411	.0868	.0026	.45	.1723	1.1410	.0650	.78	.3967	3.3200	.2314
.13	.0446	.1016	.0032	.46	.1769	1.1899	.0684	.79	.4085	3.4112	.2380
.14	.0482	.1176	.0038	.47	.1817	1.2399	.0720	.80	.4210	3.5050	.2447
.15	.0517	.1347	.0045	.48	.1865	1.2908	.0757	.81	.4343	3.6019	.2515
.16	.0553	.1530	.0053	.49	.1914	1.3427	.0795	.82	.4485	3.7021	.2584
.17	.0589	.1724	.0061	.50	.1964	1.3955	.0833	.83	.4638	3.8061	.2653
.18	.0626	.1928	.0070	.51	.2014	1.4493	.0873	.84	.4803	3.9144	.2723
.19	.0662	.2144	.0080	.52	.2065	1.5041	.0914	.85	.4982	4.0276	.2794
.20	.0699	.2371	.0091	.53	.2117	1.5598	.0956	.86	.5177	4.1465	.2865
.21	.0736	.2609	.0103	.54	.2170	1.6164	.0998	.87	.5392	4.2721	.2938
.22	.0773	.2857	.0115	.55	.2224	1.6735	.1042	.88	.5632	4.4056	.3011
.23	.0811	.3116	.0128	.56	.2279	1.7327	.1087	.89	.5900	4.5486	.3084
.24	.0848	.3386	.0143	.57	.2335	1.7923	.1133	.90	.6204	4.7033	.3158
.25	.0887	.3667	.0157	.58	.2393	1.8530	.1179	.91	.6555	4.8725	.3233
.26	.0925	.3957	.0173	.59	.2451	1.9146	.1227	.92	.6966	5.0603	.3308
.27	.0963	.4259	.0190	.60	.2511	1.9773	.1276	.93	.7459	5.2726	.3384
.28	.1002	.4571	.0207	.61	.2572	2.0409	.1326	.94	.8065	5.5183	.3460
.29	.1042	.4893	.0226	.62	.2635	2.1057	.1376	.95	.8841	5.8118	.3537
.30	.1081	.5225	.0255	.63	.2699	2.1716	.1428	.96	.9885	6.1787	.3615
.31	.1121	.5568	.0266	.64	.2765	2.2386	.1481	.97	1.1410	6.6692	.3692
.32	.1161	.5921	.0287	.65	.2833	2.3067	.1534	.98	1.3958	7.4063	.3770
.33	.1202	.6284	.0309	.66	.2902	2.3760	.1589	.99	1.9700	8.8263	.3848
								1.00	------	------	.3927

$$\frac{Qn}{D^{8/3}s^{1/2}} = k_6 \frac{1.486\pi}{4}(k_7)^{2/3} = k_8 \quad (39)$$

Values of k_8, for various flow depths, are tabulated in column 4. If $D = k_9 d$, equation (39) can be written:

$$\frac{Qn}{d^{8/3}s^{1/2}} = \frac{1.486\pi}{4}k_6(k_7)^{2/3}(k_9)^{8/3} = k_{10} \quad (40)$$

TABLE F-3.—*Uniform flow in circular sections flowing partly full.*—288-D-3196

d = Depth of flow.
D = Diameter of pipe.
A = Area of flow.
r = Hydraulic radius.

Q = Discharge in c.f.s. by Manning's formula.
n = Manning's coefficient.
s = Slope of the channel bottom and of the water surface.

$\dfrac{d}{D}$	$\dfrac{A}{D^2}$	$\dfrac{r}{D}$	$\dfrac{Qn}{D^{8/3}s^{1/2}}$	$\dfrac{Qn}{d^{8/3}s^{1/2}}$	$\dfrac{d}{D}$	$\dfrac{A}{D^2}$	$\dfrac{r}{D}$	$\dfrac{Qn}{D^{8/3}s^{1/2}}$	$\dfrac{Qn}{d^{8/3}s^{1/2}}$
1	2	3	4	5	1	2	3	4	5
0.01	0.0013	0.0066	0.00007	15.04	0.51	0.4027	0.2531	0.239	1.442
.02	.0037	.0132	.00031	10.57	.52	.4127	.2562	.247	1.415
.03	.0069	.0197	.00074	8.56	.53	.4227	.2592	.255	1.388
.04	.0105	.0262	.00138	7.38	.54	.4327	.2621	.263	1.362
.05	.0147	0325	.00222	6.55	.55	.4426	.2649	.271	1.336
.06	.0192	.0389	.00328	5.95	.56	.4526	.2676	.279	1.311
.07	.0242	.0451	.00455	5.47	.57	.4625	.2703	.287	1.286
.08	.0294	.0513	.00604	5.09	.58	.4724	.2728	.295	1.262
.09	.0350	.0575	.00775	4.76	.59	.4822	.2753	.303	1.238
.10	.0409	.0635	.00967	4.49	.60	.4920	.2776	.311	1.215
.11	.0470	.0695	.01181	4.25	.61	.5018	.2799	.319	1.192
.12	.0534	.0755	.01417	4.04	.62	.5115	.2821	.327	1.170
.13	.0600	.0813	.01674	3.86	.63	.5212	.2842	.335	1.148
.14	.0668	.0871	.01952	3.69	.64	.5308	.2862	.343	1.126
.15	.0739	.0929	.0225	3.54	.65	.5404	.2882	.350	1.105
.16	.0811	.0985	.0257	3.41	.66	.5499	.2900	.358	1.084
.17	.0885	.1042	.0291	3.28	.67	.5594	.2917	.366	1.064
.18	.0961	.1097	.0327	3.17	.68	.5687	.2933	.373	1.044
.19	.1039	.1152	.0365	3.06	.69	.5780	.2948	.380	1.024
.20	.1118	.1206	.0406	2.96	.70	.5872	.2962	.388	1.004
.21	.1199	.1259	.0448	2.87	.71	.5964	.2975	.395	0.985
.22	.1281	.1312	.0492	2.79	.72	.6054	.2987	.402	.965
.23	.1365	.1364	.0537	2.71	.73	.6143	.2998	.409	.947
.24	.1449	.1416	.0585	2.63	.74	.6231	.3008	.416	.928
.25	.1535	.1466	.0634	2.56	.75	.6319	.3017	.422	.910
.26	.1623	.1516	.0686	2.49	.76	.6405	.3024	.429	.891
.27	.1711	.1566	.0739	2.42	.77	.6489	.3031	435	.873
.28	.1800	.1614	.0793	2.36	.78	.6573	.3036	.441	.856
.29	.1890	.1662	.0849	2.30	.79	.6655	.3039	.447	.838
.30	.1982	.1709	.0907	2.25	.80	.6736	.3042	.453	.821
.31	.2074	.1756	.0966	2.20	.81	.6815	.3043	.458	.804
.32	.2167	.1802	.1027	2.14	.82	.6893	.3043	.463	.787
.33	.2260	.1847	.1089	2.09	.83	.6969	.3041	.468	.770
.34	.2355	.1891	.1153	2.05	.84	.7043	.3038	.473	.753
.35	.2450	.1935	.1218	2.00	.85	.7115	.3033	.477	.736
.36	.2546	.1978	.1284	1.958	.86	.7186	.3026	.481	.720
.37	.2642	.2020	.1351	1.915	.87	.7254	.3018	.485	.703
.38	.2739	.2062	.1420	1.875	.88	.7320	.3007	.488	.687
.39	.2836	.2102	.1490	1.835	.89	.7384	.2995	.491	.670
.40	.2934	.2142	.1561	1.797	.90	.7445	.2980	.494	.654
.41	.3032	.2182	.1633	1.760	.91	.7504	.2963	.496	.637
.42	.3130	.2220	.1705	1.724	.92	.7560	.2944	.497	.621
.43	.3229	.2258	.1779	1.689	.93	.7612	.2921	.498	.604
.44	.3328	.2295	.1854	1.655	.94	.7662	.2895	.498	.588
.45	.3428	.2331	.1929	1.622	.95	.7707	.2865	.498	.571
.46	.3527	.2366	.201	1.590	.96	.7749	.2829	.496	.553
.47	.3627	.2401	.208	1.559	.97	.7785	.2787	.494	.535
.48	.3727	.2435	.216	1.530	.98	.7817	.2735	.489	.517
.49	.3827	.2468	.224	1.500	.909	.7841	.2666	.483	.496
.50	.3927	.2500	.232	1.471	1.00	.7854	.2500	.463	.463

TABLE F-4.—*Velocity head and discharge at critical depths and static pressures in horseshoe conduits partly full.*—288-D-3197

D = Diameter of horseshoe.
d = Depth of flow.
h_{v_c} = Velocity head for a critical depth of d.
Q_c = Discharge when the critical depth is d.
P = Pressure on cross section of water prism in cubic units of water. To get P in pounds, when d and D are in feet, multiply by 62.5.

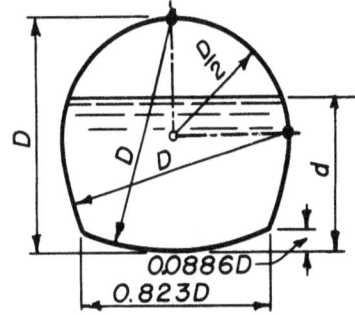

$\frac{d}{D}$	$\frac{h_{v_c}}{D}$	$\frac{Q_c}{D^{5/2}}$	$\frac{P}{D^3}$	$\frac{d}{D}$	$\frac{h_{v_c}}{D}$	$\frac{Q_c}{D^{5/2}}$	$\frac{P}{D^3}$	$\frac{d}{D}$	$\frac{h_{v_c}}{D}$	$\frac{Q_c}{D^{5/2}}$	$\frac{P}{D^3}$
1	2	3	4	1	2	3	4	1	2	3	4
0.01	0.0033	0.0009	0.0000	0.35	0.1472	0.8854	0.0449	0.69	0.3362	2.8922	0.1999
.02	.0067	.0035	.0000	.36	.1518	.9296	.0478	.70	.3443	2.9702	.2062
.03	.0100	.0079	.0001	.37	.1563	.9746	.0508	.71	.3528	3.0499	.2125
.04	.0134	.0139	.0002	.38	.1609	1.0205	.0540	.72	.3615	3.1311	.2190
.05	.0168	.0217	.0004	.39	.1655	1.0673	.0572	.73	.3707	3.2140	.2255
.06	.0201	.0312	.0007	.40	.1702	1.1148	.0605	.74	.3802	3.2987	.2321
.07	.0235	.0425	.0010	.41	.1749	1.1633	.0639	.75	.3902	3.3853	.2385
.08	.0269	.0554	.0014	.42	.1795	1.2125	.0675	.76	.4006	3.4740	.2457
.09	.0305	.0703	.0018	.43	.1843	1.2626	.0711	.77	.4116	3.5650	.2525
.10	.0351	.0879	.0024	.44	.1890	1.3135	.0748	.78	.4232	3.6584	.2595
.11	.0397	.1069	.0030	.45	.1938	1.3652	.0786	.79	.4354	3.7544	.2666
.12	.0443	.1272	.0037	.46	.1986	1.4178	.0825	.80	.4484	3.8534	.2737
.13	.0489	.1487	.0045	.47	.2035	1.4712	.0865	.81	.4623	3.9557	.2809
.14	.0534	.1714	.0054	.48	.2084	1.5253	.0907	.82	.4771	4.0616	.2882
.15	.0579	.1953	.0063	.49	.2133	1.5803	.0949	.83	.4930	4.1716	.2956
.16	.0624	.2203	.0074	.50	.2183	1.6361	.0992	.84	.5102	4.2863	.3030
.17	.0669	.2465	.0085	.51	.2234	1.6928	.1036	.85	.5289	4.4063	.3105
.18	.0714	.2736	.0098	.52	.2285	1.7505	.1081	.86	.5494	4.5325	.3181
.19	.0758	.3019	.0111	.53	.2337	1.8092	.1127	.87	.5719	4.6660	.3258
.20	.0803	.3312	.0125	.54	.2391	1.8688	.1174	.88	.5969	4.8080	.3335
.21	.0847	.3615	.0140	.55	.2445	1.9294	.1223	.89	.6251	4.9605	.3413
.22	.0891	.3928	.0156	.56	.2500	1.9911	.1272	.90	.6570	5.1256	.3492
.23	.0936	.4251	.0173	.57	.2557	2.0537	.1322	.91	.6939	5.3065	.3572
.24	.0980	.4583	.0191	.58	.2615	2.1174	.1373	.92	.7371	5.5077	.3653
.25	.1024	.4926	.0210	.59	.2674	2.1821	.1425	.93	.7889	5.7354	.3733
.26	.1069	.5277	.0229	.60	.2735	2.2479	.1478	.94	.8528	5.9996	.3813
.27	.1113	.5638	.0250	.61	.2797	2.3148	.1532	.95	.9345	6.3157	.3894
.28	.1158	.6009	0271	.62	.2861	2.3828	.1587	.96	1.0446	6.7114	.3976
.29	.1202	.6389	.0294	.63	.2926	2.4519	.1643	.97	1.2053	7.2417	.4058
.30	.1247	.6777	.0317	.64	.2994	2.5221	.1700	.98	1.4742	8.0892	4140
.31	.1292	.7175	.0342	.65	.3063	2.5936	.1758	.99	2.0804	9.5780	.4223
.32	.1337	.7582	.0367	.66	.3134	2.6663	.1817	1.00	--------	--------	.4306
.33	.1382	.7997	.0393	.67	.3208	2.7402	.1877				
.34	.1427	.8421	.0421	.68	.3283	2.8155	.1937				

Values of k_{10}, for various flow depths, are tabulated in column 5.

(b) *Pressure Flow in Conduits.*—Since factors affecting head losses in conduits are independent of pressure, the same laws apply to flow in both closed conduits and open channels, and the formulas for each take the same general form. Thus, the equation of continuity, equation (33), $Q = a_1 v_1 = a_2 v_2$, also applies to pressure flow in conduits.

TABLE F-5.—*Uniform flow in horseshoe sections flowing partly full.*—288-D-3198

d = Depth of flow.
D = Diameter.
A = Area of flow.
r = Hydraulic radius.

Q = Discharge in c.f.s. by Manning's formula.
n = Manning's coefficient.
s = Slope of the channel bottom and of the water surface.

$\dfrac{d}{D}$	$\dfrac{A}{D^2}$	$\dfrac{r}{D}$	$\dfrac{Qn}{D^{8/3}s^{1/2}}$	$\dfrac{Qn}{d^{8/3}s^{1/2}}$	$\dfrac{d}{D}$	$\dfrac{A}{D^2}$	$\dfrac{r}{D}$	$\dfrac{Qn}{D^{8/3}s^{1/2}}$	$\dfrac{Qn}{d^{8/3}s^{1/2}}$
1	2	3	4	5	1	2	3	4	5
0.01	0.0019	0.0066	0.00010	21.40	0.51	0.4466	0.2602	0.2705	1.629
.02	.0053	.0132	.00044	14.93	.52	.4566	.2630	.2785	1.593
.03	.0097	.0198	.00105	12.14	.53	.4666	.2657	.2866	1.558
.04	.0150	.0264	.00198	10.56	.54	.4766	.2683	.2946	1.524
.05	.0209	.0329	.00319	9.40	.55	.4865	.2707	.303	1.490
.06	.0275	.0394	.00473	8.58	.56	.4965	.2733	.311	1.458
.07	.0346	.0459	.00659	7.92	.57	.5064	.2757	.319	1.427
.08	.0421	.0524	.00876	7.37	.58	.5163	.2781	.327	1.397
.09	.0502	.0590	.01131	6.95	.59	.5261	.2804	.335	1.368
.10	.0585	.0670	.01434	6.66	.60	.5359	.2824	.343	1.339
.11	.0670	.0748	.01768	6.36	.61	.5457	.2844	.351	1.310
.12	.0753	.0823	.02117	6.04	.62	.5555	.2864	.359	1.283
.13	.0839	.0895	.02495	5.75	.63	.5651	.2884	.367	1.257
.14	.0925	.0964	.02890	5.47	.64	.5748	.2902	.374	1.231
.15	.1012	.1031	.0331	5.21	.65	.5843	.2920	.382	1.206
.16	.1100	.1097	.0375	4.96	.66	.5938	.2937	.390	1.181
.17	.1188	.1161	.0420	4.74	.67	.6033	.2953	.398	1.157
.18	.1277	.1222	.0467	4.52	.68	.6126	.2967	.405	1.133
.19	.1367	.1282	.0516	4.33	.69	.6219	.2981	.412	1.109
.20	.1457	.1341	.0567	4.15	.70	.6312	.2994	.420	1.087
.21	.1549	.1398	.0620	3.98	.71	.6403	.3006	.427	1.064
.22	.1640	.1454	.0674	3.82	.72	.6493	.3018	.434	1.042
.23	.1733	.1508	.0730	3.68	.73	.6582	.3028	.441	1.021
.24	.1825	.1560	.0786	3.53	.74	.6671	.3036	.448	1.000
.25	.1919	.1611	.0844	3.40	.75	.6758	.3044	.454	0.979
.26	.2013	.1662	.0904	3.28	.76	.6844	.3050	.461	.958
.27	.2107	.1710	.0965	3.17	.77	.6929	.3055	.467	.938
.28	.2202	.1758	.1027	3.06	.78	.7012	.3060	.473	.918
.29	.2297	.1804	.1090	2.96	.79	.7094	.3064	.479	.898
.30	.2393	.1850	.1155	2.86	.80	.7175	.3067	.485	.879
.31	.2489	.1895	.1220	2.77	.81	.7254	.3067	.490	.860
.32	.2586	.1938	.1287	2.69	.82	.7332	.3066	.495	.841
.33	.2683	.1981	.1355	2.61	.83	.7408	.3064	.500	.822
.34	.2780	.2023	.1424	2.53	.84	.7482	.3061	.505	.804
.35	.2878	.2063	.1493	2.45	.85	.7554	.3056	.509	.786
.36	.2975	.2103	.1563	2.38	.86	.7625	.3050	.513	.768
.37	.3074	.2142	.1635	2.32	.87	.7693	.3042	.517	.750
.38	.3172	.2181	.1708	2.25	.88	.7759	.3032	.520	.732
.39	.3271	.2217	.1781	2.19	.89	.7823	.3020	.523	.714
.40	.3370	.2252	.1854	2.13	.90	.7884	.3005	.526	.696
.41	.3469	.2287	.1928	2.08	.91	.7943	.2988	.528	.678
.42	.3568	.2322	.2003	2.02	.92	.7999	.2969	.529	.661
.43	.3667	.2356	.2079	1.973	.93	.8052	.2947	.530	.643
.44	.3767	.2390	.2156	1.925	.94	.8101	.2922	.530	.625
.45	.3867	.2422	.2233	1.878	.95	.8146	.2893	.529	.607
.46	.3966	.2454	.2310	1.832	.96	.8188	.2858	.528	.589
.47	.4066	.2484	.2388	1.788	.97	.8224	.2816	.525	.569
.48	.4166	.2514	.2466	1.746	.98	.8256	.2766	.521	.550
.49	.4266	.2544	.2545	1.705	.99	.8280	.2696	.513	.527
.50	.4366	.2574	.2625	1.667	1.00	.8293	.2538	.494	.494

A mass of water, as such, does not have pressure energy. Pressure energy is acquired by contact with other masses and is, therefore, transmitted to or through the mass under consideration. The pressure head $\frac{p}{w}$ (where p is the pressure intensity in pounds per square foot and w is unit weight in pounds per cubic foot), like velocity and elevation heads, also expresses energy. Thus, to be applicable to pressure flow in a conduit, the Bernoulli equation for flow in open channels, equation (3), can be rewritten:

$$H_A = \frac{p}{w} + Z + \frac{v^2}{2g} \qquad (41)$$

The Bernoulli theorem for flow in a reach of pressure conduit (as shown on fig. F-5) is:

$$\frac{p_1}{w} + Z_1 + h_{v_1} = \frac{p_2}{w} + Z_2 + h_{v_2} + \Delta h_L \quad (42)$$

where Δh_L represents the head losses within the reach from all causes. If H_T is the total head and v is the velocity at the outlet, Bernoulli's equation for the entire length is:

$$H_T = \Sigma(\Delta h_L) + h_v \qquad (43)$$

As in open channel flow, the Bernoulli theorem and the continuity equation are the basic formulas used in solving problems of pressure conduit flow.

(c) *Energy and Pressure Gradients.*—If piezometer standpipes were to be inserted at various points along the length of a conduit flowing under pressure, as illustrated on figure F-5, water would rise in each standpipe to a level equal to the pressure head in the conduit at those points. The pressure at any point may be equal to, greater than, or less than the local atmospheric pressure. The height to which the water would rise in a piezometer is termed the pressure gradient. The energy gradient is above the pressure gradient a distance equal to the velocity head. The fall of the energy gradient for a given length of conduit represents the loss of energy, either from friction or from friction and other influences. The relationship of the energy gradient to the pressure gradient reflects the variations between kinetic energy and pressure head.

(d) *Friction Losses.*—Many empirical formulas have been developed for evaluating the flow of fluids in conduits. Those in most common use are the Manning equation and the Darcy-Weisbach equation, previously given in this appendix and further discussed in chapter X.

The Manning equation assumes that the energy loss depends only on the velocity, the dimensions of the conduit, and the magnitude of wall roughness as defined by the friction coefficient n. The n value is related to the physical roughness of the conduit wall and is independent of the size of the conduit or of the density and viscosity of the water.

The Darcy-Weisbach equation assumes the loss to be related to the velocity, the dimensions of the conduit, and the friction factor f. The factor f is a dimensionless variable based on the viscosity and density of the fluid and on the roughness of the conduit walls as it relates to the size of the conduit.

Data and criteria for determining f values for large pipe are given in a Bureau of Reclamation engineering monograph [6].

F-4. Hydraulic Jump.—The hydraulic jump is an abrupt rise in water surface which may occur in an open channel when water flowing at high velocity is retarded. The formula for the hydraulic jump is obtained by equating the unbalanced forces acting to retard the mass of flow to the rate of change of the momentum of flow. The general formula for this relationship is:

$$v_1{}^2 = g \frac{a_2 \bar{y}_2 - a_1 \bar{y}_1}{a_1 \left(1 - \frac{a_1}{a_2}\right)} \qquad (44)$$

where:

v_1 = the velocity before the jump,

a_1 and a_2 = the areas before and after the jump, respectively, and

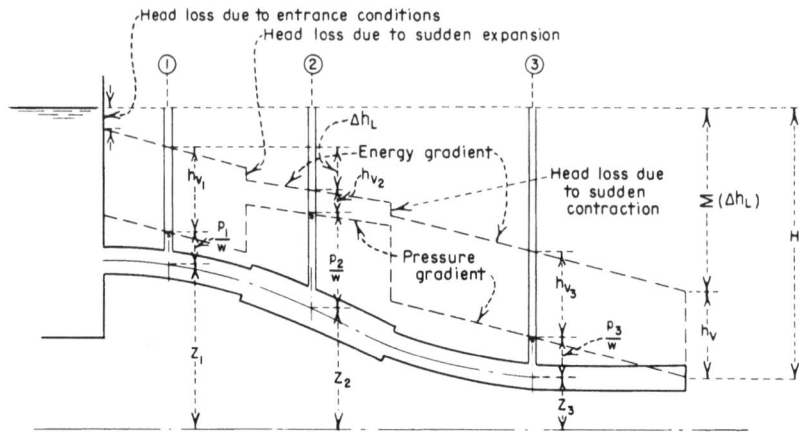

Figure F-5. Characteristics of pressure flow in conduits.—288-D-2555

$\bar{y}_1$ and $\bar{y}_2$ = the corresponding depths from the water surface to the center of gravity of the cross section.

The general formula expressed in terms of discharge is:

$$Q^2 = g \frac{a_2 \bar{y}_2 - a_1 \bar{y}_1}{\dfrac{1}{a_1} - \dfrac{1}{a_2}} \qquad (45)$$

or:

$$\frac{Q^2}{ga_1} + a_1 \bar{y}_1 = \frac{Q^2}{ga_2} + a_2 \bar{y}_2 \qquad (46)$$

For a rectangular channel, equation (44) can be reduced to $v_1{}^2 = \dfrac{gd_2}{2d_1}(d_2 + d_1)$, where d_1 and d_2 are the flow depths before and after the jump, respectively. Solving for d_2:

$$d_2 = -\frac{d_1}{2} + \sqrt{\frac{2v_1{}^2 d_1}{g} + \frac{d_1{}^2}{4}} \qquad (47)$$

Similarly, expressing d_1 in terms of d_2 and v_2:

$$d_1 = -\frac{d_2}{2} + \sqrt{\frac{2v_2{}^2 d_2}{g} + \frac{d_2{}^2}{4}}. \qquad (48)$$

A graphic solution of equation (47) is shown on figure F-8.

If the Froude number $F_1 = \dfrac{v_1}{\sqrt{gd_1}}$ is substituted in the equation (47):

$$\frac{d_2}{d_1} = \frac{1}{2}\left(\sqrt{8F_1{}^2 + 1} - 1\right) \qquad (49)$$

Figure F-6 shows a graphical representation of the characteristics of the hydraulic jump. Figure F-7 shows the hydraulic properties of the jump in relation to the Froude number, as determined from experimental data [7]. And figure F-8 is a nomograph showing the relation between variables in the hydraulic jump.

Data are for jumps on a flat floor with no chute blocks, baffle piers, or end sills. Ordinarily, the jump length can be shortened by incorporation of such devices in the designs of a specific stilling basin.

F-5. *Bibliography*.

[1] King, H. W., revised by E. F. Brater, "Handbook of Hydraulics," fourth edition, McGraw-Hill Book Co., Inc., New York, N.Y., 1954.

[2] Woodward, S. B., and Posey, C. J., "Steady Flow in Open Channels," John Wiley & Sons, Inc., fourth printing, September 1949.

[3] Bakhmeteff, B. A., "Hydraulics of Open Channels," McGraw-Hill Book Co., Inc., New York, N.Y., 1932.

[4] Binder, R. C., "Fluid Mechanics," Prentice-Hall, Inc., Englewood Cliffs, N.J., third edition, 1955.

[5] Rouse, Hunter, "Engineering Hydraulics," John Wiley & Sons, Inc., New York, N.Y., 1950.

[6] Bradley, J. N., and Thompson, L. R., "Friction Factors for Large Conduits Flowing Full," Engineering Monograph No. 7, U.S. Department of the Interior, Bureau of Reclamation, March 1951.

[7] Bradley, J. N., and Peterka, A. J., "The Hydraulic Design of Stilling Basins," ASCE Proceedings, vol. 83, October 1957, Journal of Hydraulics Division, No. HY5, Papers No. 1401 to 1404, inclusive.

q = Discharge in c.f.s. per foot of width.
H_{E_1} = Energy entering jump.
H_{E_2} = Energy leaving jump.
F_1 = Froude number at section ① $= \frac{v_1}{\sqrt{gd_1}}$
d_j = $d_2 - d_1$ = Height of jump
L = Length of jump.

$H_{E_1} - H_{E_2}$ = Energy loss in jump

$H_{E_2} = d_2 + \frac{v_2^2}{2g}$

$H_{E_1} = d_1 + \frac{v_1^2}{2g}$

(A) HYDRAULIC JUMP - ON HORIZONTAL FLOOR

(B) RELATION OF SPECIFIC ENERGY TO DEPTH OF FLOW

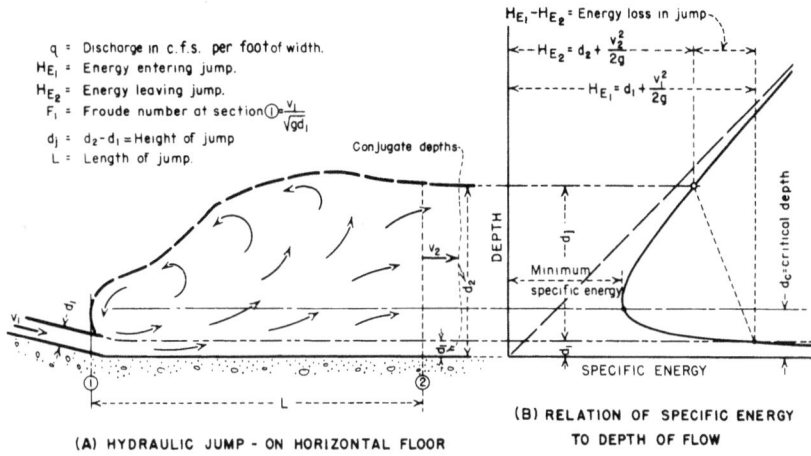

Figure F-6. Hydraulic jump symbols and characteristics.—288-D-3190

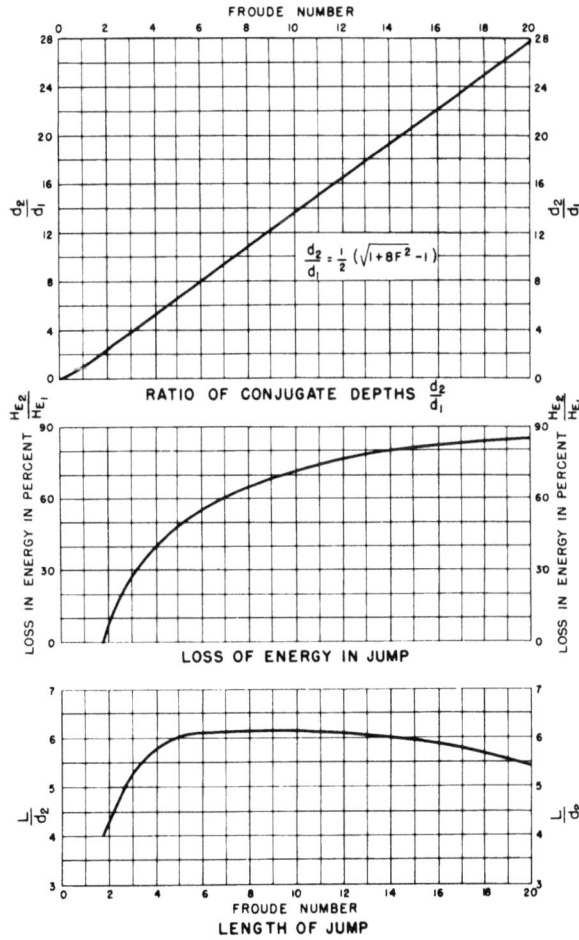

$$\frac{d_2}{d_1} = \frac{1}{2} \left(\sqrt{1 + 8F^2} - 1 \right)$$

RATIO OF CONJUGATE DEPTHS $\frac{d_2}{d_1}$

LOSS OF ENERGY IN JUMP

LENGTH OF JUMP

Figure F-7. Hydraulic jump properties in relation to Froude number.—288-D-2558

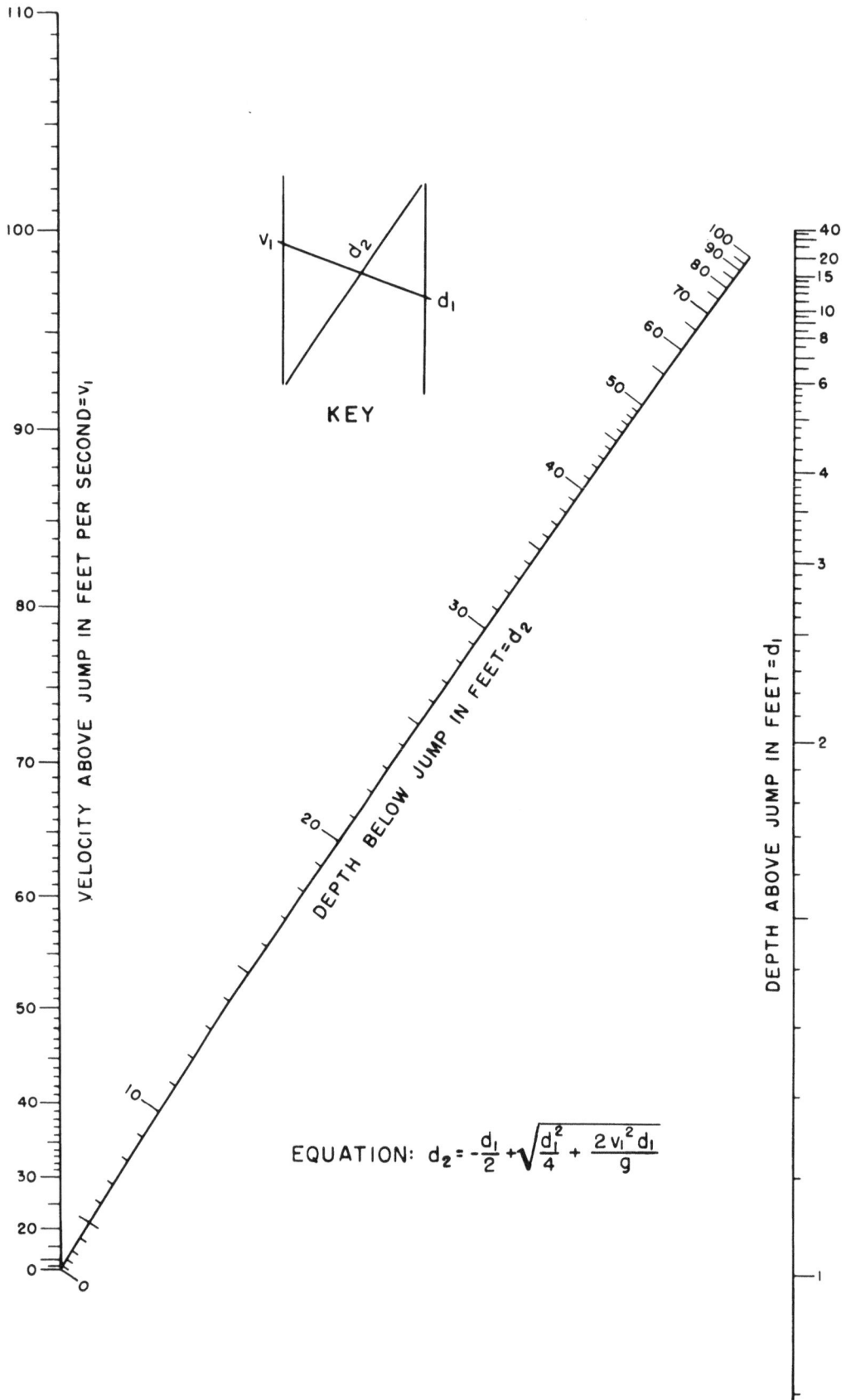

Figure F-8. Relation between variables in the hydraulic jump.—288-D-2559

Inflow Design Flood Studies

G-1. *Introduction.*—A 1970 report of the United States Committee on Large Dams (USCOLD) [1][1] gives a definition of an inflow design flood (IDF) as:

"The reservoir inflow-discharge hydrograph used in estimating the maximum spillway discharge capacity and maximum surcharge elevation finally adopted as a basis for project design . . . ."

An inflow design flood selected for design of a dam impounding considerable storage located where partial or total failure would cause sudden release of water and create major hazards to life or property downstream should be equal to a probable maximum flood (PMF). The USCOLD report defines a probable maximum flood as:

"Estimates of hypothetical flood characteristics (peak discharge, volume and hydrograph shape) that are considered to be the most severe *reasonably possible* at a particular location, based on relatively comprehensive hydrometeorological analyses of critical runoff producing precipitation (and snowmelt, if pertinent) and hydrologic factors favorable for maximum flood runoff."

This appendix discusses flood hydrology studies relating to estimates of an inflow design flood equal to a probable maximum flood, as defined in the USCOLD report. The phrase "relatively comprehensive hydrometeorological analyses" in the preceding definition refers to studies by hydrometeorologists directed towards estimation of *the physical upper limits of storm rainfall and maximum snow accumulation and melt rates.* The resulting estimates of the physical upper limits to storm rainfall in a basin or region are usually called the "probable maximum storm" or "probable maximum precipitation" [2]. Both of these terms are used in this text but with more precise meanings attached to each term as discussed in sections G-14 through G-17 on design storm studies.

Bureau of Reclamation policy in design of dams located where failure might create major hazards requires an inflow design flood estimated by evaluating the runoff from the most critical of the following situations:

(1) A probable maximum storm in conjunction with severe, but not uncommon, antecedent conditions.

(2) A probable maximum storm for the season of heavy snowmelt, in conjunction with a major snowmelt flood somewhat smaller than the probable maximum.

(3) A probable maximum snowmelt flood in conjunction with a major rainstorm less severe than the probable maximum storm for that season.

(a) *Items to be Evaluated.*—Depending on meteorological conditions for the basin above a damsite, on the size of the drainage area and, to a lesser extent, on the proposed size of reservoir and type of dam, it may be necessary to evaluate:

(1) Each of the above assumptions.

(2) Each of the two assumptions in which snowmelt is a factor.

(3) Where snowmelt is not a factor,

---

[1]Numbers in brackets refer to items in the bibliography, sec. G-32.

two probable maximum storms—a storm causing the maximum peak inflow, and a storm causing the maximum volume of inflow.

It is beyond the scope of this text to present a complete manual of all procedures used for estimating inflow design floods, because selection of procedures is dependent on available hydrological data and individual watershed characteristics.

(b) *Discussions in This Text.*—Discussions in this text will provide design engineers information about the problems encountered and some methods for their solution. Broad discussions accompany presentation of the information which concerns:

(1) Hydrologic data for estimating floodflows and data sources in the United States.

(2) Analyses of basic data.

(3) Unit hydrograph procedures for synthesizing the distribution of runoff of a basin above a damsite.

(4) Sources of generalized probable maximum precipitation values.

(5) An example of computation of a preliminary inflow design flood hydrograph and establishment of reservoir routing criteria for the flood.

Designers also need estimates of floodflows that may occur at the damsite during the construction period in order to estimate requirements for streamflow diversion. Such estimates are usually included in an inflow design flood study. Sections G-28 and G-29 discuss selected methods of estimating flood magnitudes and frequency of occurrence at the damsite.

Every damsite presents one or more unique problems to probable maximum flood estimates. An inflow design flood (IDF) used for final designs of a dam should be based on estimates by an experienced hydrometeorologist of probable maximum precipitation values *for the basin above the damsite*, not on generalized probable maximum precipitation values for a region. The methods of preparing a study which yields generalized estimates of probable maximum precipitation inherently result in values that are somewhat greater than values obtained from an individual basin study.

Sections G-14 through G-17 present a general discussion of methods and assumptions that a hydrometeorologist may use in the preparation of hydrometeorological studies for individual basins. The physical characteristics of a basin may vary as to: drainage area size, relatively small to extremely large; runoff characteristics, similar throughout the basin or including tributary areas with markedly dissimilar runoff producing conditions; contribution from snowmelt; etc. Sections G-23 through G-26 describe some methods of estimating the contribution of snowmelt runoff to inflow design floods.

The final IDF study converting probable maximum precipitation values to an IDF hydrograph should be prepared by experienced flood hydrologists. Remarks regarding considerations for development of a final IDF study are included throughout the text and a brief summary of these considerations is given in sections G-30 and G-31.

Computational procedures given in this text are oriented toward step by step "long-hand" solutions, recognizing that the ever-increasing advances in computer technology provide greatly expanded capability in all phases of flood hydrology studies. One should be mindful, though, as stated in World Meterological Organization (WMO) Technical Note No. 98 [2] that: "While the computer is a powerful tool, it must be recognized that it is simply that, and results are no better than the basic logic and methods of application."

The bibliography, section G-32, includes selected references to hydrometeorological studies in addition to those specifically referred to in the text.

A. COLLECTION OF HYDROLOGIC DATA FOR
USE IN ESTIMATING FLOODFLOWS

G-2. *General.*—For all flood studies, compilation and judgment as to quality of all available streamflow, precipitation, and watershed data are most important. Mathematical procedures cannot improve the quality of input data, and analyses procedures must be compatible with the data available.

G-3. *Streamflow Data.*—The hydrologic data most directly useful in determining floodflows are actual streamflow records of considerable length at the location of the dam. Such records are rarely available. The engineer should obtain the streamflow records available for the general region in which the dam is to be situated. Locations of stream gaging stations and precipitation stations in the United States are shown on a series of maps entitled "River Basin Maps Showing Hydrologic Stations," edition 1961,[2] prepared under the supervision of the National Weather Service. Such data collecting stations are subject to change in location, discontinuation, or initiation of new stations. These maps cannot be kept current, and information thereon must be supplemented by additional investigations in order to be sure of the location and operation of stations in a given area. The engineer should consult the water supply papers, catalogs, maps, and indexes of the U.S. Geological Survey[2] and, if possible, confer with the Survey's district engineer. He should also make a search of the records of other Federal agencies which may have collected information in the region, and the records of State water conservation agencies or State geological surveys; and he should determine whether any information may be available from other State departments, from county engineer offices, from municipalities in the vicinity, or from utility companies. Where streamflow records are not available, some agencies or inhabitants of the vicinity may have information about

high-water marks caused by specific historic floods.

With respect to the character of the streamflow data available, floodflows at the damsite may be determined under one of the following conditions:

(1) *Streamflow record at or near the damsite.*—If such a record is available and covers a period of 20 years or more, the floodflows shown by the record may be analyzed to provide flood frequency values. Hydrographs of outstanding flood events can be analyzed to provide runoff factors for use in determining the maximum probable flood.

If such a record is available but covers only a few years, it may not include any flood of great magnitude within its limits and, if used alone, it would give false indication of flood potential. Analysis may, however, give some or all of the runoff factors needed to compute the probable maximum flood. Frequency values obtained from a short record should not be used without analysis of data from nearby watersheds of comparable runoff characteristics.

(2) *Streamflow record available on the stream itself, but at a considerable distance from the damsite.*—Such a record may be analyzed to provide unitgraph characteristics and frequency data which may be transferred to the damsite by appropriate area and basin-characteristic coefficients. This transfer can be made directly from one drainage area to another if the areas have comparable characteristics. Often damsites are located within the transition zone from mountains to plains and the stream gaging stations are located well out on the plains; in such instances, special care must be exercised when using the plains record for determination of floodflows at the damsite.

(3) *No adequate streamflow data*

[2]Published by the Government Printing Office and available in libraries designated as depositories of Government publications; most important libraries in the United States are so designated.

available on the specific stream, but a satisfactory record for a drainage basin of similar characteristics in the same region.—Such a record may be analyzed for unitgraph characteristics and frequency data, and these data transferred to the damsite by appropriate area and basin-characteristic coefficients.

(4) *Streamflow records in the region, but not satisfactorily useful for application and analysis under one of the above methods.*—These records may be assembled and analyzed as reference information on general runoff characteristics.

(5) *Use of high-water marks.*—High-water marks pointed out by inhabitants of the valley should be used with caution in estimating flood magnitudes. However, where there are a number of high-water marks in the vicinity of the project, and particularly if such marks are obtained from the records of public offices (such as State highway departments or county engineers), they may be used as the basis of a separate supplemental study. These records may be used to determine the water cross-sectional area and the water surface slope for the flood to which they refer, and from these data an estimate of that particular flood peak may be prepared using the slope-area method described in appendix B of the Bureau of Reclamation publication "Design of Small Dams" [31].

Whenever it appears that there will be one or more flood seasons between the selection of the damsite and construction of the dam, facilities for securing a streamflow record for the project should be set up as promptly as possible. This is of particular importance in order to obtain watershed data directly applicable to the computation of the inflow design flood for the dam, although a record usable for frequency computations cannot be secured. The, facilities for obtaining such a record should be the best possible depending on the circumstances. A detailed discussion of these facilities, which may consist of either nonrecording or recording gages, is included in the following publications: "Equipment for Current-Meter Gaging Stations," U.S. Geological Survey Water Supply Paper 371; "Stream-Gaging Procedure," U.S. Geological Survey Water Supply Paper 888; and "Stream Flow," by Grover and Harrington, John Wiley & Sons, Inc., New York, 1943. The advice of Geological Survey engineers will be helpful in the site selection and installation, operation, and interpretation of records obtained.

A series of manuals "Techniques of Water-Resources Investigations of the United States Geological Survey," describes procedures for planning and executing specialized work in water-resources investigations. The material is grouped under major subject headings called books and further subdivided into sections and chapters; section A of book 3 is on surface water. The unit of publication, the chapter, is limited to a narrow field of subject matter. This format permits flexibility in revision and publication as the need arises.

Provisional drafts of chapters are distributed to field offices of the U.S. Geological Survey for their use. These drafts are subject to revision because of experience in use or because of advancement in knowledge, techniques, or equipment. After the technique described in a chapter is sufficiently developed, the chapter is published and is for sale by the Superintendent of Documents.[2]

The importance of utilizing records of runoff originating from the watershed above the damsite cannot be overemphasized. In the case of a damsite located on an ungaged stream, the establishment of measuring facilities as discussed above may produce basic data which would justify "eleventh hour" revision of the plans, thus improving the design of the dam.

G-4. *Precipitation Data.*—In each of the situations outlined in the preceding section, precipitation data are needed to evaluate factors for use in computing the probable maximum flood. The engineer should assemble the information with respect to precipitation

_____

[2]In loc. cit. p. 437

during the greater storms in the region, and particularly for those storms for which runoff records are available. Such information can be obtained from publications of the National Weather Service[3] and Environmental Data Service. At present (1974), daily precipitation data for each month for each State are contained in the publication "Climatological Data." Hourly data for each month for each State obtained by recording precipitation gages are contained in the publication "Hourly Precipitation Data."[4]   In areas where large storms have occurred, often precipitation data obtained by the National Weather Service precipitation stations have been supplemented by "bucket survey" data, i.e., information on rainfall amounts of unusual storms obtained from residents within the storm area by personnel of the National Weather Service and other Government agencies.

Locations of precipitation stations as of 1961 are shown on the series of maps "River Basin Maps showing Hydrologic Stations," previously referred to.

If plans are made to install streamflow measuring facilities as discussed in the preceding section, provision should also be made for obtaining precipitation records. An important item to consider is the selection of the location (or locations) of the precipitation gage, so that the catch will be a representative sample of average precipitation over the watershed. A comprehensive discussion of types of precipitation gages and observational

procedures is contained in the National Weather Service publication "Instructions for Climatological Observers," Circular B, eleventh edition, January 1962.

G-5. *Watershed Data.*—All available information concerning watershed characteristics should be assembled. A map of the area above the damsite should be prepared showing the drainage system, contours if available, drainage boundaries, and locations of any precipitation stations and streamflow gaging stations. Available data on soil types, cover, and land usage provide valuable guides to judgment of runoff potential. Soil maps prepared by the U.S. Department of Agriculture will prove helpful when the watershed lies within areas so mapped. These surveys (if in print) are available for purchase from the Superintendent of Documents, Washington, D.C. Out-of-print maps and other unpublished surveys may be available for examination from the U.S. Department of Agriculture, county extension agents, colleges, universities, and libraries.

The hydrologist preparing the flood study should make an inspection trip over the watershed to verify drainage area boundaries and soil and cover information, and to determine if any noncontributing areas are included within the drainage boundaries. The trip should also include visits to nearby watersheds if it is anticipated that records from nearby watersheds will be used in the study.

B. ANALYSES OF BASIC HYDROLOGIC DATA

G-6. *General.*—A flood hydrologist first directs attention to individual large flood events, seeking procedures whereby a good estimate may be made of the hydrograph that will result from a given amount of

precipitation. As floods which consist of combined snowmelt and rainfall runoff are difficult to separate into their two components, usually snowmelt floods and rain floods are analyzed separately. Analyses of rain floods only are discussed in these sections G-6 through G-8 with inclusion of examples of some mathematical computations. Considerations for runoff contribution from snowmelt are discussed separately in sections G-22 through G-26. Flood analyses of rainfall

---

[3]Official designation: U.S. Department of Commerce, National Oceanic and Atmospheric Administration, National Weather Service.

[4]Subscription to these publications may be made through the Superintendent of Documents, U.S. Government Printing Office, Washington, D.C. 20402.

data are interrelated to analyses of respective runoff data, so that discussions of procedures for one must include some references to the other. In the discussion that follows, analysis of storm rainfall is described first and is followed by a description of the analysis of the resulting flood runoff. Procedures used to analyze streamflow data for estimating the frequency of occurrence of flood magnitudes are discussed in sections G-28 and G-29.

G-7. *Estimating Runoff From Rainfall.*— (a) *General.*—The hydrometeorological approach to analyzing flood events and using the information obtained to estimate the magnitude of hypothetical floods requires a firm estimate of the difference between precipitation and the resulting runoff. From a flood determination point of view, this difference is considered loss, that is, loss from precipitation in the form of water over a given watershed. A simple solution to derive this loss value appears to be in finding the rate at which water will infiltrate the soil. If this infiltration rate is known, along with the amount of precipitation, a simple subtraction should give the amount of runoff. However, there are other precipitation losses in addition to infiltration, such as interception by vegetative cover, surface storage, and evaporation, that may have material effect on runoff amounts.

Various types of apparatus have been devised to test the infiltration rates of soils, and studies have been made of interception and evaporation losses. Although maps to an extremely large scale could define most of the surface storage area, it is apparent that an accurate volumetric evaluation of all the loss factors can be made only for a highly instrumented, small plot of ground and that such an evaluation is not practical for a natural watershed composed of many square miles of varying type soils, vegetative cover, and terrain features. For this reason, hydrologic literature contains arguments against the "infiltration rate approach" to determination of runoff amounts. However, the infiltration rate approach is applied on an empirical basis to obtain a practical solution to the problem of determining amounts of runoff, recognizing that the values used are of the nature of *index*

values rather than *true* values.

Natural events are studied and the difference between rainfall and runoff determined. Since this difference includes all the losses described above, it is usually called a *retention loss* or a *retention rate*. Such retention rates derived from available records may be adjusted to ungaged watersheds by analogy of soil type and cover.

The characteristics of a hydrograph must be understood so that respective amounts of runoff and precipitation are compared for estimating retention rates (and for other comparisons described later). A hydrograph of storm runoff obtained at a streamflow gaging station represents one or more of the following types of runoff from the watershed: channel runoff, surface runoff, interflow, and base flow. Brief definitions of these types are:

Channel runoff.—Caused by rain falling on the water surface of the stream. It begins with the start of precipitation and may be discernible from a slight rise of the hydrograph just after rainfall begins, but the quantity of channel runoff is so small that it is ignored in hydrograph analyses.

Surface runoff.—Occurs only when the rainfall rate is greater than the retention loss rate. This type of runoff causes most floods and the computational procedures in this text consider this type of runoff dominant.

Interflow.—Occurs when rainfall infiltrating the soil surface encounters an underground zone of lower permeability, travels above the zone to the surface downhill, and reappears to become surface runoff. This type of flow may also be called subsurface flow or *quick return flow*.

Base flow.—The fairly steady flow of a stream from natural storage as shown by hydrographs during nonstorm (or nonactive snowmelt) periods.

In flood hydrology it is customary to deal separately with base flow and to combine all other types of flow into *direct runoff* in unknown proportions as assumed in this text.

Making studies to compare rainfall with

runoff requires a knowledge of the units of measurement used and the factors for conversion to common units. These conversion factors are given in appendix F. In the United States, precipitation is measured in inches and runoff is measured in cubic feet per second (abbreviated c.f.s.).

It is necessary to know the watershed area contributing the runoff at a given measuring point, in order to express the runoff volume of inches of depth over the watershed for comparison with precipitation amounts. When making such comparisons, the amount of runoff, expressed as inches, is termed *rainfall excess*, and the difference between the rainfall excess and the total precipitation is considered retention loss as just discussed.

The following method of making a rainfall-runoff analysis has been selected for description in this text. The objectives of such analyses are: (1) the determination of a retention rate, and (2) the determination of the duration time interval of rainfall excess. A comparison of retention rates derived from several analyses leads to adoption of a rate for design flood computations. The determination of the duration of excess rainfall is necessary for the hydrograph analyses computations involving determinations of unitgraphs and lag-times, which are discussed later in this section and in sections beginning with G-9. In all such analyses, the runoff volume which is compared with precipitation amounts is that which relates directly to the rainfall under study. Therefore, the base flow of the streamflow hydrograph must be subtracted out before comparisons are made (see sec. G-8(c)).

(b) *Analysis of Observed Rainfall Data.*—

(1) *Mass curves of rainfall.*—Mass curves of cumulative rainfall during the storm period should be plotted for all precipitation stations in and near the basin as shown on figure G-1(A). To show clearly the relation of rainfall to runoff, it is sometimes desirable to plot the mass curves to the same time scale as the discharge hydrograph of storm runoff. Usually, however, the curves should be given a more expanded time scale than it is desirable to use for the hydrograph analysis. When only one recording station is located nearby, and in the absence of better information, the mass curve of precipitation at a nonrecording station is usually considered to be proportional in shape to that of the recording station, except as otherwise defined by the observer's readings and notes (fig. G-1(A)). The speed and direction of travel of the rainburst should be taken into account. Many rainfall observers enter the times of beginning and ending on the same line as the current daily reading. The notes may therefore refer to the previous day, especially when the gage is regularly read in the morning.

(2) *Isohyetal maps.*—The total amounts of rainfall occurring during the portion of the storm that produced the flood hydrograph under study should be determined from the mass curves for each station in and near the drainage area. For a flood hydrograph consisting of a single event, this will be the total depth of precipitation occurring during the storm period. For a compound hydrograph, in which individual portions of the hydrograph are studied separately, temporary cessations of rainfall will usually be indicated in the mass curves, and from inspection it usually will be apparent which of the increments of rainfall caused the runoff event under study. The appropriate depths of rainfall are then used to draw an isohyetal map, using standard procedures. A typical isohyetal map for plains-type terrain is shown on figure G-1(B). Isohyets are generally drawn smoothly, interpolating between precipitation stations. The interpolation should not be excessively mechanical.

Extreme caution should be used in drawing the isohyetal pattern in mountainous areas where the orographic effect is an important factor in the areal distribution of rainfall. For example, if there is a precipitation station in a valley on one side of a mountain range and another station in a valley on the opposite side of the range with no intervening station, it cannot be assumed that the rainfall during a storm would vary linearly between the two stations. It is likely that the rainfall would increase with increases in elevation on the windward side of the divide, whereas on the leeward side, precipitation would decrease

A, recording rain gage
B,C,D, nonrecording gages measured daily at 6 p.m.

Observer's notes:

B. Apr. 16. began 9 p.m.
    17. ended 9:30 a.m.
        began 11 a.m.
        ended 1 p.m.
        measured 6 p.m., 5.56 inches

D. Apr. 16. began 10 p.m.
    17. measured 8 a.m., 3.40 inches
        ended 1:30 p.m.
        measured 6 p.m., 4.06 inches (daily total)

C. Apr. 16. began 11 p.m.
    17. measured 6 p.m., 2.06 inches

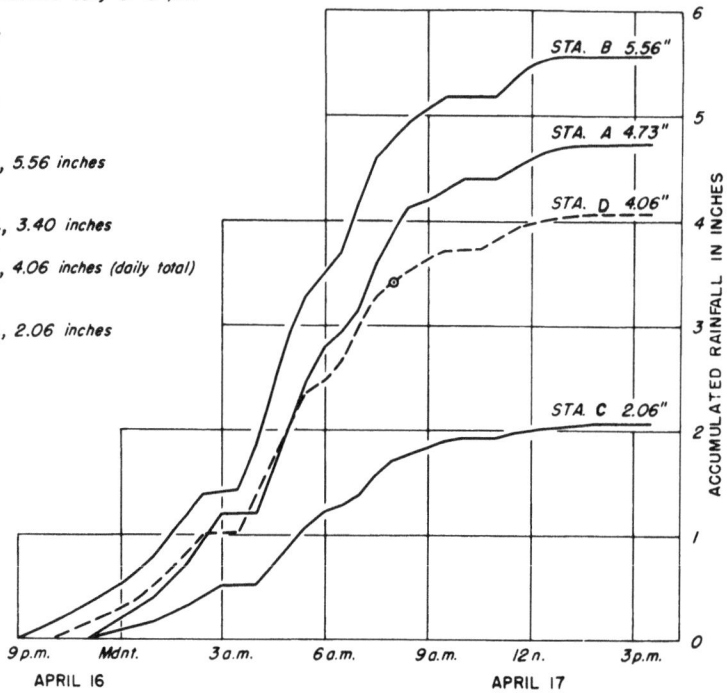

(A) MASS CURVES OF RAINFALL

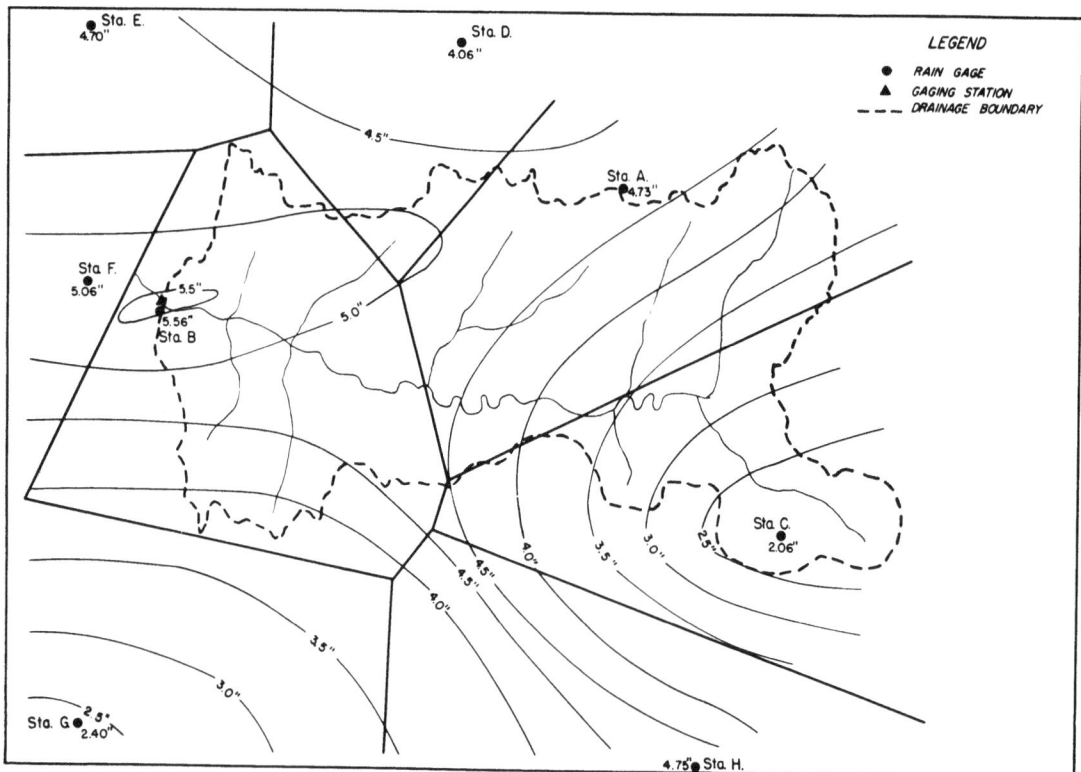

(B) ISOHYETS AND THIESSEN POLYGONS

Figure G-1. Analysis of observed rainfall data.—288-D-3158

rapidly with distance from the divide. This type of distribution can usually be verified in mountainous areas where there are sufficient precipitation stations to define the isohyetal pattern accurately.

A storm isohyetal pattern for mountainous terrain may be constructed by the isopercental technique, discussed in WMO Technical Note No. 98 [2] as follows:

"In mountainous regions the simple interpolation technique would yield unsatisfactory isohyets. Yet to prepare a valid isohyetal pattern in a mountainous region is not easy. One commonly used procedure is the isopercental technique, excellent under certain limited conditions stated in the next paragraph. This method requires a base chart of either mean annual precipitation, or preferably mean precipitation for the season of the storm, such as winter, summer, or monsoon months. In this method the ratio of the storm precipitation to the mean annual or mean seasonal precipitation (base precipitation) is plotted at each station. Isolines are drawn smoothly to these numbers. The ratios on the lines are then multiplied by the original base chart values at a large number of points to yield the storm isohyetal chart. Thus the storm isohyetal gradients and locations of centers tend to resemble the features of the base chart, which in turn is influenced by terrain.

"The first requirement for success of the isopercental technique is that a reasonably accurate mean annual or mean seasonal precipitation chart be available as a base. The base chart is of more value if it contains precipitation stations in addition to those reporting in the storm than if both charts are drawn exclusively from data observed at the same stations. The value of the base chart is also enhanced, in regions where the runoff of streams is a large percentage of the precipitation, if the precipitation shown on the chart has been adjusted not only for topographic factors, but also adjusted to agree with seasonal streamflow. In regions where a large percentage of the precipitation evaporates, adjustment to runoff volumes would be of dubious value.

"An additional requirement for success of the isopercental technique is that most of the annual or seasonal precipitation in the region result from storms with relatively the same wind direction, and from storms with minimal convective activity. Under these circumstances an individual storm will have a strong resemblance to the mean chart, as the latter is an average of kindred storms.

"In the Tropics with the dominance of convective activity and with lighter winds, the isopercental technique is of less value in analysis of an individual storm than in middle latitude locations that meet the other requirements."

After the preliminary hydrographs and the isohyetal maps have been drawn, the atypical flood events for unit hydrographs determination may readily be eliminated. *Those floods having a combination of large volume, uniform intensities, isolated periods of rainfall, and uniform areal distribution of rainfall, should be chosen for further study.*

(3) *Average rainfall by Thiessen polygons.*—The average rainfall on a drainage area can be determined from precipitation station records by the Thiessen polygon method. A sample computation of average hourly rainfall from the mass curves on figure G-1(A), using Thiessen polygons indicated on figure G-1(B), is given in table G-1.

The first step is to construct the Thiessen polygons, which are the areas bounded by the perpendicular bisectors of lines joining adjacent precipitation stations. The percentage of the drainage area controlled by each station's polygon is planimetered and entered in table G-1. Next, the average depth of rainfall over each station's polygon is determined by planimetering areas between isohyets on figure G-1(B). A factor to be used in weighing station rainfall values is obtained by multiplying the percentage of the drainage area controlled by each station's polygon by the ratio of the average depth of rainfall over each station's polygon to the observed rainfall at the station, and dividing by 100.

Hourly incremental rainfall values are determined for each precipitation station from

Table G-1.—*Computation of rainfall increments*

COMPUTATION OF STATION WEIGHTS

Station (1)	Average rainfall over Thiessen polygon (2)	Percent of basin area (3)	Rainfall at station (4)	Weight, col. (2) x col. (3) / 100 x col. (4) (5)
A	4.3	38.9	4.73	0.35
B	4.6	37.0	5.56	.31
C	2.8	21.1	2.06	.29
D	5.0	3.0	4.06	.04

COMPUTATION OF WEIGHTED AVERAGE HOURLY RAINFALL OVER BASIN

Time, hours	Station A Mass rf. (1)	Station A Δ rf. (2)	Station A 0.35×Δrf. (3)	Station B Mass rf. (1)	Station B Δ rf. (2)	Station B 0.31×Δrf. (3)	Station C Mass rf. (1)	Station C Δ rf. (2)	Station C 0.29×Δrf. (3)	Station D Mass rf. (1)	Station D Δ rf. (2)	Station D 0.04×Δrf. (3)	Weighted average, sum of cols. (3)
0				0									
1				.17	0.17	0.053				0			0.053
2	0			.33	.16	.050	0			.15	0.15	0.006	.056
3	.20	0.20	0.070	.52	.19	.059	.09	0.09	0.026	.29	.14	.006	.161
4	.40	.20	.070	.80	.28	.087	.17	.08	.023	.52	.23	.009	.189
5	.73	.33	.116	1.20	.40	.124	.32	.15	.044	.84	.32	.013	.297
6	1.20	.47	.164	1.41	.21	.065	.52	.20	.058	1.01	.17	.007	.294
7	1.20	0	0	1.85	.44	.136	.52	0	0	1.34	.33	.013	.149
8	2.05	.85	.298	2.91	1.06	.329	.89	.37	.107	2.05	.71	.028	.762
9	2.80	.75	.262	3.49	.58	.180	1.22	.33	.096	2.47	.42	.017	.555
10	3.15	.35	.122	4.19	.70	.217	1.37	.15	.044	3.00	.53	.021	.404
11	3.90	.75	.262	4.79	.60	.186	1.70	.33	.096	3.40	.40	.016	.560
12	4.20	.30	.105	5.08	.29	.090	1.83	.13	.038	3.63	.23	.009	.242
13	4.40	.20	.070	5.18	.10	.031	1.92	.09	.026	3.73	.10	.004	.131
14	4.40	0	0	5.18	0	0	1.92	0	0	3.83	.10	.004	.004
15	4.59	.19	.066	5.49	.31	.096	2.00	.08	.023	3.97	.14	.006	.191
16	4.70	.11	.038	5.56	.07	.022	2.04	.04	.012	4.04	.07	.003	.075
17	4.73	.03	.010	5.56	0	0	2.06	.02	.006	4.06	.02	.001	.017
Total		4.73	1.653		5.56	1.725		2.06	.599		4.06	.163	4.140

the mass curves of figure G-1(A) and are multiplied by the appropriate weight factors as shown in table G-1, to obtain the total for the drainage area.

Additional information on determining average rainfall is given in "Cooperative Studies Technical Paper No. 1," published by the National Weather Service, and in references [2] and [17].

(4) *Determination of rainfall excess.*—Two methods may be used to determine rainfall excess: by assuming a constant average retention rate throughout the storm period, and by assuming a retention rate varying with time. The capacity rate of retention decreases progressively throughout the storm period until a constant minimum rate is reached if the rain is sufficiently prolonged. With dry antecedent conditions, the initial capacity rate will be greater and will decline faster. Because the use of a varying retention rate requires a complicated method of computation, it is often preferable to assume an average retention rate (sometimes referred to as infiltration index) with an estimate of initial loss being made if antecedent conditions are relatively dry.

The method of determining the period of rainfall excess, when an average retention rate is used, is a trial-and-error process in which a retention rate is assumed and subtracted from hourly rainfall increments determined as the average over the basin. Various retention rates are assumed until the total of the computed rainfall excess equals the measured storm runoff. An example of this procedure is given in table G-2. If the correct retention rate has not been assumed after two trials, a rainfall

Table G-2.—*Computation of rainfall excess*

Time, hours	Rainfall increment (basin average), inches	First trial		Second trial		Third trial	
		Assumed retention rate, inches per hour	Rainfall excess, inches	Assumed retention rate, inches per hour	Rainfall excess, inches	Assumed retention rate, inches per hour	Rainfall excess, inches
0							
1	0.05	0.25		0.15		0.17	
2	.06						
3	.16				0.01		
4	.19				.04		0.02
5	.30		0.05		.15		.13
6	.29		.04		.14		.12
7	.15		0		0		0
8	.76		.51		.61		.59
9	.56		.31		.41		.39
10	.40		.15		.25		.23
11	.56		.31		.41		.39
12	.24		0		.09		.07
13	.13				0		0
14	0				0		0
15	.19				.04		.02
16	.08						
17	.02	.25		.15		.17	
Total	4.14		1.37		2.15		1.96

Total rainfall, 4.14 inches; observed runoff, 2.0 inches; total retention in 17 hours, 2.1 inches. The average retention rate of 0.17 inch per hour assumed in the third trial gives the best agreement of computed rainfall excess with measured runoff.

excess-retention curve will facilitate the solution. In the example of table G-2, the curve could be drawn through the two points represented by the coordinates 0.25, 1.37, and 0.15, 2.15. The correct retention rate corresponding to a rainfall excess of 2.0 inches would then be taken from this curve.

The duration time of excess rainfall is that time during which rainfall increments exceed the average retention rate. In the third trial, table G-2, the duration time may be taken as either 8 or 9 hours, or as two periods, one of 2 or 3 hours, and the other of 5 hours (the final 0.02 inch of precipitation being disregarded), according to the characteristics of the hydrograph. A small amount of excess rain in a marginal period is frequently assumed to have occurred within only a small part of that period and may be neglected.

(5) *Discussion of observed rainfall analyses procedures.*—The above classic procedure of rainfall-runoff analysis is simple and satisfactory, given rainfall data such as used in the illustration and a relatively homogeneous watershed not exceeding a few hundred square miles in area. As stated earlier in section G-7(a): "A comparison of retention rates derived from several analyses leads to adoption of a rate for design flood computations." Experienced judgment is needed for such comparison with due reconsideration given to the characteristics of the data for each analysis and of the watershed. The selected rate is not necessarily the minimum rate computed. Mass curves of rainfall and isohyetal patterns should always be constructed as described in sections G-7(b)(2) and (3) to obtain good results from any rainfall-runoff analysis.

The importance in flood computations of good estimates of retention losses is evident. As the ratio of retention loss to flood causative precipitation increases, the relative effect of retention loss estimates on resulting flood magnitudes increases. Research studies directed towards improved understanding and evaluation of all processes contributing to retention losses are increasing yearly. Many complex functions are being tested by electronic computer programs to model such processes. However, the most practical approach for estimating natural watershed retention losses continues to be use of empirically derived relationships, preferably from records within the watershed.

Often, relationships as percentages of runoff to rainfall, runoff coefficients, are obtained by analyses and judicially used in flood studies. This approach may be practical in cases where basic data are meager.

The following extract from WMO Technical Note No. 98 [2] gives information of a method that may be used.

"... For a particular river basin with records of streamflow and precipitation, a common procedure is to develop multiple variable rainfall-runoff correlations. Such correlations may be derived either graphically or analytically. They usually involve at least four variables, (i) depth of storm rainfall over the basin, (ii) surface runoff volume from the storm event, (iii) an index of moisture conditions in the basin prior to the storm, and (iv) a seasonal factor. In some cases storm duration is included as a fifth variable. The methods of determining these factors from the observational records in a basin or a region and graphical and analytic procedures for multiple-variable correlation analyses are outlined in the WMO Guide to Hydrometeorological Practices, Annex A, WMO 168.TP.82."

A hydrologist making an inflow design flood study seldom finds rainfall-runoff records for the watershed above a particular damsite adequate to establish a good estimate of retention loss for the watershed. Recourse is then made to information of analyses for other watersheds having similar runoff characteristics. For example, hydrologists of the Soil Conservation Service, U.S. Department of Agriculture, have made extensive analyses of runoff from small experimental watersheds having individually homogeneous soil and cover characteristics but such characteristics differing between watersheds. A procedure was developed from these studies for estimating runoff from precipitation for any watershed for which certain soil and cover data are known; such soil and cover data are usually obtainable or subject to reasonable approximations [3].

The SCS procedure with modifications to fit specific purposes is described in appendix A of the Bureau of Reclamation publication "Design of Small Dams," second edition [31]. An abridgement of that description is given in the following subsection. (The descriptive items have been renumbered for convenience.)

(6) *Method of estimating retention losses.*—This method consists of the following steps:

(I). Classification of watershed soils into hydrologic groups A, B, C, or D, and estimation of percent of areal extent of each in the watershed.

(II). Identification of land use characteristics dominant for each hydrologic group.

(III). The combination of a hydrologic group and its land use characteristics to give a *hydrologic soil-cover complex*

identification for entering tables from which respective runoff curve numbers, CN, may be obtained.

(IV). Runoff values are obtained from a family of curves on a plot of rainfall versus runoff or by solution of the equation used to define the curves.

(V). Three antecedent moisture conditions, AMC, of a watershed are considered in relation to curve numbers; namely, AMC-I, AMC-II, AMC-III.

The mathematical procedure is given in this text with minimum definitions of the terms used in the procedure and without inclusion of a list of about 4,000 soil-type names and respective hydrologic group classifications compiled by the Soil Conservation Service. A full discussion of the procedure including the list of soil-type names is given in "Design of Small Dams" [31]. Information on the development of the runoff curves may be found in the SCS National Engineering Handbook [3].

Further explanation of each of the above steps follows.

(I) *Hydrologic soil groups.*—Four major soil groups are used. The soils are classified on the basis of intake of water at the end of long-duration storms occurring after prior wetting and opportunity for swelling, and without the protective effects of vegetation.

In the definitions that follow, the *infiltration rate* is the rate at which water enters the soil at the surface and which is controlled by surface condition, and the *transmission rate* is the rate at which the water moves in the soil and which is controlled by the soil horizons. The hydrologic soil groups, as defined by SCS soil scientists, are as follows:

Group A (low runoff potential).—Soils having high infiltration rates even when thoroughly wetted and consisting chiefly of deep, well to excessively drained sands or gravels. These soils have a high rate of water transmission.

Group B.—Soils having moderate infiltration rates when thoroughly wetted and consisting chiefly of moderately deep to deep, moderately well to well drained soils with moderately fine to moderately coarse textures. These soils have a moderate rate of water transmission.

Group C.—Soils having slow infiltration rates when thoroughly wetted and consisting chiefly of soils with a layer that impedes downward movement of water, or soils with moderately fine to fine texture. These soils have a slow rate of water transmission.

Group D (high runoff potential).—Soils having very slow infiltration rates when thoroughly wetted and consisting chiefly of clay soils with a high swelling potential, soils with a permanent high water table, soils with a claypan or clay layer at or near the surface, and shallow soils over nearly impervious material. These soils have a very slow rate of water transmission.

(II). *Land use and treatment classes.*—These classes are used in the preparation of hydrologic soil-cover complexes (identified herein as item III), which in turn are used in estimating direct runoff. Types of land use and treatment are classified on a flood runoff-producing basis. The greater the ability of a given land use or treatment to increase total retention, the lower it is on a flood runoff-production scale. Land use or treatment types not described here may be classified by interpolation.

Crop rotations.—The sequence of crops on a watershed must be evaluated on the basis of its hydrologic effects. Rotations range from *poor* (or weak) to *good* (or strong) largely in proportion to the amount of dense vegetation in the rotation. *Poor rotations* are those in which a row crop or small grain is planted in the same field year after year. A poor rotation may combine row crops, small grains, or fallow, in various ways. *Good rotations* will contain alfalfa or other close-seeded legumes or grasses, to improve tilth and increase infiltration. For example, a 2-year rotation of wheat and fallow may be a good rotation for crop production where low annual rainfall is a limiting factor, but hydrologically it is a poor rotation.

Native pasture and range.—Three conditions are used, based on hydrologic considerations, not on forage production. *Poor pasture or range* is heavily grazed, has no mulch, or has plant cover on less than about 50 percent of the area. *Fair pasture or range* has between about 50 and 75 percent of the area with plant cover and is not heavily grazed. *Good pasture or range* has more than about 75 percent of the area with plant cover, and is lightly grazed.

Farm woodlots.—The classes are based on hydrologic factors, not on timber production. *Poor woodlots* are heavily grazed and regularly burned in a manner that destroys litter, small trees, and brush. *Fair woodlots* are grazed but not burned. These woodlots may have some litter, but usually these woods are not protected. *Good woodlots* are protected from grazing so that litter and shrubs cover the soil.

Forests.—See hydrologic soil-cover complex, item III following.

Straight-row farming.—This class includes up-and-down and cross-slope farming in straight rows. In areas of 1 or 2 percent slope, cross-slope farming in straight rows is almost the same as contour farming. Where the proportion of cross-slope farming is believed to be significant, it may be classed halfway between straight-row and contour farming in the table G-3(A).

Contouring.—Contour furrows used with small grains and legumes are made while planting, are generally small, and tend to disappear due to climatic action. Contour furrows, and beds on the contour, as used with row crops are generally large. They may be made in planting and later reduced in size by cultivation, or they may be insignificant after planting and become large from cultivation. Average conditions are used in table G-3(A).

Surface runoff reductions due to contour farming are greater as land slopes decrease. The curve numbers for contouring shown in table G-3(A) were obtained using data from experimental watersheds having slopes of 3 to 8 percent.

Contour furrows in pasture or range land are usually of the permanent type. Their dimensions and spacing generally vary with climate and topography. Table G-3(A) considers average conditions in the Great Plains.

Terracing.—Terraces may be graded, open-end level, or closed-end level. The effects of graded and open-end level terraces are considered in table G-3(A), and the effects of both contouring and the grass waterway outlets are included.

When considering land use and treatment classes for hydrologic soil groups within a large watershed, the above definitions should be applied broadly, estimating percentage of land use in each group, assigning proper CN and computing a weighed CN for each particular soil group.

(III) *Hydrologic soil-cover complexes.*—Combinations of hydrologic soil groups and land use and treatment classes into hydrologic soil-cover complexes with respective curve numbers are given in table G-3(A), (B), (C). The numbers show the relative value of the complexes as direct runoff-producers. The higher the number, the greater the amount of direct runoff to be expected from a storm. Table G-3(A) is applicable to farm lands and related areas, and table G-3(B) is applicable to forested watersheds. A more detailed method of estimating curve numbers for heavy forested land in humid regions is given in appendix A of "Design of Small Dams," second edition [31].

Table G-3(C) is applicable for forest-range areas in the Western United States. Descriptions of the types of cover listed are as follows:

Herbaceous.—Grass-weed-brush mixtures with brush the minor element.

Oak-Aspen.—Mountain brush mixtures of oak, aspen, mountain mahogany, bitter brush, maple, and other brush.

Juniper-Grass.—Juniper or pinon with an understory of grass.

Sage-Grass.—Sage with an understory of grass.

(IV) *Rainfall-runoff curves for estimating*

direct runoff amounts.—The curves of figure G-2 are obtained using the equation:

$$Q = \frac{(P - 0.2S)^2}{P + 0.8S} \qquad (1)$$

where:

Q = direct runoff, in inches
P = storm rainfall, in inches, and
S = maximum potential difference between P and Q, in inches, at time of storm's beginning.

There is some loss of rainfall before runoff begins due principally to interception, infiltration, and surface storage, so provision for an initial abstraction I_a is included in the runoff equation (see diagram on figure G-2). With the condition that I_a cannot be greater than P, an empirical relationship of $I_a = 0.2S$ was adopted in developing the equation, obtaining the empirical relationship of I_a and S from data from watersheds in various parts of the country.

For convenience in interpolation, the curves of figure G-2 are numbered from 100 to zero. The numbers are related to S as follows:

$$\text{Curve number, CN} = \frac{1,000}{10 + S} \qquad (2)$$

The procedure recommended in this text for estimating incremental rainfall excesses from design storm rainfall using appropriate CN and figure G-2 or the runoff equation is given in section G-19. In the process of hydrograph analyses, preliminary estimates of curve numbers for a watershed can be quickly obtained from figure G-2 by using total storm rainfall and runoff amounts. However, such preliminary estimates have to be revised by trial computations of rainfall excesses using the procedure given later in section G-19.

(V) *Antecedent moisture conditions.*—The following generalized criteria define three antecedent moisture conditions of watersheds used in the development of the runoff curve numbers.

AMC-I.—A condition of watershed soils where the soils are dry but not to the

Table G-3.—*Hydrologic soil-cover complexes and respective curve numbers (CN)*

(A) RUNOFF CURVE NUMBERS (CN) FOR FARMLANDS AND RELATED AREAS

[FOR WATERSHED CONDITION AMC-II]

Land use or cover	Treatment or practice	Hydrologic condition for infiltrating	Hydrologic soil group			
			A	B	C	D
Fallow	SR		77	86	91	94
Row crops	SR	Poor	72	81	88	91
	SR	Good	67	78	85	89
	C	Poor	70	79	84	88
	C	Good	65	75	82	86
	C&T	Poor	66	74	80	82
	C&T	Good	62	71	78	81
Small grain	SR	Poor	65	76	84	88
	SR	Good	63	75	83	87
	C	Poor	63	74	82	85
	C	Good	61	73	81	84
	C&T	Poor	61	72	79	82
	C&T	Good	59	70	78	81
Close-seeded legumes [1] or rotation meadow.	SR	Poor	66	77	85	89
	SR	Good	58	72	81	85
	C	Poor	64	75	83	85
	C	Good	55	69	78	83
	C&T	Poor	63	73	80	83
	C&T	Good	51	67	76	80
Pasture or range		Poor	68	79	86	89
		Fair	49	69	79	84
		Good	39	61	74	80
	C	Poor	47	67	81	88
	C	Fair	25	59	75	83
	C	Good	6	35	70	79
Meadow (permanent).		do	30	58	71	78
Woods (farm woodlots).		Poor	45	66	77	83
		Fair	36	60	73	79
		Good	25	55	70	77
Farmsteads			59	74	82	86
Roads (dirt)[2] (hard surface).[2]			72	82	87	89
			74	84	90	92

[1] Close-drilled or broadcast.
[2] Including right-of-way.
 SR = Straight row.
 C = Contoured.
 T = Terraced.
 C&T = Contoured and terraced.

(U.S. Soil Conservation Service.)

(B) RUNOFF CURVE NUMBERS (CN) FOR FORESTED WATERSHEDS

COMMERCIAL OR NATIONAL FOREST, FOR WATERSHED CONDITION AMC-II

Hydrologic condition class	Hydrologic soil group			
	A	B	C	D
I. Poorest	56	75	86	91
II. Poor	46	68	78	84
III. Medium	36	60	70	76
IV. Good	26	52	62	69
V. Best	15	44	54	61

(C) RUNOFF CURVE NUMBERS (CN) FOR FOREST RANGE AREAS IN WESTERN UNITED STATES (AMC-II)

Cover	Condition	Soil groups			
		A	B	C	D
Herbaceous	Poor	---	78	85	92
	Fair	---	68	81	88
	Good	---	59	71	84
Sage-Grass	Poor	---	64	78	---
	Fair	---	46	67	---
	Good	---	35	46	---
Oak-Aspen	Poor	---	63	71	---
	Fair	---	40	54	---
	Good	---	30	40	---
Juniper-Grass	Poor	---	73	84	---
	Fair	---	54	70	---
	Good	---	40	59	---

wilting point, and when satisfactory plowing or cultivation takes place. (This condition is *not* considered applicable to the design flood computation methods presented in this text.)

AMC-II.—The average case for *annual floods*, that is, an average of the conditions which have preceded the occurrence of the maximum annual flood on numerous watersheds.

AMC-III.—Heavy rainfall has occurred during the 5 days previous to the given storm and the soil is nearly saturated.

Curve numbers in table G-3(A), (B), (C) for hydrologic soil-cover complexes all relate to AMC-II. Table G-4(A) lists curve numbers for AMC-II with respective S values (column (4)) and $0.2S$ values (column (5)) which may be used to solve the runoff equation on figure

G-2. Curve numbers for AMC-I and AMC-III respective to the CN for AMC-II in column (1) are listed in columns (2) and (3). This information is useful for estimating retention losses. If data are available for analyzing observed storms and resulting runoff, an estimate of antecedent moisture condition of a watershed may be made from table G-4(B).

G-8. *Analyses of Streamflow Data*.—Streamflow data at a given location may consist of: (1) a continuous hydrograph of discharges obtained from waterstage recording mechanisms; (2) mean (average) daily discharges computed from waterstage recorders or from once or twice daily observed water stages; or, in some instances (3) peak discharges computed from flood marks or crest stage gages. U.S. Geological Survey publications should be consulted for information about

collection and processing these data for publication. However, one should be aware that U.S.G.S. publications give for each published station record an estimate of the degree of accuracy of field data and computed results for that record as follows:

"*Excellent* means that about 95 percent of the daily discharges are within 5 percent; *good*, within 10 percent; and *fair*, within 15 percent. *Poor* means that daily discharges have less than *fair* accuracy."

Objectives of streamflow data analyses for inflow design flood computations are:

(1) Determinations of watershed retention losses (previously discussed).

(2) Determination of characteristic watershed response to precipitation; this is usually accomplished by deriving a unit hydrograph for the watershed. (Complex

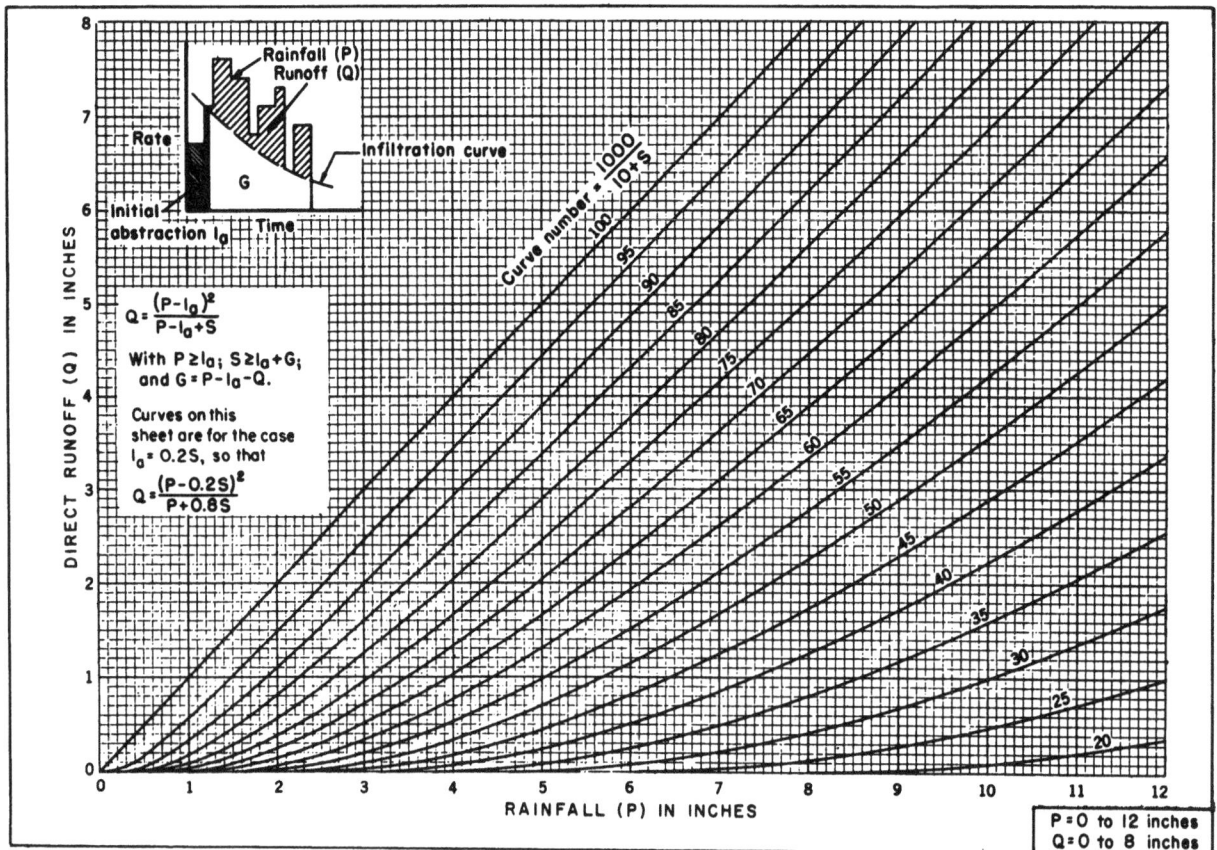

Figure G-2. Rainfall-runoff curves—solution of runoff equation, $Q = \dfrac{(P - 0.2S)^2}{P + 0.8S}$ (sheet 1 of 2) (U.S. Soil Conservation Service).—288-D-3178(1/2)

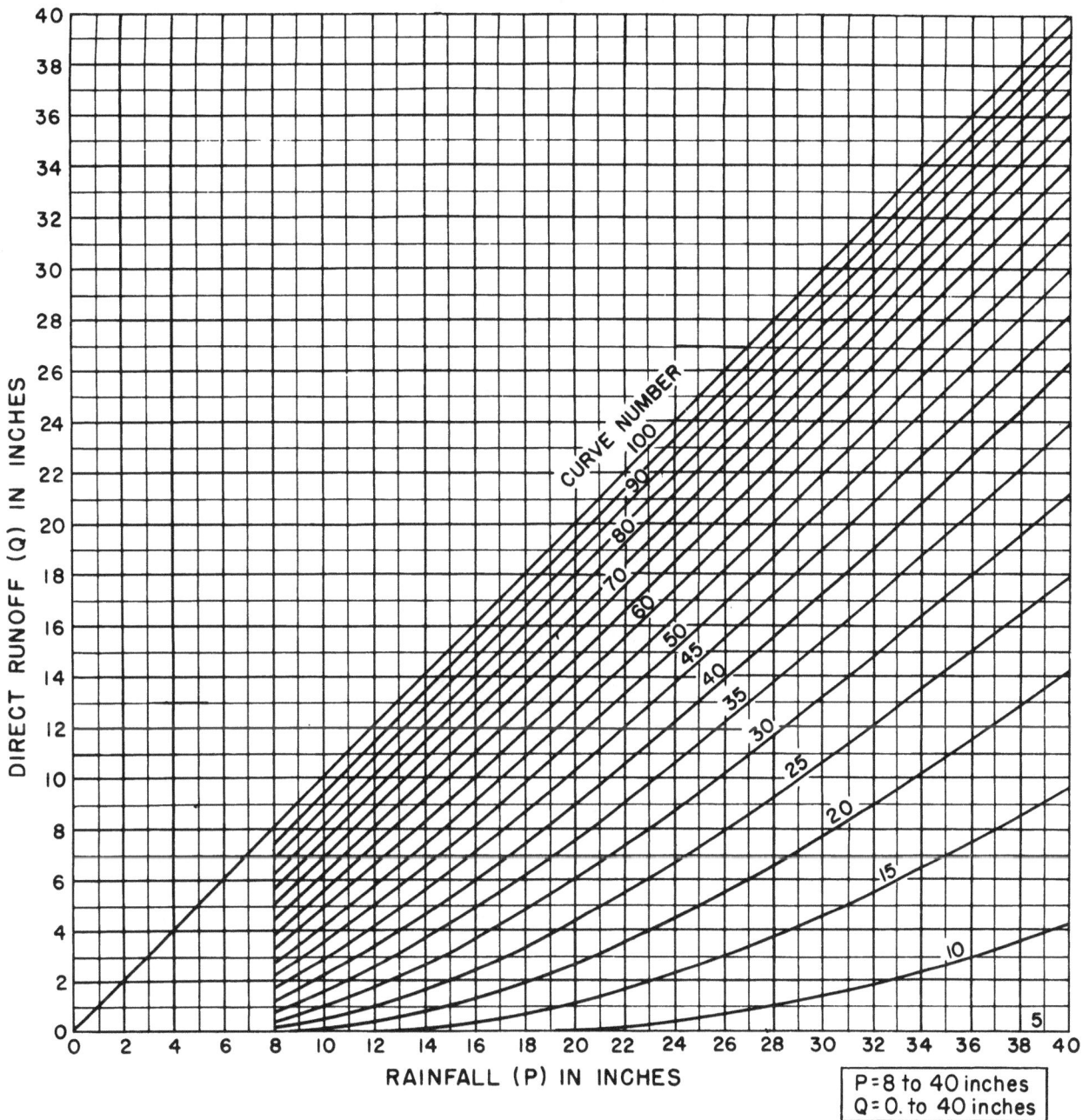

Figure G-2. Rainfall-runoff curves—solution of runoff equation, $Q = \dfrac{(P - 0.2S)^2}{P + 0.8S}$ (sheet 2 of 2) (U.S. Soil Conservation Service).—288-D-3178(2/2) (*Note*: Curve designated by number is *below* number.)

computer-programed watershed runoff models may use other means of estimating time distribution of runoff.)

Continuous hydrographs can provide for estimates of retention loss variations with time, with accumulative loss, or with accumulative precipitation. Mean daily discharges can provide ratio estimates of total retention loss to total storm precipitation.

Continuous hydrographs are essential to unit hydrograph derivations from recorded streamflow data. When mean daily discharges only are available, a continuous hydrograph is sketched for making unit hydrograph

Table G-4.—*Curve numbers, constants, and seasonal rainfall limits*

(A) CURVE NUMBERS (CN) AND CONSTANTS FOR THE CASE $I_a = 0.2S$

1	2	3	4	5	1	2	3	4	5
CN for condition II	CN for conditions		S values*	Curve* starts where $P =$	CN for condition II	CN for conditions		S values*	Curve* starts where $P =$
	I	III				I	III		
			inches	*inches*				*inches*	*inches*
100	100	100	0	0	60	40	78	6.67	1.33
99	97	100	.101	.02	59	39	77	6.95	1.39
98	94	99	.204	.04	58	38	76	7.24	1.45
97	91	99	.309	.06	57	37	75	7.54	1.51
96	89	99	.417	.08	56	36	75	7.86	1.57
95	87	98	.526	.11	55	35	74	8.18	1.64
94	85	98	.638	.13	54	34	73	8.52	1.70
93	83	98	.753	.15	53	33	72	8.87	1.77
92	81	97	.870	.17	52	32	71	9.23	1.85
91	80	97	.989	.20	51	31	70	9.61	1.92
90	78	96	1.11	.22	50	31	70	10.0	2.00
89	76	96	1.24	.25	49	30	69	10.4	2.08
88	75	95	1.36	.27	48	29	68	10.8	2.16
87	73	95	1.49	.30	47	28	67	11.3	2.26
86	72	94	1.63	.33	46	27	66	11.7	2.34
85	70	94	1.76	.35	45	26	65	12.2	2.44
84	68	93	1.90	.38	44	25	64	12.7	2.54
83	67	93	2.05	.41	43	25	63	13.2	2.64
82	66	92	2.20	.44	42	24	62	13.8	2.76
81	64	92	2.34	.47	41	23	61	14.4	2.88
80	63	91	2.50	.50	40	22	60	15.0	3.00
79	62	91	2.66	.53	39	21	59	15.6	3.12
78	60	90	2.82	.56	38	21	58	16.3	3.26
77	59	89	2.99	.60	37	20	57	17.0	3.40
76	58	89	3.16	.63	36	19	56	17.8	3.56
75	57	88	3.33	.67	35	18	55	18.6	3.72
74	55	88	3.51	.70	34	18	54	19.4	3.88
73	54	87	3.70	.74	33	17	53	20.3	4.06
72	53	86	3.89	.78	32	16	52	21.2	4.24
71	52	86	4.08	.82	31	16	51	22.2	4.44
70	51	85	4.28	.86	30	15	50	23.3	4.66
69	50	84	4.49	.90					
68	48	84	4.70	.94	25	12	43	30.0	6.00
67	47	83	4.92	.98	20	9	37	40.0	8.00
66	46	82	5.15	1.03	15	6	30	56.7	11.34
65	45	82	5.38	1.08	10	4	22	90.0	18.00
64	44	81	5.62	1.12	5	2	13	190.0	38.00
63	43	80	5.87	1.17	0	0	0	infinity	infinity
62	42	79	6.13	1.23					
61	41	78	6.39	1.28					

*For CN in column 1 (value = 0.2S)

(B) SEASONAL RAINFALL LIMITS FOR AMC

AMC group	Total 5-day antecedent rainfall, inches	
	Dormant season	Growing season
I	Less than 0.5	Less than 1.4
II	0.5 to 1.1	1.4 to 2.1
III	Over 1.1	Over 2.1

estimates; the chance of introducing considerable error is obvious. Discussions which follow assume continuous hydrographs obtained from continuous recording waterstage records converted to discharges expressed as cubic feet per second (c.f.s.), the degree of accuracy of the records being *excellent* or *good*.

(a) *Unit Hydrograph (Unitgraph) Principles.*—The 1970 USCOLD report [1] states: "In general the unit hydrograph method, in conjunction with the estimated probable maximum precipitation, is used in estimating probable maximum floods...." The unit hydrograph principle was originally developed by Sherman [4] in 1932. Although numerous refinements have been added by other investigators, the basic principles as presented by Sherman remain the same. These principles as now applied are given and illustrated on figure G-3.

Sherman's definition of unit hydrograph did not imply a specific volume of runoff, and the term was applied to the observed hydrograph as well as to a hydrograph of 1-inch volume computed from the observed graph. In present practice, observed hydrographs are usually identified as such, and the term *unitgraph* refers either to the 1-inch volume unitgraph derived from a specific observed hydrograph or to a 1-inch volume unitgraph representative of the watershed and used to compute synthetic floods from rainfall excess over the watershed. Random variations in rainfall rate in respect to time and area have a great effect on the shape of the runoff hydrograph. To minimize the effect of the time variations in rainfall rate, it has been found that the rainfall excess duration time of a basin unitgraph should not exceed one-fourth the basin lag-time as defined in section G-8(e), and the shorter the rainfall excess period with respect to lag-time, the better the unitgraph results are likely to be.

The term *unit hydrograph*, or *unitgraph*, as used in this text always means 1-inch volume of runoff; the volume notation is seldom included. The rainfall excess unit duration time is always given for a watershed representative unitgraph.

Natural flood hydrographs at a given stream gage are assumed to give integrated results of all interdependent effects on runoff such as watershed precipitation, retention losses, and routing effects of watershed vegetative cover and channel systems. A unit hydrograph which has been derived from recorded floods at a given stream location, and which will give close reconstruction of recorded flood hydrographs from recorded respective precipitation events as affected by retention losses, is considered representative of that particular watershed and also considered representative of other watersheds having similar runoff characteristics.

On this basis, synthetic unit hydrographs for ungaged basins are derived by judging comparative watershed characteristics and adjusting "representative" unit hydrographs to fit the size and lag-time of the ungaged watershed. Mathematical watershed runoff models are currently being developed by computer integration of meteorological, hydrological, and physiographical factors. Some hydrologists prefer to use these models rather than a unitgraph. However, each model includes constants related to watershed characteristics that must be empirically determined by trial analyses of recorded flows. As in the application of synthetic unitgraphs, transference of a mathematical model from a gaged to an ungaged watershed also requires experienced judgment of the effect from variations in watershed characteristics.

The use of the unit hydrograph is limited in the following ways:

(1) The principle of the unit hydrograph is applicable to basins of any size. However, it is desirable in the derivation of unitgraphs to use storms that are well distributed over the entire basin and produce runoff nearly concurrently from all parts of it. Such storms rarely occur over large areas. The extent of the basin for which a unitgraph may be derived from observed data is therefore limited in each case to the areal extent of rainfalls that have been observed.

(2) Hydrographs containing more than small amounts of snowmelt runoff are

Definitions:

Unitgraph - A hydrograph of direct runoff at a given point that will result from an isolated event of rainfall excess occurring within a unit of time and spread in an average pattern over the contributing drainage area. Identified by by the unit time and volume of the excess rainfall, that is 1-hour 1-inch unitgraph.

Rainfall excess - That portion of rainfall that enters a stream channel as direct runoff and produces the runoff hydrograph at the measuring point, base flow included.

Basic Assumptions:

(1)   The effects of all physical characteristics of a given drainage basin are reflected in the shape of the direct runoff hydrograph for that basin.

(2)   At a given point on a stream, discharge ordinates of different unitgraphs of the same unit time of rainfall excess are mutually proportional to respective volumes. See (A) at left.

(3)   A hydrograph of storm discharge that would result from a series of bursts of excess rain or from continuous excess rain of variable intensity may be constructed from a series of over-lapping unitgraphs each resulting from a single increment of excess rain of unit duration. See (B) at left.

Practical Application:

For a given runoff contributing area, a unitgraph representing exactly one inch of runoff (rainfall excess) for a selected unit time interval is computed. Increments of rainfall excess for the same unit time interval are determined for a storm. A total hydrograph of direct runoff from the storm is then computed using assumptions (2) and (3) above. See graph (B) at left.

Note: Direct runoff is defined in section G-8.

Figure G-3. Unit hydrograph principles (sheet 2 of 2).−288-D-3179(2/2)

the use of rainfall increments of measurable duration. When unitgraphs are combined they produce a regular undulation similar to a harmonic with a period equal to that of the rainfall increments, superimposed upon the fundamental hydrograph. Another obstacle to exact reproduction is the fact that the successive rainfall increments do not have the same isohyetal pattern and a single form of unitgraph is not strictly applicable to all of them. These phenomena contradict, to a certain extent, the third basic assumption of the unit hydrograph (fig. G-3). They can be disregarded in the synthesis of hydrographs, but frequently cause difficulty in the use of arithmetical procedures for analyzing them.

An engineer attempting unitgraph analyses or researching literature regarding unitgraphs soon becomes aware that the three basic

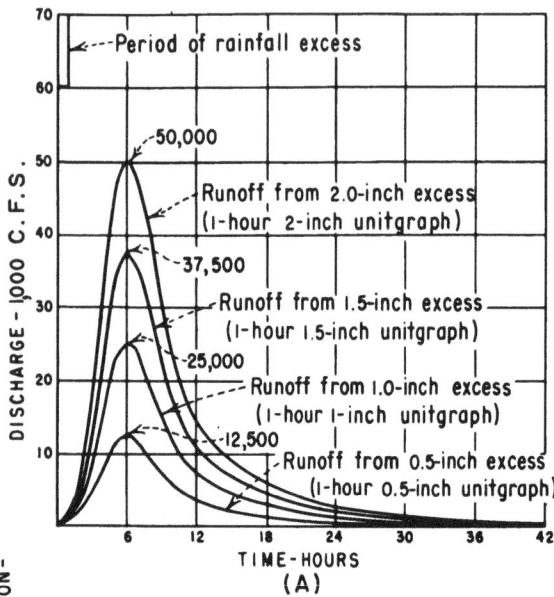

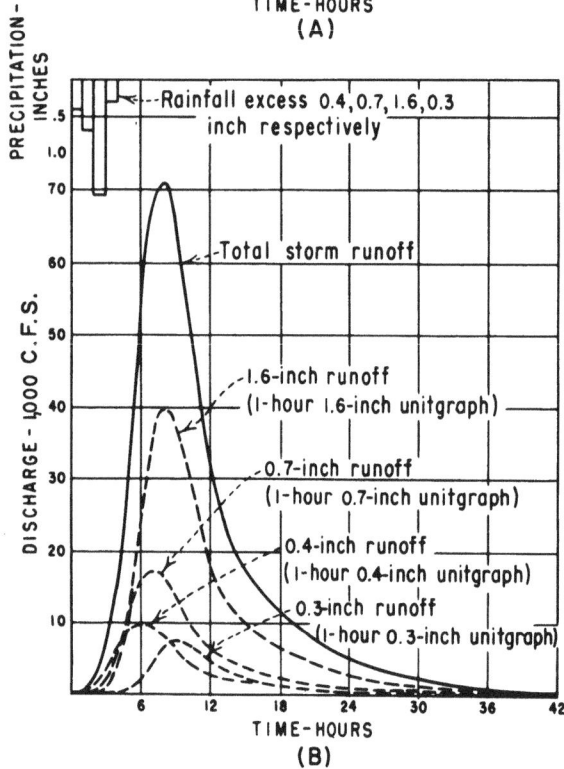

Figure G-3. Unit hydrograph principles (sheet 1 of 2).−288-D-3179(1/2)

usually unsuitable sources of unitgraphs.

(3) The observed hydrograph of storm discharge is a smooth curve, because it is actually made up of unitgraphs produced by infinitely short increments of excess rain. It cannot be reproduced perfectly by

assumptions listed on figure G-3 are not theoretically supportable. However, experience has shown that this does not negate use of the method as a practical tool.

(b) *Selection of Hydrographs to Analyze.*—The statement made in section G-7(b)(2) bears enough importance to unit hydrograph studies to be repeated: "Those floods having a combination of large volume, uniform intensity, isolated periods of rainfall, and uniform areal distribution of rainfall, should be chosen for further study."

Streamflow discharge records and basin precipitation records must be examined jointly for selection of hydrographs to analyze for unit hydrograph derivation. Isolated floods likely to merit investigation are easily identified by a rapid rise to a single peak and a smooth curve recession to low flow. Preferably, volumes of selected hydrographs should be equivalent to about one-half inch or more of runoff from the watershed. Preliminary estimates of hydrograph volumes can be made by summing the daily mean daily discharges in c.f.s.-days for the flood period. A sum of c.f.s.-days equal in number to 15 times the drainage area size in square miles is equivalent to 0.56 inch of runoff from the area. A useful equation for converting discharge volume to equivalent inches of rainfall is:

$$P_e = \frac{V}{26.89\ A} \qquad (3)$$

where:

P_e = rainfall excesses, inches, average
    depth over basin,
V = volume of runoff, c.f.s.-days, and
A = drainage area in square miles.

Hydrographs with volume sum of c.f.s.-days less than five times the drainage area size, 0.19 inch runoff, are almost always unsuitable for unit hydrograph analyses.

After noting dates of all flood hydrographs that satisfy preliminary volume criteria, rainfall records for respective flood events are examined for conformance with the ideal combination of short duration, uniform intensity, and uniform areal distribution of rainfall over the entire watershed. Those storms approaching nearest to the *ideal* criteria are analyzed as previously described in section G-7. If enough rainfall data are not available to do a good storm analysis for some of the isolated flood events having satisfactory volumes, the flood hydrographs may be analyzed for unitgraph comparisons as discussed in section G-8(e) by assuming that the beginning of rainfall excess coincides with the beginning of a sharp rise of the hydrograph, provided there is enough information available to reasonably assume the rainfall covered the total watershed.

Unit hydrograph derivations are difficult in regions where isolated flood events are rare and, instead, flood hydrographs commonly have two or more peaks caused by storms which usually persist for several days. Procedures for analyzing multipeaked flood hydrographs cannot be included in this text but can be found in publications listed in the bibliography, section G-32.

(c) *Hydrograph Analyses—Base Flow Separation.*—The purpose of flood hydrograph analyses is to determine for a watershed the time-distribution of the runoff which *quickly* reaches a particular point on a stream when rain falls on the watershed. The portion of the rainfall that infiltrates through the soil mantle into the ground-water supply will not reach the stream until days or months after the storm. Ground-water supply to a stream, base flow, may be a large proportion of that stream's total yearly discharge, but the base flow volume during an isolated flood is small in ratio to the total flood volume. However, base flow must be estimated and subtracted from the total discharge hydrograph in order to determine the direct runoff hydrograph. The schematic graphs on figure G-4 show three common approaches for estimating base flow discharges [6]. Base flow estimates are usually made graphically after plotting total flood discharges on linear or semilogarithmic graph paper.

(d) *Hydrograph Analysis of Direct Runoff—Need for Synthetic Unit Hydrographs.*—It is often necessary to use synthetic unit hydrographs for inflow design flood estimates and for obtaining indices for

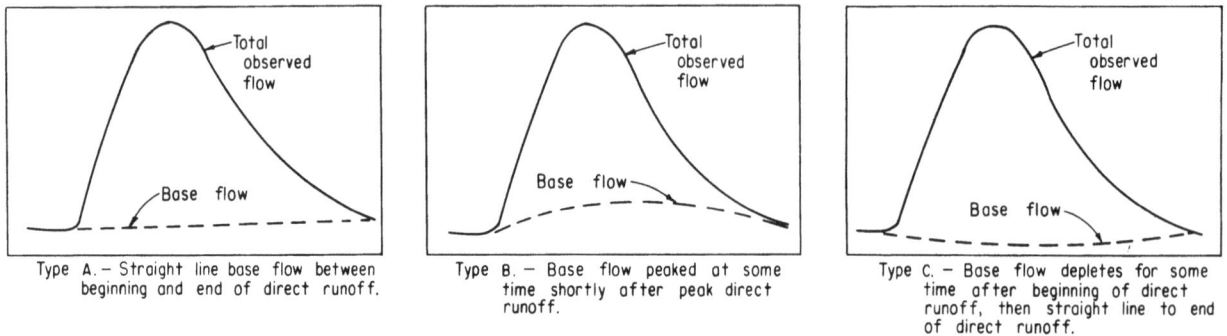

Type A. – Straight line base flow between beginning and end of direct runoff.

Type B. – Base flow peaked at some time shortly after peak direct runoff.

Type C. – Base flow depletes for some time after beginning of direct runoff, then straight line to end of direct runoff.

Figure G-4. Three common approaches for estimating base flow discharges.—288-D-3180

synthetic unitgraph estimates. Suitable records of observed discharge are seldom available at the exact stream point for which a unitgraph is needed; in this discussion, at a proposed damsite. Even if such records are available, often the proposed reservoir will be large enough to inundate several miles of stream channels above the damsite, thus causing watershed runoff to enter a full reservoir more quickly than the respective runoff would arrive at the damsite through natural channels. Therefore, a unitgraph usable for estimating floods at the damsite under natural conditions must be properly adjusted to be usable to estimate inflow to a full reservoir.

The shape of a representative watershed unitgraph can be obtained by a proper average of several unitgraphs computed from observed discharge records at a gage, or occasionally by a single unitgraph from an intense rainburst, well centered and distributed. If there are available several isolated direct runoff hydrographs suitable for simple conversion to 1-inch volume unitgraphs by multiplying the hydrograph discharge ordinates by the ratio of 1 inch to the direct runoff volume in inches, only those unitgraphs having *equal duration times of rainfall excess* can be directly averaged. Most likely, rainfall excess duration time will be different for each 1-inch unitgraph. A general similarity in shape of the unitgraphs will be recognized, but they may show pronounced differences in their relative steepness and time of peak discharge.

It is possible to eliminate these differences to a large degree by adjusting the ordinates and abscissae of each unitgraph in proportion to

some index related to both the duration of rainfall excess and to the average time interval between the rainfall excess and some representative point near the center of the respective runoff unitgraph. The index used for this purpose is known as *lag-time* which, for procedures to be described in this text is defined as: *The time interval between the mid-time of rainfall excess duration and the time of occurrence of one-half the volume of the hydrograph.*

Lag-time may be used as later described to convert each unitgraph into a dimensionless-graph form and the dimensionless-graphs can then be averaged. (*Note*: In this text, the hyphenated term dimensionless-graph refers to the particular form used within the Bureau of Reclamation. The two words, dimensionless graph(s) refer in general to graphs expressing time versus discharge as ratios.) Lag-time is also an index of time-of-concentration (time interval between end of rainfall excess and point of inflection on recession limb of direct runoff hydrograph) of runoff for a watershed, and can be correlated with certain measurable physical features common to all watersheds such as area, stream channel length, and slope. Correlations between lag-times derived from recorded floods and respective watershed features, in the form of *lag-time curves*, provide means for estimating lag-time at any desired ungaged stream point on the basis of watershed features above that point.

A synthetic unitgraph may be estimated for a watershed area, given a representative lag-time curve and dimensionless graph *based on the same lag-time definition.* Hydrology textbooks and published professional papers give many different definitions of lag-time, several different dimensionless graph forms, and many variations in correlations of basin features with lag-times.

Investigators are continually striving to improve estimates of time-distribution of runoff from rainfall. Only the lag-time versus basin factor relationships and related dimensionless-graph form used most often in Bureau of Reclamation inflow design flood studies will be described in detail in this text.

(e) *Hydrograph Analysis of Direct Runoff–Dimensionless-Graph Computations and Lag-Time Estimates.*—A direct runoff hydrograph may be converted to dimensionless-graph form using a function of lag-time. A lag-time for the flood event may also be computed if sufficient rainfall data are available to define the duration time of rainfall excess.

All hydrographs may be converted to dimensionless-graph form by the mathematical procedure to be described, but experienced judgment must be employed to select those that are suitable for further considerations. Lag-time is the basic index; however, a related value known as *lag-plus-semiduration* is the actual index used for dimensionless-graph computations. Lag-plus-semiduration is obtained by adding one-half of the duration time of rainfall excess to the lag-time. This addition provides a means of obtaining comparable dimensionless-graphs for unitgraphs of different rainfall excess durations, as, by definition, a unitgraph starts at the beginning of rainfall excess and the measurement of lag-time starts at the mid-time of rainfall excess duration. Lag-plus-semiduration is the elapsed time between the beginning of the major rise of the hydrograph and the point of 50 percent of runoff volume. Thus, in the analysis of an observed direct runoff hydrograph for which rainfall excess can be established and begins concurrently with the start of the major rise of the hydrograph, lag-time is computed as lag-plus-semiduration minus one-half of the rainfall excess duration.

When analyzing direct runoff hydrographs by the dimensionless-graph method, it is not necessary to first convert each hydrograph to a volume equivalent to 1 inch of runoff. In practice, selected observed direct runoff hydrographs are converted to dimensionless-graph form as follows. The elapsed time from the beginning of a hydrograph to the point of 50 percent volume is computed; this is the lag-plus-semiduration value for the hydrograph. The abscissae of the hydrograph is converted from actual hours into percent of the lag-plus-semiduration value. Each ordinate of the hydrograph, cubic feet per second (or c.f.s.), is multiplied by the lag-plus-semiduration value, and the product is divided by the total direct runoff hydrograph volume expressed as c.f.s.-days. The converted ordinates and abscissae are dimensionless and may be plotted for comparisons and averaging with other dimensionless-graphs similarly obtained.

The above method of eliminating the effect of rainfall excess duration time by lag-time relations is considered satisfactory in the comparison and averaging of a group of dimensionless-graphs when the maximum value of the rainfall excess duration, expressed in percent of lag-time, does not exceed about four times the minimum value found in the same group, expressed in the same way. When the duration of rainfall excess cannot be determined with reasonable accuracy, lag-plus-semiduration can frequently be measured directly from the start of rise of the direct runoff hydrograph. Thus, dimensionless-graphs may be obtained from recorded floods from watersheds where streamflows are gaged but precipitation data are meager or not collected. Use of this procedure increases the data available for synthetic unitgraph derivations.

To determine the average shape of a group of dimensionless-graphs, first determine the average of the peak ordinates and the average of the corresponding abscissae. These two values become the coordinates of the peak of the average graph. Points on the lower portions of the accession and recession are averaged *on the horizontal,* that is, an ordinate is assumed and the average of the abscissae corresponding to that ordinate is determined. If the plotting is on semilog paper and the recessions end in tangents, only two averages are needed to define the mean tangent. The *shoulder* portions of the mean graph are best sketched in by visual inspection. Arithmetical averages should not be used near the peak unless the ordinates of the points averaged are taken at a fixed percentage of the respective peak ordinates, or unless the individual peaks as plotted are at virtually the same height.

(1) *Procedures.*—A method of complete hydrograph analyses for obtaining a dimensionless-graph and lag-time estimate from a selected isolated flood event is given as a step-by-step outline with pertinent comments, graphically illustrated on figure G-5, and supplemented by a table of computation, table G-5. For illustrative purposes, computations included in table G-5 are more detailed than

necessary in practice. An outline of procedures follows:

(a) Plot recorded hydrograph on cartesian coordinate paper and on semilog paper:

① on figure G-5(A), and

① on figure G-5(B)

Hypothetical total flood discharges are listed in table G-5. A hyetograph of average hourly basin rainfall, if obtainable, plotted as shown on the same coordinate paper with the total flood hydrograph, is helpful for determining the coincidence of beginning time of rainfall excess and direct runoff. The plot on semilog paper helps in making base flow estimates.

(b) Estimate base flow, ② on figure G-5(A) and (B), by trial and error. Subtract base flow from recorded

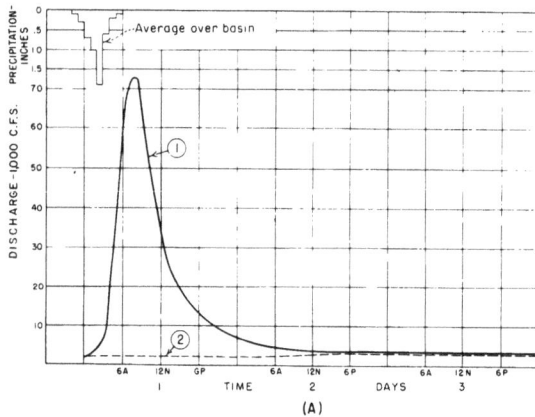

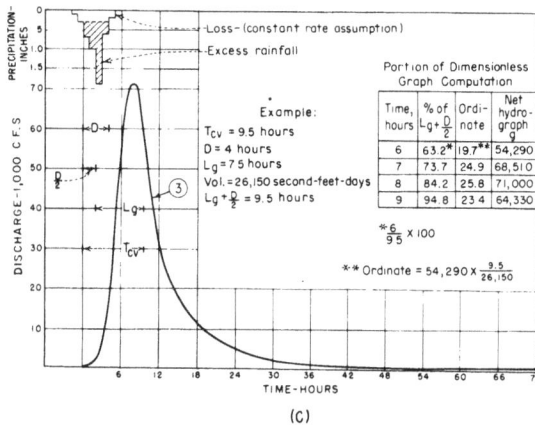

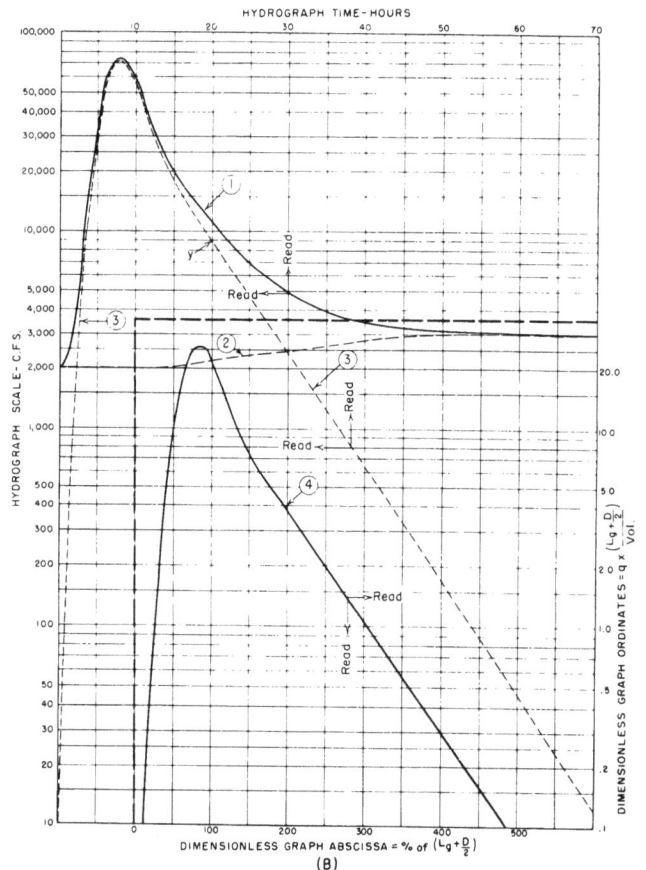

Figure G-5. **Hydrograph analysis.**—288-D-2457.

hydrograph and plot net hydrograph, ③ on figure G-5(B). If the base flow has been estimated correctly, the descending limb of hydrograph ③ on figure G-5(B) will be a straight line (exponential recession) [7]. (③ = ① minus ② on figure G-5(B).)

Large base flow discharges were used in this example to improve graphical illustration.

(c) Compute volume of net hydrograph ③ as follows (method 1, table G-5):

1. Add average hourly discharges (in c.f.s.-hours) to a point such as y on the exponential recession, ③ on figure G-5(B).

2. Compute hourly recession constant, k_{hr}, from two points on exponential recession line by use of following equation:

$$k_{hr} = \sqrt[t]{\frac{q_t}{q_o}} \qquad (4)$$

where:

q_o = discharge at first point,
q_t = discharge at second point, and
t = time interval, in hours, between points 1 and 2.

3. Storage, or volume after point y (in c.f.s.-hours) equals:

$$\frac{-q_y}{\log_e k_{hr}} \qquad (5)$$

where:

q_y = discharge in c.f.s. at point y, and
$\log_e k_{hr}$ = 2.3026 ($\log_{10} k_{hr}$).

4. Total volume is sum of volume to y plus volume after y.

(d) For comparison with rainfall data,

convert volume of ③ to inches of runoff:

Inches of runoff =

$$\frac{\text{volume in c.f.s.-hours}}{\text{(area in sq. mi.) X } 645.3*} \qquad (6)$$

*(1 inch P_{e_1}/sq. mi. = 26.888 c.f.s.-days: (26.888)(24) = 645.3 c.f.s.-hrs.)

(e) Analyze rainfall data, if available; determine period D of rainfall excess.

(f) Compute time of occurrence of one-half volume of hydrograph ③, figure G-5(C). The time to center of volume, T_{cv}, equals time from beginning of rise of net hydrograph to time one-half volume has passed measuring point.

(g) Find lag, Lg, time in hours from midpoint of excess rainfall period to time of occurrence of one-half volume.

(h) Compute dimensionless graph as follows and plot on semilog paper, ④ on figure G-5(B).

1. Abscissa—hours from beginning of excess rain expressed as percent of $(Lg + D/2)$.

2. Ordinates—discharge in c.f.s. of ③ (at respective abscissa) multiplied by $(Lg + D/2)$, all divided by net hydrograph volume expressed as

$$\text{c.f.s.-days} = \left(\frac{\text{c.f.s.-hours}}{24}\right).$$

(2) *Lag-time curves.*—Lag-time is a key function for estimating synthetic unitgraphs. An average lag-time value for a watershed is obtained by averaging the results of several good analyses of stream gage records. Such average values for different gages on a stream and/or different streams of similar runoff characteristics can be correlated empirically with certain measurable watershed features. The correlation equation most often used in the Bureau of Reclamation is of the form:

$$\text{Lag-time, hours} = C \left[\frac{LL_{ca}}{\sqrt{S}}\right]^x \qquad (7)$$

where: C and x are constants,

Table G-5.—*Hydrograph analysis computations*

BASIC DATA:

Name of streamgage = (Hypothetical for this table)   A, drainage area, sq. mi. = 319
Date of flood = (Assume May 1-3, 1970)   Volume, c.f.s.-days, net = 26,150
Time, beginning of direct runoff—net hydrograph = 12:00 p.m., 30 April
Time, point of 50 percent volume of net hydrograph, T_{cv} = 9:30 a.m., 1 May

Lag-plus-semiduration, hrs.; $\left(Lg + \dfrac{D}{2}\right) = 9.5$

Duration of rainfall excess, D, hrs. = 4 (obtained by storm analysis)

Lag-time, hrs. = $\left(Lg + \dfrac{D}{2}\right) - \left(\dfrac{D}{2}\right) = 7.5$

Q = instantaneous discharge, c.f.s.

Time		Hydrographs			Net volume		Dimensionless-graph	
Hour and day	Net Σ hr.	Total flood, Q	Base flow Q	Net Q	Increm.[2] c.f.s.-hrs.	Accum. 1,000 c.f.s.-hrs.	Abscissae, percent of $Lg + \dfrac{D}{2}$	Ordinates, net $Q \times \left[\dfrac{Lg + \dfrac{D}{2}}{\text{vol.}}\right]$
12P30	0	2,000	2,000	0	0	0	0	0
1A1	1	2,250	2,000	250	125	.12	10.5	0.09
2A1	2	3,560	2,000	1,560	905	1.03	21.1	0.57
3A1	3	8,120	2,000	6,120	3,840	4.87	31.6	2.22
4A1	4	18,640	2,000	16,640	11,380	16.25	42.1	6.0
5A1	5	36,040	2,000	34,040	25,340	41.59	52.6	12.4
6A1	6	56,290	2,000	54,290	44,165	85.76	63.2	19.7
7A1	7	70,510	2,000	68,510	61,400	147.16	73.7	24.9
8A1	8	73,000	2,000	71,000	69,755	216.91	84.2	25.8
9A1	9	66,330	2,000	64,330	67,665	284.58	94.8	23.4
10A1	10	55,360	2,000	53,360	58,845	343.42	105.8	19.4
11A1	11	43,250	2,000	41,250	47,305	390.72	115.8	15.0
12N1	12	33,520	2,000	31,520	36,385	427.11	126.4	11.4
1P1	13	26,900	2,020	24,880	28,200	455.31	136.9	9.0
2P1	14	22,830	2,050	20,780	22,830	478.14	147.4	7.5
3P1	15	19,810	2,080	17,730	19,255	497.40	158.0	6.4
4P1	16	17,230	2,100	15,310	16,520	513.92	168.5	5.6
5P1	17	15,390	2,120	13,270	14,290	528.20	179.0	4.8
6P1	18	13,780	2,150	11,630	12,450	540.66	189.5	4.2
8P1	[1]20	11,090	2,200	8,890	(20,520)	(561.18)	210.6	3.23
12P1	24	7,460	2,300	5,160	(28,100)	(589.28)		
6A2	30	4,840	2,500	2,340	(22,500)	(611.78)	[3]315.9	[3].85
12N2	36	3,700	2,650	1,050	(10,170)	(621.94)		
6P2	42	3,305	2,830	475	(4,575)	(626.52)	[3]442.3	[3].17
12P2	48	3,215	3,000	215	(2,070)	(628.59)		
6A3	54	3,100	3,000	100	(960)	(629.55)		
12N3	60	3,045	3,000	45	(420)	(629.97)		
6P3	66	3,020	3,000	20	(180)	(630.15)		
12P3	72	3,010	3,000	10	(90)	(630.24)		
6A4	78	3,000	3,000	0	(30)	(630.27)		

[1] Note variations in time intervals for listing discharges (optional).

[2] c.f.s-hrs. = $\left(\dfrac{Q_1 + Q_2}{2}\right)$ x (time interval, hrs.)

[3] For plot on semilog paper, only enough points to define a straight line need be computed.

Table G-5.–Continued

Equations for dimensionless-graph:

$$\text{Abscissae} = \frac{\text{net } \Sigma \text{ hr.}}{Lg + \dfrac{D}{2}} \times 100$$

$$\text{Ordinates} = \text{net } Q \times \frac{Lg + \dfrac{D}{2}}{\text{vol., c.f.s.-days}}$$

$$\left[\text{c.f.s.-days} = \left(\frac{\text{c.f.s.-hours}}{24} \right) \right]$$

Lag-plus-semiduration:

1/2 volume is between net Σ hrs. 9 and 10
By linear interpolation:

Volume, method 1,

$$Lg + \frac{D}{2} = 9.50 \text{ hrs.}$$

Volume, method 2,

$$Lg + \frac{D}{2} = 9.52 \text{ hrs.}$$

Except for very small watersheds, lag-plus-semiduration values are rounded to nearest 1/10 hr.

For dimensionless-graph equations:

Use: $Lg + \dfrac{D}{2} = 9.5$

Volume = 26,150 c.f.s.-days

Lag estimate:

D = 4 hrs.

$$\text{Lag} = 9.5 - \frac{D}{2} = 7.5 \text{ hrs.}$$

Net volume computations:

Method 1, by equations.

q_o: Q at net Σ hr. 20 = 8,890 c.f.s.
q_t: Q at net Σ hr. 30 = 2,340 c.f.s.
t: time interval, q_o to q_t = 10 hrs.

$$k_{hr} = \sqrt[t]{\frac{q_t}{q_o}} = \sqrt[10]{\frac{2,340}{8,890}}$$

$$k_{hr} = \sqrt[10]{0.263} = 0.875$$

$$\text{Volume after net } \Sigma \text{ hr. } 20 = \frac{-q_o}{\log_e k_{hr}}$$

$$= \frac{-8,890}{-0.1336}$$

$$= 66,540 \text{ c.f.s.-hrs.}$$

Σ net volume, hrs. 0-20 = 561,180 c.f.s.-hrs.
Total net volume = 627,720 c.f.s.-hrs.
= 26,150 c.f.s.-days
½ total net volume = 313,860 c.f.s.-hrs.

Method 2.
Ordinates of total net hydrograph used as shown in table at left.
Discharges of recession limb read at time intervals for which recession curve can be approximated as a straight line.

Total volume = 630,270 c.f.s.-hrs.
= 26,260 c.f.s.-days
½ volume = 315,140 c.f.s.-hrs.

L = length of longest watercourse from point of interest to watershed divide, measured in miles,

ca = centroid of basin—usually found by vertically suspending a cardboard cutout of basin shape successively from two or more points and finding intersection of plumb lines from each point,

L_{ca} = length of watercourse from point of interest to intersection of perpendicular from ca to stream alinement, and

S = overall slope in feet per mile of longest watercourse from point of interest to divide.

Values for the constants C and x are obtained empirically from plots on *log-log paper* of $\dfrac{LL_{ca}}{\sqrt{S}}$ values versus lag-time, hours, and fitting a straight line, either "by eye" or by least-squares computations. The lag-time indicated by the curve for an $\dfrac{LL_{ca}}{\sqrt{S}}$ value of 1.0 is the constant C, and the "slope" of the line on log-log paper is the constant x.

A lag-time curve for a watershed should be based on as many hydrograph analyses as can be obtained from the data available within the watershed and for other watersheds with similar runoff characteristics. When developing a lag-time curve, a consistent method of hydrograph analyses should be used and measurements of watercourse lengths should be made on maps of the same scale. If suitable data are limited to only one stream gage location, a lag-time curve can be constructed by drawing a line with slope of 0.33 through the point plotted on log-log paper of average lag-time versus $\dfrac{LL_{ca}}{\sqrt{S}}$ value.

In the absence of any runoff data suitable for hydrograph analyses, preliminary estimates of lag-times for *direct runoff* for watersheds having rapid runoff characteristics can be made by the following generalized equation:

$$\text{Lag-time, hours} = 1.6 \left[\frac{LL_{ca}}{\sqrt{S}} \right]^{0.33}$$

The above equation gives values acceptable as preliminary estimates of direct runoff lag-times for many streams in the plains and southwestern regions of the United States and for foothill streams of the Rocky Mountains. Certain types of watersheds have large variations in lag-times that are not adequately reflected by the generalized C value given. These include watersheds which have physical features tending to retard surface runoff such as near level terrain, dense vegetative cover, etc.; and those in which the streams extend into high, well-forested mountains or whose streamflow records show pronounced interflow contribution. Lag-time estimates for such watersheds should be made by an experienced hydrologist.

C. SYNTHETIC UNIT HYDROGRAPH

G-9. *Synthetic Unitgraphs by Lag-Time Dimensionless-Graph Method.* —Computation of a unitgraph for a watershed above a specific location by this method is done by reversing the mathematical process used to derive a dimensionless-graph. The important factors for obtaining a representative unitgraph for a given watershed are the selections of a proper lag-time curve and proper dimensionless-graph. An example of a unitgraph derivation for an ungaged watershed follows, given as a step-by-step outline with pertinent comments and graphically illustrated on figure G-6.

(1) Outline drainage boundary, determine area (fig. G-6(A)).

(2) Find basin center of area, *ca* and project to the nearest point on the longest watercourse. Measure L (to divide at head of longest watercourse) and L_{ca} miles. (Refer to sec. G-8(e)(2).) Determine S (for upper elevation, estimate average elevations along divide in vicinity of head of longest watercourse, not the specific elevation at the

point of extention of longest watercourse to divide).

(3) Compute $\dfrac{LL_{ca}}{\sqrt{S}}$.

(4) Enter graph, lag-time curve (fig. G-6(B)), with $\dfrac{LL_{ca}}{\sqrt{S}}$ value and read the corresponding lag-time. (Lag-time curve (B) represents mean curve drawn "by eye" through plotted lag-times obtained from hydrograph analyses versus respective $\dfrac{LL_{ca}}{\sqrt{S}}$ for basins of similar runoff characteristics.)

(5) Select a dimensionless-graph (fig. G-6(C)) (usually an average dimensionless-graph of a number of dimensionless-graphs derived for the same stream or for streams of similar characteristics).

(6) Select a unit rainfall duration time; this should be one-fourth or less of lag-time for basin. (Unit times are selected for

Basin Factors:
Area=300 sq mi.
L = 36 miles
L_{ca}= 16.1 miles
S = 8.8 feet per mile
$\frac{LL_{ca}}{\sqrt{S}}$ = 195

(A)

Point of interest

ca

Typical Lag Curve
1. Red Willow Creek near Mc Cook, Nebr
   (Partial Contributing Area)
2. Little Beaver Creek near Duncan, Okla
3. Cimmarron River near Boise City, Okla.

LAG - HOURS

$LL_{ca} \div \sqrt{S}$

(B)

Dimensionless graph
from figure G-5(B)

$q \times \frac{(L_g + \frac{D}{2})}{Vol.}$

% of ($L_g + \frac{D}{2}$)

(C)

DISCHARGE — 1,000 CFS

2 - hour unit graph

TIME-HOURS

(D)

Unitgraph Derivation:
Unit rainfall duration = 2 hours
Lag time = 9 hours
Area = 300 sq. mi.
Volume of 1.0 inch runoff = 300 x 26.89 =
    8,067 c f s -days
$L_g + \frac{D}{2}$ = 10 hours.

Sample Computations

Time, hours	% of $L_g+\frac{D}{2}$	* Ordinate	Unit q**
2	20	0.5	400
4	40	5.2	4190
6	60	17.0	13,710
8	80	25.8	20,810

* Read from (C) at left
** Instantaneous value at end
    of designated hour.
    q = Ordinate x $\frac{8,067}{10}$

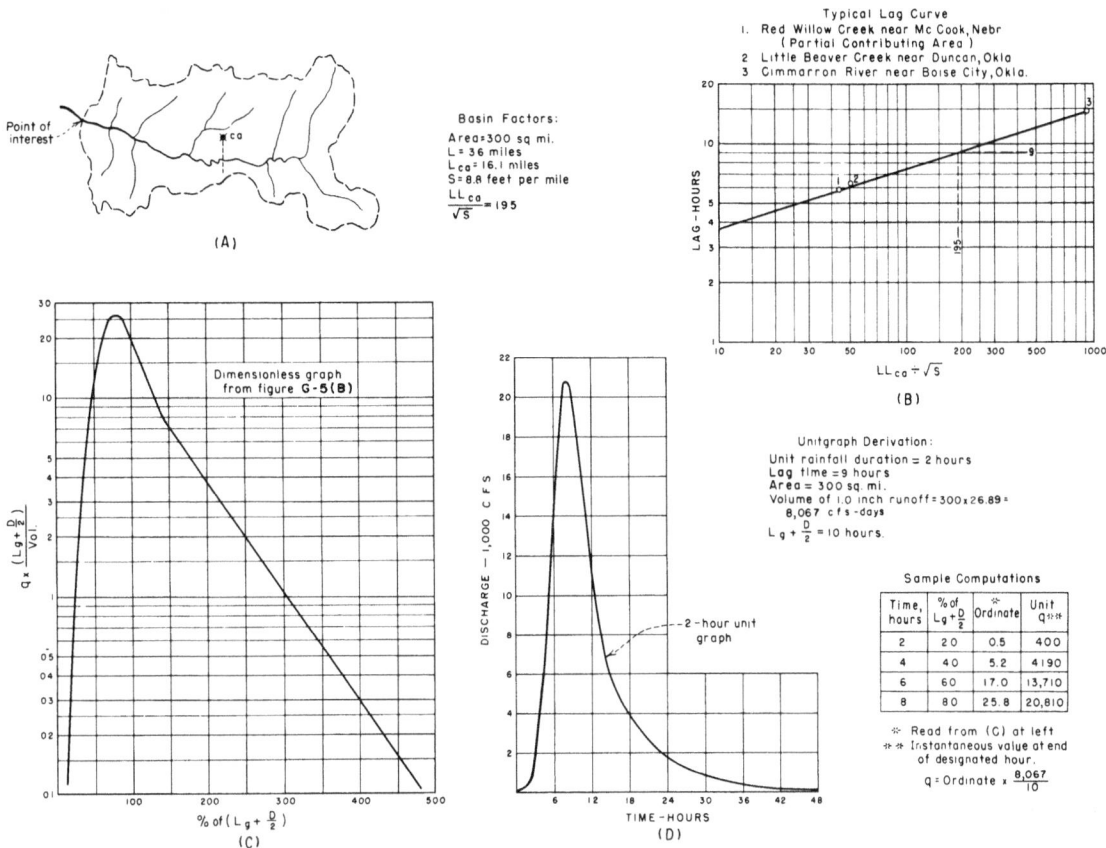

Figure G-6. Unitgraph derivation for ungaged area.—288-D-3182

computational convenience, usually 1-, 2-, 3-, 4-, or 6 hour units for lag-times of 4 hours or greater. Unit times larger than 6 hours are seldom used. Units of one-half or one-quarter hour are used for lag-times less than 4 hours.)

(7) Compute unitgraph (fig. G-6(D)) using:

(a) Basin area, square miles.

(b) Lag-time plus one-half selected unit rainfall duration time.

(c) Dimensionless-graph.

(d) Notes regarding computational procedure.

1. Equations for deriving a dimensionless-graph are given in table G-5. Unitgraph computation requires solving for instantaneous discharges at end of successive unit time intervals.

2. Time, hours, accumulative by unit time intervals are listed, and each accumulative value expressed as percent of lag-plus-semiduration.

3. Dimensionless-graph (fig. G-6(C)) is

entered with successive lag-plus-semiduration values, and respective ordinates read from the graph. Ordinates are substituted in the ordinate equation for solution of discharge values. When done by desk calculator, discharges are rounded.

(*Note*: Dimensionless-graph ordinates listed in the table of sample computations (fig. G-6) do not agree numerically at respective accumulative time values with dimensionless-graph ordinates in table G-5, because the dimensionless-graph ordinates in the table were derived at intervals of 10.5 percent of lag-plus-semiduration but the ordinates for 2-hour unitgraph derivation in figure G-6 were read at intervals of 20 percent of a different lag-plus-semiduration value.)

4. *Caution.*—The volume of a synthetic unitgraph should always be checked before being used, to be sure it has a

volume within 1 percent of 1-inch runoff volume for the watershed area. All of the ordinates of a unitgraph ((D) of fig. G-6) may be computed by reading the entire dimensionless-graph (C) and summing the ordinates to check the volume.

Another procedure may be used if the selected dimensionless-graph has an exponential recession limb such as on figure G-6(C). Unitgraph ordinates are obtained by reading the dimensionless-graph forward to an ordinate that is on the beginning portion of the exponential limb of the dimensionless-graph (see sec. G-8(e)(1)(b)). The volume of the unitgraph thus far obtained is computed and subtracted from the volume of 1 inch of runoff for the watershed area, giving the remaining volume, V_x. A recession constant, k, for the selected unit time interval can be computed by the equation,

$$\log_e k = -\frac{q}{V_x} \qquad (8)$$

where:

   q = the discharge ordinate, c.f.s., on
        the exponential limb, and
   V_x = the remaining volume expressed in
         unit time (c.f.s.-hours).

The factor k is used to compute the ordinates of the unitgraph following the last ordinate obtained by reading the

dimensionless-graph. This procedure assures correct unitgraph volume.

G-10. *Trial Reconstruction of Past Floods.*—Final decisions regarding appropriate lag-time, dimensionless-graph, and retention losses for a gaged watershed are made empirically by computing hydrographs of past recorded floods. Retention losses believed appropriate are applied to the observed storm precipitation data for each flood to be reconstructed to determine unit time increments of rainfall excess equivalent to the respective hydrograph volume. These increments are applied to a representative unitgraph according to basic assumption (3), figure G-3. The hydrograph thus computed is compared with the recorded hydrograph for *goodness of fit*; preliminary conclusions regarding appropriate factors are revised, if necessary, until an acceptable *fit* is obtained. These *test* trial reconstructions should be made for the large floods. Preferably, the largest flood of record should be excluded from the set of hydrographs selected for analyses and the parameters resulting from analyses tested by the *fit* achieved using them to reconstruct the largest flood.

G-11. *Synthetic Unitgraphs by Other Methods.*—Descriptions of several different methods of estimating synthetic unitgraphs may be found in technical publications. Among those often used are the *S*-curve hydrograph [8], Snyder's method [9], and basin routing methods based on the Clark approach [5],[10],[11],[12].

D. STREAMFLOW ROUTING

G-12. *General.*—Computation of an inflow design flood (IDF) hydrograph often requires that floodflows from several subareas within the drainage area be computed separately. Beginning with the farthest upstream subarea, hydrographs are transferred downstream by some method of streamflow routing, the flows being consecutively combined with other flood hydrographs, and the total inflow design flood

hydrograph obtained for the proposed reservoir. Watershed features above a damsite which indicate the need to subdivide the basin into subareas include:

(1) Large tributary areas which have different sizes, shapes, and cover characteristics.

(2) Existing reservoirs or natural lakes which control runoff from significant portions

of the drainage area above a proposed damsite. The flood runoff from the portion of the design storm for the total drainage area that occurs above such an existing feature should be reservoir-routed through the feature to obtain an outflow hydrograph before routing on downstream. If an existing dam impounds a large-capacity reservoir, the capability of the existing dam to safely withstand the computed inflow flood must be determined. Should the upstream dam be found to have an inadequate spillway capacity (or structural weakness), steps should be taken to get the owners of the upstream dam to make modifications as necessary to safely pass the inflow design flood. Or as an alternative, failure of the structure should be assumed and provision made at the proposed downstream dam and reservoir to safely handle the flood wave surge that might be expected with failure and an additional inflow volume equivalent to the capacity of the upstream reservoir.

(3) Drainage areas in which storm potential varies to an extent that an assumption of average precipitation over the total area during a design storm is unreasonable.

(4) Drainage areas in which during design storm conditions some streams will have snowmelt runoff in addition to rainfall runoff and other streams have only rainfall runoff.

G-13. *Practical Methods of Streamflow Routing Computations.*—Streamflow routing, the determination of a flood discharge hydrograph at any point on a stream from a discharge hydrograph at some point upstream, requires solution of the movement of flood waves in natural open channels which are extremely complex. A discussion of the theoretical and mathematical bases of flood routing methods is beyond the scope of this text. Many different methods and procedures have been described in engineering literature. If streamflow routing is necessary in the derivation of an inflow design flood hydrograph and the damsite is located on a stream that has discharge records at two or more locations, an applicable routing method may be selected from descriptions in publications, for example, "Hydrology for Engineers" [13].

Usually, inflow design flood derivations that include streamflow routing computations involve ungaged streams. Description of two practical methods of mathematical streamflow routing which can be used on the basis of an estimate of peak discharge travel time between two points on a reach of natural stream channel follows. These methods have been found to give acceptable results when tested by using recorded discharge hydrographs.

(a) *Tatum's Method* [14].—This method is also known as the *Method of Successive Averages*. Factors used when applying this method are travel time of peak discharge through the channel reach, T in hours; selected routing interval between discharges of the upstream hydrograph to be routed, t in hours; and routing constants listed in table G-6 for respective number of routing steps. Definite rules for selecting lengths of stream channel reaches for each routing computation cannot be set, but use of extremely long reaches may give very poor results. When computing an inflow design flood hydrograph, channel reaches are those on the main stream between points of inflow from subareas. Thus, inflow from a subarea can be added to the routed flow at the subarea inflow point to obtain a combined floodflow for routing through the next reach. After estimating travel time T believed applicable for a reach, a routing interval t is selected choosing an interval small enough to define well the hydrograph, and the number of routing steps for that reach computed by the equation:

$$\text{Number of routing steps} = 2T/t \qquad (9)$$

Computed fractional steps are rounded to the nearest whole number. The computational procedure is illustrated in table G-7. In actual practice when using a desk calculator, the routing constants are copied in a column on a separate sheet of paper and used as a slide beside the column of discharges to be routed. Products of the multiplications of constants and respective discharges are accumulated in the machine and only the total of each set of multiplications recorded. Constants for larger numbers of routing steps than given in table

Table G-6.—*Coefficients for floodrouting by Tatum's method*

Routing constants	Number of routing steps									
	1	2	3	4	5	6	7	8	9	10
C_1	0.5000	0.2500	0.1250	0.0625	0.0313	0.0156	0.0078	0.0039	0.0020	0.0010
C_2	.5000	.5000	.3750	.2500	.1562	.0937	.0547	.0313	.0176	.0098
C_3		.2500	.3750	.3750	.3125	.2344	.1641	.1094	.0703	.0440
C_4			.1250	.2500	.3125	.3126	.2734	.2187	.1641	.1172
C_5				.0625	.1562	.2344	.2734	.2734	.2460	.2050
C_6					.0313	.0937	.1641	.2187	.2460	.2460
C_7						.0156	.0547	.1094	.1641	.2050
C_8							.0078	.0313	.0703	.1172
C_9								.0039	.0176	.0440
C_{10}									.0020	.0098
C_{11}										.0010

G-6 may be computed from the expression $(\frac{1}{2} + \frac{1}{2})^n$ by the general equation for each term of a binomial expansion, n as the number of steps. Streamflow routing by Tatum's method using a desk calculator becomes tedious and time consuming when more than eight routing steps are used. The procedure may be easily programed for computer use.

(b) *Translation and Storage Method.*—In a paper describing a graphical reservoir-routing method, Wilson [15] also discusses streamflow routing, pointing out that it is partly analogous to reservoir routing but that natural channel storage produces less "flattening" effect on an inflow hydrograph than does reservoir storage. He suggested that in streamflow routing, the outflow (routed) hydrograph would lie between a hydrograph obtained by applying the graphical reservoir-routing method and the inflow hydrograph translated downstream a time interval equivalent to the reach travel time, and presented an example in which the routed hydrograph showed half translatory effect and half storage effect.

A report of the California Division of Water Resources [16] presented a streamflow routing method based on an adaptation of Wilson's graphical routing method showing that translation effect (travel time) and channel storage effect (attenuation) on the shape of a flood hydrograph moving downstream can be treated separately. In their studies, each effect was found to have approximately equal weight.

The translation and storage method of streamflow routing was devised[5] on the basis of evaluating separately the effects of travel time and channel storage and assuming equal weight for each effect in natural stream channels having "usual" storage characteristics. An equation for mathematical application of Wilson's graphical routing method was given in the U.S. Department of Agriculture, Soil Conservation Engineering Handbook, Supplement A, 1956. The given equation is used in the translation and storage method of streamflow routing as follows:

$$O_2 = O_1 + K(I_1 + I_2 - 2 O_1) \qquad (10)$$

where:

I_1, I_2 = inflow, consecutive incremental instantaneous discharges at the head of a stream reach, and
O_1, O_2 = outflow, successive incremental instantaneous discharges at the end of a stream reach; O_2 is the outflow resulting from I_1 and I_2 and the preceding outflow O_1.

The routing constant, K, in the above equation, is obtained as follows:

T = travel time, hours, of peak flow through the reach consisting of:

---

[5]Described in unpublished memoranda, Flood Hydrology Section, Engineering and Research Center, Bureau of Reclamation, Denver, Colo.

Table G-7.—*Illustrative example of streamflow routing by Tatum's method*

HYPOTHETICAL PROBLEM: Streamflow-route total flood hydrograph, table G-5, through channel reach having travel time of 4 hours.

If selected t = 1 hr., routing steps = $\dfrac{(2)(4)}{1}$ = 8

If selected t = 2 hrs., routing steps = $\dfrac{(2)(4)}{2}$ = 4

Hour and date	Upstream Q 1,000 c.f.s.	t = 1 hr., 8 routing steps Illustrative positioning of routing constants[2]				Routed [3]Q 1,000 c.f.s.	t = 2 hrs., 4 routing steps Illustrative positioning of routing constants[2]				Routed [3]Q 1,000 c.f.s.
4P30	[1]2.0										
5P	2.0	0.0039									
6P	2.0	.0313					0.0625				
7P	2.0	.1094									
8P	2.0	.2184					.2500				
9P	2.0	.2734									
10P	2.0	.2187					.3750				
11P	2.0	.1094									
12P30	2.0	.0313					.2500				
1A1	2.3	.0039				[4]2.0					
2A	3.6						.0625	0.0625			[4]2.1
3A	8.1		0.0039								
4A	18.6		.0313	0.0039				.2500	0.0625		
5A	36.0		.1094	.0313	0.0039						
6A	56.3		.2187	.1094	.0313			.3750	.2500	0.0625	
7A	70.5		.2734	.2187	.1094						
8A	73.0		.2187	.2734	.2187			.2500	.3750	.2500	
9A	66.3		.1094	.2187	.2734						
10A	55.4		.0313	.1094	.2187			.0625	.2500	.3750	47.7
11A	43.2		.0039	.0313	.1094	61.3					
12N1	33.5			.0039	.0313	[5]64.8			.0625	.2500	[6]58.6
1P1	26.9				.0039	61.7					
2P	22.8									.0625	37.8

[1]Constant base flow of 2,000 c.f.s. assumed to precede flood event.
[2]All routing constants are placed opposite respective Q's at t intervals.
[3]Discharge at bottom of reach; each Q is instantaneous discharge at time given in column 1.
[4]Sum of products of each constant times respective Q.
[5]Peak discharge of routed hydrograph, occurs 4 hours later than upstream peak.
[6]Peak discharge of routed hydrograph, agrees in time with routing t = 1 hr., but differs in magnitude because of longer routing interval.

T_r = translation time component, hours (when assuming equal weight to storage effect, T_r = 0.5T)

T_s = storage time component, hours (when assuming equal weight to translation effect, T_s = 0.5T)

and

$T = T_r + T_s$

Then for stream routing evaluation of storage time effect,

$$K = \frac{t}{2T_s + t}$$

where:

t = routing time interval, hours, with $t \leq 0.5T_s$.

Solving the equation for O_2 gives an instantaneous discharge value at the end of the incremental time interval designated by I_2. If I_1, I_2, etc., are designated by time at the head of a reach, the time of occurrence of O_2 at the bottom of the reach is obtained by adding the translation time component, T_r, to the time of respective I_2.

Use of the above equation with an assumption that the travel time for the reach is divided equally into translation time, T_r, and storage time, T_s, gives as acceptable results as those obtained by using Tatum's Method but requires less computational time when doing manual routing. A detailed example of application of the translation and storage method is shown in table G-8. Of course, in practice, such a detailed table is not necessary.

The translation and storage method, in addition to being easy to apply to stream reaches for which Tatum's method might be used, is also versatile enough to be applied to stream reaches having more or less storage effect than "usual." The relationship of storage time and translation time is not rigid, but may be varied depending on channel reach characteristics. If hydrographs are available at the head and bottom of a stream reach, a few trial routings will give an acceptable value for each component. Characteristics of ungaged stream channels are judged by comparison with characteristics of gaged streams when necessary to use streamflow routing methods.

(c) *Comparison of Methods.*—An illustration of results of applying the above two methods of streamflow routing is shown on figure G-7 on which the hypothetical flood hydrograph, with discharges listed in table G-5, is plotted. This hydrograph was routed downstream assuming a reach travel time of 4 hours: first, by Tatum's method assuming routing intervals of 1 hour and 2 hours; and secondly, by the translation and storage method using a routing interval of 1 hour. Routed (downstream) hydrographs are also plotted on figure G-7 (computations are not included). The two routed hydrographs obtained by Tatum's method differ because of different routing intervals; the routing by 1-hour intervals is the more representative because the upstream hydrograph is best defined in 1-hour intervals. The routed hydrograph obtained by the translation and storage method is acceptably similar to the hydrographs obtained by Tatum's method.

E. DESIGN STORM STUDIES

G-14. *General.*—Major floods, except those associated with dam failure, earthquakes, or landslides, result from a combination of severe meteorological and hydrological conditions. It follows that estimates of meteorological conditions which may approach the physical upper limits of rainfall or snow accumulation and melt rates must be considered where an inflow design flood (IDF) is required. This section is concerned only with rainfall studies. For the purpose of this text, the following terminology is used in regard to estimates of the physical upper limits of storm rainfall in a basin or region.

(a) *Probable Maximum Precipitation (PMP).*—Probable maximum precipitation values represent an envelopment of maximized intensity-duration values obtained from all types of storms. It is recognized that probable maximum precipitation values for all durations and all areas may not occur from only one type of storm. For example, a maximized thunderstorm is very likely to provide probable maximum precipitation over an area of 50 square miles for a duration of 6 hours or less, but the controlling values for longer durations or for larger areas generally will be obtained from general-type storms.

(b) *Probable Maximum Storm (PMS).*—The probable maximum storm values represent an envelopment of maximized intensity-duration values obtained from storms of a single type. Consideration is given to storm type and variations of precipitation with respect to

Table G-8.—*Translation and storage method of streamflow routing*

Equation: $O_2 = O_1 + K(I_1 + I_2 - 2 O_1)$

T = 12 hours $K = \dfrac{t}{2T_s + t}$

T_r = 6 hours $K = \dfrac{3}{12 + 3}$

T_s = 6 hours $K = 0.20$

t = 3 hours

(For definitions of symbols, see sec. G-13 (b).)

(1)	(2)	(3)	(4)	(5)	(6)	(7)	(8)
Time, hours[1]	Inflow, c.f.s.	$I_1 + I_2$, c.f.s.	$2 O_1$	(3) − (4)	(K)(5)	Outflow,[2] c.f.s.	Time, hours[4]
0	300					[3]300	6
3	300	600	600	0	0	300	9
6	415	715	600	115	23	323	12
9	1,604	2,019	646	1,373	275	598	15
12	5,458	7,062	1,196	5,866	1,173	1,771	18
15	10,093	15,551	3,542	12,009	2,402	4,173	21
18	16,567	26,660	8,346	18,314	3,663	7,836	24
21	17,924	34,491	15,672	18,819	3,764	11,600	27
24	18,608	26,532	23,200	13,332	2,666	14,266	30
27	19,244	37,852	28,532	9,320	1,864	16,130	33
30	19,772	39,016	32,260	6,756	1,351	17,481	36
33	25,913	45,685	34,962	10,723	2,145	19,626	39
36	23,499	49,412	39,252	10,160	2,032	21,658	42
39	20,552	44,051	43,316	735	147	21,805	45
42	17,377	37,929	43,610	−5,681	−1,136	20,669	48
45	14,703	32,080	41,338	−9,258	−1,852	18,817	51
48	12,054	26,757	37,634	−10,877	−2,175	16,642	54

[1] Time of instantaneous discharge at head of reach.
[2] Discharge at end of reach; (6) + preceding value in (7).
[3] Constant flow in reach assumed.
[4] Time of instantaneous discharge at end of reach. Translation time, T_r, added to time at head of reach.

location, areal coverage of a watershed, and storm duration.

(c) *Design Storm.*—The precipitation values selected for computing an inflow design flood are usually referred to as a design storm. These design storm values may or may not be equal to the PMP. The hydrometeorological report which describes the considerations and computations leading to the recommendation of a design storm for a particular watershed is usually called a "Design Storm Study."

(d) *Additional References.*—It is beyond the scope of this text to discuss in detail the meteorological considerations and computations involved in obtaining the "maximized intensity-duration values" cited in the above definitions. A comprehensive discussion of this subject is given in chapter 2, "Maximum Rainfall," of WMO Technical Note No. 98 [2]. A brief discussion on estimation of probable maximum storms is given in subsequent paragraphs. Also included in this section are generalized precipitation charts for estimating probable maximum precipitation values east of the 105° meridian and general-type design storm values west of the 105° meridian for watersheds in the 48 conterminous United States. These charts also are presented in chapter III of "Design of Small Dams," second edition [31], associated with procedures for

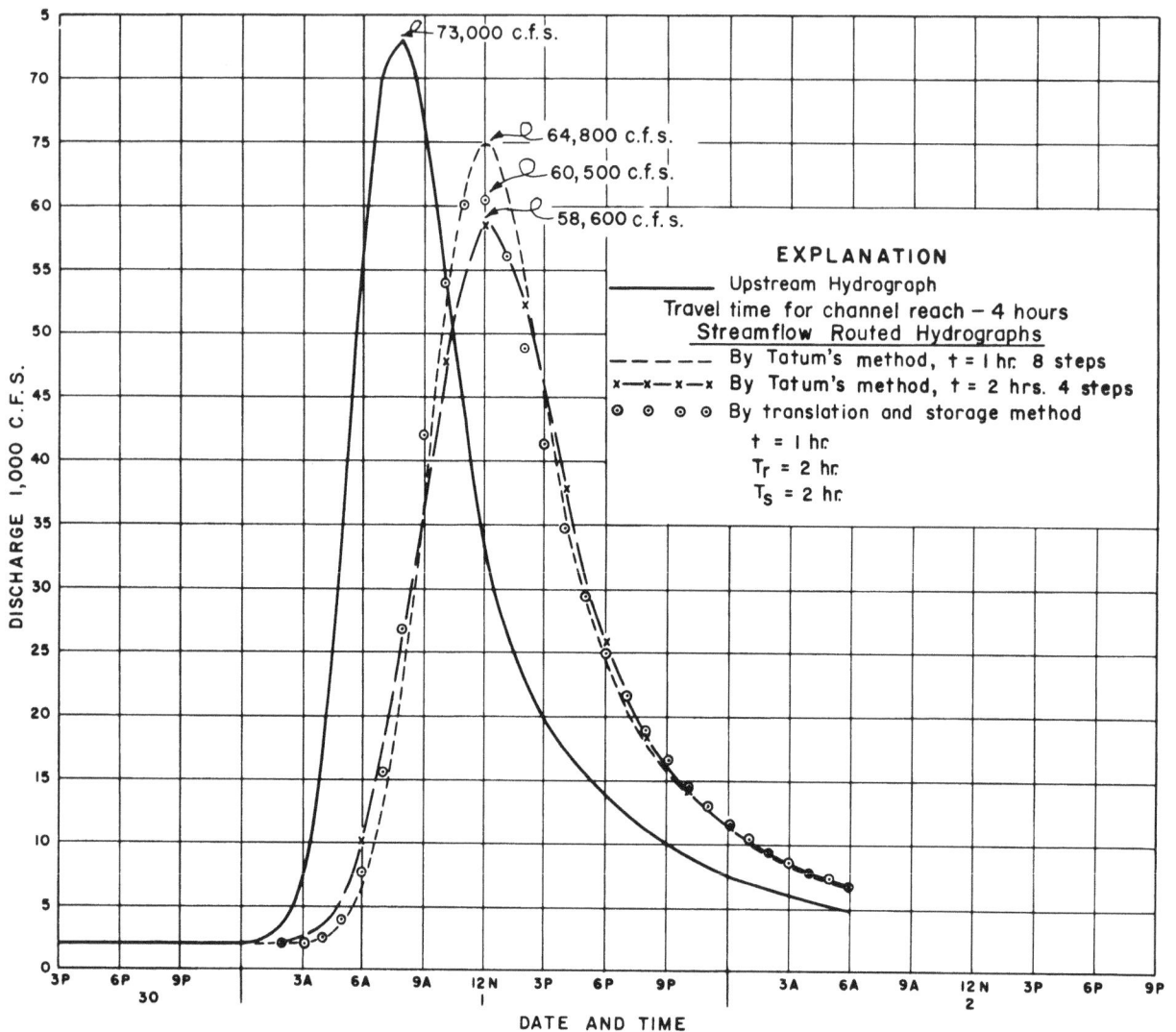

Figure G-7. Comparison of results of streamflow routings.—288-D-3183

estimating inflow design floods for small dams.

Discussion of design thunderstorm rainfall has been omitted in this text, anticipating that readers will be concerned generally with damsites controlling drainage areas large enough to preclude the use of thunderstorm rainfall. However, thunderstorm rainfall should never be ignored completely, as it may prove critical under some circumstances.

G-15. *Probable Maximum Storm Considerations.*—Estimates of probable maximum storms (PMS) are based on analyses which consist of three steps: (1) determination of the areal and temporal distribution of the larger storms of record in the general area; (2) augmentation of these observed storms through moisture adjustment; and (3) consideration of storm transposition.

One objective of the first step cited above is the determination of maximum values of storm rainfall for selected durations and area. Depth-area-duration (DAD) values of each total storm are analyzed without regard to watershed boundaries [17]. Comparison of DAD values will indicate which storms are best suited for further analysis. If hydrographs of floods for specific watersheds associated with the storms are available for analyses, determination of rainfall data for these specific watersheds can be included as a part of the analyses.

Technical literature [2] should be consulted for a detailed discussion of the theoretical assumptions included in the computational procedures for storm maximization, step (2), and storm transposition, step (3). An abridged discussion of a procedure often used for maximization and transposition of storms in plains-type terrain follows. Discussion of procedures for storm maximization and limited transposition in mountainous terrain is beyond the scope of this text.

G-16. *Procedure for Storm Maximization, Plains-Type Terrain.* —This procedure is based on assuming a saturated airmass with a pseudoadiabatic lapse rate. Moisture content under these circumstances is a unique function of surface dewpoint temperature, so that dewpoint temperatures may be used to quantitatively estimate total atmospheric water vapor or precipitable water values. Tables [18] have been published which list ambient temperatures for various elevations or pressures above a 1,000-mb. (1,000-millibar) surface, approximately equivalent to mean sea level, for selected temperatures in a saturated atmosphere with a pseudoadiabatic lapse rate.

Tables [18] also list, for each 1,000-mb. dewpoint temperature, values of precipitable water in inches for layers between the 1,000-mb. surface and various elevations to extreme heights in a saturated, pseudoadiabatic atmosphere. These precipitable water values may be used as an index to the moisture content of a unit column of air between sea level and the top of a moisture-bearing airmass. Maps with isotherms of maximum 12-hour persisting 1,000-mb. dewpoint temperatures ($^\circ$ F.) of record for each month for the 48 conterminous states are available in the "Climatic Atlas of the United States" [19].

Computational procedures for storm maximization and transposition, plains-type terrain, follow:

(a) *Maximization of a Storm in Place of Occurrence.*

(1) *Observed storm dewpoint.* —A representative 12-hour persisting surface dewpoint temperature is obtained for the storm period under study from temperature stations in the path of the inflowing moist air. If the rainfall is of a frontal type, the surface dewpoints within the rainfall area will be lower than those of the inflowing moist air, thus giving a low estimate of storm moisture content. Distance and direction from the storm center to the representative dewpoint station or stations should be recorded.

(2) *Adjustment to 1,000-mb. surface.* —Since during major storms the airmass will be saturated, the dewpoint temperature at the representative station can be adjusted to a 1,000-mb. surface temperature assuming a saturated, pseudoadiabatic lapse rate of temperature.

(3) *Precipitable water values.* —From the 1,000-mb. dewpoint temperature determined in (2) above, obtain two precipitable water values, W_p, for the observed storm:

(a) $W_{p \cdot 1}$ is the precipitable water between 1,000 mb. and the top of the moist layer for the storm system; an elevation of 40,000 feet, or pressure of 200 mb., is usually assumed.

(b) $W_{p \cdot 2}$ is the precipitable water between 1,000 mb. and the mean surface elevation of the central portion of the observed storm. If the inflowing moist air has passed over a topographical barrier with a higher elevation than at the central portion of the storm, $W_{p \cdot 2}$ is obtained using the inflow barrier elevation.

(4) *Observed storm's precipitable water, W_s.* —Compute the observed storm's moisture content or available precipitable water, W_s, as $W_{p \cdot 1}$ minus $W_{p \cdot 2}$.

(5) *Probable maximum precipitable water for the storm, W_x.* —An estimate of the probable maximum moisture content indicated for the storm is obtained as follows:

(a) From the "Climatic Atlas of the United States" [19], the maximum 12-hour persisting dewpoint temperature of record can be determined for the date of storm occurrence and the location of the representative dewpoint for the observed storm. Frequently, the maximum recorded dewpoint temperature within a period of plus or minus 15 days is used.

(b) From the maximum dewpoint of record, precipitable water is obtained for the same layers as used in W_{p-1} and W_{p-2} above. These precipitable water values are designated W_{r-1} and W_{r-2}.

(c) The estimated probable maximum precipitable water, W_x, will be W_{r-1} minus W_{r-2}.

(6) *Moisture maximization factor, M_f.*—The moisture maximization factor, M_f, is computed as the ratio of the probable maximum precipitable water to the precipitable water observed during the storm, or $M_f = W_x/W_s$.

(7) *Maximized storm values.*—Maximized storm values are computed by multiplying depth-area-duration (DAD) values of the observed storm by the maximization factor, M_f.

Note: This procedure assumes that the magnitude of rainfall in a storm is a function only of the inflow moisture charge. It also assumes that the most effective combination of storm efficiency and inflow wind has occurred or has been closely approached in the major storms of record. The procedure may not always prove adequate, particularly for regions where rainfall is strongly influenced by orographic effects [2].

(8) *Example of computations—maximization in place.*

(a) Dewpoint observation station: elevation 1000 feet.

Location: 100 miles southeast of storm center.
Representative 12-hour storm dewpoint: 69° F.
Sea level, 1,000 mb., dewpoint: 71° F.

(b) Surface elevation, storm center: 1500 feet.

W_{p-1} = 2.38 inches (at 40,000 feet)
W_{p-2} = 0.32 inch (at 1500 feet)
W_s = 2.06 inches

(c) Maximum dewpoint of record, observed 100 miles southeast of storm center: 78° F. [19].

W_{r-1} = 3.35 inches (at 40,000 feet)
W_{r-2} = 0.41 inch (at 1500 feet)
W_x = 2.94 inches

(d) Moisture maximization factor:

$$M_f = 2.94/2.06$$
$$M_f = 1.43$$

(b) *Maximization of Transposed Storm.*—When a storm is transposed and maximized for moisture content, the maximization factor is usually computed for the storm only at its transposed location. Computation of available precipitable water for the observed storm, W_s, remains the same as described above.

The moisture maximization factor is computed by determining the surface elevation at the center of the storm at its transposed position or the height of the mean inflow barrier to that location. The maximum dewpoint of record is obtained from the charts of dewpoints [19] at *the same distance from the transposed center and in the same direction as the observed storm dewpoint was obtained.*

(1) *Example of computations—moisture maximization of transposed storm.*

(a) Assume that the storm used in the previous example is transposed to a location where the elevation of the storm center is 2500 feet and that there is not a higher inflow barrier between the transposed center and the moisture source.

(b) Mark the location of the transposed center on the charts of maximum recorded dewpoint temperatures and measure 100 miles southeast to determine the maximum dewpoint of record; for example 77° F.

(c) Observed storm precipitable water remains the same; W_s = 2.06 inches.

(d) Maximum precipitable water for a dewpoint of 77° F:

W_{r-1} = 3.19 inches (at 40,000 feet)
W_{r-2} = 0.64 inch (at 2500 feet)
W_x = 2.55 inches

(e) Moisture maximization factor for the transposed storm:

$$M_f = 2.55/2.06$$
$$M_f = 1.24$$

Note: If an M_f factor greater than 2.0 is computed, reexamine the computations and all meteorological aspects of the transposed storm. An M_f factor greater than 2.0 has not been used in Bureau of Reclamation design storm studies.

(2) *Maximized transposed storm values.*—The maximized values for the transposed storm are computed by multiplying the DAD values of the observed storm by the maximization factor for the transposed location.

G-17. *Design Storm—Probable Maximum Precipitation (PMP) or Probable Maximum Storm (PMS) Estimates for a Watershed.*—Estimates of PMP or PMS, whether made by storm transposition and procedure of dewpoint adjustment described above or by more detailed theoretical computations [20][6], are based generally on the results of analyses of observed storms. In the United States, passage of the Flood Control Act of 1936 led to the development of a National Storm Study Program under the primary sponsorship of the U.S. Army Corps of Engineers. Under this program more than 600 storms throughout the United States have been analyzed in a uniform manner and summary sheets distributed to Government agencies and the engineering profession [21]. An example of a storm analysis summary sheet from the publication "Storm Rainfall in the United States" [21] is shown on figure G-8. Each storm analyzed has been assigned a designation such as MR 4-24 on the figure. Unfortunately, not all of the summary sheets have a reference to the observed storm dewpoint, such as shown on figure G-8(A). Depth-area-duration (DAD) data for each storm analyzed are given in a table,

such as the one at the bottom of figure G-8(A).

A storm location map and a few selected mass rainfall curves are given on figure G-8(B). Summaries of observed storm data such as presented in "Storm Rainfall in the United States," provide broad outlines of storm magnitudes and their seasonal and geographical variations.

A simplified example of the derivation of design storm values for a particular watershed follows. Sources of numerical values used are referenced when possible. The isohyetal patterns and watershed map are not presented. This example may provide the reader with information that will be useful in a better understanding of how preliminary design storm estimates are obtained from the generalized PMP charts given later.

(a) *Example of a Design Storm Study.*—(Final-type design storm studies should be prepared by experienced hydrometeorologists.) Let us assume that design storm values representing PMS estimates are required for a watershed with a 200-square-mile area at longitude 99°30' west, latitude 41°00' north, a region where storm transposition and maximization by dewpoint adjustment is an acceptable approach. Procedural steps are described first, then numerical computations are given.

(1) *Transposition limits of major storms.*—The broad limits within which major observed storms can be transposed should be established first. This will require consultation with an experienced hydrometeorologist. However, for the United States east of the 105° meridian, guidelines have been established in Hydrometeorological Report 33 [20].

(2) *Inventory of data of major storms.*—Referring to "Storm Rainfall in the United States" [21], rainfall depth-duration values can be obtained for an area of 200 square miles for all major storms that have been analyzed in the region for which transposition is applicable. Analysis may be required for recent major storms in the region in order to complete the inventory.

(3) *Selection of storms for further study.*—Several of the larger storms are

DEPARTMENT OF THE ARMY   CORPS OF ENGINEERS

STORM STUDIES – PERTINENT DATA SHEET

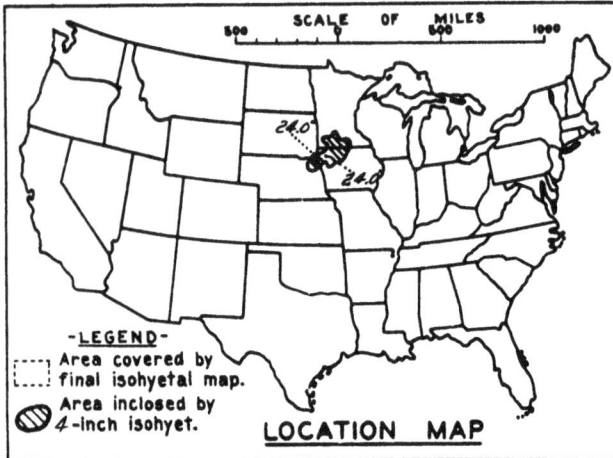

Storm of 17–19 September 1926
Assignment MR 4–24
Location Ia,Minn,Nebr.S.D. & Wisc.
Study Prepared by:
Missouri River Division
Omaha District Office

Part I Reviewed by H. M. Sec. of
  Weather Bureau, 8/5/47
Part II Approved by Office, Chief
  of Engineers for Distribution
  of Factual Data, 12/23/47
Remarks: Centers near
  Boyden & Maurice, Ia.
Dewpt. 70° – Ref. Pt. 175 SSE
           Grid C-15

LOCATION MAP

-LEGEND-
⌐ ⌐ Area covered by
└ ┘ final isohyetal map.
▨ Area inclosed by
4-inch isohyet.

DATA AND COMPUTATIONS COMPILED

PART I

Preliminary isohyetal map, in 2 sheets, scale 1:500,000
Precipitation data and mass curves: (Number of Sheets)
   Form 5001-C (Hourly precip. data)_____ 8
   Form 5001-B (24-hour " ")_____ -
   Form 5001-D (" " " ")_____ 11
   Miscl. precip. records, meteorological data, etc._____ 29
   Form 5002 (Mass rainfall curves)_____ 27

PART II

Final isohyetal maps, in 1 sheet, scale 1:1,000,000
Data and computation sheets:
   Form S-10 (Data from mass rainfall curves)_____ 3
   Form S-11 (Depth-area data from isohyetal map)_____ 2
   Form S-12 (Maximum depth-duration data)_____ 17
   Maximum duration-depth-area curves_____ 1
   Data relating to periods of maximum rainfall_____ 7

MAXIMUM AVERAGE DEPTH OF RAINFALL IN INCHES

Area in Sq. Mi.	Duration of Rainfall in Hours										
	6	12	18	24	30	36	48	54			
Max.Station	18.4	23.8	24.0	24.0	24.0	24.0	24.0	24.0			
10	15.1	20.7	21.7	21.7	21.7	21.7	21.7	21.7			
100	12.8	17.1	17.8	17.8	17.8	17.8	17.8	17.8			
200	11.7	15.8	16.6	16.6	16.6	16.6	16.6	16.6			
500	9.4	12.6	13.3	13.3	13.3	13.3	13.3	13.3			
1,000	7.5	10.1	10.4	10.6	10.6	10.6	10.6	10.6			
2,000	5.9	8.0	8.2	8.6	8.6	8.6	8.6	8.6			
5,000	4.1	6.3	6.4	6.6	6.6	6.6	6.6	6.6			
10,000	3.0	5.2	5.4	5.5	5.6	5.6	5.6	5.6			
20,000	2.1	4.1	4.3	4.4	4.6	4.8	4.9	4.9			
50,000	1.4	2.7	2.9	3.0	3.2	3.6	3.8	3.8			
63,000	1.2	2.4	2.6	2.7	2.9	3.3	3.5	3.5			

Form S-2

(A)

Figure G-8. Example of summary sheet, "Storm Rainfall in the U.S." (sheet 1 of 2).–288-D-3184(1/2)

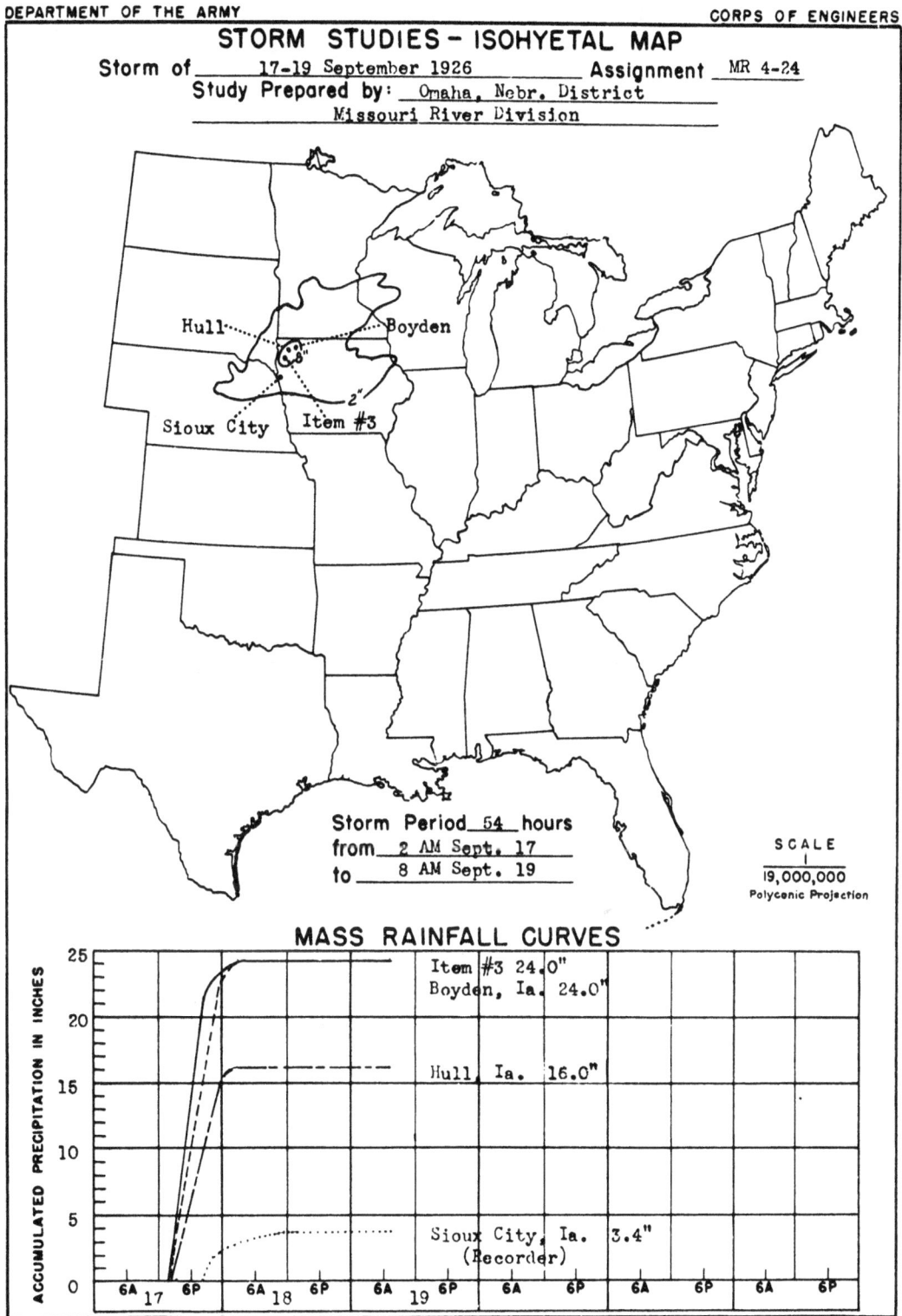

STORM STUDIES – ISOHYETAL MAP

Storm of_____17-19 September 1926_____ Assignment _MR 4-24_

Study Prepared by: _Omaha, Nebr. District_

Missouri River Division

Storm Period_54_ hours
from_2 AM Sept. 17_
to _8 AM Sept. 19_

SCALE

19,000,000
Polyconic Projection

MASS RAINFALL CURVES

Item #3 24.0"
Boyden, Ia. 24.0"

Hull, Ia. 16.0"

Sioux City, Ia. 3.4"
(Recorder)

FORM S-3E

(B)

Figure G-8. Example of summary sheet, "Storm Rainfall in the U.S." (sheet 2 of 2).–288-D-3184(2/2)

assumed transposed and the depth-duration values for 200 square miles maximized for maximum moisture charge to identify those storms that give the greatest values. Any individual storm may not yield maximum values for all durations. It may be necessary, therefore, to consider a number of storms in the final analysis.

(4) *Transposition of isohyetal patterns.* —The isohyetal patterns of the storms which yield large values should be obtained, and these patterns then overlaid individually on a map of the subject watershed. The position, within limits, that gives the greatest total basin average rainfall depth should be used. In positioning a transposed storm isohyetal pattern, the orientation of the observed storm pattern is maintained generally within limits of plus or minus 20°.

(5) *Average watershed precipitation of transposed storm.* —The average storm rainfall *within* the watershed boundaries of each transposed storm isohyetal pattern is obtained by planimetry. The depth of precipitation for a given area for the total storm was obtained from a DAD tabulation similar to that shown on figure G-8(A). These values were, of course, measured from the isohyetal pattern in the original storm without regard to any watershed boundaries. Obviously, only an assumption of a *perfect fit* of the transposed isohyetal pattern to the basin configuration would give the same total basin rainfall for the transposed storm as that listed in the DAD tables.

(6) *Fit-factor.* —A fit-factor, F_f, that is, the ratio of the watershed average rainfall depth to the storm pattern rainfall depth, for equal areas, is computed for each transposed storm. The importance of the fit-factor to PMS estimates varies depending on the size, shape, and orientation with respect to major storm patterns of each individual watershed. In the example region, watersheds are typically long and narrow with their major axis oriented generally east-west, so that a fit-factor in this region is quite important, except for extremely large drainage basins.

If $\overline{P_o}$ represents the average rainfall depth for the total observed storm for a given area and $\overline{P_{tr}}$ represents the average rainfall depth measured from the isohyetal pattern of the observed storm, as transposed, then

$$F_f = \frac{\overline{P_{tr}}}{\overline{P_o}} \qquad (11)$$

It should be obvious that $F_f \leq 1$.

(7) *Total maximization adjustment factor, Ad_f.* —The total maximization adjustment factor, Ad_f, for a storm, as transposed to a watershed, is the product of the storm moisture maximization factor, M_f, and the fit-factor, F_f, or,

$$Ad_f = (M_f)(F_f) \qquad (12)$$

(8) *Design storm values, depth-duration curve.* —The maximized depth-duration values for each storm, as transposed to a watershed, are computed by multiplying the observed storm depth-duration values by the respective maximization adjustment factor, Ad_f. The computed values for each storm should be plotted with accumulative time in hours as the abscissa versus the accumulative rainfall depths in inches as the ordinate.

A design storm depth-duration curve is obtained by drawing a smooth curve. An enveloping curve will give design storm values approaching PMP for a watershed. A curve drawn through the data for one storm only will give selected PMS values.

Since the depth-duration curve is ordered in such a manner as to show only the maximum values of rainfall for various durations, the curve does not indicate a realistic sequence of rainfall increments which might occur during the actual design storm. Incremental design storm values obtained from the smooth depth-duration curve should be arranged in realistic sequence for flood computation.

For storms of long duration (several days), the design storm depth-duration curve may not be smooth throughout but have two or more periods of intense rainfall separated by periods of little or no rainfall. Such storms are frequently critical for very large basins or basins in tropical regions. In these instances, incremental design storm values may be arranged in any realistic sequence, within the limitation that the separate periods will not be so combined as to produce a rainfall sequence that would have exceeded the recommended design storm depth-duration curve at any point.

(9) *Numerical computations.*—Table G-9 presents numerical values for procedures described in the subsections above. Maps showing the transposed storm isohyetal patterns as fitted to the watershed and the planimetry notes for determination of average basin rainfall for each transposed storm are not included. A plot of depth-duration values of the transposed storms, as maximized, and the recommended depth-duration curve of the design storm are shown on figure G-9. In this instance, the design storm duration is 17 hours and rainfall values approach PMP. The enveloping curve on figure G-9 was drawn "by eye" as adequate for a preliminary PMS estimate. Design storm values read from the curve at 1-hour intervals are listed in table G-10 because a flood hydrologist may wish to use a 1-hour unitgraph to compute an inflow design flood hydrograph for this size watershed.

(b) *Generalized Precipitation Charts.*—Maps showing smoothed isohyets of PMP for the United States east of the 105° meridian and PMS values for the United States west of the 105° meridian are presented here to provide a means of quickly obtaining *preliminary design storm values* for selected watersheds above proposed damsites. It is impossible to show on the generalized charts all of the refinements and variations that can influence the magnitude of design storm values for individual watersheds. Design storm values obtained from the generalized charts represent a reasonable upper limit and, in most instances, will exceed the values obtained for a specific watershed by a detailed hydrometeorological study, as previously discussed.

(1) *Generalized chart for the United States east of the 105° meridian.*—Figure G-10 shows probable maximum 6-hour precipitation values for any area of 10 square miles for the United States east of the 105° meridian. This chart is based on one presented in Hydrometeorological Report No. 33, prepared by the Hydrometeorological Section of the National Weather

Service in collaboration with the U.S. Army Corps of Engineers [20]. These 6-hour values for 10-square-mile areas can be modified for durations in excess of 6 hours and for larger areas up to 1,000 square miles by use of figure G-11. No variation is assumed between point and 10-square-mile precipitation. For durations shorter than 6 hours, the time distribution of precipitation can be obtained from curve C, figure G-12. Subsequent to the publication of Hydrometeorological Report No. 33, the Corps of Engineers have recommended[7] that the following adjustment percentages be applied to the depth-duration values obtained from figure G-10 in order to provide for the imperfect *fit* of the isohyetal patterns of observed storms to the shape of a particular basin.

Drainage area, square miles	Adjustment factor applicable to H.R. 33 rainfall values, percent
1,000	90
500	90
200	89
100	87
50	85
10	80

(2) *Generalized chart for the United States west of the 105° meridian.*—Figure G-13 shows probable maximum 6-hour point general-type storm values for areas of the United States west of the 105° meridian. This chart is based on the results of approximately 330 design storm analyses prepared by the Bureau of Reclamation for specific drainage basins west of the 105° meridian, as well as consideration of numerous design storm analyses made by the Special Studies and Hydrometeorological Branches of the National Weather Service.

The variable topography of this part of the United States greatly influences the storm potential and permits only limited transposition of storms. These point storm values can be applied to areas up to 1,000 square miles by use of the curves presented on figure G-14. The 6-hour general-type storm values can be extended for longer duration periods by multiplying the 6-hour value by the

[7]Engineer Circular No. 1110-2-27, dated August 1, 1966, "Policies and Procedures Pertaining to Determination of Spillway Capacities and Freeboard Allowances for Dams."

Table G-9.—*Example of design storm derivation for area east of 105° meridian*

BASIC DATA:
   Watershed location: 99°30' W, 41°00' N
   Drainage area: 200 sq. mi.
   Inflow barrier: 2,500 feet

(A) MAJOR STORMS SELECTED FOR TRANSPOSITION

Designation No.	Approximate geographic location–name	Date of storm	Inflow barrier, feet	Observed storm dewpoint		Total storm		Reference
				°F.[1]	Ref. pt.	Duration, hrs.	[2]$\overline{P}_o$	
MR4-24	Boyden, Iowa	9/17-19/26	1,200	70	175 mi. SSE	54	16.6	Fig. G-8A
MR4-5	Grant Township, Nebr.	6/3-4/40	1,200	66[3]	120 mi. S	20	11.2	[21]
MR6-15	Stanton, Nebr.	6/10-13/44	1,500	70	125 mi. SSE	78	14.4	[21]
R10-1-1[4]	Greeley, Nebr.	8/12-13/66	2,000	71	80 mi. SSE	17	13.3	[4]

[1] 1,000 millibars, or mean sea level.
[2] Average rainfall depth, 200 sq. mi.
[3] Revised value in lieu of 63° F. [21]
[4] Recent storm analysis, preliminary, Bureau of Reclamation, Engineering and Research Center, Denver, Colo.

(B) STORM TRANSPOSITION AND MAXIMIZATION

(Column heading symbols as previously defined in text.)

Storm No.	Observed storm						Transposed storms						Maximizing factors		
	Dwpt., °F.	Barrier, feet	W_{p-1}	W_{p-2}	W_s	$\overline{P}_o$	Dwpt.,[1] °F.	Barrier, feet	W_{r-1}	W_{r-3}	W_x	$\overline{P}_{tr}$	M_f	F_f	Ad_f
MR4-24	70	1,200	2.27	0.25	2.02	16.6	76	2,500	3.04	0.62	2.42	12.3	1.20	0.74	0.89
MR4-5	66	1,200	1.86	.22	1.64	11.2	76	2,500	3.04	.62	2.42	9.6	1.48	.86	1.27
MR6-15	70	1,500	2.27	.31	1.96	14.4	76	2,500	3.04	.62	2.42	13.0	1.23	.90	1.11
R10-1-1	71	2,000	2.38	.42	1.96	13.4	77	2,500	3.19	.64	2.55	12.4	1.30	.93	1.21

[1] From Climatic Atlas of United States [19].

(C) MAXIMUM OBSERVED DEPTHS, INCHES

Storm	Duration in hours											
	3	6	9	12	15	18	24	30	36	48	60	72
MR4-24		11.7		15.8		16.6	16.6	16.6	16.6	16.6	[1]16.6	
MR4-5	5.5	9.6	11.1	11.2	11.2	11.2	[2]11.2					
MR6-15		11.1		12.9		12.9	12.9	12.9	13.1	14.1	14.3	[3]14.4
R10-1-1	6.7	9.4	12.5	13.1	13.2	[4]13.4						

[1] Storm ended at 54 hrs.
[2] Storm ended at 20 hrs.
[3] Storm ended at 78 hrs., depth = 14.4 in.
[4] Storm ended at 17 hrs.

(D) MAXIMUM TRANSPOSED DEPTHS, INCHES

Storm No.	Ad_f	Duration in hours											
		3	6	9	12	15	18	24	30	36	48	60	72
MR4-24	0.89		10.4		14.1		14.8	14.8	14.8	14.8	14.8	[1]14.8	
MR4-5	1.27	7.0	12.2	14.1	14.2	14.2	14.2	[2]14.2					
MR6-15	1.11		12.3		14.3		14.3	14.3	14.3	14.5	15.7	15.9	[3]16.0
R10-1-1	1.21	8.1	11.4	15.1	15.9	16.0	[4]16.2						

[1] At 54 hrs.
[2] At 20 hrs.
[3] Also at 78 hrs.
[4] At 17 hrs.

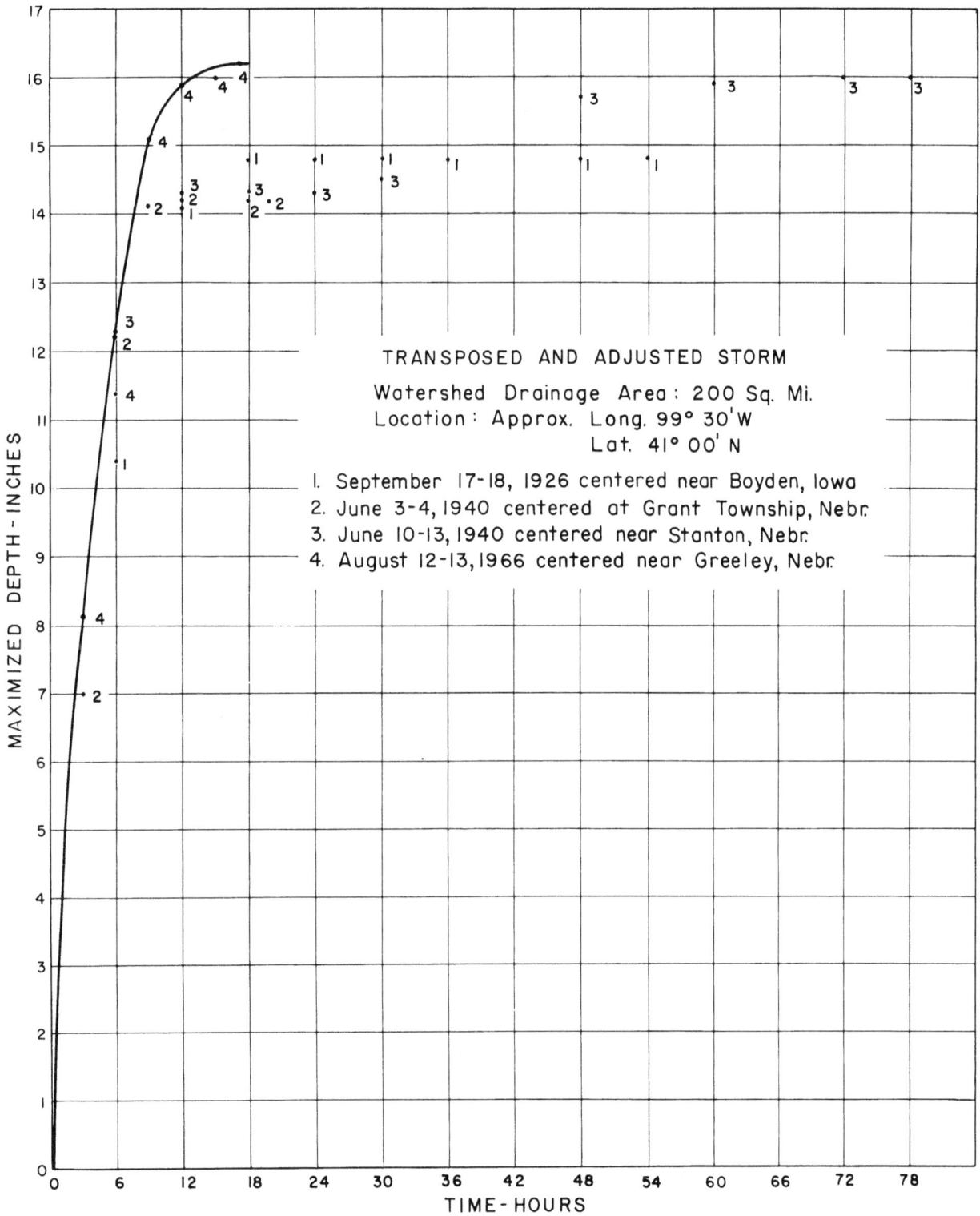

Figure G-9. Design storm—depth-duration values.—288-D-3185

Table G-10.–*Design storm depth-duration values, inches*

BASIC DATA: Hypothetical example.
Watershed area = 200 sq. mi.
Location = approximately 99°30′ W,
41°00′ N

Time, ending at hour	Accumulated depth, inches	Incremental depth, inches
0	0	0
1	4.20	4.20
2	6.40	2.20
3	8.10	1.70
4	9.70	1.60
5	11.10	1.40
6	12.30	1.20
7	13.30	1.00
8	14.30	1.00
9	15.10	.80
10	15.45	.35
11	15.70	.25
12	15.90	.20
13	16.00	.10
14	16.10	.10
15	16.15	.05
16	16.20	.05
17	16.20	0
18	16.20	0

appropriate factor shown in table G-11. Values for duration of less than 6 hours can be obtained from the appropriate curve of figure G-12.

(3) *Use of generalized charts.*–Design storm values for any watershed of a 1,000-square-mile area or less in the conterminous 48 United States may be obtained from the generalized charts, but it must be noted that such design storm values should be considered as only preliminary estimates for watersheds controlled by large dams. Design storm values obtained from figures G-10 and G-13 show considerable difference at their common boundary along the 105° meridian. This is due to the techniques used in determining the values shown on the charts.

Preliminary design storm values for a particular watershed obtained from either generalized chart should be plotted on coordinate paper and an enveloping depth-duration curve drawn. Plotting offers a method of checking the computations, as a smooth curve should be indicated, and also provides the means of obtaining hourly design storm values for the total storm period if needed. Incremental values from the depth-duration curve may be arranged in any sequence desired by a flood hydrologist for computation of a preliminary inflow design flood.

The generalization charts for estimating preliminary design storm values have been limited to an area of 1,000 square miles because generalizations of criteria become more difficult as the size of the area increases. Preliminary design storm estimates can be made for areas greater than 1,000 square miles in regions of nonorographic rainfall by the procedure described in section G-17. The step of determining a fit-factor is omitted. A depth-duration curve is drawn on the basis of information compiled in a tabulation such as table G-9(D), using the moisture maximization factor, M_f, instead of the total adjustment factor, Ad_f, to compute values for the table. Preliminary design storm estimates for large mountainous basins (with predominately orographic rainfall) should be obtained from a hydrometeorologist.

F. PRELIMINARY INFLOW DESIGN FLOOD, RAINFALL ONLY

G-18. *General.*–This subchapter outlines procedures for estimating preliminary inflow design flood (IDF) hydrographs using: (1) design storm values from the generalized precipitation charts, figures G-10 and G-13; (2) an estimation of incremental rainfall excesses from runoff curves, section G-7(b)(6); and (3) the lag-time dimensionless-graph method of obtaining unitgraphs, section G-9. An example is given of computation of preliminary inflow design flood hydrographs for a watershed east of the 105° meridian, with accompanying

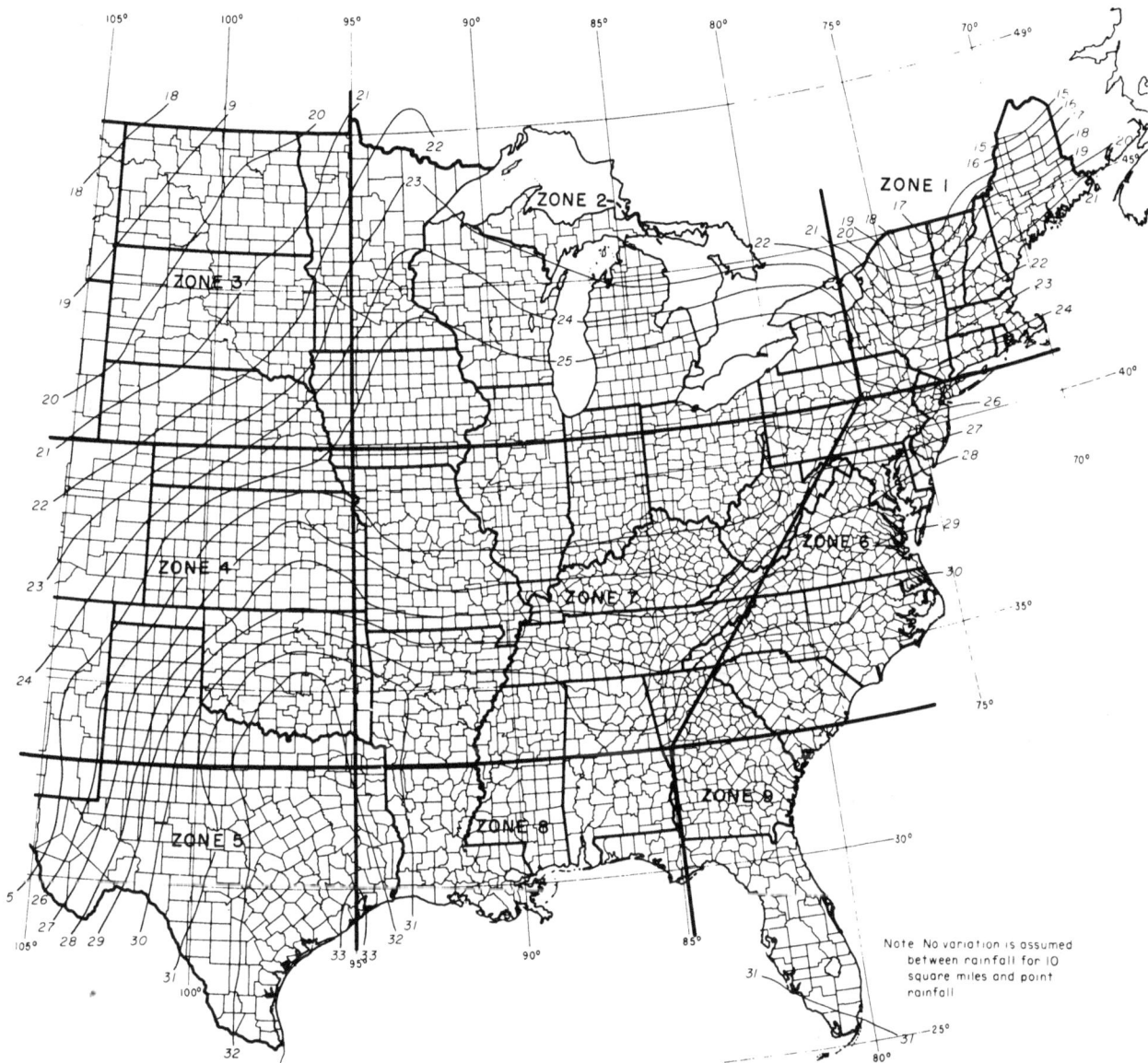

Figure G-10. Probable maximum precipitation (inches) east of the 105° meridian for an area of 10 square miles and 6 hours' duration.—288-D-3191

discussions directed toward considerations applicable to all inflow design flood studies. Procedures applicable to watersheds west of the 105° are outlined. A discussion of preparing recommendations for routing preliminary inflow design flood hydrographs through proposed reservoirs concludes this presentation.

G-19. *Example—Preliminary Inflow Design Flood Hydrographs, Watersheds East of 105° Meridian.*—A hypothetical watershed in a

general location east of the 105° meridian has been assumed in order to illustrate several of the problems encountered in IDF computations, all of which would not likely be presented by a specifically located watershed.

(a) *Basin Description.*—A map of the assumed watershed above a proposed damsite is shown on figure G-15. The center of the basin is assumed to be located in zone 4 somewhere along the 30-inch, 6-hour PMP for 10-square-mile isohyet, figure G-10. An outline

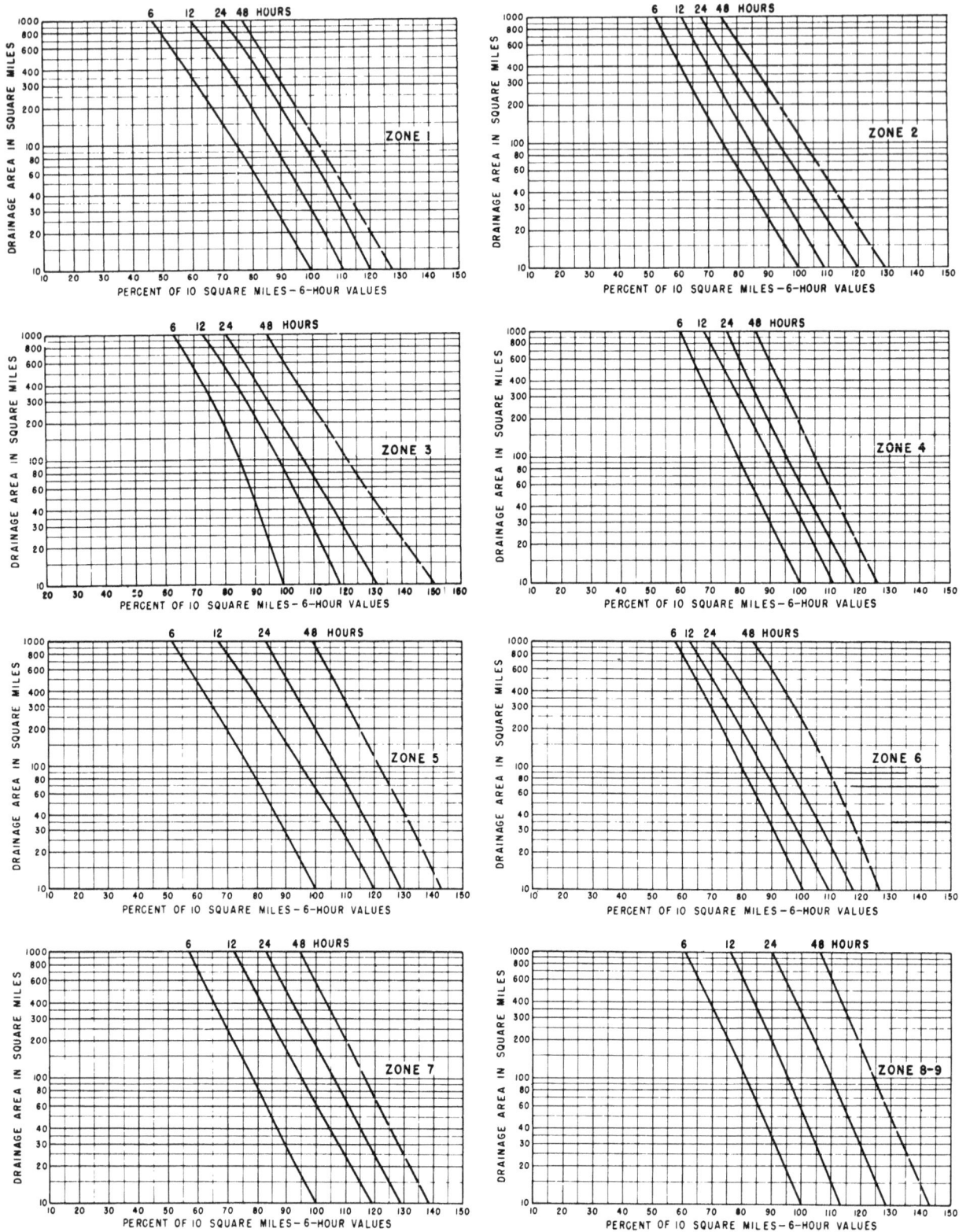

Figure G-11. Depth-area-duration relationships–percentage to be applied to 10 square miles, 6-hour probable maximum precipitation values.–288-D-2450

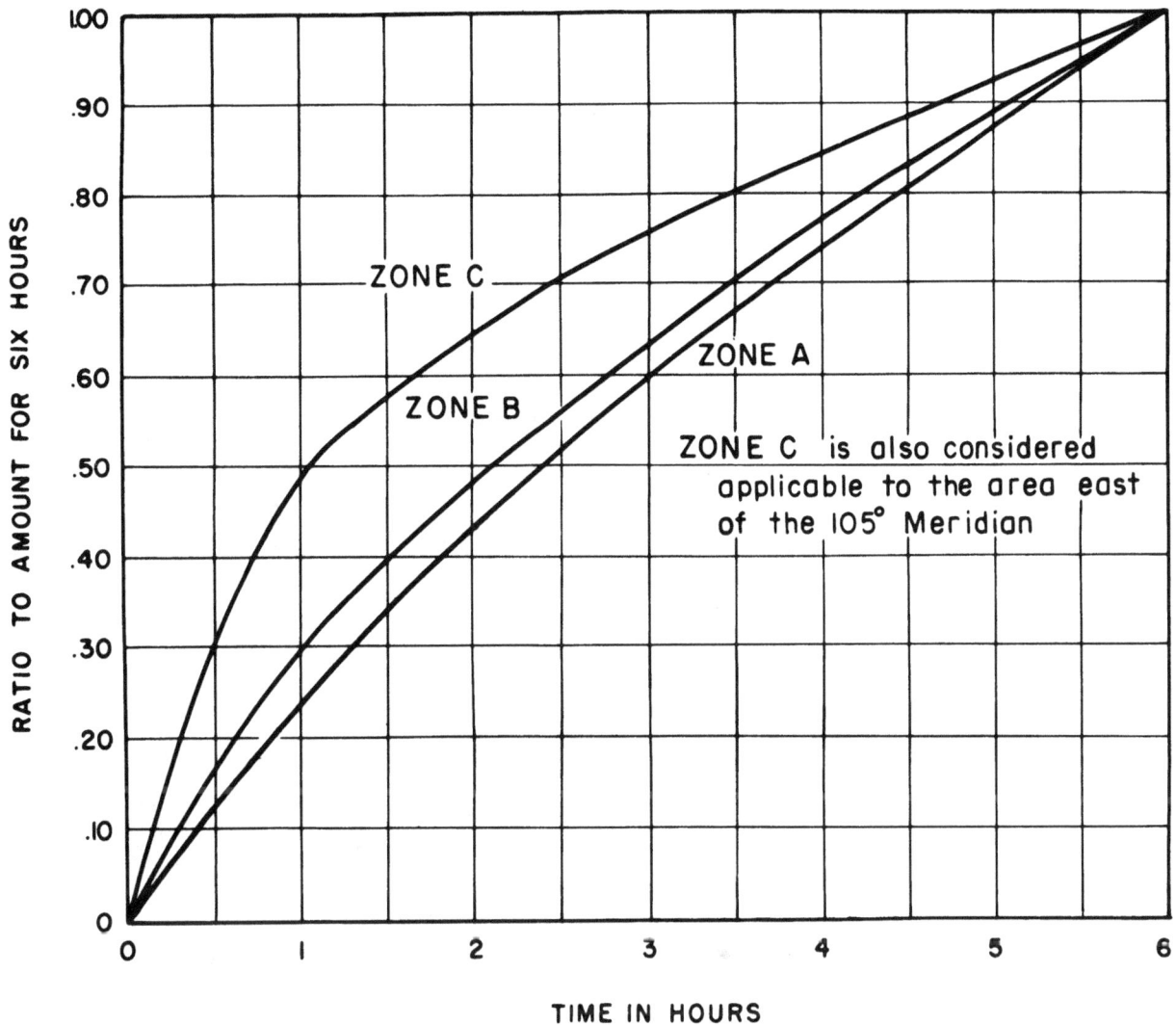

Figure G-12. Distribution of 6-hour rainfall for area west of 105° meridian (see fig. G-13 for area included in each zone).—288-D-2758

of the proposed reservoir surface at normal water storage capacity is shown, because the length of natural stream channels to be submerged influences lag-time calculations. It is assumed that runoff characteristics of the areas drained by the two main tributaries differ enough to warrant consideration of dividing the watershed into two subareas, A and B, as there is information available indicating that subarea A definitely has rapid runoff characteristics. All of the area enclosed by the natural divides contributes runoff.

(1) *Drainage areas* are:

Total basin	800 square miles
Subarea A	240 square miles
Subarea B	560 square miles
Reservoir surface	26 square miles

As the reservoir surface area is about 3 percent of the total basin area in this example, reservoir surface may be considered as land area, except for lag-time computations. Whenever there is found a reservoir surface area of about 10 percent or more of total contributing drainage area, computations should be made separately of the runoff originating from the land area, to which

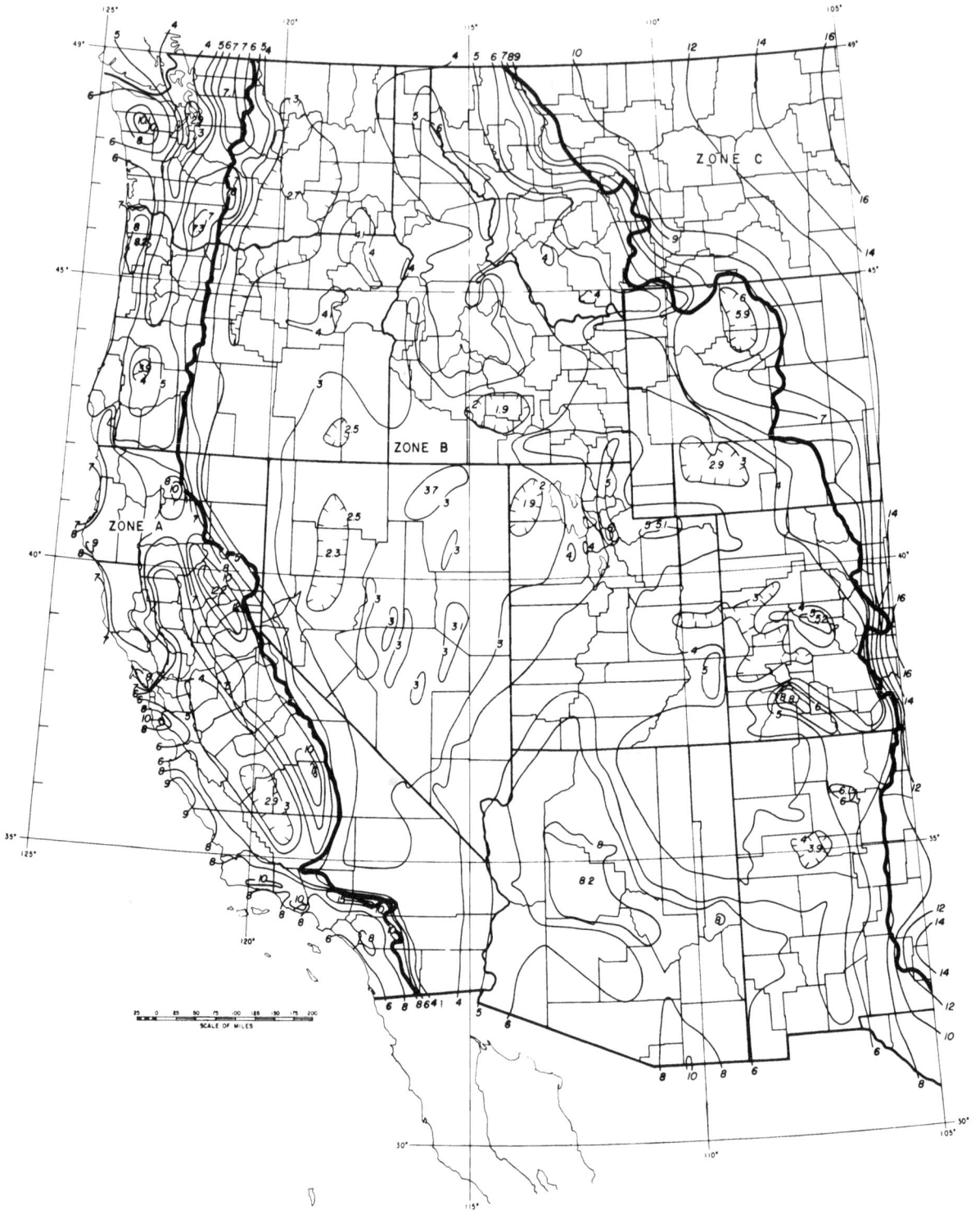

Figure G-13. Probable maximum 6-hour point precipitation values in inches for general-type storms west of the 105° meridian.—288-D-3192

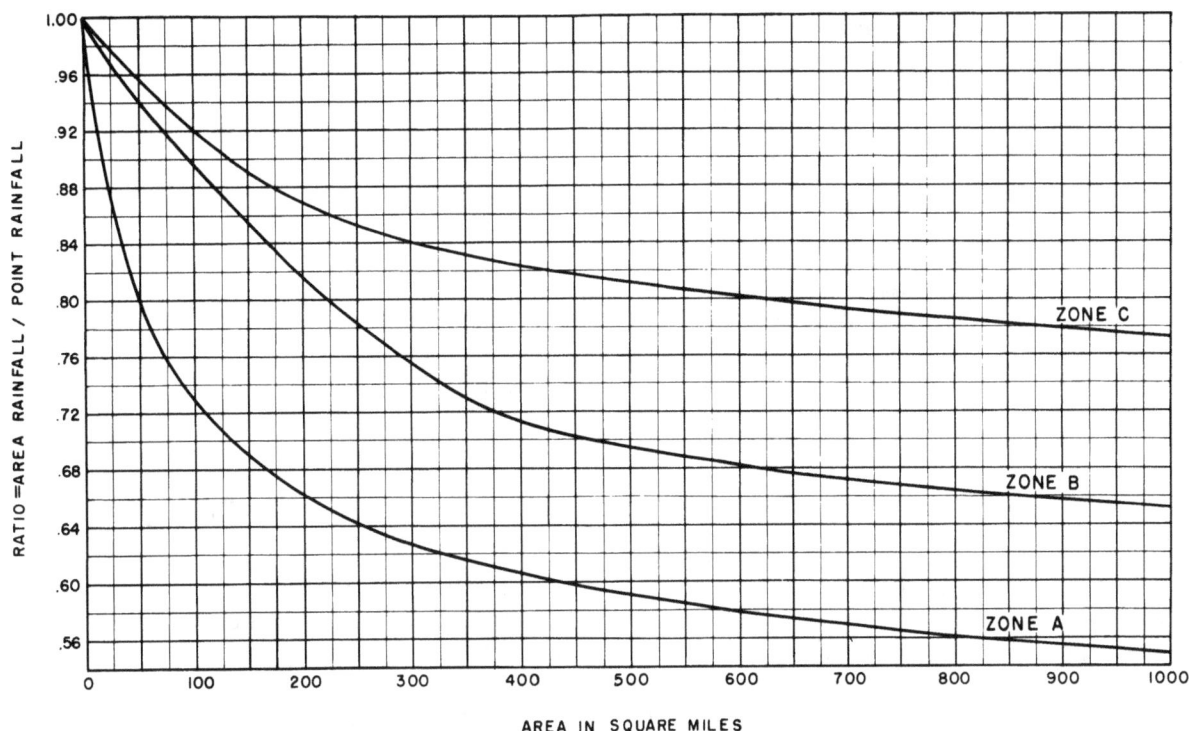

Figure G-14. General-type storm—conversion ratio from 6-hour point rainfall to area rainfall for area west of 105° meridian.—288-D-2759

Table G-11.—*Constants for extending 6-hour general-type design-storm values west of 105° meridian to longer duration periods*[1]

Duration, hours[2]	Constants		
	Zone A	Zone B	Zone C
8	1.20	1.18	1.14
10	1.39	1.36	1.26
12	1.58	1.53	1.36
14	1.76	1.66	1.43
16	1.93	1.77	1.50
18	2.10	1.87	1.57
20	2.26	1.95	1.64
22	2.42	2.03	1.71
24	2.57	2.10	1.78
30	2.95	2.28	1.97
36	3.26	2.38	2.15
42	3.55	2.40	2.25
48	3.79	2.41	2.28
60	4.14		
72	4.34		

[1]Multiply 6-hour point rainfall from figure G-13 by indicated constant.
[2]For durations shorter than 6 hours, the time distribution of storm values can be obtained from the appropriate curve presented on figure G-12.

retention losses are applicable to design storm rainfall, and the increased inflow to the reservoir due to design rainfall on the reservoir surface area where retention losses are zero. There are instances where rain falling on reservoir surfaces supplies the major portion of inflow. When rain falling on a reservoir surface must be considered, rainfall increments in inches are converted to equivalent incremental flow in cubic feet per second and combined with respectively timed increments of inflow from the land area. Watersheds in which a reservoir will submerge miles of mainstream channel, and numerous side tributaries flow directly into the reservoir, the watershed should be divided into at least two subareas, the subarea above the head of the reservoir and the area directly tributary to the reservoir. Subarea B, figure G-15, approaches this situation. If a final-type IDF study were made for the example watershed, a better evaluation of a final-type IDF would be obtained by dividing subarea B into two subbasins and

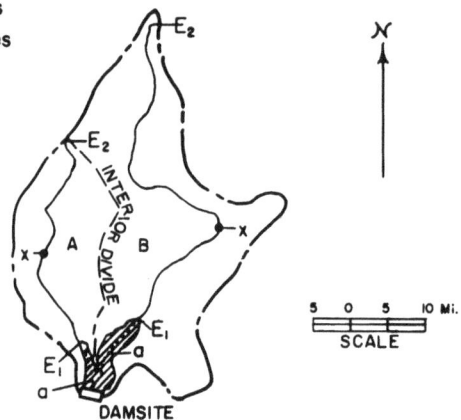

1. Measure stream length E_1 to E_2; L, miles
2. Measure stream length E_1 to x; L_{ca}, miles
    Note: Do not include "a" stream length that will be submerged.

3. $S = \dfrac{\text{Elevation } E_2 \text{ minus elevation } E_1.}{L \text{ , miles}}$

In the above, x = center of area projected.

Damsites with 2(or more) markedly different tributaries require 2 (or more) unitgraphs.

Figure G-15. Basin map—example of preliminary inflow design flood computation.—288-D-3186

deriving a unitgraph for each; the subbasins would be above and below the head of the reservoir, point E_1, figure G-15.

(2) *Streamflow records.*—Two assumptions are made for lag-time illustrating purposes: first, that there are no streamflow records available for analysis; second, that tributary B has been gaged at the mouth near the damsite, and hydrograph analyses have indicated a lag-time of 22 hours for subarea B.

(3) *Soils and cover.*—Use of runoff curves requires hydrologic classification of watershed soils and cover, discussed in section G-7(b)(6), for selection of applicable runoff curve number. These classifications are made by field inspections, examination of soils maps, etc. For this example, it is assumed available information indicates:

Subarea A:

    Soils, hydrologic group C
    Land use, mostly poor pasture
    Runoff curve, AMC-II CN86 (table G-3(A))

Subarea B:

    Soils, hydrologic group B
    Land use, mostly small grain, contour terraced
    Runoff curve, AMC-II CN70 (table G-3(A))

(b) *Dimensionless-Graph Selection.*—As hydrograph analyses cannot be made in the first instance because of lack of streamflow records, a dimensionless-graph must be selected from other sources. The dimensionless-graph shown as (C), figure G-6, which was derived from a flood hydrograph in the general region of the assumed location of the watershed, has been selected as applicable to both subareas of the watershed. It is also used in the second example, where streamflow records are available.

(c) *Lag-Times.*—A cutout of each subarea, including the respective reservoir portion in each, was made, the center of area of each determined and projected to the main streams at the points marked x on the stream channels as shown on figure G-15 (see sec. G-9(2)). Longest watercourse lengths listed below were measured from the map. Slope values for this example, S in feet per mile, were selected from general data. In the usual study, elevations for computing slope values for a given watershed are obtained from topographic maps.

Subarea A:

    L = 29.0 miles from head of reservoir to divide, E_1 to E_2, figure G-15.
    L_{ca} = 12.7 miles from head of reservoir to center of area projected, E_1 to x, figure G-15.
    S = 23.2 feet per mile (assumed in this example).

Subarea B: (Assumption of no streamflow records.)

L = 48.9 miles from head of reservoir to divide, E_1 to E_2, figure G-15.
L_{ca} = 15.4 miles from head of reservoir to center of area (projected), E_1 to x, figure G-15.
S = 12.6 feet per mile (assumed for this example).

For use in assumption that streamflow records have indicated a lag-time of 22 hours for tributary B:

L = 59.8 miles from mouth (gage) to divide.
L_{ca} = 26.3 miles from mouth (gage) to center of area, x.
S = 16.5 feet per mile (assumed for this example).

Two sets of lag-times are estimated for this example on the basis of the two assumptions regarding available streamflow records. Under the assumption that no streamflow records are available, the generalized lag-time equation is considered applicable.

$$\text{Lag-time hours} = 1.6 \left[\frac{LL_{ca}}{\sqrt{S}} \right]^{0.33} \quad \text{(Sec. G-8(e)(2).)}$$

Estimated lag-times are:

Subarea A:

$$\frac{LL_{ca}}{\sqrt{S}} = \frac{(29.4)(12.7)}{\sqrt{23.2}} = 77.5$$

Lag-time = 6.7 hours.

Subarea B:

$$\frac{LL_{ca}}{\sqrt{S}} = \frac{(48.9)(15.4)}{\sqrt{12.6}} = 212.2$$

Lag-time = 9.4 hours.

Under the assumption that hydrograph analyses for streamflow gaged near the mouth of tributary B indicates a lag-time of 22 hours for subarea B, the following lag-times are estimated:

Subarea A:

No change, lag-time = 6.7 hours.

Subarea B:

Referring to section G-8(e)(2), if a reliable lag-time for a basin is found by hydrograph analyses at a gaging station, a lag-time for an ungaged portion of the basin may be obtained by passing a curve with slope 0.33 through the point plotted on log-log paper, $\frac{LL_{ca}}{\sqrt{S}}$ versus lag hours. An $\frac{LL_{ca}}{\sqrt{S}}$ value for subarea B above the assumed gaging station is:

$$\frac{(59.8)(26.3)}{\sqrt{16.5}} = 386.7$$

If the generalized lag-time curve has been plotted on log-log paper, plot 387 versus the lag-time of 22 hours and draw a line through the plotted point parallel to the generalized lag-time curve. Read a lag-time of 18 hours for the $\frac{LL_{ca}}{\sqrt{S}}$ value of 212 from the constructed curve. In this example, the proposed reservoir has the effect of reducing the lag-time for subarea B from 22 hours for natural conditions to 18 hours after the dam is built. The effect of a proposed reservoir on natural lag-times should not be overlooked in the preparation of inflow design flood hydrographs.

Of course, the lag-time of 18.0 hours can also be obtained without plotting the curves, by solving the equation,

$$\text{Lag-time} = C \left[\frac{LL_{ca}}{\sqrt{S}} \right]^{0.33}$$

for C, substituting 22 hours for lag-time and 386.7 for $\dfrac{LL_{ca}}{\sqrt{S}}$; this gives $C = 3.08$. Then, using this computed value for C, and 212.2 for $\dfrac{LL_{ca}}{\sqrt{S}}$, lag-time in hours equals 18.0.

(d) *Preliminary Design Storm Values.*—A specific watershed location is identified on the generalized charts, figures G-10 and G-13, by county boundaries within the States and reading the zone and 6-hour PMP values applicable to the watershed. A specific location for the watershed for this example has not been designated other than it is assumed to be in zone 4 where 6-hour probable maximum precipitation (PMP) for 10 square miles is 30 inches (figure G-10). Computation of preliminary design storm values are shown in table G-12. The design storm is assumed to cover the entire watershed area of 800 square miles. Percentages of the 6-hour PMP for 10 square miles applicable to 800 square miles were read from the depth-area-duration relationships on the chart for zone 4, figure G-11, and PMP values for 6, 12, 24, and 48 hours for 800 square miles computed. These values were adjusted to 90 percent of the computed values in accordance with the fit adjustment factors given in section G-17(b)(1). Hourly depth-duration values for the maximum 6-hour period of the storm were computed by percentages read from curve C on figure G-12. Depth-duration values, line 5 of table G-12, were plotted and a preliminary design storm depth-duration curve drawn as shown on figure G-16.

(e) *Arrangement of Design Storm Rainfall Increments and Computation of Increments of Rainfall Excess.*—Arrangement of increments of rainfall of a preliminary design storm estimated from figure G-10 is illustrated in table G-13, along with the computation of respective increments of excess rainfall. Computation of table G-13 is explained in the following paragraphs. General comments on design storm arrangements are included.

(1) *Selection of design storm unit time interval.*—Design storm increments and respective rainfall excesses obtained therefrom must be for the same unit time interval as the unitgraph to which the excesses will be applied to compute an inflow design flood (IDF) hydrograph. Unit time of a unitgraph is related to the lag-time of a basin, being one-fourth or less of the lag-time (sec. G-9(6)). In this example, a 1-hour unitgraph is required for subarea A because a lag-time of 6.7 hours has been estimated for that subarea. A 2-hour unitgraph could be used for subarea B, lag-time 9.4 hours. However, the computed hydrographs for the two subareas must be combined to give the preliminary inflow design flood hydrograph. A better definition of the IDF hydrograph will be obtained if the unitgraphs for the two subareas have the same unit time interval. A 1-hour unitgraph for each subarea was used in this example. Hourly values of preliminary design storm rainfall were read to the nearest tenth inch from the depth-duration curve, figure G-16, from 1 to 24 hours and tabulated in column 2 of table G-13. Hourly increments of rainfall are listed in column 3 of table G-13.

Table G-12.—*Preliminary design storm estimate for hypothetical watershed, east of 105° meridian*

BASIC DATA:
Location: Hypothetical
Reference: Figure G-10, zone 4, 6-hr. PMP[1], 10 sq. mi.: 30 inches
Areas: Total basin, 800 sq. mi.; subarea A, 240 sq. mi.; subarea B, 560 sq. mi.

	Item	Time in hours									Text reference
		1	2	3	4	5	6	12	24	48	
1.	Percent of 6-hr. PMP[1] for 800 sq. mi.						62	70	77	87	Fig. G-11
2.	Computed PMP, 800 sq. mi., inches						18.6	21.0	23.1	26.1	
3.	PMP, adjusted to 90 percent						16.7	18.9	20.8	23.5	Sec. G-17(b)(1)
4.	Ratios to 6-hr. rainfall	0.49	0.64	0.75	0.85	0.93	1.00				Fig. G-12, zone C
5.	Design PMP, 800 sq. mi., inches	8.2	10.7	12.5	14.2	15.5	16.7	18.9	20.8	23.5	Fig. G-16

[1] PMP = probable maximum precipitation.

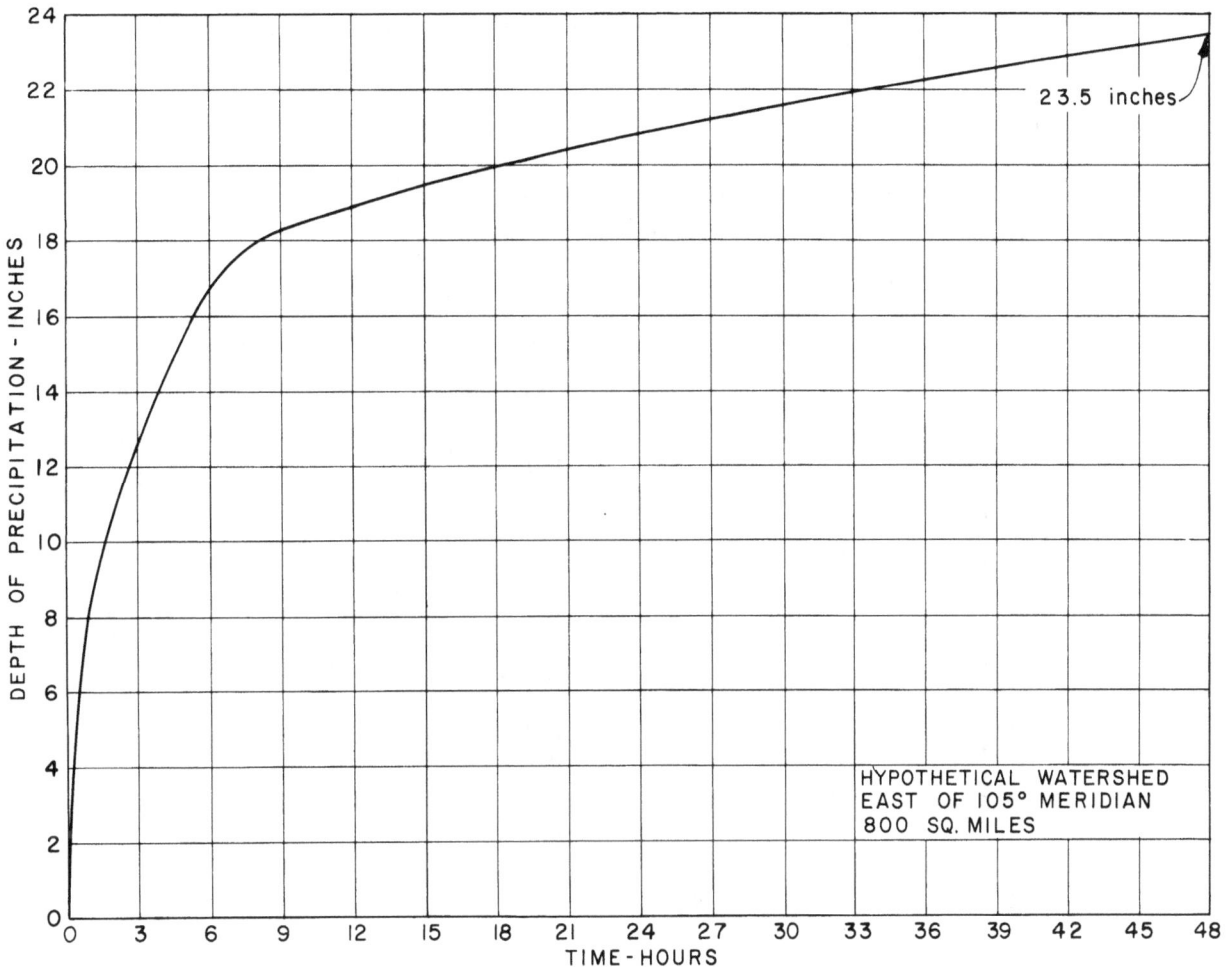

Figure G-16. Preliminary design storm—depth-duration curve.—288-D-3187

(2) *Arrangement of design storm incremental rainfall.*—Normally, the arrangement with respect to time of increments of design storm rainfall is not established in a design storm study (sec. G-17(a)(8)). Flood hydrologists arrange design storm increments to give rainfall excesses that produce the most critical inflow design flood hydrograph. Except for basins having several thousands of square miles of drainage area, design storm rainfall is assumed to occur with the same time sequence over the total watershed area. If a constant retention loss rate is used to compute rainfall excesses, a critical arrangement may be easily found by arranging design storm increments opposite the ordinates of the unitgraph for the basin, so that the largest rainfall increment (which would give the largest excess increment) is opposite the largest ordinate; and the second largest rainfall increment is opposite the second largest ordinate, etc.

This arrangement is then reversed to give the design storm arrangement in correct time sequence, because rainfall excesses are reversed in sequence of natural occurrence when being applied to unitgraph ordinates by calculators. Otherwise, much additional work must be done: (1) computing discharges for each ordinate of the unitgraph for each excess increment; (2) tabulating the individual discharges in correct time sequence; and (3) adding respectively timed incremental discharges to get the total flood hydrograph. If a retention loss rate which varies with time is used, a critical design storm arrangement is found by trial.

Table G-13.—*Preliminary design storm east of 105° meridian—arrangement of incremental rainfall; computation of incremental excesses, ΔP_e, for subareas A and B*

BASIC DATA:

Total area (for design storm estimate)—800 sq. mi.
Subarea size and retention data:
Subarea A: 240 sq. mi.; CN 86, selected minimum loss rate, 0.12 in./hr.
Subarea B: 560 sq. mi.; CN 70, selected minimum loss rate, 0.24 in./hr.

1	2	3	4	5	6	7	8	9	10	11
Time, ending hour	Design rainfall depth-duration		Arrangement of design rainfall		Rainfall excesses, P_e					
					Subarea A			Subarea B		
	ΣP, inches	ΔP, inches	ΔP, inches	ΣP, inches	$^2\Sigma P_e$, inches	ΔP_e, inches	Δ loss, inches	$^3\Sigma P_e$, inches	ΔP_e, inches	Δ loss, inches
1	8.2	8.2	1.2	1.2	0.30	0.30	0.90	0.02	0.02	1.18
2	10.7	2.5	1.7	2.9	1.57	1.27	.43	.66	.64	1.06
3	12.5	1.8	1.8	4.7	3.18	1.61	.19	1.82	1.16	.64
4	14.2	1.7	8.2	12.9	11.13	7.95	.25	8.88	7.06	1.14
5	15.5	1.3	2.5	15.4	13.51	2.38	^{4}.12	11.14	2.26	^{5}.24
6	16.7	1.2	1.3	16.7	14.69	1.18	.12	12.20	1.06	.24
7	17.4	.7	.7	17.4	15.27	.58	.12	12.66	.46	.24
8	17.9	.5	.5	17.9	15.65	.38	.12	12.92	.26	.24
9	18.2	.3	.3	18.2	15.83	.18	.12	12.98	.06	.24
10	18.5	.3	.3	18.5	16.01	.18	.12	13.04	.06	.24
11	18.7	.2	.2	18.7	16.09	.08	.12	13.04	0	.24
12	18.9	.2	.2	18.9	16.17	.08	.12			
13	19.1	.2	.2	19.1	16.25	.08	.12			
14	19.3	.2	.2	19.3	16.33	.08	.12			
15	19.5	.2	.2	19.5	16.41	.08	.10			
16	19.6	.1	.1	19.6	16.41	0	.12			
17	19.8	.2	.2	19.8	16.49	.08	.12			
18	20.0	.2	.2	20.0	16.57	.08	.12			
19	20.1	.1	.1	20.1	^{6}	^{6}				
20	20.2	.1	.1	20.2						
21	20.4	.2	.2	20.4						
22	20.6	.2	.2	20.6						
23	20.7	.1	.1	20.7						
24	^{1}20.8	.1	.1	20.8						

[1] Balance of design rainfall considered lost to retention.

[2] By equation $\Sigma P_e = \dfrac{(P - 0.2S)^2}{(P + 0.8S)}$ for CN 86, $S = 1.63$; $0.2S = 0.33$, $0.8S = 1.30$ (table G-4).

[3] By above equation, for CN 70, $S = 4.28$; $0.2S = 0.86$, $0.8S = 3.42$ (table G-4).
[4] ΔP_e by CN 86 indicates Δ loss = 0.03 in., which is less than 0.12 in. Use 0.12 in. loss/hr.
[5] ΔP_e by CN 70 indicates Δ loss = 0.15 in., which is less than 0.24 in. Use 0.24 in. loss/hr.
[6] Total of remaining excess not significant for preliminary IDF.

A definite arrangement of design storm increments has been specified for preliminary design storm values obtained from each generalized precipitation chart, figures G-10 and G-13, because the selected general method of computing rainfall excesses using rainfall runoff curves has "built-in" varying retention loss rates. The arrangement specified for preliminary design storm values east of the 105° meridian is illustrated by the arrangement of rainfall increments in column 4, table G-13. The maximum 6-hour period of design rainfall is assumed to occur during the first 6-hour period of the design storm. Hourly precipitation amounts within the maximum 6-hour period are arranged in the following order of magnitude: 6, 4, 3, 1, 2, 5. Increments of design storm rainfall after the sixth hour decrease and are taken directly from the design storm depth-duration curve.

(3) *Computation of increments of rainfall excess.*—The method of estimating excess rainfall increments given in section G-7(b)(6) has been taken from the SCS National Engineering Handbook [3] with the following modifications introduced to give a procedure applicable to preliminary design storm rainfall obtained from generalized precipitation charts.

The rainfall-runoff relationships shown by the curves of figure G-2 were developed by Soil Conservation Service hydrologists from analyses of rainfall and respective runoff records at numerous small area experimental watersheds. The relationships were developed for use with daily nonrecording rainfall data, which are more plentiful in the United States than are recording rainfall data. Data used in the development are totals for one or more storms occurring in a calendar day and nothing is known about their time distributions. The relationships developed, therefore, exclude time as an explicit variable which means that rainfall intensity is ignored.

Strict adherence to use of the runoff curves on figure G-2 results in hourly runoff increments almost equal to hourly precipitation increments after a few hours for many of the design storm values obtained from generalized precipitation charts. Infiltrometer studies indicate that all but impervious clay soils have minimum constant infiltration rates after saturation that may range from 0.05 inch per hour to greater than 1.00 inch per hour, depending on the type of soil. Therefore, to utilize the rainfall-runoff relationships in the computational procedures given in this text, time-sequences of incremental rainfall for a design storm are specified and precipitation excesses are then computed using the runoff curve relationships, with the provision that hourly retention rates indicated by use of the runoff curves be tabulated for each hourly rainfall increment. Progressively through the arranged precipitation sequence, these hourly retention rates are compared with the tabulated minimum retention rates assigned to the four hydrologic soil groups (see table G-14). When the retention rate given by use of a runoff curve becomes less than an assigned minimum retention rate, the minimum rate is used to compute excesses thereafter for the remainder of the storm.

For this example, determination of applicable runoff curve numbers, AMC-II, for subareas A and B has been assumed as described earlier in section G-19(a)(3) on soils and cover. East of the 105^0 meridian, soil moisture within a watershed which has similar to average conditions present before occurrence of the maximum annual flood (AMC-II) is considered a reasonable assumption for occurrence of a design storm. Therefore, the curve numbers referred to above were obtained from table G-3(A), which lists curve numbers for AMC-II; CN 86 was selected for subarea A and CN 70 for subarea B, to compute rainfall excesses. Minimum retention rates selected are those for general cases, table G-14: 0.12 inch per hour for subarea A, hydrologic soil group C; and 0.24 inch per hour for subarea B, hydrologic soil group B.

Computations of rainfall excesses are made to hundredths of an inch, as shown in table G-13. Runoff curves, figure G-2, cannot be accurately read to hundredths unless plotted to a large scale, so it is recommended that rainfall excesses be computed by the equation shown on figure G-2. The symbol P_e is used in this text to designate direct runoff values, rainfall excesses, in lieu of Q shown on figure G-2. Values of S and $0.2S$ in inches for each curve number are listed in table G-4. Referring to table G-13, computations of hourly rainfall excesses for subarea A are described. This procedure applies to all such computations.

(1) Obtain S and $0.2S$ values from table G-4 for CN 86. Compute $0.8S$ value.

(2) Fill in column 5, ΣP, by summing the arranged design storm increments.

Table G-14.—*Minimum retention rates for hydrologic soil groups*

Hydrologic soil group	Range of minimum retention rates, inches per hour	Recommended rate for use in general case, inches per hour
A	0.30-0.45	0.40
B	0.15-0.30	0.24
C	0.08-0.15	0.12
D	0.02-0.08	0.04

(3) To obtain column 6, begin with the first ΣP value that exceeds the applicable $0.2S$ value and, successively by hours, compute ΣP_e by the equation:

$$\Sigma P_e = \frac{(P - 0.2S)^2}{(P + 0.8S)} \qquad (13)$$

Each successive ΣP value in column 5 of table G-13 becomes the P for the equation, and the values of $0.2S$ and $0.8S$ are those obtained as in (1) above.

(4) Determine increment of excess rain, ΔP_e for each hour, and tabulate in column 7, then substract ΔP_e from respective ΔP, column 4, and enter Δ loss value thus obtained in column 8.

(5) As successively computed, compare Δ loss value with assigned minimum retention rate: 0.12 inch per hour for subarea A. If loss is greater than 0.12 inch per hour, proceed to next hour and repeat procedure; if loss is less than 0.12 inch, do not use the computed ΔP_e value. Drop use of runoff equation and use the constant hourly loss rate of 0.12 inch per hour to compute that hour's excess and the rest of the hourly increments of excess rainfall. This change occurred at hour 5 in the example in table G-13.

The hourly increments of excess rainfall listed in column 7 will be applied to a 1-hour unitgraph for subarea A.

In all cases when the generalized precipitation charts are used to estimate preliminary design storm values for a watershed, hourly increments of excess rainfall should be obtained by the above procedure. If a 2-, 3-, or 4-hour unitgraph is to be used for the watershed, the computed hourly rainfall excesses are grouped into respective 2-, 3-, or 4-hour sums and applied to the chosen unitgraph.

(f) *Computation of Preliminary Inflow Design Flood Hydrographs.*—Computation of an inflow design flood (IDF) hydrograph is a routine mathematical process after decisions are made regarding selection of dimensionless-graph, lag-time, retention rate, and design storm values and arrangement. Procedural steps for obtaining a synthetic

unitgraph for a watershed have been given in section G-9(7). The principle of obtaining a total flood hydrograph resulting from successive increments of excess rainfall is illustrated on figure G-3. Therefore, detailed tables showing computation of unitgraphs for subareas A and B and the application of respective sets of rainfall excesses to respective unitgraphs are omitted. In lieu thereof, copies of the printouts from the Bureau's Automatic Data Processing (ADP) program for application of the dimensionless-graph lag-time method of computing flood hydrographs are included as tables G-15 and G-16. Table G-15 is a simulated printout of the computed preliminary design flood contribution from subarea A resulting from the incremental rainfall excesses listed in column 7 of table G-13. The program is designed to compute discharges to the nearest cubic foot per second (c.f.s.) so the ordinates of the 1-hour unitgraph for a lag-time of 6.7 hours, listed in the third column of table G-15, are more exact than warranted by the basic data. (The same comment applies also to the computed flood hydrograph discharges.) Table G-16 is a similar printout for subarea B.

(1) *Preliminary inflow design flood hydrograph using generalized lag-time curve for both subareas.*—Design flood contributions for each subarea are tabulated, combined, and total preliminary IDF discharges listed in table G-17. Subarea hydrographs and the total hydrograph are shown on figure G-17. (In usual practice, only the total flood hydrograph is plotted.) A base flow has not been added to computed flood discharges, because base flow discharges are insignificant in relation to the computed flood discharges in this example. A method of obtaining the volume of the IDF hydrograph is detailed in table G-17.

(2) *Preliminary inflow design flood hydrograph, watershed not divided into subareas.*—Under the assumption that no streamflow records are available within the watershed and that the same dimensionless-graph, lag-time curve, and preliminary design storm values are to be used for both subareas, a preliminary inflow design flood hydrograph may be computed using one unitgraph for the

Table G-15.—*Simulated automatic data processing printout—preliminary inflow design flood (IDF) contribution, subarea A*

EXAMPLE PRELIMINARY IDF SUBAREA A

UNIT GRAPH DEVELOPED FROM DIMENSIONLESS GRAPH

DIMENSIONLESS FIGURE G-5 DESIGN OF GRAVITY DAMS   TUN = 1.00

LAG DATA GENERAL CURVE SUBAREA A   LAG = 6.70

AREA = 240.000 SQ MI UNITGRAPH RECES COEF = 0.828781 AT 18.00 HRS

EXCESS OR STORM VOLUME = 16.570 INCHES

HYDROGRAPH VOLUME IN INCHES = 16.573 AND IN AC FT = 212132.7

HOURS	EXCESSES INCHES	UNITGRAPH CFS	HYDROGRAPH CFS
.00	.000	0	0
1.00	.300	174	52
2.00	1.270	1247	595
3.00	1.610	4988	3361
4.00	7.950	12887	13591
5.00	2.380	20571	40900
6.00	1.180	23131	96644
7.00	.580	20431	184518
8.00	.380	15042	268585
9.00	.180	10787	306688
10.00	.180	7795	291631
11.00	.080	6238	242607
12.00	.080	5107	192229
13.00	.080	4217	150359
14.00	.080	3575	121417
15.00	.080	2980	99867
16.00	.000	2502	83365
17.00	.080	2140	70906
18.00	.080	1900	60650
19.00	.000	1575	52146
20.00	.000	1305	45261
21.00	.000	1082	39936
22.00	.000	897	34392
23.00	.000	743	29058
24.00	.000	616	24031
25.00	.000	510	19680
26.00	.000	423	16123
27.00	.000	351	13290
28.00	.000	291	11018
29.00	.000	241	9144
30.00	.000	200	7602
31.00	.000	165	6324
32.00	.000	137	5255
33.00	.000	114	4363
34.00	.000	94	3631
35.00	.000	78	3020
36.00	.000	65	2503
37.00	.000	54	2074
38.00	.000	44	1719
39.00	.000	37	1425
40.00	.000	31	1181
41.00	.000	25	979
42.00	.000	21	811
43.00	.000	17	672
44.00	.000	14	557
45.00	.000	12	462
46.00	.000	10	383
47.00	.000	8	317
48.00	.000	7	263
49.00	.000	6	218
50.00	.000	5	181
51.00	.000	4	150

Table G-16.—*Simulated automatic data processing printout—preliminary inflow design flood (IDF) contribution, subarea B*

EXAMPLE PRELIMINARY IDF SUBAREA B

UNIT GRAPH DEVELOPED FROM DIMENSIONLESS GRAPH

DIMENSIONLESS FIGURE G-5 DESIGN OF GRAVITY DAMS   TUN = 1.00

LAG DATA GENERAL CURVE SUBAREA B   LAG = 9.40

AREA = 560.000 SQ MI UNITGRAPH RECES COEF = 0.872335 AT 24.00 HRS

EXCESS OR STORM VOLUME = 13.040 INCHES

HYDROGRAPH VOLUME IN INCHES = 13.041 AND IN AC FT = 389484.2

HOURS	EXCESSES INCHES	UNITGRAPH CFS	HYDROGRAPH CFS
.00	.000	0	0
1.00	.020	140	3
2.00	.640	842	106
3.00	1.160	2824	757
4.00	7.060	7558	3921
5.00	2.260	16848	14710
6.00	1.060	26830	42072
7.00	.460	35618	98125
8.00	.260	38938	194141
9.00	.060	38289	304039
10.00	.060	32412	404552
11.00	.000	25421	459442
12.00	.000	19932	467687
13.00	.000	15560	424020
14.00	.000	12603	356437
15.00	.000	10754	290340
16.00	.000	9518	232873
17.00	.000	8093	188750
18.00	.000	6930	157275
19.00	.000	6232	134614
20.00	.000	5464	114743
21.00	.000	4806	98479
22.00	.000	4229	86598
23.00	.000	3721	75960
24.00	.000	3558	66750
25.00	.000	3103	58891
26.00	.000	2707	52122
27.00	.000	2362	47810
28.00	.000	2060	42485
29.00	.000	1797	37420
30.00	.000	1568	32799
31.00	.000	1368	28697
32.00	.000	1193	25054
33.00	.000	1041	21874
34.00	.000	908	19081
35.00	.000	792	16645
36.00	.000	691	14520
37.00	.000	603	12667
38.00	.000	526	11050
39.00	.000	459	9639
40.00	.000	400	8408
41.00	.000	349	7335
42.00	.000	304	6399
43.00	.000	266	5582
44.00	.000	232	4869
45.00	.000	202	4247
46.00	.000	176	3705
47.00	.000	154	3232
48.00	.000	134	2820
49.00	.000	117	2460
50.00	.000	102	2146
51.00	.000	89	1872
52.00	.000	78	1633
53.00	.000	68	1424
54.00	.000	59	1242
55.00	.000	52	1084
56.00	.000	45	945
57.00	.000	39	825
58.00	.000	34	719
59.00	.000	30	628
60.00	.000	26	547
61.00	.000	23	478
62.00	.000	20	417
63.00	.000	17	363
64.00	.000	15	317
65.00	.000	13	277
66.00	.000	11	241
67.00	.000	10	210

Table G-17.—*Preliminary inflow design flood hydrograph, east of 105°*
meridian—same lag-time curve for both subareas

Time, ending at hour	Discharges, 1,000 c.f.s.[1]			Time, ending at hour	Discharges, 1,000 c.f.s.		
	Subarea A	Subarea B	Prelim. IDF		Subarea A	Subarea B	Prelim. IDF
0	0.00	0.0	0.0	[2]33	4.4	21.9	26.3
1	.05	.0	.1	36	2.5	14.5	17.0
2	.6	.1	.7	39	1.4	9.6	11.0
3	3.4	.8	4.2	42	.8	6.4	7.2
4	13.6	3.9	17.5	45	.5	4.2	4.7
5	40.9	14.7	55.6				
				48	.3	2.8	3.1
6	96.6	42.1	138.7	51	.2	1.9	2.1
7	184.5	98.1	282.6	54	[3].1	1.2	1.3
8	268.6	194.1	462.7	57	<.1	.8	.8
9	306.7	304.0	610.7	60		.5	.5
10	291.6	404.6	696.2				
				63		.4	.4
11	242.6	459.4	702.0	66		.2	.2
12	192.2	467.7	659.9				
13	150.4	424.0	574.4				
14	121.4	356.4	477.8				
15	99.9	290.3	390.2				
16	83.4	232.9	316.3				
17	70.9	188.8	259.7				
18	60.7	157.3	218.0				
19	52.1	134.6	186.7				
20	45.3	114.7	160.0				
21	39.9	98.5	138.4				
22	34.4	86.6	121.0				
23	29.1	76.0	105.1				
24	24.0	66.8	90.8				
25	19.7	58.9	78.6				
26	16.1	52.1	68.2				
27	13.3	47.8	61.1				
28	11.0	42.5	53.5				
29	9.1	37.4	46.5				
30	7.6	32.8	40.4				

Computation of IDF volume:

Sum, discharges, 0-29 hrs.	6,977,200
½ discharge, hr. 30	20,200
Volume, 0-30 hrs.	6,997,400 c.f.s.-hrs.
½ discharge, hr. 30	20,200
Sum, discharges, 33-63 hrs.	74,400
½ discharge, hr. 66	100
Sum	94,700
Volume, 30-66 hrs., (3 times 94,700)	284,100 c.f.s.-hrs.
Total IDF volume	7,281,500 c.f.s.-hrs.
Equivalent to	303,400 c.f.s.-24 hrs.
Equivalent to	600,800 ac.-ft.

For a check, compare with the sum of volumes in tables G-15 and G-16, or 601,600 ac.-ft.

[1]Instantaneous at designated hour.
[2]Larger time intervals may be used for lower portions of hydrograph recessions.
[3]If needed, discharges "cut off" to shorten computations (see table G-15) may be extended using the hydrograph's recession coefficient.

total watershed area. Estimating a total basin lag-time by weighting subarea lag-time proportional to the areas of 240 and 560 square miles gives a lag-time of 8.6 hours. A weighted runoff curve number, CN 75, and weighted minimum retention rate, 0.20 inch per hour, are obtained as shown in table G-18. The calculations are shown because this method of weighting curve numbers is used to obtain a weighted CN for a basin (or subbasin) which contains various areas of different soil and cover complexes. Table G-18 shows the computation of incremental rainfall excesses which were applied to a 1-hour unitgraph for the watershed, lag-time 8.6 hours, area 800 square miles. Ordinates of the computed preliminary IDF hydrograph, peak discharge 768,600 c.f.s., volume 597,700 acre-feet, are plotted on figure G-17.

Either of the preliminary IDF hydrographs shown on figure G-17 could be recommended for use for preliminary designs. Under the assumptions made for computing these hydrographs, an acceptable result is obtained by considering the basin as a whole or by dividing the basin into two subareas.

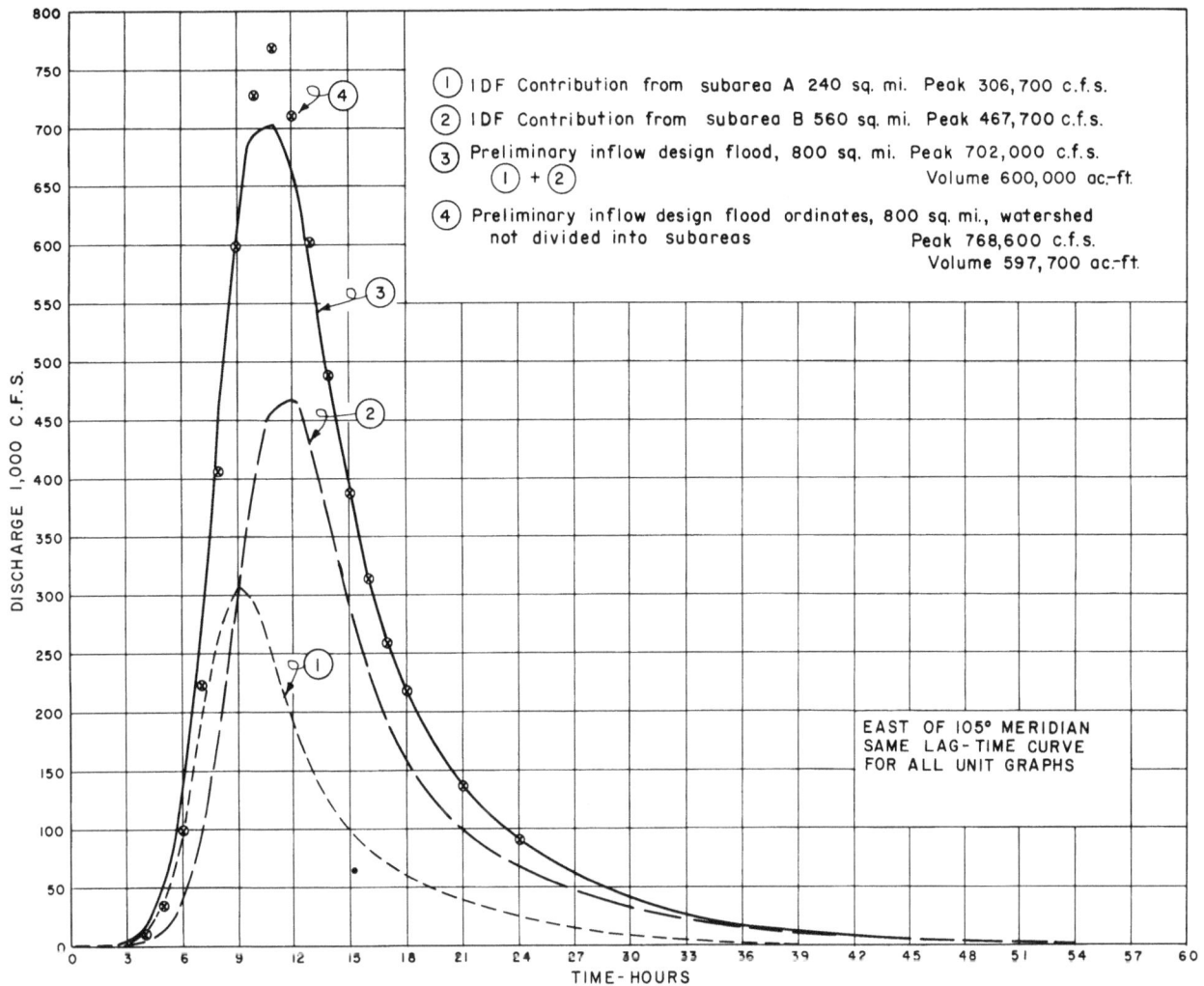

Figure G-17. Example of preliminary inflow design flood hydrographs—same lag-time curve for all unitgraphs.—288-D-3188

(3) *Preliminary inflow design flood hydrograph using a different lag-time curve for each subarea.* —As lag-time differences between subarea drainage systems within a basin increase, added consideration needs to be given to dividing the basin into subareas and obtaining the design flood contribution from each subarea for combination to form the inflow design flood. This is demonstrated by the hydrographs shown on figure G-18. Using the assumption given in section G-19(a)(2) that tributary B had streamflow records giving a lag-time of 22.0 hours from which a lag-time of 18.0 hours is obtained for subarea B for inflow to the proposed reservoir (sec. G-19(c)), a

1-hour unitgraph for subarea B was computed. The design flood contribution from subarea A shown on figure G-17 ① is not changed and is replotted on figure G-18 ① .

The increment of rainfall excesses for subarea B, table G-13, column 10, applied to the new unitgraph for subarea B gives the flood contribution shown on figure G-18 ② . Combining the hydrographs from the two subareas, table G-19, gives a preliminary inflow design flood hydrograph, figure G-18 ③ , having two peaks, the maximum of which is a peak discharge of 332,500 c.f.s. (as estimated when plotting the graphs) and a 72-hour volume of 597,000

Table G-18.—*Preliminary inflow design flood, east of 105° meridian—computation of incremental excesses, ΔP_e, considering basin as a whole, and using an areal weighted CN and minimum loss rate.*

BASIC DATA:
    Subarea A: AMC-II CN 86; min. loss, 0.12 in./hr.; area, 240 sq. mi.
    Subarea B: AMC-II CN 70; min. loss, 0.24 in./hr.; area, 560 sq. mi.
WEIGHTED VALUES FOR USE:

$$\frac{(86)(240) + (70)(560)}{800} = 74.8; \text{ use AMC-II CN 75}$$

$$\frac{(0.12)(240) + (0.24)(560)}{800} = 0.204; \text{ use 0.20 in./hr.}$$

Time, ending at hour	ΔP,[1] inches	ΣP, inches	Rainfall excesses, P_e		
			ΣP_e,[3] inches	ΔP_e, inches	Δ loss, inches
1	1.2	1.2	0.07	0.07	1.13
2	1.7	2.9	.89	.82	.88
3	1.8	4.7	2.21	1.32	.48
4	8.2	12.9	9.61	7.40	.80
5	2.5	15.4	11.91	2.30	[4].20
6	1.3	16.7	13.01	1.10	.20
7	.7	17.4	13.51	.50	.20
8	.5	17.9	13.81	.30	.20
9	.3	18.2	13.91	.10	.20
10	.3	18.5	14.01	.10	.20
11	.2	18.7	14.01	0	.20
12	.2	18.9			

[1] Arranged design rainfall, see column 4, table G-13.
[2] Balance of rainfall less than retention loss in this approach.

[3] By equation, $P_e = \dfrac{(P - 0.2S)^2}{(P + 0.8S)}$; for CN 75, $S = 3.33$, $0.2S = 0.67$, $0.8S = 2.66$ (table G-4).

[4] ΔP_e by equation indicates Δ loss of 0.10 in., less than 0.20 in.; use 0.20 in./hr.

acre-feet. Ordinates of a flood hydrograph computed using a 1-hour unitgraph having a basin weighted lag-time of 14.6 hours and incremental rainfall excesses listed in table G-18 are shown as ④ on figure G-18. This flood hydrograph has a peak of 492,000 c.f.s., excessively high in comparison with the flood hydrograph obtained by combining the two subarea flood hydrographs. The procedure of considering the watershed as a whole does not give an acceptable preliminary IDF hydrograph in this instance.

G-20. Preliminary Inflow Design Flood Estimates, Watersheds West of 105° Meridian.—It is very likely that runoff from snowmelt will contribute a portion of the discharges of an inflow design flood (IDF) hydrograph for large dams at sites west of the 105° meridian. In many instances though,

design rainstorm potential is so great that runoff from a design rainstorm gives the major portion of an inflow design flood. Preliminary inflow design flood estimates for many areas west of the 105° meridian can be made using preliminary design storm values obtained from figure G-13 and associated procedures, the methods of arranging design storm incremental rainfall and computing rainfall excesses given in this section, and adding appropriate base flows to the computed rain flood hydrograph. In general, for western mountainous watersheds having seasonal snowmelt runoff which reaches a maximum after mid-May, base flows for addition to the hydrograph computed from a preliminary design rainstorm may be estimated as those discharges likely to occur during the last 5 days of the maximum 15-day period of a 1 percent chance maximum annual 15-day

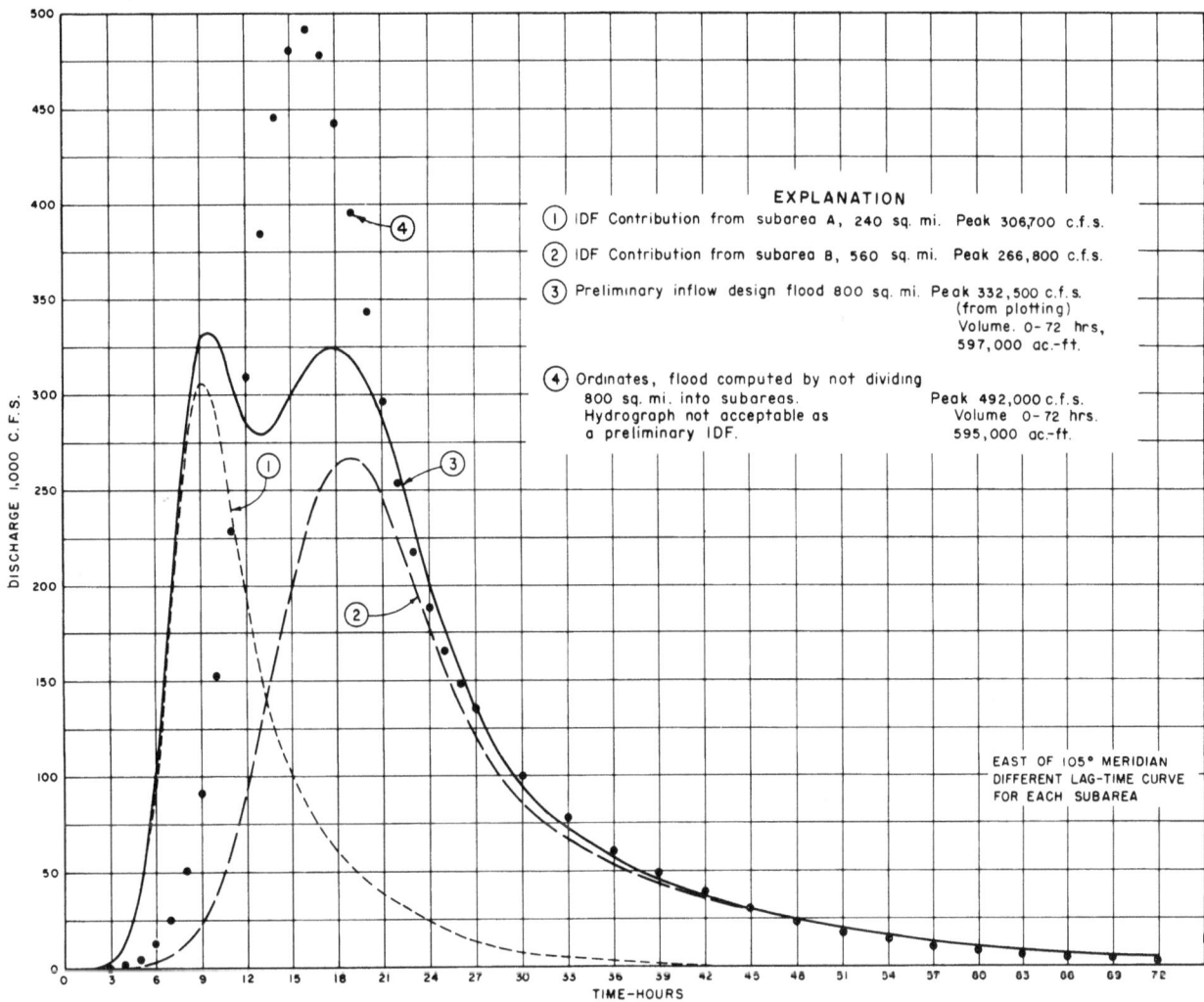

Figure G-18. Example of preliminary inflow design flood hydrograph—different lag-time curve for each subarea.—288-D-3189

seasonal snowmelt runoff flood. (See secs. G-28 and G-29 for a discussion of statistical analyses—frequency studies.) However, this general approach cannot be used for mountainous watersheds where maximum storm potential occurs during the winter months October through April. Examples are: Sierra Nevada Mountains in California and Nevada, Cascade Range in Oregon and Washington, and Mogollon Rim in Arizona. Extreme floods on streams in these regions result from rain falling on snow-covered watersheds. Estimation of rain-on-snow floods requires special procedures as discussed in sections G-22 through G-26. Exception also

must include those watersheds having a large percentage of total basin drainage area at relatively low elevations where the ground may be frozen and winter rain falling on a light snow cover can cause large floods.

Procedures for estimating the rain-flood portion of a preliminary inflow design flood hydrograph from preliminary general-type design storm values for a watershed west of the 105° meridian differ in two respects from the procedures which have been given for watersheds east of the 105° meridian; namely, arrangement of design storm rainfall increments, and assignment of appropriate runoff curve number, CN.

Table G-19.—*Preliminary inflow design flood hydrograph, east of 105°*
meridian—different lag-time curve for each subarea

Time, ending at hour	Discharges, 1,000 c.f.s.[1]			Time, ending at hour	Discharges 1,000 c.f.s.		
	Subarea [2]A	Subarea [3]B	Prelim. IDF		Subarea A	Subarea B	Prelim. IDF
0	0.00	0.0	0.0	33	4.4	67.6	72.0
1	.05	<.1	.1	36	2.5	53.2	55.7
2	.6	<.1	.6	39	1.4	43.2	44.6
3	3.4	.1	3.5	42	.8	35.2	36.0
4	13.6	.5	14.1	45	.5	28.7	29.2
5	40.9	1.3	42.2				
				48	.3	24.8	25.1
6	96.6	3.2	99.8	51	.2	20.3	20.5
7	184.5	6.9	191.4	54	.1	16.4	16.5
8	268.6	13.1	281.7	57	<.1	13.2	13.2
9	306.7	23.5	330.2	60		10.6	10.6
10	291.6	39.5	331.1				8.5
11	242.6	62.8	305.4	63		8.5	
12	192.2	94.1	286.3	66		6.8	6.8
13	150.4	129.1	279.5	69		5.5	5.5
14	121.4	165.3	286.7	72		4.4	4.4
15	99.9	200.7	300.6			*	
16	83.4	231.0	314.4				
17	70.9	252.5	323.4				
18	60.7	263.5	324.2				
19	52.1	266.8	318.9				
20	45.3	260.5	305.8				
21	39.9	244.4	284.3				
22	34.4	224.4	258.8				
23	29.1	200.8	229.9				
24	24.0	177.5	201.5				
25	19.7	156.4	176.1				
26	16.1	137.6	153.7				
27	13.3	120.8	134.1				
28	11.0	105.9	116.9				
29	9.1	94.9	104.0				
30	7.6	86.0	93.6				

*Continuing discharges may be computed at 3-hour intervals using recession coefficient of 0.8031. Volume after hour 72:

$$\text{Vol.} = \frac{-q}{\log_e k_3}$$

$$\text{Vol.} = \frac{-4,400}{-0.21928}$$

Vol. = 20,060 c.f.s.-3 hrs.
2,508 c.f.s.-24 hrs.
4,970 ac.-ft.

Vol. (0-72 hrs.), 301,050 c.f.s.-hrs.
597,100 ac.-ft.

[1]Instantaneous at designated hour.
[2]Same discharges as for subarea A, table G-17.
[3]1-hr. unitgraph, lag-time 18.0 hrs., used to compute discharges. Excesses column 10, table G-13.

(a) *Preliminary Design Storm Values, Watersheds West of 105° Meridian.*—By geographical location (county) obtain probable maximum 6-hour point rainfall value from figure G-13. Note zone designation, A, B, or C, in which watershed is located.

(1) Compute 6-hour basin rainfall by multiplying 6-hour point rainfall by ratio obtained from applicable zone curve, figure G-14, for watershed drainage area, square miles.

(2) Make a tabulation of design storm depth-duration values at 1-hour intervals for a design storm duration extending to the hour beyond which hourly rainfall increments are equal to or less than the minimum hourly retention loss rate for the watershed. Hourly distribution of maximum 6-hour rainfall is obtained from the applicable curve of figure G-12. Design storm values beyond 6 hours are computed at 2-hour intervals by appropriate constants listed in table G-11. From 6 to 24 hours, use average of even-numbered 2-hour accumulative rainfall for the intervening odd-numbered hour. If hourly rainfall increments are needed after 24 hours, draw depth-duration curve for rainfall amounts computed by constants in table G-11 and read

hourly values. Compute depth-duration rainfall values to nearest hundredth of inch.

(b) *Arrangement of Design Storm Increments of Rainfall.*—Beginning with the second largest 6-hour design storm rainfall amount, hours 6-12 of depth-duration values, arrange hourly increments of design rainfall in ascending order of magnitude for the first 6 hours of arranged design storm values. For hours 7 through 12, arrange hourly increments of maximum 6-hour rainfall in the following order of magnitude: 6, 4, 3, 1, 2, 5. Hourly rainfall amounts after the 12th hour are arranged in descending order of magnitude.

(c) *Assignment of Runoff Curve Number, CN, and Computation of Increments of Excess Rainfall.*—Watershed soils, cover and land use data are used to estimate an applicable runoff curve number from the information given in section G-7(b)(6). The estimated curve number, CN, is for antecedent moisture condition II, AMC-II. This number is then coverted to the respective AMC-III CN listed in table G-4 and the AMC-III CN used to compute hourly rainfall excesses by the method illustrated in table G-13. Antecedent moisture condition III is assumed for watersheds west of the 105° meridian, because late May and June design storm potential is likely to be concurrent with, or immediately after, snowmelt runoff while watershed soil moisture is high.

If a unit time period longer than 1 hour is used for obtaining a unitgraph, the two largest increments of rainfall excesses should be grouped together. If such grouping of hourly excesses results in only 1 hourly excess increment in a unit time period at the beginning and/or end of excess rainfall period, the 1-hour increment of excess is assumed as total excess for the unit time period.

(d) *Floods From Design Thunderstorm Rainfall.*—Data for estimating design thunderstorm rainfall have not been included in this text. If an estimate of a preliminary inflow design flood (IDF) caused by design thunderstorm rainfall is required, preliminary design thunderstorm rainfall estimates for watersheds west of the 105° meridian may be obtained from generalized data in the publication "Design of Small Dams," second edition, [31] along with data for estimating increments of excess rainfall to be applied to a unitgraph. The procedures which have been described in this text for developing a unitgraph can be used to obtain a unitgraph *for that portion of a watershed* over which a design thunderstorm might occur. In the event that this type of preliminary IDF estimate proves critical for design, a hydrometeorologist should be consulted for an estimate of design thunderstorm rainfall for the specific watershed.

G-21. *Recommendations for Routing Preliminary Inflow Design Floods Through a Proposed Reservoir.*—It is necessary for designers to assume an elevation of the reservoir pool at the start of an inflow design flood for reservoir routing studies to determine required spillway capacity. Normally, the reservoir pool is assumed to be full to the top of planned conservation storage capacity or, when either inviolate or joint use flood control capacity is proposed, full to the top of either type of flood control capacity at the beginning of a preliminary inflow design flood. If large capacities of flood control space are being considered in preliminary planning, criteria for routing a final-type IDF as discussed in sections G-30 and G-31 should be established to the extent possible with information available.

G. SNOWMELT RUNOFF CONTRIBUTIONS TO INFLOW DESIGN FLOODS

G-22. *General.*—"Hydraulic engineers responsible for planning and designing multiple-purpose storage reservoirs recognize snow as a form of precipitation possessing certain characteristics which can be evaluated and applied to advantage, both from a hydrologic and an economic viewpoint, in the planning and design of multipurpose storage reservoirs. In northern latitudes and at high elevations, snow falls and accumulates on the earth's surface in frozen crystalline form and usually remains until a proper sequence of meteorologic events provides the thermodynamic conditions essential for either evaporation or melting. Periodic snow surveys provide a reliable index of the relative snow accumulation. With knowledge of the processes of storage, evaporation, and melting, the engineer can predict, with reasonable accuracy (for normal climatic conditions and for known snowpack) the characteristics and amount of streamflow to be expected * * * In the Western United States, the economy of the arid and semiarid lands lying between the mountain ranges is increasingly dependent on development of multiple-purpose storage reservoirs to utilize the streamflow originating in the high mountain snow packs. Engineers of the Western States accept as a blessing the fact that *the predictable characteristics* (italics added) of this streamflow enable economies in planning and designing multiple-purpose reservoirs by the joint use of space allocated to the various functions and by reduction of spillway capacities."

The above extract from Mr. H. S. Riesbol's paper "Snow Hydrology for Multiple-Purpose Reservoirs" [22] is quoted to point out the importance of snow in hydrologic studies and the predictable characteristics of streamflow originating from snowpacks. These predictable characteristics often make possible employment of simple empirical correlations which give acceptable estimates of snowmelt runoff, although this runoff results from a complex thermodynamic process. Discussion of empirical methods of estimating snowmelt runoff as related to inflow design flood estimates is the main objective in these sections. Readers interested in more information about the physical and thermodynamic characteristics of snow and snowmelt processes may consult "Snow Hydrology" [23] and "Handbook of Applied Hydrology" [24].

As previously stated in section G-1, Bureau of Reclamation policy does not provide for combining probable maximum snowmelt runoff with probable maximum rainfall runoff for estimation of an inflow design flood. It is believed that such combinations are unreasonably severe. It is considered more reasonable to combine runoff from a probable maximum rainstorm that could occur during the snowmelt season with a major snowmelt flood, or to combine runoff from a major rainstorm that could occur during the snowmelt season with probable maximum snowmelt runoff. In regions where maximum probable rainstorms can occur during winter months when watersheds may have a large amount of snow on the ground, the amount of snow melted during the design rainstorm must be estimated and runoff calculated from the total combined rain and melted snow water available on the ground surface. Procedures have been developed for computing this type of rain-on-snow floods, utilizing data and analyses described in detail in the report "Snow Hydrology" [23]. One should be mindful that each individual IDF study requires some variations within the framework of a general approach, depending upon watershed characteristics, location, basic data available, and proposed operational capacity of the future reservoir.

G-23. *Major Snowmelt Runoff During Seasonal Melt Period for Combination With Probable Maximum Storm Runoff.* —A method of estimating snowmelt runoff contribution for this type of combination has been described briefly in connection with preliminary IDF estimates for watersheds west of the 105° meridian. Additional items need be considered when making "best possible" preliminary IDF or final-type IDF estimates. Inclusion of flood control capacity and its amount in a proposed reservoir may have a direct bearing on the time duration of flow required in estimation of an inflow design flood hydrograph.

(a) *Damsites for Reservoirs With no Flood Control Capacity Proposed.* —These projects are intended to store seasonal snowmelt runoff as rapidly as possible, allowing only minimum required releases until reservoir capacity becomes full to top of conservation storage. A duration time of 15 days is usually adequate for an inflow design flood hydrograph for this type of structure, as a reservoir may be assumed full to top of conservation capacity at the beginning of the 15-day period. A 1 percent chance (100 year) 15-day volume of snowmelt runoff is usually considered as a major snowmelt flood. It is obtained from a frequency study of maximum annual 15-day snowmelt runoff volumes using runoff records for the contributing watershed, if available, or records for similar nearby watersheds. The 15-day volume indicated by the frequency computations (secs. G-28 and G-29) is adjusted to the specific watershed above a damsite by area relationships.

Caution: Occasionally there will be found references or data of an extremely large snowmelt flood exceeding all recently recorded floods and, perhaps, exceeding the 1 percent chance value indicated by frequency analyses of more recent records. These data should not be ignored without making full effort to incorporate the data into the snowmelt flood estimate.

(1) *Assembly of basic streamflow data for frequency analyses.* —Concurrently with tabulation of maximum annual 15-day seasonal snowmelt runoff values from streamflow records, climatological data should be examined to determine if each year's 15-day runoff volume was snowmelt runoff or was increased by rainfall amounts large enough to cause runoff during that period (small rainfall events may be ignored). If a large snowmelt volume is indicated, an estimate of the rain-flood portion can be made and subtracted by plotting the daily discharge values on semilogarithmic paper and sketching an estimated snowmelt recession (due to lower temperatures accompanying rainfall) under the obvious rain-flood portion. This procedure may have to be used in a few regions where almost every year some rainfall runoff is concurrent with snowmelt runoff.

(2) *Daily distribution of 1 percent chance 15-day snowmelt runoff volume.* —Springtime snowmelt runoff coordinates closely with temperature fluctuations. Large areas usually have about the same daily temperature sequence. Usually snow-fed streams in a given vicinity have similar daily distribution patterns of runoff, magnitudes of discharges reflecting individual watershed snowmelt contributing areas. These distribution patterns will also be similar year to year. Therefore, a distribution pattern for one of the larger 15-day volumes recorded for the stream where a damsite is located, or for a nearby similar watershed, can be selected and the 1 percent chance 15-day snowmelt runoff volume for the damsite distributed into daily discharges proportional to the selected recorded flood. An approximately symmetrical 15-day pattern with the maximum daily discharge occurring within the 7th to 10th day of the 15-day period is usually selected. An additional refinement may be included in selecting the distribution pattern, if by chance climatological records show that a small rain event occurred a day or two after the maximum daily discharge of a large recorded 15-day volume and discharges decreased due to lowered temperatures associated with the rain event. This sequence of events agrees with the pattern of natural conditions assumed by the occurrence of a probable maximum rainstorm a day or two after the maximum day of snowmelt runoff.

(3) *Combination of probable maximum rain flood with 1 percent chance 15-day snowmelt flood.* —Selection of an appropriate day within a 15-day period of snowmelt runoff as a beginning time of design rain-flood runoff is a matter of engineering judgment. One reasonable assumption is a 2-day interval between the day of maximum temperature and the beginning of runoff caused by a design storm. Under this assumption, the apparent lag-time in days between maximum temperature and maximum daily snowmelt discharge from a watershed should be considered. The lag-time may be quickly determined by plotting a few of the larger annual maximum 15-day mean daily discharges and respective daily maximum temperatures from an 'index temperature record. Depending on size and runoff characteristics of a watershed, the time interval between maximum temperature and resulting daily maximum snowmelt discharges at a damsite may vary from zero to 3 or more days. If the time interval is zero days, design rain-flood runoff is added to the snowmelt runoff, beginning on the third day after the peak of the snowmelt flood. As the lag-time interval between

maximum temperature and peak of snowmelt runoff increases, the beginning time for a design rain-flood hydrograph is advanced closer to the peak of the snowmelt flood by 1-day intervals. Thus, for large watersheds, it may be reasonable to combine a design rain flood with the maximum daily discharges of a snowmelt flood.

(b) *Damsites for Reservoirs With Proposed Joint Use Flood Control Capacity.*—A reservoir which has a joint use flood control capacity allocation is intended to control seasonal snowmelt discharges downstream from the dam to a limit of safe channel capacity throughout the entire snowmelt season, and also to store enough water to assure that the reservoir is full to the top of the joint use capacity at the end of each snowmelt season. Forecasts of seasonal snowmelt runoff volumes are a necessary part of this kind of operation.

A seasonal major snowmelt flood as a part of an inflow design flood (IDF) hydrograph usually is required when joint use flood control capacity is proposed. However, if planned joint use capacity is small and there is a likelihood that snowmelt discharges preceding the maximum 15-day period of a 1 percent chance snowmelt flood may fill the joint use pool, a 15-day IDF hydrograph will be adequate. When a seasonal major snowmelt flood hydrograph for combination with a probable maximum rain-flood hydrograph is needed, first consideration is given to the use of streamflow data.

The duration period of a seasonal IDF corresponds with the seasonal duration of the largest snowmelt floods which have occurred in the vicinity. Frequency analyses include annual maximum 30-day, 60-day, and if needed 90-day periods of snowmelt volumes in addition to analysis of the annual maximum 15-day discharge period. A recorded seasonal snowmelt flood is selected as a pattern for runoff distribution. The design rain flood is combined with the estimated snowmelt runoff hydrograph according to the criteria previously discussed.

If available streamflow data are not suitable for satisfactory results using the above approach, one of the methods of temperature-runoff correlations described in the referenced publications may be found adaptable to the situation.

G-24. *Probable Maximum Snowmelt Floods to be Combined With Major Rain Floods.*—(a) *General.*—An estimate of probable maximum snowmelt runoff may be necessary when making an inflow design flood (IDF) study for a watershed where snowmelt runoff causes the major portion of yearly flow. The degree of refinement needed in making this type of estimate may vary from preliminary comparisons to computation by detailed procedures depending on factors such as the following: storage capacity, space allocations, and operational plans of the proposed reservoir; snowmelt runoff characteristics of the watershed; and difference in magnitudes of probable maximum rainstorm and major rainstorm potentials for the watershed. For some watersheds, a few preliminary computations may show an IDF combination of major snowmelt runoff and probable maximum rain runoff to be definitely critical for design. In other instances detailed computations of each type IDF consisting of combined snowmelt and rain runoff have to be made and both types of IDF hydrographs prepared for use in design of a dam.

Studies prepared by the Bureau of Reclamation show that usually a critical inflow design flood results from a combination of runoff of a major snowmelt flood and a probable maximum rainstorm. In most instances, an approximation of probable maximum snowmelt flood magnitude by simple correlations shows that it will not be critical for design. Development of a *best estimate* of probable maximum snowmelt runoff is a complex procedure and requires special treatment for each site. Therefore, this discussion is limited to general aspects of the problem, with references to publications containing more detailed information.

(b) *Considerations for Estimates of Probable Maximum Snowmelt Floods.*—Estimating probable maximum snowmelt contribution to an inflow design flood can be thought of as requiring three steps: (1) estimating maximum seasonal accumulation of snow on a watershed, (2) estimating critical melt rates of the snow pack, and (3) estimating the amount of snowmelt runoff and its timing at the reservoir. The probable maximum seasonal accumulation of snow on a mountainous watershed drained by one main stream can be adequately estimated by a study of winter season precipitation records in and near the watershed, supplemented by snow survey data. Special studies are required for probable maximum seasonal snow accumulation estimates for large multitributary river systems such as the Colorado River above Glen Canyon Dam. One of two basic approaches can be taken to estimate critical snowmelt rates; namely, calculation of snowmelt runoff by means of an air temperature index, or calculation of melt using generalized snowmelt equations based on energy balance considerations. Methods using some form of an air temperature index have given good results for many watersheds. There is some physical basis for using a snowmelt air temperature index. Air temperature is reasonably well correlated, at a particular time and place, with the atmospheric factors which affect melt rates, such as solar radiation and vapor pressure, although it is by no means a perfect index of these factors.

Snowmelt equations which consider energy balance are used to evaluate short-wave radiation melt, long-wave radiation melt, melt due to convective heat transfer from the atmosphere and to latent heat of water vapor condensing into the snow surface, melt due to heat of rain drops, and melt by heat conduction from the ground. The Corps of Engineers report "Snow Hydrology" [23] presents detailed information regarding both approaches. A Corps manual, "Runoff from Snowmelt," EM 1110-2-1406 [25], presents synopses of investigations of melting relationships, generalized basin snowmelt equations and their application in methods of computing maximum snowmelt floods. Selection of an approach to be used depends on the basic data available and the importance of snowmelt runoff contribution to an inflow design flood. Whichever approach is taken, it is necessary to test the snowmelt computation procedures for the basin in question in order to determine basin values of the coefficients involved.

Approximation of a maximum probable snowmelt flood for a period of 10 to 20 days usually is directed toward determination of volume. This volume is then distributed in time by using a large recorded snowmelt runoff hydrograph as a pattern, as previously described in section G-23(a)(2). If a temperature index has been used directly in the computations, the volume may be distributed by a synthetic temperature sequence.

(c) *Springtime Seasonal Probable Maximum Snowmelt Flood Estimates.*—General procedures for estimating total seasonal probable maximum snowmelt runoff are not outlined in detail in this text. Brief statements about some approaches which may be considered for use, and reference to respective specific descriptions, are given below.

(1) *Hydrothermogram approach.*—The paper, "Snow Hydrology for Multiple-Purpose Reservoirs" [22], includes a description of an approach in which during the melting season daily temperatures above a base temperature are directly related to resulting direct runoff by a device referred to as a *hydrothermogram.* A hydrothermogram is a hypothetical discharge hydrograph computed on the assumption that each effective degree of temperature above a base temperature will generate the same amount of runoff volume. This procedure, adjusted to fit individual basin problems, has been found useful in several Bureau of Reclamation IDF studies (unpublished) where probable maximum snowmelt flood estimates were important.

(2) *Generalized melt equations for springtime snowmelt floods.*—The Corps of Engineers Manual, "Runoff from Snowmelt" [25], includes a chapter describing probable maximum snowmelt flood derivation using generalized melt equations. The Salmon River Basin which drains 14,100 square miles of rugged, mountainous regions of central Idaho is cited as an example in the discussion.

(3) *Correlations.*—Correlations between temperature and runoff, snowcover and runoff,

etc., are usually evidenced . because of the predictable nature of snowmelt runoff. Hydrologists knowledgeable in the use of correlation studies may find this type of approach useful.

(d) *Major Rain-Flood Estimates for Combination With Probable Maximum Snowmelt Runoff.*—

(1) *Major rainstorm and runoff.*—Design storm studies for watersheds where snowmelt runoff contributes to inflow design floods should also include a hydrometeorological estimate of a major rainstorm that could occur during the snowmelt season. For areas where major rainstorms have often occurred in the vicinity of the watershed during the snowmelt season, the largest rainstorm of record within the area of transposability is fitted to the basin. In areas where major rainstorm occurrences during the spring snowmelt season are infrequent, watershed design storm values without maximization for moisture adjustment may be considered. A hydrograph of runoff from the major rainstorm is computed by the dimensionless-graph lag-time procedures previously discussed, but special attention is given to effects of snowmelt on retention losses applicable to the major rainstorm. The portion of the watershed covered by a melting snowpack will have little or no retention capacity for rainfall, and the portion recently denuded of snow will have high moisture content, hence low retention capacity during rainfall. Guide criteria for combining rain-flood hydrographs and snowmelt flood hydrographs have been discussed in section G-23(a)(3).

(2) *Observed rain floods.*—Occasionally, streamflow data used for snowmelt runoff analyses will include a major rain flood during a snowmelt season. In these instances, special studies are made to separate the rain-flood hydrograph from the snowmelt runoff, and the separated rain-flood hydrograph is used for combination with the estimated probable maximum snowmelt flood hydrograph.

G-25. *Probable Maximum Rain-On-Snow IDF Estimates.* —There are many watersheds along or near the coasts of the United States where major rainstorms or probable maximum rainstorms can occur during the winter months

while the watersheds are partially or completely covered with snow. In many areas, storm systems may consist of precipitation beginning as snow then changing to rain or closely spaced successive storm systems, the first system occurring as snow, the second as rain accompanied by warm temperatures. Devastating floods have resulted from certain rain-on-snow combinations; in other instances, apparently similar conditions have produced only high flows causing little damage. Detailed investigations of differences between rain-on-snow flood magnitudes point toward the following two items as the main contributors to these differences: density conditions of the snowpack at the time of rain occurrence, and convective condensation melt related to wind velocities during the rainstorm. Generalized equations for estimating snowmelt during rainfall, developed as described in "Snow Hydrology" [23], have proved very useful in procedures for estimating runoff due to rainfall on snow.

In addition to estimates of snowpack melting rates, procedures for estimating runoff caused by rain-on-snow conditions include evaluations of snowpack release of free water to the ground surface, retention losses, and distribution in time of the runoff at the point of interest. A procedure used by the Corps of Engineers is given in the manual, "Runoff from Snowmelt" [25]. The procedure used in Bureau of Reclamation studies is described in Engineering Monograph No. 35, "Effect of Snow Compaction on Runoff from Rain on Snow" [26]. In both procedures snow melting rates during rainfall are computed by the same melting equations and water released at ground surface is determined. Excesses are computed by subtracting retention losses, and are distributed in time by a basin unitgraph. Differences between the procedures lie in estimations of snowpack free-water holding capacities.

The Corps procedure establishes a limit of liquid water holding capacity of a snowpack as a percentage of snowpack water content. Nearly all data considered when developing the limit of water holding capacity were obtained from spring snowpack of densities above 35 percent. The procedure in Engineering Monograph No. 35 relates snowpack liquid water holding capacity to snowpack densities just preceding the start of rainfall, and to increases in snowpack density due to melting and added rainfall until the pack attains a density of 40 or 45 percent when release of liquid water to the ground surface is assumed to begin. Development of the procedure was directed primarily for use for evaluating wintertime conditions where a rainstorm system closely follows a snowstorm and the newly deposited snowpack has had little time to change in structure. Topics of discussion in Engineering Monograph No. 35 are a development of the procedure and reconstitution of the December 1955 flood on South Yuba River near Cisco, Calif. Estimation of a probable maximum rain-on-snow flood is not discussed in the monograph. Data required for use of the procedure for IDF computations are: (1) estimates of watershed snowcover depth and water content antecedent to a design storm occurrence; and (2) hydrometeorological data of temperatures and wind velocities concurrent with design storm rainfall increments.

G-26. *Special Situations.* —(a) *Frozen Ground.* —Frozen ground conditions seldom occur in well-forested areas or under deep snowpacks. On the other hand, open areas where periods of subfreezing temperatures and light snowfall are normal can develop frozen soil conditions such that retention losses are practically nil. These areas may experience severe winter floods due to combinations of shallow snowcover, rising temperature, and relatively minor rainfall. Frozen ground conditions may also reduce lag-time. Analyses for this type of condition require individual watershed study.

(b) *Snowmelt in the Great Plains Region of the United States.* —Probable maximum precipitation potential is so great in the Great Plains region that snowmelt runoff is not usually considered in inflow design flood studies except for large drainage areas with headwaters in the Rocky Mountains. In the northern Great Plains, major floods have resulted from rapid spring snowmelt and frozen ground conditions. Consideration of this type of flood may be necessary for large drainage areas near the northern border of the United States.

H. ENVELOPE CURVES

G-27. *General.* —Peak discharge envelope curves and flood volume envelope curves can be prepared by drawing curves enveloping plotted points representing maximum recorded values for various drainage areas. The values plotted should represent similar type floods (rain floods or snowmelt floods) that have occurred within the broad geographical subdivision within which the subject watershed lies, and should not be limited to events of a single small river system. Preparation of envelope curves for a general area provides an engineer with valuable information on past flood history and an indication of the flood of record comparable to the subject area. However, they should not be relied upon as a means of estimating probable maximum flood values. Design flood values purporting to be the probable maximum should be higher than those obtained from envelope curves. Only in specific instances where a watershed has definitely lower flood potential than neighboring watersheds due to soil type, surface storage, etc., would it be good judgment to adopt an inflow design flood of smaller magnitude than that of a flood which has occurred nearby.

A simple method of preparation of envelope curves is to tabulate maximum peak discharges (or volumes of a selected duration) and respective drainage areas prior to plotting points. In most instances, the drainage area above a stream gaging station or the point of a large flood discharge measurement is given in the U.S. Geological Survey water supply paper listing the flood. When it is known that only a portion of the drainage area above a point of measurement contributed to a flood, the size of that contributing portion should be used in the envelope curve analysis. Discharges or volumes are plotted versus respective drainage areas using log-log paper. Data thus plotted usually indicate a curved line envelopment on log-log paper which may be approximated by a straight line for small ranges in areas. High discharges from local thunderstorms may suggest consideration of two curves—one for smaller areas subject to such occurrences and another for larger areas where maximum discharges originate from general storms.

I. STATISTICAL ANALYSES—ESTIMATES OF FREQUENCY OF OCCURRENCE OF FLOODS

G-28. *General.*—Estimates of the magnitude of floods which have frequencies of 1 in 5, 1 in 10, or 1 in 25 years are helpful in estimating requirements for stream diversion during construction. These floods are often termed the "5-, 10-, or 25-year flood." The magnitude of more rare events such as the 50- or 100-year flood may be required for reasons such as to establish sill location of emergency spillways, etc. The usual term of expression, "*x*-year flood," should not lead to the wrong conclusion that the event indicated can happen only once in *x* years, and having occurred, will not happen again for another period of *x* years. It does mean that over a long span of years we can expect as many *x*-year floods (or larger) as there are *x*-year-long periods within that span. Floods occur randomly and may be bunched or spread out unevenly with respect to time. No predictions are possible for determining their distribution; the probable maximum flood *can* occur the first year after the project is built, though of course, the odds are heavily against it.

The frequency of a flood should be considered as the chances of occurrence of a flood of that size (or one larger) in any one year. Stated another way, the chances of the flood in any one year being equaled or exceeded by floods of the magnitudes indicated as the 5-, 10-, 25-, or 100-year floods have ratios of 20:100, 10:100, 4:100, and 1:100, respectively.

Many methods of flood frequency determinations based on streamflow data have been published. Excellent summaries of these methods, along with comments on factors affecting their accuracy and limitations, are contained in the papers entitled "Review of Flood Frequency Methods" [27] and "Methods of Flow Frequency Analysis" [28]. While the many methods of flood frequency determinations made from streamflow data are all based on acceptable statistical procedures, the difference in methodology can give appreciably different results when extensions are made beyond the range of adequate data. To provide for a uniformity in Federal water resources planning, the Water Resources Council has recommended that all Government agencies use the Log-Pearson type III distribution as a base method. The method is described in the publication "A Uniform Technique for Determining Flood Flow Frequencies" [29]. Hazen's method [30] gives results that are comparable to those obtained with the Log-Pearson type III method and is easier to use when computations are made by hand with or without the aid of mechanical calculating machines. A procedural outline for Hazen computations is presented in section 59 of "Design of Small Dams," second edition [31].

If streamflow data for a period of 20 years or more are available for the subject watershed or comparable watersheds, frequency curve computations yield acceptable results for estimates up to the 25-year flood and may be extrapolated to indicate the 100-year flood with a fair assurance of obtaining acceptable values.

G-29. *Hydrographs for Estimating Diversion Requirements During Construction.*—Usually, inflow design flood (IDF) studies include hydrographs of floods for different frequencies of occurrence to be used for estimation diversion requirements during construction of a dam.

The hydrograph of a particular frequency flood is usually sketched to conventional shape using the peak discharge value and corresponding volume value obtained from computed frequency curves. In some instances, a peak discharge and associated volume of a recorded flood will correspond closely with a particular frequency value, in which case the recorded flood hydrograph is used.

J. FINAL-TYPE INFLOW DESIGN FLOOD STUDIES

G-30. *General.*—Preparations of final-type inflow design flood (IDF) studies differ from preliminary studies only in the degree of refinement used to estimate each variable causing flood runoff. For example, a basin unitgraph may be derived from a single large flood hydrograph in a preliminary study, whereas in a final-type study several flood hydrographs are analyzed and a selected basin unitgraph tested by reproduction of recorded flood hydrographs. Perhaps the most important consideration in the preparation of final-type studies is making certain that all available hydrological and meteorological data available, including historical and recent events, have been considered properly. A hydrometeorologist prepares the design storm study for the basin, including therein design temperatures and wind velocities if rain-on-snow floods are to be considered. Preliminary estimates of each flood-producing variable are reviewed and revised if additional data so indicate. Preliminary dam and reservoir operation plans are examined for certainty that the critical IDF situation for the chosen type of design and operation has been used.

Hydrologists and hydrometeorologists must estimate effects of ever-varying natural phenomena. Studies of these phenomena as related to a particular watershed begin with the inception of a project and continue thereafter, unless the project is determined infeasible and not built.

G-31. *Flood Routing Criteria.*—Normally, the reservoir pool is assumed to be full to the top of conservation storage at the start of the routing of the inflow design flood (IDF). However, when either inviolate or joint use flood control space is provided, the determination of space available at the beginning of the inflow design flood will depend upon the spacing of preceding storms, the relative magnitude of snowmelt contribution to the design flood, and the operational criteria proposed for the reservoir.

(a) *Preceding Storms.*—In some areas of the west, for example areas for which the Gulf of Mexico is the moisture source, the meteorological situation is such that a major storm could occur a few days prior to the maximum possible storm. In these areas, the flood control pool is assumed to be partially or completely occupied at the start of the inflow design flood. The determination of the portion of flood control pool that is occupied depends upon the distance of the area from the moisture source and a study of historical flood events in the area.

(b) *Seasonal Flood Hydrograph.*—For those areas in which floods occur on a fixed seasonal basis, largely as the result of snowmelt, it is frequently desirable to prepare a flood-season hydrograph including the inflow design flood and maximum antecedent and supervening flows that could reasonably be expected to occur with the inflow design flood. This hydrograph is then routed through the reservoir with the conservation pool full at the beginning of the season inflow, if that assumption can be justified on the basis of carryover storage. Otherwise, the minimum drawdown for the beginning date of seasonal inflow is selected from project operation studies.

(c) *Operational Criteria.*—The assumed reservoir elevation at the start of the inflow design flood will also be dependent upon the type of flood control space, which may be a fixed inviolate amount or a varying amount, normally referred to as joint use storage space.

The varying amount of flood control storage required will be based on operational parameters which show the needed amount of flood control storage based on antecedent precipitation, or the needed amount of storage based on forecasts of the seasonal runoff expected from the snowcover measurements.

K. BIBLIOGRAPHY

G-32. *Bibliography*.

[1] "Criteria and Practice Utilized in Determining the Required Capacity of Spillways," USCOLD Committee on "Failures and Accidents to Large Dams, Other than in Connection with the Foundations," United States Committee on Large Dams, C/O Engineers Joint Council, 345 East 47th Street, New York, N.Y., 1970.

[2] "Estimation of Maximum Floods," Technical Note No. 98, Report of a working group of the Commission for Hydrometeorology, World Meteorological Organization, Secretariat of the World Meteorological Organization, Geneva, Switzerland, 1969.

[3] U.S. Department of Agriculture, Soil Conservation Service National Engineering Handbook, Section 4, Hydrology, January 1971 or most recent publication. (For sale by Superintendent of Documents, U.S. Government Printing Office, Washington, D.C. 20402.)

[4] Sherman, L. K., "Streamflow from Rainfall by the Unit-Graph Method," Engineering News-Record, vol. 108, pp. 501-505, 1932.

[5] Johnstone, D., and Cross, W. P., "Elements of Applied Hydrology," The Ronald Press Co., New York, N.Y., 1949.

[6] Federal Inter-Agency River Basin Committee, Bulletin No. 1, "Instructions for Compilation of Unit Hydrograph Data," attached to the minutes of the 28th meeting of the Subcommittee on Hydrologic Data, March 1948.

[7] Barnes, B. S., "Discussion of Analysis of Runoff Characteristics," Trans. ASCE, vol. 105, 1940, p. 106.

[8] Langbein, W. B., "Channel Storage and Unit Hydrograph Studies," Trans. American Geophysical Union, 1940, Part II, pp. 620-627.

[9] Snyder, F. F., "Synthetic Unit-Graphs," Trans. American Geophysical Union, vol. 19, 1938, pp. 447-454.

[10] Clark, C. O., "Storage and the Unit Hydrograph," Trans. ASCE, vol. 110, 1945, pp. 1419-1488.

[11] Crawford, N. H., and Linsley, R. K., "Digital Simulation in Hydrology: Stanford Watershed Model IV," Technical Report No. 39, July 1966, Department of Civil Engineering, Stanford University, Stanford, Calif.

[12] Meserve, E. C., "Use of Clark Unit Graphs and Application of Clark Method to Pond Creek Study," April 1952, AWR Joint Study on Pond Creek, Little Rock District, Corps of Engineers.

[13] Linsley, R. K., Kohler, M. A., and Paulhus, J.L.H., "Hydrology for Engineers," McGraw-Hill Book Co., Inc., New York, N.Y., 1958.

[14] Tatum, F. E., "A Simplified Method of Routing Flood Flows through Natural Valley Storage," unpublished memorandum, U.S. Engineers Office, Rock Island, Ill., May 29, 1940.

[15] Wilson, W. T., "A Graphical Flood-Routing Method," Trans. American Geophysical Union, Part III, 1941.

[16] California Division of Water Resources, "Report on Control of Floods, San Joaquin River and Tributaries between Friant Dam and Merced River," July 1954.

[17] "Manual for Depth-Area-Duration Analysis of Storm Precipitation," WMO No. 237, TP. 129, Secretariat of the World Meteorological Organization, Geneva, Switzerland, 1969.

[18] "Tables of Precipitable Water and Other Factors for a Saturated Pseudo-Adiabatic Atmosphere," Technical Paper No. 14, 1951. U.S. Department of Commerce, National Oceanic and Atmospheric Administration, National Weather Service, Washington, D.C.

[19] "Climatic Atlas of the United States," U.S. Department of Commerce, Environmental Science Service Administration, Environmental Data Service, Washington, D.C. 20402, June 1968.

[20] U.S. Department of Commerce, National Oceanic and Atmospheric Administration, National Weather Service, Hydrometeorological Reports (selected):

Report No. 2, "Maximum Possible Precipitation over the Ohio River Basin above Pittsburgh, Pennsylvania," 1942.

Report No. 3, "Maximum Possible Precipitation over the Sacramento Basin of California," 1943.

Report No. 20, "An Estimate of Maximum Possible Flood-Producing Meteorological Conditions in the Missouri River Basin Above Garrison Dam Site," 1945.

Report No. 21, "A Hydrometeorological Study of the Los Angeles Area," 1939.

Report No. 21A, "Preliminary Report on Maximum Possible Precipitation, Los Angeles Area, California," 1944.

Report No. 21B, "Revised Report on Maximum Possible Precipitation, Los Angeles Area, California," 1945.

Report No. 22, "An Estimate of Maximum Possible Flood-Producing Meteorological Conditions in the Missouri River Basin Between Garrison and Fort Randall," 1946.

Report No. 23, "Generalized Estimates of Maximum Possible Precipitation Over the United States East of the 105th Meridian, for Areas of 10, 200, and 500 Square Miles," 1947.

Report No. 24, "Maximum Possible Precipitation Over the San Joaquin Basin, California," 1947.

Report No. 25, "Representative 12-Hour Dewpoints in Major United States Storms East of the Continental Divide," 1947.

Report No. 25A, "Representative 12-Hour Dewpoints in Major United States Storms East of the Continental Divide," 2d edition, 1949.

Report No. 28, "Generalized Estimate of Maximum Possible Precipitation Over New England and New York," 1952.

Report No. 33, "Seasonal Variation of the Probable Maximum Precipitation East of the 105th Meridian for Areas from 10 to 1,000 Square Miles and Durations of 6, 12, 24, and 48 Hours," 1956.

Report No. 36, "Interim Report—Probable Maximum Precipitation in California," Washington, D.C., 1961.

Report No. 39, "Probable Maximum Precipitation in the Hawaiian Islands," Washington, D.C., 1963.

Report No. 40, "Probable Maximum Precipitation, Susquehanna River Drainage above Harrisburg, Pennsylvania," Washington, D.C., 1965.

Report No. 41, "Probable Maximum and TVA Precipitation over the Tennessee River Basin above Chattanooga," Washington, D.C., 1965.

Report No. 42, "Meteorological Conditions for the Probable Maximum Flood on the Yukon River above Rampart, Alaska," Washington, D.C., 1966.

Report No. 43, "Probable Maximum Precipitation, Northwest States," Washington, D.C., 1966.

Report No. 44, "Probable Maximum Precipitation over the South Platte River, Colorado, and Minnesota River, Minnesota," Washington, D.C., 1969.

Cooperative Studies Reports, Cooperative Studies Section, Division of Climatological and Hydrologic Services, National Weather Service, in cooperation with the Bureau of Reclamation:

Report No. 9, "Maximum Possible Flood-Producing Meteorological Conditions." (1) Colorado River Basin above Glen Canyon Damsite, (2) Colorado River Basin above Bridge Canyon Damsite, (3) San Juan River Basin above Bluff Damsite, (4) Little Colorado River Basin above Coconino Damsite. June 1949.

Report No. 11, "Critical Meteorological Conditions for Design Floods in the Snake River Basin," February 1953.

Report No. 12, "Probable Maximum Precipitation on Sierra Slopes of the Central Valley of California," Washington, D.C., March 1954.

[21] "Storm Rainfall in the United States, Depth-Area-Duration Data," Department of the Army, Office of the Chief of Engineers, Washington, D.C., 1945.

[22] Riesbol, H. S., "Snow Hydrology for Multiple-Purpose Reservoirs," Trans. ASCE, vol. 119, 1954, pp. 595-627.

[23] "Snow Hydrology," Summary Report of Snow Investigations, U.S. Corps of Engineers, June 1956.

[24] "Handbook of Applied Hydrology," A Compendium of Water-Resources Technology, Ven Te Chow (Editor-in-Chief), McGraw-Hill Book Co., Inc., New York, N.Y., 1964.

[25] "Runoff from Snowmelt," EM 1110-2-1406, U.S. Corps of Engineers, 1960.

[26] Bertle, F. A., "Effect of Snow Compaction on Runoff From Rain on Snow," Engineering Monograph No. 35, Bureau of Reclamation, 1966.

[27] "Review of Flood Frequency Methods," Final Report of the Subcommittee of the Joint Division Committee on Floods, Trans. ASCE, vol. 118, 1953, pp. 1220-1231.

[28] "Methods of Flow Frequency Analysis," Bulletin No. 13, Subcommittee on Hydrology, Inter-Agency Committee on Water Resources (now the Hydrology Committee, Water Resources Council), Washington, D.C., April 1966.

[29] "A Uniform Technique for Determining Flood Flow Frequencies," Bulletin No. 15, Hydrology Committee, Water Resources Council, Washington, D.C., December 1967.

[30] Hazen, A., "Flood Flows," John Wiley & Sons, Inc., New York, N.Y., 1930.

[31] "Design of Small Dams," second edition, Bureau of Reclamation, 1973.

Sample Specifications for Concrete

H-1. *Introduction.* —Designs of any structure are based on assumptions regarding the quality of work which will be obtained during construction. It is through the means of specifications that the assumed quality is described, and it is important that conformance to the specifications be obtained for all work.

This appendix includes sample specifications for concrete in the dam and its appurtenances. For the construction of a particular dam, these specifications will be supplemented by local conditions, selected provisions, and special measures required for the construction of the structure.

The sample specifications are written on the basis that the concrete mixes to be used in the work will be designed and controlled by the purchaser (referred to in the specifications as the Contracting Authority or simply as the Authority) within the maximum water to cement or water to cement plus pozzolan ratio and slump limitations specified, the limitations for quality and grading of aggregates, and the limitations for the other materials as specified. Also, the specifications are written on the basis that the quantity of sand and each size of coarse aggregate to be used in the concrete mixes will be determined by the purchaser. The quality limitations shown in the specifications for sand and coarse aggregate are considered as standard limits. These limits may be reduced when only substandard materials are available within economical hauling distance, and provided it has been determined by tests of concrete made with such aggregates that durable concrete meeting the design strength criteria can be produced.

Under these specifications the purchaser's own engineering force or an engineering organization retained by the purchaser would accomplish testing of proposed aggregates and other materials, perform the design of mixes, and handle the inspection and quality testing throughout the contract. If the purchaser will require the contractor to provide such mix design, inspection and control, the specifications should so provide and should include specific design compressive strength(s) at designated age(s) for the concrete. The concrete mixes should be designed to provide compressive strengths of test cylinders such that 80 percent of the cylinders will have compressive strength(s) at the specified age(s) greater than the design compressive strength [1].[1]

References to "designations" in the sample specifications refer to designations in the appendix of the Bureau of Reclamation Concrete Manual, eighth edition [1]. Where materials or other requirements are to conform to Federal specifications, or other standard specifications such as ASTM, the construction specifications for specific work should provide that the specifications for the materials or requirements concerned should be in compliance with the latest editions or revisions thereof in effect on the date bids are received or award of contract is made, whichever is appropriate.

H-2. *Contractor's Plants, Equipment, and Construction Procedures.* —Prior to the installation of the contractor's plants and

---

[1] Numbers in brackets refer to items in the bibliography, sec. H-25.

equipment for processing, handling, transporting, storing, and proportioning concrete ingredients, and for mixing, transporting, and placing concrete, the contractor shall submit drawings covering his plans for approval by the Contracting Authority, showing proposed plant arrangement, including plans of locations and description of facilities for sampling of concrete and concrete materials as hereinafter provided. Included with the plans shall be a description of the equipment the contractor proposes to use in sufficient detail that an adequate review can be accomplished. The drawings and description of plant, equipment, and sampling and testing facilities shall be submitted at least 60 days prior to plant erection.

After completion of installation, the operation of the plant and equipment shall be subject to the approval of the Contracting Authority.

Sampling and testing facilities for use by the Authority shall be provided by the contractor and shall include power-driven mechanical sampling devices, satisfactory to the Authority, as may be necessary for procuring and handling representative test samples of aggregates and other concrete materials during batching; and for obtaining samples of concrete as discharged from the mixers, for mixer efficiency, slump, and other tests, except that power-driven mechanical sampling devices will not be required for sampling concrete from truck mixers if and when the use of truck mixers is permitted by these specifications. The concrete sampling device shall be capable of procuring samples of concrete from any point in the discharge stream as the concrete is being discharged from the mixer.

After completion of the plant installation, the operation of the sample taking facilities shall be demonstrated to the satisfaction of the Authority that they are suitable for the purpose intended. If truck mixers are used where permitted by these specifications, the contractor shall provide a stable, level platform with adequate shelter, satisfactory to the Authority, for concrete tests at the point of discharge from the truck mixers. The

contractor shall also provide ample and protected working space adjacent to the batching and mixing plants, free from plant vibration; and shall furnish necessary utilities such as compressed air, water, heat, and electrical power for operation of the Authority's testing equipment and for execution of tests by Authority personnel of concrete and concrete materials at the batching and mixing plants.

Where these specifications require specific types of equipment to be used or specific procedures to be followed, such requirements are not to be construed as prohibiting use by the contractor of alternative types of equipment or procedures if it can be demonstrated to the satisfaction of the Authority that equal results will be obtained by the use of such alternatives. Approval of plants and equipment or their operation, or of any construction procedure, shall not operate to waive or modify any provisions or requirement contained in these specifications governing the quality of the materials or of the finished work.

The cost of providing facilities and working space for procuring and handling representative test samples of concrete and concrete materials at the batching and mixing plants shall be included in the prices bid in the schedule for concrete.

The contractor shall keep the Authority advised as to when batching and mixing of concrete, installation of reinforcement and forming, preparations for placing and placing of concrete, finishing, and repair of concrete will be performed. Unless inspection is waived in each specific case, these construction activities shall be performed only in the presence of a duly authorized Authority inspector.

H-3. *Composition.* —(a) *General.* —Concrete shall be composed of cement, pozzolan, sand, coarse aggregate, water, and admixtures as specified, all well mixed and brought to the proper consistency. It is contemplated that pozzolan will be used in all concrete except for miscellaneous items of concrete where elimination of pozzolan is directed by the Contracting Authority.

(b) *Maximum Size of Aggregate.*—The maximum size of coarse aggregate in concrete for any part of the work shall be the largest of the specified sizes, the use of which is practicable from the standpoint of satisfactory consolidation of the concrete by vibration.

Except where it is determined by the Authority that, owing to closely spaced reinforcement or other reasons, the use of a smaller maximum size of aggregate is necessary to obtain satisfactory placement of the concrete, the maximum size of aggregate shall be as follows:

(1) Six-inch maximum-size aggregate shall, in general, be used in concrete for the dam, stilling basins, gravity walls, and elsewhere in other equally massive portions of structures where concrete containing the 6-inch maximum-size aggregate can be properly placed.

(2) Three-inch maximum-size aggregate shall be used in concrete for walls that are 15 inches or more in thickness and in slabs that are 8 inches or more in thickness, such as in massive floors and walls, and elsewhere where concrete containing 6-inch maximum-size aggregate cannot be placed, except that the requirements of subsection (3) below shall apply for tunnels, and for structures under conditions indicated.

(3) Three-inch maximum-size aggregate shall be used in concrete in tunnels where the concrete is 12 inches or more in thickness and the reinforcement, if any, consists of only one row or will not otherwise prevent satisfactory placement of the concrete, as determined by the Authority: *Provided*, that the contractor may use 2½-inch maximum-size aggregate to facilitate pumping: *Provided further*, that the contractor may use 2½-inch maximum-size aggregate in concrete that would otherwise contain 3-inch maximum-size aggregate whenever concrete containing 2½-inch maximum-size aggregate is being used at that time in work requiring pumping. One and one-half-inch maximum-size aggregate shall be used in concrete in tunnels where the concrete is less than 12 inches in thickness and for greater thicknesses when it is determined by the Authority that concrete containing a larger maximum size of aggregate cannot be properly placed.

(4) One and one-half-inch maximum-size aggregate shall be used in concrete for walls (except tunnel walls) that are less than 15 inches in thickness and in slabs that are less than 8 inches in thickness. However, where the walls or slabs are so heavily reinforced that 1½-inch size aggregate cannot be properly placed, as determined by the Authority, ¾-inch maximum-size aggregate may be permitted.

(5) In locations where concrete is to be placed against excavated surfaces and the thickness of concrete to be placed is greater than that shown on the drawings, correspondingly larger maximum size aggregate from that specified for the thickness of concrete shown on the drawings shall be used: *Provided*, that aggregate with a maximum size greater than that indicated above will not be required.

(c) *Mix Proportions.*—The proportions in which the various ingredients are to be used for different parts of the work and the appropriate water to portland cement plus pozzolan ratio will be determined by the Authority. Adjustments in the mix proportions and water to portland cement plus pozzolan ratio will be made by the Authority from time to time during the progress of the work, as tests are made of samples of the aggregates and the resulting concrete. These adjustments will have the objective of procuring concrete having suitable workability, density, impermeability, durability, and required strength, without the use of an excessive amount of cement.

It is contemplated that the composition of the concrete will be within the ranges given in the accompanying tabulation.

The proportions shown in the referenced tabulation may be modified by the Authority to suit the work or the nature of the materials, or to comply with limitations on the water to portland cement plus pozzolan ratio, and the contractor shall be entitled to no extra compensation by reason of such modification.

The net water to portland cement plus pozzolan ratio of the concrete (exclusive of water absorbed by the aggregates) shall not

Maximum size of aggregate (inches)	Cementing materials, portland cement plus pozzolan (approximate)		Sand, percent of total aggregate, by weight	Coarse aggregate, percent of total coarse aggregate only, by weight			
	Total pounds per cubic yard of concrete	Percent pozzolan (by weight of portland cement plus pozzolan)		3/16 to ¾ inch	¾ to 1½ inches	1½ to 3 inches	3 to 6 inches
6 3 1½ ¾			(Values to be determined by laboratory tests and inserted here for specifications.)				

exceed 0.47, by weight, for concrete in thin sections of structures which will be exposed to frequent alternations of freezing and thawing, such as curbs, gutters, sills, the top 2 feet of walls, piers, and parapets; and walls of structures in the range of fluctuating water levels or subject to spray. The net water to portland cement plus pozzolan ratio shall not exceed 0.53, by weight, for other concrete in structures which will be exposed to freezing and thawing. The net water to portland cement plus pozzolan ratio shall not exceed 0.60, by weight, for mass concrete in the dam, stilling basin, gravity walls, and elsewhere in other equally massive portions of structures; and for concrete in structures that will be covered with fill material or be continually submerged or otherwise protected from freezing and thawing.

(d) *Consistency.* —The amount of water used in the concrete shall be regulated as required to secure concrete of the proper consistency and to adjust for any variation in the moisture content or grading of the aggregates as they enter the mixer. Addition of water to compensate for stiffening of the concrete before placing will not be permitted. Uniformity in concrete consistency from batch to batch will be required.

The slump of the concrete, after the concrete has been deposited but before it has been consolidated, shall not exceed 2 inches for mass concrete; for concrete in the tops of walls, piers, parapets, and curbs; and for concrete in slabs that are horizontal or nearly horizontal. Similarly, the slump shall not exceed 4 inches for concrete in sidewalls and arch of tunnel lining; and 3 inches for all other concrete. The Authority reserves the right to require a lesser slump whenever concrete of such lesser slump can be consolidated readily

into place by means of the vibration specified in section H-18(c) (Consolidation). The use of buckets, chutes, hoppers, or other equipment which will not readily handle and place concrete of such lesser slump will not be permitted.

(e) *Tests.* —The compressive strength of the concrete will be determined by the Authority through the medium of tests of 6- by 12-inch cylinders made and tested in accordance with designations 29 to 33, inclusive, of the eighth edition of the Bureau of Reclamation Concrete Manual [1], except that, for all concrete samples from which cylinders are to be cast, the pieces of coarse aggregate larger than 1½ inches will be removed by screening or hand picking. Slump tests will be made by the Authority in accordance with designation 22.

H-4. Cement. —(a) *General.* —Cement for concrete, mortar, and grout shall be furnished by the contractor. The cement shall be free from lumps, unground clinker, tramp metal, and other foreign material, and shall be otherwise undamaged when used in concrete. If the cement is delivered in paper bags, empty paper bags shall be disposed of as directed. The contractor shall inform the Contracting Authority in writing, at least 60 days before first shipments are required, concerning the mill or mills from which the cement is to be shipped; whether cement will be ordered in bulk or in bags; and the purchase order number, contract number, or other designation that will identify the cement to be used by the contractor.

When bulk cement is not unloaded from the primary carriers directly into weathertight hoppers at the batching plant, transportation from the mill, railhead, or intermediate storage to the batching plant shall be accomplished in adequately weathertight trucks, conveyors, or other means which will protect the cement completely from exposure to moisture. Separate facilities, other than those provided for pozzolan, shall be provided for unloading, transporting, storing, and handling bulk cement. Locked unloading facilities shall be provided, and unloading of cement shall be performed only in the presence of the Authority or his representative. Immediately upon receipt at the jobsite, bulk cement shall be stored in dry, weathertight, and properly ventilated bins which shall be constructed so that there will be no dead storage. All storage facilities shall be subject to approval and shall be such as to permit easy access for inspection and identification.

The bins shall be emptied and cleaned by the contractor when so directed; however, the intervals between required cleanings will normally be not less than 4 months. If cement is obtained from more than one cement plant, shipments from each plant shall be blended with those from the other plant or plants by placing the cement from the different plants in alternate layers when unloading into silos at the railhead or at the jobsite, or by any other method satisfactory to the Authority. To prevent undue aging of cement furnished in bags, after delivery, the contractor shall use the bagged cement in the chronological order in which it was delivered to the jobsite. Each shipment of cement in bags shall be stored so that it may readily be distinguished from other shipments.

The cement shall meet the requirements of Federal Specification SS-C-192G [9], including Amendment 3 for type II, low-alkali cement, and shall meet the false-set limitation specified therein. In addition, cement for contraction joint grouting shall be air separated, and 100 percent of the finished product, after processing at the cement plant, shall pass a No. 30 United States standard sieve and 97.7 percent shall pass a No. 100 United States standard sieve. Cement for contraction joint grouting shall also be screened at the jobsite through a No. 16 crimped screen which shall be installed by the contractor between the mixer and agitator in the grout plant. The cement for contraction joint grouting shall be furnished in waterproof bags which will prevent hydration of the cement from exposure and also prevent lumping of the cement due to warehouse set for a minimum of 90 days. Cement for foundation grouting shall be furnished in bags: *Provided*, that bulk cement may be used for such grouting if a suitable method, satisfactory to the Authority, is used for weighing and accounting for the cement used.

(b) *Inspection.*—Except for sieve fineness of cement for contraction joint grouting, the cement will be sampled and tested by the Authority in accordance with Federal Test Method Standard No. 158A [11], including Change Notice 1 thereto, except that for initial penetration under method 2501.1 the rod shall be released 20 seconds after completion of mixing, and except that the note at the end of method 2501.1 concerning variations in initial penetration will be disregarded.

Fineness tests of the cement for contraction joint grouting will be made by the Authority in accordance with ASTM Designation C 184 [5], except that the tests will be performed on No. 30 and No. 100 sieves.

Acceptance tests, except for false set but including fineness tests, will be made on samples taken as bins of cement are filled and reserved for exclusive Authority use. Acceptance tests for false set will be made on samples taken from the cement at the latest time, prior to shipment in cars or trucks, that the cement is still in possession of the cement company. Cement not meeting test requirements will be rejected, and the contractor shall be entitled to no adjustments in price or completion time by reason of any delays occasioned thereby.

The contractor will be charged the cost of testing of all Authority-tested cement which has been ordered in excess of the amount of cement used for the work under these specifications. The charges to be made for the cost of testing excess cement will be at the rate of 3.5 cents per hundredweight (cwt.), which charge includes the Authority overhead, and will be deducted from payments due the contractor.

(c) *Measurement and Payment.*—

Measurement, for payment, of cement furnished in bags will be on the basis of the number of bags of cement used at the mixer. Measurement, for payment, of bulk cement will be on the basis of batch weights at the batching plant. Any cement, either bulk or in bags, used for grouting, finishing, or other miscellaneous work will be measured for payment in the most practicable manner. One bag of cement shall be considered as 0.94 hundredweight.

Payment will be made for cement used in concrete placed within the pay lines for concrete; and for cement used in concrete placed outside the concrete pay lines, unless the requirement for such concrete is determined by the Authority to be the result of careless excavation, or excavation intentionally performed by the contractor to facilitate his operations. No payment will be made for cement used as follows: cement used in wasted concrete, mortar, or grout; cement used in the replacement of damaged or defective concrete; cement used in extra concrete required as a result of careless excavation; and cement used in concrete placed by the contractor in excavation intentionally performed by the contractor to facilitate his operations. As determined by the Authority, payment will be made for a reasonable amount of cement used in grout required to keep the pipelines full during the grouting operations.

Payment for furnishing and handling cement will be made at the applicable unit prices per hundredweight or bag bid therefor in the schedule, which unit prices shall include the cost of rail and truck transportation of the cement from the mill to the jobsite and the cost of storing the cement.

H-5. *Pozzolan.* —(a) *General.* —Pozzolan for concrete shall be furnished by the contractor. The contractor shall use pozzolan concrete as provided in section H-3 (Composition). The pozzolan shall be in accordance with Federal Specification SS-P-570B [10].

When bulk pozzolan is not unloaded from primary carriers directly into weathertight hoppers at the batching plant, transportation from the source railhead or intermediate storage to the batching plant shall be accomplished in adequately designed trucks, conveyors, or other means which will protect the pozzolan completely from exposure to moisture. Separate facilities, other than those for cement, shall be provided for unloading, transporting, storing, and handling bulk pozzolan. Locked unloading facilities shall be provided and unloading of pozzolan shall be performed only in the presence of the Contracting Authority or his representative.

Immediately upon receipt at the jobsite, bulk pozzolan shall be stored in dry, weathertight, and properly ventilated bins. All storage facilities shall be subject to approval and shall be such as to permit easy access for inspection and identification. Sufficient pozzolan shall be in storage at all times to complete any concrete lift or placement started. The bins shall be emptied and cleaned by the contractor when so directed; however, the intervals between required cleanings will normally be not less than 4 months. The pozzolan shall be free from lumps and shall be otherwise undamaged when used in concrete.

The contractor shall inform the Authority in writing, within 60 days after date of notice to proceed, concerning the source or sources from which he proposes to obtain the pozzolan; together with information as to location, shipping point or points, purchase order number, contract number, or other designation and information that will identify the pozzolan to be used by the contractor.

(b) *Inspection.* —The pozzolan will be sampled and tested by the Authority in accordance with Federal Specification SS-P-570B [10]. Acceptance tests will be made on a lot or lots of pozzolan, which lot or lots shall be reserved in bulk storage in sealed bins at the source for exclusive Authority use. Untested lots shall not be intermingled or combined with tested and approved lots until such lots have been tested and approved. Pozzolan will also be sampled at the jobsite when determined necessary. Release for shipment and approval for use will be based on compliance with 7-day lime-pozzolan strength requirements and other physical and chemical and uniformity requirements for which tests can be completed by the time the 7-day lime-pozzolan strength test is completed. Release for shipment and approval for use on the above basis will be contingent on continuing compliance with the other requirements of the specifications. No pozzolan shall be shipped until notice has been given that the test results are satisfactory and all shipments will be made under supervision of the Authority. Any lot or lots of pozzolan not meeting test requirements will be rejected. Rejected pozzolan shall be replaced with acceptable pozzolan, and the contractor shall be entitled to no adjustments in price or completion time by reason of any delays occasioned thereby.

The contractor will be charged the cost of testing of all Authority-tested pozzolan which has been ordered in excess of the amount of pozzolan used for the work under these specifications. The charges to be made for the cost of testing excess pozzolan will be at the testing rate per ton plus overhead cost to the Authority and will be deducted from payments due the contractor.

(c) *Measurement and Payment.*— Measurement, for payment, of pozzolan will be made on the basis of batch weights at the batching plant with deductions made for the percentage of moisture in the pozzolan. The moisture content will be determined by heating a 500-gram sample to constant weight in an oven at 105° C. The percentage of moisture will be 100 times the quantity obtained by dividing the loss in weight, in grams, by the weight in grams of the moist sample. Any pozzolan used for miscellaneous work will be measured in the most practicable manner.

Pozzolan will be paid for on the basis of the number of tons (2,000 pounds net dry weight) used in the work covered by these specifications. No payment will be made for pozzolan used as follows: pozzolan used in wasted concrete; pozzolan used in the replacement of damaged or defective concrete; pozzolan used in extra concrete required as a result of careless excavation; and pozzolan used in concrete placed by the contractor in excavation intentionally performed by the contractor to facilitate his operations.

Payment for furnishing and handling pozzolan will be made at the unit price per ton bid therefor in the schedule, which unit price shall include the cost of rail and truck transportation of the pozzolan from the mill to the jobsite and the cost of storing the pozzolan.

H-6. *Admixtures.*—(a) *Accelerator.*— Calcium chloride shall not be used in concrete in which aluminum or galvanized metalwork is to be embedded or in concrete where it may come in contact with prestressed steel. The contractor shall use 1 percent of calcium chloride, by weight of the cement, in all other concrete placed when the mean daily temperature in the vicinity of the worksite is lower than 40° F. Calcium chloride shall not be used otherwise, except upon written approval of the Contracting Authority. Request for such approval shall state the reason for using calcium chloride and the percentage of calcium chloride to be used and the location of the concrete in which the contractor desires to use the calcium chloride. Calcium chloride shall not be used in excess of 2 percent, by weight of the cement. Calcium chloride shall be measured accurately and shall be added to the batch in solution in a portion of the mixing water. Use of calcium chloride in the concrete shall in no way relieve the contractor of responsibility for compliance with the requirements of these specifications governing protection and curing of the concrete.

(b) *Air-Entraining Agents.*—An air-entraining agent shall be used in all concrete. The agent used shall conform to ASTM Designation C 260 [6], except that the limitation and test on bleeding by concrete containing the agent and the requirement relating to time of setting shall not apply. The agent shall be of uniform consistency and quality within each container and from shipment to shipment. Agents will be accepted on manufacturer's certification of compliance with specifications: *Provided*, that the Authority reserves the right to require submission of and to perform tests on samples of the agent prior to shipment and use in the work and to sample and test the agent after delivery at the jobsite.

The amount of air-entraining agent used in

each concrete mix shall be such as will effect the entrainment of the percentage of air shown in the following tabulation in the concrete as discharged from the mixer:

Maximum size of coarse aggregate in inches	Total air, percent by volume of concrete
¾	6.0 plus or minus 1
1½	4.5 plus or minus 1
3	3.5 plus or minus 1
6	3.0 plus or minus 1

The agent in solution shall be maintained at uniform strength and shall be added to the batch in a portion of the mixing water. This solution shall be accurately batched by means of a reliable mechanical batcher which shall be so constructed that the full measure of solution added to each batch of concrete can be observed in a sight gage by the plant operator prior to discharge of the solution into the mixer. When calcium chloride is being used in the concrete, the portion of the mixing water containing the air-entraining agent shall be introduced separately into the mixer.

(c) *Water-Reducing, Set-Controlling Admixture.*—The contractor shall, except as hereinafter provided, use a water-reducing, set-controlling admixture, referred to herein as WRA, in all concrete. The WRA used shall be either a suitable lignosulfonic-acid or hydroxylated-carboxylic-acid type.

The WRA shall be of uniform consistency and quality within each container and from shipment to shipment. WRA will be accepted on manufacturer's certification of conformance to Bureau of Reclamation "Specifications and Method of Test for Water-Reducing, Set-Controlling Admixtures for Concrete," dated August 1, 1971: *Provided*, that the Authority reserves the right to require submission of and to perform tests on samples of the agent prior to shipment and use in the work and to sample and test the agent after delivery at the jobsite.

If Authority testing of the WRA is required, the contractor shall submit a sample of the WRA and five bags (94 pounds each) of the cement proposed for use in the work at least

90 days before use is expected. The size of the sample of WRA to be submitted shall be 1 liquid gallon.

The quantity of WRA to be used in each concrete batch shall be determined by the Authority and for the lignosulfonic-acid type shall not exceed 0.40 percent, by weight of cement plus pozzolan, of solid crystalline lignin, and for the hydroxylated-carboxylic-acid type shall not exceed 0.50 percent, by weight of cement plus pozzolan, of liquid.

Since the quantity of WRA required will vary with changing atmospheric conditions, the quantity used shall be commensurate with the prevailing conditions. The Authority reserves the right to use lesser quantities or no WRA in concrete for any part of the work, depending on climatic or other job conditions, and the contractor shall be entitled to no additional compensation by reason of reduction in or elimination of WRA in any concrete to be placed under these specifications.

The WRA solution shall be measured for each batch by means of a reliable visual mechanical dispenser. The WRA, in a suitably dilute form, may be added to water containing air-entraining agent for the batch if the materials are compatible with each other, or shall be introduced separately to the batch in a portion of the mixing water if the two are incompatible.

When requested, the contractor shall submit test data by the manufacturer showing effects of the WRA on mixing water requirements, setting time of concrete, and compressive strength at various ages up to 1 year.

The contractor shall be responsible for any difficulties arising or damages occurring as a result of the selection and use of WRA, such as delay or difficulty in concrete placing or damage to the concrete during form removal. The contractor shall be entitled to no additional compensation above the unit prices bid in the schedule for concrete by reason of such difficulties.

(d) *Furnishing Admixtures.*—Air-entraining agent, accelerator, and WRA, as required, shall be furnished by the contractor, and the cost of the materials and all costs incidental to their use shall be included in the applicable prices bid in the schedule for concrete in which the materials are used.

H-7. *Water.*—The water used in concrete, mortar, and grout shall be free from objectionable quantities of silt, organic matter, alkali, salts, and other impurities.

H-8. *Sand.*—(a) *General.*—The term "sand" is used to designate aggregate in which the maximum size of particles is 3/16 of an inch. Sand for concrete, mortar, and grout shall be furnished by the contractor and shall be natural sand, except that crushed sand may be used to make up deficiencies in the natural sand grading. The contractor shall maintain at least three separate stockpiles of processed sand; one to receive wet sand, one in the process of draining, and one that is drained and ready for use. Sand to be used in concrete shall be drawn from the stockpile of drained sand which shall have been allowed to drain for a minimum of 48 hours. Sand, as delivered to the batching plant, shall have a uniform and stable moisture content, which shall be less than 6 percent free moisture.

(b) *Quality.*—The sand shall consist of clean, hard, dense, durable, uncoated rock fragments. The maximum percentages of deleterious substances in the sand, as delivered to the mixer, shall not exceed the following values:

Deleterious substance	Percent, by weight
Material passing No. 200 screen (designation 16)	3
Lightweight material (designation 17)	2
Clay lumps (designation 13)	1
Total of other deleterious substances (such as alkali, mica, coated grains, soft flaky particles, and loam)	2

The sum of the percentages of all deleterious substances shall not exceed 5 percent, by weight. Sand producing a color darker than the standard in the colorimetric test for organic impurities (designation 14) may be rejected. Sand having a specific gravity (designation 9), saturated surface-dry basis, of less than 2.60 may be rejected. The sand may be rejected if the portion retained on a No. 50 screen, when subjected to 5 cycles of the sodium sulfate test for soundness (designation 19), shows a weighted average loss of more than 8 percent, by weight. The designations in parentheses refer to methods of tests described in the eighth edition of the Bureau of Reclamation Concrete Manual [1].

(c) *Grading.*—The sand as batched shall be well graded, and when tested by means of standard screens (designation 4) shall conform to the following limits:

Screen No.	Individual percent, by weight, retained on screen
4	0 to 5
8	* 5 to 15
16	*10 to 25
30	10 to 30
50	15 to 35
100	12 to 20
Pan	3 to 7

*If the individual percent retained on the No. 16 screen is 20 percent or less, the maximum limit for the individual percent retained on the No. 8 screen may be increased to 20 percent.

The grading of the sand shall be controlled so that at any time the fineness moduli (designation 4) of at least 9 out of 10 consecutive test samples of finished sand will not vary more than 0.20 from the average fineness modulus of the 10 test samples.

H-9. *Coarse Aggregate.*—(a) *General.*—The term "coarse aggregate," for the purpose of these specifications, designates aggregate of sizes within the range of 3/16 of an inch to 6 inches or any size or range of sizes within such limits. The coarse aggregate shall be reasonably well graded within the nominal size ranges hereinafter specified. Coarse aggregate for concrete shall be furnished by the contractor and shall consist of natural gravel or crushed rock or a mixture of natural gravel and crushed rock.

Coarse aggregate, as delivered to the batching plant, shall have a uniform and stable moisture content.

(b) *Quality.*—The coarse aggregate shall consist of clean, hard, dense, durable, uncoated rock fragments. The percentages of deleterious substances in any size of coarse aggregate, as delivered to the mixer, shall not exceed the following values:

	Percent, by weight
Material passing No. 200 screen (designation 16)	½
Lightweight material (designation 18)	2
Clay lumps (designation 13)	½
Other deleterious substances	1

The sum of the percentages of all deleterious substances in any size, as delivered to the mixer, shall not exceed 3 percent, by weight. Coarse aggregate may be rejected if it fails to meet the following test requirements:

(1) Los Angeles rattler test (designation 21).—If the loss, using grading A, exceeds 10 percent, by weight, at 100 revolutions or 40 percent, by weight, at 500 revolutions.

(2) Sodium sulfate test for soundness (designation 19).—If the weighted average loss after 5 cycles is more than 10 percent by weight.

(3) Specific gravity (designation 10).—If the specific gravity (saturated surface-dry basis) is less than 2.60.

The designations in parentheses refer to methods of test described in the eighth edition of the Bureau of Reclamation Concrete Manual [1].

(c) *Separation.*—The coarse aggregate shall be separated into nominal sizes and shall be graded as follows:

Designation of size (inches)	Nominal size range (inches)	Minimum percent retained on screens indicated	
		Percent	Size of screen (inches)
¾	3/16 to ¾	50	3/8
1½	¾ to 1½	25	1¼
3	1½ to 3	20	2½
6	3 to 6	20	5

Coarse aggregate shall be finished screened on vibrating screens mounted over the batching plant, or at the option of the contractor, the screens may be mounted on the ground adjacent to the batching plant. The finish screens, if installed over the batching plant, shall be so mounted that the vibration of the screens will not be transmitted to, or affect the accuracy of the batching scales. The sequence of coarse aggregate handling and plant

management shall be such that, if final and/or submerged cooling are used, excessive free moisture shall be removed and diverted outside of the plant by dewatering screens prior to finish screening so that a uniform and stable moisture content is maintained in the plant storage and batching bins. The method and rate of feed shall be such that the screens will not be overloaded and will operate properly in a manner that will result in a finished product which consistently meets the grading requirements of these specifications. The finished products shall pass directly to the individual batching bins. Material passing the 3/16-inch screen that is removed from the coarse aggregate as a result of the finished screening operation shall be wasted.

Separation of the coarse aggregate into the specified sizes, after finish screening, shall be such that, when the aggregate, as batched, is tested by screening on the screens designated in the following tabulation, the material passing the undersize test screen (significant undersize) shall not exceed 2 percent, by weight, and all material shall pass the oversize test screen:

Aggregate size designation (inches)	Size of square opening in screen (inches)	
	For undersize test	For oversize test
¾	No. 5 mesh (U.S. standard screen)	7/8
1½	5/8	1¾
3	1¼	3½
6	2½	7

Screens used in making the tests for undersize and oversize will conform to ASTM Designation E 11 [7], with respect to permissible variations in average openings.

H-10. *Production of Sand and Coarse Aggregate.*—(a) *Source of Aggregate.*—Sand and coarse aggregate for concrete, and sand for mortar and grout may be obtained by the contractor from any approved source as hereinafter provided.

If sand and coarse aggregate are to be obtained from a deposit not previously tested and approved by the Contracting Authority, the contractor shall submit representative samples for preconstruction test and approval at least 60 days after date of notice to proceed. The samples shall consist of approximately 200 pounds each of sand and 3/16- to 3/4-inch size of coarse aggregate, and 100 pounds of each of the other sizes of coarse aggregate.

The approval of deposits by the Authority shall not be construed as constituting the approval of all or any specific materials taken from the deposits, and the contractor will be held responsible for the specified quality of all such materials used in the work.

In addition to preconstruction test and approval of the deposit, the Authority will test the sand and coarse aggregate during the progress of the work and the contractor shall provide such facilities as may be necessary for procuring representative samples.

If any deposit used by the contractor is located within an approved area owned or controlled by the Authority, no charge will be made to the contractor for materials taken from such deposit and used in the work covered by these specifications. Any royalties or other charges required to be paid for materials taken from deposits not owned or controlled by the Authority shall be paid by the contractor.

(b) *Developing Aggregate Deposit.* —The contractor shall carefully clear the area of the deposit, from which aggregates are to be produced, of trees, roots, brush, sod, soil, unsuitable sand and gravel, and other objectionable matter. If the deposit is owned or controlled by the Authority, the portion of the deposit used shall be located and operated so as not to detract from the usefulness of the deposit or of any other property of the Authority and so as to preserve, insofar as practicable, the future usefulness or value of the deposit. Materials, including stripping, removed from deposits owned or controlled by the Authority and not used in the work covered by these specifications shall be disposed of as directed.

The contractor's operations in and around aggregate deposits shall be in accordance with the provisions of the specifications sections on environmental protection.

(c) *Processing Raw Materials.* —Processing of the raw materials shall include screening, and washing as necessary, to produce sand and coarse aggregate conforming to the requirements of sections H-8 (Sand) and H-9 (Coarse Aggregate). Processing of aggregates produced from any source owned or controlled by the Authority shall be done at an approved site. Water used for washing aggregates shall be free from objectionable quantities of silt, organic matter, alkali, salts, and other impurities. To utilize the greatest practicable yield of suitable materials in the portion of the deposit being worked, the contractor may crush oversize material and any excess material of the sizes of coarse aggregate to be furnished, until the required quantity of each size has been secured: *Provided*, that crusher fines produced in manufacturing coarse aggregate that will pass a screen having 3/16-inch square openings shall be wasted or rerouted through the sand manufacturing plant. Crushed sand, if used to make up deficiencies in the natural sand grading, shall be produced by a suitable ball or rod mill, disk or cone crusher, or other approved equipment so that the sand particles shall be predominately cubical in shape and free from objectionable quantities of flat or elongated particles.

The crushed sand and coarse aggregate shall be blended uniformly with the uncrushed sand and coarse aggregate, respectively. Crushing and blending operations shall at all times be subject to approval by the Authority. The handling, transporting, and stockpiling of aggregates shall be such that there will be a minimum amount of fines resulting from breakage and abrasion of material caused by free fall and improper handling. Where excesses in any of the sand and coarse aggregate sizes occur, the contractor shall dispose of the excess material as directed by the Authority.

(d) *Furnishing Aggregates.* —The cost of producing aggregates required for work under these specifications and the cost of aggregates not obtained from a source owned or controlled by the Authority shall be included in the unit prices bid in the schedule for concrete in which the aggregates are used, which unit prices shall also include all expenses of the contractor in stripping, transporting, and storing the materials. The contractor shall be entitled to no additional compensation for materials wasted from a deposit, including crusher fines, excess material of any of the sizes into which the aggregates are required to be separated by the contractor, and materials which have been discarded by reason of being above the maximum sizes specified for use.

H-11. *Batching.*—(a) *General.*—The contractor shall provide equipment and shall maintain and operate the equipment as required to accurately determine and control the prescribed amounts of the various materials, including water, cement, pozzolan, admixtures, sand, and each individual size of coarse aggregate entering the concrete. The amounts of bulk cement, pozzolan, sand, and each size of coarse aggregate entering each batch of concrete shall be determined by separate weighing, and the amounts of water and each admixture shall be determined by separate weighing or volumetric measurement. Where bagged cement is used, the concrete shall be porportioned on the basis of integral bags of cement unless the cement is weighed.

When bulk cement, pozzolan, and aggregates are hauled from a central batching plant to the mixers, the cement and pozzolan for each batch shall either be placed in an individual compartment which during transit will prevent the cement and pozzolan from intermingling with each other and with the aggregates and will prevent loss of cement and pozzolan; or the cement and pozzolan shall be completely enfolded in and covered by the aggregates by loading the cement, pozzolan, and aggregates for each batch simultaneously into the batch compartment. The bins of batch trucks shall be provided with suitable covers to protect the materials therein from wind or wet weather. Each batch compartment shall be of sufficient capacity to prevent loss in transit and to prevent spilling and intermingling of batches as compartments are being emptied. If the cement and pozzolan are enfolded in aggregates containing moisture, and delays occur between filling and emptying the compartments the contractor shall, at his own expense, add extra cement to each batch in accordance with the following schedule:

*Hours of contact between cement and wet aggregate	Additional cement required
0 to 2	0 percent
2 to 3	5 percent
3 to 4	10 percent
4 to 5	15 percent
5 to 6	20 percent
Over 6	Batch will be rejected.

*The Contracting Authority reserves the right to require the addition of cement for shorter periods of contact during periods of hot weather and the contractor shall be entitled to no additional compensation by reason of the shortened period of contact.

Batch bins shall be constructed so as to be self-cleaning during drawdown and the bins shall be drawn down until they are practically empty at least three times per week. Materials shall be deposited in the batch bins directly over the discharge gates. The 1½-, 3-, and 6-inch coarse aggregates shall be deposited in the batcher bins through effective rock ladders, or other approved means. To minimize breakage, the method used in transporting the aggregates from one elevation to a lower elevation shall be such that the aggregates will roll and slide with a minimum amount of free fall.

Equipment for conveying batched materials from the batch hopper or hoppers to and into the mixer shall be so constructed, maintained, and operated that there will be no spillage of the batched materials or overlap of batches. Equipment for handling portland cement and pozzolan in the batching plant shall be constructed and operated so as to prevent noticeable increase of dust in the plant during the measuring and discharging of each batch of material. If the batching and mixing plant is enclosed, the contractor shall install exhaust fans or other suitable equipment for removing dust.

(b) *Equipment.*—The weighing and measuring equipment shall conform to the following requirements:

(1) The construction and accuracy of the equipment shall conform to the applicable requirements of Federal

Specification AAA-S-121d [8] for such equipment, except that an accuracy of 0.4 percent over the entire range of the equipment will be required.

The contractor shall provide standard test weights and any other equipment required for checking the operating performance of each scale or other measuring device and shall make periodic tests over the ranges of measurements involved in the batching operations. The tests shall be made in the presence of an Authority inspector, and shall be adequate to prove the accuracy of the measuring devices. Unless otherwise directed, tests of weighing equipment in operation shall be made at least once every month. The contractor shall make such adjustments, repairs, or replacements as may be necessary to meet the specified requirements for accuracy of measurement.

(2) Each weighing unit shall include a visible springless dial which will register the scale load at any stage of the weighing operation from zero to full capacity. The minimum clear interval for dial scale graduations shall be not less than 0.03 inch. The scales shall be direct reading to within 5 pounds for cement and 20 pounds for aggregate. The weighing hoppers shall be constructed so as to permit the convenient removal of overweight materials in excess of the prescribed tolerances. The scales shall be interlocked so that a new batch cannot be started until the weighing hoppers have been completely emptied of the last batch and the scales are in balance. Each scale dial shall be in full view of the operator.

(3) The equipment shall be capable of ready adjustment for compensating for the varying weight of any moisture contained in the aggregates and for changing the mix proportions.

(4) The equipment shall be capable of controlling the delivery of material for weighing or volumetric measurement so that the combined inaccuracies in feeding and measuring during normal operation will not exceed 1 percent for water; 1½ percent for cement and pozzolan; 3 percent for admixtures; 2 percent for sand, ¾-inch aggregate, and 1½-inch aggregate; and 3 percent for 3- and 6-inch coarse aggregate.

(5) Convenient facilities shall be provided for readily obtaining representative samples of cement, pozzolan, admixtures, sand, and each size of coarse aggregate from the discharge streams between bins and the batch hoppers or between the batch hoppers and the mixers.

(6) The operating mechanism in the water-measuring device shall be such that leakage will not occur when the valves are closed. The water-measuring device shall be constructed so that the water will be discharged quickly and freely into the mixer without objectionable dribble from the end of the discharge pipe. In addition to the water-measuring device, there shall be supplemental means for measuring and introducing small increments of water into each mixer when required for final tempering of the concrete. This equipment shall introduce the added water well into the batch. Each water-measuring device shall be in full view of the operator.

(7) Dispensers for air-entraining agents, calcium chloride solutions, and WRA shall have sufficient capacity to measure at one time the full quantity of the properly diluted solution required for each batch, and shall be maintained in a clean and freely operating condition. Equipment for measuring shall be designed for convenient confirmation by the plant operator of the accuracy of the measurement for each batch and shall be so constructed that the required quantity can be added only once to each batch.

(8) The mixing plant shall be arranged so that the mixing action in at least one of the mixers can be conveniently observed from its control station. Provisions shall be made so that the mixing action of each of the other mixers can be observed from

a safe location which can be easily reached from the control station. Provisions shall also be made so that the operator can observe the concrete in the receiving hopper or buckets after being dumped from the mixers.

(9) Equipment that fails to conform to the requirements of this section shall be effectively repaired or satisfactorily replaced.

H-12. *Mixing.*—(a) *General.*—The concrete ingredients shall be mixed thoroughly in batch mixers of approved type and size and designed so as to positively ensure uniform distribution of all of the component materials throughout the mass at the end of the mixing period. The adequacy of mixing will be determined by the method of "Variability of Constituents in Concrete" in accordance with the provisions of designation 26 of the eighth edition of the Bureau of Reclamation Concrete Manual [1]. Mixers when tested shall meet the following criteria:

(1) The unit weight of air-free mortar in samples taken from the first and last portions of the batch as discharged from the mixer shall not vary more than 0.8 percent from the average of the two mortar weights.

(2) For any one mix, the average variability for more than one batch shall not exceed the following limits:

Number of tests	Average variability (percent based on average mortar weight of all tests)
3	0.6
6	.5
20	.4
90	.3

(3) The weight of coarse aggregate per cubic foot in samples taken from the first and last portions of the batch as discharged from the mixer shall not vary more than 5.0 percent from the average of the two weights of coarse aggregate.

The Contracting Authority reserves the right to either reduce the size of batch to be mixed or to increase the mixing time when the charging and mixing operations fail to produce a concrete batch which conforms throughout

to the above-numbered criteria and in which the ingredients are uniformly distributed and the consistency is uniform. Water shall be added prior to, during, and following the mixer-charging operations. Overmixing, requiring addition of water to preserve the required consistency, will not be permitted. Any concrete retained in mixers so long as to require additional water in excess of 3 percent of the design mix water (net water-cement plus pozzolan ratio water, not including water absorbed by aggregates) to permit satisfactory placing shall be wasted. Any mixer that at any time produces unsatisfactory results shall be repaired promptly and effectively or shall be replaced.

Use of truck mixers in accordance with subsection (c) below will be permitted only for miscellaneous items of concrete work where and as approved by the Authority.

(b) *Central Mixers.*—Mixers shall not be loaded in excess of their rated capacity unless specifically authorized. The concrete ingredients shall be mixed in a batch mixer for not less than the period of time indicated in the following tabulation for various mixer capacities after all of the ingredients except the full amount of water are in the mixer, except that the mixing time may be reduced if, as determined by the Authority, thorough mixing conforming to subsections (a) (1) and (2) above can be obtained in less time.

Capacity of mixer	Time of mixing
2 cubic yards or less	1½ minutes
3 cubic yards	2 minutes
4 cubic yards	2½ minutes
Larger than 4 cubic yards	To be determined by tests performed by the Authority

(c) *Truck Mixers.*—Use of truck mixers will be permitted only when the mixers and their operation are such that the concrete throughout the mixed batch and from batch to batch is uniform with respect to consistency and grading. Any concrete retained in truck mixers sufficiently long as to require additional water to permit placing shall be wasted.

Each truck mixer shall be equipped with (1) an accurate watermeter between supply tank and mixer, the meter to have indicating dials and totalizer, and (2) a reliable revolution counter, which can be readily reset to zero for indicating the total number of revolutions of the drum for each batch. Each mixer shall have affixed thereto a metal plate on which the drum capacities for both mixing and agitating are plainly marked in terms of volume of concrete in cubic yards and the maximum and minimum speeds of rotation of the drum in revolutions per minute.

Mixing shall be continued for not less than 50 nor more than 100 revolutions of the drum at the manufacturer's rated mixing speed after all the ingredients, except approximately 5 percent of the water which may be withheld, are in the drum. The mixing speed shall be not less than 5 nor more than 20 revolutions per minute. Thereafter, additional mixing, if any, shall be at the speed designated by the manufacturer of the equipment as agitating speed; except that after the addition of withheld water, mixing shall be continued at the specified mixing speed until the water is dispersed throughout the mix. After a period of agitation a few revolutions of the drum at mixing speed will be required just prior to discharging. In no case shall the specified maximum net water-cement plus pozzolan ratio be exceeded.

When a truck mixer or agitator is used for transporting concrete, the concrete shall be delivered to the site and the discharge completed within 1½ hours after the introduction of the cement into the mixer. Each batch of concrete, when delivered at the jobsite from commercial ready-mix plants, shall be accompanied by a written certificate of batch weights and time of batching.

Mixers shall be examined daily for changes in condition due to accumulation of hard concrete or mortar or to wear of blades. No mixer shall be charged in excess of its rated capacity for mixing or agitating; however, if any mixer cannot produce concrete meeting the requirements heretofore specified when mixing at rated capacity, within the specified limitation on the number of revolutions of the mixing drum at mixing speed, the size of batch mixed in that mixer may be reduced until, upon testing, a uniformly mixed batch,

conforming to the mixer performance tests as provided in subsection (a) above, is obtained.

H-13. *Temperature of Concrete.*—The temperature of mass concrete for the dam shall, when concrete is being placed, be not more than 50° F. and not less than 40° F. For all other concrete, the temperature of concrete when it is being placed shall be not more than 90° F. and not less than 40° F. in moderate weather or not less than 50° F. in weather during which the mean daily temperature drops below 40° F. Concrete ingredients shall not be heated to a temperature higher than that necessary to keep the temperature of the mixed concrete, as placed, from falling below the specified minimum temperature. Methods of heating concrete ingredients shall be subject to approval by the Contracting Authority.

If concrete is placed when the weather is such that the temperature of the concrete would exceed the maximum placing temperatures specified, as determined by the Authority, the contractor shall employ effective means as necessary to maintain the temperature of the concrete, as it is placed, below the maximum temperatures specified. These means may include placing at night; precooling the aggregates by cool airblast, immersion in cold water, vacuum processing, or other suitable method; refrigerating the mixing water; adding chip or flake ice to the mixing water; or a combination of these or other approved means. The contractor shall be entitled to no additional compensation on account of the foregoing requirements.

H-14. *Forms.*—(a) *General.*—Forms shall be used, wherever necessary, to confine the concrete and shape it to the required lines. Forms shall have sufficient strength to withstand the pressure resulting from placement and vibration of the concrete, and shall be maintained rigidly in position. Forms shall be sufficiently tight to prevent loss of mortar from the concrete. Chamfer strips shall be placed in the corners of forms so as to produce beveled edges on permanently exposed concrete surfaces. Interior angles on such surfaces and edges at formed joints will not require beveling unless requirement for beveling is indicated on the drawings. Inside forms for nearly horizontal circular tunnels having an inside diameter of 12 feet or more shall be constructed to cover only the arch and sides. The bottom 60° of the inside circumference shall be placed without forming: *Provided,* that the contractor may increase the angle of the inside circumference to be placed without forming on written approval of the Contracting Authority. Request for approval shall be accompanied by complete plans and description of the placing methods proposed to be used.

Forms for tunnel lining shall be provided with openings along each sidewall and in each arch, each opening to be not less than 2 by 2 feet. The openings shall be located in the crown and along each sidewall, as follows:

(1) Openings in the crown shall be spaced at not more than 8 feet on centers and shall be located alternately on each side of the tunnel centerline.

(2) Openings in sidewall forms for tunnels having an inside diameter less than 12 feet shall be located at midheight of the tunnel in each sidewall and shall be spaced at not more than 8 feet on centers along each sidewall.

(3) Openings in sidewall forms for tunnels having an inside diameter of 12 feet or more shall be located along two longitudinal lines in each sidewall, the locations of which are satisfactory to the Authority. The openings along the two selected longitudinal lines in each sidewall shall be staggered and shall be spaced at not more than 8 feet on centers along each longitudinal line.

The cost of all labor and materials for forms and for any necessary treatment or coating of forms shall be included in the unit prices bid in the schedule for the concrete for which the forms are used.

(b) *Form Sheathing and Lining.*—Wood sheathing or lining shall be of such kind and quality or shall be so treated or coated that there will be no chemical deterioration or discoloration of the formed concrete surfaces. The type and condition of form sheathing and lining, and the fabrication of forms for finishes F2, F3, and F4 shall be such that the form

surfaces will be even and uniform. The ability of forms to withstand distortion caused by placement and vibration of concrete shall be such that formed surfaces will conform with applicable requirements of these specifications pertaining to finish of formed surfaces. Where finish F3 is specified, the sheathing or lining shall be placed so that the joint marks on the concrete surfaces will be in general alinement both horizontally and vertically. Where pine is used for form sheathing, the lumber shall be pinus ponderosa in accordance with the Standard Grading Rules of the Western Wood Products Association or shall be other lumber of a grading equivalent to that specified for pine. Plywood used for form sheathing or lining shall be concrete form, class I, grade B-B exterior, mill oiled and edge sealed, in accordance with Product Standard PS 1-66 of the Bureau of Standards [12]. Materials used for form sheathing or lining shall conform with the following requirements, or may be other materials producing equivalent results:

Required finish of formed surface	Wood sheathing or lining	Steel sheathing or lining*
F1	Any grade—S2E	Steel sheathing permitted. Steel lining permitted.
F2	No. 2 common or better, pine shiplap, or plywood sheathing or lining.	Steel sheathing permitted. Steel lining permitted if approved.
F3	No. 2 common or better pine tongue-and-groove or plywood sheathing or lining, except where special form material is prescribed.	Steel sheathing not permitted. Steel lining not permitted.
F4	For plane surfaces, No. 1 common or better pine tongue-and-groove or shiplap or plywood. For warped surfaces, lumber which is free from knots and other imperfections and which can be cut and bent accurately to the required curvatures without splintering or splitting.	Steel sheathing permitted. Steel lining not permitted.

*Steel "sheathing" denotes steel sheets not supported by a backing of wood boards. Steel "lining" denotes thin steel sheets supported by a backing of wood boards.

(c) *Form Ties.*—Embedded ties for holding forms shall remain embedded and, except where F1 finish is permitted, shall terminate not less than two diameters or twice the minimum dimension of the tie in the clear of the formed faces of the concrete. Where F1 finish is permitted, ties may be cut off flush with the formed surfaces. The ties shall be constructed so that removal of the ends or end fasteners can be accomplished without causing appreciable spalling at the faces of the concrete. Recesses resulting from removal of the ends of form ties shall be filled in accordance with section H-19 (Repair of Concrete).

(d) *Cleaning and Oiling of Forms.*—At the time the concrete is placed in the forms, the surfaces of the forms shall be free from encrustations of mortar, grout, or other foreign material. Before concrete is placed, the surfaces of the forms shall be oiled with a commercial form oil that will effectively prevent sticking and will not soften or stain the concrete surfaces, or cause the surfaces to become chalky or dust producing. For wood forms, form oil shall consist of straight, refined, pale, paraffin base mineral oil. For steel forms, form oil shall consist of refined mineral oil suitably compounded with one or more ingredients which are appropriate for the purpose. The contractor shall furnish certification of compliance with these specifications for form oil.

(e) *Removal of Forms.*—To facilitate satisfactory progress with the specified curing and enable earliest practicable repair of surface imperfections, forms shall be removed as soon as the concrete has hardened sufficiently to prevent damage by careful form removal. Forms on upper sloping faces of concrete, such as forms on the watersides of warped transitions, shall be removed as soon as the concrete has attained sufficient stiffness to prevent sagging. Any needed repairs or treatment required on such sloping surfaces shall be performed at once and be followed immediately by the specified curing.

To avoid excessive stresses in the concrete that might result from swelling of the forms, wood forms for wall openings shall be loosened as soon as this can be accomplished without damage to the concrete. Forms for the openings shall be constructed so as to facilitate such loosening. Forms for conduits and tunnel lining shall not be removed until the strength of the concrete is such that form removal will not result in perceptible cracking, spalling, or breaking of edges or surfaces, or other damage to the concrete. Forms shall be removed with care so as to avoid injury to the concrete and any concrete so damaged shall be repaired in accordance with section H-19 (Repair of Concrete).

H-15. *Tolerances for Concrete Construction.*—(a) *General.*—Permissible surface irregularities for the various classes of concrete surface finish as specified in section H-20 (Finishes) are defined as "finishes," and are to be distinguished from tolerances as described herein. The intent of this section is to establish tolerances that are consistent with modern construction practice, yet are governed by the effect that permissible deviations will have upon the structural action or operational function of the structure. Deviations from the established lines, grades, and dimensions will be permitted to the extent set forth herein: *Provided*, that the Contracting Authority reserves the right to diminish the tolerances set forth herein if such tolerances impair the structural action or operational function of a structure or portion thereof.

Where specific tolerances are not stated in these specifications or shown on the drawings for a structure, portion of a structure, or other feature of the work, permissible deviations will be interpreted conformably to the tolerances stated in this section for similar work. Specific maximum or minimum tolerances shown on the drawings in connection with any dimension shall be considered as supplemental to the tolerances specified in this section, and shall govern. The contractor shall be responsible for setting and maintaining concrete forms within the tolerance limits necessary to insure that the completed work will be within the tolerances specified. Concrete work that exceeds the tolerance limits specified in these specifications or shown on the drawings shall be remedied or removed and replaced at the expense of and by the contractor.

(b) *Tolerances for Dam Structures.*—

    (1) Variation of constructed linear outline from established position in plan

        In any length of 20 feet, except in buried construction . . . . ½ inch
        Maximum for entire length, except in buried construction . . . . ¾ inch
        In buried construction . . . . twice the above amounts

    (2) Variation of dimensions to individual structure features from established positions

        Maximum for overall dimension, except in buried construction . . . . 1¼ inches
        In buried construction . . . . 2½ inches

    (3) Variation from plumb, specified batter, or curved surfaces for all structures, including lines and surfaces of columns, walls, piers, buttresses, arch sections, vertical joint grooves, and visible arrises

        In any length of 10 feet, except in buried construction . . . . ½ inch
        In any length of 20 feet, except in buried construction . . . . ¾ inch
        Maximum for entire length, except in buried construction . . . . 1¼ inches
        In buried construction . . . . twice the above amounts

    (4) Variation from level or from grades indicated on the drawings for slabs, beams, soffits, horizontal joint grooves, and visible arrises

        In any length of 10 feet, except in buried construction . . . . ¼ inch
        Maximum for entire length, except in buried construction . . . . ½ inch
        In buried construction . . . . twice the above amounts

    (5) Variation in cross-sectional dimensions of columns, beams, buttresses, piers, and similar members

        Minus . . . . ¼ inch
        Plus . . . . ½ inch

    (6) Variation in the thickness of slabs, walls, arch sections, and similar members

        Minus . . . . ¼ inch
        Plus . . . . ½ inch

    (7) Footings for columns, piers, walls, buttresses, and similar members:

        (a) Variation of dimensions in plan

            Minus . . . . ½ inch
            Plus . . . . 2 inches

        (b) Misplacement or eccentricity

            2 percent of the footing width in the direction of misplacement but not more than . . . . 2 inches

        (c) Reduction in thickness

            . . . . 5 percent of specified thickness

    (8) Variation from plumb or level for sills and sidewalls for radial gates and similar watertight joints*

        . . . . Not greater than a rate of $^1/_8$ inch in 10 feet

    (9) Variation in locations of sleeves, floor openings, and wall openings

        . . . . ½ inch

---

*Dimensions between sidewalls for radial gates shall be not more than shown on the drawings at the sills and not less than shown on the drawings at the top of the walls.

(10) Variation in sizes of sleeves, floor
       openings, and wall openings .... ¼ inch

(c) *Tolerances for Tunnel Lining.*—

(1) Departure from established alinement or from   Free-flow tunnels and conduits .... 1 inch
    established grade   High-velocity tunnels and
                                                       conduits .... ½ inch

(2) Variation in thickness, at any point   Tunnel lining .... minus 0
                                                       Conduits .... minus 2½ percent
                                                                                    or ¼ inch, whichever
                                                                                    is greater
                                                       Conduits .... plus 5 percent or
                                                                                    ½ inch, which-
                                                                                    ever is greater

(3) Variation from inside dimensions .... ½ of 1 percent

(d) *Tolerances for Placing Reinforcing Bars and Fabric.*—

(1) Reinforcing steel, except for bridges:

    (a) Variation of protective covering   With cover of 2½ inches
                                                       or less .... ¼ inch
                                                       With cover of more than
                                                       2½ inches .... ½ inch

    (b) Variation from indicated spacing .... 1 inch

(2) Reinforcing steel for bridges:

    (a) Variation of protective covering   With cover of 2½ inches
                                                       or less .... ⅛ inch
                                                       With cover of more than
                                                       2½ inches .... ¼ inch

    (b) Variation from indicated spacing .... 1 inch

H-16. *Reinforcing Bars and Fabric.*—(a) *Furnishing.*—The contractor shall furnish all the reinforcing bars and fabric required for completion of the work. Reinforcing bars shall conform to ASTM Designation A 615, grade 40 or 60, or ASTM Designation A 617, grade 40 or 60. (See reference [3] or [4].) Fabric shall be electrically welded-wire fabric and shall conform to ASTM Designation A 185 [2].

(b) *Placing.*—Reinforcing bars and fabric shall be placed in the concrete where shown on the drawings or where directed. Splices shall be located where shown on the drawings: *Provided*, that the location of splices may be altered subject to the written approval of the Contracting Authority, and *Provided further*, that, subject to the written approval of the Authority, the contractor may splice bars at additional locations other than those shown on the drawings. Reinforcing bars in splices located where shown on the drawings, in relocated splices approved by the Authority, or in additional splices approved by the Authority, will be included in the measurement, for payment, of reinforcing bars.

Unless otherwise prescribed, placement dimensions shall be to the centerlines of the bars. Reinforcement will be inspected for compliance with requirements as to size, shape, length, splicing, position, and amount after it has been placed.

Before the reinforcement is embedded in concrete, the surfaces of the bars and the surfaces of any bar supports shall be cleaned of heavy flaky rust, loose mill scale, dirt, grease, or other foreign substances which, in the opinion of the Authority, are objectionable.

Heavy flaky rust that can be removed by firm rubbing with burlap or equivalent treatment is considered objectionable.

Reinforcement shall be accurately placed and secured in position so that it will not be displaced during the placing of the concrete, and special care shall be exercised to prevent any disturbance of the reinforcement in concrete that has already been placed. Welding or tack welding of grade 60 or grade 75 reinforcing bars will not be permitted except at locations shown on the drawings. Chairs, hangers, spacers, and other supports for reinforcement may be of concrete, metal, or other approved material. Where portions of such supports will be exposed on concrete surfaces designated to receive F2 or F3 finish, the exposed portion of the supports shall be of galvanized or other corrosion-resistant material, except that concrete supports will not be permitted. Such supports shall not be exposed on surfaces designated to receive an F4 finish. Unless otherwise shown on the drawings, the reinforcement in structures shall be so placed that there will be a clear distance of at least 1 inch between the reinforcement and any anchor bolts, form ties, or other embedded metalwork.

(c) *Reinforcement Drawings to be Prepared by the Contractor.*—The contractor shall prepare and submit for approval of the Authority reinforcement detail drawings for all structures including bar-placing drawings, bar-bending diagrams, and bar lists.

The contractor's reinforcement detail drawings shall be prepared from reinforcement design drawings included with these specifications and from supplemental reinforcement design drawings to be furnished by the Authority. The position, size, and shape of reinforcing bars are not shown in all cases on the drawings included with these specifications. Supplemental reinforcement design drawings in sufficient detail to permit the contractor to prepare his reinforcement detail drawings will be furnished to the contractor by the Authority after final designs have been completed and after equipment data are received from equipment manufacturers. As the supplemental reinforcement design

drawings may not be available in time to enable the contractor to purchase prefabricated reinforcing bars, it may be necessary for the contractor to purchase bars in stock lengths, and to cut and bend the bars in the field.

At least _____ days before scheduled concrete placement, the contractor shall submit to the Authority for approval three prints of each of his reinforcement detail drawings. The contractor's reinforcement detail drawings shall be prepared following the recommendations established by the American Concrete Institute's "Manual of Standard Practice for Detailing Reinforced Concrete Structures" (ACI 315-65) unless otherwise shown on the reinforcement design drawings. The contractor's drawings shall show necessary details for checking the bars during placement and for use in establishing payment quantities. Reinforcement shall conform to the requirements shown on the reinforcement design drawings.

The contractor's reinforcement detail drawings shall be clear, legible, and accurate and checked by the contractor before submittal. If any reinforcement detail drawing or group of drawings is not of a quality acceptable to the Authority, the entire set or group of drawings will be returned to the contractor, without approval, to be corrected and resubmitted. Acceptable reinforcement detail drawings will be reviewed by the Contracting Authority for adequacy of general design and controlling dimensions. Errors, omissions, or corrections will be marked on the prints, or otherwise relayed to the contractor, and one print of each drawing will be returned to the contractor for correction. The contractor shall make all necessary corrections shown on the returned prints. The corrected drawings need not be resubmitted unless the corrections are extensive enough, as determined by the Authority, to warrant resubmittal. Such Authority review and approval shall not relieve the contractor of his responsibility for the correctness of details or for conformance with the requirements of these specifications.

(d) *Measurement and Payment.*—Measurement, for payment, of reinforcing bars and fabric will be made only of the weight of the bars and fabric placed in the concrete in accordance with the drawings or as directed.

Payment for furnishing and placing reinforcing bars will be made at the applicable unit price per pound bid in the schedule for the various sizes of reinforcing bars and fabric, which unit prices shall include the cost of preparing reinforcement detail drawings, including bar-placing drawings and bar-bending diagrams; of submitting the drawings to the Authority; of preparing all necessary bar lists and cutting lists; of furnishing and attaching wire ties and metal or other approved supports, if used; and of cutting, bending, cleaning, and securing and maintaining in position, all reinforcing bars and fabric as shown on the drawings.

H-17. *Preparations for Placing.*— (a) *General.*—No concrete shall be placed until all formwork, installation of parts to be embedded, and preparation of surfaces involved in the placing have been approved. No concrete shall be placed in water except with the written permission of the Contracting Authority, and the method of depositing the concrete shall be subject to his approval. Concrete shall not be placed in running water and shall not be subjected to the action of running water until after the concrete has hardened. All surfaces of forms and embedded materials that have become encrusted with dried mortar or grout from concrete previously placed shall be cleaned of all such mortar or grout before the surrounding or adjacent concrete is placed.

(b) *Foundation Surfaces.*—Immediately before placing concrete, all surfaces of foundations upon or against which the concrete is to be placed shall be free from standing water, mud, and debris. All surfaces of rock upon or against which concrete is to be placed shall, in addition to the foregoing requirements, be clean and free from oil, objectionable coatings, and loose, semidetached, or unsound fragments. Earth foundations shall be free from frost or ice when concrete is placed upon or against them. The surfaces of absorptive foundations against which concrete is to be placed shall be moistened thoroughly so that moisture will not be drawn from the freshly placed concrete.

(c) *Surfaces of Construction and Contraction Joints.*—Concrete surfaces upon or against which concrete is to be placed and to which new concrete is to adhere, that have become so rigid that the new concrete cannot be incorporated integrally with that previously placed, are defined as construction joints.

All construction joints shall be cured by water curing or by application of wax base curing compound in accordance with the provisions of section H-22 (Curing). Wax base curing compound, if used on these joints, shall be removed in the process of preparing the joints to receive fresh concrete. The surfaces of the construction joints shall be clean, rough, and surface dry when covered with fresh concrete. Cleaning shall consist of the removal of all laitance, loose or defective concrete, coatings, sand, curing compound if used, and other foreign material. The cleaning and roughening shall be accomplished by wet sandblasting, washing thoroughly with air-water jets, and surface drying prior to placement of adjoining concrete: *Provided*, that high-pressure water blasting utilizing pressures not less than 6,000 pounds per square inch may be used in lieu of wet sandblasting for preparing the joint surfaces if it is demonstrated to the satisfaction of the Authority that the equipment proposed for use will produce equivalent results to those obtainable by wet sandblasting. High-pressure water blasting equipment, if used, shall be equipped with suitable safety devices for controlling pressures, including shutoff switches at the nozzle that will shut off the pressure if the nozzle is dropped. The sandblasting (or high-pressure water blasting if approved), washing, and surface drying shall be performed at the last opportunity prior to placing of concrete. Drying of the surface shall be complete and may be accomplished by air jet. In the process of wet sandblasting construction joints, care shall be taken to prevent undercutting of aggregate in the concrete.

The surfaces of all contraction joints shall be cleaned thoroughly of accretions of concrete or other foreign material by scraping, chipping, or other means approved by the Authority.

H-18. *Placing.*—(a) *Transporting.*—The methods and equipment used for transporting concrete and the time that elapses during transportation shall be such as will not cause appreciable segregation of coarse aggregate, or slump loss in excess of 1 inch, in the concrete as it is delivered into the work. The use of aluminum pipe for delivery of pumped concrete will not be permitted.

(b) *Placing.*—The contractor shall keep the Contracting Authority advised as to when placing of concrete will be performed. Unless inspection is waived in each specific case, placing of concrete shall be performed only in the presence of a duly authorized Authority inspector.

The surfaces of all rock against which concrete is to be placed shall be cleaned and, except in those cases where seepage or other water precludes drying of the rock face, shall be dampened and brought to a surface-dry condition. Except for tunnels, surfaces of highly porous or absorptive horizontal or nearly horizontal rock foundations to which concrete is to be bonded shall be covered with a layer of mortar approximately three-eighths of an inch thick prior to placement of the concrete. The mortar shall have the same proportions of water, air-entraining agent, cement, pozzolan, and sand as the regular concrete mixture, unless otherwise directed. The water-cement plus pozzolan ratio of the mortar in place shall not exceed that of the concrete to be placed upon it, and the consistency of the mortar shall be suitable for placing and working in the manner hereinafter specified. The mortar shall be spread and shall be worked thoroughly into all irregularities of the surface. Concrete shall be placed immediately upon the fresh mortar.

A mortar layer shall not be used on concrete construction joints. Unless otherwise directed in formed work, structural concrete placements shall be started with an oversanded mix containing ¾-inch maximum-size aggregate; a maximum net water-cement plus pozzolan ratio of 0.47, by weight; 6 percent air, by volume of concrete; and having a maximum

slump of 4 inches. This mix shall be placed approximately 3 inches deep on the joint at the bottom of the placement.

Retempering of concrete will not be permitted. Any concrete which has become so stiff that proper placing cannot be assured shall be wasted. Concrete shall be deposited in all cases as nearly as practicable directly in its final position and shall not be caused to flow such that the lateral movement will permit or cause segregation of the coarse aggregate from the concrete mass. Methods and equipment employed in depositing concrete in forms shall be such as will not result in clusters or groups of coarse aggregate particles being separated from the concrete mass, but if clusters do occur they shall be scattered before the concrete is vibrated. Where there are a few scattered individual pieces of coarse aggregate that can be restored into the mass by vibration, this will not be objectionable and should be done.

Concrete in tunnel lining may be placed by pumping or any other approved method. Where the concrete in the invert is placed separately from the concrete in the arch and without inside forms, it shall not be placed by pneumatic placing equipment unless an approved type of discharge box which prevents segregation is provided and used. The equipment used in placing the concrete and the method of its operation shall be such as will permit introduction of the concrete into the forms without high-velocity discharge and resultant separation. After the concrete has been built up over the arch at the start of a placement, the end of the discharge line shall be kept well buried in the concrete during placement of the arch and sidewalls to assure complete filling. The end of the discharge line shall be marked so as to indicate the depth of burial at any time. Special care shall be taken to force concrete into all irregularities in the rock surfaces and to completely fill the tunnel arch. Placing equipment shall be operated by experienced operators only.

Where tunnel lining placements are terminated with sloping joints, the contractor shall thoroughly consolidate the concrete at such joints to a reasonably uniform and stable slope while the concrete is plastic. If thorough consolidation at the sloping joints is not obtained, as determined by the Authority, the Authority reserves the right to require the use of bulkheaded construction joints. The concrete at the surface of such sloping joints shall be clean and surface dry before being covered with fresh concrete. The cleaning of such sloping joints shall consist of the removal of all loose and foreign material.

Except as intercepted by joints, all formed concrete other than concrete in tunnel lining, including mass concrete in the dam, shall be placed in continuous approximately horizontal layers. The depth of layers for mass concrete shall generally not exceed 18 inches, and the depth for all other concrete shall generally not exceed 20 inches. The Authority reserves the right to require lesser depths of layers where concrete in 20-inch layers cannot be placed in accordance with the requirements of these specifications. Except where joints are specified herein or on the drawings, care shall be taken to prevent cold joints when placing concrete in any portion of the work. The concrete placing rate shall be such as to ensure that each layer is placed while the previous layer is soft or plastic, so that the two layers can be made monolithic by penetration of the vibrators. To prevent featheredges, construction joints that are located at the tops of horizontal lifts near sloping exposed concrete surfaces shall be inclined near the exposed surface, so that the angle between such inclined surfaces and the exposed concrete surface will be not less than 50°.

In placing unformed concrete on slopes so steep as to make internal vibration of the concrete impracticable without forming, the concrete shall be placed ahead of a nonvibrated slip-form screed extending approximately 2½ feet back from its leading edge. Concrete ahead of the slip-form screed shall be consolidated by internal vibrators so as to ensure complete filling under the slip-form.

In placing mass concrete in the dam, the contractor shall, when required, maintain the exposed area of fresh concrete at the practical minimum, by first building up the concrete in successive approximately horizontal layers to the full width of the block and to full height of the lift over a restricted area at the downstream end of the block, and then continuing upstream in similar progressive stages to the full area of the block. The slope formed by the unconfined upstream edges of the successive layers of concrete shall be kept as steep as practicable in order to keep its area to a minimum. Concrete along these edges shall not be vibrated until adjacent concrete in the layer is placed, except that it shall be vibrated immediately when weather conditions are such that the concrete will harden to the extent that it is doubtful whether later vibration will fully consolidate and integrate it with more recently placed adjacent concrete. Clusters of large aggregate shall be scattered before new concrete is placed over them. Each deposit of concrete shall be vibrated completely before another deposit of concrete is placed over it.

Concrete shall not be placed during rains sufficiently heavy or prolonged to wash mortar from coarse aggregate on the forward slopes of the placement. Once placement of concrete has commenced in a block, placement shall not be interrupted by diverting the placing equipment to other uses.

Concrete buckets shall be capable of promptly discharging the low slump, 6-inch mass concrete mixes specified, and the dumping mechanism shall be designed to permit the discharge of as little as a ½-cubic-yard portion of the load in one place. Buckets shall be suitable for attachment and use of drop chutes where required in confined locations.

Construction joints shall be approximately horizontal unless otherwise shown on the drawings or prescribed by the Authority, and shall be given the prescribed shape by the use of forms, where required, or other means that will ensure suitable joining with subsequent work. All intersections of construction joints with concrete surfaces which will be exposed to view shall be made straight and level or plumb.

If concrete is placed monolithically around openings having vertical dimensions greater than 2 feet, or if concrete in decks, top slabs, beams, or other similar parts of structures is placed monolithically with supporting concrete, the following instructions shall be strictly observed:

(1) Placing of concrete shall be delayed from 1 to 3 hours at the top of openings and at the bottoms of bevels under decks, top slabs, beams, or other similar parts of structures when bevels are specified, and at the bottom of such structure members when bevels are not specified; but in no case shall the placing be delayed so long that the vibrating unit will not readily penetrate of its own weight the concrete placed before the delay. When consolidating concrete placed after the delay, the vibrating unit shall penetrate and revibrate the concrete placed before the delay.

(2) The last 2 feet or more of concrete placed immediately before the delay shall be placed with as low a slump as practicable, and special care shall be exercised to effect thorough consolidation of the concrete.

(3) The surfaces of concrete where delays are made shall be clean and free from loose and foreign material when concrete placing is started after the delay.

(4) Concrete placed over openings and in decks, top slabs, beams, and other similar parts of structures shall be placed with as low a slump as practicable and special care shall be exercised to effect thorough consolidation of the concrete.

(c) *Consolidation.*—Concrete shall be consolidated to the maximum practicable density, so that it is free from pockets of coarse aggregate and entrapped air, and closes snugly against all surfaces of forms and embedded materials. Consolidation of concrete in structures shall be by electric- or pneumatic-drive, immersion-type vibrators. Vibrators having vibrating heads 4 inches or more in diameter shall be operated at speeds of at least 6,000 revolutions per minute when immersed in the concrete. Vibrators having vibrating heads less than 4 inches in diameter shall be operated at speeds of at least 7,000 revolutions per minute when immersed in the concrete. Immersion-type vibrators used in mass concrete shall be heavy duty, two-man vibrators capable of readily consolidating mass concrete of the consistency specified:

Provided, that heavy-duty, one-man vibrators may be used if they are operated in sufficient number, and in a manner and under conditions as to produce equivalent results to that specified for two-man vibrators: *Provided further*, that where practicable in vibrating mass concrete, the contractor may employ gang vibrators, satisfactory to the Authority, mounted on self-propelled equipment in such a manner that they can be readily raised and lowered to eliminate dragging through the fresh concrete, and provided all other requirements of these specifications with respect to placing and control of concrete are met.

Consolidation of concrete in the sidewalls and arch of tunnel lining shall be by electric- or pneumatic-driven form vibrators supplemented where practicable by immersion-type vibrators. Form vibrators shall be rigidly attached to the forms and shall operate at speeds of at least 8,000 revolutions per minute when vibrating concrete.

In consolidating each layer of concrete the vibrator shall be operated in a near-vertical position and the vibrating head shall be allowed to penetrate and revibrate the concrete in the upper portion of the underlying layer. In the area where newly placed concrete in each layer joins previously placed concrete, particularly in mass concrete, more than usual vibration shall be performed, the vibrator penetrating deeply and at close intervals into the upper portion of the previously placed layer along these contacts. In all vibration of mass concrete, vibration shall continue until bubbles of entrapped air have generally ceased to escape. Additional layers of concrete shall not be superimposed on concrete previously placed until the previously placed concrete has been vibrated thoroughly as specified. Care shall be exercised to avoid contact of the vibrating head with surfaces of the forms.

H-19. *Repair of Concrete.*—Concrete shall be repaired in accordance with the Bureau of Reclamation "Standard Specifications for Repair of Concrete," dated November 15, 1970. Imperfections and irregularities on concrete surfaces shall be corrected in accordance with section H-20 (Finishes and Finishing).

H-20. *Finishes and Finishing.*—
(a) *General.*—Allowable deviations from plumb or level and from the alinement, profile grades, and dimensions shown on the drawings are specified in section H-15 (Tolerances for Concrete Construction): these are defined as "tolerances" and are to be distinguished from irregularities in finish as described herein. The classes of finish and the requirements for finishing of concrete surfaces shall be as specified in this section or as indicated on the drawings. The contractor shall keep the Contracting Authority advised as to when finishing of concrete will be performed. Unless inspection is waived in each specific case, finishing of concrete shall be performed only in the presence of an Authority inspector. Concrete surfaces will be tested by the Authority where necessary to determine whether surface irregularities are within the limits hereinafter specified.

Surface irregularities are classified as "abrupt" or "gradual." Offsets caused by displaced or misplaced form sheathing or lining or form sections, or by loose knots in forms or otherwise defective form lumber, will be considered as abrupt irregularities and will be tested by direct measurements. All other irregularities will be considered as gradual irregularities and will be tested by use of a template, consisting of a straightedge or the equivalent thereof for curved surfaces. The length of the template will be 5 feet for testing of formed surfaces and 10 feet for testing of unformed surfaces.

(b) *Formed Surfaces.*—The classes of finish for formed concrete surfaces are designated by use of symbols F1, F2, F3, and F4. No sack rubbing or sandblasting will be required on formed surfaces. No grinding will be required on formed surfaces, other than that necessary for repair of surface imperfections. Unless otherwise specified or indicated on the drawings, the classes of finish shall apply as follows:

F1.—Finish F1 applies to formed surfaces upon or against which fill material or concrete is to be placed, to formed surfaces of contraction joints, and to the upstream face of the dam below the minimum water pool elevation. The surfaces require no treatment after form removal except for repair of defective concrete and filling of holes left by the removal of fasteners from the ends of tie rods as required in section H-19 (Repair of Concrete), and the specified curing. Correction of surface irregularities will be required for depressions only, and only for those which, when measured as described in subsection (a) above, exceed 1 inch.

F2.—Finish F2 applies to all formed surfaces not permanently concealed by fill material or concrete, or not required to receive finishes F1, F3, or F4. Surface irregularities, measured as described in subsection (a) above, shall not exceed one-fourth of an inch for abrupt irregularities and one-half of an inch for gradual irregularities: *Provided*, that surfaces over which radial gate seals will operate without sill or wall plates shall be free from abrupt irregularities.

F3.—Finish F3 applies to formed surfaces, the appearance of which is considered by the Authority to be of special importance, such as surfaces of structures prominently exposed to public inspection. Included in this category are superstructures of large powerplants and pumping plants, parapets, railings, and decorative features on dams and bridges and permanent buildings. Surface irregularities, measured as described in subsection (a) above shall not exceed one-fourth of an inch for gradual irregularities and one-eighth of an inch for abrupt irregularities, except that abrupt irregularities will not be permitted at construction joints.

F4.—Finish F4 applies to formed surfaces for which accurate alinement and evenness of surface are of paramount importance from the standpoint of eliminating destructive effects of water action. When measured as described in subsection (a) above, abrupt irregularities shall not exceed one-fourth of an inch for irregularities parallel to the direction of flow, and one-eighth of an inch for irregularities not parallel to the direction of flow. Gradual irregularities shall not exceed one-fourth of an inch. (*Note*: When waterflow velocities on formed concrete surfaces of outlet works, spillways, etc., are calculated to exceed 40 feet per second, further limitations should be considered for the allowable irregularities to prevent cavitation.)

(c) *Unformed Surfaces.* —The classes of finish for unformed concrete surfaces are designated by the symbols U1, U2, and U3. Interior surfaces shall be sloped for drainage where shown on the drawings or directed. Surfaces which will be exposed to the weather and which would normally be level, shall be sloped for drainage. Unless the use of other slopes or level surfaces is indicated on the drawings or directed, narrow surfaces such as tops of walls and curbs, shall be sloped approximately three-eighths of an inch per foot of width; broader surfaces such as walks, roadways, platforms, and decks shall be sloped approximately one-fourth of an inch per foot. Unless otherwise specified or indicated on the drawings, these classes of finish shall apply as follows:

U1. —Finish U1 (screeded finish) applies to unformed surfaces that will be covered by fill material or by concrete. Finish U1 is also used as the first stage of finishes U2 and U3. Finishing operations shall consist of sufficient leveling and screeding to produce even, uniform surfaces. Surface irregularities measured as described in subsection (a) above, shall not exceed three-eighths of an inch.

U2. —Finish U2 (floated finish) applies to unformed surfaces not permanently concealed by fill material or concrete, or not required to receive finish U1 or U3. U2 is also used as the second stage of finish U3. Floating may be performed by use of hand- or power-driven equipment. Floating shall be started as soon as the screeded surface has stiffened sufficiently, and shall be the minimum necessary to produce a surface that is free from screed marks and is uniform in texture. If finish U3 is to be applied, floating shall be continued until a small amount of mortar without excess water is brought to the surface, so as to permit effective troweling. Surface irregularities, measured as described in subsection (a) above, shall not exceed one-fourth of an inch. Joints and edges of gutters, sidewalks, and entrance slabs, and other joints and edges shall be tooled where shown on the drawings or directed.

U3. —Finish U3 (troweled finish) applies to the inside floors of buildings, except floors requiring a bonded-concrete finish or a terrazzo finish, and to inverts of draft tubes and tunnel spillways. When the floated surface has hardened sufficiently to prevent an excess of fine material from being drawn to the surface, steel troweling shall be started. Steel troweling shall be performed with firm pressure so as to flatten the sandy texture of the floated surface and produce a dense uniform surface, free from blemishes and trowel marks. Surface irregularities, measured as described in subsection (a) above, shall not exceed one-fourth of an inch.

(*Note*: When waterflow velocities on unformed concrete surfaces of outlet works, spillways, etc., are calculated to exceed 40 feet per second, further limitations on U2 and/or U3 finishes should be considered for the allowable irregularities to prevent cavitation.)

H-21. *Protection.* —The contractor shall protect all concrete against injury until final acceptance by the Contracting Authority. Fresh concrete shall be protected from damage due to rain, hail, sleet, or snow. The contractor shall provide such protection while the concrete is still plastic and whenever such precipitation, either periodic or sustaining, is imminent or occurring, as determined by the Authority.

Immediately following the first frost in the fall the contractor shall be prepared to protect all concrete against freezing. After the first frost, and until the mean daily temperature in the vicinity of the worksite falls below 40° F. for more than 1 day, the concrete shall be protected against freezing temperatures for not less than 48 hours after it is placed.

After the mean daily temperature in the vicinity of the worksite falls below 40° F. for more than 1 day, the following requirements shall apply:

(a) *Mass Concrete.* —Mass concrete shall be maintained at a temperature not lower than 40° F. for at least 96 hours after it is placed. Mass concrete cured by application of curing compound will require no additional protection from freezing if the protection at 40° F. for 96 hours is obtained by means of approved insulation in contact with the forms or concrete surfaces; otherwise, the concrete shall be protected against freezing temperatures for 96 hours immediately following the 96 hours protection at 40° F. Mass concrete cured by water curing shall be protected against freezing temperatures for 96 hours immediately following the 96 hours of protection at 40° F. Discontinuance of protection of mass concrete against freezing temperatures shall be such that the drop in temperature of any portion of the concrete will be gradual and will not exceed 20° F. in 24 hours. After March 15, when the mean daily temperature rises above 40° F. for more than 3 successive days, the specified 96-hour protection at a temperature not lower than 40° F. for mass concrete may be discontinued for as long as the mean daily temperature remains above 40° F.: *Provided,* that the specified drop in temperature limitation is met, and that the concrete is protected against freezing temperatures for not less than 48 hours after placement.

(b) *Concrete Other Than Mass Concrete.*—All concrete other than mass concrete shall be maintained at a temperature not lower than 50° F. for at least 72 hours after it is placed. Such concrete cured by application of curing compound will require no additional protection from freezing if the protection at 50° F. for 72 hours is obtained by means of approved insulation in contact with the forms of concrete surfaces; otherwise, the concrete shall be protected against freezing temperatures for 72 hours immediately following the 72 hours protection at 50° F. Concrete other than mass concrete cured by water curing shall be protected against freezing temperatures for 72 hours immediately following the 72 hours protection at 50° F. Discontinuance of protection of such concrete against freezing temperatures shall be such that the drop in temperature of any portion of the concrete will be gradual and will not exceed 40° F. in 24 hours. After March 15, when the mean daily temperature rises above 40° F. for more than 3 successive days, the specified 72-hour protection at a temperature not lower than 50° F. may be discontinued for as long as

the mean daily temperature remains above 40° F.: *Provided,* that the specified drop in temperature limitation is met, and that the concrete is protected against freezing temperatures for not less than 48 hours after placement.

(c) *Use of Unvented Heaters.*—Where artificial heat is employed, special care shall be taken to prevent the concrete from drying. Use of unvented heaters will be permitted only when unformed surfaces of concrete adjacent to the heaters are protected for the first 24 hours from an excessive carbon dioxide atmosphere by application of curing compound: *Provided,* that the use of curing compound on such surfaces for curing of the concrete is permitted by and the compound is applied in accordance with section H-22 (Curing). (Include this proviso only when the use of sealing compound is not permitted on some concrete surfaces.)

H-22. *Curing.*—(a) *General.*—Concrete shall be cured either by water curing in accordance with subsection (b) or by application of wax base curing compound in accordance with subsection (c), except as otherwise hereinafter provided.

The unformed top surfaces of walls and piers shall be moistened by covering with water-saturated material or by other effective means as soon as the concrete has hardened sufficiently to prevent damage by water. These surfaces and steeply sloping and vertical formed surfaces shall be kept completely and continually moist, prior to and during form removal, by water applied on the unformed top surfaces and allowed to pass down between the forms and the formed concrete faces. This procedure shall be followed by the specified water curing or by application of curing compound.

(b) *Water Curing.*—Concrete cured with water shall be kept wet for at least 21 days for concrete containing pozzolan and for at least 14 days for concrete not containing pozzolan. Water curing shall start as soon as the concrete has hardened sufficiently to prevent damage by moistening the surface, and shall continue until completion of the specified curing period or until covered with fresh concrete: *Provided,* that water curing of concrete may be reduced to 6 days during periods when the mean daily temperature in the vicinity of the worksite is less than 40° F.: *Provided further*, that during the prescribed period of water curing, when temperatures are such that concrete surfaces may freeze, water curing shall be temporarily discontinued. The concrete shall be kept wet by covering with water-saturated material or by a system of perforated pipes, mechanical sprinklers, or porous hose, or by any other approved method which will keep all surfaces to be cured continuously (not periodically) wet. Water used for curing shall be furnished by the contractor and shall meet the requirements of these specifications for water used for mixing concrete in accordance with section H-7 (Water).

(c) *Wax Base Curing Compound.* —Wax base curing compound shall be applied to surfaces to form a water-retaining film on exposed surfaces of concrete, on concrete joints, and where specified, to prevent bonding of concrete placed on or against such joints. The curing compound shall be white pigmented and shall conform to Bureau of Reclamation "Specifications for Wax-Base Curing Compound," dated May 1, 1973. The compound shall be of uniform consistency and quality within each container and from shipment to shipment.

Curing compound shall be mixed thoroughly and applied to the concrete surfaces by spraying in one coat to provide a continuous, uniform membrane over all areas. Coverage shall not exceed 150 square feet per gallon, and on rough surfaces coverage shall be decreased as necessary to obtain the required continuous membrane. Mortar encrustations and fins on surfaces designated to receive finish F3 or F4 shall be removed prior to application of curing compound. The repair of all other surface imperfections shall not be made until after application of curing compound.

When curing compound is used on unformed concrete surfaces, application of the compound shall commence immediately after finishing operations are completed. When curing compound is to be used on formed concrete surfaces, the surfaces shall be moistened with a light spray of water immediately after the forms are removed and shall be kept wet until the surfaces will not absorb more moisture. As soon as the surface film of moisture disappears but while the surface still has a damp appearance, the curing compound shall be applied. Special care shall be taken to insure ample coverage with the compound at edges, corners, and rough spots of formed surfaces. After application of the curing compound has been completed and the coating is dry to touch, any required repair of concrete surfaces shall be performed. Each repair, after being finished, shall be moistened and coated with curing compound in accordance with the foregoing requirements.

Equipment for applying curing compound and the method of application shall be in accordance with the provisions of chapter VI of the eighth edition of the Bureau of Reclamation Concrete Manual [1]. Traffic and other operations by the contractor shall be such as to avoid damage to coatings of curing compound for a period of not less than 28 days. Where it is impossible because of construction operations to avoid traffic over surfaces coated with curing compound, the film shall be protected by a covering of sand or earth not less than 1 inch in thickness or by other effective means. The protective covering shall not be placed until the applied compound is completely dry. Before final acceptance of the work, the contractor shall remove all sand or earth covering in an approved manner. Any curing compound that is damaged or that peels from concrete surfaces within 28 days after application, shall be repaired without delay and in an approved manner.

(d) *Costs.* —The costs of furnishing and applying all materials used for curing concrete shall be included in the price bid in the schedule for the concrete on which the curing materials are used.

H-23. *Measurement of Concrete.* —Measurement, for payment, of concrete required to be placed directly upon or against surfaces of excavation will be made to the lines for which payment for excavation is made. Measurement, for payment, of all other concrete will be made to the neatlines of the structures, unless otherwise specifically shown on the drawings or prescribed in these specifications. In the event cavities resulting from careless excavation, as determined by the Contracting Authority, are required to be filled with concrete, the materials furnished by the Authority and used for such refilling will be charged to the contractor at their cost to the Authority at the point of delivery to the contractor. In measuring concrete for payment, the volume of all openings, recesses, ducts, embedded pipes, woodwork, and metalwork, each of which is larger than 100 square inches in cross section will be deducted.

H-24. *Payment for Concrete*.—Payment for concrete in the various parts of the work will be made at the unit prices per cubic yard bid therefor in the schedule, which unit prices shall include the cost of all labor and materials required in the concrete construction, except that payment for furnishing and handling cement, and payment for furnishing and placing reinforcing bars will be made at the unit prices bid therefor in the schedule.

H-25. *Bibliography*.

Bureau of Reclamation

[1] "Concrete Manual," eighth edition, 1975.

American Society for Testing and Materials

[2] ASTM Designation: A 185, "Welded Steel Wire Fabric for Concrete Reinforcement."
[3] ASTM Designation: A 615, "Deformed Billet-Steel Bars for Concrete Reinforcement."
[4] ASTM Designation: A 617, "Axle-Steel Deformed Bars for Concrete Reinforcement."
[5] ASTM Designation: C 184, "Standard Method of Test for Fineness of Hydraulic Cement by the No. 100 and 200 Sieves."
[6] ASTM Designation: C 260, "Standard Specifications for Air-Entraining Admixtures for Concrete."
[7] ASTM Designation: E-11, "Standard Specifications for Wire-Cloth Sieves for Testing Purposes."

*General Services Administration
(Federal Supply Service)*

[8] Federal Specification AAA-S-121d, "Scale (weighing; General Specifications for)."
[9] Federal Specification SS-C-192G (Including Amendment 3), "Portland Cement."
[10] Federal Specification SS-P-570B, "Pozzolan (for Use in Portland Cement Concrete)."
[11] Federal Test Method Standard No. 158A, "Cements, Hydraulic; Sampling, Inspection, and Testing."

U.S. Department of Commerce, Bureau of Standards

[12] Product Standard PS 1-66, "Softwood Plywood, Construction and Industrial."

Sample Specifications for Controlling Water and Air Pollution

I-1. *Scope.*—The following sample specifications prescribe water quality controls and preventive measures for discharge of wastes and/or pollution into a river, lake, or estuary due to construction operations; and the prevention and control of air pollution. They are written in the form of mandatory provisions which should be required of the contractor.

A. PREVENTION OF WATER POLLUTION

I-2. *General.*—The contractor shall comply with applicable Federal and State laws, orders, and regulations concerning the prevention, control, and abatement of water pollution. Permits to discharge wastes into receiving waters shall be obtained by the contractor either from the State water pollution control agency or from the Environmental Protection Agency.

The contractor's construction activities shall be performed by methods that will prevent entrance or accidental spillage of solid matter, contaminants, debris, and other objectionable pollutants and wastes into streams, flowing or dry watercourses, lakes, and underground water sources. Such pollutants and wastes include but are not restricted to refuse, garbage, cement, concrete, sewage effluent, industrial waste, radioactive substances, mercury, oil and other petroleum products, aggregate processing tailings, mineral salts, and thermal pollution. Pollutants and wastes shall be disposed of at sites approved by the Contracting Authority.

The contractor shall control his construction activities so that turbidity resulting from his operations shall not exist in concentrations that will impair natural or developed water supplies, fisheries, or recreational facilities downstream from the construction area.

At least 40 days prior to beginning of construction of each phase of work, the contractor shall submit for approval two copies of his plans for the treatment and disposal of all waste and for control of turbidity in the _____ River which may result from his operations. The plans shall be submitted to the Construction Engineer, Post Office Box_____, _____, _____. The plans shall include complete design and construction details of turbidity control features. Such plans shall also show the methods of handling and disposal of oils or other petroleum products, chemicals, and similar industrial wastes.

Except as otherwise provided in section I-4(a) below, approval of the contractor's plans shall not relieve the contractor of the responsibility for designing, constructing, operating, and maintaining pollution and turbidity control features in a safe and systematic manner, and for repairing at his expense any damage to, or failure of, the pollution and turbidity control structures and equipment caused by floods or storm runoff.

I-3. *Control of Turbidity.*—Turbidity increases above the natural turbidities in the _____River that are caused by construction activities shall be limited to those increases resulting from performance of required construction work in the river channel and will be permitted only for the shortest practicable period required to complete such work and as approved by the Contracting Authority. This required construction work will include such work as diversion of the river, construction or removal of cofferdams and other specified earthwork in or adjacent to the river channel, pile driving, and construction of turbidity control structures.

The spawning period for trout (or other game fish) in the _____River is normally during the period_____through _____. Accordingly, no change in the diversion or channelization of the river will be permitted during this particularly sensitive period.

Mechanized equipment shall not be operated in flowing water except as necessary to construct approved crossings or to perform the required construction, as outlined above.

The contractor's methods of unwatering, of excavating foundations, of operating in the borrow areas, and of stockpiling earth and rock materials shall include preventive measures to control siltation and erosion, and to intercept and settle any runoff of muddy waters. Waste waters from construction of dam and appurtenances, aggregate processing, concrete batching and curing, drilling, grouting, and similar construction operations shall not enter flowing or dry watercourses without the use of special approved turbidity control methods.

I-4. *Turbidity Control Methods.*—(a) *General.*—Turbidity control shall be accomplished through the use of plans approved by the Contracting Authority in accordance with section N-2 above.

The Bureau of Reclamation's methods for control of turbidity during construction at the damsite as set forth in (c) below are acceptable

methods. The contractor may adopt these methods or he may submit for approval alternative methods of equivalent adequacy. If the contractor elects to utilize the Bureau's methods and his plans for implementation are approved by the Contracting Authority, and if such approved plans do not effectively control turbidity due to no fault of the contractor, additional work will be directed for which payment will be made in accordance with the "General Provisions" portion of the specifications. If the contractor elects to propose for approval different methods of turbidity control, the contractor shall bear the full responsibility for their satisfactory operation in controlling turbidity. The approval of the contractor's alternate proposals by the Contracting Authority shall not be construed to relieve the contractor from his responsibility.

The contractor's plans, submitted in accordance with section I-2 above, shall show complete design and construction details for implementing either the Bureau's methods or the contractor's alternative methods.

(b) *Requirements for Turbidity Control During Construction at the Damsite.*—The turbidity control method to be used during construction at the damsite shall: (1) Provide for treatment of all turbid water at the damsite resulting from construction of dam and appurtenances; washing of aggregate obtained from approved sources, if such washing is performed at the damsite; drilling; grouting; or similar construction operations: Provided, that the Contracting Authority may direct that clear water removed from foundations be discharged directly to the river without treatment. The treatment plant shall have a capacity to treat 0 to_____gallons of turbid water per minute so that the turbidity of any effluent discharged to the river does not exceed ____Jackson turbidity units.

(2) Include bypass and control equipment suitable for blending treated and untreated waste waters and obtaining effluents of varying degrees of turbidity. The decision to discharge to the river completely treated effluent or a blend of treated and untreated effluent will be the responsibility of the Contracting

Authority, and will depend on the natural turbidity existing in the river at any particular time.

(3) Have a capability of adjusting the *pH* and alkalinity values of any effluent discharged to the river.

(4) Use only chemicals which have been approved by the Environmental Protection Agency for use in potable water and which have been proven to be harmless to terrestrial wildlife and aquatic life.

(5) Have provisions for accumulating, transporting, and depositing sludge in disposal areas so that the material will not wash into the river by high flows or storm runoff, as approved by the Contracting Authority.

(6) Provide for removal of the treatment plant, cleanup of the site, and restoration of the site to its original condition as approved by the Contracting Authority. All materials, plant, and appurtenances used for turbidity control shall remain the property of the contractor.

(c) *Bureau's Methods of Turbidity Control at the Damsite.*—The Bureau of Reclamation's methods for controlling turbidity during construction at the damsite are based on collecting turbid waters in sumps, and pumping from the sumps to: (1) A water clarification plant, Dorr-Oliver[1] Pretreater (__-foot diameter by__-foot water depth), or equal, with automatic chemical dosage feeders for hydrated lime, alum, and an acid or coagulant aid if needed; or

(2) A treatment plant consisting of equalizing tanks, sedimentation flumes, settling tanks, and ponds combined with innocuous stabilizing and flocculating chemicals as required. Such a treatment plant shall be the Dow Turbidity Control System, as proposed by Dow Chemical U.S.A.,[1] or equal.

(d) *Sampling and Testing of Water Quality.*—The Contracting Authority will do such water quality sampling and testing in connection with construction operations as is necessary to insure compliance with the water quality standards of the State of _____ and the Environmental Protection Agency.

---

[1]Mention of these firms should not be construed as an indication that they are the only suppliers of these or similar products nor as an endorsement by the Bureau of Reclamation.

Turbidities of all effluents discharged to the river from the contractor's construction operations shall be monitored by continuous recorders such as the HACH 6491 or 7855 strip chart recorder provided with CR Surface Scatter Turbidimeter Model 2411 or 2426,[1] or equal, which shall be furnished, installed, and operated by the contractor. Locations of the recorders shall be as approved by the Contracting Authority.

Copies of the recordings shall be submitted daily to the Contracting Authority and shall include the date, time of day, and name of person or persons responsible for operation of the equipment and recorder.

Sampling and testing by the Contracting Authority in no way relieves the contractor of the responsibility for doing such monitoring as is necessary for the controlling of his operations to prevent violation of the water quality standards.

I-5. *Payment.*—Payment for control of turbidity during construction at the damsite will be made at the applicable lump-sum price bid therefor in the schedule, which lump-sum price shall include the cost of furnishing all labor, equipment, and materials for designing, constructing, operating, maintaining, and removing all features necessary for control of turbidity in accordance with these sections.

Payment of percentages of the lump-sum price for control of turbidity during construction at the damsite will be made as follows:

(1) Fifty percent of the lump sum in the first monthly progress estimate after completion of the initial installation of the approved plant for treatment of the turbid water.

(2) Twenty-five percent of the lump sum in the first monthly progress estimate after completion of all concrete placement in the dam.

(3) Twenty-five percent of the lump sum in the first monthly progress estimate after completion of the turbidity control operation at the damsite, and removal of equipment.

The costs of all other labor, equipment, and materials necessary for control of turbidity at locations other than the damsite and for prevention of water pollution for compliance with these sections shall be included in the prices bid in the schedule for other items of work.

B. ABATEMENT OF AIR POLLUTION

I-6. *General.*—The contractor shall comply with applicable Federal, State, and local laws and regulations concerning the prevention and control of air pollution.

In his conduct of construction activities and operation of equipment, the contractor shall utilize such practicable methods and devices as are reasonably available to control, prevent, and otherwise minimize atmospheric emissions or discharges of air contaminants.

The emission of dust into the atmosphere will not be permitted during the manufacture, handling, and storage of concrete aggregates, and the contractor shall use such methods and equipment as are necessary for the collection and disposal, or prevention, of dust during these operations. The contractor's methods of storing and handling cement and pozzolans shall also include means of eliminating atmospheric discharges of dust.

Equipment and vehicles that show excessive emissions of exhaust gases due to poor engine adjustments, or other inefficient operating conditions, shall not be operated until corrective repairs or adjustments are made.

Burning shall be accomplished only at times and at locations approved by the Contracting Authority. Burning of materials resulting from clearing of trees and brush, combustible construction materials, and rubbish will be permitted only when atmospheric conditions for burning are considered favorable by appropriate State or local air pollution or fire authorities. In lieu of burning, such combustible materials may be removed from the site, chipped, or buried as provided in section _____.

Where open burning is permitted, the burn piles shall be properly constructed to minimize smoke, and in no case shall unapproved materials such as tires, plastics, rubber products, asphalt products, or other materials that create heavy black smoke or nuisance odors be burned.

Storage and handling of flammable and combustible materials, provisions for fire prevention, and control of dust resulting from drilling operations shall be done in accordance with the applicable provisions of the Department of Labor "Safety and Health Regulation for Construction" and the Bureau of Reclamation Supplement thereto.

Dust nuisance resulting from construction activities shall be prevented in accordance with section _____.

The costs of complying with this section shall be included in the prices bid in the schedule for the various items of work.

I-7. *Dust Abatement.*—During the performance of the work required by these specifications or any operations appurtenant thereto, whether on right-of-way provided by the Contracting Authority or elsewhere, the contractor shall furnish all the labor, equipment, materials, and means required, and shall carry out proper and efficient measures wherever and as often as necessary to reduce the dust nuisance, and to prevent dust which has originated from his operations from damaging crops, orchards, cultivated fields, and dwellings, or causing a nuisance to persons. The contractor will be held liable for any damage resulting from dust originating from his operations under these specifications on Authority right-of-way or elsewhere.

The cost of sprinkling or of other methods of reducing formation of dust shall be included in the prices bid in the schedule for other items of work.

INDEX